Boris Miller • Evgeny Rubinovich

Optimization of Dynamical Systems with Impulse Controls and Shocks

Birkhäuser

Boris Miller
Laboratory No. 2
A. A. Kharkevich Institute for Information Transmission Problems
Moscow, Russia

Evgeny Rubinovich
Laboratory No. 38
V. A. Trapeznikov Institute of Control Sciences
Moscow, Russia

ISSN 2324-9749 ISSN 2324-9757 (electronic)
Systems & Control: Foundations & Applications
ISBN 978-3-031-64123-7 ISBN 978-3-031-64124-4 (eBook)
https://doi.org/10.1007/978-3-031-64124-4

This book is published under the imprint Birkhäuser, www.birkhauser-science.com by the registered company Springer Nature Switzerland AG
The registered company address is: Gewerbestrasse 11, 6330 Cham, Switzerland

TO OLGA AND IRINA

Preface

This monograph deals with dynamical systems with impulsive controls and external impact actions. Although a relative long time passed since the publication of the monograph [292] and its extended version in Russian [294], this continues to be an active topic of research. The relevance of this research is explained, in particular, by its applications to various microelectromechanical systems (MEMS) and measuring tools in nanotechnologies. For example, in many vibroimpact actuators of MEMs as well as in atomic-force microscopy, the measurements are realized via impact interaction of a probe and the surface being analyzed. Moreover, since the methodology of discontinuous time change became rather popular in the analysis of systems exhibiting discontinuous behavior, most studies simply use it as a common and natural approach for the determination of the reaction of a system to shocks or impulsive actions. As it was shown already in the first edition, such a transformation is natural and perhaps the only possible way to ensure the correctness of the statement of the optimization problem. At the same time, the analog of this transformation allows one to exhibit the peculiarities in the behavior of mechanical systems subject to shocks, where it is sometimes necessary to take the fine time-spatial structure of the interaction into account.

In recent years, the method of the discontinuous time change was developed into a methodology of the so-called singular space-time transformations, which are mainly inherent to the analysis of mechanical shocks, where they enable one to obtain results in the long-standing Painlevé problems, including the case of impacts in the presence of friction. This book builds the methodology of time-spatial transformations from scratch, from external impulsive actions to mechanical shocks. The former type of actions may be described by a discontinuous time change where the control impulse is an idealization of very brief actions of very high intensity. That leads to the use of generalized controls and differential equations with measure. Such actions are inherent to various controlled objects in wide areas of human activity, from chemotherapy to the correction of trajectories of space vehicles. Mechanical systems demonstrate another nature of impulse action, where significant changes of velocities occur during very short interaction times. The nature of such changes involves rather small surface deformations, which however produce significant impulsive forces. In general, impulsive controls and shocks give rise to singular behaviors in a system, such as discontinuities of velocities and phase coordinates. In nonlinear systems, the mathematically correct description of such behaviors is nontrivial due to the fact that the reaction of a system depends on the physical properties and "fine structure" of the impulsive action. The space-time singular transformation provides an effective tool for uncovering singularities and helps solving various optimal control problems in systems involving such interactions. The mathematical basis used for the description of such systems and the associated optimization problems is the theory of differential equations with control measures, in which the supports of these measures are determined by the moments of time of shocks and/or of the impulse action application. The monograph discusses various aspects of the application of singular space-time transformations, in particular, traditional topics of impulse control theory:

- Representation of solutions with the aid of a discontinuous time change
- Description of the discontinuous solutions by means of differential equations with measure
- Existence of optimal solutions for optimal control problems
- Derivation of the optimality conditions for this class of the optimal control problems

as well as novel results concerning mechanical systems with shocks. In mechanical systems, such shocks occur usually when bodies collide. The consequences of such an interaction depend on the shape and physical properties of the surfaces involved. It should be noted that during the brief interaction of bodies during impact, microstrains create large forces that modify considerably the velocities of the bodies. The analysis of such interactions by classical methods in the presence of friction leads sometimes to incorrect and paradoxical effects [335]. However, the "uncovering" of singularities with the aid of space-time transformations helps one to obtain a correct description of the impact phenomena. Other areas of applications are also considered:

- Observation control problems, where the simultaneous use of continuous and discrete observations leads to impulsive control problems
- "Fast" control in the shock phase
- Control of hybrid systems with unilateral constraints

The contents of the book are original and based mainly on the authors' works. We distinguish two main approaches in the theory of impulse control. The first, more traditional one, defines the impulsive action as a discontinuity of the phase coordinates depending on current time, the state preceding the action, and its magnitude [235, 454]. The second one considers an impulse as an idealization of a very brief action of high intensity, which produces almost abrupt changes in the phase coordinates [128, 292, 294, 333, 464, 465]. The monograph treats both approaches, as well as the relationship between them. This relationship is based on an extension of the variational problems with unbounded derivatives of the phase coordinates of V. Krotov [219] and V. Gurman [157].

Before this monograph, there were three editions: the first one [292] in English, the others in Russian [294], and most recently [298] that is the closest to this translation. We claim that the present book presents the state of the art in this research area, with the most important results being related to control in the shock phase. The book can be used by people studying the theory of optimal control for hybrid systems, and by researchers in control engineering, operation research, and applied control science. No particular background in optimal control theory is required; however, some basic notions from the theory of ordinary differential equations and optimal control theory, as well as introductory notions from real analysis and some functional analysis would be useful. All the necessary notions are provided in the special mathematical Appendix.

The authors express their gratitude to Professors Alexandre Sesekin, Stanislav Zavalishchin (Ekaterinburg, Russia), Vladimir Dykhta (Irkutsk, Russia), Franco Rampazzo (Padova, Italy), Joseph Bentsman (Urbana-Champaign, USA), Bernard Brogliato (Grenoble, France), Fernando Lobo Pereira (Porto, Portugal) for their fruitful and motivating discussions on the subject; and to Professor Fima Klebaner (Monash University) for the opportunity to continue this work in the School of Mathematics of Monash University, Melbourne, Australia.

We gratefully acknowledge the valuable support from the Russian Foundation for Basic Research, the Russian Science Foundation, and the Australian Research Council in various stages of this research.

We are grateful to our spouses Olga and Irina and to our families for their patience and support.

Moscow, Russia
2024

Boris Miller
Evgeny Rubinovich

Contents

Chapter 1
Introduction

1.1 Discrete-Continuous (Hybrid) Systems and Problems Arising in Optimal Control

The concept of discrete-continuous or hybrid system (DCS or HS) emerged in the scientific practice at the frontier between the fifties and sixties of the last century, in connection with the appearance of novel classes of automatic control systems for continuous plants [430, 431]. These systems usually have continuous and discrete (impulsive) parts which are operating together, and hence have new properties. At that time, it became clear that the presence of elements operating in a discrete or an impulsive mode may lead to substantial changes in the characteristics and properties of a system. This made it necessary to create new directions and branches in automatic control theory, such as the theory of dynamical systems with delay, the theory of discrete deterministic and stochastic processes, as well as to develop special stability, controllability, and optimization methods [53, 198, 372, 430, 431]. During further research on various engineering, economic, and biological systems, the concept of the impulse control , regarded as an action causing instant changes in the state of a system, was developed. Impulse controls in such systems produce state changes that occur much faster than the dynamic processes inherent to the system itself. Moreover, the systems behave in such a manner that on the natural time scale these changes look as if they are instantaneous. In some areas of applications it was found useful to substitute these rapid changes with jumps, especially in mathematical models of such systems, which allows one to simplify their description. This simplification is obtained at the expense of reducing the dimension of control actions by replacing the functional description of controls with a parametric one.

Various examples of systems of the type indicated above were considered in flight dynamics [192, 193, 218, 222, 236, 251, 329–331, 378, 438, 461, 462], in optimizing protocols for chemo- and radio-therapy for cancer and other diseases [191, 366], in

B. Miller, E. Rubinovich, *Optimization of Dynamical Systems with Impulse Controls and Shocks*, Systems & Control: Foundations & Applications,
https://doi.org/10.1007/978-3-031-64124-4_1

economic analysis [32, 111, 149, 150], in control of information processes [76, 87, 153, 231, 246, 249], and in queuing systems [232, 250]. Further areas of application of impulsive control are presented in the monographs [43, 128], for instance, related to energy production, as well as stocks and finance management.

The history and evolution of the concept of DCS and its relation with variational problems and impulse control are presented in the survey [296]. The rather natural—and in some sense "naive"—interpretation of a DCS as a system subjected to jumps in its behavior at the time instants when the impulse control is applied leads to the following mathematical model .

Assume the evolution of a system is described by an $\mathbb{R}^n$-valued variable $X(t)$ which satisfies, on some interval $[0, T]$, the differential equation

$$\dot{X}(t) = F(X(t), t), \tag{1.1}$$

with the initial condition $X(0) = x_0 \in \mathbb{R}^n$, and with intermediate conditions of the form

$$X(\tau_i) = X(\tau_i-) + \Psi(X(\tau_i-), \tau_i, w_i), \tag{1.2}$$

prescribed for some sequence of times $\{\tau_i,\ i = 1, \ldots, N \leq \infty\}$ such that

$$0 \leq \tau_1 < \tau_2 < \cdots < \tau_i < \cdots < \tau_N \leq T. \tag{1.3}$$

In (1.2), $X(\tau_i-) = \lim\limits_{t \uparrow \tau_i} X(t)$ is the state of the system preceding the jump at time τ_i, and $w_i \in W \subset \mathbb{R}^m$ is the intensity of corresponding impulse control.

Thus, the state of the system changes continuously on the semi-open intervals $[0, \tau_1), \ldots, [\tau_{i-1}, \tau_i), \ldots, [\tau_N, T]$ and jumps at each time τ_i. According to Eq. (1.2), these jumping changes depend on the state preceding the jump, the time of the jump, and the intensity of the impulse control. Upon combining the Eqs. (1.1) and (1.2) one obtains the following generalized equation

$$\dot{X}(t) = F(X(t), t) + \sum_{\tau_i \leq t} \Psi(X(\tau_i-), \tau_i, w_i)\delta(t - \tau_i), \tag{1.4}$$

where $X(t)$ is right-continuous and the action of the δ-function is described by relation (1.2).

Remark 1.1 In the sequel this is precisely the way the action of the δ-function on discontinuous functions that are right-continuous and admit left limits will be understood.

The optimal control problem for DCS is to find a choice of instants and magnitudes of impulsive actions that minimizes a given objective function (quality criterion). If the number of impulses is fixed or at least finite, the impulsive optimal control problem reduces to a problem of mathematical programming. This problem,

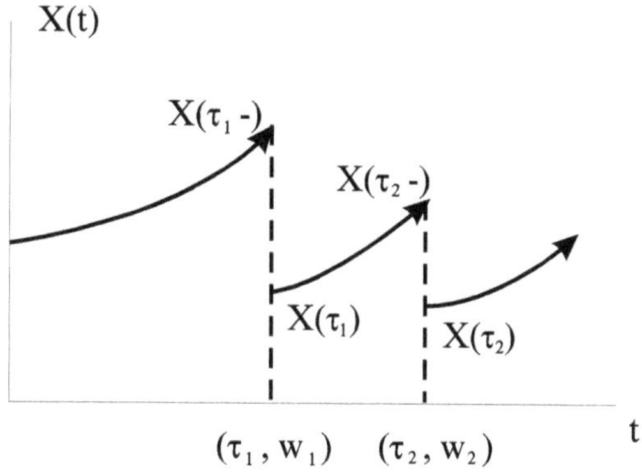

Fig. 1.1 Solution of a discrete-continuous system $X(t)$; t represents time

as a rule, has a solution obtained by solving a minimization problem for a specific function of a finite number of variables. Therefore, the original impulse control problem admits a solution within the class of purely impulsive controls. On the contrary, if the number of impulses and their localization are not fixed in advance, the optimization problem becomes nontrivial and displays specific features that make the solution more complicated (see Fig. 1.1).

1.1.1 A Typical Impulse Control Problem: Chemotherapy and Radiotherapy of Malignant Diseases

Various models of impulse control arise when dealing with the treatment of malignant diseases. Impulse control is natural for this class of processes. Here the dynamics inherent to the tumor growth is relatively slow (months or years) compared with the treatment actions, which are very fast and have characteristic times of seconds for radiotherapy [177] and hours for chemotherapy [191, 258, 425].

Let us demonstrate the specificity of the impulse control on a simple model that describes the simultaneous action of the therapy on the tumor and on healthy tissue [43, 258–260]. Various antimalignant therapies can and often lead to irreversible as well as reversible damage to normal (i.e., healthy) tissue, which must be accounted for in designing treatment protocols. This class of problems has an old history [51, 97, 319, 320]. The necessity of accounting for the aforementioned damaging action leads to optimal control problems with phase constraints [424]. We assume that the therapy process is described by a sequence of pairs

$$\mathscr{U} = \{(w_i, \tau_i),\ \ i = 1, 2, \ldots\},$$

where $\tau_1 < \tau_2 < \cdots$ are the times at which the impulses are applied and w_i are the impulse intensities. Then the process of tumor growth under the therapy $\mathscr{U}$ can be described by a differential equation with impulsive right-hand side [43, 191, 198],

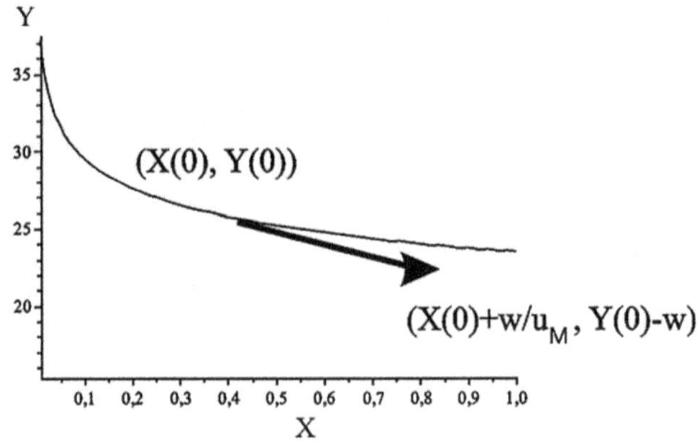

Fig. 1.2 Single-dose action. Here Y represents the "tumor size" and X is the level of normal-tissue damage

$$\dot{Y}(t) = f(Y(t)) - \sum_{i=1}^{N_t} w_i \delta(t - \tau_i), \tag{1.5}$$

where

$$N_t = \max\{i : \tau_i \leq t\},$$

and the function f is such that $f(y) \geq 0$, $f(0) = 0$ and f has a derivative for all $y > 0$ with $f'(y) \geq 0$. The control problem is to make $Y(t)$ zero without damaging healthy tissue more than some indicator value which, by convention, is set equal to 1. We assume that the damage level of healthy tissue is described by the equation

$$\dot{X}(t) = -\phi(X(t)) + \frac{1}{u_{\mathrm{M}}} \sum_{i=1}^{N_t} w_i \delta(t - \tau_i), \tag{1.6}$$

where the function ϕ has the same properties as f. Thus, by using the action w we reduce the tumor size Y by w and increase the damage level X by w/u_{M}. As a consequence, the point (X, Y) in the phase surface of the system moves to $(X + w/u_{\mathrm{M}}, Y - w)$, as shown on Fig. 1.2. Therefore, in this model

$$F(X, Y) = \begin{pmatrix} -\phi(X) \\ f(Y) \end{pmatrix} \quad \text{and} \quad \Psi(X, Y, \tau, w) = \begin{pmatrix} w/u_{\mathrm{M}} \\ -w \end{pmatrix}.$$

Let us give some examples of control problems for the system (1.5) and (1.6).

The problem of reducing the tumor size to zero (determination of a "treatability state") is formulated as follows: *determine the set of initial conditions*

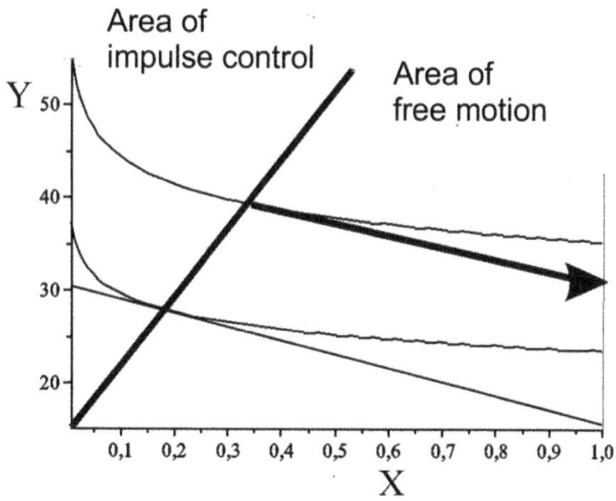

Fig. 1.3 Application of the optimal impulse control. Here Y represents the "tumor size" and X is the level of normal-tissue damage

$\{(X(0), Y(0)) : X(0) \leq 1, Y(0) > 0\}$ *for which there exists an impulse control* $\mathscr{U}$, *such that, for some* $T > 0$,

$$Y(T) = 0 \quad and \quad X(t) \leq 1\ ontheinterval \quad [0, T].$$

The complete solution of this problem was obtained by B. Miller in [191, 258]. Here we present the solution without a detailed proof. One is interested in *achieving the maximal reduction of* Y *by a single action that is admissible under the constraint* $X(t) \leq 1$. Since during the application of the control w at time t the value $Z(t) = Y(t) + u_M X(t)$ remains constant, the instant t of the impulse control application is such that $Z(t)$ achieves its minimum at t, and the corresponding intensity of the impulse action is equal to

$$w = u_M(1 - X(t)).$$

As shown in Fig. 1.3, in the "free-motion area" $Z(t)$ decreases until the moment the derivative of $Z(t)$ vanishes, while in the "impulse control area" the maximal admissible impulse control is used.

So the "treatment area" is the set of initial states (X, Y) that can be transferred by an admissible impulsive control into the area where $Z(t)$ decreases along the paths of the system

$$\dot{X}(t) = -\phi(X(t)), \quad \dot{Y}(t) = f(Y(t)).$$

This area is bounded from above by the line

$$Y + u_M X = Y_c + u_M,$$

where Y_c is the solution of equation $f(Y_c) = u_M \phi(1)$ (see Fig. 1.4).

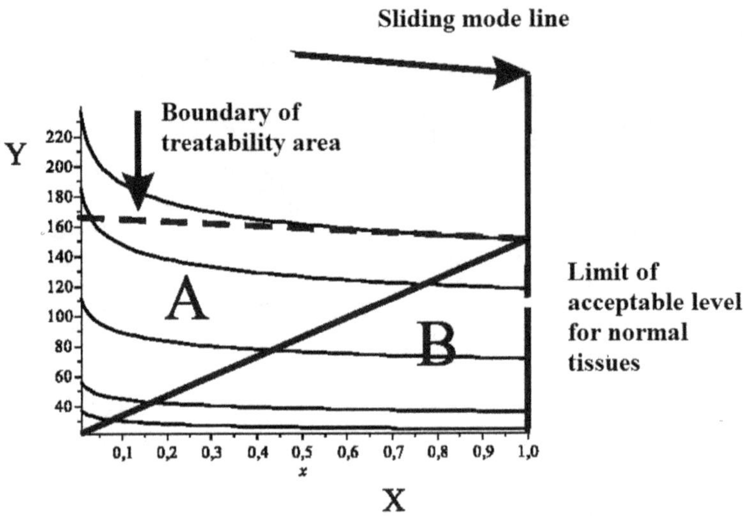

Fig. 1.4 Phase portrait of the system with optimal control. The "treatment area" is bounded above by the line $Y + u_M X = Y_c + u_M$; A is the area of impulse control, B is the area of free motion until the switching line is crossed. The impulse sliding mode is on the line $X = 1$ and can be described by the equation $\dot{Y}(t) = f(Y(t)) - u_M\phi(1)$

Tumor Confinement Problem If the initial state is above the boundary of the "treatment area", there is no admissible impulse control that can bring the tumor size to zero. In this case one can state the problem of *keeping* $Y(t)$ *below some prescribed level* $Y_M > Y_c$ *for a maximal time.* As shown in [258], the optimal control first transfers the system to the phase constraint, i.e., to the line $X(t) = 1$, and then keeps the system on the constraint with the aid of an impulsive sliding mode. During this mode the frequency of impulses tends to infinity, and their magnitudes go to zero. The equivalent continuous control is equal to $u_M\phi(1)$. The phase portrait of the system under such an optimal control is shown in Fig. 1.4.

In this model, the impulse control action is described by a linear relation, though for radiotherapy a model that is linear in state and nonlinear in action is more appropriate. For example, one can take

$$\Psi(Y, \tau, w) = Y[e^{-\alpha w} - 1],$$

where $w = w_i$ is the magnitude of the action and $\alpha > 0$ is some coefficient [191]. This model can be reduced to the one considered above by the change of variable $Z(t) = \ln Y(t)$. Then the impulsive action is given by $Z(\tau_i) = Z(\tau_i-) - \alpha w_i$. Moreover, one can consider a model in which the tumor consists of different types of cells (for example, on different steps of the cell cycle), and hence exhibits different sensitivities to radiotherapy [177].

1.1.2 Control of Bioreactors

In the kind of systems considered here the control consists in adding some reagent into the bioreactor [136]. A single addition results in an abrupt change of concentrations of the various liquids in the tank, which corresponds to the impulse control; adding a reagent continuously means that one is dealing with a continuous control problem. Sequential batch reactors are quite frequently used in biotechnology, for example, in sewage treatment. A reservoir with contaminated water is filled by an active reagent, such as silt or microorganisms, which is selected so as to destroy undesirable components. Usually, the cleaning process consists of three phases:

- filling the reactor with the liquid that needs to be cleaned,
- waiting till the concentration of undesirable components decreases to a prescribed (lower) level,
- releasing the cleaned liquid from the reactor, while the reagent remains inside.

The duration of each phase is sufficiently long and determines the effectiveness of the process as a whole. The control of the flow during the first phase determines the duration of the full process (more precisely, the duration of the first two phases listed above). However, the nonlinear kinetics of bioreactors is complicated enough, so determining the minimizing strategy is not an easy task. Similar problems arise in the production of various biomaterials under time constraints [107, 167, 174].

The dynamics of a sequential batch reactor containing several undesired components can be described by a system of differential equations of the form

$$\begin{cases} \dot{x}_i(t) = \mu_i(s(t))x_i(t) - \dfrac{F(t)}{v(t)}x_i(t), \\ \dot{s}(t) = -\displaystyle\sum_{j=1}^{n} \mu_j(s(t))x_j(t) + \dfrac{F(t)}{v(t)}(s_{\text{in}} - s(t)), \\ \dot{v}(t) = F(t), \end{cases} \tag{1.7}$$

where $x_i(t)$, $s(t)$ and $v(t)$ are the concentrations of the i-th undesired component, the concentration of the reagent, and the volume of liquid in the tank, respectively.

The constant s_{in} is the concentration of reagent in the incoming flow and the non-negative functions $\mu_i(s)$ satisfy the conditions $\mu_i(0) = 0$. The incoming flow F is the control, which of course satisfies $F \geq 0$. Such a system is related to the class of systems with unbounded controls, which may be impulsive, as in the case when fluid is added instantaneously. To include this type of controls into an optimization scheme one can assume that the control $F(t)$ is of general form, namely, it is the sum of an absolutely continuous component and a singular component with respect to the Lebesgue measure: i.e.,

$$dF(t) = u(t)dt + dF^{\text{s}}(t), \tag{1.8}$$

where the singular component $F^{s}(t)$ contains impulsive (jump like) and continuous components.

When the control $F^{s}(\cdot)$ includes an impulse $\Delta F(\tau)$ at the time instant τ, the corresponding jumps of concentration are given by the expressions

$$\begin{cases} \Delta x_i(\tau) = x_i(\tau) - x_i(\tau-) = x_i(\tau-)\left(\dfrac{v(\tau-)}{v(\tau)} - 1\right), \\ \Delta s(\tau) = s(\tau-)\left(\dfrac{v(\tau-)}{v(\tau)} - 1\right) + s_{\text{in}}\left(1 - \dfrac{v(\tau-)}{v(\tau)}\right), \\ \Delta v(\tau) = v(\tau) - v(\tau-) = \Delta F(\tau). \end{cases} \tag{1.9}$$

The expressions (1.9) may be readily obtained by considering the composition of the liquids in the tank before and after a volume of liquid of $\Delta v(\tau)$ is added. However, it is more interesting that these expressions can be derived by means of a regular procedure that determines the jump of a generalized solution (see Theorem 2.11 below).

As shown in [136], in the case of a single undesirable component, a single-impulse control at the initial time followed by an interval of continuous control is optimal. In this case, one can obtain the optimal control in feedback form. In the case of multiple undesirable components the optimal control has a more complicated structure, with more intervals of continuous controls.

1.1.3 Mechanics of Reactive Motion

Our next example concerns the mechanics of reactive motion with pulsed thrust. The ordinary reactive motion is governed by the *Meshchersky equation* [3],

$$M(t)\frac{V(t)}{dt} = -C\frac{dM(t)}{dt},$$

where $M(t)$ is the current value of the rocket mass, $V(t)$ is the velocity of the rocket, and C is the velocity of the ejected reactive flow. Assume that the rocket engine ejects a mass ΔM during the brief time interval $[\tau - \varepsilon, \tau]$. Then one can solve the Meshchersky equation and obtain Tsiolkovsky's well-known formula for the velocity increment:

$$V(\tau) - V(\tau - \varepsilon) = C \ln \frac{M(\tau - \varepsilon)}{M(\tau)} = C \ln \frac{M(\tau - \varepsilon)}{M(\tau - \varepsilon) - \Delta M}. \tag{1.10}$$

With a pulsed ejection of the working fluid, the mass of the rocket changes abruptly, and in the limit as $\varepsilon \to 0$ one obtains the following expressions for the velocity and mass jumps:

$$\begin{cases} M(\tau) - M(\tau - \varepsilon) \to \Delta M(\tau) = -\Delta M, \\ V(\tau) - V(\tau - \varepsilon) \to \Delta V^{\mathrm{c}}(\tau) = \\ \qquad = C \ln \dfrac{M(\tau-)}{M(\tau-) + \Delta M(\tau)} = C \ln \dfrac{M(\tau-)}{M(\tau-) - \Delta M}. \end{cases}$$

If the change of mass is considered as an impulse control, then when a series of impulse controls $\{(\Delta M_i, \tau_i), i = 1, 2, \ldots\}$ is used, the reactive motion with pulsed thrust is governed by the equation

$$\begin{pmatrix} \dot{V}(t) \\ \dot{M}(t) \end{pmatrix} = \sum_{\tau_i \le t} \begin{pmatrix} C \ln \dfrac{M(\tau_i-)}{M(\tau_i-) - \Delta M_i} \\ -\Delta M_i \end{pmatrix} \delta(t - \tau_i). \tag{1.11}$$

Thus, the state variables $\{V(t), M(t)\}$ obey an equation of the type (1.38), with corresponding function

$$\Psi(V, M, w) = \begin{pmatrix} C \ln \dfrac{M}{M - w} \\ -w \end{pmatrix},$$

where the magnitude of the impulse control w is the mass of working fluid ejected or the change of the rocket's mass. Since the equation of motion is integrable in closed form, it follows that the function Ψ obtained as a description of the impulse control does not depend on the approximation procedure. In other words, the increment of the velocity does not depend on the temporal profile of the ejected mass of working fluid during the brief time corresponding to the impulse action.

On the other hand, one can exhibit the specifics of this model by calculating the velocity increment directly from the momentum conservation law. Consider the case of genuine impulsive ejection of the working fluid in which the reactive motion is assumed to occur as a result of firing bullets of mass ΔM ejected with velocity C. The model obtained looks like the previous one, but simple calculations lead to the different expression

$$(M(\tau-) - \Delta M)V(\tau) + (V(\tau-) - C)\Delta M = M(\tau-)V(\tau-),$$

and consequently

$$V(\tau) - V(\tau-) = \Delta V^{\mathrm{d}}(\tau) = C \frac{\Delta M}{M(\tau-) - \Delta M}.$$

Since for any $0 < w < M$ we have

$$C \ln \frac{M}{M-w} = C \ln \left(1 + \frac{w}{M-w}\right) < C \frac{w}{M-w},$$

we conclude that

$$\Delta V^{\mathrm{c}}(\tau) = C \ln \frac{M}{M - \Delta M} < C \frac{\Delta M}{M - \Delta M} = \Delta V^{\mathrm{d}}(\tau).$$

This result becomes clear if one looks at the distribution of mass immediately after a pulsed action. In the first case, after continuous but very fast ejection of working fluid, we shall see the vehicle moving with velocity $V^{\mathrm{c}}(\tau)$ and the working fluid of mass ΔM, continuously distributed in space and moving with different velocities from $V(\tau-) - C$ to $V^{\mathrm{c}}(\tau) - C$. In the second case, of impulsive thrust of mass ΔM (single, for simplicity), we shall see the vehicle moving with velocity $V^{\mathrm{d}}(\tau)$ and the consolidated mass of working fluid moving with velocity $V(\tau-) - C$.

This argument also shows that by dividing the working fluid mass ΔM into a large number of small fragments ejected with velocity C one can achieve an arbitrary velocity increment from ΔV^{c} to ΔV^{d}.

This example shows the potential ill-posedness of the impulse control problems: indeed, the approximation of the impulse control by continuous impacts and vice versa could give essentially different results.

1.1.4 Control of Discrete-Continuous Observations

Observation control problems provide an excellent example of impulse control. In general, these problems are not well posed; however, one can choose an appropriate representation of the control as a signal to noise ratio, and in this way make the control problem well posed with respect to the approximation of the impulse control (see Chap. 7). Indeed, this type of problems motivate the research of ill-posed impulse control problems. The next example of an observation control problem shows how ill-posednedness arises.

Example 1.1 Suppose the evolution of the unobservable state $x(t) = (x^1(t), x^2(t))$ of some stochastic system is described by the system of stochastic differential equations

$$\begin{cases} dx^1(t) = x^2(t)dt, \\ dx^2(t) = x^2(t)dt + dW(t), \end{cases} \tag{1.12}$$

with a Gaussian initial state $x(0) = (x^1(0), x^2(0))$ with parameters

$$m(0) = E\{x(0)\} = E\{(x^1(0), x^2(0))\} = (m^1, m^2)$$

and

$$\gamma(0) = \operatorname{cov}\{x(0)\} = \operatorname{cov}\{(x^1(0), x^2(0))\} =$$
$$= \begin{pmatrix} E(x^1(0) - m^1)(x^1(0) - m^1) & E(x^1(0) - m^1)(x^2(0) - m^2) \\ E(x^1(0) - m^1)(x^2(0) - m^2) & E(x^2(0) - m^2)(x^2(0) - m^2) \end{pmatrix}.$$

Here E stands for mathematical expectation. Suppose the observer can choose the observation times $\{\tau_i,\ i = 1, \dots, N\}$ as well as one of the coordinates x^1 or x^2, which can be measured with a controlled signal-to-noise ratio equal to $u_i^1 > 0$ or $u_i^2 > 0$, respectively, and such that the total value obeys the constraint

$$\sum_{i=1}^{n} \left(u_i^1 + u_i^2\right) \le M < \infty. \tag{1.13}$$

The impossibility of simultaneous measurement of the two components of the state may be expressed as the suplementary constraints

$$u_i^1 u_i^2 = 0, \quad u_i^1 \ge 0, \quad u_i^2 \ge 0. \tag{1.14}$$

Therefore, the observation equations have the form

$$y(\tau_i) = u_i^1 x^1(\tau_i) + u_i^2 x^2(\tau_i) + \varepsilon_i^1 (u_i^1)^{1/2} + \varepsilon_i^2 (u_i^2)^{1/2}, \tag{1.15}$$

where $\varepsilon_i^{1,2}$, $i = 1, \dots, n$ are independent Gaussian random variables with zero mean and unit dispersion $E(\varepsilon_i^{1,2})^2 = 1$, and which are mutually independent and independent from $x(0)$ and from the Wiener process, which does not depend on $x(0)$ also. The best mean-square estimate of $x(t)$ by the observations $\{y(\tau_i), \tau_i \le t\}$ is the conditional expectation $m(t) = (m^1(t), m^2(t)) = E\{x(t)|y(\tau_i), \tau_i \le t\}$, which satisfies the equations of the generalized Kalman filter [291],

$$\begin{cases} \dot m^1(t) = m^2(t) + \sum\limits_{\tau_i \le t} \Gamma^1(\alpha_i, u_i, \tau_i)\Delta\bar y(\tau_i)\delta(t - \tau_i), \\ \dot m^2(t) = m^2(t) + \sum\limits_{\tau_i \le t} \Gamma^2(\alpha_i, u_i, \tau_i)\Delta\bar y(\tau_i)\delta(t - \tau_i), \end{cases} \tag{1.16}$$

where

$$\Delta\bar y(\tau_i) = y(\tau_i) - u_i^1 m^1(\tau_i-) - u_i^2 m^2(\tau_i-)$$

and

$$\begin{cases} \Gamma^1(u_i^1, u_i^2, \tau_i) = \dfrac{u_i^1\gamma_{11}(\tau_i-) + u_i^2\gamma_{12}(\tau_i-)}{u_i^1 + u_i^2 + (u_i^1)^2\gamma_{11}(\tau_i-) + (u_i^2)^2\gamma_{22}(\tau_i-)}, \\ \Gamma^2(u_i^1, u_i^2, \tau_i) = \dfrac{u_i^1\gamma_{12}(\tau_i-) + u_i^2\gamma_{22}(\tau_i-)}{u_i^1 + u_i^2 + (u_i^1)^2\gamma_{11}(\tau_i-) + (u_i^2)^2\gamma_{22}(\tau_i-)}. \end{cases}$$

Here $\gamma_{kl}(t)$, $k, l = 1, 2$ are the components of the covariance matrix, that is,

$$\gamma_{kl}(t) = E\big\{(x^k(t) - m^k(t))(x^l(t) - m^l(t))\big|y(\tau_i), \tau_i \leq t\big\}$$

and satisfy the system of deterministic differential equations

$$\begin{cases} \dot\gamma_{11}(t) = 2\gamma_{12}(t) + \sum\limits_{\tau_i \leq t} \Delta\gamma_{11}(\tau_i)\delta(t - \tau_i), \\ \dot\gamma_{12}(t) = \gamma_{12}(t) + \gamma_{22}(t) + \sum\limits_{\tau_i \leq t} \Delta\gamma_{12}(\tau_i)\delta(t - \tau_i), \\ \dot\gamma_{22}(t) = 2\gamma_{22}(t) + 1 + \sum\limits_{\tau_i \leq t} \Delta\gamma_{22}(\tau_i)\delta(t - \tau_i), \end{cases} \tag{1.17}$$

where

$$\begin{cases} \Delta\gamma_{11}(\tau_i) = -\dfrac{\big(u_i^1\gamma_{11}(\tau_i-) + u_i^2\gamma_{12}(\tau_i-)\big)^2}{u_i^1 + u_i^2 + (u_i^1)^2\gamma_{11}(\tau_i-) + (u_i^2)^2\gamma_{22}(\tau_i-)}, \\ \Delta\gamma_{12}(\tau_i) = -\dfrac{\big(u_i^1\gamma_{11}(\tau_i-) + u_i^2\gamma_{12}(\tau_i-)\big)\big(u_i^1\gamma_{12}(\tau_i-) + u_i^2\gamma_{22}(\tau_i-)\big)}{u_i^1 + u_i^2 + (u_i^1)^2\gamma_{11}(\tau_i-) + (u_i^2)^2\gamma_{22}(\tau_i-)}, \\ \Delta\gamma_{22}(\tau_i) = -\dfrac{\big(u_i^1\gamma_{12}(\tau_i-) + u_i^2\gamma_{22}(\tau_i-)\big)^2}{u_i^1 + u_i^2 + (u_i^1)^2\gamma_{11}(\tau_i-) + (u_i^2)^2\gamma_{22}(\tau_i-)}. \end{cases} \tag{1.18}$$

Upon denoting

$$w = (u^1, u^2) \in \mathbb{R}^2, \quad \|w\| = |u^1| + |u^2|$$

and using (1.14), the system (1.18) can be rewritten in the matrix form

$$\Delta\gamma(\tau_i) = -\gamma(\tau_i-)w_i^*\big[\|w_i\| + w_i\gamma(\tau_i-)w_i^*\big]^{-1}w_i\gamma(\tau_i-).$$

Here, * stands for transposition.

Thus, in terms of the matrix-valued variable $\gamma(t)$ the equation for our DCS takes on the form (1.38) with

$$F(\gamma) = \begin{pmatrix} 2\gamma_{12} \\ \gamma_{12} + \gamma_{22} \\ 2\gamma_{22} + 1 \end{pmatrix} \quad \text{and} \quad \Psi(\gamma, w) = \gamma w^* \big[\|w\| + w\gamma w^*\big]^{-1} w\gamma.$$

In the general case, the equation for an unobservable process that describes the evolution of the phase vector $x(t)$ of a linear stochastic system has a form

$$dx(t) = A(t)x(t)dt + C(t)dW(t),$$

with observations

$$y(\tau_i) = H(\tau_i, w_i)x(\tau_i) + G(\tau_i, w_i)\varepsilon_i,$$

in which the noise intensities are allowed to depend on time. Here the matrices $H(t, w)$ and $G(t, w)$ characterize the structure of the measurements and noises in observations. Then the $n \times n$ covariance matrix $\gamma(t)$ of estimation errors is the solution of the equation (see [15, 87, 153, 265, 267] and [291])

$$\begin{aligned} \dot\gamma(t) = {} & A(t)\gamma(t) + \gamma(t)A^*(t) + C(t)C^*(t) - \\ & - \sum_{\tau_i \le t} \gamma(\tau_i)H^*(\tau_i, w_i)\big[G(\tau_i, w_i)G^*(\tau_i, w_i) + \\ & + H(\tau_i, w_i)\gamma(\tau_i)H^*(\tau_i, w_i)\big]^{-1} H(\tau_i, w_i)\gamma(\tau_i)\delta(t - \tau_i). \end{aligned} \tag{1.19}$$

The controls $w_i \in U \subset \mathbb{R}^m$ determine the structure and performance of observations. Therefore, the observation control problem is stated as the problem of choosing optimal times τ_i and controls w_i so as to minimize the some criterion $J(\gamma(\cdot))$ that characteries the performance of the observation procedure. In this way we arrive at a standard control problem for DCS (1.19). One should mention here that the described mathematical problem has various applications in hydro-, optical- and radio-location.

In Sects. 7.5, 7.5.1, and 7.5.2 below it will be shown how to solve the observation control problem with various constraints by exploiting the well-posedness property. However, in the general case well-posedness does not hold and the system (1.19) does not satisfy the necessary robustness conditions (1.48). Therefore, for this class of problems, one can encounter all the specific peculiarities such as the absence of an optimal solution and even the impossibility of realization of the suboptimal solution. In Sect. 7.5.1 we will present a general approach to discrete-continuous observations that is free from the aforementioned difficulties. The main idea is to consider the discrete observations as a result of some continuous, but very fast single-kind measurements that realize an "almost instantaneous" observation compared with the "slow dynamics" of the unobservable process. It turns out that then the filtering equations remain the same as in the case of purely-discrete observations, but in the more general form of differential equations with measures. These equations include also the continuous part and therefore can describe the continuous observations as well. This approach is well motivated from a physical viewpoint because many physical measurement operations in radio- and optical locators and even in sampling from a various set of sensors are indeed of the described kind. Of course, the measurement operation, even though very fast, is not instantaneous due to the bounded frequency band of the measuring device, and therefore its dynamics must be taken into account. Moreover, modern sensors permit to compose the set of measurement parameters even inside the brief measurement time-interval and therefore one needs to develop more detailed observation models.

The approach sketched above shows that the derivation of the jump representation from the"fast" continuous motion may be useful and constructive. It enables us to describe all possible jumps allowed by constraints and, by that, to include the jump approximation procedure into the optimization scheme as a kind of additional control. In this maner the ambiguity in the solution of the optimal control problem can be removed. This approximation procedure is based on a special transformation, the so-called *discontinuous time change method*. The idea is to stretch the brief time interval of the impulse action into a finite interval, where the motion of the system is described by ordinary differential equations with bounded controls. This transformation is of course singular; however, the existing theory allows one to derive the motion on the original time scale and to construct the sequence of controls approximating this motion.

1.2 Specific Features of Nonlinear Systems with Impulse Controls

As shown above, many problems can be solved in the framework of simple models; nevertheless, in the nonlinear case, optimal impulse control problems exhibit specific features, such as ill-posedness and incompleteness of the description in terms of ordinary differential equations with intermediate conditions. Typical occurrences are the absence of the optimal controls, the unfeasibility of the impulse control approximation, lack of stability of solutions with respect to small variations of impulse controls, and appearance of sliding impulse modes in solutions of the optimal impulse control problems. The following examples illustrate these features of DCS and outline the circle of problems considered in this monograph.

1.2.1 Nonlinear Systems: Absence of the Optimal Control

Example 1.2 Consider the system governed by the Eqs. (1.1) and (1.2) with

$$F(x,t) = 0, \qquad \Psi(x,t,w) = xw.$$

Assume that an impulse control of intensity w is applied at $\tau \in (0, T)$. Then the solution with initial condition $X(0) > 0$ at $t = 0$ can be represented in the form

$$X(t) = \begin{cases} X(0), & \text{for} \quad t < \tau, \\ X(0)(1+w), & \text{for} \quad t \geq \tau. \end{cases} \tag{1.20}$$

For a given sequence of impulse controls $\{w_i\}$ applied at time instants $\{\tau_i\} \in (0, T)$, the corresponding solution is the product

$$X(t) = X(0) \prod_{\tau_i \leq t} (1 + w_i). \tag{1.21}$$

Suppose the magnitudes of the impulses are bounded, i.e.,

$$w_i > 0, \quad \sum_i w_i \leq M. \tag{1.22}$$

The aim of the optimal control is to maximize the criterion

$$J(X(\cdot)) = X(T). \tag{1.23}$$

Since

$$X(T) = X(0) \prod_{\tau_i \leq T} (1 + w_i),$$

to maximize $X(T)$ we need to maximize $\prod_i (1 + w_i)$ under the constraints (1.22).

Since for any $w > 0$ and $w_1, w_2 > 0$ such that $w_1 + w_2 = w$ it holds that $(1 + w_1)(1 + w_2) = 1 + w_1 + w_2 + w_1 w_2 > 1 + w$, we see that for any sequence of controls $\{w_i\}$ one can increase the value (1.23) by splitting each impulse w_i into two impulses, and so on. This shows that there is no optimal control by means of a sequence of separate impulses, applied at some time instants $\tau_i \in [0, T]$.

At the same time, the criterion $X(T)$ is bounded, therefore, the supremum of the value (1.23) exists. Indeed,

$$1 + w < e^w, \quad \text{for} \quad w > 0$$

and so

$$\prod_i (1 + w_i) < \prod_i e^{w_i} \le e^M.$$

Thus,

$$\sup_{\{w_i\}} J(X(\cdot)) \le X(0)e^M.$$

It is readily verified that in this inequality is in fact an equality. Indeed, consider the sequence of impulse controls with magnitudes $w_i^n = M/n,\ i = 1, \ldots, n$. The moments of application of impulses are irelevant, it is only required that the impulses are distinct. This sequence satisfies the constraint (1.22), and for the corresponding solutions we have that

$$X^n(T) = X(0)\left(1 + \frac{M}{n}\right)^n \to X(0)e^M, \quad n \to \infty,$$

and consequently

$$\sup_{\{w_i\}} J(X(\cdot)) = X(0)e^M.$$

Thus, this example demonstrates that there is no optimal impulse control, though the maximal value of the criterion may be achieved as a limit of some sequence of impulse controls.

From a practical point of view it would be useful to describe in some way or another the optimal path, but even if the sequence of solutions converges to some function, the latter cannot be described by the Eqs. (1.1) and (1.2) with some impulse controls, as the next example shows.

Example 1.3 Consider the impulse control problem with

$$F(x, t) = x, \quad \Psi(x, w) = xw,$$

with initial condition $X(0) > 0$, with impulse controls subject to the constraints

$$w_i \ge 0, \quad \sum_{i=1}^{N} w_i \le M,$$

and with the objective function

$$J(\{(\tau_i, w_i),\ i = 1, \ldots, N\}) = x(T) \to \max.$$

Choose again the sequence of impulse controls with intensities $w_i^n = M/n$, applied at the time instants

$$\tau_i^n = T - \frac{n-i}{n^2}, \quad i = 1, \ldots, n.$$

It is readily seen that

$$J(\{(\tau_i^n, w_i^n),\ i = 1, \ldots, n\}) = X^n(T) = X(0)e^T \left(1 + \frac{M}{n}\right)^n \uparrow X(0)e^T e^M.$$

Moreover, for any set of controls w_i,

$$x(T) = X(0)e^T \prod_i (1 + w_i) < X(0)e^T \prod_i e^{w_i} = X(0)e^T e^{\sum_i w_i} \leq X(0)e^T e^M.$$

Therefore, for the sequence of impulse controls under consideration, which localizes around the point $t = T$ when $n \to \infty$, the supremum of the objective function is achieved. Moreover, the sequence of solutions $X^n(\cdot)$ converges to some "optimal" solution

$$X_0(t) = \begin{cases} X(0)e^t, & \text{for} \quad t < T, \\ X(0)e^T e^M, & \text{for} \quad t = T. \end{cases}$$

At the same, there is no impulse control that realizes such a solution, since it would have to be localized at $t = T$, but the application of the impulse control using the full control resource M at $t = T$ gives

$$X(T) = X(0)e^T (1 + M) < X(0)e^T e^M.$$

The next example demonstrates the absence of the optimal control due to the fact that the "optimal path" is realized by a so-called sliding impulse mode.

Example 1.4 Consider the same DCS as in Example 1.2, but with the criterion

$$J(X(\cdot)) = \int_0^T \left| X(t) - X(0)e^{Mt/T} \right| dt, \tag{1.24}$$

to be minimized. Since any trajectory, corresponding to some admissible impulse control, satisfies (1.21) as well, it is a piecewise-constant function and so the criterion is positive for any admissible impulse control. However, there exists a sequence of controls $\{(w_i^n, \tau_i^n),\ i = 1, \ldots, n\}$, such that the corresponding solution $X^n(t)$ converges to

$$X(0)e^{Mt/T}$$

everywhere on $[0, T]$, and consequently $J(X^n(\cdot)) \to 0$ as $n \to \infty$. Indeed, take

$$w_i^n = \frac{M}{n}, \quad \tau_i = T\frac{i}{n}, \quad i = 1, \dots, n.$$

Then the number of impulses located on the interval $[0, t]$ is equal to $[tn/N]$, where $[a]$ denotes the integer part of a real number a. Therefore,

$$X^n(t) = X(0) \prod_{i \le [tn/T]} \left(1 + w_i^n\right) = X(0)\left(1 + \frac{M}{n}\right)^{[tn/T]}.$$

On the other hand, since

$$\frac{tn}{T} - 1 < \left[\frac{tn}{T}\right] \le \frac{tn}{T},$$

we have that

$$X(0)\left(1 + \frac{M}{n}\right)^{(tn/T)-1} < X^n(t) \le X(0)\left(1 + \frac{M}{n}\right)^{tn/T},$$

and so

$$X^n(t) \to X(0)e^{Mt/T}, \quad \text{as} \quad n \to \infty.$$

It follows that the sequence of paths $X^n(\cdot)$ converges to the "optimal path", namely, the function $X(0)e^{mt/T}$, which satisfies the differential equation

$$\dot{X}(t) = X(t)w(t),$$

with the control

$$w(t) = \frac{M}{T}$$

such that

$$\int_0^T w(t)dt = M.$$

We see that even if we seek the optimal control in a purely impulsive form, we may arrive at a solution that is a continuous function which does not satisfy the Eqs. (1.1) and (1.2) with some impulsive controls. This means that to guarantee the existence of the optimal control one needs to extend the concept of solution so as to include possible continuous paths that correspond to so-called "impulsive sliding modes" [157], arising as a result of splitting a single impulse into a sequence of impulses with infinitely decreasing magnitudes. The natural extension of purely impulsive controls consists of the control measures and the differential equations with measures, which provide a universal tool for the description of such paths.

To explain the necessity of introducing such equations, consider the function

$$U(t) = \sum_{\tau_i \le t} w_i,$$

which quantifies the control resource used up to time t.

Then for the solution of the system in Example 1.2 we get

$$\Delta X(\tau_i) = X(\tau_i) - X(\tau_i-) = X(\tau_i-)w_i = X(\tau_i-)\Delta U(\tau_i).$$

Thus,

$$X(t) = X(0) + \sum_{\tau_i \le t} \Delta X(\tau_i),$$

and by writing the sum as a Stieltjes integral with respect to measure induced by the function of bounded variation $U(t)$, we obtain the expression

$$X(t) = X(0) + \sum_{\tau_i \le t} X(\tau_i-)\Delta U(\tau_i) = X(0) + \int_0^t X(s-)dU(s).$$

In differential form, $X(t)$ the following differential equation with control measure $U(t)$:

$$dX(t) = X(t-)dU(t).$$

Equations of this type, which contain integrals with respect to a measure, are usually called *differential equations with measure* [99, 165, 377, 395]. They represent a natural extension of DCS, since they provide a universal description for the continuous as well as the discrete components. General properties of such equations are presented in the Appendix.

Earlier, such equations were successfully used to study linear systems, where the control measures appeared naturally upon taking the closure of the sets of trajectories, corresponding to ordinary controls, bounded in the L_1-norm.

1.2.2 Differential Equations with Control Measures

Consider the linear system described by an equation of the form

$$\dot{X}(t) = A(t)X(t) + B(t)u(t), \tag{1.25}$$

where $X(t) \in \mathbb{R}^n$, $u(t) \in \mathbb{R}^m$ and $A(t)$, $B(t)$ are continuous matrix-valued functions of the appropriate dimensions. The system is defined on the interval $[0, T]$ with the constraint $u(\cdot) \in L_1[0, T]$ imposed on the control, i.e.,

$$\int_0^T \|u(t)\| dt \leq M, \tag{1.26}$$

where $\|u\| = \sum_{i=1}^{m} |u_i|$. If the control aim is to minimize the terminal criterion

$$J(X(\cdot)) = \phi(X(T)) \tag{1.27}$$

(with a smooth function $\phi(X)$), then for the existence of the optimal control, one needs to guarantee the compactness in $\mathbb{R}^n$ of the set of terminal points $\{X(T)\}$ (that is, the attainable set at time $t = T$). On the other hand, the set of admissible (i.e., satisfying (1.26)) controls is not closed in the weak-$*$ topology [204], and consequently the set of terminal points that correspond to various controls is also not closed, though it is bounded. The reason is that if the set of controls is not bounded in norm (in magnitude), then it contains functions that are close to controls of δ-function type. As a result, the set of paths contains functions that are arbitrarily close to discontinuous functions.

To guarantee the existence of the optimal solution, the following extension of the set of possible paths was proposed [218, 328, 329]:

- Equation (1.25) is rewritten in a more general form as a differential equation with control measure $U(t)$, where $U(t)$ is a right-continuous function of bounded variation:

$$dX(t) = A(t)X(t)dt + B(t)dU(t); \tag{1.28}$$

- the constraint (1.26) is given in terms of the variation of the control $U(t)$ as

$$U(0) = 0, \qquad \operatorname*{Var}_{[0,T]} U(t) \leq M < \infty; \tag{1.29}$$

- finally, the original optimal control problem with ordinary control $u(t)$ [172] is generalized by replacing the system (1.25) and the constraint (1.26) with the system (1.28) and the constraint (1.29), respectively, replacing the control $u(t)$ with $U(t)$, and retaing the terminal criterion (1.27).

This new formulation of the problem is a natural extension of the original one, since for any admissible control $u(\cdot)$ obeying (1.29), there exists a corresponding control $U(\cdot)$, defined by

$$U(t) = \int_0^t u(s)ds,$$

such that $dU(t) = u(t)dt$, and for any measurable and integrable control $u(\cdot)$ Eq. (1.25) is equivalent to (1.28). Moreover, since

$$\operatorname*{Var}_{[0,T]} U(t) = \int_0^T \|u(s)\|ds \le M$$

the control $U(\cdot)$ satisfies constraint (1.26).

The optimal control problem with control measures has at least two advantages:

- thanks to Helly's theorem [322, 409] (see Theorem A.5), the set of functions $U(\cdot)$ of uniformly bounded variation is compact in the weak-$*$ topology, i.e., the topology where convergence of functions implies the convergence at all points of continuity. Consequently, the set of terminal points, defined by the formula

$$X(T) = \Phi(T,0)X(0) + \int_0^T \Phi(T,s)B(s)dU(s),$$

  where $X(0)$ is the initial condition and $\Phi(t,s)$ is the fundamental solution of corresponding linear system of differential equations, is a compact subset of $\mathbb{R}^n$ (see also Sect. A.2.3 and Theorem A.18). It follows that there exists the optimal control $U^0(\cdot)$ and the corresponding optimal path $X^0(\cdot)$;
- for any function of bounded variation $U(\cdot)$, there exists a sequence of absolutely continuous functions $U^n(\cdot)$, such that

$$\operatorname*{Var}_{[0,T]} U^n(t) \le \operatorname*{Var}_{[0,T]} U(t) \quad \text{and} \quad \lim_n U^n(t) = U(t)$$

  at all points of continuity and also for $t \ge T$, because $U^0(\cdot)$ can be continued for $t \ge T$ as a right-continuous function and even as a constant one (see Theorem A.6).

We claim that the extended problem has an optimal solution which provides the infimum in the original problem, that is

$$\min_{U(\cdot)} J = \inf_{u(\cdot)} J.$$

Indeed, since $U^n(t)$ converges to $U^0(t)$ at all points of continuity as well as for $t \ge T$, we have

$$\int_0^T f(s)dU^n(s) \to \int_0^T f(s)dU^0(s), \quad n \to \infty$$

for any continuous function $f(t)$; therefore, $X^n(T)$, which is equal to

$$X^n(T) = \Phi(T,0)X(0) + \int_0^T \Phi(T,s)B(s)dU^n(s) \to X^0(T),$$

converges to $X^0(T)$ as $n \to \infty$.

Since each of the functions $U^n(\cdot)$ admits a representation of the form

$$U^n(t) = \int_0^t u^n(s)ds,$$

the function $X^n(t)$ satisfies the equation

$$\dot{X}^n(t) = A(t)X^n(t) + B(t)u^n(t),$$

with controls $u^n(t)$ that meet the constraint

$$\operatorname*{Var}_{[0,T]} U^n(t) = \int_0^T \|u^n(t)\|dt \leq \operatorname*{Var}_{[0,T]} U^0(t) \leq M.$$

The optimal control problem stated in the class of control measures $U(\cdot)$ has the optimal solution $U^0(\cdot)$, and so

$$\min_{U(\cdot)} J = \phi(X^0(T)) = \lim_n \phi(X^n(T)) = \inf_{u(\cdot)} J.$$

The last relation means that the optimal control problem posed in the class of control measures is indeed an extension of the original problem and allows one to circumvent the issue of lack of a solution. At the same time, since the optimal control $U^0(t)$ in the class of extended solutions is a function of bounded variation, possibly discontinuous, the optimal path $X^0(t)$ in the extended problem may be discontinous as well. By the Lebesgue decomposition for functions of bounded variation (Sect. A.1), we can write

$$U^0(t) = (U^0)^c(t) + \sum_{\tau \leq t} \Delta U^0(\tau),$$

where $(U^0)^c(t)$ is the continuous component in the decomposition and the sum runs over the set of all points of discontinuity of the function $U^0(\cdot)$. The corresponding solution $X^0(t)$ has the form

$$X^0(t) = \Phi(t, 0)X(0) + \int_0^t \Phi(t, s)B(s)d(U^0)^c(s) + \sum_{\tau \leq t} \Phi(t, \tau)B(\tau)\Delta U^0(\tau),$$

where the integral over the continuous component of $U^0(t)$ is understood in the sense of Lebesgue-Stieltjes [204].

Thus, the solution of the problem under consideration may contain impulsive control components. These impulses correspond to the points of discontinuity of the control $U^0(t)$ and, correspondingly, of the trajectory $X^0(t)$. The relation between this problem and the purely impulsive control problem for DCS is made clearer by examining the following example.

Consider the DCS system with

$$F(x,t) = A(t)x, \qquad \Psi(x,t,w) = B(t)w.$$

For a given sequence of impulses $\{w_i, \tau_i\}$, the solution of the system (1.1) and (1.2) has the form

$$X(t) = \Phi(t,0)X(0) + \sum_{\tau_i \le t} \Phi(t,\tau_i)B(\tau_i)w_i.$$

If $U(t)$ is the piecewise-constant function $U(t) = \sum_{\tau_i \le t} w_i$, then

$$X(t) = \Phi(t,0)X(0) + \int_0^t \Phi(t,s)B(s)dU(s). \tag{1.30}$$

To solve this problem with the criterion (1.27) and subject to the constraint

$$\sum_{\tau_i \le T} \|w_i\| \le M, \tag{1.31}$$

one resorts to the following analogous extension:

- the system satisfies Eq. (1.28),
- the constraint (1.31) is expressed in terms of the function $U(\cdot)$ as

$$\operatorname*{Var}_{[0,T]} U(t) = \sum_{\tau_i \le T} \|U(\tau_i) - U(\tau_i-)\| = \sum_{\tau_i \le T} \|w_i\| \le M.$$

Thus, the optimal control problem under consideration can be formulated on the set of control measures given by functions $U(t)$ of bounded variation. This novel problem is an extension of the original purely impulsive control problem. The equivalence of the two problems is a consequence of the fact that any function of bounded variation $U(\cdot)$ can be approximated by a sequence of piecewise-constant functions $U^n(\cdot)$ which converges to $U(t)$ at all points of continuity and such that

$$\operatorname*{Var}_{[0,T]} U^n(t) \le \operatorname*{Var}_{[0,T]} U(t) \le M,$$

(see Theorem A.6). Therefore, as follows from the formula (1.30),

$$X^n(T) \to X(T), \quad n \to \infty,$$

and so $\phi(X^n(T)) \to \phi(X(T))$, and finally

$$\inf_{\{w_i,\tau_i\}} J(X(\cdot)) = \min_{U(\cdot)} J(X(\cdot)).$$

The above discussion explains the connection between problems with control measures and those with purely impulsive controls in DCS. Notice that the problems with control measures do sometimes have solutions with measures localized on finite sets. This means that there exist purely impulsive optimal controls [5, 218, 328]; however, this usually concerns linear systems with a convex terminal criterion. One should emphasize that the success of the extension based on replacing ordinary controls with control measures is a consequence of Helly's theorem.

In the nonlinear case, the situation is considerably more complex.

Example 1.5 Consider a system described by the differential equation with scalar measure

$$dX(t) = X(t-)dU(t), \tag{1.32}$$

with initial condition $X(0)$ and the following constraints imposed on the measure $dU(t)$:

$$U(0) = 0, \quad dU(t) \geq 0, \quad U(T) \leq M, \tag{1.33}$$

which correspond to the constraints (1.22). The aim of the control is to minimize the objective function

$$J(X(\cdot)) = \int_0^T |X(t) - X(0)|dt + |X(T) - X(0)e^M| \to \min. \tag{1.34}$$

In this case there exists the minimizing sequence of impulsive controls $\{(\tau_i^n, w_i^n)\}$, $i = 1, \ldots, n$, given by

$$\tau_i^n = T - \frac{n-i+1}{n^2} \quad \text{and} \quad w_i^n = \frac{M}{n} > 0, \quad i = 1, \ldots, n.$$

For this choice of the sequence $\{(\tau_i^n, w_i^n)|]$, we have

$$X^n(t) = X(0) \quad \text{for} \quad t \in [0, T - (1/n))$$

and

$$X^n(T) = X^n\left(T - \frac{1}{n^2}\right) = X(0)\prod_{i=1}^n (1 + w_i) =$$

$$= X(0)\left(1 + \frac{M}{n}\right)^n \to X(0)e^M, \quad n \to \infty,$$

and consequently

$$J(X^n) \to 0 = \inf_{\{w_i, \tau_i\}} J(X), \quad n \to \infty.$$

Despite the fact that X^n converges to the "optimal" solution

$$X(t) = \begin{cases} X(0), & \text{for } t \in [0, T), \\ X(0)e^M, & \text{for } t = T, \end{cases}$$

there is no control measure that realizes such an "optimal path." The reason is that such a measure must vanish on $[0, T)$ and may contain a purely discrete component at $t = T$ only. However, the application of this kind of control gives

$$X(T) = X(0)(1 + \Delta U(T)) \leq X(0)(1 + M),$$

and therefore the equality $X(T) = X(0)e^M$ is impossible, since $1 + M < e^M$ for $M > 0$.

We see that for nonlinear DCS, even when differential equations with measures are used, the optimal control problem may not admit a solution. Moreover, a gap may exists between the values of the optimal criterion in the original problem and in the extended problem. The following example illustrates this last possibility.

Example 1.6 Consider the optimal control problem for a system of the form

$$\begin{cases} dX(t) = X(t)u(t)dt, & X(0) = 1, \\ dU(t) = u(t)dt, & U(0) = 0, \quad u(t) \geq 0, \end{cases}$$

with the criterion

$$J(X(\cdot), U(\cdot)) = \int_0^1 |X(t) - 1|dt + |X(1) - 2| + |U(1) - 1| \to \min.$$

The embedding of this problem in the class of problems with control measures gives the system

$$\begin{cases} dX(t) = X(t-)dU(t), & X(0-) = 1 \\ dU(t) \geq 0, & U(0-) = 0. \end{cases}$$

The *infimum* of the criterion J exists in both problems; however, the minimum of J is really achievable only in the second problem, by means of the control

$$U^*(t) = \begin{cases} 0, & \text{for } t < 1, \\ 1, & \text{for } t = 1, \end{cases} \qquad X^*(t) = \begin{cases} 1, & \text{for } t < 1, \\ 2, & \text{for } t = 1, \end{cases}$$

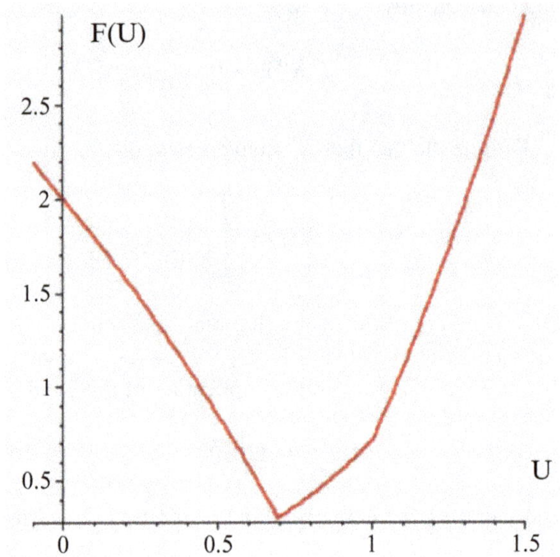

Fig. 1.5 Graph of the function $|e^U - 2| + |U - 1|$ for $U \geq 0$

for which

$$J(X^*(\cdot), U^*(\cdot)) = 0.$$

At the same time, in the first problem we have

$$X(1) = e^{U(1)}$$

and

$$J(X(\cdot), U(\cdot)) \geq |e^{U(1)} - 2| + |U(1) - 1| \geq C > 0$$

(see the Fig. 1.5).

This example shows that there is a real gap between the problem with unbounded controls and the extended problem with control measures. The reason is that for the correct extension of the optimal control problems for DCS, one requires a special type of nonlinear equations with measure based on the approximation of impulse control by ordinary ones.

1.2.3 Approximation of Impulsive Controls by Ordinary Controls

As shown in the above example of the system (1.28) with the constraint (1.29), the extension of the optimal control problem by using control measures is natural in the case where the function $B(t)$ is continuous. However, this extension does not work when $B(t)$ is discontinuous. Indeed, in this case Helly's theorem is not applicable and the sequence of integrals $\int_0^t B(s)dU^n(s)$ does not converge to $\int_0^t B(s)dU(s)$ even if $U^n(t) \to U(t)$ at all points of continuity of $U(t)$. Moreover, in this case there exists an entire set of possible limits, depending on the choice of the sequence $U^n(t)$. This fact is related to the well-known ambiguity concerning the product of a discontinuous function and a generalized function [464–466].

Thus, when $B(t)$ is discontinuous it is impossible to determine the optimal discontinuous trajectory by means of the Eq. (1.28) with some regular measure $U(dt)$. In this case the extension problem may be solved in the class of finitely-additive measures [84, 85, 312]; however, the resulting so-called optimal solution has rather peculiar properties, as displayed by the next example.

Example 1.7 Consider the system described by the equation

$$\dot{X}(t) = b(t)dU(t), \quad X(t), b(t), U(t) \in \mathbb{R}^1, \quad X(0-) = 0, \tag{1.35}$$

with

$$b(t) = \begin{cases} 1, & \text{for } t = \frac{1}{2}, \\ 0, & \text{otherwise}, \end{cases}$$

under the constraints

$$U(0-) = 0, \quad U(1) \leq 1, \quad dU(t) \geq 0,$$

and with the criterion

$$J[X(\cdot)] = X(1) \to \max .$$

The obvious optimal solution in the class of regular and finitely-additive measures is obtained for the choice

$$U^0(t) = \begin{cases} 0, & \text{for } t < \frac{1}{2}, \\ 1, & \text{for } t \geq \frac{1}{2}, \end{cases}$$

for which

$$J[X^0(\cdot)] = \int_0^1 b(t)dU^0(t) = 1.$$

However, for any admissible continuous control $U(t)$ the corresponding integral vanishes:

$$\int_0^1 b(t)dU(t) = 0.$$

This means that the supremum of the criterion on the set of continuous controls also vanishes, so a real gap exists between the supremum achievable on the set of continuous (physically realizable) controls, on one hand, and the one achievable on the set of controls measures in the extended problem, on the other.

The importance of well-posedness (correctness) of the transition to control measures was emphasized by R. Rishel in [377]. This work treats the optimal control problem for the nonlinear system described by an equation of the form

$$dX(t) = F(X(t), u(t), t)dt + B(u(t), t)\, d\mu(t) \tag{1.36}$$

with controls of two types: $u(\cdot)$, with values in some compact set, and a scalar non-negative regular measure $\mu(dt)$, subject to the constraint

$$\mu\{[0, T]\} \leq M < \infty. \tag{1.37}$$

This problem is regarded as an extension of the problem with two standard controls $\{u(t), w(t)\}$, where the $u(t)$ is the same as above, but $w(t)$ is subject to the constraint

$$w(t) \geq 0, \qquad \int_0^T w(t)dt \leq M$$

instead of (1.37). Then Eq. (1.36) must be replaced by the ordinary differential equation

$$\dot{X}(t) = F(X(t), u(t), t) + B(u(t), t)w(t).$$

To establish the correspondence between the original problem with standard controls and the problem with control measures $B(u, t)$ a special requirement, namely, "the constancy condition," to [377], must be fulfilled. This condition ensures the well-posedness of the transition to control measures and allows to approximate the optimal control measure by ordinary controls. Indeed, as demonstrated by Example 1.7, if the control $u(t)$ is discontinuous, one should not expect that the minimum values in the problem with standard controls and the problem with control measures will coincide. Moreover, if $u(t)$ is merely Lebesgue measurable, the integral of $B(t, u(t))$ with respect to an arbitrary measure $\mu(dt)$ may not exist. This issue may be settled if the measure μ and the corresponding control u are constructed simultaneously during the same approximation procedure, which enables one to prove the existence of the optimal control and to obtain the necessary optimality conditions [266, 269, 441].

The problem of correct transition from standard to impulse controls is the key issue in the optimal control of DCS. We note that a DCS that satisfies Eq. (1.1) with the intermediate conditions (1.2) can be described by a generalized differential equation of the form

$$\dot{X}(t) = F(X(t), t) + \sum_{\tau_i \le t} \Psi(X(\tau_i-), \tau_i, w_i)\delta(t - \tau_i), \tag{1.38}$$

where $\delta(\cdot)$ is the standard Dirac δ-function, characterized by the property that

$$\int_{-\infty}^{\infty} f(s)\delta(s - \tau)ds = f(\tau)$$

for any continuous function $f(\cdot)$.

In engineering applications, the δ-function is just a useful tool for describing inputs of very large magnitude which cause very fast, "almost jump-like" changes of the trajectory of a system. Hence, when using a well-posed transition one can expect that if we replace δ-functions with some continuous "almost impulse" functions we must reproduce a similar almost jumping behavior of the path described by Eq. (1.38). Thus, the *well-posedness question* may be formulated as follows:

- Can Eq. (1.38) be regarded as the result of approximating impulse inputs by ordinary inputs?
- if so, does the result of the approximation depend on the approximating sequence? (If it does not, the DCS under consideration will be said to have the *well-posedness property*);
- if a system does not have the well-posedness property, how to describe the results of all possible approximations of impulsive controls?

The well-posedness problem for differential equations with generalized coefficients was probably first formulated by Kurzwail [229, 230], who noted the fact that it is non-trivial. This problem remained open over for many years, and, as indicated in [365], the solutions of such equations cannot be defined unambiguously without taking into account the limit procedure that generates the equations in question. At the same time, if generalized coefficients describe the impulsive inputs in control problems, then settling the well-posedness problem is very important. Indeed, well-posedness ensures that many useful properties hold, for instance:

- robustness with respect to small variation of control measures ;
- the above property ensures, in its turn, the continuity of the mapping from the set of control measures to the set of discontinuous trajectories, which enables one to establish the existence of the optimal controls for a sufficiently wide class of problems;
- moreover, such continuity allows us to obtain the necessary optimality conditions, derived by means of control variations, and even the maximum principle.

We now provide an example that exhibits some specific aspects of the well-posedness problem.

Fig. 1.6 Approximation of $\delta(t-\tau)$ by a sequence $\phi^n(t,\tau)$

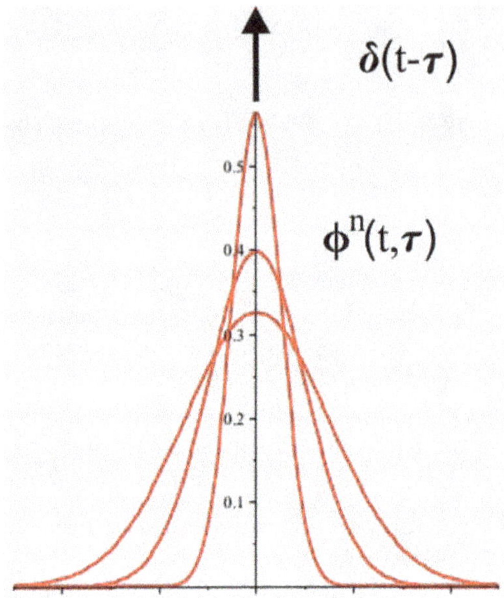

Example 1.8 Consider a DCS described by the equation

$$\dot{x}(t) = x(t-)w\delta(t-\tau), \quad x(0) = x_0 \in \mathbb{R}^1, \tag{1.39}$$

where $\tau > 0$. The equation

$$\dot{x}^n(t) = x^n(t-)w\phi^n(t,\tau), \quad x(0) = x_0. \tag{1.40}$$

where $\phi^n(t,\tau)$ is a sequence of continuous functions that approximates the δ-function in (1.39), may be regarded as a continuous approximation of (1.39).

Remark 1.2 The sequence $\{\phi^n(t,\tau)\}$ is said to approximate the δ-function concentrated at τ in the weak-$*$ topology if $\phi^n(t,\tau) \geq 0$ and

$$\int_{-\infty}^{\infty} f(t)\phi^n(t,\tau)dt \to f(\tau), \quad n \to \infty$$

for any continuous function $f(t)$.

A typical approximating sequence of functions $\phi^n(s,\tau)$ is depicted in Fig. 1.6 Equation (1.39) has the solution

$$x(t) = \begin{cases} x_0, & \text{for } t < \tau, \\ x_0(1+w), & \text{for } t \geq \tau. \end{cases} \tag{1.41}$$

The sequence of solutions corresponding to the approximations $\phi^n(t, \tau)$ is given by

$$x^n(t) = x_0 \exp\left\{\int_0^t w\,\phi^n(s, \tau)ds\right\}$$

and converges to the function

$$x(t) = \begin{cases} x_0, & \text{for } t < \tau, \\ x_0 \exp\{w\}, & \text{for } t \geq \tau, \end{cases} \tag{1.42}$$

which however differs from (1.41).

This example shows that what one should think as the "natural" approximation does not give the expected result, even in very simple cases. Incidentally, the function

$$\Psi(x, w) = x[\exp\{w\} - 1]$$

obtained as a result of the approximation does not depend on the choice of the approximating sequence $\phi^n(t, \tau)$.

The analogous approximation in the problem of impulse control of the motion of a rocket yields the well-known Tsiolkovsky formula (see Sect. 1.1.3). However, in the case of multi-dimensional impulse control a well-posed approximation may not exist (see Example 1.9).

These examples and further analysis show (see Sect. 1.3) that in the case of scalar inputs one can guarantee well-posedness by choosing the "appropriate" passage to the limit that defines the "right" expression of the function Ψ. Even in this case, a difference exists between the "natural" DCS model and the model obtained via approximation. This difference may lead to a special type of lack of robustness related to small variations of impulse controls.

1.2.4 *Small Variations of Impulse Controls: Example of Unstable DCS*

Example 1.9 Let us consider make the previous example (see Sect. 1.1.3) more complicated. We will now look at a rocket vehicle with two identical jets, with thrusters directed along the same axis, but in opposite directions. In this case the flow velocities are equal to

$$C = C_1 = -C_2.$$

Assume that the jets are used for accelleration and braking, respectively. Then the equations of motion take on the form

$$\begin{cases} M(t)\dfrac{dV(t)}{dt} = -C\dfrac{dM_1(t)}{dt} + C\dfrac{dM_2(t)}{dt}, \\ \dfrac{dM(t)}{dt} = -\left(\dfrac{dM_(t)}{dt} + \dfrac{dM_(t)}{dt}\right), \end{cases}$$

where $M(t)$ is the rocket's mass, $M_1(t)$ and $M_2(t)$ are the masses of working fluid spent until time t by the first and the second jet, respectively, $V(t)$ is the rocket's velocity, and C is the speed of the reactive stream.

We assume that the jets operate synchronously, so that

$$\frac{dM_1(t)}{dt} = \frac{dM_2(t)}{dt},$$

and therefore

$$\frac{dV(t)}{dt} = 0.$$

In this case, assuming a synchronous impulsive switching-on of the two jets at some instant $t = \tau$, with ejected masses $\Delta M_1(\tau) = \Delta M_2(\tau)$, we obtain the equalities

$$\Delta M(\tau) = -(\Delta M_1(\tau) + \Delta M_2(\tau)), \quad \Delta V(\tau) = 0. \tag{1.43}$$

This idealized model is however not realistic, since in practice an exact synchronization of the operation of the two jets is not possible.

If there is a small time interval between the operation of the two jets, so that $\tau_1 \neq \tau_2$, then the dynamic equations after two impulse controls $(\Delta M_1, \tau_1)$ and $(\Delta M_2, \tau_2)$ have the form

$$\begin{aligned} \begin{pmatrix} \dot{V}(t) \\ \dot{M}(t) \end{pmatrix} &= \begin{pmatrix} C \ln \dfrac{M(\tau_1-)}{M(\tau_1-) - \Delta M_1} \\ -\Delta M_1 \end{pmatrix} \delta(t-\tau_1) - \\ &- \begin{pmatrix} C \ln \dfrac{M(\tau_2-)}{M(\tau_2-) - \Delta M_2} \\ -\Delta M_2 \end{pmatrix} \delta(t-\tau_2). \end{aligned} \tag{1.44}$$

Let us try to obtain (1.43) by assuming that

$$\Delta M_1 = \Delta M_2 = \Delta M, \quad \tau_1, \tau_2 \to \tau$$

However, there is no limit for the velocity increment. Indeed, if during the passage to limit the inequality $\tau_1 < \tau_2$, we have that

$$\Delta V^1(\tau) = C \ln \frac{M(\tau-)[M(\tau-) - 2\Delta M(\tau)]}{(M(\tau-) - \Delta M(\tau))^2} < 0.$$

If, conversely, $\tau_1 > \tau_2$, then

$$\Delta V^2(\tau) = -\Delta V^1(\tau) = C \ln \frac{(M(\tau-) - \Delta M(\tau))^2}{M(\tau-)[M(\tau-) - 2\Delta M(\tau)]} > 0.$$

Moreover, if there is no strict ordering of τ_1, τ_2, then the resulting velocity increment $\Delta V(\tau)$ may take an arbitrary value in the interval

$$\Delta V^1(\tau) \leq \Delta V(\tau) \leq \Delta V^2(\tau).$$

As further considerations make clear, for the system at hand there is no unique description of the limit motion under the impulsive actions: this motion depends on the way one passess to the limit. Thus, the system does not have the well-posedness property.

Moreover, one can see the connection between lack of well-posedness and lack of robustness with respect to small desynchronization. This type of lack of robustness is usual for nonlinear impulsive systems. However, there exists a special class of impulse systems that are stable with respect to small variations of the impulsive controls [261], namely, the so-called robust DCS.

1.3 Stability of Controlled DCS with Respect to Small Variations of Impulse Controls: Robustness

1.3.1 Robustness of DCS: Frobenius Condition

In the previous section we saw that defining the concept of a solution for DCS is a nontrivial task from the point of view of the existence of the optimal control, as well as from the point of view of the unambiguity of the representation of the system's response to impulse actions. Both issues are related to well-posedness in the definition of an impulse control (see Sect. 1.2.3). The next example shows how lack of well-posedness arises in a general case.

Example 1.10 Consider an impulse control $\{\tau, w\}$ and a sequence of impulse controls $\{(\tau_1^n, w_1^n), (\tau_2^n, w_2^n)\}$, such that $\tau_1^n, \tau_2^n \to \tau$, $w_1^n \to w_1'$, $w_2^n \to w_2'$ as $n \to \infty$, where $w_1' + w_2' = w$.

Remark 1.3 This sequence approximates the impulse control $\{\tau, w\}$ in the sense that

$$\lim_n \int_{-\infty}^{\infty} f(t)\big[w_1^n \delta(t-\tau_1^n) + w_2^n \delta(t-\tau_2^n)\big]dt = \lim_n \big[f(\tau_1^n)w_1^n + f(\tau_2^n)w_2^n\big] =$$

$$= \int_{-\infty}^{\infty} f(t) w \delta(t-\tau)dt = f(\tau)w$$

for any continuous function $f(t)$.

In other words, this sequence represents a decomposition of the original impulse control into two impulse control actions with almost equal total amplitude and very close moments of application. It is natural to expect that the response to the original impulse control and the response to the approximating sequence will be close. However, this is not true even in rather simple cases.

The DCS in Example 1.2 under the control $\{\tau, w\}$ satisfies the equation

$$\dot{x}(t) = x(t-) w \delta(t-\tau), \quad x(0) = x_0 \in \mathbb{R}^1, \tag{1.45}$$

which has the solution

$$x(t) = \begin{cases} x_0, & \text{for } t < \tau, \\ x_0(1+w), & \text{for } t \geq \tau. \end{cases} \tag{1.46}$$

However, the sequence of solutions corresponding to the chosen approximation of this impulse control does not converge to (1.46), whatever the meaning of convergence. Indeed, the solution corresponding to the impulse control $\{(\tau_1^n, w_1^n), (\tau_2^n, w_2^n)\}$ has a form

$$x^n(t) = \begin{cases} x_0, & \text{for } t < \tau_1^n, \\ x_0(1+w_1^n), & \text{for } t \in [\tau_1^n, \tau_2^n), \\ x_0(1+w_1^n)(1+w_2^n), & \text{for } t \geq \tau_2^n, \end{cases}$$

and converges to the function

$$x(t) = \begin{cases} x_0, & \text{for } t < \tau, \\ x_0(1+w_1')(1+w_2'), & \text{for } t \geq \tau, \end{cases}$$

which differs from (1.46), since

$$(1+w_1')(1+w_2') = 1 + w_1' + w_2' + w_1' w_2' = 1 + w + w_1' w_2'.$$

This example shows that despite the fact that the solution of Eq. (1.45) is well defined, it is not robust with respect to the decomposition of the single-impulse control into two sequences of impulses. This is typical for *non-robust* DCS with impulse controls. At the same time, Example 1.8 and relation (1.42) shows that the solution of the dynamic equation of the DCS, obtained as a result of approximating the δ-function in the right-hand side, namely

$$\dot{x}(t) = x(t-)[\exp\{w\} - 1]\delta(t - \tau), \quad x(0) = x_0 \in \mathbb{R}^1, \tag{1.47}$$

is robust with respect to the decomposition of the single-impulse control into two impulses thanks to the equality

$$\exp\{w_1\}\exp\{w_2\} = \exp\{w_1 + w_2\}.$$

Next, we show that in the general case the stability condition has the form

$$\Psi(x, t, w_1 + w_2) = \Psi(x, t, w_1) + \Psi(x + \Psi(x, t, w_1), t, w_2), \tag{1.48}$$

for any $x \in \mathbb{R}^n$, $t \in [0, T]$ and $w_1, w_2 \in W$; it expresses a necessary and sufficient condition for robustness with respect to small variations of the impulse controls (see Theorem 2.2).

For scalar impulse controls this robustness property is directly connected with the correctness of the approximation of the impulsive control; however, in the multi-dimensional case this is no longer true. The analytical condition for the well-posed approximability of the impulse controls (*robustness*) for general systems of the type

$$\dot{X}(t) = F(X(t), t) + B(X(t), t)u(t) \tag{1.49}$$

is the so-called Frobenius condition [79]

$$\sum_{k=1}^{n} \frac{\partial b_{ij}(X, t)}{\partial x_k} b_{km}(X, t) = \sum_{k=1}^{n} \frac{\partial b_{im}(X, t)}{\partial x_k} b_{kj}(X, t), \tag{1.50}$$

where b_{ij} are the components of the matrix function B and x_i are the components of a vector X. In Sect. 2.2 we will show that (1.50) is equivalent to the condition (1.48) provided the matrix function $B(x, t)$ satisfies the equality

$$B(x, t) = \frac{\partial}{\partial w} \Psi(x, t, w)\Big|_{w=0}. \tag{1.51}$$

These conditions were first obtained in [258–262] as robustness (stability) conditions of solutions of nonlinear differential equations with measures with respect to small variations of the measure Later, the Frobenius condition was derived in [398–400] as a necessary condition for the correct definition of the product of a generalized function and a discontinuous function in a controlled dynamical system with impulsive controls. This condition appeared also in works related to impulse-trajectory extension of optimal control problems [122], in works related to *vibrostability* of solutions of differential equations [216, 217, 333], and in [168–170] as a condition allowing to express solutions of differential equations with multi-dimensional inputs by means of integrals. Thus, the Frobenius condition specifies an important class of *robust discrete-continuous systems* (DCS) and plays a key role in the general theory of DCS.

We note that the assimilation of the well-posedness conditions was a highly important step in the creation of the optimal control theory for DCS with impulse controls: it is precisely these conditions that allowed to establish a complete such theory for the class of robust DCS. Using these conditions it became possible:

- to introduce and describe the class of generalized (discontinuous) solutions, which also includes the impulse sliding modes, and to derive for them nonlinear differential equations with measure [64, 262, 398, 399];
- to prove the existence of solutions for a wide class of optimal control problems with various non-standard constraints [119, 159, 262, 333, 399–401, 464];
- to obtain first-order optimality conditions in the form of the maximum principle for optimal control problems with phase and/or terminal constraints [64, 120, 205, 263, 264, 333, 411, 464];
- to obtain stronger optimality conditions in the form of a variational maximum principle [120, 123, 126, 128];
- to prove the sufficiency of the maximum principle as the optimality condition in the class of optimal control problems with a convex structure [264, 269, 290];
- to derive the dynamic programming equation for optimal control problems with unbounded controls [111, 315] and for certain differential games [375];
- finally, to establish second-order optimality conditions and investigate in detail the types of extrema in optimization problems with impulse controls [110, 118, 120, 127, 205, 206].

We should emphasize again that, without ignoring its importance, the class of robust DCS is characterized (limited) by the conditions (1.48) and (1.50).

However, there exists numerous problems of practical importance where such conditions are not satisfied. Hence, to develop further the theory of nonlinear DCS theory, one needs to circumvent somehow the restrictions imposed by the Frobenius condition. This is necessary for the completeness of the mathematical theory, but even more for the solution of practical problems where well-posedness may fail.

Remark 1.4 A very important consequence of rejecting the Frobenius condition is the ambiguous response of a system to impulsive control, which makes the way in which the impulse action is approximated an additional component of the control. In other words, the approximation recipe must be included in the optimization scheme.

A typical system where the Frobenius conditions fail is given by Example 1.9, where different approximations of the impulsive controls yield different motions. In what follows we will consider some concrete examples of such non-robust systems.

1.3.2 Mechanical Systems with Impacts: Additional Examples of Non-robustness

The dynamics of mechanical systems with impacts is of great interest in research. On the one hand, it provides a very simple class of phenomena where solutions that are well known from the school physics, suddenly become very complicated and intricate with small changes of the conditions of the problems. On the other hand, the mechanics of systems with impacts has numerous engineering applications, in areas such as vibrotechnique [17], robotics [155, 253, 254, 289], or micro- and nano-mechanics, etc. [166, 238]. Real mechanical shock is a rather complicated process, which depends on properties and shape of the surface of the bodies involved, external conditions, and other factors. It is the more so tempting to try describing the impact in simple terms, for example, in the form of a relation that connects the bodies' velocities before and after impact. In the relevant literature, such a relation is often referred to as the *restitution law* [70]. This law may be derived with the aid of various hypotheses characterizing the collision process, for example, whether the impact is elastic or anelastic, or by introducing the restitution coefficient for the normal velocity component when *the Newton's hypothesis* is in force [215]. However, in order to account not only for the nonuniqueness of the shock model , but also for the possibility that following the shock the bodies may undergo for a time a coupled motion before they separate, a more detailed description of the phenomenon is required. Thus, the shock itself is described as a phase of fast motion during which the impulse force produces a jump-like change of velocities. Numerous works concerned with mathematical models of the impact phenomena appeared recently. A short list includes [18, 54, 81, 143, 179–185, 194, 199, 213, 308, 310, 311, 336, 346–353, 362, 363, 370, 384, 391–393, 453].

It is clear that the description of shocks and jump-like changes in velocities in terms of some fast motion is very close to the concept of the impulse control. There are numerous works devoted to control in mechanical systems with velocity discontinuities that are anayzed from the point of view of impulse control [58, 59]. There are also recent works using the idea of transformation of some phase coordinates of a mechanical system into controls [60–62]. This approach models the discontinuities of the velocities as jumps of certain control variables and is very close to the methodology suggested by V. I. Gurman and V. A. Dykhta [118, 158, 160].

The papers [57, 66] provide examples from impact mechanics where the idea of *singular space-time transformation* is used. This transformation may be considered as a further development of the discontinuous time transformation method. This approach allow one to obtain a correct description of a single shock; however, the case of multiple impacts and the possibility of accumulation points of the impacts and of "sticking" cause serious difficulties. The investigation of such phenomena became very popular recently in relation to robotics, precision mechanics and control of motion. Many authors use the concept of unilateral constraint as a supplementary control tool which allows to ensure controllability and stability

[69, 71, 72, 86, 256, 413, 429]. However, one can go even farther by regarding the impact phase as a controllable process. We do not consider here the position and velocity control of the obstacle's surface. This aspect does not concern the impact itself, but rather the external conditions, since for the short duration of the impact these factors remain unchanged. When we talk about control in the impact phase we mean control actions that occur on the same time scale as the shock itself, that is, the control action is assumed to be fast enough. A typical example can be observed in a ping-pong game, when a good player is able to turn the racket during the brief time of contact with the ball, giving the ball additional spin (this type of impact is called *top-spin*). Another example is either the selective capture or, on the contrary, repulsion of metallic objects by an electromagnet, which needs to be switched on or off when the object touches its surface. In such considerations, the control in mechanical systems with unilateral constraints is similar to impulse control. One of the main difficulties in the analysis of control problems in mechanics is the absence of robustness. To show the typical issues that arrise in this connection, consider the following classical example.

Example 1.11 Suppose a ball of unit mass moves along the vertical axis due to the force of gravity until it hits (partially elastictically) the horizontal surface, for example, a tennis racket which moves vertically with some velocity. Since during the impact the ball's velocity changes almost instantaneously, we consider its motion under the action of the impulsive force arising during the impact and causing the jump-like change of the ball's velocity. Let us describe the appropriate model. Let $y(t)$ and $x(t)$ be the coordinates of the ball and of the surface of the racket, respectively.

The free motion of the ball is described by its coordinate-velocity pair $(y(t), v(t))$, which satisfies the system of equations

$$\dot{v}(t) = -g, \quad \dot{y}(t) = v(t).$$

Similarly, the motion of the racket in terms its coordinate-velocity pair $(x(t), w(t))$ is governed by the equations

$$M\dot{w}(t) = F(t), \quad \dot{x}(t) = w(t),$$

where g is the gravitational acceleration, $M \gg 1$ is the mass of the racket, and $F(t)$ is the external force, subject to the constraints

$$|F(t)| \leq F_0 < \infty.$$

Suppose that $y(0) > x(0)$, so that the ball is above the racket. If at some instant $\tau > 0$ we have $y(\tau) = x(\tau)$, then adopting Newton's law of partial restitution of the normal component of the velocity we have

$$v(\tau) = -v(\tau-)k + w(\tau)(1 + k), \tag{1.52}$$

where $0 \leq k < 1$ is the restitution coefficient [70]. Since $M \gg 1$, one can neglect the change in the racket's velocity. So we have

$$\Delta v(\tau) = v(\tau) - v(\tau-) = (w(\tau) - v(\tau-))(1+k),$$

and therefore, the motion of the ball is governed by the equations

$$\begin{cases} \dot{v}(t) = -g + \sum_{\tau \leq t} (w(\tau) - v(\tau-))(1+k)\delta(t-\tau), \\ \dot{y}(t) = v(t), \end{cases} \tag{1.53}$$

where summation is over all the times τ at which the racket hits the ball during the time interval $[0, t]$.

The optimal control problem may be formulated as follows: choose a control force $F(t)$ which ensures that a specific control objective is achieved, for instance, the minimization of $|v(T)|$ in the problem of effective braking (dampening), or vice versa, the maximization of $v(T)$ in the acceleration problem, etc.

Example 1.12 Consider case when $x(0) = 0$, $w(0) = 0$, $F(t) = 0$, $y(0) = h$, $v(0) = 0$. The first shock occurs at time $t_1 = \sqrt{2h/g}$, and the after-shock velocity of the ball equals $v(t_1) = k\sqrt{2gh}$. The next shock occurs at the time

$$t_2 = t_1 + 2k\sqrt{\frac{2h}{g}}.$$

The time interval between n-th and $(n+1)$-th shocks is

$$t_{n+1} - t_n = 2k^n\sqrt{\frac{2h}{g}},$$

and the velocity satisfies

$$v(t_n) = k^n\sqrt{2gh} \to 0, \quad n \to \infty.$$

Moreover,

$$t_{n+1} = t_1 + 2t_1 \sum_{l=1}^{n} k^l \to t_1 + \frac{2t_1 k}{1-k} := T^*.$$

Thus, for $t \geq T^*$ the ball attains the equilibrium with $y(t) = 0$, $v(t) = 0$, as shown in Fig. 1.7. However, this state cannot be described by the Eq. (1.53), even if we put $x(t) = y(t) = 0$.

Fig. 1.7 Bouncing ball tending to equilibrium

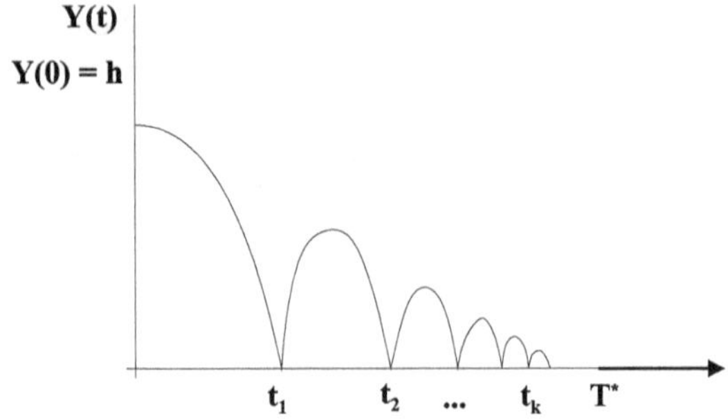

The issue here is that for the equilibrium one needs to introduce an additional, gravity-compensating force. But the existence of such force does not follow from the Eq. (1.53). The situation could be made even more complicated if the equality $x(t) = y(t) \neq 0$ is allowed to hold on some non-zero interval. This situations requires an additional non-impulsive force and the corresponding extension of the Eq. (1.53). These equations can usually be derived from a complete set of equations describing the collision dynamics for a finite value of the rigidity by subsequently letting the rigidity go to infinity (*"penalization" method*) [346, 349]. In this case, the equations of the limit motion can be derived in the form of complementarity equations or the form of differential equations with measures, localized on the boundary [72, 73].

However, in these equations one needs to determine simultaneously the motion and a corresponding measure, which sometimes results in a lack of uniqueness of the solution. Known uniqueness results require the analyticity in time of the external control action [18, 58], but this is very strong requirement. Moreover, lack of uniqueness of the solution is almost generic for such differential equations with measures [18, 56].

Some researchers believe that standard methods of classical mechanics known for systems with unilateral constraints are applicable also for controllable systems [73]. This is of course true for problems in which the shocks are separated in time [71, 72]. However, some concrete examples show that complementarity gives results that are far from realistic [37].

Example 1.13 Consider the following system of equations that describes the interaction of a body with a surface with large rigidity coefficient μ; here (x_p, x_v) are the position and the velocity of the body and the elastic obstacle is bounded by the surface of equation $x_p = 0$:

$$\begin{cases} \dot{x}_p^\mu(t) = x_v^\mu(t), \\ \dot{x}_v^\mu(t) = -f(t) - \mu x_p^\mu(t)\mathbf{I}\{x_p^\mu(t) \leq 0\}, \end{cases} \tag{1.54}$$

where $\mathbf{I}\{A\}$ denotes the indicator function of a set A. The external force $f(t)$ is assumed to satisfy the constraint $0 \leq f(t) \leq F < \infty$.

In [346] and [353] such systems are used as an approximation of the following complementarity model of a bouncing ball:

$$\begin{cases} \dot{\bar{x}}_p(t) = \bar{x}_v(t), \\ \dot{\bar{x}}_v(t) = -f(t), \end{cases} \tag{1.55}$$

in the domain $\bar{x}_p(t) > 0$, with the restitution law assumed to hold for the absolutely elastic shock occurring at time τ:

$$\bar{x}_v(\tau+) = -\bar{x}_v(\tau-), \tag{1.56}$$

where $\bar{x}_p(\tau) = 0$, $\bar{x}_v(\tau) < 0$.

To determine the limit motion of the system (1.54) when $\mu \to \infty$, note that at the time τ where $x_p^\mu(\tau) = 0$, $x_v^\mu(\tau) < 0$, the system (1.54) gives the following relation for single shock: if $f(t) = 0$ for $t \geq \tau$, then

$$x_p^\mu\left(\tau + \frac{\pi}{\mu^{1/2}}\right) = 0, \quad x_v^\mu\left(\tau + \frac{\pi}{\mu^{1/2}}\right) = -x_v^\mu(\tau). \tag{1.57}$$

If $f(t) \neq 0$ and is bounded, then relations (1.57) are valid up to order $\mu^{-1/2}$ [346]. Therefore, in the case of a bounded external force, when $\mu \to \infty$ the motion of the system converges to an instantaneous jump at the time of the absolutely elastic shock. Hence, we can say that the correct complementarity equations are the result of passing to the limit when the rigidity goes to infinity.

Let us change this example by assuming that $x_p^\mu(\tau) = 0$, $x_v^\mu(\tau) = 0$ and an impulsive control is applied during the contact with the obstacle. Now the external force is not bounded, but obeys the integral constraint

$$f_{\Delta t}(t) \geq 0, \quad \int_\tau^{\tau+\Delta t} f_{\Delta t}(t)dt \leq M < \infty.$$

Let us fit this force with the elastic deformation, so that

$$f_{\Delta t}(t) = \begin{cases} M/\Delta t, & \text{for } \tau \leq t \leq \tau + \Delta t, \\ 0, & \text{otherwise.} \end{cases}$$

This gives the solution of the system (1.54), which at the first time of exit t^* from the region bounded by $x_p = 0$ satisfies the relations

$$x_p^\mu(t^*) = 0, \quad x_v^\mu(t^*) = \frac{M}{(t^* - \tau)\mu^{1/2}}\sqrt{2(1 - \cos\left((t^* - \tau)\mu^{1/2}\right))},$$

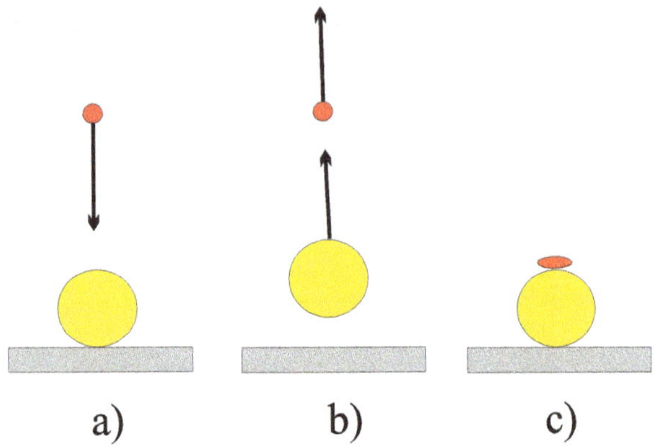

Fig. 1.8 Various motions corresponding to an impulsive external action

where t^* obeys the constraints

$$\frac{\pi}{2\mu^{1/2}} \leq t^* - \tau \leq \frac{\pi}{\mu^{1/2}}.$$

Then by passing to the limit $\mu \to \infty,\ \Delta t \to 0$, if this limit exists, we get

$$\bar{x}_v(0+) = \lim_{\Delta t \to 0, \mu \to \infty} x_v^{\mu}(t^*).$$

Two possibilities arise:

- if the external action is like a hammer hit or a bullet impact, as shown in Fig. 1.8a, and we assume that $\lim_{\Delta t \to 0, \mu \to \infty} \mu^{1/2}\Delta t < \infty$, then the system will execute various motions with a positive velocity $\bar{x}_v(t) \in [0, M]$, as shown on Fig. 1.8b;
- if, however, $\lim_{\Delta t \to 0, \mu \to \infty} \mu^{1/2}\Delta t = \infty$, then in the limit we get only the zero solution $\bar{x}_v(t) = 0$, as shown in Fig. 1.8c.

Let us emphasize that the complementarity equations corresponding to the two cases are the same; they have the form

$$\begin{cases} \dot{\bar{x}}_p(t) = \bar{x}_v(t), \\ d\bar{x}_v(t) = -M\delta(t)dt + dv(t), \end{cases} \tag{1.58}$$

and together with the restitution law (1.56) admit only the zero solution, since in the case of a bounded external force $\bar{x}_v(0+) = \bar{x}_v(0-) = 0$, with corresponding measure $dv(t) = M\delta(t)dt \geq 0$. This is an obvious conclusion when the complementarity approach takes into account the geometry of the obstacle and the kinematics in the form of the restitution law (1.56), but does not take into account the impulsive character of the external forces in the shock phase. Our approach based on a space-time transformation [37] allows to go further and to "uncover" the singularity at the impact, and thus to adequately determine the motion corresponding to large values of the rigidity. It is worth emphasizing that the space-time transformation approach, which will be discussed in Sect. 5.4, not only enables us to solve the control problems in the impact phase, but also provides an approach to classical mechanics problems such as the Painlevé paradox, where the complementarity approach gives a solution only in cases corresponding to a small friction coefficient [143]. In these problems the external force is the impulsive friction force, which is correlated with the duration of the contact.

In Sect. 8.2.2, we shall present a formalization method for such problems. This method is a generalization of the discontinuous time change, namely, the *singular space-space transformation*. General control problems for hybrid systems and dynamical systems with unilateral constraints were treated based on this method in [33–37, 283–285].

So, we will again focus the reader's attention on the method of discontinuous time change again. This approach, based on a singular space-time transformation in the vicinity of shock instances, is a constructive and in some sense universal tool for solving various problems with impulse controls and discontinuous trajectories.

1.3.3 Method of Discontinuous Time Change in Impulse Control Problems

As already mentioned, in this monograph the main emphasis is on the systematic use of singular space-time transformations. In a few words, the idea of this method is to stretch very small intervals of impulse action into intervals whose lengths are proportional to the impulse magnitudes. This procedure does automatically eable a transition to impulse controls after solving a certain auxiliary control problem with bounded controls and leads to substantial progress in the theory of impulse control. The idea of applying a time transformation for nonlinear control problems was first used by J. Warga [445–447]. These works dealt with systems with an arbitrary dependence on an unbounded control. Warga's approach became an effective tool for the study of optimization problems and provided the general scheme of the method of discontinuous time transformation.

For robust DCS the applicability of such a transformation is ensured by the relation (1.48), which says that the function

$$\Phi(x, t, w, 1) = x + \Psi(x, t, w)$$

is the solution of the ordinary differential equation

$$\frac{d}{ds}\Phi(x, t, w, s) = B(\Phi(x, t, w, s), t)w, \tag{1.59}$$

where $B(x, t) = \frac{\partial}{\partial w}\Psi(x, t, w)\Big|_{w=0}$, with the initial condition $\Phi(x, t, 0) = x$. Hence, one can introduce the new controllable system

$$\dot{y}(s) = \alpha(s)F(y(s), \eta(s)) + (1 - \alpha(s))B(y(s), \eta(s))l(s), \quad y(0) = x, \tag{1.60}$$

with the "new time" $\eta(s)$ that obeys the equation

$$\dot{\eta}(s) = \alpha(s), \quad \eta(0) = 0,$$

and with the two new bounded controls $\{\alpha(s), l(s)\}$, defined as follows:

$$\alpha(s) = \begin{cases} 1, & \text{if } s < \tau, \quad s > \tau + \|w\|, \\ 0, & \text{otherwise,} \end{cases}$$

and

$$l(s) = \begin{cases} 0, & \text{if } s < \tau, \quad s > \tau + \|w\|, \\ w/\|w\|, & \text{if } s \in [\tau, \tau + \|w\|]. \end{cases}$$

Let us compare the solutions of the Eq. (1.60) and

$$\dot{x}(t) = F(x(t), t) + \Psi(x(\tau-), \tau, w)\delta(t - \tau), \quad x(0) = x.$$

It is readily seen that on the intervals and $s \in [0, \tau)$ and $t \in [0, \tau)$ these equations for $y(s)$ and for $x(t)$ are identical, and with the same initial conditions; moreover, $\eta(s) = s$. Consequently,

$$x(t) = y(t) \quad \text{and} \quad x(\tau-) = y(\tau).$$

Equation (1.59) and the definition of the control $\alpha(s)$ imply that

$$y(\tau + \|w\|) = y(\tau) + \Psi(y(\tau), \tau, w) = x(\tau-) + \Psi(x(\tau-), \tau, w) = x(\tau)$$

and

$$\eta(\tau) = \eta(\tau + \|w\|).$$

Furthermore, for $t \geq \tau$ and correspondingly for $s \geq \tau + \|w\|$ we again have identical equations for $x(t)$ and $y(s)$, with the same initial conditions, but posed at different initial points, that is, at τ and $\tau + \|w\|$ respectively. Therefore, $x(t) = y(t + \|w\|)$. Thus, we arrived at two systems of equations, and the correspondence between their solutions can be established with the aid of a discontinuous

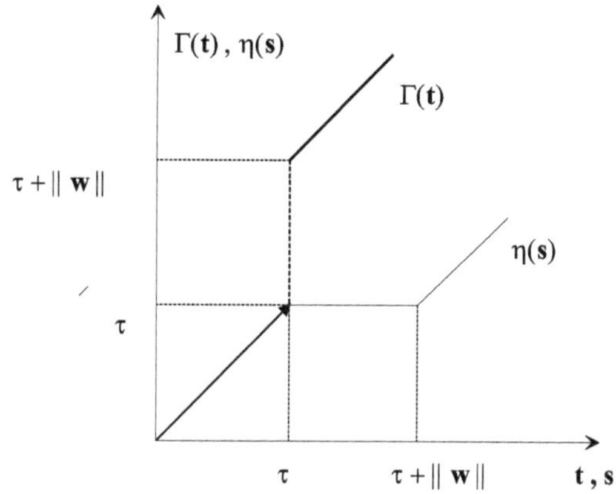

Fig. 1.9 Change of time: $\eta(s)$ and the inverse function $\Gamma(t)$

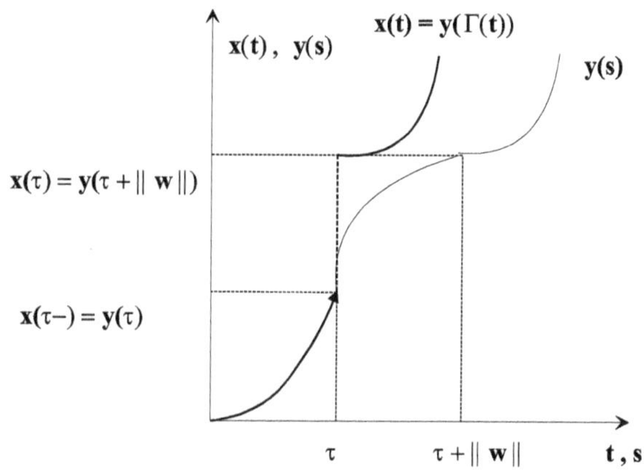

Fig. 1.10 Change of time: $x(t) = y(\Gamma(t))$

time transformation. To define this time transformation more precisely, we introduce the function

$$\Gamma(t) = t + \|w\|\mathbf{I}\{t \geq \tau\},$$

where recall that $\mathbf{I}\{A\}$ denotes the indicator function of the set A. The function $\Gamma(t)$ is the inverse to $\eta(s)$ (see Fig. 1.9), and for all t we have that $x(t) = y(\Gamma(t))$ takes place (see Fig. 1.10).

The idea of usings the method of discontinuous time transformation for the analysis of DCS was a principal motivation for this monograph. We tried to include results that demonstrate the most effective applications of this method. We also tried to give a description of the method in the most general form and to cast it as a powerful tool in the analysis of various problems concerned with systems with discontinuous paths.

Most of the results concerned with the theory of optimal control in nonlinear impulse systems were published mainly in periodicals. Among the books discussing themes that closely parallel the ones in the present monograph we emphasise those of Yu. Orlov [333] and of S. Zavalishin and A. Sesekin [465]. The first of these two books is based on the methodology of vibrosolutions introduced by M. Krasnosel′skii and A. Pokrovskii [216, 217], generally for another class of problems, and which, broadly speaking, leads to some incorrect conclusions in the analysis of optimal control problems with phase constraints. The reason for the latter is that the equation for the conjugate variable in the case of constraints is not vibro-well-posed, even though it is so for the main variable. The second book mentioned above considers mainly problems related to differential equations with impulse inputs. The closest to our book is the monograph [128], which considers optimality conditions in various impulse control problems. Some new results concerning optimal impulsive actions in feedback form are presented in the works of A. Kurzhanski and A. Daryin [103–105, 226–228]. A short presentation of the impulse control problems based on the usage of the discontinuous time change may be found also in the recent monograph of A. Arutyunov, D. Karamzin and F.-L. Pereira [11], which basically repeats known results from our earlied book [292]. The present monograph continues the series of texts where we analyze impulse controls in DCS [292, 294, 298] and in mechanical systems with shocks. We designed it as a textbook, so we try to explain the emergence of our ideas from scratch, rather than pulled out instantly like a rabbit from a magician's hat. We hope that this book is the most complete state-of-art presentation in this area of optimal control and provides a comprehensive completion of the existing monographs. Moreover, we are focusing on the method of singular space-time transformations, which allows us to discover the peculiar aspects of and develop effective tools for optimal control issues.

In the present text, the class of nonlinear systems with impulse controls described by differential equations with measure is considered for the first time in full detail in book form. As applications, we consider two important classes of problems: control by discrete-continuous observations and control of hybrid systems with unilateral constraints.

Here we do not touch upon other results obtained with the aid of the discontinuous time change method in dynamical programming, differential games, and stochastic systems, providing instead relevant references (see, for example, [78, 111, 114, 300, 315, 375]).

1.4 Structure of the Book

The text consists of eight chapters and the Appendix.

Chapter 1 is the Introduction.

In Chap. 2, we consider DCS with impulse controls. We give the general description of a nonlinear system with impulse control satisfying the robustness condition (self-consistency in terms of [262]), which is the synonym of *robustness*. We present analytic robustness conditions and define the discontinuous time change for robust DCS. The use of this method permits the representation of *generalized* or discontinuous solutions such as *impulsive sliding modes*, and subsequently the derivation of differential equations with a measure for this kind of solutions.

In Chap. 3, we continue the investigation of robust DCS and we solve the impulse control problem with restrictions on the number of admissible impulses and their duty cycle. In this case, the method of discontinuous time change allows us to derive the optimality conditions in the maximum principle form.

In Chap. 4, we present a general concept of the generalized (discontinuous) solutions in optimal control problems. We describe the discontinuous time-change method for general nonlinear control systems with ordinary and impulse controls, for both robust and non-robust DCS with impulse control. We review systems with linear dependence on an unbounded control and derive a representation of the jumps of trajectories for this class of systems in full detail. This chapter establishes a bridge between the "naive" description of DCS and systems with impulsive controls.

Chapter 5 is devoted to various optimal control problems for systems admitting generalized (discontinuous) solutions. The main purpose is to define a class of trajectories (generally speaking, discontinuous ones) wide enough to include the optimal solution. The main result of this chapter shows that the class of *generalized solutions* introduced for the description of discontinuous trajectories does indeed contain the optimal path. In other words, this class serves well as a class of *generalized curves* [456] (by analogy with the same concept in classical optimal control theory). Later, we consider the *generalized solutions* in problems with state constraints. Two concepts of admissible solution are introduced, namely those of *strong* solution and *weak* solution.

Strong solutions satisfy the state constraints "strictly," that is, constraint violations during the action of the impulse control are not allowed. On the contrary, the concept of weak solution admits constraint violations, but on the infinitesimally brief time interval when the impulse control acts. These two concepts lead to different types of trajectories, which require different tools of investigation. It is shown that by using the discontinuous time change it is possible to formulate two different equivalent auxiliary problems with ordinary controls. With the aid of the solutions of these auxiliary problems one can easily derive the solutions for the original problems. Moreover, in Chap. 5 we consider the optimal control problem for a hybrid system with possible state discontinuities. Here we again express a jump in the state by means of the discontinuous time change and also for the case of unilateral constraints inherent to mechanical systems.

In Chap. 6, we derive a series of necessary and sufficient optimality conditions for generalized solutions. We use the following general methodology, first suggested in [263]:

- the original optimal control problem with unbounded controls and which admits generalized (discontinuous) solutions is reformulated in the equivalent form of an auxiliary optimization problem with bounded controls; the correspondence between these problems is established with the aid of the discontinuous time change;

- in the new auxiliary problem the necessary optimality conditions are formulated in the form of the maximum principle;
- finally, these conditions are transformed into optimality conditions for the original problem by means of the inverse time transformation.

This methodology was used for the derivation of sufficient optimality conditions in the maximum principle form [264, 269] and for the derivation of dynamic programming equations [315]. We use it to obtain optimality conditions in optimal control problems with state constraints, in hybrid systems, and in linear-convex problems with state constraints. In linear and linear-convex problems this generalized maximum principle gives necessary and sufficient optimality conditions. Moreover, we use the presented methodology to derive necessary optimality conditions in hybrid systems with unilateral constraints.

In Chap. 7, the general discrete-continuous observation control problem is formulated. In this class of problems, a single discrete observation plays the role of an impulse control and induces a discontinuity of the estimate and its covariance. We prove existence theorems of the optimal control and give a representation of the solution in the form of a nonlinear differential equation with a measure. We establish convexity properties of this problem with constraints and then derive necessary and sufficient optimality conditions. We give examples of various observation control problems and show how the methods of impulse control help to solve various problems, for example, determining of power and composition of the sounding impulse in radar systems.

Chapter 8 is devoted to a novel application of the impulse control, namely to the control of mechanical systems in the shock phase. This phase is considered as a motion parametrized by certain parameters μ and in the limit $\mu \to \infty$ converges to a discontinuous function. Various applications of this approach have been considered, in particular, to the long-standing P. Painlevé problem, where using classical mechanics in the analysis of shock interaction with dry friction leads sometimes to the impossibility of continuation of the solution. We consider systems with a series of shocks and with controls during the impact phase of motion.

In the Appendix, we give some necessary concepts from measure theory and analysis of functions of a real variable, and discuss elements of the theory of linear and nonlinear differential equations with measures: existence and uniqueness of solutions, dependence of solutions on initial conditions and parameters. We believe that the choice of material in the Appendix makes the monograph self-contained. Some fundamental results that are necessary for the understanding of various parts of the text are given directly in the corresponding chapters. The book is supplied with a Subject Index.

Chapter 2
Discrete-Continuous Systems with Impulse Control

2.1 Definition of the Notions of Impulse Control and the Solution of a DCS

Let us consider the evolution of a dynamical system whose state is described by the variable $X(t) \in \mathbb{R}^n$ defined on some interval $[0, T]$. Suppose that $X(t)$ satisfies the differential equation

$$\dot{X}(t) = F(X(t), t) \tag{2.1}$$

with a given initial condition $X(0) = x_0 \in \mathbb{R}^n$ and intermediate conditions

$$X(\tau_i) = X(\tau_i -) + \Psi(X(\tau_i -), \tau_i, w_i), \tag{2.2}$$

which are defined for some sequence of instants $\{\tau_i,\ i = 1, \ldots, N \leq \infty\}$ such that

$$0 \leq \tau_1 < \tau_2 < \cdots < \tau_i < \cdots < \tau_N \leq T. \tag{2.3}$$

In Eq. (2.2), $X(\tau_i -) = \lim_{t \uparrow \tau_i} X(t)$ and $w_i \in W \subset \mathbb{R}^m$ are the intensities of the impulse actions.

Thus, the state of the system changes continuously in the half-intervals $[0, \tau_1), \ldots, [\tau_{i-1}, \tau_i), \ldots, [\tau_N, T]$, and it undergoes a sudden change at each of the instants τ_i. According to the Eq. (2.2), the magnitudes of these changes depend on the state immediately preceding the jump $X(\tau_i -)$, the time τ_i of the jump, and the intensity w_i of the impulse action.

Definition 2.1 The sequence of pairs $\mathscr{I} = \{(\tau_i, w_i),\ i = 1, \ldots, N\}$, where τ_i satisfy (2.3) and $N \leq \infty$, will be referred to as the *impulse control*.

B. Miller, E. Rubinovich, *Optimization of Dynamical Systems with Impulse Controls and Shocks*, Systems & Control: Foundations & Applications,
https://doi.org/10.1007/978-3-031-64124-4_2

Definition 2.2 A right-continuous function $X(t)$ that satisfies the integral equation

$$X(t) = x_0 + \int_0^t F(X(s), s)ds + \sum_{\tau_i \le t} \Psi(X(\tau_i-), \tau_i, w_i) \tag{2.4}$$

is called a *solution of the DCS with impulse control.*

The typical behaviour of such a solution was depicted above in Fig. 1.1.

It readily seen that the Eq. (2.4) incorporates the differential equation (2.1) and the intermediate conditions (2.2). In what follows we will use also the Eq. (2.4) in its differential form, i.e., in the form of a generalized differential equation

$$\dot{X}(t) = F(X(t), t) + \sum_{\tau_i \le t} \Psi(X(\tau_i-), \tau_i, w_i)\delta(t - \tau_i) \tag{2.5}$$

with the initial condition $X(0) = x_0$, where $\delta(t - \tau)$ is the standard delta-function [204], i.e., $\int_{-\infty}^{+\infty} f(t)\delta(t - \tau)dt = f(\tau)$ for any continuous function $f(t)$.

Remark 2.1 These two forms of dynamic equations for DCS are equivalent because, formally,

$$\int_{-\infty}^{t} \delta(s - \tau)ds = \mathbf{I}\{t : t \ge \tau\} = \begin{cases} 1, & \text{if } t \ge \tau \\ 0, & \text{if } t < \tau, \end{cases}$$

and

$$\sum_{\tau_i \le t} \Psi(X(\tau_i-), \tau_i, w_i) = \sum_{\tau_i \le T} \Psi(X(\tau_i-), \tau_i, w_i)\mathbf{I}\{t : t \ge \tau_i\}.$$

A formal differentiation of Eq. (2.4) yields Eq. (2.5).

To ensure the existence and uniqueness of the solution, a number of assumptions are needed.

Assumption 2.1 *Let the functions $F(X, t)$, $\Psi(X, t, w)$ and the set W satisfy the following conditions:*

1. *$F(X, t)$ and $\Psi(X, t, w)$ are continuous with respect to all the variables.*
2. *There exist positive constants L_1 and K_1 such that, for any $x, y \in \mathbb{R}^n$ and any $t \in [0, T]$,*

$$\|F(x, t) - F(y, t)\| \le L_1\|x - y\| \quad \text{and} \quad \|F(x, t)\| \le K_1(\|x\| + 1).$$

3. *There exist positive constants L_2 and K_2 such that, for any $x, y \in \mathbb{R}^n$, any $w \in W$ and any $t \in [0, T]$,*

$$\|\Psi(x, t.w) - \Psi(y, t, w)\| \le L_2\|x - y\|\|w\| \text{ and } \|\Psi(x, t.w)\| \le K_2(\|x\| + 1)\|w\|.$$

4. *The set* $W \subset \mathbb{R}^m$ *contains the point* 0 *and*

$$\Psi(x, t, 0) = 0$$

for any $x \in \mathbb{R}^n$ *and any* $t \in [0, T]$.

Now one can state an existence and uniqueness theorem.

Theorem 2.1 ([262]) *Suppose Assumption 2.1 holds. Then for any initial condition* $X(0) = x_0 \in \mathbb{R}^n$ *and any impulse control* $\mathcal{I} = \{(\tau_i, w_i),\ i = 1, \dots, N\}$ *satisfying the condition*

$$\sum_{i=1}^{N} \|w_i\| = \sum_{\tau_i \le T} \|w_i\| < \infty, \tag{2.6}$$

Eq. (2.4) or (2.5)) have a unique solution $X(t)$ *in the class of right-continuous functions of bounded variation.*

Remark 2.2 This theorem is a particular case of the general existence and uniqueness theorem (see Theorem A.24) for nonlinear differential equations with a measure. Nevertheless, it's proof is interesting and very useful for the understanding and illustration of some features of DCS.

Proof The assertion of the theorem is obvious if $N < \infty$, since the solution of (2.4) satisfies in each of the half-intervals $[\tau_i, \tau_{i+1})$ the ordinary differential equation $\dot{X}(t) = F(X(t), t)$, with the initial condition $X(\tau_i) = X(\tau_i-)+\Psi(X(\tau_i-), \tau_i, w_i)$. Therefore, the Lipschitz property of F (see Condition 2 of Assumption 1, and [96, 365]) guarantees that the solution exists and is unique in each of the half-intervals $[\tau_i, \tau_{i+1})$, and can be continued to the whole interval $[0, T]$ by concatenation.

If $N = \infty$, then the increasing and bounded sequence $\{\tau_i\}$ has a limit

$$\lim_{i \to \infty} \tau_i = T_1 \le T,$$

and to establish the existence of the solution, one needs to continue $X(t)$ to the point T_1 and then in half-interval $(T_1, T]$.

Each of the half-intervals $[0, T_1 - \varepsilon)$, where $\varepsilon > 0$, contains only finitely many points τ_i; hence, in each of these intervals, the solution of Eq. (2.4) exists and is unique. For $t \in [T_1 - \varepsilon, T_1]$, the solution of (2.4) must satisfy the integral equation

$$X(t) = X(T_1 - \varepsilon) + \int_{T_1-\varepsilon}^{t} F(X(s), s)ds + \sum_{T_1-\varepsilon<\tau_i \le t} \Psi(X(\tau_i-), \tau_i, w_i).$$

Thus, to prove the theorem it suffices to establish the existence and uniquenessof a solution in some interval $[T_1 - \varepsilon, T_1]$ with a given initial condition y_0 at the point $t = T_1 - \varepsilon$.

Fix $\varepsilon > 0$ and $y_0 \in \mathbb{R}^n$. Let us introduce a subspace $BV\{[T_1 - \varepsilon, T_1], y_0\}$ of the space of functions of bounded variations defined in $[T_1 - \varepsilon, T_1]$ with the initial condition y_0.

Definition 2.3 A function $y(\cdot)$ belongs to $BV\{[T_1 - \varepsilon, T_1], y_0\}$ if

$$y(T_1 - \varepsilon) = y_0 \quad \text{and} \quad \operatorname*{Var}_{[T_1-\varepsilon,T_1]} y(s) < \infty.$$

The subspace $BV\{[T_1 - \varepsilon, T_1], y_0\}$ thus defined is a complete metric space with the metric [198] (see also Lemma A.1)

$$\rho[\, y_1(\cdot),\ y_2(\cdot)\,] = \operatorname*{Var}_{[T_1-\varepsilon,T_1]} [y_1(s) - y_2(s)].$$

We will prove that (for an appropriate choice of $\varepsilon > 0$) the maping $\mathscr{A}_\varepsilon$ acting as

$$\mathscr{A}_\varepsilon[x(\cdot)] = y_0 + \int_{T_1-\varepsilon}^{t} F(x(s), s)ds + \sum_{T_1-\varepsilon<\tau_i\leq t} \Psi(x(\tau_i-), \tau_i, w_i)$$

is a contraction in $BV\{[T_1 - \varepsilon, T_1], y_0\}$ for any y_0. Then, by the Contraction Mapping Theorem [204], Eq. (2.4) admits one, and exactly one solution in the interval $[0, T_1]$. Thus, $X(t)$ with a given initial condition at the point $T_1 - \varepsilon$ can be continued to the point T_1, and so the solution of (2.4) will be defined and unique in the whole interval $[0, T]$.

First let us show that, for any $\varepsilon > 0$, $\mathscr{A}_\varepsilon[x(\cdot)]$ maps the space $BV\{[T_1-\varepsilon, T_1], y_0\}$ into itself. Let $y(t) = \mathscr{A}_\varepsilon[x(\cdot)](t)$, where $x(\cdot) \in BV\{[T_1 - \varepsilon, T_1], y_0\}$. Then

$$y(T_1 - \varepsilon) = y_0$$

and

$$\operatorname*{Var}_{[T_1-\varepsilon,T_1]} y(s) \leq \|y_0\| + \int_{T_1-\varepsilon}^{T_1} \|F(x(t), t)\|dt + \sum_{\tau_i\in(T_1-\varepsilon,T_1]} \|\Psi(x(\tau_i-), \tau_i, w_i)\|.$$

Conditions 2 and 3 of Assumption 2.1, the boundedness of $x(t)$ and the relation (2.6) imply that

$$\begin{aligned}\operatorname*{Var}_{[T_1-\varepsilon,T_1]} y(s) \leq \|y_0\| + \int_{T_1-\varepsilon}^{T_1} K_1(\|x(t)\| + 1)dt+ \\ + \sum_{\tau_i\in(T_1-\varepsilon,T_1]} K_2\|w_i\|(\|x(\tau_i)\| + 1) < \infty,\end{aligned}$$

Therefore, $y(t) = \mathscr{A}_\varepsilon[x(\cdot)](t) \in BV\{[T_1 - \varepsilon, T_1], y_0\}$.

Now let

$$y_1(t) = \mathscr{A}_\varepsilon[x_1(\cdot)](t) \quad \text{and} \quad y_2(t) = \mathscr{A}_\varepsilon[x_2(\cdot)](t),$$

where $x_1(\cdot), x_2(\cdot) \in BV\{[T_1 - \varepsilon, T_1], y_0\}$. Then, by using again conditions 2 and 3 of Assumption 2.1, we have

$$\begin{aligned}
&\rho[\, y_1(\cdot),\ y_2(\cdot)\,] = \\
&= \operatorname*{Var}_{[T_1-\varepsilon,T_1]} [y_1(s) - y_2(s)] \leq \int_{T_1-\varepsilon}^{T_1} \|F(x_1(t), t) - F(x_2(t), t)\| dt + \\
&\quad + \sum_{\tau_i \in (T_1-\varepsilon, T_1]} \|\Psi(x_1(\tau_i-), \tau_1, w_i) - \Psi(x_2(\tau_i-), \tau_1, w_i)\| \leq \\
&\leq \int_{T_1-\varepsilon}^{T_1} L_1\|x_1(t) - x_2(t)\| dt + \sum_{\tau_i \in [T_1-\varepsilon, T_1]} L_2\|x_1(\tau_i-) - x_2(\tau_i-)\| \, \|w_i)\| \leq \\
&\leq \left\{ L_1\varepsilon + L_2 \sum_{\tau_i \in (T_1-\varepsilon, T_1]} \|w_i\| \right\} \sup_{[T_1-\varepsilon, T_1]} \|x_1(t) - x_2(t)\| \leq \\
&\leq \left\{ L_1\varepsilon + L_2 \sum_{\tau_i \in (T_1-\varepsilon, T_1]} \|w_i\| \right\} \operatorname*{Var}_{[T_1-\varepsilon, T_1]} \|x_1(t) - x_2(t)\| \leq \\
&\leq \left\{ L_1\varepsilon + L_2 \sum_{\tau_i \in (T_1-\varepsilon, T_1]} \|w_i\| \right\} \rho[\, x_1(\cdot),\ x_2(\cdot)\,].
\end{aligned}$$

By (2.6), one can choose $\varepsilon > 0$ such that

$$L_1\varepsilon + L_2 \sum_{\tau_i \in (T_1-\varepsilon, T_1]} \|w_i\| < 1.$$

Therefore, the mapping $\mathscr{A}_\varepsilon[x(\cdot)]$ is a contraction.

We conclude that Eq. (2.4) has a unique solution in the class of functions of bounded variation and, as it follows from the integral equation itself, this solution is right-continuous. □

Next we consider the robustness properties of solutions to DCS.

2.2 Stability (Robustness) Conditions for Systems with Impulse Control

2.2.1 *Necessary Stability Conditions*

In the preceding section, we introduced the concept of DCS with impulse control. However, the example from Sect. 1.1.3 and Example 1.9 show that this concept requires further investigation to determine its properties concerning stability of solutions with respect to small variations of the impulse controls as well as of the behavior of solutions when the impulse controls are approximated by ordinary controls. To treat the stability aspect, one should define the concept of the *nearness* in the set of impulse controls in such a manner that it will be compatible with its intuitive meaning. A routine way to define nearness is to introduce a metric on the space of impulse controls, and in this way equip that space with a topological structure of complete metric space [204]. However, in our considerations, when the instants of the impulse actions are not fixed and their number is unbounded, we need to introduce a much weaker topology than the topology of complete metric space.

For any impulse control $\mathcal{I} = \{(\tau_i, w_i),\ i = 1, \dots, N\}$, define a function

$$U(t) = \sum_{\tau_i \le t} w_i,$$

which measures the total control resource spent before time t. If the impulse control satisfies the condition

$$\sum_{\tau_i \le T} \|w_i\| < \infty,$$

or if the number of impulses N is finite, then $U(t)$ is a function of bounded variation and also right-continuous.

Definition 2.4 We say that the sequence of impulse controls $\mathcal{I}^n = \{(\tau_i^n, w_i^n),\ i = 1, \dots, N^n\}$ *converges to the impulse control* $\mathcal{I} = \{(\tau_i, w_i),\ i = 1, \dots, N\}$, if the corresponding sequence of functions $U^n(t)$ has uniformly bounded variation, i.e.,

$$\sup_n \operatorname*{Var}_{[0,T]} U^n(t) = \sup_n \sum_{\tau_i \le T} \|w_i^n\| < \infty,$$

and $U^n(t)$ converges to $U(t)$ on $t \in [0, T]$ at all points of continuity.

Remark 2.3 Definition 2.4 corresponds to convergence in the weak-$*$ topology on the space of functions of bounded variation [204]. Indeed, if $U^n(\cdot)$ converges to $U(\cdot)$ according to this definition, then, by Helly's theorem (see Sect. A.1, Theorem A.12),

$$\lim_n \int_0^T f(t)dU^n(t) = \int_0^T f(t)dU(t)$$

for any continuous function $f(t)$.

Let $X(t)$ be a solution of Eq. (2.4) with the initial condition x_0 and corresponding to the impulse control $\mathcal{I}$, and let $X^n(t)$ be a sequence of solutions of (2.4) with the same initial condition and corresponding to some sequence of impulse controls $\mathscr{I}^n$ that converges to $\mathscr{I}$.

Definition 2.5 A DCS with impulse control is said to be *stable with respect to variations of the impulse control*, or *robust in the sense of* [411], if for any initial condition x_0 and any impulse control $\mathscr{I}$, the convergence of the sequence $\mathscr{I}^n$ to $\mathscr{I}$ implies the convergence of $X^n(t)$ to $X(t)$ at all points of continuity.

The next result provides a necessary condition for stability.

Theorem 2.2 ([261]) *If a DCS with impulse control is stable with respect to variations of the impulse control, then for any $x \in \mathbb{R}^n$, any $t \in [0, T]$, and any pair $w_1, w_2 \in W$,*

$$\Psi(x, t, w_1 + w_2) = \Psi(x, t, w_1) + \Psi(x + \Psi(x, t, w_1), t, w_2). \tag{2.7}$$

Proof Arguing by contradiction, suppose that the relation (2.7) is violated at some point, i.e., there exist $\bar{x} \in \mathbb{R}^n$, $\bar{t} \in [0, T]$ and $\bar{w_1}, \bar{w_2} \in W$ such that

$$\Psi(\bar{x}, \bar{t}, \bar{w_1} + \bar{w_2}) \neq \Psi(\bar{x}, \bar{t}, \bar{w_1}) + \Psi(\bar{x} + \Psi(\bar{x}, \bar{t}, \bar{w_1}), \bar{t}, \bar{w_2}). \tag{2.8}$$

Choose the initial condition $x_0 \in \mathbb{R}^n$ such that the solution of the equation $\dot{X}(t) = F(X(t), t)$ with $X(0) = x_0$ will satisfy $X(\bar{t}) = \bar{x}$. (Thanks to the Lipschitz properties of $F(x, t)$ (see Assumption 2.1), any solution of (2.1) can be continuously extended in both directions of time [365]).

Now define the impulse control $\mathscr{I} = \{(\bar{t}, \bar{w_1} + \bar{w_2})\}$ and the sequence of impulse controls $\mathscr{I}^n = \left\{\left(\bar{t} - (1/n), \bar{w_1}\right), (\bar{t}, \bar{w_2})\right\}$, which converges to $\mathscr{I}$ according to Definition 2.4.

Let $\bar{X}(t)$ and $\bar{X}^n(t)$ be the solutions of Eq. (2.4) with the initial condition x_0 and corresponding to the impulse controls $\mathscr{I}$ and $\mathscr{I}^n$, respectively.

By the definition of what a solution to Eq. (2.4) means,

$$\bar{X}(\bar{t}) = \bar{x} + \Psi(\bar{x}, \bar{t}, \bar{w_1} + \bar{w_2})$$

and

$$\bar{X}^n(\bar{t}) = \bar{X}^n(\bar{t}-) + \Psi(\bar{X}^n(\bar{t}-), \bar{t}, \bar{w_2});$$

here $\bar{X}^n(\bar{t}-)$ is the solution of the equation

$$\dot{\bar{X}}^n(t) = F(\bar{X}^n(t), t) \tag{2.9}$$

with the initial condition

$$\bar{X}^n\left(\bar{t} - \frac{1}{n}\right) = \bar{X}^n\left(\left[\bar{t} - \frac{1}{n}\right] -\right) + \Psi\left(\bar{X}^n\left(\left[\bar{t} - \frac{1}{n}\right] -\right), \bar{t}, \bar{w_1}\right),$$

where $\bar{X}^n\left(\left[\bar{t}-\frac{1}{n}\right]-\right)$ is the solution of Eq. (2.9) with the initial condition $\bar{X}^n(0) = x_0$. However,

$$\bar{X}^n\left(\left[\bar{t}-\frac{1}{n}\right]-\right) = \bar{X}\left(\left[\bar{t}-\frac{1}{n}\right]-\right),$$

and consequently

$$\bar{X}^n\left(\left[\bar{t}-\frac{1}{n}\right]-\right) \to \bar{x}, \quad \text{if} \quad n \to \infty.$$

Therefore, by the continuity of $\Psi(x, t, w)$ and the equicontinuity property of solutions of (2.9) [365], the sequence $\bar{X}^n(\bar{t})$ satisfies the relation

$$\lim_n \bar{X}^n(\bar{t}) = \bar{x} + \Psi(\bar{x}, \bar{t}, \bar{w}_1) + \Psi(\bar{x} + \Psi(\bar{x}, \bar{t}, \bar{w}_1), \bar{t}, \bar{w}_2).$$

Hence, since we assumed that (2.8) holds, $\bar{X}^n(\bar{t})$ does not converge to $\bar{X}(\bar{t})$, which contradicts the convergence of $\bar{X}^n(t)$ to $\bar{X}(t)$ for $t > \bar{t}$. This in turn contradicts the stability assumption of the the theorem, and so the proof is complete. □

Another stability condition is obtained by looking at the approximation of impulse actions. Consider a system described by an ordinary differential equation of the form

$$\dot{X}(t) = F(X(t), t) + B(X(t), t)w(t) \tag{2.10}$$

where $F(x, t)$ is the same as in (2.4), and $B(x, t)$ is a continuous $(n \times m)$-matrix valued function which satisfies the Lipschitz condition

$$\|B(x, t) - B(y, t)\| \leq L\|x - y\| \quad \text{for all } x, y \in \mathbb{R}^n,$$

with some constant $L > 0$. The input $w(t) \in \mathbb{R}^m$ is a control. We consider this system as a *continuous approximation* of the DCS with impulse control, namely, we consider controls of impulse type that approximate generalized controls of δ-function type. Our aim is to define the response of the DCS to this kind of impulse actions.

So, let us approximate the impulse action $w(t) = w_0\delta(t - \tau)$ by a sequence of ordinary controls $\{w^n(\cdot)\}$. For example, we can take

$$w^n(t) = w_0\phi^n(t, \tau), \qquad n = 1, 2, \ldots,$$

where the sequence $\{\phi^n(t, \tau)\}$ satisfies the conditions

$$\phi^n(t,\tau) \geq 0, \quad \phi^n(t,\tau) \to 0 \quad \text{almost everywhere on} \quad [0,T],$$

$$\lim_n \int_0^{\tau-\varepsilon} \phi^n(t,\tau)dt = 0, \qquad \lim_n \int_{\tau+\varepsilon}^{T} \phi^n(t,\tau)dt = 0, \tag{2.11}$$

$$\lim_n \int_{\tau-\varepsilon}^{\tau+\varepsilon} \phi^n(t,\tau)dt = 1,$$

for any $\varepsilon > 0$.

The typical behavior of such sequence is displayed in Fig. 1.6.

Example 2.1 As examples of such a function $\phi^n(t,\tau)$ one can take

$$\phi^n(t,\tau) = \begin{cases} n, & \text{if } \tau - \dfrac{1}{2n} \leq t \leq \tau + \dfrac{1}{2n}, \\ 0, & \text{otherwise,} \end{cases}$$

or

$$\phi^n(t,\tau) = \frac{n}{\sqrt{2\pi}} \exp\left\{ -\frac{n^2(t-\tau)^2}{2} \right\}.$$

Let $\{X^n(t)\}$ be the solution of Eq. (2.10) with some initial condition x_0, and corresponding to the control $w^n(t) = w_0\phi^n(t,\tau)$. To describe the limit of the sequence $\{X^n(t)\}$, we need the following definition.

Definition 2.6 Let $\Phi(x,t,w,s)$ be a solution of the differential equation

$$\dot{Y}(s) = B(Y(s),t)w$$

with the initial condition $Y(0) = x \in \mathbb{R}^n$. Define the *function* Ψ in Eq. (2.4) by the formula

$$\Psi(x,t,w) = \Phi(x,t,w,1) - x.$$

Thanks to the continuity and the Lipschitz property of $B(x,t)$ (see Assumption 2.1), the functions $\Phi(x,t,w,s)$ and $\Psi(x,t,w)$ are continuous in all the variables [96, 365].

Consider the solution $\bar{X}(t)$ of Eq. (2.4) with the function Ψ introduced above, the impulse control $\mathscr{I} = \{(\tau, w_0)\}$, and the initial condition x_0. That is,

$$\dot{\bar{X}}(t) = F(\bar{X}(t),t) + \Psi(\bar{X}(\tau-),\tau,w_0)\delta(t-\tau) \quad \text{and} \quad \bar{X}(0) = X^n(0) = x_0.$$

The next theorem shows that $\bar{X}(t)$ is the limit of the sequence $\{X^n(t)\}$ at all points of continuity.

Theorem 2.3 *The sequence* $\{X^n(t)\}$ *converges to* $\bar{X}(t)$ *everywhere on* $[0, T]$, *except possibly at the point* τ.

To proof the theorem, we need several lemmas.

Lemma 2.1 *The sequence* $\{X^n(t)\}$ *is uniformly bounded on* $[0, T]$, *i.e., there exists a constant* $C > 0$, *such that* $\|X^n(t)\| \leq C$ *for all* n *and all* $t \in [0, T]$.

Proof By the continuity and the Lipschitz property of the functions F and B, there exists a constant $K > 0$ such that

$$\|F(x, t)\| + \|B(x, t)\| \leq K(1 + \|x\|)$$

for any $x \in \mathbb{R}^n$ and all $t \in [0, T]$. Consequently, $X^n(t)$ satisfies the inequality

$$\|X^n(t)\| \leq \|x_0\| + \int_0^t K(1 + \|X^n(s))\|(1 + \|w_0\|\phi^n(s, \tau))ds \leq$$

$$\leq \|x_0\| + \int_0^T K(1 + \|w_0\|\phi^n(s, \tau))ds + K\int_0^t \|X^n(s)\|(1 + \|w_0\|\phi^n(s, \tau))ds$$

where, by (2.11),

$$\sup_n \int_0^T K(1 + \|w_0\|\phi^n(s, \tau))ds \leq K_1 < \infty.$$

By the Gronwall-Bellman inequality [237],

$$\|X^n(t)\| \leq [\|x_0\| + K_1]\exp\{K_1\},$$

which completes the proof. □

Now choose some $\varepsilon > 0$ and consider the functions $\{X^n(t)\}$ and $\bar{X}(t)$ on the interval $[0, \tau - \varepsilon]$.

Lemma 2.2 *The sequence* $\{X^n(t)\}$ *converges to* $\bar{X}(t)$ *uniformly on* $[0, \tau - \varepsilon]$, *for any* $0 < \varepsilon < \tau$.

Proof By the Lipschitz property of the function F,

$$\|X^n(t) - \bar{X}(t)\| \leq \int_0^t L_1\|X^n(s) - \bar{X}(s)\| + \int_0^{\tau-\varepsilon} \|B(X^n(s), s)\| \, \|w_0\|\phi^n(s, \tau)ds.$$

By Lemma 2.1 and (2.11), we have

$$C(n) = \int_0^{\tau-\varepsilon} \|B(X^n(s), s)\| \, \|w_0\|\phi^n(s, \tau)ds \to 0, \quad \text{if} \quad n \to \infty.$$

Therefore, by the Gronwall-Bellman inequality [237], we conclude that

$$\|X^n(t) - \bar{X}(t)\| \leq C(n) \exp\{L_1[\tau - \varepsilon]\}$$

for $t \in [0, \tau - \varepsilon]$, which completes the proof. □

Proof of Theorem 2.3 Fix some $\varepsilon > 0$ and introduce the change of variables

$$t^{\varepsilon}(s) = \tau - \varepsilon + 2\varepsilon s,$$

which maps the interval [0, 1] to $[\tau - \varepsilon, \tau + \varepsilon]$. Set $Y^{n,\varepsilon}(s) = X^n(t^{\varepsilon}(s))$. Then $Y^{n,\varepsilon}(s)$ satisfies on [0, 1] the integral equation

$$\begin{aligned} Y^{n,\varepsilon}(s) = X^n(\tau - \varepsilon) + 2\varepsilon \int_0^s F(Y^{n,\varepsilon}(u), t^{\varepsilon}(u))du + \\ + 2\varepsilon \int_0^s B(Y^{n,\varepsilon}(u), t^{\varepsilon}(u))w_0\phi^n(t^{\varepsilon}(u), \tau)du. \end{aligned} \tag{2.12}$$

Denote by $\bar{Y}^{n,\varepsilon}(s)$ the solution of the equation

$$\bar{Y}^{n,\varepsilon}(s) = \bar{X}(\tau - \varepsilon) + 2\varepsilon \int_0^s B(\bar{Y}^{n,\varepsilon}(u), \tau)w_0\phi^n(t^{\varepsilon}(u), \tau)du. \tag{2.13}$$

Subtracting (2.12) from (2.13) we obtain the estimate

$$\begin{aligned} &\|Y^{n,\varepsilon}(s) - \bar{Y}^{n,\varepsilon}(s)\| \leq \|X^n(\tau - \varepsilon) - \bar{X}(\tau - \varepsilon)\| + \\ &+ 2\varepsilon \int_0^s \|F(Y^{n,\varepsilon}(u), t^{\varepsilon}(u))\|du, + \\ &+ 2\varepsilon \int_0^s \|B(Y^{n,\varepsilon}(u), t^{\varepsilon}(u)) - B(\bar{Y}^{n,\varepsilon}(u), \tau)\| \, \|w_0\|\phi^n(t^{\varepsilon}(u), \tau)du. \end{aligned} \tag{2.14}$$

By Lemma 2.1, the sequences $\{Y^{n,\varepsilon}(s)\}$ and $\{\bar{Y}^{n,\varepsilon}(s)\}$ are uniformly bounded for $|t^{\varepsilon}(s) - \tau| \leq \varepsilon$. Therefore, thanks to the continuity of the functions F and B, for any $\delta > 0$ there exists $\varepsilon(\delta)$, such that for any $\varepsilon < \varepsilon(\delta)$ we have

$$2\varepsilon \int_0^s \|F(Y^{n,\varepsilon}(u), t^{\varepsilon}(u))\|du < \frac{\delta}{2}, \tag{2.15}$$

and

$$2\varepsilon \int_0^s \|B(\bar{Y}^{n,\varepsilon}(u), t^{\varepsilon}(u)) - B(\bar{Y}^{n,\varepsilon}(u), \tau)\| \|w_0\|\phi^n(t^{\varepsilon}(u), \tau)du < \frac{\delta}{2}. \tag{2.16}$$

Now substituting the bounds (2.15) and (2.16) into the inequality (2.14) and using the Lipschitz property of B, we obtain

$$\begin{aligned}\|Y^{n,\varepsilon}(s) - \bar{Y}^{n,\varepsilon}(s)\| &\le \|X^n(\tau-\varepsilon) - \bar{X}(\tau-\varepsilon)\| + \delta + \\ &+ 2\varepsilon \int_0^s \|B(Y^{n,\varepsilon}(u),\tau) - B(\bar{Y}^{n,\varepsilon}(u),\tau)\| \, \|w_0\| \phi^n(t^{\varepsilon}(u),\tau)du \le \\ &\le \|X^n(\tau-\varepsilon) - \bar{X}(\tau-\varepsilon)\| + \delta + \\ &+ 2\varepsilon \int_0^s L\|Y^{n,\varepsilon}(u) - \bar{Y}^{n,\varepsilon}(u)\| \, \|w_0\| \phi^n(t^{\varepsilon}(u),\tau)du.\end{aligned} \tag{2.17}$$

Applying again the Gronwall-Bellman inequality [237], we deduce from (2.17) that

$$\begin{aligned}&\|Y^{n,\varepsilon}(s) - \bar{Y}^{n,\varepsilon}(s)\| \le \\ &\le \left[\|X^n(\tau-\varepsilon) - \bar{X}(\tau-\varepsilon)\| + \delta\right] \exp\left\{2\varepsilon \int_0^1 L\|w_0\|\phi^n(t^{\varepsilon}(u),\tau)du\right\}.\end{aligned} \tag{2.18}$$

Next, the change-of-variables formula and conditions (2.11) give

$$2\varepsilon \int_0^1 \phi^n(t^{\varepsilon}(u),\tau)du = \int_{\tau-\varepsilon}^{\tau+\varepsilon} \phi^n(t,\tau)dt \le L_1 < \infty, \tag{2.19}$$

which in conjunction with (2.18) yields the inequality

$$\|Y^{n,\varepsilon}(1) - \bar{Y}^{n,\varepsilon}(1)\| \le \left[\|X^n(\tau-\varepsilon) - \bar{X}(\tau-\varepsilon)\| + \delta\right] \exp\left\{LL_1\|w_0\|\right\}. \tag{2.20}$$

By the definition of $t^{\varepsilon}(s)$, we have $Y^{n,\varepsilon}(1) = X^n(t+\varepsilon)$. Moreover, $\bar{Y}^{n,\varepsilon}(s)$, as a solution of the differential equation

$$\dot{\bar{Y}}^{n,\varepsilon}(s) = 2\varepsilon B(\bar{Y}^{n,\varepsilon}(s),\tau)w_0\phi^n(t^{\varepsilon}(s),\tau)$$

with the initial condition

$$\bar{Y}^{n,\varepsilon}(0) = \bar{X}(\tau-\varepsilon),$$

can be represented in the form

$$\bar{Y}^{n,\varepsilon}(s) = \Phi\big(\bar{X}(\tau-\varepsilon), \tau, w_0, 2\varepsilon \int_0^s \phi^n(t^{\varepsilon}(u),\tau)du\big),$$

which follows directly from Definition 2.6 of the function Φ.

Using (2.19) in the representation for $\bar{Y}^{n,\varepsilon}(1)$, we get

$$\bar{Y}^{n,\varepsilon}(1) = \Phi\big(\bar{X}(\tau - \varepsilon), \tau, w_0, \int_{\tau-\varepsilon}^{\tau+\varepsilon} \phi^n(t,\tau)du\big).$$

Letting $n \to \infty$, we obtain

$$\lim_n \bar{Y}^{n,\varepsilon}(1) = \Phi(\bar{X}(\tau - \varepsilon), \tau, w_0, 1). \tag{2.21}$$

The equicontinuity property of $\bar{X}(t)$ in the half-interval $[0, \tau)$ implies that

$$\lim_n \bar{Y}^{n,\varepsilon}(1) = \Phi(\bar{X}(\tau-), \tau, w_0, 1) + o(\varepsilon) = \bar{X}(\tau + \varepsilon) + o(\varepsilon), \tag{2.22}$$

where $o(\varepsilon) \to 0$ when $\varepsilon \to 0$. Combining (2.20) and (2.22) we conclude that for any $\delta > 0$ there exist $N(\delta)$ and $\varepsilon(\delta)$, such that

$$\|X^n(\tau + \varepsilon) - \bar{X}(\tau + \varepsilon)\| < \delta \tag{2.23}$$

if $n > N(\delta)$ and $\varepsilon < \varepsilon(\delta)$.

An argument similar to that in the proof of Lemma 2.1 shows that estimate (2.23) implies the uniform convergence of $X^n(t)$ to $\bar{X}(t)$ on any interval $[\tau + \varepsilon, T]$, which in conjunction with Lemma 2.1 completes the proof. □

As a corollary of Theorem 2.3 we can prove the following approximation result for any general impulse control $\mathscr{I} = \{(\tau_i, w_i),\ i = 1, \ldots, N < \infty\}$.

Consider the sequence

$$w^n(t) = \sum_{i=1}^{N} w_i \phi^n(t, \tau_i),$$

where the functions $\phi^n(t, \tau_i)$ satisfy (2.11). This sequence approximates the impulse control

$$w(t) = \sum_{i=1}^{N} w_i \delta(t - \tau_i).$$

Denote by $X^n(t)$ the solution of Eq. (2.10) corresponding to the initial condition x_0 and to the control $w^n(t)$, and denote by $\bar{X}(t)$ the solution of (2.4), i.e.,

$$\dot{\bar{X}}(t) = F(\bar{X}(t), t) + \sum_{\tau_i \le t} \Psi(\bar{X}(\tau_i-), \tau_i, w_i)\delta(t - \tau_i),$$

with the initial condition x_0 and with the function $\Psi(x, t, w) = \Phi(x, t, w, 1) - x$ introduced in Definition 2.6.

Corollary 2.1 *The sequence* $\{X^n(t)\}$ *converges to the function* $\bar{X}(t)$ *everywhere on* $[0, T]$, *except possibly on the set of points* $\{\tau_i, \; i = 1, \ldots, N\}$.

The proof is carried out by the successive application of Theorem 2.3 on the half-intervals $[\tau_i, \tau_{i+1})$. □

The next result, which is a direct consequence of the above theorem, is the stability condition expressed in terms of the function Φ. We first need to introduce a type of approximation.

Definition 2.7 We say that a sequence of functions $\{w^n(t)\}$ approximates the impulse control

$$w(t) = \sum_{i=1}^{N} w_i \delta(t - \tau_i)$$

if the following conditions are satisfied:

$$\sup_n \int_0^T \|w^n(t)\| dt < \infty,$$

$$\lim_n \int_0^T f(t) w^n(t) dt = \int_0^T f(t) w(t) dt = \sum_{i=1}^{N} f(\tau_i) w_i \quad \text{as} \quad n \to \infty, \tag{2.24}$$

for any continuous function $f(t)$.

Remark 2.4 Here and below the integrals with respect to δ-functions will be understood in the sense of Remark 2.1

Remark 2.5 In other words, the sequence $\{w^n(\cdot)\}$ approximates the control $w(\cdot)$ in the weak-$*$ topology of the space of functions of bounded variation [204].

Example 2.2 It is easy to see that the sequence

$$w^n(t) = \sum_{i=1}^{N} w_i \phi^n(t, \tau_i),$$

where the functions $\phi^n(t, \tau_i)$ satisfy (2.11), meets the conditions (2.24). Moreover, if we introduce the functions

$$U(t) = \sum_{\tau_i \le t} w_i = \int_0^t w(s) ds, \quad U^n(t) = \int_0^t w^n(s) ds = \sum_{\tau_i \le t} w_i \int_0^t \phi^n(s, \tau_i) ds,$$

then the relations (2.11) ensure that $U^n(t) \to U(t)$ at all points of continuity and, by Helly's theorem (see Sect. A.1, Theorem A.12),

$$\lim_n \int_0^T f(t)dU^n(t) = \int_0^T f(t)dU(t), \quad \text{as} \quad n \to \infty,$$

for any continuous function $f(t)$.

A natural requirement in various applications is that *the result of an approximation should not depend on the choice of the approximating sequence*, i.e., it would be natural to assume that stability with respect to the impulse control approximation holds.

Further consideration shows that, in genera,l this requirement is not fulfilled. The next theorem provides a criterion for the stability of the solution to a DCS with respect to various approximations of the impulse control.

Theorem 2.4 (Necessary Stability Condition) *Suppose the sequence of controls* $\{w^n(\cdot)\}$ *approximates the impulse control*

$$w(t) = \sum_{i=1}^{N} w_i \delta(t - \tau_i)$$

in the sense of Definition 2.7.

Let $X^n(t)$ *be a solution of Eq. (2.10) with the initial condition* x_0 *and corresponding to the control* $w^n(t)$. *Suppose that the limit of the sequence* $\{X^n(t)\}$ *exists, for any initial condition* x_0 *and any impulse control* $\mathscr{I}$, *everywhere on* $[0, T]$ *(except possibly on the set of points* $\{\tau_i,\ i = 1, \ldots, N\}$*) and is equal to* $X(t)$. *If this limit does not depend on the choice of the sequence* $\{w^n(\cdot)\}$, *then the function* $\Phi(x, t, w, 1)$ *introduced in Definition 2.6 satisfies for any* $x \in \mathbb{R}^n$, *any* $t \in [0, T]$, *and any* $w_1, w_2 \in W$, *the following condition* ***semigroup property with respect to*** w*]:*

$$\Phi(x, t, w_1 + w_2, 1) = \Phi(\Phi(x, t, w_1, 1), t, w_2, 1); \tag{2.25}$$

moreover, the function $X(t)$ *satisfies the equation*

$$\dot{X}(t) = F(X(t), t) + \sum_{\tau_i \le t} \Psi(X(\tau_i-), \tau_i, w_i)\delta(t - \tau_i)$$

with the initial condition x_0 *with* $\Psi(x, t, w) = \Phi(x, t, w, 1) - x$.

Proof Consider the impulse control $\mathscr{I} = \{(\tau, w_1 + w_2)\}$ and the sequence of impulse controls $\mathscr{I}^m = \left\{(\tau - \frac{1}{m}, w_1), (\tau, w_2)\right\}$. Then, by Definition 2.4, the sequence $\mathscr{I}^n$ converges to $\mathscr{I}$.

Consider also two sequences of ordinary controls

$$w^n(t) = (w_1 + w_2)\phi^n(t, \tau)$$

and

$$w^{n,m}(t) = w_1 \phi^n \left(t, \tau - \frac{1}{m} \right) + w_2 \phi^n(t, \tau)$$

that approximate $\mathscr{I}$ and $\mathscr{I}^n$, respectively, in the sense of Definition 2.7. Moreover, the sequence $\{w^{n,m}(\cdot)\}$ approximates the impulse control $\mathscr{I}$ as $n, m \to \infty$. Denote the solutions of Eq. (2.10) corresponding to the controls $w^n(t)$ and $w^{n,m}(t)$ by $X^n(t)$ and $X^{n,m}(t)$, respectively. When $n \to \infty$, they converge, by Theorem 2.3, to functions $\bar{X}(t)$ and $\bar{X}^m$ that satisfy the respective equations

$$\dot{\bar{X}}(t) = F(\bar{X}(t), t) + \Psi(\bar{X}(\tau-), \tau, w_1 + w_2)\delta(t - \tau),$$

and

$$\begin{aligned} \dot{\bar{X}}^m(t) = F(\bar{X}^m(t), t) + \\ &+ \Psi\left(\bar{X}^m\left(\tau - \frac{1}{m}\right), \tau - \frac{1}{m}, w_1\right)\delta\left(t - \tau + \frac{1}{m}\right) \\ &+ \Psi(\bar{X}^m(\tau-), \tau, w_2)\delta(t - \tau), \end{aligned}$$

everywhere on $[0, T]$, except possibly at the points τ and $\tau-(1/m)$, respectively. An argument similar to that used in the proof of Theorem 2.2 shows that the sequence $\{\bar{X}^m(\cdot)\}$ converges to the solution of the equation

$$\begin{aligned} \dot{\widetilde{X}}(t) = F(\widetilde{X}(t), t) + \\ &+ \left[\Psi(\widetilde{X}(\tau-), \tau, w_1) + \Psi(\widetilde{X}(\tau-) + \Psi(\widetilde{X}(\tau-), \tau, w_1), \tau, w_2)\right]\delta(t - \tau). \end{aligned}$$

Therefore, the sequence $X^{n,n}(t)$, corresponding to the sequence of controls $w^{n,n}(t)$ converges to $\widetilde{X}(t)$ everywhere on $[0, T]$, except possibly at the point τ.

The sequences $w^n(t)$ and $w^{n,n}(t)$ approximate the same impulse control $\mathscr{I}$, yet the corresponding limits $\bar{X}(t)$ and $\widetilde{X}(t)$ satisfy different equations. They coincide for all initial conditions and for all admissible impulse controls if and only if, for any $x \in \mathbb{R}^n$, any $t \in [0, T]$, and any $w_1, w_2 \in W$, the function Ψ satisfies the condition (2.7).

Substituting the expression $\Psi(x, t, w) = \Phi(x, t, w, 1) - x$, with $\Phi(x, t, w, s)$ as in Definition 2.6, in (2.7), we obtain the relation (2.25), which completes the proof. □

Relation (2.25) can be rewritten in the equivalent form

$$\begin{aligned} \Phi(\Phi(x, t, w_2, 1), t, w_1, 1) = \Phi(\Phi(x, t, w_1, 1), t, w_2, 1) = \\ = \Phi(x, t, w_1 + w_2, 1), \end{aligned} \tag{2.26}$$

for any $w_1, w_2 \in W$, any $x \in \mathbb{R}^n$, and any $t \in [0, T]$, which is a form of the *commutativity property* (*commutativity condition*).

The analytical relation (2.26) is extremely important for the further understanding of the properties of the function $\Phi(x, t, w, 1)$. It follows from (2.26) that any function satisfying the commutativity property on some convex set W that contains the point $0 \in \mathbb{R}^m$ can be extended to the minimal convex cone containing W, namely,

$$\text{con}(W) = \{l \in \mathbb{R}^n \, : l = \lambda w \mid \lambda \leq 0, \; w \in W\},$$

with preservation of condition (2.26) for all $w_1, w_2 \in \text{con}(W)$.

Definition 2.8 The *extension* $\widetilde{\Phi}(x, t, w, 1)$ of the function $\Phi(x, t, w, 1)$ is defined as follows:

1. if $w \in W$, then

$$\widetilde{\Phi}(x, t, w, 1) = \Phi(x, t, w, 1);$$

2. if $w \notin W$, but $w \in \text{con}(W)$, then find a point $w_0 \in W$ and an integer k, such that $w = kw_0$, and define $\widetilde{\Phi}(x, t, w, 1)$ recursively as

$$\widetilde{\Phi}(x, t, (i+1)w_0, 1) = \Phi(\widetilde{\Phi}(x, t, iw_0, 1), t, w_0, 1), \quad \text{for} \quad i = 1, \ldots, k-1.$$

The next theorem establishes the correctness of this definition.

Theorem 2.5 *The function* $\widetilde{\Phi}(x, t, w, 1)$ *is uniquely determined and satisfies the condition (2.26) for all* $w_1, w_2 \in \text{con}(W)$.

Proof First, we prove the uniqueness. Let $w \in \text{con}(W) \setminus W$, and suppose we have two different representations of w as $w = k_1 w_1$ and $w = k_2 w_2$, where $w_1, w_2 \in W$ and $k_1 \neq k_2$ are integers. By the convexity of W and as $0 \in W$, we have $w_0 = w/(1/k_1 k_2) \in W$. By Definition 2.8, three different representations of $\widetilde{\Phi}(x, t, w, 1)$ are available, namely:

$$\widetilde{\Phi}_1(x, t, w, 1) = \underbrace{\Phi(\Phi \ldots \Phi(x, t, w/k_1, 1), \ldots, t, w/k_1, 1) \ldots)}_{k_1\text{-times}},$$

$$\widetilde{\Phi}_2(x, t, w, 1) = \underbrace{\Phi(\Phi \ldots \Phi(x, t, w/k_2, 1), \ldots, t, w/k_2, 1) \ldots)}_{k_2\text{-times}},$$

and

$$\widetilde{\Phi}_3(x, t, w, 1) = \underbrace{\Phi(\Phi \ldots \Phi(x, t, w/(k_1 k_2), 1), \ldots, t, w/(k_1 k_2), 1) \ldots)}_{k_1 k_2\text{-times}}.$$

However, the identity (2.26), which is valid for $w/k_1 \in W$ and $w/(k_1k_2) \in W$ implies that

$$\Phi(x, t, w/k_1, 1) = \underbrace{\Phi(\Phi \ldots \Phi(x, t, w/(k_1k_2), 1), \ldots, t, w/(k_1k_2), 1) \ldots)}_{k_2\text{-times}},$$

and if we substitute this relation in the representation for $\widetilde{\Phi}_1(x, t, w, 1)$, we immediately conclude that $\widetilde{\Phi}_1(x, t, w, 1) = \widetilde{\Phi}_3(x, t, w, 1)$. The same argument applied to the representation of $\widetilde{\Phi}_2(x, t, w, 1)$ gives $\widetilde{\Phi}_2(x, t, w, 1) = \widetilde{\Phi}_3(x, t, w, 1)$, which proves the uniqueness and establishes the correctness of the definition.

To prove the relation

$$\widetilde{\Phi}(x, t, w_1 + w_2, 1) = \widetilde{\Phi}(\widetilde{\Phi}(x, t, w_1, 1), t, w_2, 1) \tag{2.27}$$

we need the following analogue of the commutativity condition. Let $\widetilde{\Phi}(x, t, w_1, 1)$ be defined for $w_1 \in \mathrm{con}(W)$. Then for any $w_2 \in W$ it holds that

$$\Phi(\widetilde{\Phi}(x, t, w_1, 1), t, w_2, 1) = \widetilde{\Phi}(\Phi(x, t, w_2, 1), t, w_1, 1). \tag{2.28}$$

Applying Definition 2.8 and the commutativity condition (2.26), we obtain the chain of equalities

$$\begin{aligned}
&\Phi(\widetilde{\Phi}(x, t, w_1, 1), t, w_2, 1) = \\
&\quad = \Phi(\underbrace{\Phi(\Phi \ldots \Phi(x, t, w_1/k, 1), \ldots, t, w_1/k, 1) \ldots)}_{k\text{-times}}, t, w_2, 1) = \\
&\quad = \underbrace{\Phi(\Phi \ldots \Phi(x, t, w_2, 1), \ldots, t, w_1/k, 1) \ldots)}_{k\text{-times}} = \\
&\quad = \Phi(\underbrace{\Phi(\Phi \ldots \Phi(x, t, w_2, 1), \ldots, t, w_1/k, t, 1) \ldots)}_{(k-1)\text{-times}}, t, w_1k, 1) = \\
&\quad = \Phi(\underbrace{\Phi(\Phi \ldots \Phi(x, t, w_2, 1), \ldots, t, 2w_1/k, t, 1) \ldots)}_{(k-2)\text{-times}}, t, w_1/k, 1) = \\
&\quad \ldots\ldots\ldots\ldots\ldots\ldots\ldots\ldots\ldots\ldots\ldots\ldots\ldots\ldots \\
&\quad = \widetilde{\Phi}(\Phi(x, t, w_2, 1), t, w_1, 1).
\end{aligned}$$

Now we are ready to prove the relation (2.27).

Let $w_1, w_2 \in \mathrm{con}(W)$. Then there exists an integer $k \geq 1$, such that all the three points w_1/k, w_2/k and $(w_1 + w_2)/k$ lie in W.

By Definition 2.8, $\widetilde{\Phi}(x, t, w_1 + w_2, 1)$, $\widetilde{\Phi}(x, t, w_1, 1)$ and $\widetilde{\Phi}(x, t, w_2, 1)$ are defined by the recurrence relations

$$\widetilde{\Phi}(x,t,(i+1)(w_1+w_2)/k,1)=\Phi(\widetilde{\Phi}(x,t,i(w_1+w_2)/k,1),t,(w_1+w_2)/k,1),$$

$$\widetilde{\Phi}(x,t,(i+1)w_1/k,1)=\Phi(\widetilde{\Phi}(x,t,iw_1/k,1),t,w_1/k,1),$$

$$\widetilde{\Phi}(x,t,(i+1)w_2/k,1)=\Phi(\widetilde{\Phi}(x,t,iw_1/k,1),t,w_1/k,1),$$

for $i=1,\dots,k-1$. Therefore, to prove (2.27) it suffices to verify that if the relation

$$\widetilde{\Phi}(x,t,i(w_1+w_2)/k,1)=\widetilde{\Phi}(\widetilde{\Phi}(x,t,iw_1/k,1),t,iw_2/k,1), \tag{2.29}$$

holds for some $i\geq 1$, then it holds for $i'=i+1$.

Definition 2.8, the commutativity property (2.28), and our assumption (2.29) imply that

$$\begin{aligned}
&\widetilde{\Phi}(x,t,(i+1)(w_1+w_2)/k,1)\\
&=\Phi(\widetilde{\Phi}(x,t,i(w_1+w_2)/k,1),t,(w_1+w_2)/k,1)=\\
&=\widetilde{\Phi}(\widetilde{\Phi}(\Phi(x,t,(w_1+w_2)/k,1),t,iw_1/k,1),t,iw_2/k,1)=\\
&=\widetilde{\Phi}(\Phi(\widetilde{\Phi}(x,t,iw_1/k,1),t,(w_1+w_2)/k,1),t,iw_2/k,1)=\\
&=\widetilde{\Phi}(\Phi(\Phi((\widetilde{\Phi}(x,t,iw_1/k,1),t,w_1/k,1),t,w_2/k,1),t,iw_2/k,1)=\\
&=\widetilde{\Phi}(\widetilde{\Phi}(x,t,(i+1)w_1/k,1),t,(i+1)w_2/k,1).
\end{aligned}$$

Since (2.29) holds for $i=1$, it follows that it holds for all $i=1,\dots,k-1$, which establishes (2.27) and completes the proof of the theorem. □

In what follows, for the sake of convenience we will use the same notation for the extended function $\widetilde{\Phi}(x,t,w,1)$, namely, we will assume that $\Phi(x,t,w,1)$ is defined for all $w\in \mathrm{con}(W)$.

Conditions (2.25) and (2.27) express the semigroup property of the functions $\Phi(x,t,w,1)$ and $\widetilde{\Phi}(x,t,w,1)$ with respect to the vectorial parameter $w\in \mathrm{con}(W)$ [371]. Generally speaking, the function $\Phi(x,t,w,1)$, introduced in Definition 2.6 as a solution of the differential equation

$$\dot{Y}(s)=B(Y(s),t)w$$

with the initial condition $Y(0) = x \in \mathbb{R}^n$, does not satisfy semigroup property with respect to w. However, one can find a condition which ensures that the function $B(x, t)$ satisfies the semigroup condition (2.25). This condition, as well as our previous results, are also necessary for stability, however, it can be expressed directly in terms of the function $B(x, t)$. A similar condition is known as the Frobenius complete integrability condition of in the theory of differential forms [79], but here, for the reader's convenience, we derive this condition directly from the stability (robustness) requirement.

Assumption 2.2

(1) *The function $B(x, t)$ is continuous with respect to (x, t) and admits partial derivatives with respect to x,*

$$B'_x(x, t) = \frac{\partial B(x, t)}{\partial x},$$

that are uniformly bounded and continuous with respect to (x, t); moreover, $B'_x(x, t)$ satisfies a local Lipschitz condition, i.e., if C is any bounded set in $\mathbb{R}^n$, then there exists a constant $L(C) > 0$ such that, for any $x, y \in C$,

$$\|B'_x(x, t) - B'_x(y, t)\| \leq L(C)\|x - y\|.$$

(2) *For any two vectors $h, g \in K$, where K is some convex cone in $\mathbb{R}^m$, and for any $x \in \mathbb{R}^n$ and any $t \in [0, T]$,*

$$[B(x, t)h]'_x[B(x, t)g] = [B(x, t)g]'_x[B(x, t)h],$$

where we use the notation

$$[B(x, t)h]'_x = \frac{\partial}{\partial x}[B(x, t)h].$$

If condition (2) of Assumption 2.2 is valid for $K = \mathbb{R}^m$, then it can be expressed in components as

$$\sum_{k=1}^{n} \frac{\partial b_{ij}(x, t)}{\partial x_k} b_{kl}(x, t) = \sum_{k=1}^{n} \frac{\partial b_{il}(x, t)}{\partial x_k} b_{kj}(x, t), \quad j, l = 1, \dots, m,$$

for all $x \in \mathbb{R}^n$ and all $t \in [0, T]$, where $b_{ij}(x, t)$, $i = 1, \dots, n$, $j = 1, \dots, m$ are the components of the matrix-valued function $B(x, t)$.

Theorem 2.6 *Let the matrix-valued function $B(x, t)$ satisfy Assumption 2.2. Then:*

(i) *the function $\Phi(x, t, w, 1)$ introduced in Definition 2.6 satisfies the semigroup property (2.25) for any two vectors $w_1, w_2 \in K$, and*

(ii) $\Phi(x, t, w, 1)$ *is the unique solution of the partial differential equation*

$$\begin{aligned} \frac{\partial}{\partial \vec{e}} \Phi(x, t, w, 1) &= B(\Phi(x, t, w, 1), t)\vec{e}, \quad \vec{e} \in K, \\ \Phi(x, t, 0, 1) &= x, \end{aligned} \tag{2.30}$$

where $\frac{\partial}{\partial \vec{e}}$ *stands for the directional derivative with respect to the unit vector* $\vec{e} \in K$.

Definition 2.9 The directional derivative of Φ at $(x, t, w, 1)$ with respect to the direction $\vec{e} \in K$, with $\|\vec{e}\| = 1$, is defined as

$$\begin{aligned} \frac{\partial}{\partial \vec{e}} \Phi(x, t, w, 1) &= \lim_{s \downarrow 0} \frac{\Phi(x, t, w + s\vec{e}, 1) - \Phi(x, t, w, 1)}{s} \\ &= \frac{d}{ds} \Phi(x, t, w + s\vec{e}, 1)\Big|_{s=0+}. \end{aligned}$$

Proof of Item (i) of Theorem 2.6 First we show that if $B(x, t)$ satisfies Assumption 2.2, then the function $\Phi(x,t,w,1)$ introduced in Definition 2.6 satisfies the semigroup property . Thus, we have to verify that

$$\Phi(x, t, w_1 + w_2, s) = \Phi(\Phi(x, t, w_1, s), t, w_2, s),$$

for any $s \geq 0$. Denote

$$H(s) = \Phi(x, t, w_1 + w_2, s) - \Phi(\Phi(x, t, w_1, s), t, w_2, s).$$

To verify that $H(s) \equiv 0$, it suffices to prove that the derivative $H'(s) = 0$, because $H(0) = 0$.

By the properties of solutions to ordinary differential equations (see [79, 96]), the function $\Phi(x, t, w, s)$ admits a partial derivative $\Phi'_x(x, t, w, s)$ with respect to the first argument x, and this derivative is the unique solution of the differential equation

$$\frac{d}{ds} \Phi'_x(x, t, w, s) = [B(\Phi(x, t, w, s), t)w]'_x \Phi'_x(x, t, w, s), \tag{2.31}$$

with the initial condition $\Phi'_x(x, t, w, 0) = I_{(n \times n)}$ where $I_{(n \times n)}$ is the $n \times n$ identity matrix [79].

Remark 2.6 We will use the following notation for the partial derivative of the function $B(x, t)$ with respect to the first argument:

$$[B(z, t)w]'_x = \frac{\partial}{\partial x} [B(x, t)w]\Big|_{x=z}.$$

Thus, $H(s)$ is differentiable with respect to s and, by Definition 2.6,

$$
\begin{aligned}
H'(s) &= \frac{d}{ds}\Phi(x,t,w_1+w_2,s) - \frac{d}{ds}\Phi(\Phi(x,t,w_1,s),t,w_2,s) = \\
&= B(\Phi(x,t,w_1+w_2,s),t)(w_1+w_2) - \\
&\quad - B(\Phi(\Phi(x,t,w_1,s),t,w_2,s),t)w_2 - \\
&\quad - \Phi'_x(\Phi(x,t,w_1,s),t,w_2,s)B(\Phi(x,t,w_1,s),t)w_1 = \\
&= \big[B(\Phi(x,t,w_1+w_2,s),t) - \\
&\quad - B(\Phi(\Phi(x,t,w_1,s),t,w_2,s),t)\big](w_1+w_2) + L(s),
\end{aligned}
\tag{2.32}
$$

where

$$
\begin{aligned}
L(s) = {} & B(\Phi(\Phi(x,t,w_1,s),t,w_2,s),t)w_1 - \\
& - \Phi'_x(\Phi(x,t,w_1,s),t,w_2,s)B(\Phi(x,t,w_1,s),t)w_1.
\end{aligned}
$$

Next, for any $y \in \mathbb{R}^n$ we set

$$
\bar{L}(y,s) = B(\Phi(y,t,w_2,s),t)w_1 - \Phi'_x(y,t,w_2,s)B(y,t)w_1
$$

and prove that $\bar{L}(y,s) = 0$.

Since $\bar{L}(y,0) = 0$, it suffices to show that $\bar{L}(y,s)$ satisfies a homogeneous linear differential equation. Differentiating $\bar{L}(y,s)$ with respect to s and taking into account (2.31), we get

$$
\begin{aligned}
\frac{d}{ds}\bar{L}(y,s) = {} & [B(\Phi(y,t,w_2,s),t)w_1]'_x B(\Phi(y,t,w_2,s),t)w_2 - \\
& - [B(\Phi(y,t,w_2,s),t)w_2]';_x \Phi'_x(y,t,w_2,s)B(y,t)w_1.
\end{aligned}
$$

In view of Assumption 2.2, the first term on the right-hand side of this equation is identical to

$$
[B(\Phi(y,t,w_2,s),t)w_2]'_x B(\Phi(y,t,w_2,s),t)w_1,
$$

and consequently

$$
\frac{d}{ds}\bar{L}(y,s) = [B(\Phi(y,t,w_2,s),t)w_2]'_x \bar{L}(y,s).
$$

This is a homogeneous linear differential equation with zero as initial condition; therefore, $\bar{L}(y, s) = 0$. Substituting $y(s) = \Phi(x, t, w_1, s)$ in this last relation yields

$$L(s) = \bar{L}(y(s), s) = 0$$

which in view of (2.32) can be rewritten as

$$\begin{aligned} H'(s) = \big\{&B(\Phi(x, t, w_1 + w_2, s), t) - \\ &- B(\Phi(\Phi(x, t, w_1, s), t, w_2, s), t)\big\}(w_1 + w_2). \end{aligned} \tag{2.33}$$

Since $H(0) = 0$, (2.33) implies that

$$\begin{aligned} &\|H(s)\| \le \\ &\le \int_0^s \|B(\Phi(x, t, w_1 + w_2, \tau), t) - B(\Phi(\Phi(x, t, w_1, \tau), t, w_2, \tau), t)\| \\ &\times \|(w_1 + w_2)\| d\tau, \end{aligned}$$

and in view od the Lipschitz property of B we obtain

$$\begin{aligned} &\|H(s)\| \le \\ &\le \int_0^s L\|\Phi(x, t, w_1 + w_2, \tau), t) - \Phi(\Phi(x, t, w_1, \tau), t, w_2, \tau), t)\| \\ &\quad \times \|(w_1 + w_2)\| d\tau = \\ &= \int_0^s L\|H(\tau)\| \|w_1 + w_2\| d\tau. \end{aligned}$$

Therefore, by the Gronwall-Bellman inequality [237], $H(s) \equiv 0$ which completes the proof of item (i) of Theorem 2.6.

To address the proof of item (ii) of the theorem, we need the following auxiliary result.

Lemma 2.3 *Let the function $\Phi(x, t, w, 1)$ be as in Definition 2.6. Then $\Phi(x, t, w, 1)$ admits at any point (x, t, w) a directional derivative with respect to the direction $\vec{e} = w/\|w\| \in K$, and*

$$\frac{\partial}{\partial \vec{e}} \Phi(x, t, w, 1) = B(\Phi(x, t, w, 1), t)\, \vec{e}. \tag{2.34}$$

Proof of Lemma 2.3 First let us show that

$$\Phi(x, t, w, s) = \Phi(x, t, sw, 1), \quad \text{for all } s \in [0, 1]. \tag{2.35}$$

By Definition 2.6, $\Phi(x, t, w, s)$ is the unique solution of the equation

$$\frac{d}{ds}\Phi(x, t, w, s) = B(\Phi(x, t, w, s), t)w \tag{2.36}$$

with the initial condition $\Phi(x, t, w, 0) = x \in \mathbb{R}^n$. Consider the equation

$$\dot{Y}(s) = B(Y(s), t)w\tau$$

with some $\tau > 0$ and with the same initial condition $\Phi(x, t, w, 0) = Y(0) = x \in \mathbb{R}^n$. Then, by Definition 2.6,

$$Y(s) = \Phi(x, t, \tau w, s),$$

while making the time change $s' = s\tau$ in Eq. (2.36) we obtain

$$Y(s) = \Phi(x, t, w, s\tau).$$

Thus, for any $s, \tau > 0$, we have

$$\Phi(x, t, \tau w, s) = \Phi(x, t, w, s\tau)$$

as we needed to show.

Next, differentiating the equality (2.35) with respect to the positive direction of s we obtain

$$\frac{d}{ds}\Phi(x, t, sw, 1) = \frac{d}{d}\Phi(x, t, w, s) = B(\Phi(x, t, w, s), t)w.$$

But

$$\frac{d}{ds}\Phi(x, t, sw, 1) = \frac{\partial}{\partial \vec{e}}\Phi(x, t, sw, 1)\|w\|,$$

which establishes the relation

$$\frac{\partial}{\partial \vec{e}}\Phi(x, t, sw, 1) = B(\Phi(x, t, sw, 1), t)\,\vec{e}.$$

Setting here of $s = 1$ completes the proof. □

Proof of Item (ii) of Theorem 2.6 By Definition 2.9, using Eq. (2.25) and the identity (2.36) in the relation

$$\frac{\partial}{\partial \vec{e}}\Phi(x, t, w, 1) = \frac{d}{ds}\Phi(x, t, w + s\vec{e}, 1)\Big|_{s=0+} \tag{2.37}$$

yields

$$\frac{\partial}{\partial \vec{e}} \Phi(x,t,w,1) = \frac{d}{ds}\Phi(\Phi(x,t,w,1),t,s\vec{e},1)\Big|_{s=0+} =$$
$$= \frac{d}{ds}\Phi(\Phi(x,t,w,1),t,\vec{e},s)\big|_{s=0+} =$$
$$= B(\Phi(\Phi(x,t,w,1),t,\vec{e},s),t)\,\vec{e}\big|_{s=0+} = B(\Phi(x,t,w,1),t)\vec{e},$$

which establishes (2.30) for all $\vec{e} \in K$.

To prove the uniqueness of the solution of (2.30), suppose that there exists another function $\overline{\Phi}(x,t,w)$ thath satisfies the equation

$$\begin{aligned} &\frac{\partial}{\partial \vec{e}} \overline{\Phi}(x,t,w) = B(\bar{\Phi}(x,t,w),t)\vec{e}, \quad \text{for all } \vec{e} \in K, \\ &\overline{\Phi}(x,t,0) = x. \end{aligned} \tag{2.38}$$

Then

$$\frac{d}{ds}\overline{\Phi}(x,t,\tau\vec{e}+s\vec{e})\Big|_{s=0+} = \frac{d}{ds}\bar{\Phi}(x,t,(\tau+s)\vec{e})\Big|_{s=0+} =$$
$$= \frac{d}{d\tau}\bar{\Phi}(x,t,(\tau+s)\vec{e})\Big|_{s=0+} = \frac{d}{d\tau}\bar{\Phi}(x,t,\tau\vec{e}).$$

On the other hand, using Lemma 2.3 implies that

$$\frac{d}{ds}\bar{\Phi}(x,t,\tau\vec{e}+s\vec{e})\Big|_{s=0+} = B(\bar{\Phi}(x,t,\tau\vec{e}),t)\,\vec{e}.$$

Therefore, the function $\bar{\Phi}(x,t,\tau\,\vec{e}\tau)$ satisfies the ordinary differential equation

$$\frac{d}{d\tau}\bar{\Phi}(x,t,\tau\vec{e}) = B(\bar{\Phi}(x,t,\tau\vec{e}),t)\vec{e}$$

with the initial condition $\bar{\Phi}(x,t,0) = x$.

The uniqueness of the solution of this equation (see Definition 2.6) and relation (2.35) imply that

$$\bar{\Phi}(x,t,\tau\vec{e}) = \Phi(x,t,\tau\vec{e},1), \quad \text{for all } \vec{e} \in K$$

Substituting here $\tau = \|w\|$ and $\vec{e} = w/\|w\|$ for any $w \in K$ completes the proof. □

Corollary 2.2 *If $K = \mathbb{R}^m$, then $\Phi(x, t, w, 1)$ is the unique solution of the partial differential equation*

$$\frac{\partial \Phi(x, t, w, 1)}{\partial w} = B(\Phi(x, t, w, 1), t), \qquad \text{with the initial condition} \quad \Phi(x, t, 0, 1) = x. \tag{2.39}$$

Therefore, the equation

$$\frac{\partial \Phi}{\partial w} = B(\Phi, t)$$

is completely integrable [79].

Corollary 2.3 *If $K \in \mathbb{R}^1$, i.e., $m = 1$, then the equation*

$$\frac{\partial \Phi}{\partial w} = B(\Phi, t)$$

is completely integrable for any function $B(x, t)$ that satisfies condition (1) of Assumption 2.2, because condition (2) is satisfied identically.

Theorem 2.7 *Suppose the function $\Psi(x, t, w)$ satisfies the condition*

$$\Psi(x, t, w_1 + w_2) = \Psi(x, t, w_1) + \Psi(x + \Psi(x, t, w_1), t, w_2)$$

for any $x \in \mathbb{R}^n$, any $t \in [0, T]$ and any $w_1, w_2 \in W$, where $\operatorname{con}(W) = \mathbb{R}^m$, and admits a derivative with respect to w at $w = 0$,

$$B(x, t) = \Psi'_w(x, t, w)\big|_{w=0}, \tag{2.40}$$

such that

$B(x, t)$ satisfies the Lipschitz condition with respect to x,
$B'_x(x, t)$ satisfies the local Lipschitz condition with respect to x.

Then $B(x, t)$ satisfies the Frobenius condition (see Assumption 2.2).

Proof The function $\Phi(x, t, w, 1) = x + \Psi(x, t, w)$ satisfies semigroup condition (2.25) and therefore, by Theorem 2.5, it can be extended to $\operatorname{con}(W)$ with preservation of the semigroup property. It follows that the function $\Psi(x, t, w)$ is defined for all $w \in \operatorname{con}(W) = \mathbb{R}^m$. By the assumption of the theorem, the function $\Phi(x, t, w, 1) = \Psi(x, t, w) + x$ has a derivative with respect to w at $w = 0$ and

$$\Phi'_w(x, t, w, 1)\big|_{w=0} = \Psi'_w(x, t, w)\big|_{w=0}.$$

Take for $B(x, t)$ the function

$$B(x,t) = \Psi'_w(x,t,w)\big|_{w=0}.$$

Then for any $w \in \mathbb{R}^m$

$$\Phi'_w(x,t,w,1) = \Psi'_w(x,t,w) = B(x + \Psi(x,t,w),t) = B(\Phi(x,t,w,1),t),$$

which in conjunction with (2.35) implies that

$$\Phi'_w(x,t,w,s) = B(\Phi(x,t,w,s),t). \tag{2.41}$$

Indeed, by the assumption of the theorem and relation (2.40),

$$\Psi(x,t,w+\Delta w) - \Psi(x,t,w) = \Psi(x + \Psi(x,t,w),t,\Delta w) =$$

$$= B(x + \Psi(x,t,w),t)\Delta w + O(\|\Delta w\|^2),$$

where $O(\|\Delta w\|^2)/\|\Delta w\| \to 0$ as $\|\Delta w\| \to 0$, which establishes (2.41).

To prove the Frobenius property, it is sufficient to differentiate the identity

$$\Phi(\Phi(x,t,w_1,s),t,w_2,s) = \Phi(\Phi(x,t,w_2,s),t,w_1,s).$$

This yields

$$\frac{\partial^2}{\partial w_1 \partial w_2}\Phi(\Phi(x,t,w_1,s),t,w_2,s) = \frac{\partial^2}{\partial w_2 \partial w_1}\Phi(\Phi(x,t,w_2,s),t,w_1,s)$$

where the derivatives exist thanks to the assumption of the theorem.

Applying the relation (2.41) and setting $s = 0$ we obtain

$$[B(x,t)w_1]'_x B(x,t)w_2 = [B(x,t)w_2]'_x B(x,t)w_1$$

which completes the proof. □

Let us consider some examples of functions satisfying the Frobenius condition.

Example 2.3 Let $K = \mathbb{R}^m$; then

$$B(x,t)w = \sum_{i=1}^{m} w_i B_i(x,t),$$

where w_i, $i = 1,\ldots,m$ are the components of a vector w and vectors $B_i(x,t)$, $i = 1,\ldots,m$ are the columns of the matrix-valued function $B(x,t)$. Suppose that B satisfies the Frobenius conditions, i.e.,

$$(B_i)'_x(x,t)B_j(x,t) = (B_j)'_x(x,t)B_i(x,t), \quad \forall (x,t) \in R^n \times [0,T], \tag{2.42}$$

for $i, j = 1, \ldots, m$. Each vector $w = (w_1, \ldots, w_m)$ can be expressed as

$$w = \sum_{i=1}^{m} w_i \vec{e}_i,$$

where $\vec{e}_i = (0, \ldots, 0, 1, 0, \ldots, 0)^*$, $i = 1, \ldots, m$ are the standard unit vectors in $\mathbb{R}^m$. By the semigroup property (2.25),

$$\Phi(x, t, w, 1) = \Phi(\Phi(\ldots \Phi(x, t, w_1\vec{e}_1, 1)\ldots), t, w_{m-1}\vec{e}_{m-1}, 1), t, w_m\vec{e}_m, 1),$$

and $\Phi(x, t, w_k\vec{e}_k, s)$ is the solution of the differential equation

$$\frac{d}{ds}\Phi(x, t, w_k\vec{e}_k, s) = B_k(\Phi(x, t, w_k\vec{e}_k, s), t)w_k, \quad \Phi(x, t, w_k\vec{e}_k, 0) = x.$$

Let $\Phi_k(x, t, s)$ denote the general solution of the differential equation

$$\frac{d}{ds}\Phi_k(x, t, s) = B_k(\Phi_k(x, t, s), t), \quad \Phi_k(x, t, 0) = x; \tag{2.43}$$

then

$$\Phi(x, t, w_k\vec{e}_k, s) = \Phi_k(x, t, sw_k).$$

By Lemma 2.3 and relation (2.35), we obtain

$$\Phi(x, t, w_k\vec{e}_k, 1) = \Phi_k(x, t, w_k),$$

and finally

$$\Phi(x, t, w, 1) = \Phi_m(\Phi_{m-1}(\ldots \Phi_1(x, t, w_1)\ldots), t, w_{m-1}), t, w_m). \tag{2.44}$$

This formula shows that $\Phi(x, t, w, 1)$ can be represented as a superposition of general solutions of the differential equations (2.43). It is important to note that the left-hand side of (2.44) does not depend on the order of the functions in the right-hand-side.

The next example uses specific functions to illustrate the general case.

Example 2.4 Let $x = (x_1, x_2) \in \mathbb{R}^2$, $w = (w_1, w_2) \in \mathbb{R}^2$ and

$$B_1(x_1, x_2) = \begin{pmatrix} x_2 \\ -x_1 \end{pmatrix}, \quad B_2(x_1, x_2) = \begin{pmatrix} x_1 \\ x_2 \end{pmatrix}.$$

It is easy to check directly that

$$(B_1)'_x(x_1, x_2)B_2(x_1, x_2) = (B_2)'_x(x_1, x_2)B_1(x_1, x_2) = B_1(x_1, x_2),$$

so the Frobenius condition is fulfilled. The general solutions of the corresponding Eq. (2.43) are

$$\Phi_1(x, s) = \begin{pmatrix} x_1 \cos s + x_2 \sin s \\ -x_1 \sin s + x_2 \cos s \end{pmatrix} \quad \text{and} \quad \Phi_2(x, s) = \begin{pmatrix} x_1 \exp s \\ x_2 \exp s \end{pmatrix}.$$

This yields the following representation of the function $\Phi(x, w)$:

$$\Phi(x, w) = \Phi_2(\Phi_1(x_1, w_1), w_2) = \begin{pmatrix} \exp w_2(x_1 \cos w_1 + x_2 \sin w_1) \\ \exp w_2(-x_1 \sin w_1 + x_2 \cos w_2) \end{pmatrix},$$

which of course does not depend on the order of functions Φ_1 and Φ_2 in the right-hand-side of the above relation.

Generally, in the case of linear functions $B_i(x, t)$ one has the following specific representation of the Frobenius condition.

Example 2.5 Lett $x \in \mathbb{R}^n$ and $B_i(x, t) = A_i(t)x$, where $A_i(t)$, $i = 1, \ldots, m$ are continuous $(n \times n)$-matrix valued functions of $t \in [0, T]$. Then

$$(B_i)'_x(x, t)B_j(x, t) = A_i(t)A_j(t)x,$$

hence the Frobenius condition is fulfilled if and only if

$$A_i(t)A_j(t) = A_j(t)A_i(t), \quad \forall i, j = 1, \ldots, m, \quad \forall t \in [0, T].$$

Thus, in this linear case the Frobenius condition is equivalent to the commutativity of the matrices $A_i(t)$, $A_j(t)$. The general solutions of the corresponding Eq. (2.43) are to

$$\Phi_i(x, t, s) = \Phi_i(t, s, 0)x,$$

where $\Phi_i(t, s, 0)$ is the solution of the linear differential matrix equation

$$\frac{d}{ds}\Phi_i(t, s, 0) = A_i(t)\Phi_i(t, s, 0),$$

with the initial condition $\Phi_i(t, s, 0) = I_{(n\times n)}$. Therefore,

$$\Phi(x, w) = \prod_{i=1}^{m} \Phi_i(t, w_i, 0)x,$$

and the right-hand-side does not depend on the order of the functions Φ_i.

Finally we consider an example related to observation control problems (see Example 1.1).

Example 2.6 Suppose that $x > 0$ and $w > 0$ are positive definite $n \times n$ matrices and $B(x)w$ has the form $B(x)w = -xwx$. Consider the matrix differential equation

$$\frac{d}{ds}\Phi(x, w, s) = -\Phi(x, w, s)w\Phi(x, w, s), \quad \Phi(x, w, 0) = x.$$

This equation has an explicit solution which can be obtained by using the transformation $y(s) = (\Phi(x, w, s))^{-1}$, where $y(s)$ satisfies the equation

$$\dot{y}(s) = w, \quad y(0) = x^{-1},$$

so $y(s) = x^{-1} + ws$, and

$$\Phi(x, w, 1) = (x^{-1} + w)^{-1}.$$

If K denotes the cone of positive-definite matrices, then for $w_1, w_2 \in K$ we have

$$\Phi(x, w_1 + w_2, 1) = \Phi(\Phi(x, w_1, 1), w_2, 1).$$

Thus, we obtained a matrix-valued function Φ that satisfies the semigroup condition. The relevance of this function Φ to the observation control problem becomes clear if w is represented as $w = H^*H$, where H is an $m \times n$ matrix. Then

$$\Phi(x, w) = \Phi(x, H) = -xH^*\left(I_{(m\times m)} + HxH^*\right)^{-1}Hx$$

(this result is a particular case of Lemma 7.1). A function Φ of this form arises in the Riccati equation (1.19) of the observation control problem in Example 1.1 with the observation equation

$$y(t_i) = Hx(t_i) + \varepsilon_i.$$

This semigroup property of the function that decribes the jump allows one to reduce the observation control problem for a discrete-continuous system to a standard optimal control problem and to obtain effective methods for its solution (see Chaps. 3 and 7).

2.2.2 Sufficient Stability Conditions

Theorems 2.2, 2.4, 2.6 and 2.7 provide a collection of natural and intrinsic necessary stability conditions. Now we are ready to show that under some additional assumptions they are also sufficient conditions.

Theorem 2.8 (Sufficient Stability Condition ([260], [261])) *Suppose the function $\Psi(x, t, w)$ satisfies the condition*

$$\Psi(x, t, w_1 + w_2) = \Psi(x, t, w_1) + \Psi(x + \Psi(x, t, w_1), t, w_2),$$

for any $x \in \mathbb{R}^n$, any $t \in [0, T]$ and any $w_1, w_2 \in W$, where $\mathrm{con}(W) = \mathbb{R}^m$, and admits the derivatives $\Psi'_x(x, t, w)$, $\Psi'_t(x, t, w)$ and $\Psi'_w(x, t, w)$ for any $x \in \mathbb{R}^n$, any $t \in [0, T]$, and any $w \in W$. Suppose, in addition, that these derivatives are locally Lipschitz with respect to (x, w), i.e., for any bounded set $C \subset \mathbb{R}^n \times \mathbb{R}^m$ there exists a constant $L(C) > 0$, such that for any $(x, w), (y, v) \in C$,

$$\|\Psi'_x(x, t, w) - \Psi'_x(y, t, v)\| \leq L(C)(\|x - y\| + \|w - v\|),$$

$$\|\Psi'_t(x, t, w) - \Psi'_t(y, t, v)\| \leq L(C)(\|x - y\| + \|w - v\|),$$

$$\|\Psi'_w(x, t, w) - \Psi'_w(y, t, v)\| \leq L(C)(\|x - y\| + \|w - v\|).$$

Then for any impulse control $\mathscr{I} = \{(\tau_i, w_i),\ i = 1, \dots, N\}$ and any sequence of impulse controls $\mathscr{I}^n = \{(\tau_i^n, w_i^n)\ i = 1, \dots, N^n\}$ that converges to $\mathscr{I}$ according to Definition 2.4, the corresponding sequence of solutions of the equations (2.4),

$$X^n(t) = x_0 + \int_0^t F(X^n(s), s)ds + \sum_{\tau_i^n \leq t} \Psi(X^n(\tau_i^n-), \tau_i^n, w_i^n),$$

converges to the solution of the equation

$$X(t) = x_0 + \int_0^t F(X(s), s)ds + \sum_{\tau_i \leq t} \Psi(X(\tau_i-), \tau_i, w_i)$$

at all points of continuity.

Proof By assumption, the function $\Phi(x, t, w, 1) = x + \Psi(x, t, w)$ satisfies the semigroup condition (2.25). Therefore, by Theorem 2.5, it can be extended on $\mathrm{con}(W)$ with preservation of the semigroup property. Thus $\Psi(x, t, w)$ is defined for all $w \in \mathrm{con}(W) = \mathbb{R}^m$. Also by assumption, the function $\Phi(x, t, w, 1) = \Psi(x, t, w) + x$ has derivatives with respect to all variables, namely,

$$\Phi'_x(x, t, w, 1) = \Psi'_x(x, t, w) + I_{(n \times n)}, \quad \Phi'_t(x, t, w, 1) = \Psi'_t(x, t, w)$$

and

$$\Phi'_w(x, t, w, 1) = \Psi'_w(x, t, w).$$

Moreover, for any fixed $t \in [0, T]$ and $w \in \mathrm{con}(W) = \mathbb{R}^m$, the function $x \mapsto \Phi(x, t, w, 1)$ has an inverse. The inverse function $\Phi^{-1}(z, t, w, 1)$ is in fact given by

$$\Phi^{-1}(z, t, w, 1) = \Phi(z, t, -w, 1),$$

because the relation

$$\Phi(\Phi(z, t, -w, 1), t, w, 1) = \Phi(z, t, 0, 1) = z$$

holds for any $z \in \mathbb{R}^n$, any $t \in [0, T]$, and any $w \in \mathrm{con}(W) = \mathbb{R}^m$.

Let us introduce the new variable

$$Z(t) = \Phi^{-1}(X(t), t, U(t), 1) = \Phi(X(t), t, -U(t), 1),$$

where $X(t)$ is the solution of Eq. (2.4) and

$$U(t) = \sum_{\tau_i \le t} w_i.$$

Let us prove that is continuous. We know that $Z(t)$ can have jumps only at the points τ_i; but

$$\begin{aligned}
\Delta Z(\tau_i) = Z(\tau_i) - Z(\tau_i-) = \\
&= \Phi(X(\tau_i), \tau_i, -U(\tau_i), 1) - \Phi(X(\tau_i-), \tau_i, -U(\tau_i-), 1) = \\
&= \Phi(\Phi(X(\tau_i-), \tau_i, w_i, 1), \tau_i, -U(\tau_i-) - w_i, 1) - \\
&\quad - \Phi(X(\tau_i-), \tau_i, -U(\tau_i-), 1) = \\
&= \Phi(X(\tau_i-), \tau_i, -U(\tau_i-), 1) - \Phi(X(\tau_i-), \tau_i, -U(\tau_i-), 1) = 0.
\end{aligned}$$

Hence, $Z(t)$ is differentiable with respect to t for all $t \in [0, T] \setminus \{\tau_i, i = 1, \dots, N\}$, and

$$\frac{d}{dt} Z(t) = \Phi'_x(X(t), t, -U(t), 1) F(X(t), t) + \Phi'_t(X(t), t, -U(t), 1),$$

because $dU(t)/dt = 0$ for all $t \in [0, T] \setminus \{\tau_i \ i = 1, \dots, N\}$. By the definition of the inverse function,

$$X(t) = \Phi(Z(t), t, U(t), 1),$$

and the substitution of this expression in the formula for $dZ(t)/dt$ yields the following differential equation for the variable $Z(t)$:

$$\begin{aligned}\frac{d}{dt}Z(t) &= \Phi'_x(\Phi(Z(t),t,U(t),1),t,-U(t),1)F(\Phi(Z(t),t,U(t),1),t)+\\ &+\Phi'_t(\Phi(Z(t),t,U(t),1),t,-U(t),1)=\widetilde{F}(Z(t),t,U(t)),\end{aligned} \tag{2.45}$$

with the initial condition

$$Z(0)=\Phi(X(0),0,-U(0),1)=X(0).$$

Being a superposition of locally-Lipschitz functions (see the assumptions of the theorem), the function $\widetilde{F}(Z,t,U)$ is locally Lipschitz with respect to the variables (Z,U). Consequently, the differential equation (2.45) has a unique solution in the class of bounded functions.

Now consider the sequence of functions

$$Z^n(t)=\Phi(X^n(t),t,-U^n(t),1),$$

corresponding to the sequence of impulse controls $\mathscr{I}^n$.

The sequence $\{X^n(t)\}$ is uniformly bounded. This follows from the properties of the functions F and Ψ (see Assumption 2.1) and the uniform boundedness of the sequence $\{U^n(t)\}$ (see Definition 2.4). Here, we omit the proof of this assertion, which however is a direct consequence of the generalized Gronwall-Bellman inequality (see Theorem A.21).

Using the uniform boundedness of the sequences $X^n(t)$ and $U^n(t)$ we obtain the inequality

$$\|Z^n(t)-Z(t)\|\le\int_0^t L\|Z^n(s)-Z(s)\|ds+\int_0^t L\|U^n(s)-U(s)\|ds,$$

where the constant $L>0$ may depend on the sequence $\{X^n,U^n\}$.

The sequence $\{U^n(\cdot)\}$ converges to $U(t)$ almost everywhere on $[0,T]$ and is uniformly bounded. Hence, by the Lebesgue theorem [322],

$$\lim_n\int_0^T L\|U^n(s)-U(s)\|ds=0.$$

Using this fact in the above inequality and applying the Gronwall-Bellman inequality [237], we conclude that $\|Z^n(t)-Z(t)\|\to 0$ uniformly on $[0,T]$. Therefore, $X^n(t)=\Phi(Z^n(t),t,U^n(t),1)$ converges to $X(t)=\Phi(Z(t),t,U(t),1)$ at all the points of continuity, which completes the proof. □

Another sufficient stability condition concerns the case of approximation of an impulse control by ordinary controls.

Theorem 2.9 *Suppose the function $B(x,t)$ in Eq. (2.10) satisfies the Frobenius condition (see Assumption 2.2), $K = \mathbb{R}^m$, and $B(x,t)$ admits a derivative with respect to t that is locally Lipschitz with respect to x.*

Then for any impulse control $\mathscr{I} = \{(\tau_i, w_i)\ , i = 1, \dots, N\}$ and for any sequence of ordinary controls $w^n(\cdot)$ that approximates $\mathscr{I}$ according to Definition 2.7, the corresponding sequence of solutions of the Eq. (2.10), i.e.,

$$\dot{X}^n(t) = F(X^n(t), t) + B(X^n(t), t)w^n(t)$$

with the initial condition $X^n(0) = x_0$ converges to the solution of the equation

$$X(t) = x_0 + \int_0^t F(X(s), s)ds + \sum_{\tau_i \le t} \Psi(X(\tau_i-), \tau_i, w_i)$$

at all points of continuity. Here $\Psi(x,t,w) = \Phi(x,t,w,1) - x$ and $\Phi(x,t,w,1)$ is the function introduced in Definition 2.6.

Proof By Theorem 2.6, the function $\Phi(x,t,w,1)$ satisfies the semigroup condition (2.25). By Corollary 2.2, $\Phi(x,t,w,1)$ is the unique solution of the partial differential equation

$$\frac{\partial \Phi(x,t,w,1)}{\partial w} = B(\Phi(x,t,w,1),t), \quad \Phi(x,t,0,1) = x.$$

As solution of an ordinary differential equation (see Definition 2.6) with differentiable right-hand side, $\Phi(x,t,w,s)$ has derivatives with respect to both x and t, which satisfy the differential equations

$$\frac{d}{ds}\Phi'_x(x,t,w,s) = [B(\Phi(x,t,w,s),t)w]'_x \Phi'_x(x,t,w,s)$$

with the initial condition $\Phi'_x(x,t,w,0) = I_{(n\times n)}$, and

$$\frac{d}{ds}\Phi'_t(x,t,w,s) = [B(\Phi(x,t,w,s),t)w]'_x \Phi'_t(x,t,w,s) + B'_t(\Phi(x,t,w,s),t)$$

with the initial condition $\Phi'_t(x,t,w,0) = 0$, respectively [79].

As in the proof of Theorem 2.8, one concludes that, for any fixed $t \in [0,T]$ and $w \in \mathbb{R}^m$, the function $x \mapsto \Phi(x,t,w,1)$ has an inverse $\Phi^{-1}(z,t,w,1)$, given by

$$\Phi^{-1}(z,t,w,1) = \Phi(z,t,-w,1).$$

Let us introduce the new variable

$$Z(t) = \Phi^{-1}(X(t),t,U(t),1) = \Phi(X(t),t,-U(t),1),$$

where $X(t)$ is the solution of Eq. (2.4) with $\Psi(x, t, w) = \Phi(x, t, w, 1) - x$, and

$$U(t) = \sum_{\tau_i \le t} w_i.$$

Arguing in much the same way as in the proof of Theorem 2.8, one finds that $Z(t)$ satisfies the differential equation

$$\frac{d}{dt}Z(t) = \Phi'_x(\Phi(Z(t), t, U(t), 1), t, -U(t), 1)F(\Phi(Z(t), t, U(t), 1), t) + \\ + \Phi'_t(\Phi(Z(t), t, U(t), 1), t, -U(t), 1) = \widetilde{F}(Z(t), t, U(t)) \tag{2.46}$$

with the initial condition

$$Z(0) = \Phi(X(0), 0, -U(0), 1) = X(0).$$

Similarly, consider the sequence of variables

$$Z^n(t) = \Phi^{-1}(X^n(t), t, U^n(t), 1) = \Phi(X^n(t), t, -U^n(t), 1),$$

where $X^n(t)$ is the solution of the equation

$$\dot{X}^n(t) = F(X^n(t), t) + B(X^n(t), t)w^n(t)$$

with the initial condition $X^n(0) = x_0$, and

$$U^n(t) = \int_0^t w^n(s)ds.$$

By the definition of the inverse function,

$$X^n(t) = \Phi(Z^n(t), t, U^n(t), 1),$$

and substitution of this expression in the formula for $dZ^n(t)/ds$ yields the following differential equation for $Z^n(t)$:

$$\frac{d}{dt}Z^n(t) = \Phi'_x(\Phi(Z^n(t), t, U^n(t), 1), t, -U^n(t), 1)\times \\ \times (F(\Phi(Z^n(t), t, U^n(t), 1), t) + B(\Phi(Z^n(t), t, U^n(t), 1), t)w^n(t)) + \\ + \Phi'_t(\Phi(Z^n(t), t, U^n(t), 1), t, -U^n(t), 1) - \\ - \Phi'_w(\Phi(Z^n(t), t, U^n(t), 1), t, -U^n(t), 1)w^n(t) \tag{2.47}$$

with the initial condition

$$Z^n(0) = \Phi(X^n(0), 0, -U^n(0), 1) = X(0) = x_0.$$

However, for any (x, t, w) we have the identity

$$\Phi'_x(x, t, -w, 1)B(x, t) - \Phi'_w(x, t, -w, 1) = 0. \tag{2.48}$$

To deduce (2.48) from the semigroup property, take the identity

$$\Phi(\Phi(z, t, w, 1), t, -w, 1) = z$$

and differentiate it with respect to w to obtain

$$\begin{aligned}
&\frac{\partial}{\partial w}\Phi(\Phi(z, t, w, 1), t, -w, 1) = \\
&\quad = \Phi'_x(\Phi(z, t, w, 1), t, -w, 1)\Phi'_w(z, t, w, 1) + \\
&\quad\quad + \Phi'_w(\Phi(z, t, w, 1), t, -w, 1) = \\
&\quad = \Phi'_x(\Phi(z, t, w, 1), t, -w, 1)B(\Phi(z, t, w, 1), t) + \\
&\quad\quad + \Phi'_w(\Phi(z, t, w, 1), t, -w, 1) = 0.
\end{aligned} \tag{2.49}$$

Replacing here x by $\Phi(z, t, w, 1)$ we obtain (2.48).

Now using the identity (2.48) in (2.47), we conclude that $Z^n(t)$ satisfies the differential equation

$$\frac{d}{dt}Z^n(t) = \widetilde{F}(Z^n(t), t, U^n(t)), \tag{2.50}$$

with the initial condition

$$Z^n(0) = x_0.$$

The sequences $\{X^n(t)\}$ and $\{Z^n(t)\}$ are uniformly bounded: this is established by the same argument as in the proof of Lemma 2.1 (by using the uniform boundedness of $\{U^n(t)\}$ and the Gronwall-Bellman inequality [237]). The rest of the proof runs as before in the proof of Theorem 2.8. □

Theorems 2.8 and 2.9 provide sufficient stability conditions. Let us emphasize that these conditions are very close to being necessary as well, so that under higher smoothness assumptions on the functions F, Ψ, B and their derivatives, the necessary and sufficient stability conditions can be expressed in the form of the relations (2.7) and Assumption 2.2.

Corollary 2.4 *Under the hypotheses of Theorems 2.8 and 2.9, the relation*

$$\Psi(x, t, w_1 + w_2) = \Psi(x, t, w_1) + \Psi(x + \Psi(x, t, w_1), t, w_2)$$

which holds for any $x \in \mathbb{R}^n$, any $t \in [0, T]$, and any $w_1, w_2 \in \mathbb{R}^m$, and the equality

$$[B(x,t)g]'_x B(x,t)h = [B(x,t)h]'_x B(x,t)g$$

which holds for any $x \in \mathbb{R}^n$, any $t \in [0, T]$, and any $g, h \in \mathbb{R}^m$, respectively, are necessary and sufficient stability conditions in the following sense:

For any impulse control $\mathscr{I}$ and the corresponding solution $X(t)$ of Eq. (2.4), the convergence of the impulse control sequence $\mathscr{I}^n$ to $\mathscr{I}$ in the sense of Definition 2.4, as well as the convergence of the ordinary control sequence $w^n(\cdot)$ to the impulse control $\mathscr{I}$ in the sense of Definition 2.7, imply that the sequences $X_1^n(t)$ and $X_2^n(t)$ of solutions of Eqs. (2.4) and (2.10), respectively, with the same initial condition $X_1^n(0-) = X_2^n(0) = X(0-)$, converge to $X(t)$ at all points of continuity.

In the next section, we will consider the concept of the generalized solution of a DCS with impulse control, defined as a pointwise limit of solutions of the system (2.4). As demonstrated by the examples given in the Introduction, this limit does not satisfy the original Eq. (2.4). Moreover, to obtain a correct description, it is necessary to introduce the new class of nonlinear *differential equations with measure*, which provide a universal description for systems with impulse and ordinary controls.

2.3 Generalized Solutions of Systems with Impulse Control and Their Representations

2.3.1 Definition of the Generalized Solution and its Representation by Change of Variables

This book aims to elaborate an effective mathematical tool for the investigation of various optimization problems for DCS with impulse control. In the Introduction and the preceding section we considered the approximation problem for an arbitrary impulse control. This problem is closely connected with the optimal control problem. Indeed, when we investigate an optimization problem, for example, when we prove an existence theorem or construct a procedure for the computation of the optimal solution, we usually deal with some sequence of controls (ordinary or impulsive) and the corresponding sequence of solutions, which minimize (in the limit) the value of the performance criterion. These sequences may be convergent or not, and the same is true for the corresponding sequences of solutions of the system. However, if the set of admissible solutions is compact, then one can extract a subsequence that converges to some function, and the latter can be treated as the optimal solution [456]. In optimal control theory such solutions are called *weak solutions* [237], or *generalized curves* [456]. In the optimal control for DCS, we have to introduce an analog of this concept, and indeed our previous results allow us to provide a mathematically correct description of such *generalized solutions of DCS*.

Definition 2.10 A right-continuous function $X(t)$ of bounded variation, defined on the interval $[0, T]$, is said to be a *generalized solution* of the DCS (2.4) if there exists a sequence of impulse controls $\mathscr{I}^n = \{(\tau_i^n, w_i^n),\ i = 1, \dots, N^n\}$, satisfying the conditions

$$w_i^n \in W, \quad \sup_n \sum_{i=1}^{N^n} \|w_i^n\| < \infty,$$

such that the corresponding sequence $\{X^n(t)\}$ of solutions of the system (2.4) with the initial condition $X^n(0) = X(0-)$ converges to $X(t)$ at all points of continuity.

Our next task is to obtain a representation for generalized solutions. This problem has an explicit answer for systems satisfying the stability conditions (or for robust DCS), though it does not have one for general DCS with impulse control. To illustrate the complexity of this task in the general case, we look again at the DCS of Example 1.2.

Example 2.7 Consider a system with impulse control described by the equation

$$\dot{x}^n(t) = \sum_{\tau_i^n \le t} x^n(\tau_i^n -) w_i^n \delta(t - \tau_i^n), \quad x^n(0) = x_0 \in \mathbb{R}^1, \tag{2.51}$$

with the impulse control $\mathscr{I}^n = \{(\tau_i^n, w_i^n),\ i = 1, \dots, n\}$, where $\tau_i^n = \tau - i/(n^2)$, $i = 1, \dots, n,\ w_i^n = w_i > 0$, and $\sum\limits_{i=1}^{\infty} w_i = M < \infty$. It is readily verified that

$$x^n\left(\tau - (1/n^2)\right) = x_0 \prod_{i=1}^{n} (1 + w_i)$$

and for any $\varepsilon > 0$

$$\lim_n x^n(\tau + \varepsilon) = x_0 \prod_{i=1}^{\infty} (1 + w_i),$$

where the limit of the infinite product exists because $\sum\limits_{i=1}^{\infty} w_i < \infty$ and $w_i \ge 0$ [388].

While the sequence $\mathscr{I}^n$ converges to the unique optimal impulse control $\mathscr{I} = \{(\tau, M)\}$, the corresponding sequence of solutions $X^n(\cdot)$ does not have limit that depends only on M. Indeed, using different ways of distributing the resource control M, one can obtain different values of $\lim\limits_n X^n(\tau)$, ranging from the minimal to the maximal value of the product $x_0 \prod\limits_{i=1}^{\infty} (1 + w_i)$, namely,

$$x_0(1 + M) \le X(\tau) \le x_0 \exp\{M\}.$$

This example shows that, generally speaking, one cannot obtain a description of the generalized solution as a function of the limit impulse control, because a jump of the limit solution depends on the "fine" structure of the approximating sequence in the neighborhood of the jump. Moreover, using various approximations of the same impulse control one can obtain different generalized solutions.

The situation is considerably simpler for DCS satisfying the stability (robustness) conditions.

Theorem 2.10 *Suppose the hypotheses of Theorem 2.8 are in force. Then for any generalized solution $X(t)$ of the system (2.4) with a closed set W, there exists a right-continuous function $U(t)$ of bounded variation on $[0, T]$ satisfying*

$$U(t_1) - U(t_2) \in \text{con}(W), \quad \textit{for all} \quad t_1 \ge t_2,$$

such that $X(t)$ admits the representation

$$X(t) = \Phi(Z(t), t, U(t), 1), \tag{2.52}$$

where $\Phi(x, t, w, 1) = \Psi(x, t, w) + x$ and $Z(t)$ is the solution of the differential equation (2.45),

$$\frac{d}{dt} Z(t) = \widetilde{F}(Z(t), t, U(t)), \qquad Z(0) = X(0-).$$

Proof Let $X(t)$ be a generalized solution and $\mathscr{I}^n = \{(\tau_i^n, w_i^n),\ i = 1, \dots, N^n\}$ be a sequence of impulse controls such that the corresponding sequence of solutions $\{X^n(t)\}$ of the system (2.4) approximates $X(t)$ in the sense of Definition 2.10.

By the assumptions of the theorem, the functions

$$U^n(t) = \sum_{\tau_i^n \le t} w_i^n, \qquad n \ge 0$$

are uniformly bounded and have uniformly bounded variation on $[0, T]$.

By Helly's theorem (see Theorem A.5), one can extract from the sequence $\{U^n(\cdot)\}$ a subsequence $\{U^{n_k}(\cdot)\}$ that converges to a function $U(t)$ of bounded variation for any $t \in [0, T]$.

As in the proof of Theorem 2.8, the convergence of the sequence $\{U^{n_k}(\cdot)\}$ to $U(t)$ implies the uniform convergence of the sequence of functions $Z^{n_k}(t) = \Phi(X^{n_k}(t), t, -U^{n_k}(t), 1)$ to the function $Z(t)$, which is the solution of Eq. (2.45) with the initial condition $Z(0) = X(0-)$.

Therefore,

$$\lim_k X^{n_k}(t) = \lim_k \Phi(Z^{n_k}(t), t, U^{n_k}(t), 1) = \Phi(Z(t), t, U(t), 1) \text{ for all } t \in [0, T].$$

By the Definition 2.10 of $X(t)$,

$$\lim_k X^{n_k}(t) = X(t)$$

and, therefore,

$$X(t) = \Phi(Z(t), t, U(t), 1)$$

at all points of continuity.

Since $U(t)$ is a function of bounded variation, it has a right limit $U(t+)$ at each point $t \in [0, T]$ [322], and so, due to the right-continuity $X(t)$ we have

$$X(t) = X(t+) = \Phi(Z(t), t, U(t+), 1).$$

Moreover, $U(t+) = U(t)$ almost everywhere on $[0, T]$, except, possibly, on a countable set of points of discontinuity. Hence $Z(t)$, as a solution of the equation (2.45), does not change when $U(t)$ is replaced by $U(t+)$.

This is almost the assertion of the theorem. It only remains to verify that $U(t+)$ satisfies the condition

$$U(t_1+) - U(t_2+) \in \text{con}(W), \quad \text{for any} \quad t_1 \geq t_2.$$

This is done by letting $\varepsilon \downarrow 0$ in the relation

$$U(t_1 + \varepsilon) - U(t_2 + \varepsilon) \in \text{con}(W), \quad \text{for any} \quad t_1 \geq t_2,$$

and using the fact that $\text{con}(W)$ is a closed set. The proof is complete. □

If in the representation of a generalized solution the function $U(t)$ has only a finite number of jumps and is differentiable everywhere, except at these jumps, i.e.,

$$U(t) = \int_0^t \dot{U}(t)dt + \sum_{\tau_i \leq t} \Delta U(\tau_i), \tag{2.53}$$

where $\Delta U(\tau_i) = U(\tau_i) - U(\tau_i-)$, then one can obtain an explicit representation of the generalized solution with the aid of a generalized differential equation, as follows:

Theorem 2.11 *Under the hypotheses of Theorem 2.8, suppose that the function $U(t)$ in the representation (2.52) of a generalized solution $X(t)$ admits the representation (2.53) with a finite number of discontinuity points τ_i. Then $X(t)$ satisfies the equation*

$$\dot{X}(t) = F(X(t), t) + B(X(t), t)\dot{U}(t) + \sum_{\tau_i \le t} \Psi(X(\tau_i -), \tau_i, \Delta U(\tau_i))\delta(t - \tau_i), \tag{2.54}$$

where

$$B(x, t) = \Psi'_w(x, t, w)\big|_{w=0}.$$

Proof In view of the representation (2.52), the function $X(t)$ is differentiable for any $t \in [0, T] \setminus \{\tau_i\}$ and has jumps at the points τ_i. To prove the theorem, it is sufficient to show that

$$\dot{X}(t) = F(X(t), t) + B(X(t), t)\dot{U}(t) \qquad \text{for all } t \in [0, T] \setminus \{\tau_i\} \tag{2.55}$$

and

$$\Delta X(\tau_i) = X(\tau_i) - X(\tau_i -) = \Psi(X(\tau_i -), \tau_i, \Delta U(\tau_i)). \tag{2.56}$$

Since the function $\Phi(x, t, w, 1) = \Psi(x, t, w) + x$ admits partial derivatives with respect to all its arguments, we have

$$\begin{aligned}\dot{X}(t) = \Phi'_x(Z(t), t, U(t), 1)\tilde{F}(Z(t), t, U(t)) + \\ + \Phi'_t(Z(t), t, U(t), 1) + \Phi'_w(Z(t), t, U(t), 1)\dot{U}(t).\end{aligned}$$

for all $t \in [0, T] \setminus \{\tau_i\}$ Substituting here the expressions (2.45) and (2.52) we obtain

$$\begin{aligned}&\dot{X}(t) = \\ &= \Phi'_x(\Phi(X(t), t, -U(t), 1), t, U(t), 1)\Phi'_x(X(t), t, -U(t), 1)F(X(t), t) + \\ &\quad + \Phi'_x(\Phi(X(t), t, -U(t), 1), t, U(t), 1)\Phi'_t(X(t), t, -U(t), 1) + \\ &\quad + \Phi'_t(\Phi(X(t), t, -U(t), 1), t, U(t), 1) + \\ &\quad + \Phi'_w(\Phi(X(t), t, -U(t), 1), t, U(t), 1)\dot{U}(t).\end{aligned} \tag{2.57}$$

On the other hand, for any (x, t, w) we have the identities

$$\begin{aligned}&\Phi'_x(\Phi(x, t, -w, 1), t, w, 1)\Phi'_x(x, t, -w, 1) = \\ &\qquad = \frac{\partial}{\partial x}\Phi(\Phi(x, t, -w, 1), t, w, 1) = \frac{\partial x}{\partial x} = I_{(n \times n)},\end{aligned}$$

$$\begin{aligned}&\Phi'_x(\Phi(x, t, -w, 1), t, w, 1)\Phi'_t(x, t, -w, 1) + \Phi'_t(\Phi(x, t, -w, 1), t, w, 1) = \\ &\qquad = \frac{\partial}{\partial t}\Phi(\Phi(x, t, -w, 1), t, w, 1) = \frac{\partial x}{\partial t} = 0,\end{aligned}$$

and

$$\Phi'_w(\Phi(x,t,-w,1),t,w,1) = B(\Phi(\Phi(x,t,-w,1),t,w,1),t) = B(x,t).$$

Substituting these identities in (2.57) we get (2.55).

Calculating $\Delta X(\tau_i)$ by using the semigroup property (2.25) one obtains

$$\begin{aligned} l\Delta X(\tau_i) &= X(\tau_i) - X(\tau_i-) = \\ &= \Phi(Z(\tau_i),\tau_i,U(\tau_i),1) - \Phi(Z(\tau_i),\tau_i,U(\tau_i-),1) = \\ &= \Phi(\Phi(X(\tau_i),\tau_i,-U(\tau_i),1),\tau_i,U(\tau_i),1) - \\ &\quad - \Phi(\Phi(X(\tau_i),\tau_i,-U(\tau_i),1),\tau_i,U(\tau_i-),1) = \\ &= X(\tau_i) - \Phi(X(\tau_i),\tau_i,-\Delta U(\tau_i),1). \end{aligned}$$

Thus,

$$X(\tau_i-) = \Phi(X(\tau_i),\tau_i,-\Delta U(\tau_i),1),$$

and finally

$$X(\tau_i) = \Phi(X(\tau_i-),\tau_i,\Delta U(\tau_i),1)$$

which establishes (2.56) and completes the proof. □

However, in the general case $U(t)$, as a function of bounded variation, admits only a representation of the form (see Theorem A.11)

$$U(t) = U^{c}(t) + \sum_{\tau\le t} \Delta U(\tau), \tag{2.58}$$

where $U^{c}(t)$ is a continuous function of bounded variation, the set of discontinuity points $\{\tau_i\}$ is at most countable, and

$$\sum_{\tau\le T} \|\Delta U(\tau)\| < \infty.$$

A formal application of Theorem 2.11 leads to the following equation for the generalized solution $X(t)$:

$$\begin{aligned} dX(t) &= F(X(t),t)dt + B(X(t),t)dU^{c}(t) + \\ &+ \sum_{\tau\le t} \Psi(X(\tau-),\tau,\Delta U(\tau)\delta(t-\tau)dt, \end{aligned} \tag{2.59}$$

which is a differential form of the integral equation

$$X(t) = X(0-) + \int_0^t F(X(s), s)ds + \int_0^t B(X(s), s)U^c(s) +$$
$$+ \sum_{\tau \le t} \Psi(X(\tau-), \tau, \Delta U(\tau)),$$

where the second integral is understood as the Lebesgue-Stieltjes integral with respect to the vector-valued measure corresponding to the $U^c(t)$ (see Sect. A.1.3).

To prove (2.59), one needs to derive a special representation of the generalized solution via *the discontinuous time change*, to which we devote the remaining part of this section.

2.3.2 Discontinuous Time Change and Representation of Generalized Solutions

The idea of using a discontinuous time change was apparently originally suggested by J. Warga [445, 446], and R. Rishel [377, 378] for the investigation of optimization problems with unbounded or impulse controls. Their approach was based on a special transformation of the time scale which reduces an impulse-driven dynamical equation to an ordinary differential equation with bounded right-hand side. As a result, the original impulsive optimal control problem can be transformed into a standard problem with bounded controls. The analogous approach to DCS with impulse control allows one to derive the representation of a generalized solution by applying the discontinuous time transformation to the solution of a certain auxiliary system of ordinary differential equations. The idea of the method is to stretch the jump points into intervals of lengths proportional to the magnitudes (norms) of the corresponding impulse actions, so that the values of a solution at the left and right endpoints of these intervals are equal to the left and right limits of the the discontinuous solution, respectively. In this manner a discontinuous trajectory is transformed into an absolutely continuous trajectory which satisfies an auxiliary system of differential equations with (bounded) ordinary controls. If subsequently one applies the inverse time transformation, each such interval shrinks to the point, and a continuous trajectory of this auxiliary system goes into a discontinuous trajectory of the original system. Thus, the idea of the method just sketched is that, after the inverse transformation, we obtain a generalized solution of the original DCS with impulse control. Moreover, every generalized solution can be represented with the aid of such a time transformation, applied to a trajectory of the auxiliary system.

The method is really effective if after such an inverse transformation one does indeed get a generalized solution, that is, exactly one that can be obtained by passing to the limit along a sequence of purely impulse or continuous controls and, moreover, all generalized solutions of the given DCS with impulse controls can be obtained in this manner. In this chapter, we consider only stable DCS, while unstable DCS and more general systems admitting generalized controls will be discussed later. The approach described below for nonlinear dynamical systems was first proposed in [258] for problems with scalar impulse actions, and was then extended to vector-valued controls in [262]. Further generalizations of the method were obtained in [399, 401, 405, 406].

Thus, let us consider a DCS described by the Eq. (2.4), where the function Ψ satisfies the assumptions of Theorem 2.8.

Next, consider the auxiliary system described by the equations

$$\begin{cases} \dot{Y}(s) = \alpha(s)F(Y(s), \eta(s)) + (1 - \alpha(s))B(Y(s), \eta(s))l(s), \\ \dot{\eta}(s) = \alpha(s), \end{cases} \tag{2.60}$$

with the initial conditions

$$\begin{cases} Y(0) = X(0-), \\ \eta(0) = 0, \end{cases}$$

where $B(x, t) = \Psi'_w(x, t, w)\big|_{w=0}$. Here $\{\alpha(\cdot), l(\cdot)\}$ are admissible controls that satisfy the constraints

$$\alpha(s) \in [0, 1], \qquad l(s) \in \text{con}(W) \cap \{l : \|l\| \leq 1\}. \tag{2.61}$$

In (2.60) and (2.61), the set W is the same as for the original system (2.4).

The following theorem describes the connection between the auxiliary system (2.60) and the original system (2.4).

Theorem 2.12 *Let $\mathscr{I} = \{(\tau_i, w_i),\ i = 1, \ldots, N\}$ be an impulse control and let $X(t)$ be the corresponding solution of the system (2.4).*

Then there exist admissible controls $\{\alpha(\cdot), l(\cdot)\}$ defined on the interval $[0, T_1]$, where

$$T_1 = T + \sum_{i=1}^{N} \|w_i\|,$$

such that the corresponding solution $\{Y(s), \eta(s)\}$ of the system (2.60) is connected with $X(t)$ by the relation

$$X(t) = Y(\Gamma(t)), \qquad \text{where } \Gamma(t) = \inf\{s : \eta(s) > t\}. \tag{2.62}$$

Remark 2.7 Relation (2.62) is the fundamental relation of the discontinuous time change method.

Proof Define a sequence $\{s_k,\ k = 1, \dots, 2N\}$ by the recursion relations

$$s_1 = \tau_1, \quad s_2 = s_1 + \|w_1\|,$$

$$\dots\dots\dots\dots\dots\dots\dots\dots\dots\dots\dots\dots$$

$$s_{2k+1} = s_{2k} + \tau_{k+1} - \tau_k, \quad s_{2k+2} = s_{2k+1} + \|w_{k+1}\|,$$

$$\dots\dots\dots\dots\dots\dots\dots\dots\dots\dots\dots\dots$$

$$s_{2N-1} = s_{2N-2} + \tau_N - \tau_{N-1}, \quad s_{2N} = s_{2N-1} + \|w_N\|.$$

Let $T_1 = T + \sum_{i=1}^{N} \|w_i\|$ and define the functions $\alpha(s)$ and $l(s)$ on $[0, T_1]$ by

$$\alpha(s) = \begin{cases} 1, & \text{if } s \in [0, s_1] \cup [s_{2N}, T_1] \cup \{\bigcup_{k=1}^{N} [s_{2k}, s_{2k+1})\}, \\ 0, & \text{otherwise,} \end{cases}$$

and

$$l(s) = \begin{cases} w_k / \|w_k\|, & \text{if } s \in [s_{2k-1}, s_{2k}) \\ \text{arbitrary, but in } \mathrm{con}(W) \cap \{l : \|l\| \le 1\}, & \text{otherwise.} \end{cases}$$

It is readily seen that the controls $\alpha(s)$ and $l(s)$ satisfy the constraints (2.61).

We claim that the solution of (2.60) corresponding to these controls does indeed satisfy (2.62).

Indeed, by the definition of $\alpha(\cdot)$,

$$\eta(s) = \begin{cases} \tau_k, & \text{if } s \in [s_{2k-1}, s_{2k}), \\ \tau_k + s - s_{2k}, & \text{if } \quad s \in [s_{2k}, s_{2k+1}), \end{cases}$$

and

$$\eta(T_1) = T.$$

Typical graphs of the functions $\eta(s)$ and $\Gamma(t)$ are shown in Fig. 2.1.

The next lemma describes some general properties of the function $\Gamma(\cdot)$.

Lemma 2.4 *Suppose the function $\eta(\cdot)$ is absolutely continuous with respect to the Lebesgue measure, i.e., it has a derivative $\dot{\eta}(s)$ almost everywhere on $[0, T]$ (see Sect. A.1.1), and this derivative satisfies inequality*

$$0 \le \dot{\eta}(s) \le 1,$$

almost everywhere in $[0, T_1]$.

Fig. 2.1 Change of time: $\eta(s)$ and its inverse $\Gamma(t)$

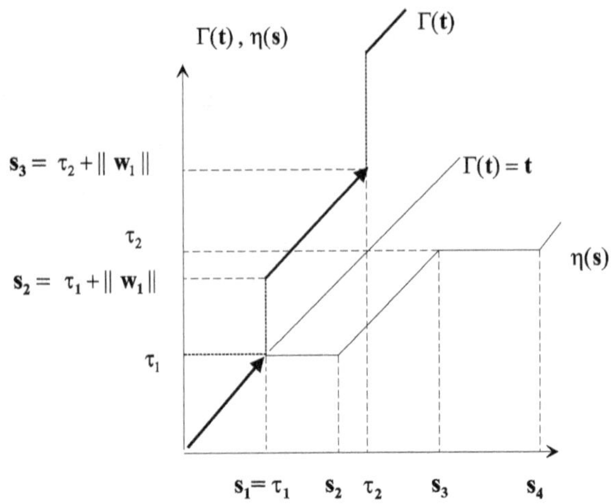

Moreover, suppose that

$$\eta(0) = 0, \qquad \eta(T_1) = T.$$

Then the function $\Gamma(\cdot)$ *defined as*

$$\Gamma(t) = \begin{cases} \inf\{s : \eta(s) > t\}, & \text{for } t \in [0, T) \\ T_1, & \text{for } t \geq T, \end{cases}$$

is non-decreasing, right-continuous, and satisfies the relations

$$\eta(\Gamma(t)) = t \quad \text{for any} \qquad t \in [0, T],$$

$$\Gamma(\eta(s)) = s, \quad \text{if } t = \eta(s) \text{ is a point of continuity of } \Gamma(\cdot).$$

Proof of Lemma 2.4 Since the function $\eta(\cdot)$ is non-decreasing, it is clear that

$$\{s : \eta(s) > t_1\} \subseteq \{s : \eta(s) > t_2\},$$

and if $t_1 \geq t_2$, then

$$\Gamma(t_1) = \inf\{s : \eta(s) > t_1\} \geq \inf\{s : \eta(s) > t_2\} = \Gamma(t_2).$$

To prove the right-continuity, note that thanks to the continuity of $\eta(\cdot)$ for any $t \in [0, T]$, there exists $s \in [0, T_1]$ such that $\eta(s) = t$, and so

$$\Gamma(t) = \sup\{s : \eta(s) = t\}.$$

Moreover, since the set $\{s : \eta(s) = t\}$ is closed, there exists s such that

$$\Gamma(t) = s \quad \text{and} \quad \eta(s) = t.$$

Consider a sequence $\{t_n\}$ with $t_n \downarrow t$. Then for each t_n there exists s_n, such that

$$\Gamma(t_n) = s_n \quad \text{and} \quad \eta(s_n) = t_n.$$

The sequence s_n is monotonically non-increasing, and so it has a limit

$$\lim_n s_n = \overline{s},$$

such that $\eta(\overline{s}) = t$. Moreover, for any s with $\eta(s) = t$, we have $\overline{s} \geq s$. We conclude that

$$\overline{s} = \sup\{s : \eta(s) = t\} = \Gamma(t) \quad \text{and} \quad \lim_n \Gamma(t_n) = \lim_n s_n = \overline{s} = \Gamma(t),$$

which proves the right continuity of $\Gamma(\cdot)$.

Let us extend the definition of $\eta(\cdot)$ as $\eta(s) = 0$ for $s \leq 0$, and $\eta(s) = T$ for $s \geq T_1$. Then, that for any $t \in [0, T)$,

$$\Gamma(t) = \sup\{s : \eta(s) = t\}, \quad \Gamma(t-) = \inf\{s : \eta(s) = t\} \tag{2.63}$$

and

$$\Gamma(0) = 0, \qquad \Gamma(T) = T_1,$$

by definition. Therefore, the relation $\eta(\Gamma(t)) = t$ is an immediate consequence of (2.63).

Further, by the definition of $\Gamma(\cdot)$,

$$\Gamma(\eta(s)) = \inf\{\tau : \eta(\tau) > \eta(s)\} = \sup\{\tau : \eta(\tau) = \eta(s)\}.$$

If the function $\Gamma(\cdot)$ is continuous for $t = \eta(s)$ then, as it follows from (2.63),

$$\{\tau : \eta(\tau) = \eta(s)\} = \{s\}$$

and

$$\Gamma(\eta(s)) = \Gamma(\eta(s)-) = s.$$

This completes the proof of the lemma. □

Remark 2.8 If $t = \eta(s)$ is a point of discontinuity of $\Gamma(\cdot)$, then

$$\Gamma(\eta(s)) = \sup\{\tau : \eta(\tau) = \eta(s) = t\} \geq s \geq \inf\{\tau : \eta(\tau) = \eta(s) = t\},$$

and the relation $\Gamma(\eta(s)) = \Gamma(\eta(s)-) = s$ may be violated.

Remark 2.9 Lemma 2.4 shows that the function $\Gamma(t)$ has properties of an inverse for the function $\eta(s)$.

For every interval of constancy of $\eta(s)$, i.e., $[s_{2k-1}, s_{2k})$ we have that $s_{2k} - s_{2k-1} = \|w_i\|$. The function $\Gamma(t)$, as an inverse to $\eta(s)$, is defined on $[0, \eta(T_1)] = [0, T]$ and is equal to

$$\Gamma(t) = t + \sum_{\tau_i \leq t} \|w_i\|.$$

By Lemma 2.4, $\Gamma(t)$ satisfies the relation

$$\eta(\Gamma(t)) = t. \tag{2.64}$$

Moreover, $\Gamma(t)$ has the same jump points as $X(t)$.

Let us derive the relation between $Y(\Gamma(\tau_k-))$ and $Y(\Gamma(\tau_k))$. By the definition of $\alpha(s)$ and $l(s)$, the function $Y(s)$ satisfies on the interval $[s_{2k-1}, s_{2k})$ the differential equation

$$\dot{Y}(s) = B(Y(s), \tau_k)\frac{w_k}{\|w_k\|}$$

with the initial condition $Y(s_{2k-1})$. Hence, by Definition 2.6 and Lemma 2.3,

$$Y(s_{2k}) = \Phi(Y(s_{2k-1}), \tau_k, w_k, 1) = \Psi(Y(s_{2k-1}), \tau_k, w_k) + Y(s_{2k-1}),$$

and therefore

$$Y(\Gamma(\tau_k)) = \Psi(Y(\Gamma(\tau_k-)), \tau_k, w_k) + Y(\Gamma(\tau_k-)). \tag{2.65}$$

If $t \in (\tau_k, \tau_{k+1})$, then $\dot{\Gamma}(t) = 1$ and $\alpha(\Gamma(t)) = 1$. The substitution of these relations and (2.64) in Eq. (2.60) yields

$$\begin{aligned}\dot{Y}(\Gamma(t)) &= [\alpha(\Gamma(t))F(Y(\Gamma(t)), \eta(\Gamma(t)) + \\ &\quad + (1 - \alpha(\Gamma(t)))B(Y(\Gamma(t)), \eta(\Gamma(t)))l(\Gamma(t))]\dot{\Gamma}(t) = \\ &= F(Y(\Gamma(t)), t).\end{aligned} \tag{2.66}$$

Combining (2.65) and (2.66) we conclude that $Y(\Gamma(t))$ satisfies (2.4) with the same impulse control and initial condition as $X(t)$. In view of the uniqueness of the solution of this equation, $Y(\Gamma(t)) = X(t)$, which proves (2.62).

An analogous result is valid for any generalized solution.

Theorem 2.13 *Let $X(t)$ be a generalized solution of the DCS (2.4).*

Then there exist controls $\{\alpha(\cdot), l(\cdot)\}$, satisfying the conditions (2.61) and defined on some interval $[0, T_1]$, where $T_1 \geq T$, such that the corresponding solution $\{Y(\cdot), \eta(\cdot)\}$ of the system (2.60) is connected with $X(t)$ by the relation (2.62).

Proof According to Definition 2.10, for any generalized solution $X(t)$ there exists a sequence $\mathscr{I}^n = \{(\tau_i^n, w_i^n), i = 1, \ldots, N^n\}$, such that the corresponding sequence $X^n(t)$ of solutions of (2.4) with the initial condition $X^n(0) = X(0-)$ converges to $X(t)$ at all points of continuity; moreover, there exists a constant $M > 0$ such that

$$\sum_{i=1}^{N^n} \|w_i^n\| \leq M.$$

By Theorem 2.12, for any n there exist controls $\{\alpha^n(\cdot), l^n(\cdot)\}$, defined on the interval $[0, T_1^n]$, where

$$T_1^n = T + \sum_{i=1}^{N^n} \|w_i^n\|,$$

and satisfying (2.61), such that the corresponding solution $\{Y^n(s), \eta^n(s)\}$ of the system (2.60) is connected with $X^n(t)$ by the relation

$$X^n(t) = Y^n(\Gamma^n(t)), \qquad \text{where } \Gamma^n(t) = \inf\{s : \eta^n(s) > t\}. \tag{2.67}$$

Let $T_1 = T + M \geq T_1^n$ and extend the definition of the functions $\{\alpha^n(\cdot), l^n(\cdot)\}$ by zero on the interval $(T_1^n, T_1]$. Then the functions $\{Y^n(s), \eta^n(s)\}$ are still constant on these semi-intervals, as it follows from Eq. (2.60).

Since the control set (2.61) is uniformly bounded and the functions F and B satisfy linear growth conditions, the the sequence $\{Y^n(s), \eta^n(s)\}$ is uniformly bounded and equicontinuous on $[0, T_1]$ [237]. Therefore, by the Arzèla-Ascoli theorem [204], there exists a subsequence $\{Y^{n_k}(s), \eta^{n_k}(s)\}$ that converges uniformly on $[0, T_1]$ to some functions $\{Y(s), \eta(s)\}$.

Since the right-hand side of the system (2.60), under the constraints (2.61), is a convex closed set for any $\{y, \eta\}$, and the set (2.61) is closed and bounded, one concludes that the functions $\{Y(s), \eta(s)\}$ satisfy the system (2.60) with some admissible controls $\{\alpha(s), l(s)\}$ [237].

To complete the proof it remains to show that the subsequence $\{Y^{n_k}(\Gamma^{n_k}(\cdot))\}$ that corresponds to the controls $\{\alpha^{n_k}(\cdot), l^{n_k}(\cdot)\}$ converges to $Y(\Gamma(t))$ at all points of continuity. This will imply that

$$Y^{n_k}(\Gamma^{n_k}(t)) = X^{n_k}(t) \to Y(\Gamma(t)), \qquad k \to \infty$$

at all points of continuity. However, in view of the definition of a generalized solution, also

$$X^{n_k}(t) \to X(t)$$

at all points of continuity. Therefore, $X(t) = Y(\Gamma(t))$ on the set of points where both functions $X(t)$ and $Y(\Gamma(t))$ are continuous.

Now, since $X(t)$ and $Y(\Gamma(t))$ are the right-continuous functions of bounded variations, their sets of points of continuity are dense in $[0, T]$, and at any point of discontinuity t we have

$$X(t) = \lim_{\tau^n \downarrow t} X(\tau) = \lim_{\tau^n \downarrow t} Y(\Gamma(\tau)) = Y(\Gamma(t)),$$

where τ^n is any sequence of points of continuity points. Hence, $X(t) = Y(\Gamma(t))$ everywhere on $[0, T]$.

To prove the convergence of $\{Y^{n_k}(\Gamma^{n_k}(t))\}$ to $Y(\Gamma(t))$ we need the following general result, which will be frequently used in our further considerations.

Lemma 2.5 *Suppose that the sequence of monotonically increasing functions* $\{\eta^n(\cdot)\}$, *defined on an interval* $[0, T_1]$, *satisfies the conditions*

$$\eta^n(0) = 0, \quad \eta^n(T_1) = T,$$
$$0 < \eta^n(s_2) - \eta^n(s_1) \le s_2 - s_1, \quad \text{if } s_1 < s_2$$

and converges to a function $\eta(s)$ *uniformly on* $[0, T_1]$.

Then the sequence of inverse functions

$$\Gamma^n(t) = \inf\{s : \eta^n(s) > t\}, \quad \Gamma^n(T) = T_1,$$

defined on the interval $[0, T]$, *converges to the function* $\Gamma(t) = \inf\{s : \eta(s) > t\}$ *with* $\Gamma(T) = T_1$, *at all points of continuity.*

Proof of Lemma 2.5 The limit function $\eta(\cdot)$ is non-decreasing and absolutely continuous, so it admits a derivative $\alpha(s) = \dot{\eta}(s) \le 1$, almost everywhere on $[0, T]$ (see Theorem A.7). By Lemma 2.4, for any $t \in [0, T]$ there exists $s \in [0, T_1]$, such that $\Gamma(t) = s$ and $\eta(s) = t$. Let t be point of continuity of $\Gamma(\cdot)$.

Consider the sequence $\{s^n : \eta^n(s^n) = t\}$. By the definition of the inverse function, $\Gamma^n(t) = s^n$. We claim that $\{s^n\}$ converges to some s^* such that $\eta(s^*) = t$.

Indeed, for any n we have the relation

$$\eta(s^n) - t = \eta(s^n) - \eta^n(s^n),$$

the right-hand side of which goes to zero due to the uniform convergence of $\eta^n(\cdot)$ to $\eta(\cdot)$. Hence,

$$\lim_n \eta(s^n) = t.$$

The sequence $\{s^n\}$ is bounded, and if $\overline{s}$ is one of its partial limits, then $\eta(\overline{s}) = t$. Since $\Gamma(\cdot)$ is continuous at the point t, there is exactly one point s^* where $\eta(s^*) = t$.

Therefore,

$$\lim_n \Gamma^n(t) = \lim_n s^n = s^*;$$

but if $t = \eta(s^*)$ is a point of continuity of $\Gamma(\cdot)$, then, by Lemma 2.4,

$$\Gamma(t) = \Gamma(\eta(s^*)) = s^* = \lim_n \Gamma^n(t),$$

and this is precisely the assertion of Lemma 2.5. □

To complete the proof of the theorem it remains to show that the sequence $\{Y^{n_k}(\Gamma^{n_k}(t))\}$ converges to $Y(\Gamma(t))$.

By Lemma 2.5, $\Gamma^{n_k}(t)$ converges to $\Gamma(t)$ at all points of continuity. But

$$\begin{aligned} &\|Y^{n_k}(\Gamma^{n_k}(t)) - Y(\Gamma(t))\| \leq \\ &\quad \leq \|Y^{n_k}(\Gamma^{n_k}(t)) - Y(\Gamma^{n_k}(t))\| + \|Y(\Gamma^{n_k}(t)) - Y(\Gamma(t))\|. \end{aligned} \tag{2.68}$$

Thanks to the uniform convergence of Y^{n_k} to Y, the first term in the right-hand side of inequality (2.68) goes to zero everywhere. Also, thanks to the continuity of Y and the convergence of $\Gamma^{n_k}(t)$ to $\Gamma(t)$ at all points of continuity, the second term also goes to zero at all points of continuity of the function Γ, as we desired. □

Remark 2.10 Thus, according to Theorem 2.4 any generalized solution of a robust DCS with impulse control can be represented, via the time transformation, by means of some solution of the auxiliary system (2.60). In the next subsection we will show how this result enables us to represent a generalized solution by means of a nonlinear differential equation with measure.

We will also show, in Chap. 4, that the set of generalized solutions coincides with the set of functions that can be obtained with the aid of a time transformation similar to (2.62). This coincidence plays a key role in the proof of the existence of the optimal solution as well as in the derivation of optimality conditions.

2.3.3 Representation of Generalized Solutions by Means of Differential Equations with a Measure

In this subsection we introduce a special class of differential equation with a measure, which provide a natural description of the generalized solutions of the DCS (2.4). Our approach is based on a change of the independent variable in the Lebesgue-Stieltjes integral. For the reader's convenience we present the proof of this result, thereby making our exposition self-contained. However, for an extended exposition of the theory of measures and integrals we refer to the Appendix and to [204, 322, 376, 388], and [409].

The Lebesgue-Stieltjes integral with respect to a measure $\mu(dt)$ can be reduced to the integral with respect to the Lebesgue measure dt which assigns the measure $\mu\{(a, b]\} = b - a$ to any half-interval $(a, b]$.

Lemma 2.6 ([257]) *Let $\mu(dt)$ be a regular non-negative measure on the interval $[0, T]$, let $P(t) = \mu\{[0, t]\}$ be its distribution function, and let $\nu(s) = \inf\{t : P(t) > s\}$. If $f(t)$ is a Borel-measurable function such that the Lebesgue-Stieltjes integral $\int_a^b f(t)\mu(dt)$ exists, then*

$$\int_a^b f(t)dP(t) = \int_{P(a)}^{P(b)} f(\nu(s))ds. \tag{2.69}$$

Remark 2.11 Equality (2.69) is actually valid for *any* Borel-measurable function. Indeed, it is obviously valid for any linear combination of indicator functions of intervals, and any Borel-measurable function f can be approximated everywhere on $[a, b]$ by a sequence of such linear combinations and the sequence of their Lebesgue-Stieltjes integrals converges to the integral of f [322].

Proof It is sufficient to prove (2.69) for $f(t) = \mathbf{I}(c, d](t)$, where $(c, d] \subset [a, b]$ and $\mathbf{I}(c, d](t)$ is the indicator function of the set $(c, d]$.

By the definition of the indicator function, we have

$$\int_a^b \mathbf{I}(c, d](t)dP(t) = P(d) - P(c). \tag{2.70}$$

To deal with the right-hand side of (2.69), we note that

$$\mathbf{I}(c, d](\nu(s)) = \mathbf{I}(c', d'](s),$$

where $c' = P(c)$ and $d' = P(d)$. Therefore,

$$\int_{P(a)}^{P(b)} \mathbf{I}(c, d]\nu(s)ds = \int_{P(a)}^{P(b)} \mathbf{I}(P(c), P(d)](s)ds = P(d) - P(c),$$

which coincides with the right-hand side of (2.70). This completes the proof of (2.69). □

Theorem 2.14 *Let the assumptions of Theorem 2.8 be satisfied. Let $X(t)$ be an arbitrary generalized solution of the DCS (2.4). Then there exists a vector-valued measure $\mathscr{U}(dt)$ taking values in* $\mathrm{con}(W)$ *(i.e., $\mathscr{U}(A) \in \mathrm{con}(W)$ for any Borel subset $A \subset [0, T]$), such that $X(t)$ satisfies the integral equation*

$$X(t) = X(0-) + \int_0^t F(X(s), s)ds + \int_0^t B(X(s), s)dU^{\mathrm{c}}(s) + \\ + \sum_{\tau \le t} \Psi(X(\tau-), \tau, \Delta U(\tau)), \tag{2.71}$$

Here $U(t)$ is the distribution function of the measure $\mathscr{U}(dt)$, $U^{\mathrm{c}}(t)$ is the continuous component of $\mathscr{U}(dt)$, and $\{\Delta U(\tau)\}$ is the set of jumps of $U(t)$, so that $U(t)$ admits the Lebesgue decomposition

$$U(t) = U^{\mathrm{c}}(t) + \sum_{\tau \le t} \Delta U(\tau).$$

We begin with a general result concerning the functions $\Psi(x, t, w)$ and $B(x, t)$.

Lemma 2.7 *Under the assumptions of Theorem 2.8, the differential equation*

$$\dot{Y}(s) = B(Y(s), t)w(s), \quad Y(a) = y, \tag{2.72}$$

has exactly one solution for any integrable $\mathrm{con}(W)$*-valued function $w(s)$ on $[a, b]$, and*

$$Y(b) = y + \Psi\left(y, t, \int_a^b w(s)ds\right). \tag{2.73}$$

Remark 2.12 By Definition 2.6 and Lemma 2.3, the lemma is valid for any constant function $w(\cdot)$.

Proof of Lemma 2.7 First we show that (2.73) holds if $w(\cdot)$ is a step function, i.e.,

$$w(s) = \sum_{k=0}^{N} w_k \mathbf{I}(s_k, s_{k+1}](s)$$

where $s_0 = a$, $s_{N+1} = b$, $s_k < s_{k+1}$, and $w_k \in \mathrm{con}(K)$.

By the semigroup property of the function $\Phi(x, t, w, 1) = x + \Psi(x, t, w)$, we have the relations

$$Y(s_1) = \Phi(y, t, w_1(s_1 - a), 1) = y + \Psi(y, t, w_1(s_1 - a)) =$$

$$= y + \Psi\left(y, t, \int_a^{s_1} w(s)ds\right)$$

$$Y(s_2) = \Phi(Y(s_1), t, w_2(s_2 - s_1), 1) = Y(s_1) + \Psi(Y(s_1), t, w_2(s_2 - s_1)) =$$

$$= Y(s_1) + \Psi\left(Y(s_1), t, \int_{s_1}^{s_2} w(s)ds\right) = y + \Psi\left(y, t, \int_a^{s_2} w(s)ds\right),$$

....

$$Y(s_{k+1}) = y + \Psi\left(y, t, \int_a^{s_{k+1}} w(s)ds\right),$$

which is precisely the assertion of the lemma for a step function.

For any integrable function $w(\cdot)$ there exists a sequence of step functions $w^n(\cdot)$ that approximates $w(\cdot)$ in the space $L_1[a, b]$, that is,

$$\int_a^b \|w^n(s) - w(s)\|ds \to 0.$$

By the Lipschitz property of $B(x, t)$ and the linearity in w of the right-hand side of Eq. (2.72), the sequence of solutions $\{Y^n(\cdot)\}$ generated by the functions w^n converges to the unique solution of (2.72), uniformly on $[a, b]$. Thanks to the continuity of Ψ,

$$Y(b) = \lim_n Y^n(b) = \lim_n \left[y + \Psi\left(y, t, \int_a^b w^n(s)ds\right)\right] = y + \Psi\left(y, t, \int_a^b w(s)ds\right),$$

which completes the proof of the lemma. □

Proof of Theorem 2.14 By Theorem 2.13, if $X(t)$ is a generalized solution of (2.4), then there exists admissible controls $\{\alpha(\cdot), l(\cdot)\}$, defined on an interval $[0, T_1]$ with $T_1 \geq T$, such that the solution of the system

$$\begin{cases} \dot{Y}(s) = \alpha(s)F(Y(s), \eta(s)) + (1 - \alpha(s))B(Y(s), \eta(s))l(s), \\ \dot{\eta}(s) = \alpha(s), \end{cases}$$

with the initial conditions

$$\begin{cases} Y(0) = X(0-), \\ \eta(0) = 0, \end{cases}$$

satisfies the relation

$$X(t) = Y(\Gamma(t)), \quad \text{where} \quad \Gamma(t) = \inf\{s : \eta(s) > t\}.$$

Define $D_\Gamma = \{\tau \in [0, T] : \Delta\Gamma(\tau) > 0\}$. Since $\Gamma(\cdot)$ is non-decreasing and bounded, the set D_Γ is at most countable.

Therefore, $X(t)$ admits the representation

$$
\begin{aligned}
X(t) &= Y(\Gamma(t)) = X(0-) + \int_0^{\Gamma(t)} \alpha(s)F(Y(s), \eta(s))ds + \\
&+ \int_0^{\Gamma(t)} (1-\alpha(s))B(Y(s), \eta(s))l(s)ds = \\
&= X(0-) + \int_0^{\Gamma(t)} \alpha(s)F(Y(s), \eta(s))ds + \\
&+ \int_0^{\Gamma(t)} (1-\alpha(s))\mathbf{I}\{s : \eta(s) \in [0, T] \setminus D_\Gamma\}(s)B(Y(s), \eta(s))l(s)ds + \\
&+ \sum_{\tau \in D_\Gamma \cup \{\tau \le t\}} \int_{\Gamma(\tau-)}^{\Gamma(\tau)} B(Y(s), \tau)l(s)ds = \\
&= X(0-) + X^{\mathrm{a}}(t) + X^{\mathrm{c}}(t) + X^{\mathrm{d}}(t),
\end{aligned}
\tag{2.74}
$$

where

$$
X^{\mathrm{a}}(t) = \int_0^{\Gamma(t)} \alpha(s)F(Y(s), \eta(s))ds, \tag{2.75}
$$

$$
X^{\mathrm{c}}(t) = \int_0^{\Gamma(t)} (1-\alpha(s))\mathbf{I}\{s : \eta(s) \in [0, T] \setminus D_\Gamma\}(s)B(Y(s), \eta(s))l(s)ds, \tag{2.76}
$$

and

$$
X^{\mathrm{d}}(t) = \sum_{\tau \in D_\Gamma \cup (\tau \le t)} \Delta X(\tau) = \sum_{\tau \in D_\Gamma \cup \{\tau \le t\}} \int_{\Gamma(\tau-)}^{\Gamma(\tau)} B(Y(s), \tau)l(s)ds. \tag{2.77}
$$

Now define a vector-valued measure $\mathscr{U}(dt)$ by the formula

$$
\mathscr{U}(A) = \int_{\{s:\eta(s)\in A\}} (1-\alpha(s))l(s)ds.
$$

Then $\mathscr{U}(dt)$ has the properties required in the statement of the theorem and its distribution function is

$$
\begin{aligned}
U(t) = \mathscr{U}([0, t]) &= \int_{\{s:\eta(s)\in[0,t]\}} (1-\alpha(s))l(s)ds = \\
&= \int_{\{s\in[0,\Gamma(t)]\}} (1-\alpha(s))l(s)ds = \int_0^{\Gamma(t)} (1-\alpha(s))l(s)ds.
\end{aligned}
$$

The set of jump points of the function $U(t)$ lies in D_Γ, so $U(t)$ admits the decomposition

$$U(t) = U^{\mathrm{c}}(t) + U^{\mathrm{d}}(t),$$

where

$$U^{\mathrm{c}}(t) = \int_0^{\Gamma(t)} \mathbf{I}\{s : \eta(s) \in [0,T] \setminus D_\Gamma\}(s)(1-\alpha(s))l(s)ds$$

and

$$U^{\mathrm{d}}(t) = \sum_{\tau \in D_\Gamma \cup \{\tau \le t\}} \Delta U(\tau) = \sum_{\tau \in D_\Gamma \cup \{\tau \le t\}} \int_{\Gamma(\tau-)}^{\Gamma(\tau)} (1-\alpha(s))l(s)ds.$$

Using the change-of-variable formula (2.69) one can rewrite the relations (2.75)–(2.77) as follows.

Setting $\eta(s)=\tau$ in (2.75) and taking into account that $\alpha(s)ds=d\eta(s)$, $\eta(\Gamma(t))=t$ and $Y(\Gamma(t)) = X(t)$, we have

$$\begin{aligned} X^{\mathrm{a}}(t) &= \int_0^{\Gamma(t)} \alpha(s)F(Y(s),\eta(s))ds = \int_0^{\Gamma(t)} F(Y(s),\eta(s))d\eta(s) = \\ &= \int_0^{\eta(\Gamma(t))} \alpha(s)F(Y(\Gamma(\tau)),\eta(\Gamma(\tau))d\tau = \int_0^t F(X(\tau),\tau))d\tau. \end{aligned} \tag{2.78}$$

In much the same way,

$$\begin{aligned} X^{\mathrm{c}}(t) &= \int_0^{\Gamma(t)} (1-\alpha(s))\mathbf{I}\{s : \eta(s) \in [0,T] \setminus D_\Gamma\}(s)B(Y(s),\eta(s))l(s)ds = \\ &= \int_0^{\Gamma(t)} B(Y(s),\eta(s))dU^{\mathrm{c}}(\eta(s)) = \int_0^t B(X(\tau),\tau)dU^{\mathrm{c}}(\tau). \end{aligned} \tag{2.79}$$

Our final objective is to derive an expression for the jumps of the function $X(t)$,

$$\Delta X(\tau) = Y(\Gamma(\tau)) - Y(\Gamma(\tau-)) = \int_{\Gamma(\tau-)}^{\Gamma(\tau)} B(Y(s),\tau)l(s)ds.$$

Note that if $s \in [\Gamma(\tau-), \Gamma(\tau)]$, then $\eta(s) = \tau$, $\alpha(s) = 0$, and so $Y(s)$ satisfies the equation

$$\dot{Y}(s) = B(Y(s),\tau)l(s).$$

By Lemma 2.7

$$
\begin{aligned}
Y(\Gamma(\tau)) &= Y(\Gamma(\tau-)) + \Psi\left(Y(\Gamma(\tau-)), \tau, \int_{\Gamma(\tau-)}^{\Gamma(\tau)} l(s)ds\right) = \\
&= Y(\Gamma(\tau-)) + \Psi(Y(\Gamma(\tau-)), \tau, \Delta U(\tau)).
\end{aligned}
$$

Substituting here $X(t) = Y(\Gamma(t))$ we obtain

$$
\Delta X(\tau) = \Psi(X(\tau-), \tau, \Delta U(\tau)). \tag{2.80}
$$

Now combining (2.78)–(2.80) with (2.74) we arrive at the Eq. (2.71), which completes the proof. □

Equations of the form (2.71) provide a universal description of trajectories in DCS with impulse controls. If the measure $\mathscr{U}(dt)$ has only a finite numbers of atoms, that is, a finite number of points τ, where $\mathscr{U}(\{\tau\}) \neq 0$, and $U^c(t) = 0$, then it represents a standard impulse control $\mathscr{I} = \{(\tau, \mathscr{U}(\{\tau\})$. If the measure $\mathscr{U}(dt)$ is absolutely continuous with respect to the Lebesgue measure [322], then it represents an ordinary control $w(t) = \dot{U}(t)$, where $U(t)$ is the distribution function of $\mathscr{U}(dt)$. In the general case we will say that the measure $\mathscr{U}(dt)$ defines a *generalized control.*

We develop the general theory of generalized solutions in nonlinear differential equations with impulse controls in Chap. 4. There it will be shown that a differential equation with measure of a more general type than those considered here provides a general description of generalized solutions. Some basic properties of differential equations with measure are presented in the Appendix.

In Chap. 3 we consider an optimal control problem for DCS with a restricted number of impulses. This is a direct application of the method of discontinuous time change in optimal control. By means of this method, we reduce the original impulse control problem to a standard optimal control problem with bounded control, establish the equivalence of these problems, derive a necessary optimality condition, and demonstrate an example of solution of the observation control problem.

Chapter 3
Optimal Impulse Control with Constrained Number of Impulses

3.1 Introduction and Problem Statement

The classical statement of the impulse control problems [159, 218, 262, 263, 467] assumes, as a rule, that energy-type constraints are imposed on the total intensity of control actions. If no restrictions are imposed on the total number of impulses or/and on the impulse repetition rate, then impulse sliding modes may arise as optimal solutions (see [159, 263] and Examples 1.4 and 1.5). The realization of sliding modes requires an extremely high impulse repetition rate, which for technical considerations is not achievable in some real systems. One of the available methods that take such constraints into account is to restate the optimal control problem as a problem of mathematical programming. In this way one can obtain optimality conditions of Kuhn-Tucker type and solve the problem with the aid of numerical procedures. On the other hand, in optimal control problems there are some more powerful tools, such as the Pontryagin maximum principle [53, 133, 372], which is much more effective than mathematical programming methods. However, to derive the maximum principle for DCS, one has to justify the convexity of the attainability set for the state of the system after the application of each impulse action [12, 13, 439]. Generally, one is dealing with a rather complicated problem that can be effectively solved only for linear systems.

One of the advantages of our approach to DCS is that no such convexity conditions are needed. The idea is that the dynamics of robust DCS can be represented as the result of approximation of impulse controls by continuous controls, for which the maximum principle is valid without any convexity requirements. The remarkable result is that robust DCS also have this property.

In this chapter, we derive optimality conditions in the maximum-principle form for problems in which the restrictions imposed on the number of impulses and on the impulse repetition rate are of critical importance. Since we deal with the robust DCS, one can resort to the method of discontinuous time change and by that reduce

B. Miller, E. Rubinovich, *Optimization of Dynamical Systems with Impulse Controls and Shocks*, Systems & Control: Foundations & Applications,
https://doi.org/10.1007/978-3-031-64124-4_3

the original impulse control problem to an equivalent auxiliary non-impulse optimal control problem in a conventional form.

The solution of the observation control problem is given as an illustration in Sect. 7.5.

Let the dynamics of the DCS under consideration be governed by the equation

$$X(t) = X_0 + \int_0^t F(X(s), s)ds + \sum_{\tau_i \le t} \Psi(X(\tau_i-), \tau_i, w_i), \tag{3.1}$$

where $X(t) \in \mathbb{R}^n$ and $t \in [0, T]$.

We define the impulse control as a set of pairs $\mathscr{I} = \{(w_i, \tau_i) : i = 1, \dots, N\}$, where $N < \infty$ is the (fixed) number of impulses. The intensities of the impulse actions are subject to the constrains

$$w_i \in W \subset \mathbb{R}^m, \tag{3.2}$$

$$\sum_{i=1}^{N} \|w_i\| \le M, \tag{3.3}$$

where W is a bounded closed convex set in $\mathbb{R}^m$, $\|w_i\| = \sum_{k=1}^{m} |w_i^k|$, w_i^k is the k-th component of the vector w_i.

The instants τ_i of impulse actions satisfy the monotonicity condition (2.3), i.e.,

$$0 = \tau_0 < \tau_1 < \dots < \tau_N < \tau_{N+1} = T,$$

where T is the terminal time of the controlled process. The functions F and Ψ satisfy the Assumption 2.1, which guarantee the existence and uniqueness of the solution to Eq. (3.1) (see Theorem 2.1). We also require that Ψ satisfies the robustness conditions (see Theorem 2.2 and (2.7)).

Moreover, we assume the existence of derivatives

$$\Psi_t'(x, t, w) \quad \text{and} \quad \Psi_w'(x, t, w)\big|_{w=0} = B(x, t) \tag{3.4}$$

which satisfy the Lipschitz condition in the variables (x, t). They are an n-dimensional vector-function and $n \times m$ matrix-valued function, respectively.

The admissible trajectory $X(t)$ and the appropriate impulse control $\mathscr{I}$ are subject to the following mixed-type constraints of inequality type ($j = 1, \dots, k_1$):

$$\begin{gathered} g_j\big(X(\tau_1-), \dots, X(\tau_N-), X(\tau_1), \dots, X(\tau_N), X(T), w_1, \dots, w_N, \\ \tau_1, \dots, \tau_N, T\big) \le 0 \end{gathered} \tag{3.5}$$

where the functions g_j are continuous and differentiable with respect to all arguments.

In addition to (3.5), there is an explicit constraint on the impulse repetition rate, namely,

$$\tau_{i-1} + T_i \leq \tau_i \leq T - \sum_{k=i+1}^{N+1} T_k, \quad i = 1, \dots, N \tag{3.6}$$

where $\tau_0 = 0$ and $T_i, i = 1, \dots, N+1$ are given non-negative constants.

We call $\mathscr{I}$ an *admissible control* if $\mathscr{I}$ and the corresponding path $X(t)$ satisfy the constraints (3.2), (3.3), (3.5), and (3.6).

The optimal control problem asks for minimization of the performance criterion

$$J(X(\cdot)) = G(X(T), T) \tag{3.7}$$

over the set of admissible controls, where $G(x, T)$ is a smooth function.

3.2 Auxiliary Optimal Control Problem

3.2.1 Reduction to the Auxiliary Problem

Let us reduce the original impulse control problem to an auxiliary problem with ordinary controls. To this end we introduce the change of time $t = t(\theta)$, $\theta \in [0, 2N+1]$, so that

$$\overset{\circ}{t}(\theta) = \beta(\theta)v(\theta), \quad t(0) = t_0, \quad t(2N+1) = T \tag{3.8}$$

where $\circ$ stands for differentiation with respect to θ, the function $v(\theta) \in [0, \infty)$ is a new scalar control, and $\beta(\theta)$ is an indicator of "odd" subintervals of the segment $[0, 2N+1]$ (see Fig. 3.1):

$$\beta(\theta) = \begin{cases} 1, & \theta \in [2i-2,\ 2i-1], \quad \text{in "odd" intervals,} \\ 0, & \theta \in (2i-1,\ 2i), \quad \text{in "even" intervals.} \end{cases} \tag{3.9}$$

Consider the auxiliary optimal control problem for the dynamical system in the "new" time variable $\theta \in [0, 2N+1]$, $i = 1, \dots, N$

$$\overset{\circ}{y}_i(\theta) = \alpha_i(\theta) f(z(\theta), t(\theta), v(\theta), l(\theta), \theta), \tag{3.10}$$

$$\overset{\circ}{z}_i(\theta) = \beta_i(\theta) f(z(\theta), t(\theta), v(\theta), l(\theta), \theta), \tag{3.11}$$

$$\overset{\circ}{z}(\theta) = f(z(\theta), t(\theta), v(\theta), l(\theta), \theta), \tag{3.12}$$

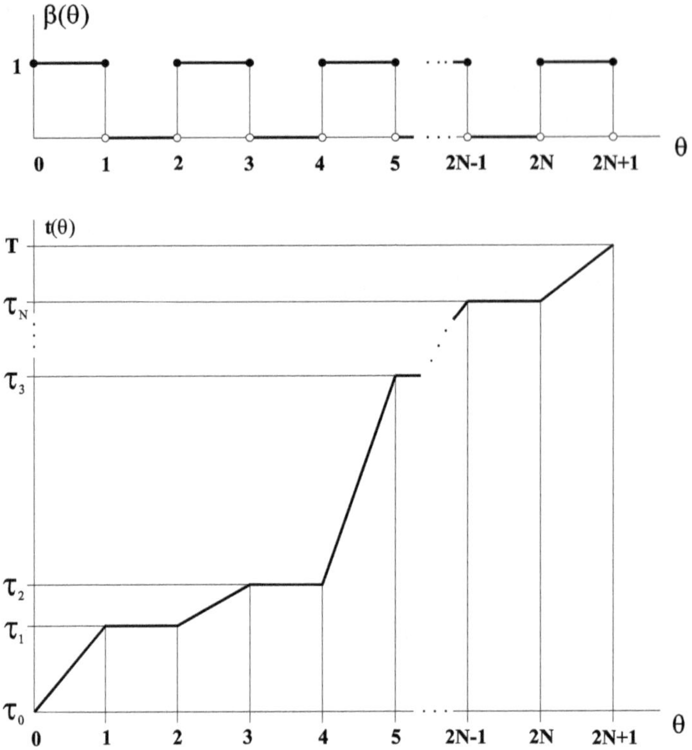

Fig. 3.1 Discontinuous time change in the problem with a constrained number of impulses

$$\overset{\circ}{t}(\theta) = \beta(\theta)v(\theta), \tag{3.13}$$

$$\overset{\circ}{t}_i(\theta) = \alpha_i(\theta)\beta(\theta)v(\theta), \tag{3.14}$$

$$\overset{\circ}{\eta}_i(\theta) = (\beta_i(\theta) - \alpha_i(\theta))l(\theta), \tag{3.15}$$

with the boundary conditions

$$z(0) = z_i(0) = y_i(0) = X_0, \tag{3.16}$$

$$t(0) = t_i(0) = t_0, \tag{3.17}$$

$$\eta_i(0) = 0, \tag{3.18}$$

where

$$
\begin{aligned}
&f(z(\theta), t(\theta), v(\theta), l(\theta), \theta) = \\
&\quad = \beta(\theta)v(\theta)F(z(\theta), t(\theta)) - (1-\beta(\theta))B(z(\theta), t(\theta))l(\theta).
\end{aligned} \tag{3.19}
$$

Here the functions $\alpha_i(\theta)$ and $\beta_i(\theta)$ are defined as

$$
\begin{aligned}
\alpha_i(\theta) &= \begin{cases} 1, & \text{if } \theta \le 2i-1, \\ 0, & \text{if } \theta > 2i-1, \end{cases} \quad i = 1, \dots, N, \\
\beta_i(\theta) &= \begin{cases} 1, & \text{if } \theta \le 2i, \\ 0, & \text{if } \theta > 2i, \end{cases} \quad i = 1, \dots, N.
\end{aligned} \tag{3.20}
$$

In this auxiliary problem the role of admissible controls is played by the functions $v(\theta)$ and $l(\theta)$, defined for $\theta \in [0, 2N+1]$ and satisfying the constraints

$$
0 \le v(\theta) < \infty, \tag{3.21}
$$

$$
l(\theta) \in W \in \mathbb{R}^m, \tag{3.22}
$$

$$
\sum_{i=1}^{N} \sum_{k=1}^{m} \left| \int_{2i-1}^{2i} l^k(\theta) d\theta \right| \le M, \tag{3.23}
$$

where l^k is the k-th component of the vector l. The quality of the control process is defined by the terminal criterion

$$
\widetilde{J}(z(\cdot),\ t(\cdot)) = G(z(2N+1),\ t(2N+1)), \tag{3.24}
$$

to be minimized on the set of admissible controls under the following terminal conditions:

$$
\begin{aligned}
&g_j(y_1(2N+1), \dots, y_N(2N+1),\ z_1(2N+1), \dots, z_N(2N+1), \\
&\quad z(2N+1),\ \eta_1(2N+1), \dots, \eta_N(2N+1),\ t_1(2N+1), \dots \\
&\quad \dots, t_N(2N+1),\ t(2N+1)) \le 0, \quad j = 1, \dots, k_1,
\end{aligned} \tag{3.25}
$$

$$
\begin{aligned}
T_i \le & t_i(2N+1) - t_{i-1}(2N+1) \le \\
& \le t(2N+1) - t_{i-1}(2N+1) - \sum_{k=i+1}^{N+1} T_k, \quad i = 1, \dots, N,
\end{aligned} \tag{3.26}
$$

where $t_0(2N+1) = 0$.

3.2.2 Equivalence of the Primary and Auxiliary Problems

The equivalence of the primary impulse control problem (3.1)–(3.7) and the auxiliary problem (3.10)–(3.26) is established by the following result, which is the extension of Theorem 2.12 to problems with constraints (3.25) and (3.26).

Lemma 3.1 *For a given admissible impulse control* $\mathscr{I} = \{(w_i, \tau_i), i = 1, \ldots, N\}$*, there exists an admissible control* $\{v(\theta), l(\theta)\}$ *in the auxiliary control problem, such that the corresponding solution* $X(t)$*,* $t \in [0, T]$*, of (3.1) and the solution of the auxiliary problem* $\{z(\theta), t(\theta)\}$ *satisfy the following discontinuous time change relations*

$$X(t) = z(\mu(t)) \quad \text{and} \quad X(T) = z(2N + 1), \tag{3.27}$$

where $\mu(t) = \sup\{\theta : t(\theta) = t\}$*.*

The corresponding phase variables $\{z_i(\theta), y_i(\theta), t_i(\theta)\}$ *of the auxiliary problem satisfy the relations*

$$X(\tau_i -) = y_i(2N + 1), \quad X(\tau_i) = z_i(2N + 1), \tag{3.28}$$

$$w_i = \eta_i(2N + 1), \tag{3.29}$$

$$\tau_i = t_i(2N + 1). \tag{3.30}$$

Moreover, the controls $\{v(\theta), l(\theta)\}$ *can be chosen as step functions on the corresponding intervals, for example, as follows:*

$$v(\theta) = \begin{cases} \tau_i - \tau_{i-1}, & \text{if } \theta \in [2i - 2,\ 2i - 1],\ i = 1, \ldots, N + 1, \\ \text{an arbitrary non-negative function,} & \\ \text{if } \theta \in (2i - 1,\ 2i),\ i = 1, \ldots, N, & \end{cases} \tag{3.31}$$

$$l(\theta) = \begin{cases} w_i, & \text{if } \theta \in (2i - 1,\ 2i),\ i = 1, \ldots, N, \\ \text{an arbitrary } W\text{-valued function} & \\ \text{if } \theta \in [2i - 2,\ 2i - 1],\ i = 1, \ldots, N + 1. & \end{cases} \tag{3.32}$$

Proof of Lemma 3.1 Consider the intervals

$$t \in [\tau_{i-1}, \tau_i], \quad i = 1, \ldots, N + 1, \quad \tau_0 = 0,\ \tau_{N+1} = T,$$

and formally set $w_{N+1} = 0$. It follows from the definition of $\mu(t)$ (see also Fig. 3.2) that

$$\mu(\tau_i -) = 2i - 1, \quad \mu(\tau_i) = 2i.$$

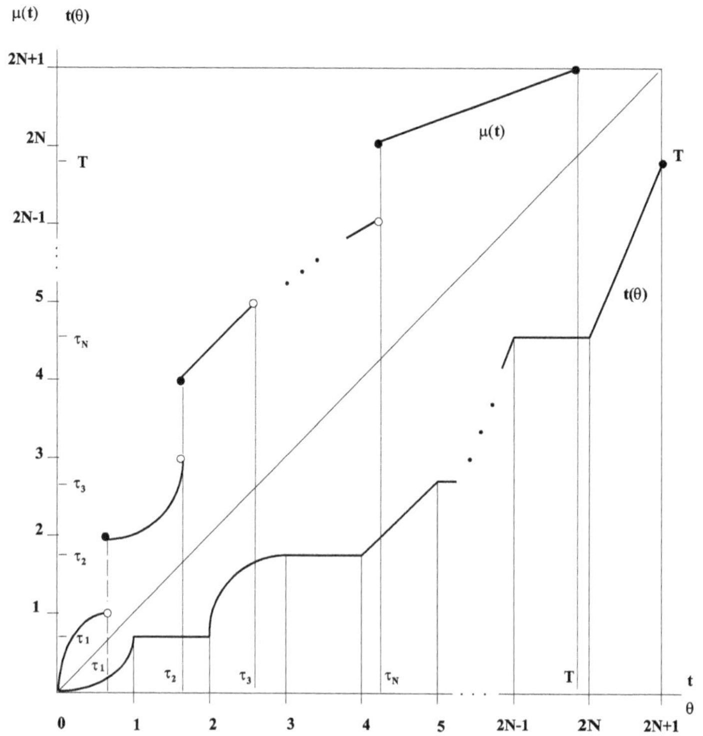

Fig. 3.2 Discontinuous time change in the problem with constrained number of impulses. Direct and inverse transformation

In addition, for any $s \in [0, T]$ we have, by Lemma 2.4,

$$t(\mu(s)) = s. \tag{3.33}$$

Let us take the controls $\{v(\theta), l(\theta)\}$ given by the formulas (3.31) and (3.32) and derive the relation between $z(\mu(\tau_i))$ and $z(\mu(\tau_i-))$. In view of (3.12), (3.16), and (3.13), the function $z(\theta)$ satisfies for $\theta \in [\mu(\tau_i-), \mu(\tau_i)] = [2i-1, 2i]$ the equation

$$z(\theta) = z(2i-1) + \int_{2i-1}^{\theta} B(z(\vartheta), t(\vartheta))w_i \, d\vartheta, \tag{3.34}$$

where $t(\vartheta) \equiv t(2i-1) = \tau_i$ for $\vartheta \in [2i-1, 2i]$. Now, by Definition 2.6 and Lemma 2.7, Eq. (3.34) implies that

$$z(2i) = z(2i-1) + \Psi(z(\mu(2i-1)), \tau_i, w_i),$$

and

$$z(\mu(\tau_i)) = z(\mu(\tau_i-)) + \Psi(z(\mu(\tau_i-1)), \tau_i, w_i). \tag{3.35}$$

Hence, for $\theta \in [\mu(\tau_{i-1}), \mu(\tau-)] = [2i-2,\ 2i-1],\ i = 1, \dots, N+1$, we have

$$z(\theta) = z(2i-2) + \int_{2i-2}^{\theta} v(\vartheta)F(z(\vartheta), t(\vartheta))d\vartheta,$$

or

$$z(\mu(t)) = z(\mu(\tau_{i-1})) + \int_{\mu(\tau_{i-1})}^{\mu(t)} v(\theta)F(z(\theta), t(\theta))d\theta. \tag{3.36}$$

By (3.8) and (3.33), for any admissible control $v(\theta)$ we have $v(\theta)d\theta = dt(\theta)$, and hence

$$v(\mu(s))\, d\mu(s) = dt(\mu(s)) = ds. \tag{3.37}$$

The change variable formula (2.69) (see Lemma 2.6) yields

$$z(\mu(t)) = z(\mu(\tau_{i-1})) + \int_{\tau_{i-1}}^{t} F(z(\mu(s)), t(\mu(s)))v(\mu(s))\, d\mu(s)$$

or

$$z(\mu(t)) = z(\mu(\tau_{i-1})) + \int_{\tau_{i-1}}^{t} F(z(\mu(s)), s)\, ds \tag{3.38}$$

where $t \in [\tau_{i-1}, \tau_i]$. By combining (3.35) and (3.38), we obtain for $t \in [\tau_{i-1}, \tau_i]$ the condition

$$\begin{aligned} z(\mu(t)) = z(\mu(\tau_{i-1})) + \int_{\tau_{i-1}}^{t} F(z(\mu(s)), s)\, ds + \\ + \Psi(z(\mu(\tau_i-)), \tau_i, w_i)\mathbf{I}\{t = \tau_i\} \end{aligned} \tag{3.39}$$

(recall that $\mathbf{I}\{A\}$ denotes the indicator function of the set A) and the inclusion $t \in [\tau_{i-1}, \tau_i]$ is guaranteed (due to (3.8)) by any admissible control $v(\theta)$ such that

$$\int_{2i-2}^{2i-1} v(\theta)\, d\theta = \tau_i - \tau_{i-1}, \quad i = 1, \dots, N+1. \tag{3.40}$$

In particular, this last condition is satisfied by step control functions of the form (3.31).

Combining the Eq. (3.39) for all intervals $[\tau_{i-1}, \tau_i]$, $i = 1, \dots, N+1$, we conclude that the function $z(\mu(t))$ satisfies the equation

$$z(\mu(t)) = z(\mu(\tau_0)) + \int_{\tau_0}^{t} F(z(\mu(s)), s)\, ds + \sum_{\tau_i \le t} \Psi(z(\mu(\tau_i-)), \tau_i, w_i).$$

Since $z(\mu(\tau_0)) = z(0) = X(0)$, the uniqueness of the solution of Eq. (3.1) implies that $z(\mu(t)) = X(t)$ for all $t \in [0, T]$, where, by (3.8), $T = t(2N + 1)$. This completes the proof of the first statement of Lemma 3.1.

Finally, we note that the initial conditions and the right-hand sides of Eqs. (3.10) and (3.12) coincide for $\theta \in [0, 2i - 1]$, $i = 1, \ldots, N$, and that $\alpha_i(\theta) = 0$ for $\theta > 2i - 1$. Therefore,

$$y_i(2N + 1) = y_i(2i - 1) = z(2i - 1) = X(\tau_i -),$$

so the first equality in (3.28) is proved. The remaining equalities in (3.28)–(3.30) can be obtained by the same arguments. □

Theorem 3.1 *Let the set of variables*

$$\left\{l^0(\theta), v^0(\theta), t^0(\theta), y_i^0(\theta), z_i^0(\theta), z^0(\theta), t_i^0(\theta), \eta_i^0(\theta), i = 1, \ldots, N\right\},$$

consist of the optimal controls and trajectories in the auxiliary problem and let $G(z^0(2N + 1),\ t^0(2N + 1))$ *be the corresponding value of the performance criterion.*

Then the impulse control $\mathscr{I} = \{(w_i^0,\ \tau_i^0),\ i = 1, \ldots, N\}$, *where*

$$w_i^0 = \int_{2i-1}^{2i} l^0(\theta)\, d\theta \quad \text{and} \quad \tau_i^0 = t_i^0(2N + 1) = t^0(2i), \tag{3.41}$$

is optimal in the original problem.

Moreover, the corresponding optimal trajectory $X^0(t)$ *with the terminal time* T^0 *and the pair* $\{z^0(\theta), t^0(\theta)\}$ *satisfy the relations*

$$\begin{aligned} &X^0(t) = z^0(\mu^0(t)),\\ &G(X^0(T^0),\ T^0) = G(z^0(2N + 1),\ t^0(2N + 1)), \end{aligned} \tag{3.42}$$

where $\mu^0(t) = \sup\{\theta :\ t^0(\theta) = t\}$.

Remark 3.1 This theorem establishes the equivalence of the original impulse control problem and the auxiliary problem.

Proof of Theorem 3.1

(i) Let us define w_i^0 by formula (3.41) and put

$$\widetilde{l}(\theta) = \begin{cases} w_i^0, & \text{if } \theta \in (2i - 1,\ 2i),\ i = 1, \ldots, N,\\ \text{an arbitrary } W\text{-valued function,} & \\ \text{if } \theta \in [2i - 2,\ 2i - 1],\ i = 1, \ldots, N + 1, & \end{cases} \tag{3.43}$$

$$\tau_i^0 = t_i^0(2N + 1), \quad i = 1, \ldots, N, \quad \tau_0^0 = t_0,\ \tau_{N+1}^0 = t^0(2N + 1),$$

$$\widetilde{v}_i = \int_{2i-2}^{2i-1} v^0(\theta)\, d\theta = \tau_i^0 - \tau_{i-1}^0, \quad i = 1, \ldots, N+1,$$

$$\widetilde{v}(\theta) = \begin{cases} \widetilde{v}_i, & \text{if } \theta \in [2i-2,\ 2i-1],\ i = 1, \ldots, N+1, \\ \text{an arbitrary non-negative function,} \\ \text{if } \theta \in (2i-1,\ 2i),\ i = 1, \ldots, N. \end{cases}$$

Then, by Lemma 3.1,

$$\begin{aligned} X^0(t) &= \widetilde{z}(\widetilde{\mu}(t)), \quad T^0 = \widetilde{t}(2N+1), \\ \tau_i^0 &= \widetilde{t}_i(2N+1), \quad u_i^0 = \widetilde{\eta}_i(2N+1), \end{aligned} \tag{3.44}$$

and

$$G(X^0(T^0), T^0) = G(\widetilde{z}(2N+1), \widetilde{t}(2N+1)), \tag{3.45}$$

where the "tilde" refers to a solution of (3.10)–(3.18) with the controls $\widetilde{l}(\theta),\ \widetilde{v}(\theta)$.

Integrating Eqs. (3.13) and (3.14) with the controls $v^0(\theta)$ and $\widetilde{v}(\theta)$, we obtain

$$\widetilde{t}(\theta) = t^0(\theta), \quad \widetilde{t}_i(\theta) = t_i^0(\theta), \quad i = 1, \ldots, N, \tag{3.46}$$

and so $\widetilde{\mu}(t) = \mu^0(t) = \sup\{\theta : t^0(\theta) = t\}$.

Next, we have to show that

$$\widetilde{z}(\widetilde{\mu}(t)) = \widetilde{z}(\mu^0(t)) = z^0(\mu^0(t)). \tag{3.47}$$

From (3.12) and (3.16) it follows that

$$z^0(\theta) = z^0(2i-1) + \int_{2i-1}^{\theta} B(z^0(\vartheta), t^0(\vartheta)) l^0(\vartheta)\, d\vartheta$$

for $\theta \in [2i-1, 2i]$, $i = 1, \ldots, N$, and, by Lemma 2.7,

$$z^0(2i) = z^0(2i-1) + \Psi\big(z^0(2i-1), \tau_i^0, \int_{2i-1}^{2i} l^0(\theta)\, d\theta\big).$$

In view of the relations (3.41) we finally obtain

$$z^0(\mu^0(\tau_i^0)) = z^0(\mu^0(\tau_i^0-)) + \Psi(z^0(\mu^0(\tau_i^0-)), \tau_i^0, w_i^0). \tag{3.48}$$

Similarly, (3.43) and (3.46) imply that

$$\widetilde{z}(\theta) = \widetilde{z}(2i-1) + \int_{2i-1}^{\theta} B(\widetilde{z}(\vartheta), t^0(\vartheta))u^0(\vartheta)\, d\vartheta,$$

whence

$$\widetilde{z}(\mu^0(\tau_i^0)) = \widetilde{z}(\mu^0(\tau_i^0-)) + \Psi(z^0(\mu^0(\tau_i^0-)),\ \tau_i^0,\ w_i^0). \tag{3.49}$$

Then, as in the proof of Lemma 3.1, we deduce that

$$\widetilde{z}(\mu^0(t)) = \widetilde{z}(\mu^0(\tau_{i-1}^0)) + \int_{\tau_{i-1}^0}^{t} F(\widetilde{z}^{\,0}(\mu^0(s)), s)\, ds \tag{3.50}$$

and

$$z(\mu^0(t)) = z(\mu^0(\tau_{i-1}^0)) + \int_{\tau_{i-1}^0}^{t} F(z^0(\mu^0(s)), s)\, ds \tag{3.51}$$

for $t \in [\tau_{i-1}^0,\ \tau_i^0-]$. Combining (3.49), (3.50) and (3.48), (3.51), we obtain for $t \in [\tau_{i-1}^0, \tau_i^0]$ the two equations

$$\begin{aligned} \widetilde{z}(\mu^0(t)) &= \widetilde{z}(\mu^0(\tau_{i-1}^0)) + \int_{\tau_{i-1}^0}^{t} F(\widetilde{z}(\mu^0(s)), s) ds + \\ &\quad + \Psi(\widetilde{z}(\mu^0(\tau_i^0-)), u_i^0, \tau_i^0)\mathbf{I}\{t = \tau_i^0\}, \\ z^0(\mu^0(t)) &= z^0(\mu^0(\tau_{i-1}^0)) + \int_{\tau_{i-1}^0}^{t} F(z^0(\mu^0(s)), s)\, ds + \\ &\quad + \Psi(z^0(\mu^0(\tau_i^0-)), u_i^0, \tau_i^0)\mathbf{I}\{t = \tau_i^0\} \end{aligned} \tag{3.52}$$

which differ from (3.39) only by the notation of arguments. These equations involve the same endpoints of the intervals $[\tau_{i-1}^0, \tau_i^0]$, $i = 1, \ldots, N+1$, and the initial conditions coincide:

$$\widetilde{z}(0) = \widetilde{z}(\mu^0(\tau_0^0)) = X_0, \quad z^0(0) = z(\mu^0(\tau_0^0)) = X_0.$$

Consequently, (3.47) holds. Now (3.47), (3.44) and (3.45) imply that

$$\begin{aligned} &X^0(t) = z^0(\mu^0(t)), \\ &G(X^0(T^0), T^0) = G(z^0(2N+1), t^0(2N+1)). \end{aligned} \tag{3.53}$$

(ii) To prove the optimality of the control $\mathscr{I}^0$ we assume that there exists an impulse control $\widehat{\mathscr{I}} = \{(\widehat{w}_i,\ \widehat{\tau}_i),\ i = 1, \ldots, N\}$, different from $\{(w_i^0,\ \tau_i^0)\}$, such the value of the performance criterion corresponding to this control satisfies the inequality

$$G(\widehat{X}(\widehat{T}), \widehat{T}) < G(X^0(T^0), T^0), \tag{3.54}$$

where $\widehat{X}(t)$ and $\widehat{T}$ are the trajectory and the terminal time corresponding to the control $\widehat{\mathscr{I}}$.

Let $(\widehat{z}(\theta),\ \widehat{t}(\theta))$ be the solutions of Eqs. (3.12) and (3.13) with the initial conditions (3.16) and (3.17), and with the control

$$l(\theta) = \begin{cases} \widehat{u}_i, & \text{if } \theta \in (2i-1,\ 2i),\ i = 1, \ldots, N, \\ \text{an arbitrary } W - \text{valued function}, \\ \text{if } \theta \in [2i-2,\ 2i-1],\ i = 1, \ldots, N+1, \end{cases}$$

and the arbitrary admissible control $\widehat{v}(\theta)$ which satisfies (3.40), i.e.,

$$\int_{2i-2}^{2i-1} \widehat{v}(\theta)\, d\theta = \widehat{\tau}_i - \widehat{\tau}_{i-1}, \quad 1, \ldots, N+1, \quad \widehat{\tau}_0 = t_0, \widehat{\tau}_{N+1} = \widehat{t}(2N+1).$$

Then, thanks to the optimality of the controls $\{v^0, l^0\}$, we obviously have

$$G(z^0(2N+1),\ t^0(2N+1)) \le G(\widehat{z}(2N+1),\ \widehat{t}(2N+1)). \tag{3.55}$$

At the same time, by Lemma 3.1, $\widehat{X}(t) = \widehat{z}(\widehat{\mu}(t))$, where $\widehat{\mu}(t) = \sup\{\theta : \widehat{t}(\theta) = t\}$ and $\widehat{T} = \widehat{t}(2N+1)$, $\widehat{X}(\widehat{T}) = \widehat{z}(2N+1)$. Therefore,

$$G(\widehat{X}(\widehat{T}), \widehat{T}) = G\left(\widehat{z}(2N+1),\ \widehat{t}(2N+1)\right). \tag{3.56}$$

Relations (3.53)–(3.56) imply the contradictory chain of inequalities

$$G\left(X^0(T^0), T^0\right) = G\left(z^0(2N+1), t^0(2N+1)\right) \le$$
$$\le G(\widehat{z}(2N+1), \widehat{t}(2N+1)) \le G\left(\widehat{X}(\widehat{T}), \widehat{T}\right) < G\left(X^0(T^0), T^0\right),$$

which completes the proof of the theorem. □

Theorem 3.1 and the representation (3.41) admit the following

Corollary 3.1 *Let* $\{v^0(\theta), l^0(\theta)\}$ *be an optimal control and let*

$$\widetilde{v}(\theta) = \begin{cases} \widetilde{v}_i = \tau_i^0 - \tau_{i-1}^0, & \text{if } \theta \in [2i-2,\ 2i-1],\ i = 1, \dots, N+1, \\ \text{an arbitrary non-negative function,} \\ \text{if } \theta \in (2i-1,\ 2i),\ i = 1, \dots, N, \end{cases}$$

$$\widetilde{l}(\theta) = \begin{cases} l_i = \int_{2i-1}^{2i} l^0(\theta)\, d\theta, & \text{if } \theta \in (2i-1,\ 2i), \quad i = 1, \dots, N, \\ \text{an arbitrary } W\text{-valued function,} \\ \text{if } \theta \in [2i-2,\ 2i-1], \quad i = 1, \dots, N+1. \end{cases} \tag{3.57}$$

Then the new control $\{\widetilde{v}(\theta), \widetilde{l}(\theta)\}$ *thus defined will also be optimal in the auxiliary problem.*

3.3 Necessary and Sufficient Optimality Conditions

3.3.1 Necessary Optimality Conditions in the Auxiliary Problem

Lemma 3.1 and Corollary 3.1 show that one can search for the optimal solution in the auxiliary problem within the class of step control functions $\{v(\theta), l(\theta\})$ that are constant in "odd" and "even" intervals and have the form (3.31) and (3.57), respectively.

To derive necessary optimality conditions in the auxiliary problem, we use the classical maximum principle for an optimal control problem with terminal constraints [173]. First, we reduce the auxiliary problem to standard form by introducing a new phase variable and changing the control variables. Specifically, we put

$$\overset{\circ}{\zeta}(\theta) = (1 - \beta(\theta))\, \|l(\theta)\|, \quad \zeta(0) = 0, \tag{3.58}$$

where $\|l(\theta)\| = \sum_{k=1}^{m} |l^k(\theta)|$, with l^k the k-th component of the m-dimensional piecewise-constant vector $l(\theta)$. In terms of the variable $\zeta(\theta)$, the constraint (3.23) has the form $\zeta(2N+1) \le M$, or

$$g_{k_1+1}(\zeta(2N+1)) = \zeta(2N+1) - M \le 0. \tag{3.59}$$

We also introduce the variables $\rho_i(\theta)$, $i = 1, \dots, N$, subject to the differential equations

$$\overset{\circ}{\rho}_i(\theta) = 0 \tag{3.60}$$

with the boundary conditions

$$\begin{cases} \rho_i(2N+1) = t(2N+1) - t_{i-1}(2N+1) - \sum\limits_{k=i+1}^{N+1} T_k, \\ \rho_{N+1}(2N+1) = t(2N+1) - t_N(2N+1), \end{cases} \tag{3.61}$$

where $t_0(2N+1) = t_0$. Later we will use these conditions in the form of equality constraints

$$h_i(\rho_i(2N+1), t_{i-1}(2N+1), t(2N+1)) = 0, \quad i = 1, \dots, N+1. \tag{3.62}$$

We include these new variables in the set of phase variables, with corresponding Eqs. (3.58) and (3.60) and constraints (3.59) and (3.62).

Since $\rho_i(\theta) = \text{const}$, one can rewrite inequality (3.26) in the form

$$T_i \le v_i \le \rho_i(\theta). \tag{3.63}$$

Now let us replace the control variable, which is constant in each "odd" interval and satisfies the inequality (3.63), by the new control $u_i \in [-1,\ 1]$, as follows:

$$v_i = \frac{1}{2}(\rho_i - T_i)u_i + \frac{1}{2}(\rho_i + T_i), \quad i = 1, \dots, N+1, \tag{3.64}$$

where, by definition, $u_{N+1} = 1$.

In other words, we replaced the step control function $v(\theta) \ge 0$ with the new step control function $u(\theta) \in [-1,\ 1]$ given by

$$u(\theta) = \begin{cases} u_i \in [-1, 1], \quad \text{if } \theta \in [2i-2, 2i-1],\ i = 1, \dots, N+1, \\ \text{an arbitrary } [-1, 1]\text{-valued function,} \\ \text{if } \theta \in (2i-1, 2i),\ i = 1, \dots, N. \end{cases} \tag{3.65}$$

Hence, one can use the function

$$v(\rho, u, \theta) = \sum_{i=1}^{N} \left[\frac{1}{2}(\rho_i - T_i)u + \frac{1}{2}(\rho_i + T_i) \right] \mathbf{I}\{\theta \in [2i-2, 2i-1]\}, \tag{3.66}$$

which is continuous with respect to the phase variables and a step function with respect to the "new" time variable θ.

In terms of the new variables, the auxiliary problem is the standard optimal control problem with terminal constraints and where the right-hand sides of the differential equations describing the dynamical system are step functions. This is a class of control problems for which the Pontryagin maximum principle is valid [173]. The method applied above was known for the optimal control problems with intermediate conditions [236, 251, 439], and was subsequently extended by the authors [290] to systems with impulse controls. To formulate the maximum principle we introduce the following notations:

1. $\mathscr{X} = \{t, y_i, z_i, z, t_i, \eta_i, \zeta, \rho_i, \rho_{N+1}, \quad i = 1, \ldots, N,\}$ is the set of phase variables in the auxiliary problem;
2. $\mathscr{P} = \{p_t, p_{y_i}, p_{z_i}, p_z, p_{t_i}, p_{\eta_i}, p_\zeta, p_{\rho_i}, p_{\rho_{N+1}}, \quad i = 1, \ldots, N\}$ is the set of conjugate variables;
3. $\Lambda = \{\lambda_0, \ldots, \lambda_{k_1+1}\}$ is a vector of non-negative numbers;
4. $\mathscr{V} = \{\nu_1, \ldots, \nu_{N+1}\}$ is a vector of dimension $N + 1$.

Next, we define the associated Hamilton and Lagrange functions by (the argument θ is omitted):

$$
\begin{aligned}
\mathscr{H}(\mathscr{X}, \mathscr{P}, \theta) = & \Big\langle \sum_{i=1}^{N} (\alpha_i p_{y_i} + \beta_i p_{z_i}) + p_z, f(z, t, v(\rho, u), l) \Big\rangle + \\
& + \sum_{i=1}^{N} \Big(\alpha_i \beta v(\rho, u) p_{t_i} + (\beta_i - \alpha_i) p^*_{\eta_i} l \Big) + \\
& + \beta v(\rho, u) p_t + (1 - \beta) \, \|l\| \, p_\zeta
\end{aligned}
\tag{3.67}
$$

and

$$
\mathscr{L}(\mathscr{X}, \Lambda, \mathscr{V}) = \lambda_0 G(\mathscr{X}) + \sum_{j=1}^{k_1+1} \lambda_j g_j(\mathscr{X}) + \sum_{i=1}^{N+1} \nu_i h_i(\mathscr{X}),
\tag{3.68}
$$

respectively. Recall that the symbol $\langle a, b \rangle$ stands for the scalar product $\sum_{i=1}^{n} a_i b_i$ of two vectors $a, b \in \mathbb{R}^n$.

The maximum principle for the auxiliary problem is stated as follows (see [173]).

Theorem 3.2 *Let $\{l^0(\theta), u^0(\theta), \mathscr{X}^0(\theta)\}$ be the optimal controls and the corresponding trajectory in the auxiliary problem. Then there exist a vector-function $\mathscr{P}(\theta)$ of the conjugate variables, and constant vectors Λ and $\mathscr{V}$, with the following properties:*

(1) *These variables and constants do not vanish simultaneously and satisfy the following system of differential equations (where the argument θ is omitted, and the superscript ^{0} indicates that the functions are computed for the optimal values of variables):*

$$
\overset{\circ}{p}_{y_i} = 0, \quad \overset{\circ}{p}_{z_i} = 0, \quad \overset{\circ}{p}_{t_i} = 0, \quad \overset{\circ}{p}_{\eta_i} = 0, \quad \overset{\circ}{p}_{\zeta} = 0,
\tag{3.69}
$$

i.e., p_{t_i}, p_{z_i}, p_{t_i}, p_{η_i}, p_ς *are constant functions, and*

$$\overset{\circ}{p}_z = -\left[f_z'(z^0, t^0, v(\rho^0, u^0), l^0)\right]^* \left(p_z + \sum_{j=1}^{N}\left(\alpha_j p_{y_j} + \beta_j p_{z_j}\right)\right), \tag{3.70}$$

$$\overset{\circ}{p}_t = -\left[f_t'(z^0, t^0, v(\rho^0, u^0), l^0)\right]^* \left(p_z + \sum_{j=1}^{N}\left(\alpha_j p_{y_j} + \beta_j p_{z_j}\right)\right), \tag{3.71}$$

$$\overset{\circ}{p}_{\rho_i} = -\mathbf{I}\{\theta \in [2i-2, 2i-1]\}\frac{1}{2}(u_i^0 + 1) \times$$
$$\times \left(p_t + \sum_{j=1}^{N} p_{t_j} + \left\langle p_z + \sum_{j=i-1}^{N}(p_{y_j} + p_{z_j}), F(z^0, t^0)\right\rangle\right), \tag{3.72}$$

$$\overset{\circ}{p}_{\rho_{N+1}} = -\mathbf{I}_{\{\theta\in[2N,2N+1]\}}[p_t + p_z^* F(z^0, t^0)] \tag{3.73}$$

where the notation $v(\rho^0, u^0)$ *corresponds to the expression (3.66),* f_z' *and* f_t' *are the partial derivatives of the function* f *with respect to* z *and* t*, respectively, and* $*$ *stands for transposition.*

(2) *When* $\theta = 2N + 1$*, the following transversality conditions are satisfied (the arguments of the functions* g *and* G *are omitted),* $i = 1, \ldots, N$*:*

$$\begin{aligned}
p_{y_i} &= -\sum_{j=1}^{k_1}\lambda_j\frac{\partial g_j^0}{\partial y_i}, \quad p_{z_i} = -\sum_{j=1}^{k_1}\lambda_j\frac{\partial g_j^0}{\partial z_i},\\
p_z &= -\sum_{j=1}^{k_1}\lambda_j\frac{\partial g_j^0}{\partial z} - \lambda_0\frac{\partial G^0}{\partial z},\\
p_t &= -\sum_{j=1}^{k_1}\lambda_j\frac{\partial g_j^0}{\partial t} - \lambda_0\frac{\partial G^0}{\partial t} + \sum_{k=1}^{N+1}\nu_k,\\
p_{t_i} &= -\sum_{j=1}^{k_1}\lambda_j\frac{\partial g_j^0}{\partial t_i} - \nu_{i+1},\\
p_{\eta_i} &= -\sum_{j=1}^{k_1}\lambda_j\frac{\partial g_j^0}{\partial \eta_i}, \quad p_\varsigma = -\lambda_{k_1+1},\\
p_{\rho_i} &= -\nu_i, \quad p_{\rho_{N+1}} = -\nu_{N+1}.
\end{aligned} \tag{3.74}$$

(3) *In addition, when* $\theta = 2N + 1$ *the following complementary slackness conditions are satisfied:*

$$\begin{cases} \lambda_j g_j^0 = 0, \quad j = 1, \ldots, k_1, \\ \lambda_{k_1+1}(\zeta^0 - M) = 0. \end{cases} \tag{3.75}$$

(4) *The Hamiltonian*

$$\mathscr{H}(z, t, \rho_i, p_{y_i}, p_{z_i}, p_z, p_{t_i}, p_{\eta_i}, p_\zeta, l, u) =$$

$$= \Big\langle \sum_{i=1}^{N} (\alpha_i p_{y_i} + \beta_i p_{z_i}) + p_z \, , \, f(z, t, v(\rho, u), l) \Big\rangle +$$

$$+ \sum_{i=1}^{N} \Big(\alpha_i \beta v(\rho, u) p_{t_i} + (\beta_i - \alpha_i) p_{\eta_i}^{*} l \Big) + \beta v(\rho, u), \, p_t + (1 - \beta) \|l\| p_\zeta,$$

satisfies for any admissible controls $\{l, u\}$ *the inequality*

$$\begin{aligned} &\mathscr{H}(z^0, t^0, \rho_i^0, p_{y_i}, p_{z_i}, p_z, p_{t_i}, p_{\eta_i}, p_\zeta, l^0, u^0) \geq \\ &\qquad \geq \mathscr{H}(z^0, t^0, \rho_i^0, p_{y_i}, p_{z_i}, p_z, p_{t_i}, p_{\eta_i}, p_\zeta, l, u). \end{aligned} \tag{3.76}$$

Here and in what follows the dependence of the Hamiltonian on the i*-th component of a vector-function means its dependence on all the components of that vector-function.*

Corollary 3.2 *In view of (3.61) and (3.64), the constants* u_i^0*, which define the piecewise-constant optimal control* $u^0(\theta)$*, have the property that, if one denotes* $i_0 = \min\{i : u_i^0 = 1\}$*, then*

$$u_i^0 = 1 \quad for \ all \ \ i \geq i_0.$$

Remark 3.2 This means that, starting with the index i_0, all the subsequent control impulses (in the primary problem) must be taken with the maximal repetition rate at the end of the interval $[t_0, T^0]$.

3.3.2 Necessary Optimality Conditions in the Primary Problem

In the preceding section we obtained the maximum principle for the auxiliary control problem. Here we derive necessary optimality conditions for the primary problem (3.1)–(3.7). For this purpose we consider the system (3.69)–(3.73) of joint variables in the auxiliary problem and perform the inverse time transformation

$\theta = \mu^0(t) = \sup\{\vartheta : t^0(\vartheta) = t\}$ in the "even" and "odd" intervals. We assume that

$$B(X, t) = \Psi'_w(X, t, w)\big|_{w=0}.$$

is a smooth function. First let us derive a relation between the functions $p_z(\mu^0(\tau_i^0))$ and $p_z(\mu^0(\tau_i^0-))$.

Let $\theta \in (2i-1, 2i) = (\mu^0(\tau_i^0-), \mu^0(\tau_i^0))$. Then, in view of (3.70) and the fact that $t^0(\theta) \equiv t^0(2i-1) = \tau_i^0$, the variables $p_z(\theta)$ and $z^0(\theta)$ satisfy the relations

$$p_z(\theta) = p_z(2i-1) - \int_{2i-1}^{\theta} A^*(\vartheta)\big(p_z(\vartheta) + \sigma_i\big)\, d\vartheta \tag{3.77}$$

and

$$z^0(\theta) = z^0(2i-1) + \int_{2i-1}^{\theta} B(z^0(\vartheta), \tau_i^0) w_i^0\, d\vartheta, \tag{3.78}$$

respectively, where

$$A(\theta) = \Big[B(z, \tau_i^0) w_i^0\Big]'_{z=z^0(\theta)},$$

(see Remark 2.6 for notation) and

$$\sigma_i = \sum_{j=i}^{N} \big(\alpha_j(\theta) p_{y_j} + \beta_j(\theta) p_{z_j}\big) = p_{z_i} + \sum_{j=i+1}^{N} \big(p_{y_j} + p_{z_j}\big) = \text{const.}$$

By Lemma 2.7, the solution of Eq. (3.78) has the form

$$z^0(\theta) = z^0(2i-1) + \Psi\big(z^0(2i-1), \tau_i^0, (\theta - (2i-1)) w_i^0\big). \tag{3.79}$$

Differentiating this expression with respect to the initial condition $z_0 = z^0(2i-1)$ at the point $2i-1$ one obtains

$$Z(\theta) = I_{(n\times n)} + \Psi'_z(z^0(2i-1), \tau_i^0, (\theta - (2i-1)) w_i^0), \tag{3.80}$$

where $Z(\theta) = \partial z^0(\theta)/\partial z_0$ and, as before, $I_{(n\times n)}$ is the $n \times n$ identity matrix. As follows from (3.78),

$$\overset{\circ}{Z}(\theta) = A(\theta) Z(\theta), \quad \text{with} \quad Z(2i-1) = I_{(n\times n)}. \tag{3.81}$$

Let $Q(\theta) = Z^{-1}(\theta)$. Then from (3.80) it follows that

$$Q(\theta) = Z^{-1}(\theta) = \left[I_{(n\times n)} + \Psi'_z(z^0(2i-1), \tau_i^0, (\theta - (2i-1))w_i^0) \right]^{-1}. \tag{3.82}$$

Lemma 3.2 *The matrix function $Q(\theta)$ satisfies the linear matrix differential equation*

$$\overset{\circ}{Q}(\theta) = -Q(\theta)A(\theta) \tag{3.83}$$

with the initial condition $Q(\mu^0(\tau_i^0-)) = Q(2i-1) = I_{(n\times n)}$.

Proof of Lemma 3.2 Differentiating the identity

$$Z(\theta)Q(\theta) = I_{(n\times n)} \tag{3.84}$$

with respect to θ one obtains

$$\overset{\circ}{Z}(\theta)Q(\theta) + Z(\theta)\,\overset{\circ}{Q}(\theta) = 0, \tag{3.85}$$

which in conjunction with (3.81) and (3.84) yields

$$A(\theta) + Z(\theta)\,\overset{\circ}{Q}(\theta) = 0.$$

Therefore, in view of (3.84), $\overset{\circ}{Q}(\theta) = -Q(\theta)A(\theta)$, and this equation coincides with (3.83).

Corollary 3.3 *The matrix $Q(\theta)$ is the fundamental solution of the Eq. (3.77).*

Setting

$$q(\theta) = p_z(\theta) + \sigma_i \quad \text{for} \quad \theta \in (2i-1, 2i), \tag{3.86}$$

one can rewrite Eq. (3.77) as

$$q(\theta) = q(2i-1) - \int_{2i-1}^{\theta} A^*(\vartheta)q(\vartheta)\,d\vartheta. \tag{3.87}$$

By Corollary 3.3 and formula (3.82), the solution of Eq. (3.87) has the form

$$\begin{aligned} q(\theta) &= Q^*(\theta)q(2i-1) = \\ &= \left(\left[I_{(n\times n)} + \Psi'_z(z^0(2i-1), \tau_i^0, (\theta - (2i-1))w_i^0) \right]^{-1} \right)^* q(2i-1). \end{aligned} \tag{3.88}$$

Thanks to the continuity of the function $p_z(\theta)$ (which follows from (3.70)), formulas (3.88) and (3.86) imply that when $\theta = 2i$,

$$p_z(2i)+\sigma_i = \left(\left[I_{(n\times n)}+\Psi_z'(z^0(2i-1),\tau_i^0,w_i^0)\right]^{-1}\right)^*\left[p_z(2i-1)+\sigma_i\right]. \tag{3.89}$$

Therefore,

$$p_z(2i-1) = p_z(2i)+\left[\Psi_z'(z^0(2i-1),\tau_i^0,w_i^0)\right]^*[p_z(2i)+\sigma_i].$$

Since $\mu^0(\tau_i^0-) = 2i-1$ and $\mu^0(\tau_i^0) = 2i$, this last equation can be rewritten as

$$\begin{aligned} p_z(\mu^0(\tau_i^0-)) &= p_z(\mu^0(\tau_i^0))+ \\ &+\left[\Psi_z'(z^0(\mu^0(\tau_i^0-)),\tau_i^0,w_i^0)\right]^*\left[p_z(\mu^0(\tau_i^0))+\sigma_i\right]. \end{aligned} \tag{3.90}$$

Denoting

$$p(t) = p_z(\mu^0(t)), \tag{3.91}$$

and taking into account the relations $z^0(\mu^0(\tau_i^0-)) = X^0(\tau_i^0-)$ and (3.90), we have

$$p(\tau_i^0-) = p(\tau_i^0)+\left[\Psi_x'(X^0(\tau_i^0-),\tau_i^0,w_i^0)\right]^*\left[p(\tau_i^0)+\sigma_i\right]. \tag{3.92}$$

Now let $\theta \in [\mu^0(\tau_{i-1}^0),\mu^0(\tau_i^0-)] = [2i-2,2i-1]$. Then, by (3.64) and (3.70),

$$p_z(\theta) = p_z(2i-2) - \int_{2i-2}^{\theta} v_i^0\left(F_z'(z^0(\vartheta),t^0(\vartheta))\right)^*[p_z(\vartheta)+\widehat{\sigma}_i]\,d\vartheta, \tag{3.93}$$

where

$$\widehat{\sigma}_i = \sum_{j=i}^{N}\left(p_{y_j}+p_{z_j}\right) = \sigma_i + p_{y_i} = \text{const.} \tag{3.94}$$

Since $v(\mu(s))\,d\mu(s) = dt(\mu(s)) = ds$, relations (3.91) and (3.93) show that

$$p(t) = p(\tau_{i-1}^0) - \int_{\tau_{i-1}^0}^{t}\left(F_x'(X^0(s),s)\right)^*[p(s)+\widehat{\sigma}_i]\,ds \tag{3.95}$$

which holds for $t \in [\tau_{i-1}^0,\tau_i^0)$. Upon combining (3.95) and (3.92), we obtain for $t \in [\tau_{i-1}^0,\tau_i^0)$ that

$$
\begin{aligned}
p(t) = p(\tau_i^0) + \int_t^{\tau_i^0} \left(F_x'(X^0(s), s)\right)^* [p(s) + \widehat{\sigma}_i]\, ds + \\
+ \left[\Psi_x'(X^0(\tau_i^0-), \tau_i^0, w_i^0)\right]^* \left[p(\tau_i^0) + \sigma_i\right].
\end{aligned}
\tag{3.96}
$$

Let us introduce the set of functions $\{p_i(t)\}_{i=1,\ldots,N}$,

$$
p_i(t) = p(t) + \widehat{\sigma}_i, \tag{3.97}
$$

defined for each half-interval $t \in [\tau_{i-1}^0, \tau_i^0)$.

Then, by (3.96) and (3.94),

$$
\begin{aligned}
p_i(t) = p_i(\tau_i^0) + \int_t^{\tau_i^0} \left(F_x'(X^0(s), s)\right)^* p_i(s)\, ds + \\
+ \left[\Psi_x'(X^0(\tau_i^0-), \tau_i^0, w_i^0)\right]^* \left[p_i(\tau_i^0) - p_{y_i}\right],
\end{aligned}
\tag{3.98}
$$

where, in view of the system (3.74),

$$
p_{y_i} = -\sum_{j=1}^{k_1} \lambda_j \frac{\partial g_j^0}{\partial y_i} = \text{const}.
$$

Moreover, the boundary conditions ($i = 1, \ldots, N$) satisfy

$$
p_{i+1}(\tau_i^0) = p_i(\tau_i^0) - p_{y_i} - p_{z_i}, \tag{3.99}
$$

where, in view of the system (3.74),

$$
p_{z_i} = -\sum_{j=1}^{k_1} \lambda_j \frac{\partial g_j^0}{\partial z_i} = \text{const}.
$$

On the interval $t \in [\tau_N^0, T]$ the function $p_{N+1}(t)$ is defined by the equation

$$
p_{N+1}(t) = -\sum_{j=1}^{k_1} \lambda_j \frac{\partial g_j^0}{\partial z} - \lambda_0 \frac{\partial G^0}{\partial z} + \int_t^T \left(F_x'(X^0(s), s)\right)^* p_{N+1}(s)\, ds. \tag{3.100}
$$

Next, the conjugate variables are defined recursively.

Further, let us consider the inequality (3.76) (the maximum principle) in the "even" and "odd" intervals and perform the inverse time transformation. We shall show that, as a result, the inequality (3.76) splits into two independent inequalities that give the necessary optimality conditions for the impulse actions $w_i^0 = w^0(\tau_i^0)$ and the instants τ_i^0.

For $\theta \in (2i-1, 2i)$ ("even" interval) $\beta(\theta) = 0$ and the inequality (3.76) has the form

$$\begin{aligned}&\langle p_z(\theta) + \sigma_i, B(z^0(\theta), t^0(\theta))l^0(\theta)\rangle + p_{\eta_i}^* l^0(\theta) + p_\zeta \|l^0(\theta)\| \geq \\ &\geq \langle p_z(\theta) + \sigma_i, B(z^0(\theta), t^0(\theta))l(\theta)\rangle + p_{\eta_i}^* l(\theta) + p_\zeta \|l(\theta)\|,\end{aligned} \tag{3.101}$$

where, in view of Corollary 3.1, the control functions $l^0(\theta)$ and $l(\theta)$ can be replaced by some constant vectors $w_i^0, w \in W$. Here the constants p_{η_i} and p_ζ are taken from the system (3.74).

Since all the functions in the inequality (3.101) are continuous with respect to θ, it remains valid for $\theta = 2i = \mu^0(\tau_i^0)$ as well. Taking into account (3.89) and (3.79) and the notations (3.91), (3.97), and (3.94), we obtain the inequality

$$\begin{aligned}&\langle\left(\left[E + \Psi_x'(X^0(\tau_i^0-), \tau_i^0, w_i^0)\right]^{-1}\right)^* \left(p_i(\tau_i^0-) - p_{y_i}\right), B(X^0(\tau_i^0)), \tau_i^0)\, w_i^0\rangle + \\ &\quad + p_{\eta_i}^* w_i^0 + p_\zeta \|w_i^0\| \geq \\ &\geq \langle\left(\left[E + \Psi_x'(X^0(\tau_i^0-), \tau_i^0, w_i^0)\right]^{-1}\right)^* \left(p_i(\tau_i^0-) - p_{y_i}\right), B(X^0(\tau_i^0), \tau_i^0)\, w\rangle + \\ &\quad + p_{\eta_i}^* w + p_\zeta \|w\|,\end{aligned} \tag{3.102}$$

which holds at each instant $\tau_i^0, i = 1, \ldots, N,$ for any vector $w \in W$.

For $\theta \in [2i-2, 2i-1]$ ("odd" interval), $\beta(\theta) = 1$ and the Hamiltonian function (3.67) has the form

$$\mathscr{H}(\mathscr{Z}, \mathscr{P}, \theta) = v_i H_i(z^0(\theta), t^0(\theta), \theta),$$

where v_i =const and

$$H_i(z^0(\theta), t^0(\theta), \theta) = \langle p_z(\theta) + \widehat{\sigma}_i, F(z^0(\theta), t^0(\theta))\rangle + \sum_{j=i}^{N} p_{t_j} + p_t(\theta). \tag{3.103}$$

Note that in the interval $[2i-2, 2i-1], i = 1, \ldots, N,$ we have

$$H_i(z^0(\theta), t^0(\theta), \theta) = H_i = \text{const}.$$

Indeed, if we differentiate the function $H_i(\cdot)$ with respect to θ by virtue of the system (3.12), (3.13), and (3.74), then we easily get

$$\frac{d}{d\theta} H_i(z^0(\theta), t^0(\theta), \theta) = 0.$$

Hence, $\mathscr{H}(\mathscr{Z}, \mathscr{P}, \theta) = v_i H_i$ and the inequality (3.76) (the maximum principle) becomes

$$H_i(v_i^0 - v_i) \geq 0, \tag{3.104}$$

where, due to (3.40) and Corollary 3.1, the parameters

$$v_i^0 = \tau_i^0 - \tau_{i-1}^0, \qquad v_i = \tau_i - \tau_{i-1}$$

specify the time intervals between the impulse actions for the optimal trajectory and an arbitrary trajectory, respectively.

It follows from the condition (3.6) that

$$\begin{aligned} &v_i^0 \in [T_i,\, T^0 - \sum_{k=i+1}^{N+1} T_k], \\ &v_i \in [T_i,\, T - \sum_{k=i+1}^{N+1} T_k], \quad \text{where} \quad T \geq \sum_{k=i}^{N+1} T_k, \end{aligned} \tag{3.105}$$

which in view of (3.104) implies that

$$\begin{aligned} &\text{if } H_i < 0, \quad \text{then } v_i^0 = T_i, \\ &\text{if } H_i > 0, \quad \text{then } v_i^0 = T_i^0 - \sum_{k=i+1}^{N+1} T_k, \end{aligned} \tag{3.106}$$

i.e., all the subsequent control impulses $w_{i+1}, w_{i+2}, \ldots,$ must follow with a maximum repetition rate at the end of the interval $[t_0, T^0]$.

If $H_i = 0$, then one cannot obtain an optimal value v_i^0 from the maximum principle.

Summarizing the results obtained above, we formulate them as the maximum principle for the primary problem.

Theorem 3.3 *Let* $\{w_i^0, v_i^0, X^0(\tau_i^0-)\}$ *be the collection of the optimal impulse actions, the optimal on-off time ratios, and the optimal trajectories at the moments "immediately preceding"* τ_i^0, *respectively, where* τ_i^0 *is the optimal instant of the* i*-th impulse action,* $i = 1, \ldots, N$.

Then there exist:

1. *a set of vectors* $\{p_\zeta, p_{\eta_i}, p_{y_i}, p_{z_i}, p_{t_i}\}$, *satisfying (3.74), and a sequence of vector-valued functions* $p_i(t)$ *defined by the formulas (3.98)–(3.100) (these constants and functions, which play the role of conjugate variables, do not vanish simultaneously);*

2. *numbers* H_i *generated by the expression (3.103), where all the functions of* θ *satisfy the conditions of Theorem 3.2;*
such that the following holds:

(a) *At each instant* τ_i^0, $i = 1, \dots, N$, *the inequality (3.102) is valid for any vector* $w \in W$.
(b) *The optimal on-off time ratios* $\upsilon_i^0 = \tau_i^0 - \tau_{i-1}^0$ *satisfy the condition (3.106).*

3.3.3 Sufficient Optimality Conditions

The conditions obtained above play a key role in the solution of the optimal control problem, since they enable us to identify the controls and trajectories that could be optimal. However, there exists the class of so-called convex problems where these conditions actually ensure optimality. Here we derive such sufficient optimality conditions in the problem with a constrained number of impulses. The general problem of optimality conditions will be considered later in Sects. 6.2 and 6.4.

Theorem 3.4 *Let the dynamic equations of a controlled DCS be linear with respect to the control actions, i.e.,*

$$X(t) = X_0 + \int_{t_0}^{t} F\,(X(s), s)\,ds + \sum_{\tau_i \le t} B(\tau_i) w(\tau_i). \tag{3.107}$$

Assume that:

1. *the functions* g_j, $j = 1, \dots, k_1$, *and* G *figuring in the constraints are convex in all the variables;*
2. *the Hamiltonian* $\mathscr{H}$ *in the auxiliary problem is a concave function with respect to the phase variables* $\{z, t\}$;
3. *the pair* $\{l^0(\theta), u^0(\theta)\}$ *satisfies the maximum principle with a constant* $\lambda_0 > 0$.

Then the control $\{l^0(\theta), u^0(\theta)\}$ *is the optimal control in the auxiliary problem and the impulse control* $\mathscr{I} = \{(w_i^0, \tau_i^0), i = 1, \dots, N\}$, *the trajectory* $X^0(t)$, *and the terminal time* T^0, *defined by relations (3.41) and (3.42), are the optimal ones in the original impulse control problem.*

Proof of Theorem 3.4 Assume that in the auxiliary problem the set of phase variables

$$\mathscr{Z} = \{z, t, \zeta, \eta_j, y_j, z_j, t_j, \rho_j, \rho_{N+1}, \quad j = 1, \dots, N\}$$

(the argument θ is omitted), corresponds to an admissible control. Assume also that the set of variables

$$\mathscr{Z}^0 = \{z^0, t^0, \zeta^0, \eta_j^0, y_j^0, z_j^0, t_j^0, \rho_j^0, \rho_{N+1}^0, \quad j = 1, \dots, N\}$$

satisfies the maximum principle with the conjugate variables p_z, p_t, p_ζ, p_{η_j}, p_{z_j}, p_{t_j}, p_{ρ_j}, $p_{\rho N+1}$ and constants λ_i, $i = 0, \ldots, k_1 + 1$, and ν_k, $k = 1, \ldots, N + 1$.

Define the function $\mathscr{M}(t)$ by

$$\begin{aligned}\mathscr{M}(t) &= \langle p_z(\mu(t)), z(\mu(t)) \rangle + p_t(\mu(t))\, t + p_\zeta \zeta(\mu(t)) + \\ &+ \sum_{j=1}^{N} \Big[\langle p_{\eta_j}, \eta_j(\mu(t)) \rangle + \langle p_{y_j}, y_j(\mu(t)) \rangle + \langle p_{z_j}, z_j(\mu(t)) \rangle + p_{t_j} t_j(\mu(t)) \Big] + \\ &+ \sum_{j=1}^{N} \rho_j p_{\rho_j}(\mu(t)),\end{aligned}$$

where $\mu(t) = \sup\{\theta : t(\theta) = t\}$.

Then the differentiation formula (A.14) yields

$$\begin{aligned}d\mathscr{M}(s) &= \langle z(\mu(s-),)\, dp_z(\mu(s)) \rangle + \langle p_z(\mu(s)), dz(\mu(s)) \rangle + \\ &+ t(\mu(s-))\, dp_t(\mu(s)) + p_t(\mu(s))\, dt(\mu(s)) + p_\zeta d\zeta(\mu(s)) + \\ &+ \sum_{j=1}^{N} \Big[\langle p_{\eta_j}, d\eta_j(\mu(s)) \rangle + \langle p_{y_j}, dy_j(\mu(s)) \rangle + \\ &+ \langle p_{z_j}, dz_j(\mu(s)) \rangle + p_{t_j} dt_j(\mu(s)) \Big] + \\ &+ \sum_{j=1}^{N} \rho_j dp_{\rho_j}(\mu(s)),\end{aligned}$$

Now let us calculate

$$\Delta\mathscr{M}(T) = \mathscr{M}(T) - \mathscr{M}(0) = \sum_{i=1}^{N+1} \int_{(\tau_{i-1}, \tau_i]} d\mathscr{M}(s), \tag{3.108}$$

where $\tau_0 = 0$ and $\tau_{N+1} = T$. To this end we calculate the integrals in (3.108) for $i = 1, \ldots, N + 1$ by formally setting $w_{N+1} = 0$ and $\sum\limits_{j=N+1}^{N} (\cdot)_j = 0$. In view of the form of the right-hand side of Eq. (3.107), where the function $B(\cdot)$ does not depend on z, i.e.,

$$f(z, t, v, l) = \beta v F(z, t) + (1 - \beta) B(t) l$$

we have (see also Remark A.10)

$$J_1^1 = \int_{(\tau_{i-1},\tau_i]} \big\langle z\left(\mu(s-)\right), dp_z\left(\mu(s)\right)\big\rangle = \int_{(\tau_{i-1},\tau_i[} \big\langle z\left(\mu(s-)\right), dp_z\left(\mu(s)\right)\big\rangle +$$
$$+ \big\langle z\left(\mu(\tau_i-)\right), \left[p_z\left(\mu(\tau_i)\right) - p_z\left(\mu(\tau_i-)\right)\right]\big\rangle =$$
$$= \int_{2i-2}^{2i-1} \big\langle z(\theta), dp_z(\theta)\rangle + \langle z(2i-1), [p_z(2i) - p_z(2i-1)]\big\rangle =$$
$$= -\int_{2i-2}^{2i-1} \Big\langle z, v^0 (F_z^{0'})^* \left[p_z + \sum_{j=i}^{N} (p_{y_j} + p_{z_j}) \right] \Big\rangle d\theta$$

(the arguments of the integrands are omitted), where the superscript 0 indicates that the functions are calculated for values of variables such that the maximum principle is satisfied, and F_z' denotes the partial derivative of F with respect to z (the argument θ is omitted). Since in the "even" intervals $f_z' = 0$ and $p_z(2i) = p_z(2i-1)$ (see (3.70)), we have

$$J_2^i = \int_{(\tau_{i-1},\tau_i]} \big\langle p_z\left(\mu(s)\right), dz\left(\mu(s)\right)\big\rangle =$$
$$= \int_{(\tau_{i-1},\tau_i)} \big\langle p_z\left(\mu(s)\right), dz\left(\mu(s)\right)\big\rangle + \big\langle p_z\left(\mu(\tau_i)\right), \left[z\left(\mu(\tau_i)\right) - z\left(\mu(\tau_i-)\right)\right]\big\rangle =$$
$$= \int_{2i-2}^{2i-1} \big\langle p_z(\theta), dz(\theta)\rangle + \langle p_z(2i), [z(2i) - z(2i-1)]\big\rangle =$$
$$= \int_{2i-2}^{2i-1} \big\langle p_z, vF\big\rangle\, d\theta + \big\langle p_z(2i), B(\tau_i) w_i\big\rangle$$

(the arguments of the integrands v and f are omitted). Similar calculations based on the time transformation give

$$J_3^i = \int_{(\tau_{i-1},\tau_i]} t\left(\mu(s-)\right)\, dp_t\left(\mu(s)\right) =$$
$$= -\int_{2i-2}^{2i-1} t(\theta) \Big\langle p_z + \sum_{j=i}^{N} (p_{y_j} + p_{z_j}), v^0 F_t^{0'} \Big\rangle\, d\theta + \tau_i \left[p_t(2i) - p_t(2i-1)\right],$$

$$J_4^i = \int_{(\tau_{i-1},\tau_i]} p_t\left(\mu(s)\right)\, dt\left(\mu(s)\right) = \int_{2i-2}^{2i-1} p_t(\theta) v\, d\theta,$$

$$J_5^i = \int_{(\tau_{i-1},\tau_i]} p_\zeta\, d\zeta\left(\mu(s)\right) = p_\zeta[\zeta(2i) - \zeta(2i-1)] = p_\zeta \|w_i\|,$$

$$J_6^i = \int_{(\tau_{i-1},\tau_i]} \sum_{j=1}^{N} \langle p_{\eta_j},\, d\eta_j(\mu(s))\rangle = p_{\eta_i}^* w_i,$$

$$J_7^i = \int_{(\tau_{i-1},\tau_i]} \sum_{j=1}^{N} \langle p_{y_j},\, dy_j(\mu(s))\rangle$$

$$= \langle \sum_{j=i}^{N} p_{y_j},\, \int_{2i-2}^{2i-1} vF\, d\theta\rangle + \langle \sum_{j=i+1}^{N} p_{y_j},\, B(\tau_i)w_i\rangle,$$

$$J_8^i = \int_{(\tau_{i-1},\tau_i]} \sum_{j=1}^{N} \langle p_{z_j}, dz_j(\mu(s))\rangle$$

$$= \langle \sum_{j=i}^{N} p_{z_j},\, \int_{2i-2}^{2i-1} vF\, d\theta\rangle + \langle \sum_{j=i}^{N} p_{z_j},\, B(\tau_i)w_i\rangle$$

$$J_9^i = \int_{(\tau_{i-1},\tau_i]} \sum_{j=1}^{N} p_{t_j}\, dt_j(\mu(s)) = \left(\sum_{j=i}^{N} p_{t_j}\right) \int_{2i-2}^{2i-1} v\, d\theta,$$

$$J_{10}^i = \int_{(\tau_{i-1},\tau_i]} \sum_{j=1}^{N+1} \rho_j\, dp_{\rho_j}(\mu(s)) =$$

$$= -\rho_i \int_{2i-2}^{2i-1} \left[p_t + \sum_{j=i}^{N} p_{t_j} + \langle p_z + \sum_{j=i}^{N} (p_{y_j} + p_{z_j},\, F^0\rangle\right] \frac{1}{2}(u_i^0 + 1)\, d\theta.$$

On the other hand, by the definition of $\mathscr{M}(t)$,

$$\Delta\mathscr{M}(T) = \mathscr{M}(T) - \mathscr{M}(0) =$$

$$= \langle p_z(2N+1), z(2N+1)\rangle + p_t(2N+1)t(2N+1) + p_\zeta\zeta(2N+1) +$$

$$+ \sum_{j=1}^{N} \Big[\langle p_{\eta_j}, \eta_j(2N+1)\rangle + \langle p_{y_j}, y_j(2N+1)\rangle +$$

$$+ \langle p_{z_j}, z_j(2N+1)\rangle + p_{t_j} t_j(2N+1)\Big] +$$

$$+ \sum_{j=1}^{N+1} \rho_j p_{\rho_j}(2N+1) - \mathscr{M}(0).$$

In much the same way we calculate $\Delta\mathscr{M}^0(T^0) = \mathscr{M}^0(T^0) - \mathscr{M}(0)$, which corresponds to the set of the variables $\mathscr{Z}^0$, and then we calculate the difference

$\Delta\mathscr{M}(T)-\Delta\mathscr{M}^0(T^0)=\mathscr{M}(T)-\mathscr{M}^0(T^0)$. Our aim is to prove its non-positivity, i.e., that $\Delta\mathscr{M}(t)-\Delta\mathscr{M}^0(T^0)\le 0$. Indeed, on the one hand we have

$$\Delta\mathscr{M}(T)-\Delta\mathscr{M}^0(T^0)=\mathscr{M}(T)-\mathscr{M}^0(T^0)=$$

$$=\Bigg\{\langle p_z,(z-z^0)\rangle+p_t(t-t^0)+p_\zeta(\zeta-\zeta^0)+$$

$$+\sum_{j=1}^{N}\Big[\langle p_{\eta_j},(\eta_j-\eta_j^0)\rangle+\langle p_{y_j},(y_j-y_j^0)\rangle+\langle p^{z_j},(z_j-z_j^0)\rangle+p_{t_j}(t_j-t_j^0)\Big]+$$

$$+\sum_{j=1}^{N+1}p_{\rho_j}(\rho_j-\rho_j^0)\Bigg\}\Bigg|_{\theta=2N+1}. \tag{3.109}$$

where all the variables are evaluated at $\theta=2N+1$. On the other hand,

$$\Delta\mathscr{M}(T)-\Delta\mathscr{M}^0(T^0)=\sum_{i=1}^{N+1}\left(\Delta J_1^i+\Delta J_2^i+\cdots+\Delta J_{10}^i\right),$$

where

$$\Delta J_1^i=J_1^i-J_1^{oi}=-\int_{2i-2}^{2i-1}\Big\langle z-z^0,v^0\left(F_z^{0'}\right)^*\Big[p_z+\sum_{j=i}^{N}(p_{y_j}+p_{z_j})\Big]\Big\rangle\,d\theta.$$

Similarly,

$$\Delta J_2^i=\int_{2i-2}^{2i-1}\langle p_z,vF-v^0F^0\rangle\,d\theta+\langle p_z(2i),B(\tau_i)w_i-B(\tau_i^0)w_i^0\rangle,$$

$$\Delta J_3^i=-\int_{2i-2}^{2i-1}(t-t^0)\Big\langle p_z+\sum_{j=i}^{N}(p_{y_j}+p_{z_j}),v^0F_z^{0'}\Big\rangle\,d\theta+$$

$$+(\tau_i-\tau_i^0)[p_t(2i)-p_t(2i-1)],$$

$$\Delta J_4^i=\int_{2i-2}^{2i-1}p_t(v-v^0)\,d\theta,$$

$$\Delta J_5^i=p_\zeta(\|w_i\|-\|w_i^0\|),$$

$$\Delta J_6^i=p_{\eta_i}^*(w_i-w_i^0),$$

$$\Delta J_7^i = \Big\langle \sum_{j=i}^{N} p_{y_j}, \int_{2i-2}^{2i-1} (vF - v^0 F^0)\Big\rangle \, d\theta + \Big\langle \sum_{j=i+1}^{N} p_{y_j}, B(\tau_i)w_i - B(\tau_i^0)w_i^0 \Big\rangle,$$

$$\Delta J_8^i = \Big\langle \sum_{j=i}^{N} p_{z_j} \int_{2i-2}^{2i-1}, (vF - v^0 F^0)\Big\rangle \, d\theta + \Big\langle \sum_{j=i}^{N} p_{z_j}, B(\tau_i)w_i - B(\tau_i^0)w_i^0 \Big\rangle,$$

$$\Delta J_9^i = \left(\sum_{j=i}^{N} p_{t_j} \right) \int_{2i-2}^{2i-1} (v - v^0) \, d\theta,$$

$$\Delta J_{10}^i = -(\rho_i - \rho_i^0) \int_{2i-2}^{2i-1} \Bigg[p_t + \sum_{j=i}^{N} p_{t_j} +$$

$$+ \Big\langle p_z + \sum_{j=i}^{N} (p_{y_j} + p_{z_j}), F^0 \Big\rangle \Bigg] \frac{1}{2}(u_i^0 + 1) \, d\theta = \tag{3.110}$$

$$= -(\rho_i - \rho_i^0) \int_{2i-2}^{2i-1} \frac{\partial \mathscr{H}^0}{\partial \rho_i} \, d\theta,$$

where $\mathscr{H}^0$ denotes the Hamiltonian in the left-hand side of (3.76) (the arguments are omitted). Let us introduce the column vector κ of the variables involved in $\mathscr{H}^0$,

$$\kappa = (z, t, \rho_1, \ldots, \rho_{N+1}, u, l)$$

(where each of the components has the appropriate dimension) and calculate the functional

$$\Delta = \sum_{i=1}^{N+1} \int_{2i-2}^{2i-1} \left[\mathscr{H} - \mathscr{H}^0 - \langle \kappa - \kappa^0, \mathscr{H}_\kappa^{0'} \rangle \right] d\theta +$$

$$+ \sum_{i=1}^{N} \int_{2i-1}^{2i} \left[\mathscr{H} - \mathscr{H}^0 - \langle \kappa - \kappa^0, \mathscr{H}_\kappa^{0'} \rangle \right] d\theta =$$

$$= \int_0^{2N+1} \left[\mathscr{H} - \mathscr{H}^0 - \langle \kappa - \kappa^0, \mathscr{H}_\kappa^{0'} \rangle \right] d\theta,$$

where $\mathscr{H}$ is the Hamiltonian evaluated in the set of variables $\mathscr{Z}$ and $\mathscr{H}^0$ is evaluated in the set of variables $\mathscr{Z}^0$. To do this, we have to integrate the difference $\mathscr{H} - \mathscr{H}^0$ over the “odd” intervals, where $\beta(\theta) = 1$, and then over the “even” intervals, where $\beta(\theta) = 0$. Then we have to integrate the term $-\langle \kappa - \kappa^0, \mathscr{H}_\kappa^{0\prime} \rangle$ separately over the "odd" and "even" intervals.

The integration of $\mathscr{H} - \mathscr{H}^0$ with $\beta(\theta) = 1$ gives (the argument θ is omitted, $i = 1, \ldots, N+1$):

$$\int_{2i-2}^{2i-1} \Bigg\{ \langle \sum_{j=1}^{N} (\alpha_j p_{y_j} + \beta_j p_{z_j}) + p_z, vF - v^0 F^0 \rangle +$$

$$+ \sum_{j=1}^{N} \left[\alpha_j \beta p_{t_j} (v - v^0) + (\beta_j - \alpha_j) p_{\eta_j}^* (l - l^0) \right] + \beta p_t (v - v^0) \Bigg\} \, d\theta =$$

$$= \langle \sum_{j=i}^{N} p_{y_j}, \int_{2i-2}^{2i-1} (vF - v^0 F^0) \rangle \, d\theta + \langle \sum_{j=i}^{N} p_{z_j}, \int_{2i-2}^{2i-1} (vF - v^0 F^0) \rangle \, d\theta +$$

$$+ \int_{2i-2}^{2i-1} \langle p_z, vF - v^0 F^0 \rangle \, d\theta + \left(\sum_{j=i}^{N} p_{t_j} \right) \int_{2i-2}^{2i-1} (v - v^0) \, d\theta +$$

$$+ \int_{2i-2}^{2i-1} p_t (v - v^0) \, d\theta =$$

$$= \Delta_1 J_7^i + \Delta_1 J_8^i + \Delta_1 J_2^i + \Delta J_9^i + \Delta J_4^i, \tag{3.111}$$

where in the right-hand side of the last relation in (3.111) one substitutes the values of the appropriate integral terms (3.110), and the index "1" corresponds the integration over the interval $[2i-2, 2i-1]$.

Now integrating the difference $\mathscr{H} - \mathscr{H}^0$ with $\beta(\theta) = 0$, we get (the argument θ is omitted, $i = 1, \ldots, N$):

$$\int_{2i-1}^{2i} \langle \sum_{j=1}^{N} (\alpha_j p_{y_j} + \beta_j p_{z_j}) + p_z, B(t) l - B(t^0) l^0 \rangle \, d\theta +$$

$$+ \int_{2i-1}^{2i} \sum_{j=1}^{N} (\beta_j - \alpha_j) \langle p_{\eta_j}, l - l^0 \rangle \, d\theta + \int_{2i-1}^{2i} p_\varsigma (\|l\| - \|l^0\|) \, d\theta =$$

$$= \langle \sum_{j=i+1}^{N} p_{y_j}, B(\tau_i) w_i - B(\tau_i^0) w_i^0 \rangle + \langle p_z(2i), B(\tau_i) w_i - B(\tau_i^0) w_i^0 \rangle +$$

$$+\big\langle\sum_{j=i}^{N} p_{z_j}, B(\tau_i)w_i - B(\tau_i^0)w_i^0\big\rangle + \langle p_{\eta_i}, w_i - w_i^0\rangle + p_\zeta(\|w_i\| - \|w_i^0\|) =$$

$$= \Delta_2 J_7^i + \Delta_2 J_2^i + \Delta_2 J_8^i + \Delta J_6^i + \Delta J_5^i, \tag{3.112}$$

where in the right-hand side of the last equality in (3.112) one substitutes the values of the appropriate integral terms (3.110), and the index "2" corresponds to the integration over the interval $(2i-1, 2i)$. Since

$$\Delta J_2^i = \Delta_1 J_2^i + \Delta_2 J_2^i, \quad \Delta J_7^i = \Delta_1 J_7^i + \Delta_2 J_7^i, \quad \Delta J_8^i = \Delta_1 J_8^i + \Delta_2 J_8^i,$$

combining (3.111) and (3.112), we obtain

$$\int_0^{2N+1} (\mathscr{H} - \mathscr{H}^0)\, d\theta = \sum_{i=1}^{N+1} \big(\Delta J_2^i + \Delta J_4^i + \Delta J_5^i + \Delta J_6^i + \Delta J_7^i + \Delta J_8^i + \Delta J_9^i\big).$$

Below we are listing the different terms that arise during the integration. By integrating the term $-\langle \kappa - \kappa^0, \mathscr{H}_\kappa^{0\prime}\rangle$ when $\beta(\theta) = 1$, we have (the argument θ is omitted)

$$-\int_{2i-2}^{2i-1} \langle z - z^0, \mathscr{H}_z^{0\prime}\rangle\, d\theta =$$

$$= -\int_{2i-2}^{2i-1} \big\langle z - z^0, v^0 (F_z^{0\prime})^* \big[p_z + \sum_{j=i}^{N} (p_{y_j} + p_{z_j})\big]\big\rangle\, d\theta = \Delta J_1^i,$$

$$-\int_{2i-2}^{2i-1} (t - t^0)\mathscr{H}_t^{0\prime}\, d\theta$$

$$= -\int_{2i-2}^{2i-1} (t - t^0)\big\langle p_z + \sum_{j=i}^{N} (p_{y_j} + p_{z_j}), v^0 F_t^{0\prime}\big\rangle\, d\theta = \Delta_1 J_3^i,$$

$$-\int_{2i-2}^{2i-1} \sum_{j=i}^{N+1} (\rho_j - \rho_j^0)\mathscr{H}_{\rho_j}^{0\prime}\, d\theta = -(\rho_i - \rho_i^0)\int_{2i-2}^{2i-1} \mathscr{H}_{\rho_i}^{0\prime}\, d\theta = \Delta J_{10}^i,$$

$$-\int_{2i-2}^{2i-1} \langle u - u^0, \mathscr{H}_u^{0\prime}\rangle\, d\theta.$$

When $\beta(\theta) = 0$ we have (the argument θ is omitted):

$$-\int_{2i-1}^{2i} (t - t^0)\mathcal{H}_t^{0'}\, d\theta = -\int_{2i-1}^{2i} (t - t^0)\langle p_z + \sum_{j=i}^{N} (\alpha_j p_{y_j} + p_{z_j}\rangle,$$

$$B_t{}'(t^0)l^0\rangle\, d\theta =$$

$$= (\tau_i - \tau_i^0)[p_t(2i) - p_t(2i-1)] = \Delta_2 J_3^i.$$

Note that, by definition,

$$\Delta J_3^i = \Delta_1 J_3^i + \Delta_2 J_3^i.$$

Hence, the integration of the term $-\langle \kappa - \kappa^0, \mathcal{H}_\kappa^{0'}\rangle$ gives

$$\int_0^{2N+1} -\langle \kappa - \kappa^0, \mathcal{H}_\kappa^{0'}\rangle\, d\theta = \sum_{i=1}^{N} \Bigg[\Delta J_1^i + \Delta J_3^i + \Delta J_{10}^i -$$

$$- \int_{2i-2}^{2i-1} \langle u - u^0, \mathcal{H}_u^{0'}\rangle\, d\theta - \int_{2i-1}^{2i} \langle l - l^0, \mathcal{H}_l^{0'}\rangle\, d\theta \Bigg].$$

Combining the results of integration of the function

$$\mathcal{H} - \mathcal{H}^0 - \langle \kappa - \kappa^0, \mathcal{H}_\kappa^{0'}\rangle$$

with a direct calculation of the difference $\Delta M(T) - \Delta M^0(T^0)$, we obtain the inequality

$$\Delta M(T) - \Delta M^0(T^0) - \int_0^{2N+1} \left[\langle u - u^0, \mathcal{H}_u^{0'}\rangle + \langle l - l^0, \mathcal{H}_l^{0'}\rangle \right] d\theta =$$

$$= \int_0^{2N+1} \left[\mathcal{H} - \mathcal{H}^0 - \langle \kappa - \kappa^0, \mathcal{H}_\kappa^{0'}\rangle \right] d\theta \le 0.$$

Here the right-hand side is non-positive due to the concavity of $\mathcal{H}$ in κ, and the integral in the left-hand side is non-negative due to (3.76) (the maximum principle). Therefore, $\Delta M(T) - \Delta M^0(T^0) \le 0$. Upon substituting in this inequality (due to (3.109)) the terminal values of the conjugate variables given by formulas (3.74) we have (the arguments are omitted)

$$
\begin{aligned}
&\big\langle z - z^0, -\sum_{j=1}^{k_1} \lambda_j \frac{\partial g_j^0}{\partial z} - \lambda_0 \frac{\partial G^0}{\partial z}\big\rangle + (T - T^0)\Big(- \sum_{j=1}^{k_1} \lambda_j \frac{\partial g_j^0}{\partial t} - \lambda_0 \frac{\partial G^0}{\partial t}\Big) + \\
&+ (T - T^0)\Big[\sum_{j=1}^{N+1} \nu_j \Big] + (-\lambda_{k_1+1})(\zeta - \zeta^0) + \\
&+ \sum_{i=1}^{N} \Big[\Big(- \sum_{j=1}^{k_1} \lambda_j \frac{\partial g_j^0}{\partial \eta_j} (\eta_i - \eta_i^0) \Big) + \big\langle y_i - y_i^0, - \sum_{j=1}^{k_1} \lambda_j \frac{\partial g_j^0}{\partial y_i} \big\rangle + \\
&+ \big\langle z_i - z_i^0, \sum_{j=1}^{k_1} \lambda_j \frac{\partial g_j^0}{\partial z_i} \big\rangle + (t_i - t_i^0)\Big(\sum_{j=1}^{k_1} \lambda_j \frac{\partial g_j^0}{\partial t_i} \Big) \Big] + \\
&+ \Big[- \sum_{j=1}^{N+1} \nu_j (\rho_j - \rho_j^0) - \sum_{j=1}^{N+1} \nu_{j+1} (t_j - t_j^0) \Big] \le 0.
\end{aligned}
\tag{3.113}
$$

Here all the arguments are computed at $\theta = 2N + 1$. Let q be any of the vector arguments of the functions $g_j(\cdot)$ and $G(\cdot)$. The convexity condition of these functions yields the inequalities

$$
\big\langle - (q - q^0), \frac{\partial g_j^0}{\partial q} \big\rangle \ge g_j^0 - g_j, \quad j = 1, \dots, k_1 + 1, \quad g_{k_1+1} = \zeta - \overline{M},
$$

$$
\big\langle - (q - q^0), \frac{\partial G^0}{\partial q} \big\rangle \ge G^0 - G.
$$

Therefore, thanks to the complementary slackness condition (3.75), the non-negativity of λ_j and the non-positivity of g_j, we have that

$$
\big\langle - (q - q^0), \frac{\partial g_j^0}{\partial q} \big\rangle \lambda_j \ge \lambda_j g_j^0 - \lambda_j g_j \ge 0.
$$

Moroever, it follows from (3.75) and (3.61) that

$$
- \sum_{j=1}^{N+1} \nu_j (\rho_j - \rho_j^0) + (T - T^0) \sum_{j=1}^{N+1} \nu_j - \sum_{j=1}^{N} \nu_{j+1} (t_j - t_j^0) = 0
$$

(this is the sum of terms enclosed in square brackets in (3.113)). Therefore we have

$$
\lambda_0 (G^0 - G) \le \text{ left hand side of } ((3.113)) \le 0.
$$

But $\lambda_0 \geq 0$, and so $G^0 \leq G$. Thus, the pair $\{l^0(\theta), u^0(\theta)\}$ is optimal for the auxiliary problem, therefore, by Theorem 3.1, the impulse control $\{w_i^0, \tau_i^0\}$, $i = 1, \ldots, N$, the trajectory $X^0(t)$ and the terminal instant T^0 generated by this pair are optimal for the primary problem. Theorem 3.4 is proved. □

Remark 3.3 In Sect. 7.5, we give an example of an observation control problem with a constrained number of the observation instants. It is a typical optimal control problem where the methods developed above can be applied rather effectively.

3.4 Impulse Control with Impulse Actions of Two Types

3.4.1 Problem Statement

The problem with a constrained number of impulses considered above admits the following generalization [295].

Suppose the dynamics of a control plant is governed by the equation

$$
\begin{aligned}
X(t) = X_0 + \int_{t_0}^{t} F(X(s), s)\, ds + \sum_{\tau_{1i} \leq t} \Psi_1(X(\tau_{1i}-), u(\tau_{1i}), \tau_{1i}) \\
+ \sum_{\tau_{2j} \leq t} \Psi_2(X(\tau_{2j}-), u(\tau_{2j}), \tau_{2j}),
\end{aligned}
\tag{3.114}
$$

where $X(t) \in \mathbb{R}^n$ and $t \in [t_0,\ T]$.

The collection of pairs $\{u(\tau_{1i}), \tau_{1i}\}$, $i = 1, \ldots, N$, constitutes an impulse control of first type (referred to here as type I) with constrained number of impulses, while the collection of pairs $\{u(\tau_{2j}), \tau_{2j}\}$, $j = 1, 2, \ldots$, constitutes an impulse control of the second type (referred to here as type II), with generally unbounded number of impulses. The intensities of the impulsive actions are subject to the constraints

$$
\begin{cases}
u_{1i} = u(\tau_{1i}) \in U_1 \subset \mathbb{R}^{r_1}, \\
\sum_{i=1}^{N} \|u_{1i}\| \leq \overline{M}_1
\end{cases}
\tag{3.115}
$$

and

$$
\begin{cases}
u_{2j} = u(\tau_{2j}) \in U_2 \subset \mathbb{R}^{r_2}, \\
\sum_{j=1}^{\infty} \|u_{2j}\| \leq \overline{M}_2,
\end{cases}
\tag{3.116}
$$

respectively, where $\overline{M}_1$ and $\overline{M}_2$ are given constants, U_m, $m = 1, 2$, are bounded, closed and convex sets in $\mathbb{R}^{r_m}$ with $0 \in U_2$. Also, $\|u_{mi}\| = \sum_{k=1}^{r_m} |u^k_{mi}|$, where u^k_{mi} is the k-th component of the vector u_{mi}. The instants τ_{1i} and τ_{2j} of impulse actions satisfy the monotonicity conditions

$$t_0 = \tau_{10} < \tau_{11} < \tau_{12} < \cdots < \tau_{1N} < \tau_{1(N+1)} = \overline{T},$$

$$t_0 = \tau_{20} < \tau_{21} < \tau_{22} < \cdots \leq \overline{T},$$

where $\overline{T}$ is the terminal instant of the control; moreover $\tau_{1i} \neq \tau_{2j}$, $i = 1, \ldots, N$, $j = 1, 2, \ldots$. The functions $F(\cdot)$, $\Psi_1(\cdot)$ and $\Psi_2(\cdot)$ are assumed to be smooth in the variable $t \in [t_0, \overline{T}]$, continuous in the variable X, and to satisfy the Lipschitz condition uniformly in $t \in [t_0, T]$, that is, for any $x_1, x_2 \in \mathbb{R}^n$ and $u_1 \in U_1, u_2 \in U_2$

$$\|F(x_1, t) - F(x_2, t)\| \leq L_0 \|x_1 - x_2\|$$

and

$$\|\Psi_m(x_1, u_m, t) - \Psi_m(x_2, u_m, t)\| \leq L_m \|u_m\| \; \|x_1 - x_2\|, \qquad m = 1, 2,$$

where L_0, L_m are constants independent of t. If these conditions are satisfied, then Eq. (3.114) has a unique solution [173, 237].

As in Sect. 3, we require that the conditions for correct impulsive description (robustness) (2.7) and (3.4) be satisfied, where $\Psi(\cdot)$ is any of $\Psi_1(\cdot)$ or $\Psi_2(\cdot)$.

For type-I impulse actions, as in Sect. 3, we impose the restrictions (3.5) and (3.6), where in the present case $T = \overline{T}$, $\tau_i = \tau_{1i}$, $T_i = \overline{T}_i$, $u_i = u_{1i}$, $i = 1, \ldots, N$ (the non-negative constants $\overline{T}_i$ are fixed).

Controls that satisfy these restrictions are called *admissible*.

The problem at hand is to minimize the functional (3.7) on the set of admissible controls.

3.4.2 Reduction to an Auxiliary Problem

We convert the problem considered in Sect. 3.4.1, which we call here the *basic problem*, to two auxiliary problems with constrained control, via two consecutive time substitutions:

$$t = t(\theta), \quad \theta \in [0,\ 2N + 1], \quad \text{and} \quad \theta = \theta(\vartheta), \quad \vartheta \in [0,\ 2N + 1 + \overline{M}_2].$$

The first time substitution $t(\theta)$ is performed in the same way as when solving the problem in Sect. 3 using the formulas (3.8), (3.9). In this case, we get the *first auxiliary problem* (FAP), specified by the equations (3.10)–(3.26), where one denotes $B(x,t) = (\psi_1)'_{u_1}|_{u_1=0} = B_1(x,t)$.

Let $l(\theta)$, $v(\theta)$, $t(\theta)$, $y_i(\theta)$, $z_i(\theta)$, $z(\theta)$, $t_i(\theta)$, $\eta_i(\theta)$, $i = 1, \ldots, N$ be admissible controls and corresponding trajectories in the FAP .

To move on to the second auxiliary problem, we replace the time variable by $\theta = \theta(\vartheta)$, $\vartheta \in [0,\ 2N + 1 + \overline{M}_2]$, so that

$$\dot{\theta}(\vartheta) = \alpha(\vartheta), \quad \theta(0) = 0,$$

where $\alpha(\vartheta) \in [0,\ 1]$ plays the role of a new control.

In the new time variable, consider the *second auxiliary problem* (SAP) of controlling a dynamic system of the form ($i = 1, \ldots, N$)

$$\dot{Y}_i(\vartheta) = \alpha_i(\theta(\vartheta))\Phi(Z(\vartheta), T(\vartheta), v(\theta(\vartheta)), l(\theta(\vartheta)), \theta(\vartheta)), \tag{3.117}$$

$$\dot{Z}_i(\vartheta) = \beta_i(\theta(\vartheta))\Phi(Z(\vartheta), T(\vartheta), v(\theta(\vartheta)), l(\theta(\vartheta)), \theta(\vartheta)), \tag{3.118}$$

$$\dot{Z}(\vartheta) = \Phi(Z(\vartheta), T(\vartheta), v(\theta(\vartheta)), l(\theta(\vartheta)), \theta(\vartheta)), \tag{3.119}$$

$$\dot{T}(\vartheta) = \beta(\theta(\vartheta))v(\theta(\vartheta))\alpha(\vartheta), \tag{3.120}$$

$$\dot{T}_i(\vartheta) = \alpha_i(\theta(\vartheta))\beta(\theta(\vartheta))v(\theta(\vartheta))\alpha(\vartheta), \tag{3.121}$$

$$\dot{\Lambda}_i(\vartheta) = [\beta_i(\theta(\vartheta)) - \alpha_i(\theta(\vartheta))]\, l(\theta(\vartheta))\alpha(\vartheta), \tag{3.122}$$

$$\dot{\theta}(\vartheta) = \alpha(\vartheta)\beta(\theta(\vartheta)) + (1 - \beta(\theta(\vartheta))), \tag{3.123}$$

$$\dot{\zeta}(\vartheta) = (1 - \alpha(\vartheta))\|L(\vartheta)\|\beta(\theta(\vartheta)), \tag{3.124}$$

with boundary conditions

$$Z(0) = Z_i(0) = Y_i(0) = X_0, \tag{3.125}$$

$$T(0) = T_i(0) = t_0, \tag{3.126}$$

$$\Lambda_i(0) = 0, \tag{3.127}$$

$$\zeta(0) = 0, \tag{3.128}$$

$$\theta(0) = 0. \tag{3.129}$$

Here

$$
\begin{aligned}
&\Phi(Z(\vartheta), T(\vartheta), v(\theta(\vartheta)), l(\theta(\vartheta)), \theta(\vartheta)) = \\
&\quad = v(\theta(\vartheta)), \alpha(\vartheta), \beta(\theta(\vartheta)), F(Z(\vartheta), T(\vartheta)) + \\
&\quad + [1 - \beta(\theta(\vartheta))] B_1(Z(\vartheta), T(\vartheta)) l(\theta(\vartheta)) + \\
&\quad + [1 - \alpha(\vartheta)] \beta(\theta(\vartheta)) B_2(Z(\vartheta), T(\vartheta)) L(\vartheta).
\end{aligned}
\tag{3.130}
$$

The admissible controls in the SAP are functions $\alpha(\vartheta)$ and $L(\vartheta)$ that obey for $\vartheta \in [0,\ 2N + 1 + \overline{M}_2]$ the restrictions

$$
\begin{cases}
\alpha(\vartheta) \in [0,\ 1], \\
L(\vartheta) \in \{\|L\| \le 1\} \cap K(U_2),
\end{cases}
\tag{3.131}
$$

where $K(U_2)$ is the smallest convex cone with vertex at zero containing the set U_2.

The performance criterion in the SAP is the terminal functional

$$
J_2(Z(\cdot), T(\cdot)) = G(Z(\overline{\vartheta}),\ \overline{\vartheta}),
\tag{3.132}
$$

which must be minimized over the set of admissible controls (3.131). Here $\overline{\vartheta}$ is the instant when the control process ends, and satisfies the inequality

$$
\overline{\vartheta} \le 2N + 1 + \overline{M}_2.
\tag{3.133}
$$

3.4.3 Equivalence of the Main and Auxiliary Problems

The equivalence of the main and auxiliary problems (FAP and SAP) is established by the following results.

Lemma 3.3 *Let $\{u_{1i}, \tau_{1i}\}$, $i = 1, \dots, N$ and $\{u_{2j}, \tau_{2j}\}$, $j = 1, 2, \dots,$ be arbitrary admissible impulse controls. Let $X(t)$, $t \in [t_0,\ T]$ be the corresponding solution of Eq. (3.114) describing the dynamics of the control system in the basic problem.*

There exist admissible controls $\alpha(\vartheta) \in [0,\ 1]$ and $L(\vartheta)$ for SAP, a terminal instant $\overline{\vartheta}$, as well as an admissible control $v(\theta)$ for FAP (where $\theta \in [0,\ 2N + 1]$ satisfies the equation (3.123) with the initial condition (3.129)) such that if $Z(\vartheta)$, $\vartheta \in [0,\ \overline{\vartheta}]$, denotes the solution of Eq. (3.119) with the initial condition (3.125) with the controls $\alpha(\vartheta)$, $L(\vartheta)$, and the control

$$
l(\theta) = \begin{cases}
u_{1i}, \quad if\ \theta \in]2i - 1,\ 2i[, \quad i = 1, \dots, N, \\
arbitrary\ U_1\text{-}valued\ function, \\
if\ \theta \in [2i - 2,\ 2i - 1], \quad i = 1, \dots, N + 1,
\end{cases}
\tag{3.134}
$$

then under the time substitution $\vartheta = \mu(t)$ with

$$\mu(t) = \sup\{\vartheta : t(\theta(\vartheta)) = t\}, \tag{3.135}$$

one has that $X(t) = Z(\mu(t))$ *and* $X(\overline{T}) = Z(\overline{\vartheta})$. *Moreover, the following equalities hold:*

$$X(\tau_{1i}-) = Y_i(\overline{\vartheta}), \quad X(\tau_{1i}) = Z_i(\overline{\vartheta}), \tag{3.136}$$

$$u_{1i} = \Lambda_i(\overline{\vartheta}), \tag{3.137}$$

$$\tau_{1i} = T_i(\overline{\vartheta}). \tag{3.138}$$

Proof Denote $v_i = \tau_{1i} - \tau_{1,i-1}$, $i = 1, \ldots, N+1$, and define the function $v(\theta)$ by

$$v(\theta) = \begin{cases} v_i, & \text{for } \theta \in [2i-2, 2i-1], \; i = 1, \ldots, N+1, \\ 0, & \text{for } \theta \in \,]2i-1, 2i\,[, \;\; i = 1, \ldots, N. \end{cases} \tag{3.139}$$

In turn, $v(\theta)$ defines the piecewise-linear function $t(\theta)$ with constant values on the "even" intervals by the formula

$$t(\theta) = t_0 + \int_0^\theta \beta(s)v(s)ds = t_0 + \int_0^\theta v(s)ds. \tag{3.140}$$

Next, let us mark on the ordinate axis the time instants $\{\tau_{2j}\}$, $j = 1, 2, \ldots$, and let, at the instants $\tau_{21}, \tau_{22}, \ldots$, shift successively the segments of the graph of the function $t(\theta) > \tau_{21}$, $t(\theta) > \tau_{22}, \ldots$ upwards, parallel to the abscissa axis, by the respective amounts $\|u_{21}\|$, $\|u_{22}\|, \ldots$ (Fig. 3.3). We call the resulting piecewise-constant curve $T(\vartheta)$ with $\vartheta \in \left[0, \; 2N + 1 + \sum_{j \geq 1} \|u_{2j}\|\right]$, the *deformation* of the (graph of) the function $t(\theta)$.

The constancy segments of the function $T(\vartheta)$, with having the lengths $\|u_{2j}\|$, are called the j-th *deformation intervals*. By construction, the ordinates of these

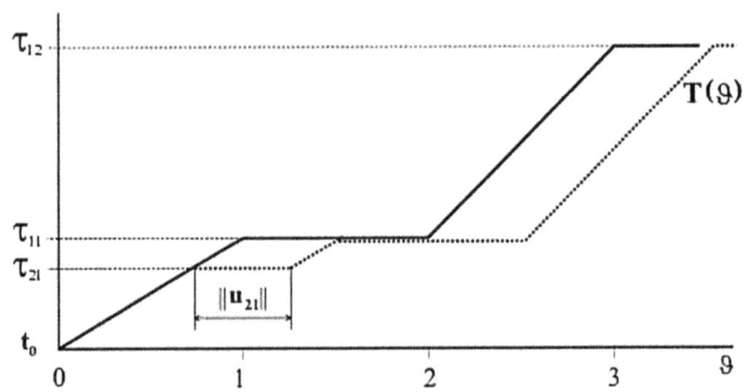

Fig. 3.3 Deformation process of the graph of $t(\vartheta)$

segments are equal to τ_{2j}, $j = 1, 2, \ldots$. As in the case of the function $t(\vartheta)$, the constancy segments of the function $T(\vartheta)$, with the length 1 (inherited from the function $t(\vartheta)$) are called the "even" intervals; their ordinates are equal to τ_{1i}, $i = 1, \ldots, N$. Finally, $T(\vartheta) = t(\theta(\vartheta))$.

We assume that

$$\alpha(\vartheta) = \begin{cases} 0, & \text{for } \vartheta \text{ belonging to deformation intervals,} \\ 1, & \text{for the remaining values of } \vartheta, \end{cases}$$

$$L(\vartheta) = \begin{cases} \dfrac{u_{2j}}{\|u_{2j}\|}, & \text{for } \vartheta \text{ in the } j\text{-th deformation interval, } j = 1, 2, \ldots, \\ \text{arbitrary value in } \{\|L\| \leq 1\} \cap K(U_2), \\ \text{for the remining values of } \vartheta. \end{cases} \tag{3.141}$$

Next, set $\overline{\beta}(\vartheta) = \beta(\theta(\vartheta))$. By (3.123), this function is given by

$$\overline{\beta}(\vartheta) = \begin{cases} 0, & \text{for } \vartheta \text{ in even intervals,} \\ 1, & \text{for the remaining values of } \vartheta. \end{cases} \tag{3.142}$$

Finally, using the right-continuous function $\mu(t)$ we obtain the inverse time transformation in the form

$$\mu(t) = \sup\{\vartheta : T(\vartheta) = t\}.$$

The instants $\{\tau_{1i}\}_{i=1,\ldots,N}$ and $\{\tau_{2j}\}_{j=1,2,\ldots}$ at which the impulses of the first and second type, respectively, are applied, are the discontinuity points of the function $\mu(t)$, and

$$T(\mu(s)) \equiv s. \tag{3.143}$$

Let τ be an arbitrary instant of impulse action. Two cases are possible:

(i) $\tau \in \{\tau_{1i}\}_{i=1,\ldots,N}$, $\quad \tau \notin \{\tau_{2j}\}_{j=1,2,\ldots}$;
(ii) $\tau \notin \{\tau_{1i}\}_{i=1,\ldots,N}$, $\quad \tau \in \{\tau_{2j}\}_{j=1,2,\ldots}$.

Let us establish the relation between $Z(\mu(\tau-))$ and $Z(\mu(\tau))$ in each of these two cases.

In case (i), $\tau = \tau_{1i}$, that is, the interval $[\mu(\tau-), \mu(\tau)]$ is even together with the number i and, consequently, the relations

$$\alpha(\vartheta) = 0, \quad \overline{\beta}(\vartheta) = 0, \quad \theta(\vartheta) \in]2i - 1,\ 2i[,$$

$$\Phi(Z(\vartheta),\ T(\vartheta),\ v(\theta(\vartheta)),\ l(\theta(\vartheta)),\ \theta(\vartheta)) = B_1(Z(\vartheta),\ T(\vartheta))\, l(\theta(\vartheta)),$$

$$l(\theta(\vartheta)) = u_{1i}, \quad \mu(\tau) - \mu(\tau-) = 1, \quad T(\vartheta) \equiv \tau_{1i}.$$

hold for $\vartheta \in [\mu(\tau-), \mu(\tau)]$. Therefore, according to (3.119), the function $Z(\vartheta)$ satisfies the equation

$$Z(\vartheta) = Z(\mu(\tau-)) + \int_{\mu(\tau-)}^{\vartheta} B_1(Z(s),\ T(s))\, l(\theta(s))\, ds.$$

By Lemma 1 in [263],

$$Z(\mu(\tau)) = Z(\mu(\tau-)) + \Psi_1(Z(\mu(\tau-)), u_{1i}, \tau) \tag{3.144}$$

where $\tau = \tau_{1i}$.

In case (ii), $\tau = \tau_{2j}$, that is, $[\mu(\tau-), \mu(\tau)]$ is the j-th deformation interval, and consequently

$$\alpha(\vartheta) = 0, \quad \overline{\beta}(\vartheta) = 0, \quad \mu(\tau) - \mu(\tau-) = \|u_{2j}\|, \quad T(\vartheta) \equiv \tau = \tau_{2j}$$

for all $\vartheta \in [\mu(\tau-), \mu(\tau)]$. According to (3.119), the function $Z(\vartheta)$ satisfies the equation

$$Z(\vartheta) = Z(\mu(\tau-)) + \int_{\mu(\tau-)}^{\vartheta} B_2(Z(s),\ T(s)) \frac{u_{2j}}{\|u_{2j}\|}\, ds.$$

By Lemma 1 [263],

$$Z(\mu(\tau)) = Z(\mu(\tau-)) + \Psi_2(Z(\mu(\tau-)), u_{2j}, \tau), \tag{3.145}$$

where $\tau = \tau_{2j}$.

Let $\widehat{\tau} < \tau$ be the first instant preceding τ at which an impulse is applied, or $\widehat{\tau} = \lim \tau_{2j}$, $j \to \infty$. Then $\beta(\theta(\vartheta)) = \overline{\beta}(\vartheta) = 1$ and $\alpha(\vartheta) = 1$ for all $\vartheta \in [\mu(\widehat{\tau}), \mu(\tau-)]$, and $\theta(\vartheta) \in [2i-2, 2i-1]$ for some $i = 1, \ldots, N+1$. Consequently, $v(\theta(\vartheta)) = v_i = \tau_{1i} - \tau_{1,i-1} = \text{const}$, and therefore

$$Z(\vartheta) = Z(\mu(\widehat{\tau})) + \int_{\mu(\widehat{\tau})}^{\vartheta} v_i F(Z(s),\ T(s))\, ds$$

or

$$Z(\mu(t)) = Z(\mu(\widehat{\tau})) + \int_{\mu(\widehat{\tau})}^{\mu(t)} v_i F(Z(s),\ T(s))\, ds.$$

Hence,

$$Z(\mu(\tau)) = Z(\mu(\tau-)) + \int_{\widehat{\tau}}^{t} F(Z(\mu(s)), s)\, ds, \tag{3.146}$$

where $t \in [\hat{\tau}, \tau-]$.

Combining (3.146) with (3.144) and (3.146) with (3.145), we see that

$$\begin{aligned} Z(\mu(\tau)) &= Z(\mu(\tau-)) + \int_{\hat{\tau}}^{t} F(Z(\mu(s)), s)\, ds + \\ &+ \Psi_1(Z(\mu(\tau-)), u_{1i}, \tau)\, \mathbf{I}\{t = \tau_{1i}\} + \\ &+ \Psi_2(Z(\mu(\tau-)), u_{2j}, \tau)\, \mathbf{I}\{t = \tau_{2j}\}, \end{aligned} \tag{3.147}$$

for all $t \in [\hat{\tau}, \tau]$. If we now take together these relations for all intervals between impulses, we conclude that the function $Z(\mu(t))$ satisfies the equation

$$\begin{aligned} Z(\mu(\tau)) &= Z(\mu(t_0)) + \int_{t_0}^{t} F(Z(\mu(s)), s)\, ds + \\ &+ \sum_{\tau_{1i} \le t} \Psi_1(Z(\mu(\tau_{1i}-)), u_{1i}, \tau_{1i}) + \\ &+ \sum_{\tau_{2j} \le t} \Psi_2(Z(\mu(\tau_{2j}-)), u_{2j}, \tau_{2j}). \end{aligned}$$

The equality of the initial conditions,

$$Z(\mu(t_0)) = Z(0) = X_0 = X(t_0),$$

and the uniqueness of the solution of equation (3.114) imply that $Z(\mu(t))$ and $X(t)$ coincide for any $t \in [t_0, \overline{T}]$, where $\overline{T} = T(\overline{\vartheta})$; this proves the first assertion of Lemma 3.3.

Now, equalities (3.136) follow from the coincidence of the initial conditions and (in view of (3.20)) the coincidence of the right-hand sides of Eqs. (3.117) and (3.119) for $\vartheta : \theta(\vartheta) \in [0, 2i-1], i = 1, \dots, N$, as well as the coincidence of the right-hand sides of Eqs. (3.118) and (3.119) for $\vartheta : \theta(\vartheta) \in [0, 2i], i = 1, \dots, N$. Next, equality (3.137) follows from (3.134) and (3.123), and the definition of the indicator functions (3.20). Finally, equality (3.138) follows from (3.20), (3.123), and (3.121). This completes the proof of the lemma. □

Theorem 3.5 *Let $l^0(\theta)$, $v^0(\theta)$ be the optimal control in FAP and let*

$$\alpha^0(\vartheta),\ L^0(\vartheta),\ Y_i^0(\vartheta),\ Z_i^0(\vartheta),\ Z^0(\vartheta),\ T^0(\vartheta),\ T_i^0(\vartheta),\ \Lambda_i^0(\vartheta),\ \theta^0(\vartheta),\ \zeta^0(\vartheta)$$

be the optimal controls and trajectories in SAP with the functions $l^0(\cdot)$ and $v^0(\cdot)$ used in the Eqs. (3.117)–(3.124).

Let $\overline{\vartheta}^0$ be the optimal terminal instant and let $G(Z^0(\overline{\vartheta}^0), T^0(\overline{\vartheta}^0))$ be the corresponding value of the performance criterion.

Then, there exist a sequence of admissible impulse controls of the second type $\{u^n_{2j},\ \tau^n_{2j}\}_{j=1,2,\dots}$ *and a collection of pairs* $\{u^0_{1i},\ \tau^0_{1i}\}_{i=1,\dots,N}$*, defining the impulsive control of the first type, with*

$$\begin{cases} \tau^0_{1i} = t^0_i(2N+1) = t^0(2i), \\ u^0_{1i} = \displaystyle\int_{2i-1}^{2i} l^0(\theta)\, d\theta, \end{cases} \tag{3.148}$$

(here $t^0_i(\theta)$ *and* $t^0(\theta)$ *are generated by the controls* $l^0(\theta)$ *and* $v^0(\theta)$ *in FAP) such that the relations*

$$\overline{T}^0 = T^0(\overline{\vartheta}^0) = T^n(\overline{\vartheta}^0) = \overline{T}^n, \quad n = 1, 2, \dots \tag{3.149}$$

$$\lim_n G\big(X^n(\overline{T}^0),\ \overline{T}^0\big) = G\big(Z^0(\overline{\vartheta}^0),\ \overline{T}^0\big). \tag{3.150}$$

hold for the corresponding sequence of solutions $\{X^n(t)\}$ *of Eq. (3.114) and the sequence of the terminal instants* $\{\overline{T}^n\}$*. Moreover,*

$$\lim_n \left[\int_0^{\overline{T}^0} \|X^n(s) - X^0(s)\|\, ds + \|X^n(\overline{T}^0) - X^0(\overline{T}^0)\| \right] = 0, \tag{3.151}$$

where $X^0(t) = Z^0(\mu^0(t))$*, with* $\mu^0(t) = \sup\{\vartheta\ :\ T^0(\vartheta) = t\}$*.*

The following lemmas are required for the proof of the theorem.

Lemma 3.4 *Suppose the collection of pairs* $\{u^0_{1i},\ \tau^0_{1i}\}_{i=1,\dots,N}$ *satisfies the conditions (3.148) and let* $\{u_{2j},\ \tau_{2j}\}_{j=1,2,\dots}$ *be the admissible control of the second type with the set* $K(U_2)$ *instead of* U_2*. Assume that* $N_T = \max\{j\ :\ \tau_{2j} \le \overline{T}\} < \infty$*. Then there exists a sequence of controls* $\{u^n_{2j},\ \tau^n_{2j}\}_{j \le N_T}$ *that are admissible with the set* U_2 *and such that*

$$\lim_n \left[\int_0^{\overline{T}} |X^n(s) - X(s)|\, ds + |X^n(\overline{T}) - X(\overline{T})| \right] = 0. \tag{3.152}$$

Proof of Lemma 3.4 For each $u_{2j} \in K(U_2)$ there exists an integer k_j such that $(1/k_j)u_{2j} \in U_2$. Pick $\epsilon > 0$ such that the ϵ-neighborhoods of the points τ_{2j} are disjoint (which is possible because by assumption $N_T < \infty$), and also disjoint with the ϵ-neighborhoods of (a finite number of) the points τ_{1i}, $i = 1, \dots, N$. Place k_j points τ in the ϵ-neighborhood of the point τ_{2j} and assign to each of them the control $(1/k_j)u_{2j} \in U_2$. Upon enumerating all these points τ in ascending order, one obtains a finite set of points $\{\tau_k\}$. Denote the controls corresponding to these points by u_k. Obviously, the impulse control $\{u_k,\ \tau_k\}_\epsilon$ will be admissible with the set U_2. The index ϵ indicates here that the performed construction of the impulse controls depends on the choice of ϵ-neighborhoods of a finite number of points τ_{2j} and τ_{1i}.

Now suppose a sequence $\epsilon(n) \to 0$ is given. Let us show that the sequence of controls $\{u_k, \tau_k\}_{\epsilon(n)}$ has the desired properties. Let $X^n(t)$ denote the solution of Eq. (3.114) corresponding to the impulsive control $\{u_k, \tau_k\}_{\epsilon(n)}$. Then, if $\tau_{21} \neq 0$, we have

$$X^n(\tau_{21} - \epsilon(n)) = X(\tau_{21} - \epsilon(n)) \quad \text{and} \quad X^n(t) = X(t)$$

for n large enough and $t \in [0, \tau_{21} - \epsilon(n)]$. Consequently, $\lim_n X^n(\tau_{21} - \epsilon(n)) = X(\tau_{21}-)$. On the other hand, thanks to (2.7),

$$X^n(\tau_{21} + \epsilon(n)) = X^n(\tau_{21} - \epsilon(n)) +$$
$$+ \Psi_2\left(X^n(\tau_{21} - \epsilon(n)), \sum_{\tau_k^n \in [\tau_{21} - \epsilon(n),\, \tau_{21} + \epsilon(n)]} u_2(\tau_k^n), \tau_{21} - \epsilon(n) \right) + O(\epsilon(n)),$$

where $O(\epsilon(n)) \to 0$ as $n \to \infty$.

Now, by construction,

$$\sum_{\tau_k^n \in [\tau_{21} - \epsilon(n),\, \tau_{21} + \epsilon(n)]} u_2(\tau_k^n) = u_2(\tau_{21}) = u_{21}.$$

Note that in this sum (for any n) k_1 terms are equal to $(1/k_1)u_{21}$. Letting $n \to \infty$, we obtain

$$\lim_n X^n(\tau_{21} + \epsilon(n)) = X(\tau_{21}-) + \Psi_2(X(\tau_{21}-), u_{21}, \tau_{21}). \tag{3.153}$$

The case when $\tau_{21} = 0$ is dealt with in much the same way. For $t \in [\tau_{21} + \epsilon(n), \tau_{22} - \epsilon(n)]$, the functions $X(t)$ and $X^n(t)$ satisfy identical equations, namely

$$\begin{aligned} X(t) &= X(\tau_{21} + \epsilon(n)) + \int_{\tau_{21}+\epsilon(n)}^t F(X(s), s)\, ds + \\ &+ \sum_{\tau_{21} < \tau_{1i} < \tau_{22}} \Psi_1(X(\tau_{1i}-), u_{1i}, \tau_{1i}), \\ X^n(t) &= X^n(\tau_{21} + \epsilon(n)) + \int_{\tau_{21}+\epsilon(n)}^t F(X^n(s), s)\, ds + \\ &+ \sum_{\tau_{21} < \tau_{1i} < \tau_{22}} \Psi_1(X^n(\tau_{1i}-), u_{1i}, \tau_{1i}). \end{aligned} \tag{3.154}$$

Let $\kappa^n(t) = |X(t) - X^n(t)|$. Since the functions $F(\cdot)$ and $\Psi_1(\cdot)$ satisfy the Lipschitz condition with respect to the variable X, we get

$$\kappa^n(t) \leq \kappa^n(\tau_{21}+\epsilon(n))+k\left[\int_{\tau_{21}+\epsilon(n)}^{t} \kappa^n(s)\,ds + \sum_{\tau_{21}<\tau_{1i}<\tau_{22}} \kappa^n(\tau_{1i}-)\right], \tag{3.155}$$

where k =const= $\max\{L_0,\ L_1,\ \overline{M}_1\}$.

Next, since $\kappa^n(\tau_{21}+\epsilon(n)) \to 0$ as $n \to \infty$ thanks to (3.153), applying the Gronwall-Bellman lemma to (3.155) we obtain that

$$\sup_{t\in[\tau_{21}+\epsilon,\ \tau_{22}-\epsilon]} \kappa^n(t) \to 0, \quad n \to \infty$$

for any $\epsilon > 0$ satisfying the aforementioned condition on disjointness of the neighborhoods of the points $\{\tau_{1i}\}$ and $\{\tau_{2j}\}$. This implies that

$$X^n(\tau_{22}-\epsilon(n)) \to X(\tau_{22}-), \quad n \to \infty.$$

By continuing this procedure over all τ_{2j}, $j = 1, \ldots, N_T$, we conclude that, for any $\epsilon > 0$,

$$\sup |X^n(t) - X(t)| \to 0, \quad n \to \infty,$$

where the supremum is taken over the set $[t_0,\ T] \setminus \bigcup_{j=1}^{N_T} O_\epsilon(\tau_{2j})$. Here, $O_\epsilon(\tau)$ denotes the $\epsilon 0$-neighborhood of the point τ.

The uniform boundedness of the sums $\sum_k \|u_k^n\| = \sum_{j=1}^{N_T} \|u_{2j}\|$ implies that the set of functions $\{X^n(t)\}$ is uniformly bounded, and consequently

$$\int_0^T |X^n(t) - X(T)|\,dt \to 0, \quad n \to \infty.$$

On the other hand,

$$X^n(\tau_{2j}-\epsilon(n)) \to X(\tau_{2j}-), \quad n \to \infty,$$

$$X^n(\tau_{2j}+\epsilon(n)) \to X(\tau_{2j}), \quad n \to \infty.$$

for any $j = 1, \ldots, N_T$. If $\tau_{2N_T} \neq \overline{T}$, then $X^n(\overline{T}) \to X(\overline{T})$ as $n \to \infty$. If $\tau_{2N_T} = \overline{T}$, then, by construction,

$$X^n(\overline{T}) = X^n(\overline{T}-\epsilon(n)) + \Psi_2(X^n(\overline{T}-\epsilon(n)), u_{2N_T}, \overline{T}-\epsilon(n)) + O(\epsilon(n)).$$

But

$$X^n(\overline{T}-\epsilon(n)) \to X(\overline{T}-), \quad n \to \infty,$$

and so, in view of the continuity of $\Psi_2(\cdot)$,

$$X^n(\overline{T}) \to X(\overline{T}), \quad n \to \infty,$$

which completes the proof of the lemma. □

Lemma 3.5 *Let $\alpha(\vartheta)$, $L(\vartheta)$ be admissible controls, and let $\overline{\vartheta}$ be the corresponding terminal instant in SAP. Then there exists a sequence $\{\alpha^n(\vartheta), L^n(\vartheta)\}$ such that the functions $\alpha^n(\vartheta)$ are piecewise-constant with values in the set $\{0, 1\}$ and the functions $\alpha^n(\vartheta)$ and $(1-\alpha^n(\vartheta))L^n(\vartheta)$ converge weakly to the functions $\alpha(\vartheta)$ and $(1-\alpha(\vartheta))L(\vartheta)$, respectively. Moreover,*

$$\int_0^{\overline{\vartheta}} \alpha^n(s)\,ds = \int_0^{\overline{\vartheta}} \alpha(s)\,ds,$$

and

$$\int_0^{\overline{\vartheta}} (1-\alpha^n(s))L^n(s)\,ds = \int_0^{\overline{\vartheta}} (1-\alpha(s))L(s)\,ds.$$

Proof of Lemma 3.5 Consider an interval $[\vartheta_0, \vartheta_1] := \Delta$ and the functions $\alpha'(\vartheta)$ and $L'(\vartheta)$ defined on Δ by

$$\begin{aligned} \alpha'(\vartheta) &= \begin{cases} 1, & \text{for } \vartheta \in \left[\vartheta_0, \vartheta_0 + \int_0^{\vartheta_1} \alpha(s)\,ds\right[, \\ 0, & \text{otherwise,} \end{cases} \\ L'(\vartheta) &= \begin{cases} 0, & \text{for } \vartheta \in \left[\vartheta_0, \vartheta_0 + \int_0^{\vartheta_1} \alpha(s)\,ds\right[, \\ \dfrac{\int_0^{\vartheta_1} (1-\alpha(s))L(s)\,ds}{\int_0^{\vartheta_1} (1-\alpha(s))\,ds}, & \text{otherwise.} \end{cases} \end{aligned} \tag{3.156}$$

Since the set $V = \{\|L\| \le 1\} \bigcap K(U_2)$ is convex, the values of the function $L'(\vartheta)$ belong to V. Moreover,

$$\int_0^{\vartheta_1} \alpha'(s)\,ds = \int_0^{\vartheta_1} \alpha(s)\,ds,$$

$$\int_0^{\vartheta_1} (1-\alpha'(s))L^n(s)\,ds = \int_0^{\vartheta_1} (1-\alpha(s))L(s)\,ds.$$

We decompose the segment $[0, \overline{\vartheta}]$ into n half-intervals Δ_k, $k = 1, \ldots, n$ of equal lengths and use the recipe (3.156) to construct on each of them the functions $\alpha'_k(\vartheta)$ and $L'_k(\vartheta)$ for $\vartheta \in \Delta_k$. Then, by joining the intervals Δ_k and the corresponding functions $\alpha'_k(\vartheta)$ and $L'_k(\vartheta)$, we construct the functions $\alpha^n(\vartheta)$ and $L^n(\vartheta)$ satisfying

$$\int_{\Delta_k} \alpha^n(s)\, ds = \int_{\Delta_k} \alpha(s)\, ds$$

and

$$\int_{\Delta_k} (1 - \alpha^n(s))L^n(s)\, ds = \int_{\Delta_k} (1 - \alpha(s))L(s)\, ds.$$

Now let $n \to \infty$. Since the lengths of half-intervals Δ_k tend to zero uniformly in k, there exist the weak limits

$$\alpha^n(\vartheta) \overset{\text{weak}}{\longrightarrow} \alpha(\vartheta), \quad n \to \infty,$$

and

$$(1 - \alpha^n(\vartheta))L^n(\vartheta) \overset{\text{weak}}{\longrightarrow} (1 - \alpha(\vartheta))L(\vartheta), \quad n \to \infty.$$

This implies that, for any function $h(\vartheta)$ that is continuous on $[0, \overline{\vartheta}]$,

$$\int_0^{\overline{\vartheta}} h(s)\alpha^n(s)\, ds \to \int_0^{\overline{\vartheta}} h(s)\alpha(s)\, ds \quad \text{as } n \to \infty, \tag{3.157}$$

and

$$\int_0^{\overline{\vartheta}} h(s)(1 - \alpha^n(s))L^n(s)\, ds \to \int_0^{\overline{\vartheta}} h(s)(1 - \alpha(s))L(s)\, ds, \quad \text{as } n \to \infty. \tag{3.158}$$

Indeed, let $\overline{h}_n(\vartheta) = \sum_{k=1}^{n} \overline{h}^k \mathbf{I}\{\Delta_k\}$, $\overline{h}^k$ =const, be step functions approximating $h(\vartheta)$ from above, that is, $h(\vartheta) \le \overline{h}_{n+1}(\vartheta) \le \overline{h}_n(\vartheta)$ and $h(\vartheta) = \inf_n \overline{h}_n(\vartheta)$, for all ϑ. Similarly, let $h(\vartheta) = \sup_n \underline{h}_n(\vartheta)$, with $\underline{h}_n(\vartheta) \le \underline{h}_{n+1}(\vartheta) \le h(\vartheta)$, for all ϑ. Then, letting $n \to \infty$, we obtain that, on the one hand,

$$\begin{aligned} &\int_0^{\overline{\vartheta}} h(s)\alpha^n(s)\, ds \le \int_0^{\overline{\vartheta}} \overline{h}_n(s)\alpha^n(s)\, ds = \\ &= \sum_{k=1}^{n} \overline{h}^k \int_{\Delta_k} \alpha^n(s)\, ds = \sum_{k=1}^{n} \overline{h}^k \int_{\Delta_k} \alpha(s)\, ds \to \int_0^{\overline{\vartheta}} h(s)\alpha(s)\, ds \end{aligned} \tag{3.159}$$

and, on the other hand,

$$\int_0^{\overline{\vartheta}} h(s)\alpha^n(s)\,ds \geq \int_0^{\overline{\vartheta}} \underline{h}_n(s)\alpha^n(s)\,ds =$$
$$= \sum_{k=1}^{n} \underline{h}^k \int_{\Delta_k} \alpha^n(s)\,ds = \sum_{k=1}^{n} \underline{h}^k \int_{\Delta_k} \alpha(s)\,ds \to \int_0^{\overline{\vartheta}} h(s)\alpha(s)\,ds. \tag{3.160}$$

Relation (3.157) follows from (3.159) and (3.160). Relation (3.158) is proved along the same lines. The lemma is proved. □

Proof of Theorem 3.5 Assuming, for definiteness, that $\alpha^n(0) = 1$, let us construct, as instructed by Lemma 3.5, sequences of functions $\alpha^n(\vartheta)$ and $L^n(\vartheta)$ for the given functions $\alpha^0(\vartheta)$ and $L^0(\vartheta)$ and the terminal instant $\overline{\vartheta}^0$. Let $\{\vartheta_k^n\}$, $k = 1, \ldots, 2n$, be the (finite number of) discontinuity points of the function $\alpha^n(\vartheta)$, and let $\{\varepsilon_n\}$ be an arbitrary numerical sequence such $\varepsilon_n \downarrow 0$ as $n \to \infty$. We define the sequence of vectors $u_{2,k}^n$ and instants θ_k^n, $\tau_{2,k}^n$ by the formulas

$$\begin{cases} u_{2,k}^n = \displaystyle\int_{\vartheta_{2k-1}^n}^{\vartheta_{2k}^n} L^n(s)\,ds, \\ \theta_k^n = \displaystyle\int_0^{\vartheta_{2k-1}^n} \alpha^n(s)\,ds, \\ \tau_{2,k}^n = \displaystyle\int_0^{\theta_k^n} \beta(s)v^0(s)\,ds. \end{cases} \tag{3.161}$$

We claim that the control $\{(u_{2,k}^n, \tau_{2,k}^n)\}$ is admissible with the set $K(U_2)$. Indeed, $u_{2,k}^n \in K(U_2)$, because

$$\sum_k \|u_{2,k}^n\| \leq \sum_k \|\int_{\vartheta_{2k-1}^n}^{\vartheta_{2k}^n} L^n(s)\,ds\| \leq \sum_k (\vartheta_{2k}^n - \vartheta_{2k-1}^n) = \overline{\vartheta}^0 - (2N+1) \leq \overline{M}_2$$

and

$$u_{2,k}^n = \sum_k \int_{\vartheta_{2k-1}^n}^{\vartheta_{2k}^n} L^n(s)\,ds = \int_0^{\overline{\vartheta}^0} (1-\alpha^n(s))L^n(s)\,ds = \int_0^{\overline{\vartheta}^0} (1-\alpha(s))L(s)\,ds.$$

If we have equality $\tau_{2,k'}^n = \tau_{1,i'}^0$ for some k' and i', then from the sequence $\{\varepsilon_n\}$ we take terms ε_{m_n} such that the $(v_*^0\varepsilon_{m_n})$-neighborhoods of the non-coinciding points from the sets $\{\tau_{1,i}^0\}_{i=1,\ldots,N}$ and $\{\tau_{2,k}^0\}_{k=1,2,\ldots}$ are disjoint. Here, $v_*^0 = \min_{1\leq i\leq N} v_i^0$. Then, instead of the function $\alpha^n(\vartheta)$, we consider a similar function $\alpha_\varepsilon^n(\vartheta)$ which differs from $\alpha^n(\vartheta)$ in that its discontinuity points $\vartheta_k^n(\varepsilon)$ indexed by $2k'-1$ and

$2k'$ are shifted to the right (or left) by $+\varepsilon_{m_n}$ (or $-\varepsilon_{m_n}$) relative to the discontinuity points $\vartheta^n_{2k'-1}$ and $\vartheta^n_{2k'}$ of the function $\alpha^n(\vartheta)$:

$$\begin{cases} \vartheta^n_{2k-1}(\varepsilon) = \vartheta^n_{2k-1}, & \vartheta^n_{2k}(\varepsilon) = \vartheta^n_{2k}, \quad \text{for } k \neq k', \\ \vartheta^n_{2k-1}(\varepsilon) = \vartheta^n_{2k-1} \pm \varepsilon_{m_n}, & \vartheta^n_{2k}(\varepsilon) = \vartheta^n_{2k} \pm \varepsilon_{m_n}, \quad \text{for } k = k', \end{cases}$$

The shift is by $+\varepsilon_{m_n}$ or $-\varepsilon_{m_n}$ according to whether the segment $[\vartheta^n_{2k'-1}, \vartheta^n_{2k'}]$ is adjacent to the right or to the left to the corresponding "even" interval $(\beta(\vartheta) = 0)$. Under the shift, the sequence of vectors $\{u^n_{2,k}\}_{k=1,2,\ldots}$ defined by (3.161) does not change, and instead of the instants θ^n_k and $\tau^n_{2,k}$ we obtain

$$\begin{cases} \theta^n_k(\varepsilon) = \displaystyle\int_0^{\vartheta^n_{2k-1}(\varepsilon)} \alpha^n_\varepsilon(s)\, ds = \theta^n_k, \quad \text{if } k \neq k', \\ \theta^n_k(\varepsilon) = \displaystyle\int_0^{\vartheta^n_{2k-1}(\varepsilon)} \alpha^n_\varepsilon(s)\, ds = \theta^n_k + \varepsilon_{m_n}, \quad \text{if } k = k', \end{cases}$$

and consequently,

$$\begin{cases} \tau^n_{2,k}(\varepsilon) = \tau^n_{2,k}, & k \neq k', \\ \tau^n_{2,k}(\varepsilon) = \tau^n_{2,k} \pm \varepsilon_{m_n} v^0_{i'=1}, & k = k', \end{cases}$$

where the indices i' are the ones figuring in the equalities $\tau^0_{2,k'}(\varepsilon) = \tau^0_{1,i'}$. The sequences $\{\tau^0_{1,i}\}_{i=1,\ldots,N}$ and $\{\tau^0_{2,k}(\varepsilon)\}_{k=1,2,\ldots}$ are already distinct (no coinciding points; see, e.g., Fig. 3.4 where $k' = 3$ and $i = 1$). The subsequence $\{\varepsilon_{m_n}\}$ is taken to be monotonically decreasing. Further, the functions $\alpha^n_\varepsilon(\vartheta)$ obviously satisfy the equalities

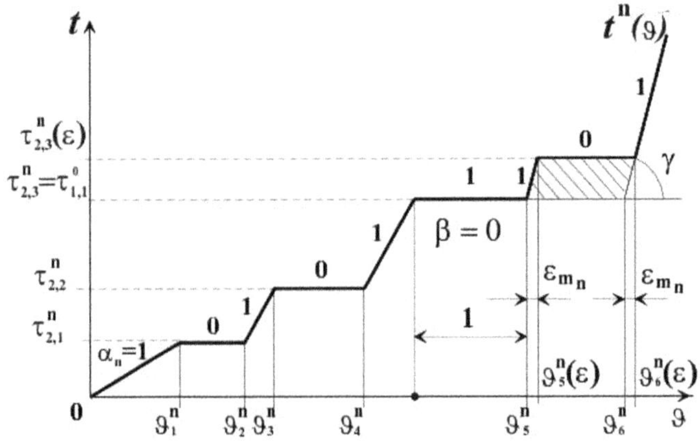

Fig. 3.4 Sequence of $\tau^n_{2,k}$ and the graph of function $t^n(\vartheta)$

$$\begin{cases} \int_0^{\overline{\vartheta}^0} \alpha_\varepsilon^n(s)\,ds = \int_0^{\overline{\vartheta}^0} \alpha^0(s)\,ds, \\ \int_0^{\overline{\vartheta}^0} (1-\alpha_\varepsilon^n(s))L^n(s)\,ds = \int_0^{\overline{\vartheta}^0} (1-\alpha^0(s))L^0(s)\,ds, \end{cases} \tag{3.162}$$

and consequently (see the proof of Lemma 3.5)

$$\begin{cases} \alpha_\varepsilon^n(\vartheta) \xrightarrow{\text{weak}} \alpha^0(\vartheta), & n \to \infty, \\ (1-\alpha_\varepsilon^n(\vartheta))L^n(\vartheta) \xrightarrow{\text{weak}} (1-\alpha^0(\vartheta))L^0(\vartheta), & n \to \infty. \end{cases}$$

Now consider the solution of (3.114) with the controls $\{u_{1i}^0,\ \tau_{1i}^0\}_{i=1,\dots,N}$ from (3.148) and the controls $\{u_{2k}^n,\ \tau_{2k}^n\}_{k=1,2,\dots}$. As follows from the proof of Lemma 3.3,

$$Z^n(\mu_n(t)) = X^n(t), \quad \text{for} \quad \mu_n(t) = \sup\{\vartheta : t^n(\vartheta) = t\}, \quad \vartheta \in [0,\ \overline{\vartheta}^0],$$

where $t^n(\vartheta) = T^n(\vartheta)$, and $T^n(\vartheta)$ satisfies Eq. (3.120) with $v(\cdot) = v^0(\cdot)$ and $\alpha(\cdot) = \alpha_\varepsilon^n(\cdot)$. The weak convergence of the functions $\alpha_\varepsilon^n(\cdot)$ and $(1-\alpha_\varepsilon^n(\cdot))L^n(\cdot)$ to $\alpha^0(\cdot)$ and $(1-\alpha^0(\cdot))L^0(\cdot)$, respectively, ensures that for $\vartheta \in [0,\ \overline{\vartheta}^0]$ of the sequence $Z^n(\vartheta)$ converges uniformly to $Z^0(\vartheta)$, and similar uniform convergence holds for the rest of the phase variables satisfying (3.117)–(3.124). In particular, $Z^n(\overline{\vartheta}^0) \to Z^0(\overline{\vartheta}^0)$. Moreover, in view of (3.162),

$$\overline{T}^0 = T^0(\overline{\vartheta}^0) = T^n(\overline{\vartheta}^0) = \overline{T}^n. \tag{3.163}$$

By definition,

$$\mu_n(T^n(\overline{\vartheta}^0)) = \sup\left\{\vartheta : T^n(\vartheta) = T^n(\overline{\vartheta}^0)\right\} = \overline{\vartheta}^0.$$

Similarly,

$$\mu_0(T^0(\overline{\vartheta}^0)) = \overline{\vartheta}^0.$$

Consequently,

$$Z^n(\overline{\vartheta}^0) = X^n(T^n(\overline{\vartheta}^0)) = X^n(\overline{T}^n) = X^n(\overline{T}^0). \tag{3.164}$$

From (3.163) and (3.164) it follows that

$$X^n(\overline{T}^0) = Z^n(\overline{\vartheta}^0) \to Z^n(\overline{\vartheta}^0), \quad n \to \infty.$$

In view of the continuity of the function $G(\cdot)$, this implies that

$$G(X^n(\overline{T}^0), \overline{T}^0) \to G(Z^0(\overline{\vartheta}^0), \overline{T}^0), \quad n \to \infty. \tag{3.165}$$

By Lemma 3.4, for any fixed n there exists a sequence (indexed by $m = 1, 2, \ldots$) of admissible controls $\{u_{2j}, \tau_{2j}\}_{j=1,2,\ldots}^{n,m}$ with the set U_2 instead of $K(U_2)$, such that

$$\lim_m X^{n,m}(\overline{T}^0) = X^n(\overline{T}^0).$$

Then, for this sequence

$$\lim_n X^n(\overline{T}^0) = Z^0(\overline{\vartheta}^0). \tag{3.166}$$

Relations (3.163), (3.165), and (3.166) imply that (3.149) and (3.150) are satisfied for the controls $\{u_{2j}, \tau_{2j}\}_{j=1,2,\ldots}^{n,m}$.

Now let us prove the equality (3.151). It follows from the definition of the functions $\alpha_\varepsilon^n(\vartheta),\ \vartheta \in [0, \overline{\vartheta}^0]$, that

$$\int_0^\vartheta \alpha_\varepsilon^n(s)\, ds \to \int_0^\vartheta \alpha^0(s)\, ds, \quad n \to \infty. \tag{3.167}$$

Therefore,

$$\begin{aligned}
&\int_0^{\overline{T}^0} |\mu_0(s) - \mu_n(s)|\, ds = \int_0^{\vartheta_*^0} |T(\vartheta) - T^n(\vartheta)|\, d\vartheta = \\
&\quad = \int_0^{\vartheta_*^0} \left| \int_0^\vartheta \big(\beta(\theta^0(s))v^0(\theta^0(s))\alpha^0(s) - \beta(\theta_n(s))v^0(\theta_n(s))\alpha_n(s)\big)\, ds \right| d\vartheta = \\
&\quad = \int_0^{\vartheta_*^0} \left| \int_0^\vartheta \big(\beta(\theta^0(s))v^0(\theta^0(s))\alpha^0(s) - \beta(\theta_n(s))v^0(\theta_n(s))\alpha^0(s) + \right. \\
&\quad\quad \left. + \beta(\theta_n(s))v^0(\theta_n(s))\alpha^0(s) - \beta(\theta_n(s))v^0(\theta_n(s))\alpha_n(s)\big)\, ds \right| d\vartheta \le \\
&\quad \le \int_0^{\vartheta_*^0} \int_0^\vartheta \left| \beta(\theta^0(s))v^0(\theta^0(s)) - \beta(\theta_n(s))v^0(\theta_n(s)) \right| ds\, d\vartheta + \\
&\quad\quad + \int_0^{\vartheta_*^0} \int_0^\vartheta \left| \beta(\theta_n(s))v^0(\theta_n(s)) \right| \left| \alpha^0(s) - \alpha_n(s) \right| ds\, d\vartheta \to 0, \quad n \to \infty,
\end{aligned}$$

thanks of (3.162) and (3.167), the weak convergence of $\alpha_\varepsilon^n(\cdot)$ to $\alpha^0(\cdot)$, and the fact that

$$\beta(\theta_n(s))v^0(\theta_n(s)) \to \beta(\theta^0(s))v^0(\theta^0(s)), \quad n \to \infty.$$

Further,

$$\int_0^{\overline{T}^0} \left|X^0(s) - X^{n,n}(s)\right| ds = \int_0^{\overline{T}^0} \left|Z^0(\mu_0(s)) - X^{n,n}(s)\right| ds \le$$

$$\le \int_0^{\overline{T}^0} \left|Z^0(\mu_0(s)) - Z^0(\mu_n(s))\right| ds + \int_0^{\overline{T}^0} \left|Z^0(\mu_n(s)) - Z^n(\mu_n(s))\right| ds +$$

$$+ \int_0^{\overline{T}^0} \left|Z^n(\mu_n(s)) - X^{n,n}(s)\right| ds.$$

Since the function $Z(\vartheta)$ is Lipschitz, the first integral above obeys the estimate

$$\int_0^{\overline{T}^0} \left|Z^0(\mu_0(s)) - Z^0(\mu_n(s))\right| ds \le \overline{K} \int_0^{\overline{T}^0} |\mu_0(s) - \mu_n(s)|\, ds,$$

and the second integral tends to zero thanks to the uniform (in $\vartheta \in [0, \overline{\vartheta}^0]$) convergence of $Z^n(\vartheta)$ to $Z^0(\vartheta)$. The third integral tends to zero by Lemma 3.4. Consequently, when $n \to \infty$,

$$X^{n,n}(s) \to Z^0(\mu_0(s)) \quad \text{in } L_1[0, \overline{T}^0].$$

The convergence of $X^{n,n}(\overline{T}^0)$ to $Z^0(\overline{\vartheta}^0) = X^0(\overline{T}^0)$ was established in (3.166). This completes the proof of the theorem. □

Theorem 3.6 *Under the assumptions of Theorem 3.5, in the basic problem the impulse controls* $\{u^0_{1i}, \tau^0_{1i}\}$, $i = 1, \dots, N$, *and the sequence* $\{u_{2j}, \tau_{2j}\}^{n,n}_{j=1,\dots}$ *are minimizing.*

Proof Let us assume that in the basic problem there exist admissible impulsive controls $\{\widehat{u}_{1i}, \widehat{\tau}_{1i}\}_{i=1,\dots,N}$ and a sequence $\{\widehat{u}_{2j}, \widehat{\tau}_{2j}\}^n_{j=1,2,\dots}$ such that the corresponding sequences of trajectories $\widehat{X}^n(t)$ and terminal instants $\{\widehat{T}^n\}$ satisfy the inequality

$$\inf_n G(\widehat{X}^n(\widehat{T}^n), \widehat{T}^n) < G(X^0(\overline{T}^0), \overline{T}^0).$$

By Lemma 3.3, for the controls $\{\widehat{u}_{1i}, \widehat{\tau}_{1i}\}_{i=1,\dots,N}$ and $\{\widehat{u}_{2j}, \widehat{\tau}_{2j}\}^n_{j=1,2,\dots}$ there exist admissible controls in the auxiliary problems such that

$$\widehat{X}^n(t) = \widehat{Z}^n(\widehat{\mu}_n(t)) \quad \text{and} \quad \widehat{X}^n(\widehat{T}^n) = \widehat{Z}^n(\overline{\vartheta}^n), \; t \in [t_0, \widehat{T}^n],$$

where the symbol "$\widehat{\ }$" indicates the variables calculated under the controls generated by $\{\widehat{u}_{1i}, \widehat{\tau}_{1i}\}_{i=1,\dots,N}$ and $\{\widehat{u}_{2j}, \widehat{\tau}_{2j}\}^n_{j=1,2,\dots}$. Consequently, also

$$\inf_n G(\widehat{X}^n(\widehat{T}^n), \widehat{T}^n) = \inf_n G(\widehat{Z}^n(\overline{\vartheta}^n), \widehat{T}^n(\overline{\vartheta}^n)).$$

These relations lead to the following contradictory chain of inequalities:

$$G(X^0(\widehat{T}^0), \widehat{T}^0) = G(Z^0(\widehat{\vartheta}^0), \widehat{T}^0) \le \inf_n G(\widehat{Z}^n(\overline{\vartheta}^n), \widehat{T}^n(\overline{\vartheta}^n)) =$$

$$= \inf_n G(\widehat{X}^n(\widehat{T}^n), \widehat{T}^n) < G(X^0(\widehat{T}^0), \widehat{T}^0).$$

This completes the proof of the theorem. □

Chapter 4
Generalized Solutions of Nonlinear Differential Equations

4.1 Impulse and Ordinary Controls in Nonlinear Differential Equations

In this chapter, we discuss the control problem of a nonlinear dynamic system whose state is governed by a nonlinear differential equation of the form

$$\dot{X}(t) = F(X(t), u(t), w(t), t), \quad X(0) = x_0, \tag{4.1}$$

where $F(x, u, w, t)$ is a given function, $x \in \mathbb{R}^n$, $t \in [0, T]$, and $u(t)$, $w(t)$ are Lebesgue-measurable controls defined on $[0, T]$. It is assumed that $u(t)$ is an ordinary bounded control such that

$$u(t) \in U \subset \mathbb{R}^k, \tag{4.2}$$

where U is a closed bounded subset of $\mathbb{R}^k$. The generalized control $w(t)$ is unbounded in norm, but restricted in the integral sense, that is

$$w(t) \in W \subset \mathbb{R}^m, \tag{4.3}$$

$$\int_0^T \|w(t)\| dt \leq M < \infty, \tag{4.4}$$

where W is a subset of $\mathbb{R}^m$ and M is a constant.

Remark 4.1 A solution of the system (4.1) is a triple $\{X(\cdot), u(\cdot), w(\cdot)\}$, where $X(\cdot)$ is absolutely continuous with respect to the Lebesgue measure, the functions $u(\cdot)$ and $w(\cdot)$ are Lebesgue measurable and satisfy the conditions (4.2) and (4.3) almost everywhere on $[0, T]$, and Eq. (4.1) is understood in the integral sense, that is

B. Miller, E. Rubinovich, *Optimization of Dynamical Systems with Impulse Controls and Shocks*, Systems & Control: Foundations & Applications,
https://doi.org/10.1007/978-3-031-64124-4_4

$$X(t) = x_0 + \int_0^t F(X(s), u(s), w(s), s)ds, \tag{4.5}$$

which means that (4.1) holds almost everywhere on $[0, T]$. This concept of solution is common for differential equations with measurable right-hand side in the sense of Carathéodory [237, 365]. Henceforth we will not mention again explictly that this is the concept of solution used if it is not necessary for understanding; moreover, since it is always possible to modify the controls ($u(\cdot)$ and $w(\cdot)$ on a set of measure zero so that they will satisfy the constraints the (4.2) and (4.3) everywhere on $[0, T]$ without changing the solution $X(\cdot)$, we will assume that these constraints are indeed satisfied everywhere on $[0, T]$.

Optimization problems with impulse and ordinary controls arise, for example, in flight dynamics [192, 193, 218, 222, 236, 378], and [461]. For instance, the direction of the thrust of a jet vehicle can be considered as bounded control, while the jet flow rate, which may be of impulsive type, can be considered as an unbounded control. Another example of such problems arises in observation control [87, 153] for discrete-continuous systems where two different types of control characterize the possibility of controlling the composition of observations, their timing, as well as their density. Condition (4.4) means that the control w can be taken as a function that is as close as desired to an impulse or a generalized function. As a result, the optimal control problem for the system (4.1) with absolutely continuous paths $\{x(\cdot)\}$ and subject to the constraints (4.2)–(4.4) may not admit a solution in the class of measurable ordinary or relaxed control functions. Thus, the concept of a solution should be extended by allowing a discontinuous behavior of the optimal trajectory under the impulse control action.

If the function $F(x, u, w, t)$ is linear in the variable w, i.e.,

$$F(x, u, w, t) = f(x, u, t) + B(t)w,$$

where the matrix-valued function $B(\cdot)$ is continuous, the concept of solution can be easily extended, interpreting (4.1) as an equation with measure of the form (see [237, 377], and [395])

$$dx(t) = f(x(t), u(t), t)dt + B(t)d\mu(t).$$

However, if the function $B(\cdot)$ is discontinuous or depends on the other control $u(\cdot)$, i.e., $B = B(t, u)$, then the representation of the optimal solution by means of the above equation can be incorrect without imposing a special "constancy" condition. For the investigation of optimization problems with impulse controls in the system

$$dx(t) = f(x(t), u(t), t)dt + B(t, u(t))d\mu(t),$$

a special method, which we refer to as the *method of a discontinuous time change*, was suggested by R. Rishel [377]. Rishel's approach is to introduce a new independent variable with respect to which the trajectories become absolutely continuous. This leads to the consideration of an auxiliary control problem with bounded controls where the time variable plays the role of one of the state variable. This approach enables one to reduce the original optimization problem to an equivalent auxiliary problem with bounded controls, and to deduce the optimality conditions for the original process.

The general approach based on the idea of replacing time by a new independent variable was applied to nonlinear systems by J. Warga [445–447], who considered nonlinear systems with an arbitrary dependence on the unbounded components of the control. Warga's approach provides an effective tool for the investigation of optimization problems and provides a general representation of optimal trajectories via the discontinuous change of time. We shall present a further extension of this method, namely, we will derive the representation of a generalized (discontinuous) solution in optimization problems in terms of differential equations with measure.

Here we consider the concept of a generalized solution as the pointwise limit of some sequence of ordinary solutions on the set of continuity of this generalized solution. The application of Warga's approach [445, 446] to the problem with constraint (4.4) gives a very natural form of the auxiliary system for the representation of the generalized solution with an additional variable that is referred to as the "new time". These ideas were later developed by B. Miller [273], where for the function φ in the equation for the "new time" one can take $\dot{\varphi} = 1 + \|w\|$. Then our next steps in deriving the representation of the generalized solution follow directly from the results presented in [273], with slight modifications. This approach enables us to represent any generalized solution by means of the discontinuous time change, in terms of the solution of an auxiliary system with ordinary (bounded) controls.

Our purpose is to study the representation problem for the right-hand side of the auxiliary system in the so-called "control form," as it is in the paper [447]. This is a key issue for the representation of the generalized solution by sampling of some control in the auxiliary system. Our main result shows that in the case when $F(x, u, w, t)$ behaves regularly at infinity, such representation of the right-hand side of the auxiliary system is achievable.

Now, in the regular case, one can derive a differential equation with measure for any generalized solution of the system (4.1). As examples of this approach we consider representations of generalized solutions for various types of continuous and discrete-continuous systems.

4.1.1 Definition of Generalized Solutions and Their Representation by Means of a Discontinuous Change of Time

Consider a controllable system described by the Eq. (4.1) with the state variable $X(t) \in \mathbb{R}^n$ and controls $\{u(t), w(t)\}$ satisfying (4.2)–(4.4).

Assumption 4.1

1. *The vector-function F is continuous with respect to $(x, u, w, t) \in \mathbb{R}^{n+k+m+1}$, and for any fixed $(u, w, t) \in U \times W \times [0, T]$, F satisfies the Lipschitz condition, i.e.,*

$$\|F(x_1, u, w, t) - F(x_2, u, w, t)\| \le L_1(1 + \|w\|)(\|x_1 - x_2\|)$$

for any $x_1, x_2 \in \mathbb{R}^n$, with a constant $L_1 > 0$.

2. *$F(x, u, w, t)$ has linear growth in both x and w, i.e.,*

$$\|F(x, u, w, t)\| \le L_2(1 + \|x\|)(1 + \|w\|)$$

for any (x, u, w, t), with a constant $L_2 > 0$.

To account for the possible impulsive behaviors of the component $w(\cdot)$ of the control, we introduce the following definition of generalized (discontinuous) solution of the system (4.1)

Definition 4.1 A right-continuous function $X(\cdot)$ of bounded variation on the interval $[0, T]$ is called a *generalized solution* of the system (4.1), if there exists a sequence of admissible controls $\{u^n(\cdot), w^n(\cdot)\}$ satisfying the constraints (4.2)–(4.4), such that the corresponding sequence $\{X^n(\cdot)\}$ of solutions of (4.1) with the initial condition $X^n(0) = x_0$ converges to $X(\cdot)$ at all the points of continuity of $X(\cdot)$.

In other words, a generalized solution is a limit of a sequence of ordinary solutions in the sense of the weak-$\star$ topology on the space of right-continuous functions with bounded variation.

Our purpose now is to characterize generalized solutions with the help of the auxiliary control system, using the method of the discontinuous time change.

Consider the following auxiliary controllable system for a pair of variables $\{Y(s), \eta(s)\} \in \mathbb{R}^n \times \mathbb{R}^1$, defined on an interval $[0, T_1]$, where $T_1 \le T + M$,

$$\begin{pmatrix} \dot{Y}(s) \\ \dot{\eta}(s) \end{pmatrix} = \begin{pmatrix} \dfrac{F(Y(s), u_1(s), w_1(s), \eta(s))}{1 + \|w_1(s)\|} \\ \dfrac{1}{1 + \|w_1(s)\|} \end{pmatrix}, \tag{4.6}$$

with the initial conditions

$$\begin{pmatrix} Y(0) \\ \eta(0) \end{pmatrix} = \begin{pmatrix} x_0 \\ 0 \end{pmatrix},$$

and with the controls $u_1(s), \; w_1(s)$ subject to the constraints

$$u_1(s) \in U, \quad w_1(s) \in W. \tag{4.7}$$

In the auxiliary system (4.6)–(4.7), the function $F(x, u, w, t)$, the sets U and W, and the constants T and M are the same as in (4.1)–(4.4).

As follows from Assumption 4.1, the right-hand side of Eq. (4.6) satisfies the Lipschitz condition and has a linear growth in the usual sense with respect to the variable Y.

The relationship between the systems (4.1) and (4.6) and their solutions is established by the following theorem.

Theorem 4.1 *Let the triplet of functions* $\{X(\cdot), u(\cdot), w(\cdot)\}$ *satisfy the system (4.1) and let the controls* $\{u(\cdot), w(\cdot)\}$ *be Lebesgue measurable and obey the constraints (4.2)–(4.4).*

Then there exists a collection of functions $\{Y(\cdot), \eta(\cdot), u_1(\cdot), w_1(\cdot)\}$ *that are defined on some interval* $[0, T_1]$, *where* $\eta(T_1) = T$, *and satisfy the system (4.6) with some Lebesgue-measurable controls* $\{u_1(\cdot), w_1(\cdot)\}$ *obeying the constraints (4.7), and with the initial condition* $Y(0) = x_0$, *such that, for any* $t \in [0, T]$,

$$X(t) = Y(\Gamma(t)), \quad \textit{where} \quad \Gamma(t) = \inf\{s : \eta(s) > t\}. \tag{4.8}$$

Remark 4.2 Recall that the same correspondence was established earlier between a robust DCS and the associated auxiliary system (see Remark 2.7).

Proof The function

$$\Gamma(t) = t + \int_0^t \|w(\tau)\| d\tau$$

is absolutely continuous and monotonically increasing. Therefore, it has an inverse function, which is defined for any s in the interval $[0, \Gamma(T)] = [0, T_1]$, where $T_1 \leq T + M$. This inverse function, denoted here by $\eta(s)$, is given by $\eta(s) = \inf\{t : \Gamma(t) > s\} = \Gamma^{-1}(s)$ and satisfies the inequalities

$$0 \leq \eta(s_1) - \eta(s_2) \leq s_1 - s_2 \quad \text{if} \quad s_1 \leq s_2. \tag{4.9}$$

The left inequality is a consequence of the monotonicity of the function $\Gamma(\cdot)$, while the right inequality follows from the relations

$$\eta(s_1) - \eta(s_2) = t_1 - t_2 \leq \Gamma(t_1) - \Gamma(t_2) = s_1 - s_2,$$

where $\Gamma(t_1) = s_1$, $\Gamma(t_2) = s_2$, and the points t_1, t_2 exist thanks to the continuity of $\Gamma(\cdot)$.

As follows from (4.9), the function $\eta(\cdot)$ is absolutely continuous and admits a derivative almost everywhere in $[0, T_1]$, given by

$$\dot{\eta}(s) = \left. \frac{1}{\dot{\Gamma}(t)} \right|_{t=\eta(s)} = \frac{1}{1 + \|w(\eta(s))\|}. \tag{4.10}$$

Define the functions $u_1(\cdot)$ and $w_1(\cdot)$ by

$$u_1(s) = u(\eta(s)), \quad w_1(s) = w(\eta(s)), \tag{4.11}$$

and consider the system (4.6) with the controls (4.11).

The functions $u_1(\cdot)$ and $w_1(\cdot)$ satisfy the constraints (4.7), so in order to regard them as controls in the system (4.6) it remains to verify that they are Lebesgue measurable. We do this by showing that if u^i is one of the components of $u(\cdot)$, and c is an arbitrary real number, then the set $N_c = \{s : u^i(\eta(s)) \leq c\}$ is Lebesgue measurable.

In view of the Lebesgue measurability of the function $u(\cdot)$, the set $L_c = \{t : u^i(t) \leq c\}$ is also measurable, and $N_c = \{s : \eta(s) \in L_c\}$. However, since

$$\eta(\Gamma(t)) = t, \qquad \Gamma(\eta(s)) = s,$$

the set N_c is the image of L_c under the mapping specified by the function $\Gamma(\cdot)$, i.e.,

$$N_c = \{s : \eta(s) \in L_c\} = \eta^{-1}(L_c) = \Gamma(L_c).$$

Since $\Gamma(\cdot)$ is absolutely continuous, it maps Lebesgue-measurable sets into Lebesgue-measurable ones [322]. Hence, as N_c is Lebesgue measurable, so is the function $u_1(s) = u(\eta(s))$, as we needed to show. The measurability of the function $w_1(s) = w(\eta(s))$ is established in much the same way.

Further, substituting the controls (4.11) in the system (4.6) we obtain

$$\begin{pmatrix} \dot{Y}(s) \\ \dot{\eta}(s) \end{pmatrix} = \begin{pmatrix} \dfrac{F(Y(s), u(\eta(s)), w(\eta(s)), \eta(s))}{1 + \|w(\eta(s))\|} \\ \dfrac{1}{1 + \|w(\eta(s))\|} \end{pmatrix}.$$

Set $Y(0) = X(0)$, $\eta(0) = 0$, and consider the functions $Y(\Gamma(t))$ and $\eta(\Gamma(t))$. This pair of functions satisfies the system

$$\begin{pmatrix} \dot{Y}(\Gamma(t)) \\ \dot{\eta}(\Gamma(t)) \end{pmatrix} = \begin{pmatrix} \dfrac{F(Y(\Gamma(t)), u(\eta(\Gamma(t)), w(\eta(\Gamma(t)), \eta(\Gamma(t))}{1 + \|w(\eta(\Gamma(t))\|} \dot{\Gamma}(t) \\ \dfrac{1}{1 + \|w(\eta(\Gamma(t))\|} \dot{\Gamma}(t), \end{pmatrix}$$

with the initial conditions

$$\begin{pmatrix} Y(\Gamma(0)) \\ \eta(\Gamma(0)) \end{pmatrix} = \begin{pmatrix} X(0) \\ 0 \end{pmatrix}.$$

By the definition of the inverse function, $\eta(\Gamma(t)) = t$; therefore, the first equation in the system (4.6) coincides with (4.1), and, by the uniqueness of its solution,

$$X(t) = Y(\Gamma(t)),$$

where $\Gamma(\cdot)$, as the inverse function to $\eta(\cdot)$ satisfies

$$\Gamma(t) = \inf\{s : \eta(s) > t\}.$$

Using again the definition of the inverse functions, we have

$$\eta(T_1) = \eta(\Gamma(T)) = T,$$

which completes the proof. □

The auxiliary system (4.6) seems more amenable to analysis than the original system (4.1), because its right-hand side is uniformly bounded for any pair of state variables (y, η), as ensured by Assumption 4.1. Moreover, it satisfies the Lipschitz condition and has a linear growth in y in the ordinary sense.

Therefore, the set of solutions of Eq. (4.6) with controls $\{u_1(\cdot), w_1(\cdot)\}$ that satisfy the constraints (4.7) is totally bounded and its closure is a compact subset of the space of continuous functions [237].

All this allows us to express any generalized solution of the original system (4.1) in terms of the solution of the auxiliary system (4.6) via a change of the independent (time) variable. However, in the general case, this change of time will be discontinuous, since the functions $\eta(\cdot)$ lying in the closure of the set of solutions of (4.6) are not necessarily strictly increasing. Hence, the inverse function $\Gamma(\cdot)$ will have jumps, corresponding to the constancy intervals of the appropriate function $\eta(\cdot)$. This means that, in the general case, $Y(\Gamma(t))$ does not satisfy the Eq. (4.1) with some measurable control. Nevertheless, any generalized solution admits the representation $X(t) = Y(\Gamma(t))$, where $\{Y(\cdot), \eta(\cdot)\}$ is one of the elements of this closure.

Theorem 4.2 *Let $X(\cdot)$ be the generalized solution of the system (4.1) with the initial condition $X(0-)$.*

Then there exists a sequence of controls $\{u_1^n(\cdot), w_1^n(\cdot)\}$, defined on some interval $[0, T_1]$, where $T_1 \leq T + M$, and satisfying (4.7), such that the corresponding sequence $\{Y^n(\cdot), \eta^n(\cdot)\}$ of solutions of the system (4.6) with the initial conditions

$$Y^n(0) = X(0-), \qquad \eta^n(0) = 0,$$

converges uniformly in $[0, T_1]$ to the pair of functions $\{Y(\cdot), \eta(\cdot)\}$, and $X(\cdot)$ can be represented in $[0, T]$ by the relation (4.8), where, by definition, $\Gamma(T) = T_1$.

Proof Since $X(\cdot)$ is the generalized solution of the system (4.1), there exists a sequence of functions $\{X^n(\cdot), u^n(\cdot), w^n(\cdot)\}$ satisfying the Eq. (4.1) with the constraints (4.2)–(4.4), such that $\{X^n(\cdot)\}$ converges to $X(\cdot)$ at all points of continuity.

By Theorem 4.1, for each n there exists the functions $\{Y^n(\cdot), \eta^n(\cdot), u_1^n(\cdot), w_1^n(\cdot)\}$, which satisfy the system (4.6) with the constraints (4.7) in some interval $[0, T_1^n]$, where $\eta(T_1^n) = T$, and such that for any $t \in [0, T]$ we have

$$X^n(t) = Y^n(\Gamma^n(t)) \tag{4.12}$$

and $X^n(0) = Y(0) = x_0$, $\eta^n(0) = 0$, where $\Gamma^n(t) = \inf\{s : \eta^n(s) > t\}$. Moreover, for each n we have $\dot{\Gamma}^n(t) = 1 + \|w_n(t)\|$, and so

$$T_1^n = \Gamma^n(T) = T + \int_0^T \|w^n(\tau)\| d\tau \le T + M.$$

Thus, there is a constant T_1 such that $T_1 \ge T_1^n$ for all n, and $T_1 \le T + M$.

Let us extend the functions $\{u_1^n(\cdot), w_1^n(\cdot)\}$ by assigning them constant values in the half-interval $(T_1^n, T_1]$, i.e.,

$$u_1^n(s) = u_1^n(T_1^n), \qquad w_1^n(s) = w_1^n(T_1^n)$$

for $s \in (T_1^n, T_1]$, and keep the same notation for the extensions.

Consider the sequence $\{Y^n(\cdot), \eta^n(\cdot), T_1^n\}$, with the functions $\{Y^n(\cdot), \eta^n(\cdot)\}$ corresponding to the extended controls on the interval $[0, T_1]$. Thanks to the Lipschitz and linear growth properties of the vector-function $\widetilde{F}(y, u_1, w_1, \eta)$ in the right-hand side of (4.6), namely,

$$\widetilde{F}(y, u_1, w_1, \eta) = \begin{pmatrix} \dfrac{F(y, u_1, w_1, \eta)}{1 + \|w_1\|} \\ \dfrac{1}{1 + \|w_1\|} \end{pmatrix}, \tag{4.13}$$

the sequence $\{Y^n(\cdot), \eta^n(\cdot), T_1^n\}$ is uniformly bounded, and the sequence of functions $\{Y^n(\cdot), \eta^n(\cdot)\}$ is equicontinuous on $[0, T_1]$ [237].

Therefore, by Arzelà's theorem (see, e.g., [204]), one can extract a subsequence $\{Y^{n_k}(\cdot), \eta^{n_k}(\cdot), T_1^{n_k}\}$, such that the sequence $\{Y^{n_k}(\cdot), \eta^{n_k}(\cdot)\}$ converges uniformly on $[0, T_1]$ to a function $\{Y(\cdot), \eta(\cdot)\}$, and $\{T_1^{n_k}\}$ converges to a value $\widetilde{T}_1 \le T_1$.

Our next objective is to prove that the subsequence of controls $\{u_1^{n_k}(\cdot), w_1^{n_k}(\cdot)\}$, defined on the interval $[0, \widetilde{T}_1]$, has the properties claimed in the statement of the theorem.

By Lemma 2.5, the sequence $\Gamma^{n_k}(t) = \inf\{s : \eta^{n_k}(s) > t\}$ converges to $\Gamma(t) = \inf\{s : \eta(s) > t\}$ at all points of continuity.

Thus, at all points of continuity of $\Gamma(\cdot)$ we have

$$\|Y^{n_k}(\Gamma^{n_k}(t)) - Y(\Gamma(t))\| \le$$
$$\le \|Y^{n_k}(\Gamma^{n_k}(t)) - Y(\Gamma^{n_k}(t))\| + \|Y(\Gamma^{n_k}(t)) - Y(\Gamma(t))\| \to 0,$$

where the first term on the right-hand side tends to zero due to the uniform convergence of $\{Y^{n_k}(\cdot)\}$ to $Y(\cdot)$, and the second terms tends to zero due to the convergence of $\Gamma^{n_k}(t)$ to $\Gamma(t)$. Consequently,

$$X(t) = \lim_{k} X^{n_k}(t) = \lim_{k} Y^{n_k}(\Gamma^{n_k}(t)) = Y(\Gamma(t)),$$

at all points of continuity of $\Gamma(\cdot)$.

By Lemma 2.4, the function $\Gamma(\cdot)$ is right-continuous and monotonically increasing, so for every $t \in (0, \widetilde{T}_1)$ we have

$$X(t) = \lim_{\tau^n \downarrow t} X(\tau^n) = \lim_{\tau^n \downarrow t} Y(\Gamma(\tau^n)) = Y(\Gamma(t)),$$

where $\tau^n \downarrow t$ is any monotonically decreasing sequence of points of continuity of $\Gamma(\cdot)$.

The equality $X(T) = Y(\Gamma(T))$ follows from the fact that $\Gamma^n(T) = T_1^n \to \widetilde{T}_1$, where $\widetilde{T}_1 = \Gamma(T)$ by definition. This completes the proof. □

Thus, Theorem 4.2 gives a representation for any generalized solution of the system (4.1) in terms of a sequence of ordinary solutions of the auxiliary system (4.1). However, for the solution of optimal control problems we need a more explicit representation of generalized solutions. Consider again a sequence of solutions of the auxiliary system (4.6) which converges to a pair of functions $\{y(\cdot), \eta(\cdot)\}$ uniformly in some interval $[0, T_1]$. Know results from the theory of differential inclusions allow us to assert that, in view of the continuity of the function $\widetilde{F}(y, u, w, \eta)$, the limit pair of functions $\{y(\cdot), \eta(\cdot)\}$ satisfies the differential inclusion [237]

$$\begin{pmatrix} \dot{y}(s) \\ \dot{\eta}(s) \end{pmatrix} \in \overline{\operatorname{conv}}\, \{\widetilde{F}(y(s), u_1, w_1, \eta(s)) : u_1 \in U, w_1 \in W\} \tag{4.14}$$

where the set on the right-hand side is the closed convex hull of the right-hand side of the system (3.1) for every (y, η).

Assumption 4.2

1. *There exists an appropriate representation of the set in the right-hand side of (4.14), i.e., there exist an $\mathbb{R}^{n+1}$-valued function $\mathcal{G}(y, \omega, \eta)$ and a closed bounded set Ω in a vector space of appropriate dimension, such that for every $(y, \eta) \in \mathbb{R}^{n+1}$*

$$\begin{aligned} \overline{\operatorname{conv}}\, \{\widetilde{F}(y, u_1, w_1, \eta) : u_1 \in U, w_1 \in W\} = \overline{\operatorname{conv}}\, \widetilde{F}(y, U, W, \eta) = \\ = \mathcal{G}(y, \Omega, \eta) = \{\mathcal{G}(y, \omega, \eta) : \omega \in \Omega\}. \end{aligned} \tag{4.15}$$

2. *The function $\mathcal{G}(y, \omega, \eta)$ is continuous with respect to (y, ω, η) and for any fixed $\omega \in \Omega$ satisfies the Lipschitz condition and has a linear growth with respect to the variables (y, η), i.e., there are constants $L_1 > 0$ and $L > 0$ such that*

$$\|\mathcal{G}(y_1, \omega, \eta_1) - \mathcal{G}(y_2, \omega, \eta_2)\| \leq L_1\{\|y_1 - y_2\| + | \eta_1 - \eta_2 |\} \tag{4.16}$$

and

$$\|\mathscr{G}(y, \omega, \eta)\| \le L(1 + \|y\| + \mid \eta \mid) \tag{4.17}$$

for any $y_1, y_2 \in \mathbb{R}^n$ *and any* $\eta_1, \eta_2 \in \mathbb{R}$.

Remark 4.3 When Assumption 4.2 holds, we will say that $\widetilde{F}(y, U, W, \eta)$ is representable in "control form" [447].

Theorem 4.3 *Suppose Assumption 4.2 holds. Then for any generalized solution* $X(t)$ *of the system (4.1) there exists a Lebesgue-measurable control* $\omega(s) \in \Omega$ *defined a.e. in some interval* $[0, T_1]$ *with* $T_1 \le T + M$*, such that* $X(t)$ *admits the representation (4.8), where the pair of functions* $\{Y(\cdot), \eta(\cdot)\}$ *satisfies the differential equation*

$$\begin{pmatrix} \dot{Y}(s) \\ \dot{\eta}(s) \end{pmatrix} = \mathscr{G}(Y(s), \omega(s), \eta(s)), \tag{4.18}$$

with the initial and terminal conditions

$$Y(0) = X(0-), \quad \eta(0) = 0 \quad \textit{and } \eta(T_1) = T,$$

respectively.

Proof By Theorem 4.2, there exists a sequence $\{Y^n(\cdot), \eta^n(\cdot)\}$ of solutions of (4.6) with the initial conditions

$$Y^n(0) = X(0-), \quad \eta^n(0) = 0$$

which converges uniformly in $[0, T_1]$ to a pair of functions $\{Y(\cdot), \eta(\cdot)\}$ that satisfy the initial and terminal conditions

$$Y(0) = X(0-), \quad \eta(0) = 0 \quad \text{and } \eta(T_1) = T,$$

respectively, such that $X(\cdot)$ admits on $[0, T]$ the representation (4.8).

It remains to prove that $\{Y(\cdot), \eta(\cdot)\}$ satisfies (4.18) with some Lebesgue-measurable control $\omega(s) \in \Omega$. By Assumption 4.2, the set of solutions of (4.18) with the fixed initial condition is compact in the topology of uniform convergence [237], [446]. Therefore, such a control function $\omega(s) \in \Omega$ exists, as claimed. □

The assumptions concerning the existence of a suitable function $\mathscr{G}(y, \omega, \eta)$ and the set Ω may seem rather artificial; however, one can show that they are fulfilled when the behavior of the function F when $\|w\| \to \infty$ is in some sense "regular."

Definition 4.2 The function $F(x, u, w, t)$ is said to be *regular at infinity* if:

1. it is continuous with respect to all the variables and Lipschitz with respect to x and t, i.e., there is a constant $L > 0$ such that

$$\|F(x', u, w, t') - F(x, u, w, t)\| \leq L(1 + \|w\|)(\|x - x'\| + |t - t'|)$$

for any $x', x \in \mathbb{R}^n$, $t', t \in [0, T]$, $u \in U$ and $w \in W$;
2. for any pair of vectors (u, e) with $u \in U$ and $e \in E = \{\|e\| = 1\}$, the limit

$$\lim_n \frac{F(x, u_n, w_n, t)}{1 + \|w_n\|} = B(x, u, e, t), \tag{4.19}$$

exists when

$$u_n \in U, \ w_n \in W, \quad \text{and} \quad u_n \to u, \quad \frac{w_n}{1 + \|w_n\|} \to e, \quad \|w_n\| \to \infty.$$

Moreover, the function $B(x, u, e, t)$ is continuous with respect to all the variables, and the limit (4.19) is independent of the choice of the sequences $\{u_n\}$ and $\{w_n\}$.

Theorem 4.4 *If $F(x, u, w, t)$ is regular at infinity, then there exist an appropriate function $\mathcal{G}(y, \omega, \eta)$ and a set Ω such that Assumption 4.2 holds.*

Proof Define the new control variable ω by

$$\omega = \omega(w) = \frac{w}{1 + \|w\|}. \tag{4.20}$$

The map (4.20) sends the unbounded set W in a one-to-one manner onto the set $\omega(W)$, which lies the interior of the unit ball in $\mathbb{R}^m$. The inverse map is given by

$$w(\omega) = \frac{\omega}{1 - \|\omega\|},$$

and is defined and continuous on the interior of the unit ball in $\mathbb{R}^m$.

Replacing the control $w \in W$ with $\omega \in \omega(W)$ in the expression (4.13) for the function $\widetilde{F}$ gives

$$\widetilde{F}(y, u_1, w_1, \eta) = \begin{pmatrix} (1 - \|\omega\|)F(y, u_1, \omega/(1 - \|\omega\|), \eta) \\ 1 - \|\omega\| \end{pmatrix} = \widetilde{\mathcal{G}}(y, u_1, \omega, \eta).$$

We claim that in the regular case the function $\widetilde{\mathcal{G}}$ "satisfies almost completely" Assumption 4.2.

First, let us prove that $\widetilde{\mathcal{G}}(y, u_1, \omega, \eta)$ is jointly continuous in the variables $(y, \eta, u_1, \omega) \in \mathbb{R}^n \times \mathbb{R}^1 \times U \times \overline{\omega(W)}$ (recall that overline stands for closure). Continuity is evident if $\|\omega\| < 1$, as a composition of continuous functions is continuous. Thus, if (y, η, u_1, ω) is an arbitrary point with $\|\omega\| = 1$ and $\omega \in \overline{\omega(W)}$, then there exists a sequence $\{(u_1^n, \omega^n)\}$ such that $u_1^n \in U$, $\omega^n \in \omega(W)$, $u_1^n \to u_1$, and $\omega^n \to \omega$. Now for each n set

$$w^n = \frac{\omega^n}{1 - \|\omega^n\|}.$$

Then $w^n \in W$ and

$$\|w^n\| \to \infty, \qquad \frac{w^n}{1 + \|w^n\|} \to \omega,$$

and

$$\widetilde{\mathscr{G}}(y, u^n, \omega^n, \eta) = \widetilde{F}(y, u^n, w^n, \eta). \tag{4.21}$$

By the regularity condition, the function in the right-hand side in (4.21) has a limit when $u^n \to u$ and $\dfrac{w^n}{1 + \|w^n\|} \to \omega$, and this limit is independent of the choice of the sequence $\{(u^n, w^n)\}$. Therefore, $\widetilde{\mathscr{G}}$ can be extended by continuity to $\overline{\omega(W)}$. We will use the same notation $\widetilde{\mathscr{G}}$ for the extended function.

We have

$$\begin{aligned} \|\widetilde{\mathscr{G}}(y', u_1', \omega', \eta') - \widetilde{\mathscr{G}}(y, u_1, \omega, \eta)\| \leq & \\ \leq \|\widetilde{\mathscr{G}}(y_1', u_1', \omega', \eta') - \widetilde{\mathscr{G}}(y, u_1', \omega', \eta)\| + & \\ + \|\widetilde{\mathscr{G}}(y, u_1', \omega', \eta) - \widetilde{\mathscr{G}}(y, u_1, \omega, \eta)\|. & \end{aligned} \tag{4.22}$$

The first term on the right-hand side is estimated by $L(\|y_m' - y\| + |\eta' - \eta\ |)$, thanks to the Lipschitz property of the original function $F(x, u, w, t)$; the second term tends to zero when $(u_1', \omega') \to (u_1, \omega)$, thanks to the regularity condition. Therefore, $\widetilde{\mathscr{G}}(y', u_1', \omega', \eta') \to \widetilde{\mathscr{G}}(y, u_1, w, \eta)$ when $(y', u_1', \omega', \eta') \to (y, u_1, w, \eta)$, and so the function $\widetilde{\mathscr{G}}$ is continuous. Moreover, it is Lipschitz and has a linear growth.

To finalize the construction of the function $\mathscr{G}(y, \overline{\omega}, \eta)$ we can use the description of the convex hull of the bounded set $\widetilde{\mathscr{G}}(y, U, \omega(W), \eta)$ provided by well-known theorem of Carathéodory [237] . Then the set Ω and the function $\mathscr{G}(y, \overline{\omega}, \eta)$ can be expressed as

$$\Omega = \Big\{\overline{\omega} = (u_1, \ldots, u_{n+2}, \omega_1, \ldots, \omega_{n+2}, \alpha_1, \ldots, \alpha_{n+2}) : \\ u_i \in U,\ \omega_i \in \overline{\omega(W)},\ \alpha_i \geq 0,\ i = 1, \ldots, n+2,\ \sum_{i=1}^{n+2} \alpha_i = 1\Big\},$$

and

$$\mathscr{G}(y, \overline{\omega}, \eta) = \sum_{i=1}^{n+2} \alpha_i \widetilde{\mathscr{G}}(y, u_i, \omega_i, \eta),$$

respectively. □

Remark 4.4 By results in [445, 446], the homeomorphism $w \to \omega(w) = w/(1 + \|w\|)$ of W onto its image $\omega(W)$ defines the *metric compactification* of the problem, in the terminology proposed by Warga in [446].

4.1.2 Differential Inclusions for Generalized Solutions of Nonlinear Differential Equations

In this section we provide a representation of generalized solutions with the aid of differential inclusions with a measure.

Suppose the function $X(\cdot)$ is a generalized solution of the system (4.1) in the interval $[0, T]$. Then $X(\cdot)$ is right-continuous and of bounded variation. Therefore, it admits a representation of the form

$$X(t) = X(0-) + X^{\mathrm{c}}(t) + \sum_{\tau \le t} \Delta X(\tau) \tag{4.23}$$

where

$$\Delta X(\tau) = X(\tau) - X(\tau-)$$

and $X^{\mathrm{c}}(\cdot)$ is a continuous function. Each summand in (4.23) admits a representation with the aid of a *differential inclusion with measure.*

Definition 4.3 ([442]) Let $\mathscr{F}(x, t) : [0, T] \times \mathbb{R}^n \to \mathbb{R}^n$ be a multifunction, i.e., $\mathscr{F}$ sends every point (x, t) into a subset $\mathscr{F}(x, t) \subset \mathbb{R}^n$.

Let $\mu(dt)$ be a non-negative scalar-valued measure on the Borel subsets of $[0, T]$.

We say that a right-continuous function $x(\cdot)$ with bounded variation in $[0, T]$ satisfies the *differential inclusion*

$$dx(t) \in \mathscr{F}(x(t), t)\mu(dt)$$

with the initial condition $x(0-) = x_0 \in \mathbb{R}^n$ if there exists a μ-measurable function $f(t)$ such that

$$f(t) \in \mathscr{F}(x(t), t) \quad \mu\text{-a.e.}$$

and

$$x(t) = x_0 + \int_0^t f(\tau)\mu(d\tau) \quad \text{for all } t \in [0, T].$$

In other words, $f(t)$ is the Radon-Nikodym derivative of the vector-valued measure $\mathscr{X}(dt)$ with distribution function $x(t) = \mathscr{X}([0, t])$ with respect to the measure $\mu(dt)$:

$$f(t) = \frac{dx(t)}{d\mu(t)},$$

where $\mu(t)$ is the distribution function of the measure $\mu(dt)$.

Suppose that there exist a function $\mathscr{G}$ and a set Ω satisfying Assumption 4.2. Define the set $\mathscr{G}^1(y, \eta)$ of vectors in $\mathbb{R}^n$ as the set of the first n components of the vectors in the set $\mathscr{G}(y, \Omega, \eta)$ whose $(n+1)$-th component is equal to zero, i.e.,

$$\mathscr{G}^1(y, \eta) = \left\{ e_1 = \begin{pmatrix} \mathscr{G}_1(y, \omega, \eta) \\ \vdots \\ \mathscr{G}_n(y, \omega, \eta) \end{pmatrix} \in \mathbb{R}^n : \omega \in \Omega,\ \mathscr{G}_{n+1}(y, \omega, \eta) = 0 \right\}. \tag{4.24}$$

This set is not empty, since it contains all partial limits of sequences of the form

$$\left\{ \frac{F(y, u^{(k)}, w^{(k)}, \eta)}{1 + \|w^{(k)}\|} \right\}$$

such that $u^{(k)} \in U$, $w^{(k)} \in W$ and $\|w^{(k)}\| \to \infty$. By Assumption 4.1, any such sequence is bounded for fixed (y, η), and therefore, the set of its partial limits is not empty. By Assumption 4.2, the set $\mathscr{G}(y, \Omega, \eta)$ contains all partial limits of the sequences with terms of the form

$$\begin{pmatrix} e_1^{(k)} \\ e_{n+1}^{(k)} \end{pmatrix} = \begin{pmatrix} \dfrac{F(y, u^{(k)}, w^{(k)}, \eta)}{1 + \|w^{(k)}\|} \\ \dfrac{1}{1 + \|w^{(k)}\|} \end{pmatrix} \in \mathbb{R}^n \times \mathbb{R}$$

where $u^{(k)} \in U$ and $w^{(k)} \in W$, hence it contains all partial limits for $\|w^{(k)}\| \to \infty$ and $e_{n+1}^{(k)} \to 0$, respectively.

The set $\mathscr{G}^1(y, \eta)$ is convex and closed for every fixed (y, η), since it is the cross-section of the convex and closed set $\mathscr{G}(y, \Omega, \eta)$.

Let us also introduce the conic hull $\operatorname{con} \mathscr{G}^1(y, \eta)$ of the set $\mathscr{G}^1(y, \eta)$ by

$$\operatorname{con} \mathscr{G}^1(y,\eta) = \Big\{ e \in \mathbb{R}^n : e = \sum_{i=1}^{k} \beta_i e_i,\ \beta_i \geq 0,\ k \geq 1,\ e_i \in \mathscr{G}^1(y,\eta) \Big\}. \tag{4.25}$$

The next theorem gives a representation of a generalized solution with the aid of a differential inclusion with measure.

Theorem 4.5 *Let Assumption 4.2 hold. Let $X(\cdot)$ be a generalized solution of the system (4.1). Then there exists a non-negative regular scalar measure $\mathscr{V}(dt)$, defined on the Borel subsets of the interval $[0, T]$, such that the following holds:*

$$\mathscr{V}([0,T]) \leq M$$

and the corresponding distribution function

$$V(t) = \mathscr{V}([0,t])$$

can be expressed as (see, e.g., [322])

$$V(t) = \int_0^t \dot{V}(s)ds + V^{\mathrm{s}}(t) + \sum_{\tau \leq t} \Delta V(\tau), \tag{4.26}$$

where

1. *$\dot{V}(t)$ is the derivative of $V(\cdot)$, defined almost everywhere on $[0, T]$,*
2. *$V^{\mathrm{s}}(t)$ is a continuous non-decreasing function whose set of growth points is of null Lebesgue measure (i.e., the corresponding measure $\mathscr{V}^{\mathrm{s}}(dt)$ is singular with respect to the Lebesgue measure),*
3. *$\Delta V(\tau) = V(\tau) - V(\tau-) = \mathscr{V}(\{\tau\}) > 0$ are the jumps of the function $V(\cdot)$, and the set of jumps is at most countable [322].*

The generalized solution $X(\cdot)$ admits the representation

$$X(t) = x_0 + X^{\mathrm{a}}(t) + X^{\mathrm{s}}(t) + \sum_{\tau \leq t} \Delta X(\tau), \tag{4.27}$$

where $X^{\mathrm{a}}(t)$ is absolutely continuous with respect to the Lebesgue measure and its derivative satisfies the differential inclusion

$$\dot{X}^{\mathrm{a}}(t) \in \overline{\operatorname{conv}}\, F(X(t), U, W, t) + \operatorname{con} \mathscr{G}^1(X(t), t), \quad \text{a.e. on } [0, T], \tag{4.28}$$

$X^{\mathrm{s}}(t)$ is absolutely continuous with respect to the measure $\mathscr{V}^{\mathrm{s}}(dt)$ and its derivative with respect to the measure $\mathscr{V}^{\mathrm{s}}(dt)$ satisfies the differential inclusion

$$\frac{dX^{\mathrm{s}}(t)}{dV^{\mathrm{s}}(t)} \in \mathscr{G}^1(X(t), t), \quad \mathscr{V}^{\mathrm{s}}(dt)\text{-a.e. in } [0, T], \tag{4.29}$$

and finally

$$\Delta X(\tau) = X(\tau) - X(\tau-) = y_\tau(\Delta V(\tau)) - X(\tau-), \tag{4.30}$$

where the function $y_\tau(\cdot)$ *satisfies on the interval* $[0, \Delta V(\tau)]$ *the differential inclusion*

$$\dot{y}_\tau(s) \in \mathscr{G}^1(y_\tau(s), \tau) \tag{4.31}$$

with the initial condition

$$y_\tau(0) = X(\tau-).$$

To prove the theorem we need the following auxiliary result.

Lemma 4.1 *Suppose the function* $\eta(\cdot)$ *is absolutely continuous with respect to the Lebesgue measure on* $[0, T_1]$, *and its derivative satisfies the inequality*

$$0 \le \dot{\eta}(s) \le 1, \quad a.e.\ on\ [0, T_1].$$

Suppose also that

$$\eta(0) = 0, \quad \eta(T_1) = T.$$

Define the function $\Gamma(t)$ *by*

$$\Gamma(t) = \inf\{s : \eta(s) > t\}$$

and associate with it the measure $\Gamma(dt)$ *by setting*

$$\Gamma((a, b]) = \Gamma(b) - \Gamma(a),$$

for any half-interval $(a, b]$.

Let $\omega(\cdot) : [0, T_1] \to \mathbb{R}^1$ *be a Lebesgue-measurable function on* $[0, T_1]$. *Then the composite function* $\omega(\Gamma(t))$ *is both Lebesgue- and* Γ*-measurable on* $[0, T]$.

Proof of Lemma 4.1 If we prove that the set $N_c = \{t : \omega(\Gamma(t)) \le c\}$, where c is any real number, is Lebesgue measurable set, the assertion follows.

Thanks to the Lebesgue-measurability of the function $\omega(\cdot)$, the set $L_c = \{s : \omega(s) \le c\}$ is also measurable and $N_c = \{t : \Gamma(t) \in L_c\}$. Let $t \in N_c$ be a continuity point of Γ. Then there exists exactly one point $s \in [0, T_1]$ such that $\eta(s) = t$ and $\Gamma(\eta(s)) = s$; moreover,

$$\omega(\Gamma(t)) = \omega(\Gamma(\eta(s))) = \omega(s) \le c$$

and so t belongs to the image $\eta(L_c)$ of the set L_c under the non-decreasing function $\eta(\cdot)$. Hence,

$$N_c \setminus D_\Gamma^d \subset \eta(L_c)$$

where D_Γ^{d} is set of discontinuity points of Γ, which is at most countable.

Further, if $s \in L_c$ and $\eta(s)$ is a continuity point of Γ, then $\Gamma(\eta(s)) = s$ and $\omega(\Gamma(\eta(s))) = \omega(s) \le c$, and consequently $\eta(s) \in N_c$. If $s \in L_c$ and $\eta(s)$ is a discontinuity point of Γ, then $\eta(s) \in D_\Gamma^{\mathrm{d}}$, and so

$$\eta(L_c) \subset N_c \cup D_\Gamma^{\mathrm{d}},$$

which in conjunction with the preceding inclusion shows that

$$N_c = D' \cup \eta(L_c),$$

where $D' \subset D_\Gamma^{\mathrm{d}}$ is at most countable. Therefore, N_c is Lebesgue measurable.

The set $\eta(L_c)$ is Lebesgue measurable as the image of the Lebesgue-measurable set L_c under the absolutely continuous map $\eta(\cdot)$ [322]. Hence, the set N_c is Lebesgue measurable as a union of two Lebesgue-measurable sets.

To prove that $\omega(\Gamma(t))$ is Γ-measurable it is sufficient to show that the indicator function $\mathbf{I}\{N_c\}(t)$ is Γ-integrable for any c. By Lemma 2.4

$$\Gamma(\eta(s)) = \begin{cases} s, & \text{if } \eta(s) \in [0, T] \setminus D_\Gamma^{\mathrm{d}}, \\ \Gamma(\tau), & \text{if } \eta(s) = \tau \in D_\Gamma^{\mathrm{d}} \end{cases}$$

and so

$$\begin{aligned} &\mathbf{I}\{s : \Gamma(\eta(s)) \in L_c\}(s)\mathbf{I}\{s : \eta(s) \in [0, T] \setminus D_\Gamma^{\mathrm{d}}\}(s) = \\ &\qquad = \mathbf{I}\{s : s \in L_c\}(s)\mathbf{I}\{s : \eta(s) \in [0, T] \setminus D_\Gamma^{\mathrm{d}}\}(s). \end{aligned}$$

Since the function on the right-hand is Lebesgue measurable, the change variable formula (see Lemma 2.6) yields

$$\begin{aligned} &\int_0^{\Gamma(T)} \mathbf{I}\{s : \Gamma(\eta(s)) \in L_c\}(s)\mathbf{I}\{s : \eta(s) \in [0, T] \setminus D_\Gamma^{d}\}(s)ds = \\ &\qquad = \int_0^T \mathbf{I}\{t : \Gamma(t) \in L_c\}(\tau)\mathbf{I}\{t : t \in [0, T] \setminus D_\Gamma^{\mathrm{d}}\}(\tau)d\Gamma(\tau). \end{aligned} \tag{4.32}$$

Note that the set N_c can be represented as the union of two disjoint sets:

$$N_c = \big\{\{t : \Gamma(t) \in L_c\} \cap \{t : t \in [0, T] \setminus D_\Gamma^{\mathrm{d}}\}\big\} \cup \big\{\{t : \Gamma(t) \in L_c\} \cap \{t : t \in D_\Gamma^{\mathrm{d}}\}\big\},$$

where the second set is at most countable as a subset of D^{d}_{Γ}, and so it is Γ-measurable. As for the first set, its indicator function is Γ-integrable by (4.32), and therefore the set itself is Γ-measurable. This completes the proof of the lemma. □

Proof of Theorem 4.5 By Theorem 4.3, $X(t) = y(\Gamma(t))$, where $\Gamma(t) = \inf\{s \in [0, T_1] : \eta(s) > t\}$ and the pair $(y(s), \eta(s))$ satisfies (4.18) with the initial conditions $y(0) = x_0$, $\eta(0) = 0$, with a measurable control $\omega(s) \in \Omega$ a.e. in $[0, T_1]$. By Lemma 2.4, the function $\Gamma(\cdot)$ is right-continuous, monotonically increasing, and admits the representation

$$\Gamma(t) = \int_0^t \dot{\Gamma}(s)ds + \Gamma^{\mathrm{s}}(t) + \sum_{\tau \le t} \Delta\Gamma(\tau),$$

where $\dot{\Gamma}(t)$ is the derivative of $\Gamma(\cdot)$, the function $\Gamma^{\mathrm{a}}(t) = \int_0^t \dot{\Gamma}(s)ds$ is the distribution function of the measure $\Gamma^{\mathrm{a}}(dt)$ in the Lebesgue decomposition of the function $\Gamma(\cdot)$, $\Gamma^{\mathrm{s}}(t)$ is a continuous non-decreasing function such that the corresponding measure $\Gamma^{\mathrm{s}}(dt)$ is singular with respect to the Lebesgue measure, and $\Delta\Gamma(\tau) = \Gamma(\tau) - \Gamma(\tau-)$.

The interval $[0, T]$ can decomposed into a union of three disjoint subsets,

$$[0, T] = D^{\mathrm{a}}_{\Gamma} \cup D^{\mathrm{s}}_{\Gamma} \cup D^{\mathrm{d}}_{\Gamma}$$

where D^{a}_{Γ} and D^{s}_{Γ} are the supports of the measures $\Gamma^{\mathrm{a}}(dt)$ and $\Gamma^{\mathrm{s}}(dt)$, respectively, and D^{d}_{Γ} is the at most countable set of points τ such that $\Gamma(\tau) - \Gamma(\tau-) > 0$. Since the sets D^{s}_{Γ} and D^{d}_{Γ} are of Lebesgue measure zero, the Lebesgue measure of D^{a}_{Γ} is equal to T.

The generalized solution $X(t)$ satisfies the integral equation

$$X(t) = y(\Gamma(t)) = x_0 + \int_0^{\Gamma(t)} \dot{y}(s)ds,$$

which can be rewritten as

$$\begin{aligned} X(t) = x_0 &+ \int_0^{\Gamma(t)} \dot{y}(s)\mathbf{I}\{s : \eta(s) \overline{\in} D^{\mathrm{d}}_{\Gamma}\}(s)ds \\ &+ \int_0^{\Gamma(t)} \dot{y}(s)\mathbf{I}\{s : \eta(s) \in D^{\mathrm{d}}_{\Gamma}\}(s)ds = \\ = x_0 &+ \int_0^{\Gamma(t)} \dot{y}(s)\mathbf{I}\{s : \eta(s) \notin D^{\mathrm{d}}_{\Gamma}\}ds + \sum_{\tau \le t} \int_{\Gamma(\tau-)}^{\Gamma(\tau)} \dot{y}(s)ds = \\ = x_0 &+ \int_0^{\Gamma(t)} \dot{y}(s)\mathbf{I}\{s : \eta(s) \notin D^{\mathrm{d}}_{\Gamma}\}ds + \sum_{\tau \le t} \Delta X(\tau). \end{aligned} \tag{4.33}$$

Making here the substitution $\eta(s) = \tau$ into (4.33) and taking into account that $\eta(\Gamma(t)) = t$ and $\Gamma(\eta(s)) = s$ if $\eta(s) \notin D_\Gamma^{\mathrm{d}}$, we obtain

$$\int_0^{\Gamma(t)} \dot{y}(s)\mathbf{I}\{s : \eta(s) \notin D_\Gamma^{\mathrm{d}}\}(s)ds = \int_0^{\Gamma(t)} \dot{y}(\Gamma(\eta(s)))\mathbf{I}\{s : \eta(s) \notin D_\Gamma^{\mathrm{d}}\}(s)ds =$$
$$= \int_0^{\eta(\Gamma(t))} \dot{y}(\Gamma(t))\mathbf{I}\{\tau : \tau \notin D_\Gamma^{\mathrm{d}}\}(s)d\Gamma(\tau) =$$
$$= \int_0^{t} \dot{y}(\Gamma(\tau))\mathbf{I}\{\tau : \tau \notin D_\Gamma^{\mathrm{d}}\}(s)d\Gamma(\tau) =$$
$$= \int_0^{t} \dot{y}(\Gamma(\tau))(\dot{\Gamma}(\tau)d\tau + d\Gamma^{\mathrm{s}}(\tau)) =$$
$$= \int_0^{t} \dot{y}(\Gamma(\tau))\dot{\Gamma}(\tau)d\tau + \int_0^{t} \dot{y}(\Gamma(\tau))d\Gamma^{\mathrm{s}}(\tau) = X^{\mathrm{a}}(t) + X^{\mathrm{s}}(t). \tag{4.34}$$

Combining (4.33) and (4.34), we obtain for $X(t)$ the representation

$$X(t) = x_0 + X^{\mathrm{a}}(t) + X^{\mathrm{s}}(t) + \sum_{\tau \le t} \Delta X(\tau).$$

Here

$$X^{\mathrm{a}}(t) = \int_0^{t} \dot{y}(\Gamma(\tau))\dot{\Gamma}(\tau)d\tau \tag{4.35}$$

and so $X^{\mathrm{a}}(\cdot)$ is absolutely continuous with respect to the Lebesgue measure:

$$X^{\mathrm{s}}(t) = \int_0^{t} \dot{y}(\Gamma(\tau))d\Gamma^{\mathrm{s}}(\tau), \tag{4.36}$$

that is, $X^{\mathrm{s}}(\cdot)$ is absolutely continuous with respect to the measure $\Gamma^{\mathrm{s}}(dt)$; and

$$\Delta X(\tau) = y(\Gamma(\tau)) - y(\Gamma(\tau-)) = \int_{\Gamma(\tau-)}^{\Gamma(\tau)} \dot{y}(s)ds \tag{4.37}$$

is the jump of the function $X(\cdot)$ at the point $\tau \in D_\Gamma^{\mathrm{d}}$.

Let us show that the functions $X^{\mathrm{a}}(\cdot)$, $X^{\mathrm{s}}(\cdot)$ and the jumps $\Delta X(\tau)$ are as claimed in the statement of the theorem.

Define the measure $\mathscr{V}(dt)$ by the formula

$$\mathscr{V}((a, b]) = \Gamma(b) - \Gamma(a) - (b - a). \tag{4.38}$$

The continuity of $\eta(\cdot)$ ensures that for any $t \in [0, T]$ there exists $s \in [0, T_1]$ such that $\eta(s) = t$. Therefore, there exist s_a and s_b such that $\eta(s_a) = a$ and $\eta(s_b) = b$. If a and b are continuity points of $\Gamma(\cdot)$, then, by Lemma 2.4,

$$\Gamma(b) - \Gamma(a) = \Gamma(\eta(s_b)) - \Gamma(\eta(s_a)) = s_b - s_a.$$

However, $\dot{\eta}(s) \leq 1$, hence $s_b - s_a \geq b - a$, implying that

$$\Gamma(b) - \Gamma(a) \geq b - a.$$

If, on the other hand, a and b are discontinuity points of $\Gamma(\cdot)$, then in view of the right-continuity of $\Gamma(\cdot)$ we have

$$\Gamma(b) = \lim_{b_n \downarrow b} \Gamma(b_n), \qquad \Gamma(a) = \lim_{a_n \downarrow a} \Gamma(a_n),$$

where $\{b_n\}$ and $\{a_n\}$ are sequences of continuity points of $\Gamma(\cdot)$, and so

$$\Gamma(b) - \Gamma(a) = \lim_n (\Gamma(b_n) - \Gamma(a_n)) \geq \lim_n (b_n - a_n) = b - a.$$

This relation and the right-continuity of the function $\Gamma(\cdot)$ show that the measure $\mathscr{V}(dt)$ is non-negative and regular [322]. Moreover,

$$\mathscr{V}\{[0, T]\} = \Gamma(T) - T \leq M,$$

and

$$\mathscr{V}^{\mathrm{s}}(dt) = \Gamma^{\mathrm{s}}(dt), \qquad \mathscr{V}(\{\tau\}) = \Delta\Gamma(\tau).$$

Consider the Eq. (4.35). On the subset D_Γ^{a} the function $\Gamma(\cdot)$ has a derivative $\dot{\Gamma}(t)$ such that

$$0 < \dot{\Gamma}(t) = \left. \frac{1}{\dot{\eta}(s)} \right|_{s=\Gamma(t))} < \infty \quad \text{a.e. on} \quad D_\Gamma^{\mathrm{a}},$$

and so this double inequality holds also a.e. on $[0, T]$.

Therefore, $\dot{\eta}(\Gamma(t)) = \dot{\eta}(s)|_{s=\Gamma(t)} > 0$ a.e.in $[0, T]$, and the derivative of $X^{\mathrm{a}}(\cdot)$ exists and satisfies the relation

$$\dot{X}^{\mathrm{a}}(t) = \left. \frac{\dot{y}(s)}{\dot{\eta}(s)} \right|_{s=\Gamma(t)} \quad \text{a.e. on } [0, T]. \tag{4.39}$$

Let us show that $\dot{X}^{\mathrm{a}}(t)$ satisfies the inclusion (4.28). By Lemma 4.1, $\omega(\Gamma(t))$ is Lebesgue measurable, hence,

$$\begin{pmatrix} \dot{y}(\Gamma(t)) \\ \dot{\eta}(\Gamma(t)) \end{pmatrix} = \mathscr{G}(y(\Gamma(t)), \omega(\Gamma(t)), \eta(\Gamma(t))) = \mathscr{G}(X(t), \omega(\Gamma(t)), t) \tag{4.40}$$

a.e. on $[0, T]$.

The set $\mathscr{G}(x, \Omega, t)$ is the convex hull of the closure of the set $\widetilde{F}(x, u, w, t)$; hence, by the Carathéodory theorem [237], , for any $\omega \in \Omega$ there exists a set $\{(\alpha_i, l_i)i = 1, \dots, n+2\}$ such that,

$$\alpha_i \geq 0, \quad \sum_{i=1}^{n+2} \alpha_i = 1, \quad l_i \in \overline{\widetilde{F}(x, u, w, t)},$$

and

$$\mathscr{G}(x, \omega, t) = \sum_{i=1}^{n+2} \alpha_i l_i = \sum_{i=1}^{n+2} \alpha_i \begin{pmatrix} l_i^1 \\ l_i^2 \end{pmatrix}, \tag{4.41}$$

where l_i^1 is the vector of the first n components of the vector l_i, and l_i^2 is the $(n+1)$-th component of l_i.

If $l_i^2 > 0$, then, due to the continuity of $\widetilde{F}(x, u, w, t)$, there exists a pair $(u_i, w_i) \in U \times W$ such that

$$\begin{pmatrix} l_i^1 \\ l_i^2 \end{pmatrix} = \widetilde{F}(x, u_i, w_i, t).$$

If $l_i^2 = 0$, then $l_i^1 \in \mathscr{G}^1(x, t)$ by the definition of the set $\mathscr{G}^1(x, t)$.

Since $\mathscr{G}(x, \omega, t)$ is continuous and $\mathscr{G}(X(t), \omega(\Gamma(t)), t)$ is Lebesgue measurable, applying the measurable selection theorem [237, 456] we obtain the representation

$$\mathscr{G}(X(t), \omega(\Gamma(t)), t) = \sum_{i=1}^{n+2} \alpha_i(t) \begin{pmatrix} l_i^1(t) \\ l_i^2(t) \end{pmatrix} =$$

$$= \sum_{i=1}^{n+2} \alpha_i(t) \begin{pmatrix} l_i^1(t) \\ l_i^2(t) \end{pmatrix} \mathbf{I}\{l_i^2(t) > 0\}(t) + \sum_{i=1}^{n+2} \alpha_i(t) \begin{pmatrix} l_i^1(t) \\ l_i^2(t) \end{pmatrix} \mathbf{I}\{l_i^2(t) = 0\}(t),$$

with some Lebesque-measurable functions $\{\alpha_i(t), l_i^1(t), l_i^2(t)\}$, $i = 1, \dots, n+2$. Hence, as the derivative of $X^{\mathrm{a}}(t)$ is equal to

$$\dot{X}^{\mathrm{a}}(t) = \frac{\sum_{i=1}^{n+2} \alpha_i(t) l_i^1(t)}{\sum_{i=1}^{n+2} \alpha_i(t) l_i^2(t)} \quad \text{a.e. on} \quad [0, T],$$

we obtain for $\dot{X}^{\mathrm{a}}(t)$ the expression

$$\dot{X}^{\mathrm{a}}(t) = \sum_{i=1}^{n+2} \beta_i^1(t) \frac{l_i^1(t)}{l_i^2(t)} \mathbf{I}\{l_i^2(t) > 0\}(t) + \sum_{i=1}^{n+2} \beta_i^2(t) l_i^1(t) \mathbf{I}\{l_i^2(t) = 0\}(t), \tag{4.42}$$

where

$$\beta_i^1(t) := \frac{\alpha_i(t) l_i^2(t)}{\sum_{i=1}^{n+2} \alpha_i(t) l_i^2(t)} \mathbf{I}\{l_i^2(t) > 0\}(t) \tag{4.43}$$

and

$$\beta_i^2(t) := \frac{\alpha_i(t)}{\sum_{i=1}^{n+2} \alpha_i(t) l_i^2(t)} \mathbf{I}\{l_i^2(t) = 0\}(t). \tag{4.44}$$

Note that

$$\beta_i^1(t) \geq 0, \quad \sum_{i=1}^{n+2} \beta_i^1(t) = 1, \tag{4.45}$$

and

$$\beta_i^2(t) \geq 0. \tag{4.46}$$

The functions $\frac{l_i^1(t)}{l_i^2(t)}$ in the first sum in (4.42) satisfy the inclusions

$$\frac{l_i^1(t)}{l_i^2(t)} = F(X(t), u_i(t), w_i(t), t) \in F(X(t), U, W, t),$$

while the functions $l_i^1(t)$ in the second sum satisfy the inclusion

$$l_i^1(t) \in \mathscr{G}^1(X(t), t).$$

Hence, the derivative $\dot{X}^{\mathrm{a}}(t)$ can be expressed as

$$\dot{X}^{\mathrm{a}}(t) = \sum_{i=1}^{n+2} \beta_i^1(t) F(X(t), u_i(t), w_i(t), t) + \sum_{i=1}^{n+2} \beta_i^2(t) l_i^1(t),$$

where $l_i^1(t) \in \mathscr{G}^1(X(t), t)$, and the functions $\beta_i^1(t)$ and $\beta_i^2(t)$ satisfy the relations (4.45) and (4.46).

This proves the claim of the theorem for the component $X^{\mathrm{a}}(t)$.

Consider now the component $X^{\mathrm{s}}(t)$ given by the formula (4.36), and note that $\dot{\eta}(\Gamma(t)) = 0$ on the subset $D_\Gamma^{\mathrm{s}} = D_{\mathscr{V}}^{\mathrm{s}}$. Then, a similar argument applied to the component $X^{\mathrm{s}}(t)$ leads to the inclusion

$$\dot{y}^{\mathrm{s}}(\Gamma(t)) \in \mathscr{G}^1(X(t), t),$$

and the relation

$$X^{\mathrm{s}}(t) = \int_0^t l^1(\tau) dV^{\mathrm{s}}(\tau), \qquad l^1(\tau) \in \mathscr{G}^1(X(\tau), \tau),$$

where, in view of (4.36), the function $l^1(\cdot)$ is both Γ^{s}- and $\mathscr{V}^{\mathrm{s}}$-measurable, proving the claim of the theorem for the component $X^{\mathrm{s}}(t)$.

Finally, since in the interval $[\Gamma(\tau-), \Gamma(\tau)]$, where $\eta(s) = \tau \in D_{\mathscr{V}}^{\mathrm{d}} = D_\Gamma^{\mathrm{d}}$, the derivative of $\eta(s)$ is equal to zero, we have

$$\dot{y}(s) \in \mathscr{G}^1(y(s), \tau).$$

Therefore, if $s \in [\Gamma(\tau-), \Gamma(\tau)]$, then $y(s)$ satisfies this differential inclusion with the initial condition $y(\Gamma(\tau-)) = x(\tau-)$. Denoting

$$y_\tau(s) = y(\Gamma(\tau-) + s)$$

for $s \in [0, \mathscr{V}(\{\tau\})] = [0, \Delta\Gamma(\tau)]$, we obtain the assertion concerning the jumps of $X(\cdot)$.

This completes the proof of the theorem. □

4.1.3 Representation of Generalized Solutions by Means of Differential Equations with Measure

In this section, we show that whenever the function $F(x, u, w, t)$ fulfills the regularity condition, one can derive a representation of the generalized solutions by means of differential equations with measure.

First, let us prove an auxiliary result concerning the description of the set $\mathcal{G}^1(x,t)$.

Lemma 4.2 *Suppose that $F(x,u,w,t)$ is regular at infinity. Then*

$$\mathcal{G}^1(x,t) = \text{conv}\, B(x,U,E,t), \tag{4.47}$$

where

$$B(x,U,E,t) = \big\{l \in \mathbb{R}^n : l = B(x,u,e,t),\ u \in U,\ e \in E = \{\|e\| = 1\}\big\}. \tag{4.48}$$

Proof If $l \in B(x,U,E,t)$, i.e,. $l = B(x,u,e,t)$ with $(u,e) \in U \times E$, then for any sequence $w_n \in W$, such that $\|w_n\| \to \infty$ and $\dfrac{w_n}{1+\|w_n\|} \to e$ we have that

$$l = \lim_n \frac{F(x,u,w_n,t)}{1+\|w_n\|} = B(x,u,e,t).$$

Moreover,

$$\lim_n \frac{1}{1+\|w_n\|} = 0.$$

Hence, the extended vector

$$\begin{pmatrix} l \\ 0 \end{pmatrix} \in \mathcal{G}(x,\Omega,t)$$

and, by the definition of the set $\mathcal{G}^1(x,t)$,

$$l \in \mathcal{G}^1(x,t).$$

Since $\mathcal{G}^1(x,t)$ is convex,

$$\text{conv}\, B(x,U,E,t) \subseteq \mathcal{G}^1(x,t). \tag{4.49}$$

Next, if $l \in \mathcal{G}^1(x,t)$, then

$$\begin{pmatrix} l \\ 0 \end{pmatrix} \in \mathcal{G}(x,\Omega,t),$$

and, by the definition of the set $\mathcal{G}(x,\Omega,t)$, this extended vector belongs to the closed convex hull of the set $\widetilde{F}(x,U,W,t)$. Hence, there exists numbers $\alpha_i > 0$, $i = 1,\ldots,n+2$, such that $\sum_{i=1}^{n+2} \alpha_i = 1$ and

$$\begin{pmatrix} l \\ 0 \end{pmatrix} = \sum_{i=1}^{n+2} \alpha_i \begin{pmatrix} l_i^1 \\ 0 \end{pmatrix}, \tag{4.50}$$

where the vectors $\begin{pmatrix} l_i^1 \\ 0 \end{pmatrix}$ belong to the closure of the set $\widetilde{F}(x, U, W, t)$. Therefore, for each i there exists a sequence $\{u_i^n, w_i^n\} \subset U \times W$, such that

$$\begin{pmatrix} l_i^1 \\ 0 \end{pmatrix} = \lim_n \begin{pmatrix} \dfrac{F(x, u_i^n, w_i^n, t)}{1 + \|w_i^n\|} \\ \dfrac{1}{1 + \|w_i^n\|} \end{pmatrix}.$$

Extracting a subsequence $\{u_i^{n_k}, w_i^{n_k}\}$ such that

$$u_i^{n_k} \to u_i \in U \quad \text{and} \quad \frac{w_i^{n_k}}{1 + \|w_i^{n_k}\|} \to e_i \in E \quad \text{as} \quad k \to \infty$$

we see that

$$l_i^1 = B(x, u_i, e_i, t),$$

which together with (4.50) gives

$$\begin{pmatrix} l \\ 0 \end{pmatrix} \in \operatorname{conv} B(x, U, E, t).$$

Hence,

$$\mathscr{G}^1(x, t) \subseteq \operatorname{conv} B(x, U, E, t), \tag{4.51}$$

which in conjunction with (4.49) completes the proof. □

The next result provides a representation of generalized solutions by means of differential equations with a measure.

Theorem 4.6 *Suppose the function $F(x, u, w, t)$ is regular at infinity and for any fixed pair (x, t) the set of vectors*

$$\mathscr{L}(x, t) = \left\{ l = \begin{pmatrix} l_1 \\ l_2 \end{pmatrix} = \begin{pmatrix} F(x, u, w, t) \\ \|w\| \end{pmatrix} \in \mathbb{R}^{n+1} : u \in U, \ w \in W \right\} \tag{4.52}$$

is convex.

Then for any generalized solution $X(\cdot)$ of the system (4.1) there exist:

1. *a non-negative regular scalar measure* $\mathscr{V}(dt)$ *whose distribution function*

$$V(t) = \mathscr{V}([0, t]),$$

is defined on the Borel subsets of the interval $[0, T]$ *and satisfies*

$$\mathscr{V}([0, T]) = V(T) \le M;$$

2. *a set of both Lebesgue- and* $\mathscr{V}$*-measurable controls*

$$\{u_0(t), w_0(t), \beta_i^0(t), \beta_i^{\mathrm{s}}(t), u_i(t), e_i(t)\}, \quad i = 1, \dots, n+2,$$

which satisfy the constraints

$$u_0(t) \in U, \quad w_0(t) \in W, \quad \beta_i^0(t) \ge 0,$$

$$u_i(t) \in U, \quad e_i(t) \in E, \quad \beta_i^{\mathrm{s}}(t) \ge 0, \quad \sum_{i=1}^{n+2} \beta_i^{\mathrm{s}}(t) = 1,$$

a.e. on the interval $[0, T]$ *with respect both to the Lebesgue and* $\mathscr{V}$ *measures;*

3. *a set of Lebesgue-measurable controls*

$$\{u_i^\tau(s), e_i^\tau(s), \beta_i^\tau(s)\}, \quad i = 1, \dots, n+2,$$

which are defined on each interval $[0, \Delta V(\tau)]$ *for all* $\tau \in D_{\mathscr{V}}^{\mathrm{d}}$ *and satisfy the constraints*

$$u_i^\tau(s) \in U, \quad e_i^\tau(s) \in E, \quad \beta_i^\tau(s) \ge 0, \quad \sum_{i=1}^{n+2} \beta_i^\tau(s) = 1,$$

a.e. on the interval $[0, \Delta V(\tau)]$*;*

such that the following holds:

The generalized solution $X(t)$ *and the function* $V(t)$ *satisfy the system consisting of the equation*

$$\begin{aligned} X(t) = x_0 + &\int_0^t F(X(\tau), u_0(\tau), w_0(\tau), \tau)\, d\tau + \\ &+ \int_0^t \sum_{i=1}^{n+2} \beta_i^0(\tau) B(X(\tau), u_i(\tau), e_i(\tau), \tau)\, d\tau + \\ &+ \int_0^t \sum_{i=1}^{n+2} \beta_i^{\mathrm{s}}(\tau) B(X(\tau), u_i(\tau), e_i(\tau), \tau)\, dV^{\mathrm{s}}(\tau) + \sum_{\tau \le t} \Delta X(\tau), \end{aligned} \tag{4.53}$$

where

$$\Delta X(\tau) = y_\tau(\Delta V(\tau)) - X(\tau-),$$

and $y_\tau(s)$ *is the solution of the differential equation*

$$\dot{y}_\tau(s) = \sum_{i=1}^{n+2} \beta_i^\tau(s) B(y_\tau(s), u_i^\tau(s), e_i^\tau(s), \tau)$$

with the initial condition $y_\tau(0) = X(\tau-)$, *and the equation*

$$V(t) = \int_0^t \|w_0(\tau)\| d\tau + \int_0^t \sum_{i=1}^{n+2} \beta_i^0(\tau) d\tau + V^s(t) + \sum_{\tau \le t} \Delta V(\tau) \tag{4.54}$$

respectively.

Proof By the definition (4.38) of the measure $\mathscr{V}(dt)$, the derivative of its distribution function is equal to

$$\dot{V}(t) = \dot{\Gamma}(t) - 1.$$

An analysis similar to that in the proof of Theorem 4.5 shows that the functions $X^{\mathrm{a}}(\cdot)$ and $V^{\mathrm{a}}(\cdot)$ satisfy the equations

$$\begin{aligned} \dot{X}^{\mathrm{a}}(t) &= \left.\frac{\dot{y}(s)}{\dot{\eta}(s)}\right|_{s=\Gamma(t)} \\ \dot{V}^{\mathrm{a}}(t) &= \left.\frac{1}{\dot{\eta}(s)}\right|_{s=\Gamma(t)} - 1 \end{aligned} \tag{4.55}$$

almost everywhere in $[0, T]$.

Moreover, the derivatives of the functions $X^{\mathrm{a}}(\cdot)$ and $V^{\mathrm{a}}(\cdot)$ can be written in the form

$$\begin{aligned} \dot{X}^{\mathrm{a}}(t) &= \sum_{i=1}^{n+2} \beta_i^1(t) \frac{l_i^1(t)}{l_i^2(t)} \mathbf{I}\{l_i^2(t) > 0\}(t) + \sum_{i=1}^{n+2} \beta_i^2(t) l_i^1(t) \mathbf{I}\{l_i^2(t) = 0\}(t), \\ \dot{V}^{\mathrm{a}}(t) &= \sum_{i=1}^{n+2} \beta_i^1(t) \frac{1 - l_i^2(t)}{l_i^2(t)} \mathbf{I}\{l_i^2(t) > 0\}(t) + \sum_{i=1}^{n+2} \beta_i^2(t), \end{aligned} \tag{4.56}$$

where $l_i^1(t)$ and $l_i^2(t)$ are the components of the vector-function $\mathscr{G}(X(t), \omega(\Gamma(t)), t)$, i.e.,

$$\begin{pmatrix} l_i^1(t) \\ l_i^2(t) \end{pmatrix} = \mathscr{G}(X(t), \omega(\Gamma(t)), t),$$

and the coefficients $\beta_i^1(t)$ and $\beta_i^2(t)$ are defined by formulas (4.43) and (4.44).

If $l_i^2(t) > 0$, then there exists a pair of vectors $u_i(t) \in U$, $w_i(t) \in W$, such that

$$\frac{l_i^1(t)}{l_i^2(t)} = F(x(t), u_i(t), w_i(t), t), \tag{4.57}$$

and

$$\frac{1 - l_i^2(t)}{l_i^2(t)} = \|w_i(t)\|. \tag{4.58}$$

If $l_i^2(t) = 0$, then

$$l_i^1(t) \in \mathscr{G}^1(x(t), t) = \text{conv}\, B(x(t), U, E, t). \tag{4.59}$$

Using the relations (4.57)–(4.59), the system (4.56) can be recast as

$$\begin{pmatrix} \dot{X}^{\text{a}}(t) \\ \dot{V}^{\text{a}}(t) \end{pmatrix} = \sum_{i=1}^{n+2} \beta_i^1(t) \begin{pmatrix} F(X(t), u_i(t), w_i(t), t) \\ \|w_i(t)\| \end{pmatrix} + \sum_{i=1}^{n+2} \beta_i^2(t) \begin{pmatrix} l_i^1(t) \\ 1 \end{pmatrix}. \tag{4.60}$$

In view of the convexity of the set (4.52) and the relations (4.45), there exists functions $u_0(t)$ and $w_0(t)$ such that $u_0(t) \in U$ and $w_0(t) \in W$ a.e. on $[0, T]$ and the first term on the right-hand side of Eq. (4.60) can be expressed as

$$\sum_{i=1}^{n+2} \beta_i^1(t) \begin{pmatrix} F(X(t), u_i(t), w_i(t), t) \\ \|w_i(t)\| \end{pmatrix} = \begin{pmatrix} F(X(t), u_0(t), w_0(t), t) \\ \|w_0(t)\| \end{pmatrix}. \tag{4.61}$$

Using the standard measurable selection procedure [456], one can assume that $u_0(t)$ and $w_0(t)$ are Lebesgue measurable .

The second term on the right-hand side of (4.61) can be expressed as

$$\sum_{i=1}^{n+2} \beta_i^2(t) \begin{pmatrix} l_i^1(t) \\ 1 \end{pmatrix} = \left(\sum_{i=1}^{n+2} \beta_i^2(t) \right) \sum_{i=1}^{n+2} \frac{\beta_i^2(t)}{\sum_{i=1}^{n+2} \beta_i^2(t)} \begin{pmatrix} l_i^1(t) \\ 1 \end{pmatrix},$$

where the summands satisfy the inclusion

$$\begin{pmatrix} l_i^1(t) \\ 1 \end{pmatrix} \in \begin{pmatrix} \mathscr{G}^1(X(t),t) \\ 1 \end{pmatrix},$$

thanks to the convexity of the set $\mathscr{G}^1(X(t),t)$. By Lemma 4.2, $\mathscr{G}^1(x,t) =$ conv $B(x,U,E,t)$, hence there exists functions

$$\gamma_k(t),\ u_k(t),\ e_k(t), \quad k = 1,\dots,n+2$$

satisfying the constraints $\gamma_k(t) \geq 0$, $\sum\limits_{k=1}^{n+2} \gamma_k(t) = 1$, $u_k(t) \in U$, $e_k(t) \in E$ a.e. in $[0,T]$, such that

$$\sum_{i=1}^{n+2} \frac{\beta_i^2(t)}{\sum\limits_{i=1}^{n+2} \beta_i^2(t)} \begin{pmatrix} l_i^1(t) \\ 1 \end{pmatrix} = \sum_{k=1}^{n+2} \gamma_k(t) \begin{pmatrix} B(x(t),u_k(t),e_k(t),t) \\ 1 \end{pmatrix},$$

and

$$\sum_{i=1}^{n+2} \beta_i^2(t) \begin{pmatrix} l_i^1(t) \\ 1 \end{pmatrix} = \sum_{i=1}^{n+2} \beta_i^0(t) \begin{pmatrix} B(X(t),u_i(t),e_i(t),t) \\ 1 \end{pmatrix}, \tag{4.62}$$

where

$$\beta_i^0(t) = \gamma_i(t) \sum_{k=1}^{n+2} \beta_k^2(t).$$

If the functions $\{\beta_i^0, u_i, e_i\}$ are selected to be Lebesgue measurable, then the substitution of expressions (4.62) and (4.61) into (4.60) establishes the representations (4.53) and (4.54) for the absolutely continuous components of the functions $X(\cdot)$ and $V(\cdot)$.

In much the same way one can obtain the representation for the singular components of the functions $X(\cdot)$ and $V(\cdot)$. From (4.29) it follows that the derivative of the singular component of the function $X^{\mathrm{s}}(t)$ with respect to $V^{\mathrm{s}}(t)$ equals

$$\frac{dX^{\mathrm{s}}(t)}{dV^{\mathrm{s}}(t)} = \sum_{i=1}^{n+2} \beta_i^{\mathrm{s}}(t) B(x(t),u_i(t),e_i(t),t),$$

where the functions $u_i(t)$ and $e_i(t)$ are $\mathscr{V}^{\mathrm{s}}$-measurable and

$$\beta_i^s(t) \geq 0, \qquad \sum_{i=1}^{n+2} \beta_i^{\mathrm{s}}(t) = 1.$$

Therefore, if in relation (4.62) we extend the the Lebesgue-measurable functions $\{u_i(t)\ e_i(t)\}$, which are defined on the set $D^{\mathrm{a}}_{\mathscr{V}}$, by respective $\mathscr{V}^{\mathrm{s}}$-measurable functions defined on the set $D^{\mathrm{s}}_{\mathscr{V}}$ of null Lebesgue measure, then we obtain the statement 2. of this theorem. Statement 3. is proved by the same arguments. □

These results were first published in [258, 270, 273]. In the next section we discuss the application of the general representation to various specific problems.

4.2 Generalized Solutions of Differential Equations with Affine Dependence on Unbounded Controls

In this section, we consider generalized solutions for some special classes of differential equations, namely, equations in which the right-hand side depends on the unbounded control in an affine manner. This class of systems satisfies Assumption 4.2 almost automatically, and the appropriate representation of the auxiliary system (4.15) can be obtained in an explicit form [274].

4.2.1 Differential Equations with Affine Dependence on Unbounded Controls

Consider a system described by a differential equation that is affine in an unbounded control,

$$\begin{cases} \dot{X}(t) = f(X(t), V(t), u(t), t) + B(X(t), V(t), u(t), t)w(t) \\ \dot{V}(t) = \|w(t)\|, \end{cases} \tag{4.63}$$

with two types of controls: $u \in U$ and $w \in K$, where $U \subset \mathbb{R}^k$ is a compact set and $K \subset \mathbb{R}^m$ is a convex cone. Thus, the component $u(t)$ is a bounded control, while the component $w(t)$ corresponds to a controls that is unbounded in norm, but restricted in the integral sense, i.e., $w(t)$ satisfies the constraint

$$\int_0^T \|w(t)\|dt \leq M < \infty \tag{4.64}$$

with some constant M.

The first equation in (4.63) is affine with respect to the control variable $w(t)$, while the second equation is only sublinear. The second equation governs the control resources spent till the current time. The dependence of the dynamic equations on the variable $V(t)$ is typical for problems with phase constraints as well as for systems admitting restoration of the control resources during the operational period.

Suppose Assumption 4.1 holds for right-hand side of the system (4.63), so that the functions $f(X, V, u, t)$ and $B(X, V, u, t)$ are continuous in all the variables and satisfy the Lipschitz condition with respect to (X, V).

Then for any e with $\|e\| = 1$ there exists the limit

$$\lim_{\|w\|\to\infty} \frac{f(X, V, u, t) + B(X, V, u, t)w}{1 + \|w\|} = B(X, V, u, t)e, \tag{4.65}$$

if $w/(1 + \|w\|) \to e$ when $\|w\| \to \infty$.

The right-hand side of the auxiliary system is described by the function

$$\begin{pmatrix} \dot{y} \\ \dot{z} \\ \dot{\eta} \end{pmatrix} = \widetilde{\mathscr{G}}(y, z, u, \eta, \omega) =$$

$$= \begin{pmatrix} (1 - \|\omega\|)\left[f(y, z, u, \eta) + B(y, z, u, \eta)\dfrac{\omega}{1 - \|\omega\|}\right] \\ = (1 - \|\omega\|)f(y, z, u, \eta) + B(y, z, u, \eta)\omega \\ \|\omega\| \\ 1 - \|\omega\| \end{pmatrix}, \tag{4.66}$$

where

$$\omega(w) = \frac{w}{1 + \|w\|}. \tag{4.67}$$

If one takes ω as a new control variable instead of w, then it should obey the constraint

$$\omega \in \{0 \le \|\omega\| < 1, \ \omega \in \omega(K) = K \cap \{\|\omega\| \le 1\}\}. \tag{4.68}$$

Therefore, the closure of the set $\widetilde{\mathscr{G}}(y, z, U, \eta, \omega(K))$ can be described as

$$\overline{\widetilde{\mathscr{G}}(y, z, U, \eta, \omega(K))} = \widetilde{\mathscr{G}}(y, z, U, \eta, \overline{\omega(K)}) =$$

$$= \left\{ \begin{pmatrix} \alpha f + (1 - \alpha)Be \\ (1 - \alpha)\|e\| \\ \alpha \end{pmatrix} : \alpha \in [0, 1], \ e \in K, \ \|e\| \le 1, \ u \in U \right\}. \tag{4.69}$$

If the set $\widetilde{\mathscr{G}}(y, z, U, \eta, \overline{\omega(K)})$ is convex for any fixed triplet (y, z, η), then the right-hand side of (4.69) gives the desired representation of the auxiliary system with the set of controls

$$\Omega = \left\{\overline{\omega} = (\alpha, e, u) \,|\, 0 \leq \alpha \leq 1,\ e \in K \cap \{\|e\| \leq 1\},\ u \in U\right\};$$

otherwise, to obtain the representation in the desired form we need to use convexification (i.e., construct the convex hull of Ω) by using the Carathéodory theorem [237].

Here we provide an approach based on the general idea of discontinuous time transformation [266]. Subsequently, similar results were obtained independently by other authors (see [66, 315, 405]).

Since the limit in (4.65) exists and the function $B(X, V, u, t)$ is continuous, the right-hand side of the system (4.63) is regular at infinity, and one can apply Theorem 4.6 to obtain the sought-for representation of a generalized solution for this class of systems.

Theorem 4.7 *Suppose the set of vectors*

$$\mathscr{L}_1(x, v, t) = f(x, v, U, t) = \{l \in \mathbb{R}^n \,:\, l = f(x, v, u, t),\ u \in U\},$$

$$\mathscr{L}_2(x, v, t) = \left\{ l = \begin{pmatrix} l_1 \\ l_2 \end{pmatrix} : \begin{array}{l} l_1 = B(x, v, u, t)w \in \mathbb{R}^n, \\ l_2 = \|w\| \in \mathbb{R}, \end{array} \quad u \in U,\ w \in K \right\} \tag{4.70}$$

is convex for any fixed triplet (x, v, t).

Then for any generalized solution $\{X(\cdot), V(t)\}$ of the system (4.63) there exist:

1. *a non-negative regular scalar measure $\mathscr{V}(dt)$ whose distribution function*

$$V(t) = \mathscr{V}([0, t]),$$

is defined on the Borel subsets of the interval $[0, T]$ and satisfies

$$\mathscr{V}([0, T]) = V(T) \leq M;$$

2. *a set of both Lebesgue- and $\mathscr{V}$-measurable controls*

$$\{u_0(t), w_0(t)\}$$

which satisfy the constraints

$$u_0(t) \in U, \quad w_0(t) \in K,$$

a.e. on the interval $[0, T]$ with respect to both the Lebesgue and $\mathscr{V}$ measures;

3. *a set of Lebesgue-measurable controls*

$$\{u^\tau(s), e^\tau(s)\}$$

that are defined in every interval $[0, \Delta V(\tau)]$ *with* $\tau \in D^{\mathrm{d}}_{\mathscr{V}}$ *and satisfy the constraints*

$$u^\tau(s) \in U, \quad e^\tau(s) \in K \cap \{\|e\| \leq 1\}$$

a.e. on the interval $[0, \Delta V(\tau)]$,

such that the generalized solution $\{X(t), V(t)\}$ *satisfies the equations*

$$\begin{aligned} X(t) = x_0 + &\int_0^t f(X(\tau), V(\tau), u_0(\tau), \tau) d\tau + \\ &+ \int_0^t B(X(\tau), V(\tau), u_0(\tau), \tau) w_0 \, dV^{\mathrm{c}}(\tau) + \sum_{\tau \leq t} \Delta X(\tau), \end{aligned} \tag{4.71}$$

where $V^{\mathrm{c}}(\cdot)$ *is the continuous component in the Lebesgue decomposition of the function* $V(\cdot)$ *of bounded variation [322], and*

$$\Delta X(\tau) = y_\tau(\Delta V(\tau)) - X(\tau-),$$

in which $y_\tau(s)$ *is the solution of the system of differential equations*

$$\begin{cases} \dot{y}_\tau(s) = B(y_\tau(s), z_\tau(s), u^\tau(s), \tau) \, e^\tau(s), \\ \dot{z}_\tau(s) = 1, \end{cases}$$

with the initial condition $y_\tau(0) = X(\tau-)$, $z_\tau(0) = V(\tau-)$.

Proof By (4.69) and the convexity of the set (4.70), the right-hand side of the auxiliary system

$$\begin{aligned} &\widetilde{\mathscr{G}}(y, z, U, \eta, \overline{\omega(K)}) = \\ &= \left\{ \begin{pmatrix} \alpha f + (1-\alpha) B e \\ (1-\alpha)\|e\| \\ \alpha \end{pmatrix} : \ \alpha \in [0, 1], \ e \in K, \ \|e\| \leq 1, \ u \in U \right\} \end{aligned} \tag{4.72}$$

is convex for any fixed triplet (y, z, η). Therefore, by Theorem 4.3, for any generalized solution $\{X(t), V(t)\}$ there exists a Lebesgue-measurable control $\overline{\omega}(s) = \{\alpha(s), e(s), u_1(s)\}$ satisfying the constraints

$$\overline{\omega}(s) \in \Omega = \big\{(\alpha, e, u) \,|\, 0 \leq \alpha \leq 1, \ e \in K, \ \|e\| \leq 1, \ u \in U\big\},$$

such that the corresponding solution $\{y(s), z(s), \eta(s)\}$ of the auxiliary system satisfies the relations

$$X(t) = y(\Gamma(t), \quad V(t) = z(\Gamma(t)), \quad \text{where } \Gamma(t) = \inf\{s : \eta(s) > t\}.$$

Now define the measure $\mathscr{V}(dt)$ by means of its distribution function as

$$\mathscr{V}\{(0, t]\} = V(t) = \int_0^{\Gamma(t)} (1 - \alpha(s))\|e(s)\|ds,$$

and define the controls $\{u_0(t), w_0(t)\}$ by

$$u_0(t) = u_1(\Gamma(t)), \quad w_0(t) = e(\Gamma(t)).$$

An analysis similar to that in the proof of Theorem 4.6 shows that $\{X(\cdot)\}$ satisfies the equations

$$\begin{aligned} X^{c}(t) = x_0 + \int_0^t f(X(\tau), V(\tau), u_0(\tau), \tau)d\tau + \\ + \int_0^t B(X(\tau), V(\tau), u_0(\tau), \tau)w_0\, dV^{c}(\tau) + \sum_{\tau \le t} \Delta X(\tau), \end{aligned} \tag{4.73}$$

where

$$\Delta X(\tau) = \int_{\Gamma(\tau-)}^{\Gamma(\tau)} B(y(s), z(s), u(s), \tau)e(s)\, ds,$$

and $\{y(s), z(s)\}$ is the solution of the system

$$\begin{cases} \dot{y}(s) = B(y(s), z(s), u_1(s), \tau)e(s), \\ \dot{z}(s) = \|e(s)\|, \end{cases} \tag{4.74}$$

with the initial conditions

$$y(\Gamma(\tau-)) = X(\tau-), \quad z(\Gamma(\tau-) = V(\tau-).$$

The substitution

$$s' = z(s) - V(\tau-)$$

transforms the interval $[\Gamma(\tau-), \Gamma(\tau)]$ into $[0, \Delta V(\tau)]$ and using the time transformation

$$z^{-1}(s') = \inf\{s : z(s) - V(\tau-) > s'\}$$

we find that the set of variables

$$\widetilde{y}(s') = y(z^{-1}(s')), \quad \widetilde{z}(s') = z(z^{-1}(s'))$$

$$\widetilde{u}_1(s') = u(z^{-1}(s')), \quad \widetilde{e}(s') = e(z^{-1}(s'))$$

satisfies the following system of differential equations:

$$\frac{d\widetilde{y}(s')}{ds'} = B(\widetilde{y}(s'), \widetilde{z}(s'), \widetilde{u}_1(s'), \tau)\widetilde{e}(s')\|\widetilde{e}(')\|^+,$$

$$\frac{d\widetilde{z}(s')}{ds'} = \|\widetilde{e}(s')\| \, \|\widetilde{e}(s')\|^+ = 1.$$

Here the superscript $^+$ stands for "pseudoinverse", i.e., $a^+ = a^{-1}$ if $a \neq 0$ and $0^+ = 0$. Defining the new control $\bar{e}(s')$ as

$$\overline{e}(s') = \|\widetilde{e}(s')\| \, \|\widetilde{e}(s')\|^+$$

completes the proof. □

Remark 4.5 The control $\overline{e}(s')$ satisfies the condition $\|\overline{e}(s')\| = 1$.

As a corollary of this general theorem we derive a representation for generalized solutions of a system described by equations of the form

$$\begin{cases} \dot{X}(t) = f(X(t), V(t), u(t), t) + B(u(t), t)w(t), \\ \dot{V}(t) = \|w(t)\|. \end{cases} \tag{4.75}$$

Corollary 4.1 *Suppose the set of vectors*

$$\begin{aligned} &\mathscr{L}_1(x, v, t) = f(x, v, U, t) = \{l \in \mathbb{R}^n, l = f(x, v, u, t) : u \in U\}, \\ &\mathscr{L}_2(x, v, t) = \left\{ l = \begin{pmatrix} l_1 \\ l_2 \end{pmatrix} : \begin{array}{l} l_1 = B(u, t)w \in \mathbb{R}^n, \\ l_2 = \|w\| \in \mathbb{R}, \end{array} \quad u \in U, \; w \in K \right\} \end{aligned} \tag{4.76}$$

is convex for any fixed triplet (x, v, t).

Then for any generalized solution $\{X(\cdot), V(t)\}$ *of the system (4.75) there exist:*

1. *a non-negative regular scalar measure* $\mathscr{V}(dt)$ *whose distribution function*

$$V(t) = \mathscr{V}([0, t]),$$

is defined on the Borel subsets of the interval $[0, T]$ *and satisfies*

$$\mathscr{V}([0, T]) = V(T) \leq M;$$

2. *a set of both Lebesgue- and* $\mathscr{V}$*- measurable controls*

$$\{u_0(t), w_0(t)\},$$

which satisfy the constraints

$$u_0(t) \in U, \quad w_0(t) \in K,$$

a.e. with respect both to the Lebesgue and $\mathscr{V}$ *measures in the interval* $[0, T]$, *such that the generalized solution* $\{X(t), V(t)\}$ *satisfies the equations*

$$X(t) = x_0 + \int_0^t f(X(\tau), V(\tau), u_0(\tau), \tau)d\tau + \\ + \int_0^t B(u_0(\tau), \tau)w_0 dV(\tau), \tag{4.77}$$

where the integral with respect to the measure $V(dt)$ *is understood in the Lebesgue-Stieltjes sense.*

Proof By Theorem 4.7, any generalized solution of system (4.75) satisfies the system of the equations

$$\begin{cases} X(t) = x_0 + \int_0^t f(X(\tau), V(\tau), u_0(\tau), \tau)d\tau + \\ \quad + \int_0^t B(u_0(\tau), \tau)w_0\, dV^c(\tau) + \sum_{\tau \leq t} \Delta X(\tau), \\ V(t) = V^c(t) + \sum_{\tau \leq t} \Delta V(\tau), \end{cases} \tag{4.78}$$

where $V^c(\cdot)$ is the continuous component in the Lebesgue decomposition of the function $V(\cdot)$ of bounded variation [322],

$$\Delta X(\tau) = y_\tau(\Delta V(\tau)) - X(\tau-),$$

and $y_\tau(s)$ is a solution of the system of the differential equation

$$\begin{cases} \dot{y}_\tau(s) = B(u^\tau(s), \tau)e^\tau(s), \\ \dot{z}_\tau(s) = 1, \end{cases}$$

with the initial conditions $y_\tau(0) = X(\tau-)$, $z_\tau(0) = V(\tau-)$. Therefore,

$$\Delta X(\tau) = \int_0^{\Delta V(\tau)} B(u^\tau(s), \tau)e^\tau(s)ds,$$

and since the set $B(U, \tau)K$ is convex, Lyapunov's theorem [173] ensures that there exist vectors $u(\tau) \in U$ and $w(\tau) \in K \cap \{\|w\| \leq 1\}$ such that

$$\Delta X(\tau) = \int_0^{\Delta V(\tau)} B(u^\tau(s), \tau)e^\tau(s)ds = B(u(\tau), \tau)w(\tau)\Delta V(t).$$

Finally, let us extend the functions $u_0(t)$ and $w_0(t)$ on the set $D_V^{\mathrm{d}} = \{t : \Delta V(t) > 0\}$ by the formulas

$$u_0(\tau) = u(\tau), \quad w_0(\tau) = w(\tau), \quad \text{for} \quad \tau \in D_V^{\mathrm{d}}.$$

This extension to the measurable and countable set D_V^{d} preserves the $\mathscr{V}$-measurability of the functions $u_0(t)$ and $w_0(t)$. Moreover, the values of $u_0(t)$ and $w_0(t)$ at the points $\tau \in D_V^{\mathrm{d}}$ belong to the sets U and K, respectively. Thus, we have shown that

$$\Delta X(\tau) = B(u_0(\tau), \tau)w_0(\tau)\Delta V(t).$$

for $\tau \in D_V^{\mathrm{d}}$, which completes the proof of Corollary 4.1. □

4.2.2 Differential Equations with a Lie-Algebra Structure

The specific properties of the function $B(x, v, u, t)$ allow us to obtain an explicit representation of the jumps of a generalized solution. We saw an example of explicit representation in Theorem 2.14, which describes the generalized solutions of DCS satisfying the robustness condition by means of a nonlinear differential equation with measure (2.71). This is a direct consequence of the Frobenius property (see Assumption 2.2), which can be alternatively expressed in algebraic terms, as follows.

Let $f(x)$ and $g(x)$ be smooth vector fields in $\mathbb{R}^n$.

Definition 4.4 The *Lie bracket* or *commutator* $[f, g]$ is given by

$$[f, g](x) = f'_x(x)g(x) - g'_x(x)f(x),$$

where $f'_x(x)$ and $g'_x(x)$ are the $(n \times n)$-matrices of partial derivatives of the components of the vector fields f and g, respectively.

If $[f, g] = 0$, then we say that the pair of vector fields f, g is *commutative*, or that f and g *commute*.

Definition 4.5 A set of the vector fields $\{f_i(x),\ i = 1, \ldots, N\}$ is called *involutive* if the Lie bracket of any pair f_i, f_j is a linear combination of the fields $\{f_i(x), i = 1, \ldots, N\}$, i.e., there exists constants $\{c^k_{i,j}, i, j, k = 1, \ldots, N\}$ such that

$$[f_i, f_j] = \sum_{k=1}^{N} c^k_{i,j} f_k. \tag{4.79}$$

Alternatively, in this case we will say that the set of functions $\{f_i,\ i = 1, \ldots, N\}$ is *in involution*. The constants $\{c^k_{i,j},\ i, j, k = 1, \ldots, N\}$ are called the *structure constants* of the Lie algebra $\mathscr{L}\{f_1, f_2, \ldots, f_N\}$ generated by the vector fields $\{f_i(x),\ i = 1, \ldots, N\}$.

It is readily seen that in the case of robust DCS, i.e., the case where the function $B(x, v, u, t) \equiv B(x, t)$ satisfies the Frobenius condition (see Assumption 2.2), the columns of the matrix $B(x, t)$, regarded as vector fields, form a commutative set for any fixed $t \in [0, T]$.

By the Frobenius theorem [68, 79] (see also Lemma 2.7), in the commutative case (i.e., when the vector fields f_i pairwise commute) the general solution of the system of differential equations

$$\dot{X}(t) = \sum_{i=1}^{N} f_i(X(t))w_i(t)$$

with the initial condition $X(0) = x_0$ can be expressed in some neighborhood of the initial condition in the form

$$X(t) = \Phi_N(\xi_N(t), \Phi_{N-1}(\xi_{N-1}(t), \ldots, \Phi(\xi_1(t), x_0)\ldots)), \tag{4.80}$$

where

$$\xi_k = \int_0^t w_k(s)ds$$

and $\Phi_k(t, y)$ is the general solution of the differential equation

$$\dot{y}(t) = f_k(y(t)), \quad y(0) = y. \tag{4.81}$$

Although the representation (4.80) holds in some neighborhood of $t = 0$, it can be continued for all $t > 0$ provided the solutions of (4.81) are defined for any initial condition for all values of t. Applying this result to the representation of the jumps of the generalized solution given by Theorem 4.7, one concludes that the jump $\Delta X(\tau) = y_\tau(\Delta V(\tau)) - X(\tau-)$, where $y_\tau(s)$, is the solution of the equation

$$\dot{y}_\tau(s) = B(y_\tau(s), \tau)e^\tau(s).$$

Consequently,

$$\Delta X(\tau) = \Phi_N(\mathscr{V}_N\{\tau\}, \tau, \Phi_{N-1}(\mathscr{V}_{N-1}\{\tau\}, \tau, \ldots, \Phi(\mathscr{V}_1\{\tau\}, \tau, X(\tau-))\ldots))- \\ - X(\tau-), \tag{4.82}$$

where $\Phi_k(s, \tau, y)$ is the solution of the differential initial value problem

$$\dot{y}(s) = B_k(y(s), \tau), \quad y(0) = y,$$

Here $B_k(x, \tau)$ is the k-th column of the matrix-valued function $B(x, t)$, and

$$\mathscr{V}_k\{\tau\} = \int_0^{|\mathscr{V}|\{\tau\}} e_k^\tau(s)ds$$

is the k-th component of the vector-valued measure $\mathscr{V}\{\tau\}$.

By Theorem 2.6, formula (4.82) provides one of possible analytical representations of the jump function, which due to the semigroup property depends on τ, on $X(\tau-)$, and on the value of the measure $\mathscr{V}\{\tau\}$, localized at the point τ.

To obtain a representation like (4.82) in the general case where the functions $\{f_i\}$ are in involution (see Definition 4.5), we need the following result.

Lemma 4.3 *Let $\Phi_i(\xi, y) \in \mathbb{R}^N$, where $\xi \in \mathbb{R}^1$, $y \in \mathbb{R}^N$, be a solution of the differential equation*

$$\frac{\partial}{\partial \xi}\Phi_i(\xi, y) = f_i(\Phi_i(\xi, y)) \tag{4.83}$$

with the initial condition

$$\Phi_i(0, y) = y.$$

Suppose that the functions $\{f_i,\ i = 1, \ldots, N\}$ are in involution, i.e., (4.79) holds. Then for any $i, j = 1, \ldots, N$,

$$\frac{\partial \Phi_i(\xi, y)}{\partial y} f_j(y) = \sum_{k=1}^N \gamma_{ij}^k(\xi) f_k(\Phi_i(\xi, y)), \tag{4.84}$$

where $\gamma_{ij}^k(\xi)$ are analytic functions that depend only on the structure constants of the Lie algebra $\mathscr{L}\{f_1, \ldots, f_N\}$.

Proof Denote the function on the left-hand side of relation (4.84) by

$$Z_{ij}(\xi, y) = \frac{\partial \Phi_i(\xi, y)}{\partial y} f_j(y). \tag{4.85}$$

In view of (4.83), $\frac{\partial \Phi_i(\xi, y)}{\partial y}$ satisfies the differential equation (see [79])

$$\frac{\partial}{\partial \xi}\frac{\partial \Phi_i(\xi, y)}{\partial y} = \frac{\partial f_i}{\partial y}(\Phi_i(\xi, y))\frac{\partial \Phi_i(\xi, y)}{\partial y}$$

with the initial condition

$$\frac{\partial \Phi_i(0, y)}{\partial y} = I_{(N\times N)}.$$

Then

$$\frac{\partial}{\partial \xi} Z_{ij}(\xi, y) = \frac{\partial f_i}{\partial y}(\Phi_i(\xi, y))\frac{\partial \Phi_i(\xi, y)}{\partial y} f_j(y) = \frac{\partial f_i}{\partial y}(\Phi_i(\xi, y)) Z_{ij}(\xi, y),$$

so $Z_{ij}(\xi, y)$ satisfies the differential equation

$$\frac{\partial}{\partial \xi} Z_{ij}(\xi, y) = \frac{\partial f_i}{\partial y}(\Phi_i(\xi, y)) Z_{ij}(\xi, y) \tag{4.86}$$

with the initial condition

$$Z_{ij}(0, y) = \frac{\partial \Phi_i(0, y)}{\partial y} f_j(y) = f_j(y).$$

Differentiating the right-hand side of (4.84) with respect to ξ and using the relation (4.79) we obtain

$$\frac{\partial}{\partial \xi}\sum_{k=1}^{N} \gamma_{ij}^k(\xi) f_k(\Phi_i(\xi, y)) =$$

$$= \sum_{k=1}^{N}\left[\frac{\partial \gamma_{ij}^k(\xi)}{\partial \xi} f_k(\Phi_i(\xi, y)) + \gamma_{ij}^k(\xi)\frac{\partial f_k}{\partial y}(\Phi_i(\xi, y)) f_i(\Phi_i(\xi, y))\right] =$$

$$= \sum_{k=1}^{N}\left[\frac{\partial \gamma_{ij}^k(\xi)}{\partial \xi} f_k(\Phi_i(\xi, y)) + \right.$$

$$\left. + \gamma_{ij}^k(\xi)\left\{\frac{\partial f_i}{\partial y}(\Phi_i(\xi, y)) f_k(\Phi_i(\xi, y)) - \sum_{l=1}^{N} c_{ik}^l f_l(\Phi_i(\xi, y))\right\}\right].$$

Changing here the order of summation we obtain

$$\frac{\partial}{\partial \xi}\sum_{k=1}^{N}\gamma_{ij}^{k}(\xi)f_k(\Phi_i(\xi,y)) = \frac{\partial f_i}{\partial y}(\Phi_i(\xi,y))\sum_{k=1}^{N}\gamma_{ij}^{k}(\xi)f_k(\Phi_i(\xi,y)) + \\ + \sum_{k=1}^{N} f_k(\Phi_i(\xi,y))\left\{\frac{\partial \gamma_{ij}^{k}(\xi)}{\partial \xi} - \sum_{l=1}^{N} c_{il}^{k}\gamma_{ij}^{l}(\xi)\right\}. \tag{4.87}$$

Now let $\gamma_{ij}^{k}(\xi)$ be the solution of the system of linear differential equations

$$\frac{d\gamma_{ij}^{k}(\xi)}{d\xi} = \sum_{l=1}^{N} c_{il}^{k}\gamma_{ij}^{l}(\xi), \tag{4.88}$$

with the initial conditions

$$\gamma_{ij}^{j} = 1 \quad \text{and} \quad \gamma_{ij}^{k} = 0, \quad \text{if} \quad k \neq j. \tag{4.89}$$

Remark 4.6 Since $\gamma_{ij}^{k}(\xi)$ are solutions of a linear differential equation with constant coefficients c_{ij}^{k}, they are analytic functions of ξ and depend only on the structure constants c_{ij}^{k}, $i, j, k = 1, \ldots, N$.

Substituting the expression (4.88) in the right-hand side of (4.87) we find that the function

$$Z_{ij}^{1}(\xi, y) = \sum_{k=1}^{N}\gamma_{ij}^{k}(\xi)f_k(\Phi_i(\xi,y))$$

satisfies the Eq. (4.86) with the initial condition

$$Z_{ij}^{1}(0, y) = \sum_{k=1}^{N}\gamma_{ij}^{k}(0)f_k(\Phi_i(0,y)) = f_j(\Phi_i(0,y) = f_j(y) = Z_{ij}(0,y).$$

Since $\frac{\partial f_i}{\partial y}(y)$ and $\Phi_i(\xi, y)$ are the locally Lipschitz with respect to y, the solution of Eq. (4.86) is unique, and therefore

$$Z_{ij}^{1}(\xi, y) = Z_{ij}(\xi, y).$$

This completes the proof. □

Now we can establish the following result.

Lemma 4.4 *Suppose the assumptions of Lemma 4.3 are in force. Let* $\{\xi_i(t) \in \mathbb{R}^1,\ i = 1, \dots, N\}$ *be a given set of differentiable functions and let*

$$Z(t) = \Phi_N(\xi_N(t), \Phi_{N-1}(\xi_{N-1}(t), \dots, \Phi_1(\xi_1(t), y), \dots,)).$$

Then

$$\dot{Z}(t) = \sum_{i=1}^{N} f_i(Z(t)) \sum_{j=1}^{N} L^{ij}(\xi_N(t), \dots, \xi_1(t))\dot{\xi}_j(t), \tag{4.90}$$

where the functions $L^{ij}(\xi_N, \dots, \xi_1)$ *are analytic and depend only on the structure constants of the Lie algebra* $\mathscr{L}\{f_1, \dots, f_N\}$.

Proof Denote

$$Z_k(t) = \Phi_N(\xi_k(t), \Phi_{k-1}(\xi_{k-1}(t), \dots, \Phi_1(\xi_1(t), y), \dots,)), \quad \text{for } k = 1, \dots, N.$$

Then

$$Z_k(t) = \Phi_k(\xi_k(t), Z_{k-1}(t)), \quad \text{for } k = 2, \dots, N \tag{4.91}$$

and $Z(t) = Z_N(t)$. We claim that $Z_k(t)$ satisfies the equation

$$\dot{Z}_k(t) = \sum_{i=1}^{N} f_i(Z_k(t)) \sum_{j=1}^{k} L_k^{i\,j}(\xi_k(t), \dots, \xi_1(t))\dot{\xi}_j(t), \tag{4.92}$$

where the functions $\{L_k^{i\,j},\ i = 1, \dots, N;\ j = 1, \dots, k\}$ are given by the recursion relations

$$L_k^{i\,j}(\xi_k, \dots, \xi_1) = \sum_{\nu=1}^{N} \gamma_{k\nu}^{i}(\xi_k) L_{k-1}^{\nu\,j}(\xi_{k-1}, \dots, \xi_1), \quad \text{if } 1 \le j \le k-1,$$

where (4.93)

$$L_k^{i\,k}(\xi_k, \dots, \xi_1) = \begin{cases} 1, & \text{if } i = k, \\ 0, & \text{if } i \neq k, \end{cases}$$

$$L_1^{i\,1} = \begin{cases} 1, & \text{if } i = 1, \\ 0, & \text{if } i \neq 1, \end{cases}$$

and the functions $\gamma_k^{i\,j}(\xi)$ are the solution of the system of Eqs. (4.88) and (4.89).

For $k = 1$ the validity of (4.92) is obvious. Assume now that relation (4.92) holds for some $k > 1$ and let us prove that it holds for $k + 1$. This will establish the assertion of the theorem with $L^{ij} = L_N^{ij}$.

So, suppose that $Z_k(t)$ satisfies (4.92) with some analytic functions L_k^{ij}. Differentiating the function $Z_{k+1}(t)$ with respect to t we obtain

$$\dot{Z}_{k+1}(t) = \frac{\partial \Phi_{k+1}}{\partial \xi}(\xi_{k+1}(t), Z_k(t))\dot{\xi}_{k+1}(t) + \frac{\partial \Phi_{k+1}}{\partial z}(\xi_{k+1}(t), Z_k(t))\dot{Z}_k(t).$$

Then, using the expression (4.92) for $\dot{Z}_k(t)$, we have

$$\begin{aligned}\dot{Z}_{k+1}(t) &= f_{k+1}(\xi_{k+1}(t), Z_k(t))\dot{\xi}_{k+1}(t) + \\ &+ \sum_{i=1}^{N} \frac{\partial \Phi_{k+1}}{\partial z}(\xi_{k+1}(t), Z_k(t)) f_i(Z_k(t)) \sum_{j=1}^{k} L_k^{ij}(\xi_k(t), \ldots, \xi_1(t))\dot{\xi}_j(t).\end{aligned} \tag{4.94}$$

Next, using the expression (4.84) provided by Lemma 4.3 for the terms

$$\frac{\partial \Phi_{k+1}}{\partial z}(\xi_{k+1}(t), Z_k(t)) f_i(Z_k(t)),$$

we obtain

$$\begin{aligned}\dot{Z}_{k+1}(t) &= f_{k+1}(\xi_{k+1}(t), Z_k(t))\dot{\xi}_{k+1}(t) + \\ &+ \sum_{i=1}^{N}\sum_{l=1}^{N} \gamma_{k+1i}^{l}(\xi_{k+1}(t)) f_l(Z_{k+1}(t)) \sum_{j=1}^{k} L_k^{ij}(\xi_k(t), \ldots, \xi_1(t))\dot{\xi}_j(t) = \\ &= f_{k+1}(\xi_{k+1}(t), Z_k(t))\dot{\xi}_{k+1}(t) + \\ &+ \sum_{i=1}^{N} f_i(Z_{k+1}(t)) \sum_{j=1}^{k} \left(\sum_{\nu=1}^{N} \gamma_{k+1\nu}^{i}(\xi_{k+1}(t)) L_k^{\nu j}(\xi_k(t), \ldots, \xi_1) \right) \dot{\xi}_j(t) = \\ &= \sum_{i=1}^{N} f_i(Z_{k+1}(t)) \sum_{j=1}^{k+1} L_{k+1}^{ij}(\xi_{k+1}(t), \ldots, \xi_k(t))\dot{\xi}_j(t),\end{aligned}$$

with L_{k+1}^{ij} satisfying the relations (4.93).

Thanks to the analyticity of γ_{jk}^{i}, the functions L_{k+1}^{ij} are also analytic. Therefore, the lemma is proved and one can pass to the proof of the representation (4.82) in the case where the functions $\{f_i,\ i = 1, \ldots, N\}$ are in involution. □

Theorem 4.8 *Let $L(\vec{\xi})$ be an $(N \times N)$- matrix valued function with the elements $L^{ij}(\vec{\xi}) = L^{ij}(\xi_N, \dots, \xi_1)$, where the functions on right-hand side are the same as in the representation (4.90). Denote*

$$\vec{\xi}(t) = (\xi_N(t), \dots, \xi_1(t)), \quad \vec{u}(t) = (u_N(t), \dots, u_1(t))$$

and consider the system of differential equations

$$L(\vec{\xi}(t))\dot{\vec{\xi}}(t) = \vec{u}(t), \tag{4.95}$$

with the initial condition $\vec{\xi}(0) = 0$.

(1) *Suppose the system (4.95) has a bounded solution $\vec{\xi}(t)$ on some interval $[0, T]$ and that the inial value problem*

$$\dot{x}(t) = \sum_{i=1}^{N} f_i(x(t))u_i(t), \quad x(0) = x_0 \in \mathbb{R}^N \tag{4.96}$$

has a unique solution on $[0, T]$. Then this solution $x(t)$ admits the representation

$$x(t) = \Phi_N(\xi_N(t), \Phi_{N-1}(\xi_{N-1}(t), \dots, \Phi_1(\xi_1(t), x_0), \dots,)). \tag{4.97}$$

(2) *Suppose the controls $u_N(t), \dots, u_1(t)$ are integrable on $[0, T]$. Then there exists an interval $[0, t^*]$, $t^* > 0$, where the system (4.95) has a unique solution, and consequently the representation (4.97) for the solution of (4.96) also holds on $[0, t^*]$.*

Proof Assertion (1) is an obvious consequence of Lemma 4.4 and can be obtained by substituting the relation (4.95) into (4.90), which gives

$$\dot{Z}(t) = \sum_{i=1}^{N} f_i(Z(t)) \sum_{j=1}^{N} L^{ij}(\xi_N(t), \dots, \xi_1(t))\dot{\xi}_j(t) = \sum_{i=1}^{N} f_i(Z(t))u_i(t).$$

Thus, $Z(t)$ satisfies the Eq. (4.96) and thanks to the uniqueness of the solution, $Z(t)$ coincides with the function $x(t)$ given by the formula (4.97).

To prove assertion (2), we will show that the matrix-valued function $L(\vec{\xi})$ is nonsingular in some neighborhood of the point $\vec{\xi} = 0$. First let us verify that $L(0) = I_{(N \times N)}$.

Indeed, each matrix $L_k(0)$ has the block form

$$L_k(0) = \begin{pmatrix} I_{(k\times k)} \\ \vdots \\ 0_{((N-k)\times k)} \end{pmatrix}, \tag{4.98}$$

where $0_{((N-k)\times k)}$ is the null matrix of the indicated dimensions.

The representation of $L_k(0)$ comes from the recursion relations for L_k, because by (4.93) we have

$$L_{k+1}^{i\,j}(0) = \sum_{\nu=1}^{N} \gamma_{k+1\nu}^{i}(0) L_k^{\nu\, j}(0), \quad \text{if} \quad j \le k,$$

and

$$L_{k+1}^{i\,k+1} = \begin{cases} 1, & \text{if } i = k+1, \\ 0 & \text{if } i \ne k+1. \end{cases}.$$

Assume for the moment that $L_k(0)$ has the form (4.98) and notice that, by (4.89),

$$\gamma_{k+1\,\nu}^{i}(0) = \begin{cases} 1, & \text{if } i = \nu, \\ 0, & \text{if } i \ne \nu, \end{cases} \tag{4.99}$$

for $i, \nu = 1, \ldots, N$. Upon substituting the expression (4.99) into the recursion relation for $L_{k+1}^{i\,j}(0)$, $j \le k$, and using the expression for $L_{k+1}^{i\,k+1}(0)$, we conclude that under this assumption the matrix $L_{k+1}^{i\,j}(0)$ is also of the form (4.98). Since $L_1^{i\,j}$ has the form (4.98), it follows that all the matrices $L_k(0)$ have the form (4.98) and $L(0) = L_N(0) = I_{(N\times N)}$. Therefore, the matrix-valued function $L(\vec{\xi})$ with the analytic coefficients is nonsingular and has an inverse in some neighborhood of the point $\vec{\xi} = 0$. Hence, Eq. (4.95) can be solved with respect to $\dot{\vec{\xi}}(t)$ and be transformed into the equivalent equation

$$\dot{\vec{\xi}}(t) = L^{-1}(\vec{\xi}(t))\vec{u}(t), \tag{4.100}$$

with the initial condition $\vec{\xi}(0) = 0$. Thanks to the analyticity of $L(\vec{\xi})$, the inverse matrix $L^{-1}(\vec{\xi})$ is also analytic in some neighborhood of the point $\vec{\xi} = 0$ and Eq. (4.100) has a unique solution in some interval $[0, t^*]$, $t^* > 0$. This solution satisfies Eq. (4.95) as well; moreover, Eq. (4.96) has a unique solution due to the integrability of $u_i(t)$. Therefore, according to assertion (1) of the theorem, the solution of (4.96) coincides on $[0, t^*]$ with the function given by the expression (4.97). This completes the proof. □

Remark 4.7 The order of the functions Φ_i in the representation (4.97) can be changed with an appropriate change of the system (4.95).

As a corollary of Theorem 4.8 we can obtain a representation for the jumps of solutions of systems with the involutive property.

Corollary 4.2 *Suppose that the columns of the matrix-valued function $B(x, t)$ are in involution for any $t \in [0, T]$, i.e., satisfy the relations*

$$[B_i(x,t), B_j(x,t)] = \sum_{k=1}^{m} c_{ij}^k(t) B_k(x,t),$$

and that the initial value problem

$$L(\vec{\xi}(s), t)\frac{d\vec{\xi}(s)}{ds} = \vec{e}(s), \quad \vec{\xi}(0) = 0, \tag{4.101}$$

has a unique solution for any function $\{\vec{e}(s)\}$ with $\|\vec{e}(s)\| \le 1\}$. Here $L(\vec{\xi}, t)$ is the same matrix as in the representation (4.90) corresponding to the set of constants $\{c_{ij}^k(t)\}$ at the point $t \in [0, T]$.

Then the jump of a generalized solution $X(t)$ at a point τ admits the representation

$$\begin{aligned}\Delta X(\tau) =& \Phi_N(\xi_N(|\mathcal{V}|\{\tau\}), \tau, \Phi_{N-1}(\xi_{N-1}(|\mathcal{V}|\{\tau\}), \tau, \ldots, \\ & \Phi_1(\xi_1(|\mathcal{V}|\{\tau\}), \tau, X(\tau-))\ldots)) - X(\tau-);\end{aligned} \tag{4.102}$$

here

$$|\mathcal{V}|\{\tau\} = \sum_{k=1}^{m} |\mathcal{V}_k|\{\tau\}$$

is the total value of the vector measure $\mathcal{V}(dt)$ at the point τ, $\Phi_k(s, \tau, y)$ is the solution of the differential initial value problem

$$\dot{y}(s) = B_k(y(s), \tau), \quad y(0) = y$$

with $B_k(x, \tau)$ the k-th column of the matrix-valued function $B(x, t)$, and $\vec{\xi}(s)$ is the solution of Eq. (4.101) in the interval $[0, |\mathcal{V}|\{\tau\}]$ with some function $\vec{e}(s)$ such that $\|\vec{e}(s)\| \le 1$.

Remark 4.8 This result is a direct consequence of Theorems 4.7 and 4.8; its proof is omitted to save space.

Remark 4.9 According to Remark 4.7, in the representation of jumps one can use any other order of the functions Φ_i, with the appropriate system (4.101).

Next we consider two examples where the representation of jumps can be obtained in explicit form.

Example 4.1 (The Case of a 2-Dimensional Lie Algebra) Suppose $B(x,t) \equiv B(x)$, where $B(x)$ is of dimension $N \times 2$ and its columns are in involution, i.e.,

$$[B_1(x), B_2(x)] = c_{12}^1 B_1(x) + c_{12}^2 B_2(x).$$

Then the corresponding system of equations for $\{\gamma_{12}^1(\xi), \gamma_{12}^2(\xi)\}$ has the form (see (4.88) and (4.89))

$$\begin{cases} \dot{\gamma}_{12}^1(\xi) = c_{12}^1 \gamma_{12}^2(\xi), \\ \dot{\gamma}_{12}^2(\xi) = c_{12}^2 \gamma_{12}^2(\xi), \end{cases} \quad \text{where} \quad \begin{aligned} \gamma_{12}^1(0) = 0, \\ \gamma_{12}^2(0) = 1. \end{aligned} \tag{4.103}$$

If $c_{12}^1 \neq 0$ and$c_{12}^2 \neq 0$, we have

$$\begin{cases} \gamma_{12}^1(\xi) = \dfrac{c_{12}^1}{c_{12}^2}\left(\exp\{c_{12}^2\xi\} - 1\right), \\ \gamma_{12}^2(\xi) = \exp\{c_{12}^2\xi\}. \end{cases}$$

If $c_{12}^1 = 0$, then

$$\begin{cases} \gamma_{12}^1(\xi) = 0, \\ \gamma_{12}^2(\xi) = \exp\{c_{12}^2\xi\}, \end{cases}$$

and if $c_{12}^2 = 0$, then

$$\begin{cases} \gamma_{12}^1(\xi) = c_{12}^1\xi, \\ \gamma_{12}^2(\xi) = 1. \end{cases}$$

The variables $\{\xi_1(s),\ \xi_2(s)\}$ satisfy the system (see (4.101))

$$\begin{cases} \dot{\xi}_1(s) + \gamma_{12}^1(\xi_1(s))\dot{\xi}_2(s) = e_1(s), & \xi_1(0) = 0, \\ \gamma_{12}^2(\xi_1(s))\dot{\xi}_2(s) = e_2(s), & \xi_2(0) = 0. \end{cases}$$

Since $\gamma_{12}^2(\xi) \neq 0$, this system can be reduced to the equivalent form

$$\begin{cases} \dot{\xi}_1(s) = e_1(s) - \dfrac{\gamma_{12}^1(\xi_1(s))}{\gamma_{12}^2(\xi_1(s))} e_2(s), \\ \dot{\xi}_2(s) = \dfrac{1}{\gamma_{12}^2(\xi_1(s))} e_2(s). \end{cases}$$

If $c_{12}^2 = 0$ the above system has the form

$$\begin{cases} \dot{\xi}_1(s) = e_1(s) - c_{12}^1 \xi_1(s) e_2(s) \\ \dot{\xi}_2(s) = e_2(s). \end{cases}$$

and admits the explicit solution

$$\begin{cases} \xi_1(s) = \displaystyle\int_0^s \exp\left(-c_{12}^1 \int_u^s e_2(v)dv\right) e_1(u)du, \\ \xi_2(s) = \displaystyle\int_0^s e_2(u)du. \end{cases} \tag{4.104}$$

Thus, the system (4.101) for $\{\xi_1(s), \xi_2(s)\}$ has a unique solution for any integrable functions $\{e_1(s), e_2(s)\}$ and, by Corollary 4.2, the jump of the generalized solution at a point τ can be expressed as

$$\Delta X(\tau) = \Phi_1(\xi_1(|\mathscr{V}|\{\tau\}), \Phi_2(\xi_2(|\mathscr{V}|\{\tau\}), X(\tau-)) - X(\tau-),$$

where $|\mathscr{V}|\{\tau\} = |\mathscr{V}_1|\{\tau\} + |\mathscr{V}_2|\{\tau\}$. Here $\Phi_k(s, y)$, $k = 1, 2$ are the solutions of differential initial value problems

$$\dot{y}(s) = B_k(y(s)), \quad y(0) = y, \quad k = 1, 2,$$

and $\{\xi_1(s), \xi_2(s)\}$ are given by formulas (4.104) with some functions $\{e_1(s), e_2(s)\}$ that satisfy the constraints

$$\begin{aligned} &|e_1(s)| + |e_2(s)| \leq 1, \\ &\int_0^{|\mathscr{V}|\{\tau\}} e_1(s)ds = \mathscr{V}_1\{\tau\}, \qquad \int_0^{|\mathscr{V}|\{\tau\}} e_2(s)ds = \mathscr{V}_2\{\tau\}. \end{aligned} \tag{4.105}$$

Formulas (4.104) exhibit the difference between the commutative and involutive cases. In the first case the values of jumps are uniquely defined by the values of the components of the vector measure, i.e., they depend only on $\{\mathscr{V}_1\{\tau\}$ and $\mathscr{V}_2\{\tau\}\}$.

In the involutive case we have a set of various jump values, corresponding to various functions $\{e_1(s), e_2(s)\}$ satisfying (4.105). Moreover, every admissible pair $\{e_1(s), e_2(s)\}$ corresponds to some method of approximation of an impulsive control localized at the point τ. For example, the pair

$$e_1(s) = \begin{cases} 1, & \text{if } s \in [0, |\mathscr{V}_1|\{\tau\}), \\ 0, & \text{if } s \in [|\mathscr{V}_1|\{\tau\}, |\mathscr{V}|\{\tau\}], \end{cases} \qquad e_2(s) = \begin{cases} 0, & \text{if } s \in [0, |\mathscr{V}_1|\{\tau\}), \\ 1, & \text{if } s \in [|\mathscr{V}_1|\{\tau\}, |\mathscr{V}|\{\tau\}], \end{cases}$$

corresponds to the method of approximation where the jump of the first component precedes the jump of the second component. On the contrary, the pair

$$e_1(s) = \begin{cases} 0, & \text{if } s \in [0, |\mathcal{V}_2|\{\tau\}), \\ 1, & \text{if } s \in [|\mathcal{V}_2|\{\tau\}, |\mathcal{V}|\{\tau\}], \end{cases} \qquad e_2(s) = \begin{cases} 1, & \text{if } s \in [0, |\mathcal{V}_2|\{\tau\}), \\ 0, & \text{if } s \in [|\mathcal{V}_2|\{\tau\}, |\mathcal{V}|\{\tau\}], \end{cases}$$

corresponds to the method of approximation where the impulse applied to the second component precedes the impulse applied to the first component.

In the commutative case the jump does not depend on the method of approximation, whereas in the involutive case we obtain different values of the jumps, namely,

$$\Delta X(\tau) = \Phi_1(\mathcal{V}_1\{\tau\}), \Phi_2(\mathcal{V}_2\{\tau\}), X(\tau-)) - X(\tau-),$$

$$\Delta X(\tau) = \Phi_1(\exp\{-c_{12}^1 \mathcal{V}_2\{\tau\}\}\mathcal{V}_1\{\tau\}), \Phi_2(\mathcal{V}_2\{\tau\}), X(\tau-)) - X(\tau-),$$

in the first and second method of approximation, respectively.

In the following example, we consider a system for which one can also obtain an explicit representation of jumps, but the original system does not possess a Lie algebra structure, and to obtain such a representation in the form (4.102), one needs to complement the system by an additional "fictitious" input with an appropriate vector of coefficients.

Example 4.2 (The Case of a 3-Dimensional Lie Algebra) Consider a system with the (3×2)-dimensional matrix $B(x)$, where $x = (x_1, x_2, x_3) \in \mathbb{R}^3$ and

$$B(x) = (B_1(x), B_2(x)) = \begin{pmatrix} f_1(x_1) & 0 \\ 0 & f_2(x_2) \\ 1 & x_3 \end{pmatrix}.$$

It is readily verified that

$$[B_1(x), B_2(x)] = \begin{pmatrix} 0 \\ 0 \\ -1 \end{pmatrix} = B_3(x),$$

and so the set of functions $\{B_1(x), B_2(x), B_3(x)\}$ has a structure of Lie algebra with the commutation relations

$$[B_1, B_2] = B_3, \quad [B_1, B_3] = 0, \quad [B_2, B_3] = B_3,$$

hence with the following structure constants:

$$c_{12}^1 = 0, \qquad c_{12}^2 = 0, \quad c_{12}^3 = 1,$$

$$c_{13}^1 = 0, \qquad c_{13}^2 = 0, \quad c_{13}^3 = 0,$$

$$c_{23}^1 = 0, \qquad c_{23}^2 = 0, \quad c_{23}^3 = 1.$$

Let us supplement the input by a third component $u_3(t) \equiv 0$. Applying the standard procedure for the derivation of the Eq. (4.102), we get

$$\Delta X(\tau) =$$
$$= \Phi_3(\xi_3(|\mathscr{V}|\{\tau\}), \Phi_2(\xi_2(|\mathscr{V}|\{\tau\}), \Phi_1(\xi_1(|\mathscr{V}|\{\tau\}), \tau, X(\tau-)))) - X(\tau-),$$

where $|\mathscr{V}|\{\tau\}$ denotes the total value of the vector measure $\mathscr{V}(dt)$ localized at the point τ and $\Phi_k(s, y)$ are the solutions of the differential initial value problems

$$\dot{y}(s) = B_k(y(s)), \quad y(0) = y, \quad k = 1, \dots, 3.$$

In this case

$$\Phi_1(s, y) = \begin{pmatrix} \Phi_{11}(s, y_1) \\ y_2 \\ y_3 + s \end{pmatrix}, \quad \Phi_2(s, y) = \begin{pmatrix} y_1 \\ \Phi_{22}(s, y_2) \\ y_3 \exp\{s\} \end{pmatrix}, \quad \Phi_3(s, y) = \begin{pmatrix} y_1 \\ y_2 \\ y_3 - s \end{pmatrix},$$

where Φ_{11} and Φ_{22} are the general solutions of the equations $\dot{y} = f_1(y)$ and $\dot{y} = f_2(y)$, respectively.

The functions $\{\xi_1(s), \xi_2(s), \xi_3(s)\}$ satisfy the system of equations

$$\begin{cases} \dot{\xi}_1(s) = e_1(s), \\ \dot{\xi}_2(s) = e_2(s), \\ \dot{\xi}_3(s) = (\exp\{\xi_2(s)\} - 1)e_1(s) + \xi_3(s)e_2(s), \end{cases} \tag{4.106}$$

where $\vec{e}(s) = \{e_1(s), e_2(s)\}$ obeys on the interval $[0, |\mathscr{V}|\{\tau\}]$ the constraints

$$\|\vec{e}(s)\| \leq 1,$$
$$\int_0^{|\mathscr{V}|\{\tau\}} e_1(s)ds = \mathscr{V}_1\{\tau\}, \quad \int_0^{|\mathscr{V}|\{\tau\}} e_2(s)ds = \mathscr{V}_2\{\tau\}. \tag{4.107}$$

Since the solution of the system (4.106) is given by the formulas

$$\begin{cases} \xi_1(s) = \int_0^s e_1(v)dv, \\ \xi_2(s) = \int_0^s e_2(v)dv, \\ \xi_3(s) = \exp\{\xi_2(s)\}\left[\xi_1(s) - \int_0^s \exp\{-\xi_2(v)\}e_1(v)dv\right], \end{cases}$$

the jumps of the generalized solution admit the representation

$$\begin{cases} \Delta X_1(\tau) = \Phi_{11}(\mathscr{V}_1\{\tau\}, x_1(\tau-)) - x_1(\tau-), \\ \Delta X_2(\tau) = \Phi_{22}(\mathscr{V}_2\{\tau\}, x_2(\tau-)) - x_2(\tau-), \\ \Delta X_3(\tau) = (\exp\{\mathscr{V}_2\} - 1)x_3(\tau-) + \exp\{\mathscr{V}_2\{\tau\}\} \int_0^{|\mathscr{V}|\{\tau\}} \exp\{-\xi_2(s)\}e_1(s)ds. \end{cases}$$

Clearly, the jumps of the first and second components do not depend on the way the impulse input is approximated i.e., on the choice of the functions e_1, e_2 subject to (4.107). By contrast, the jump of the third component does depend on the approximation and may assume different values. For example, if $0 \leq e_1, e_2 \leq 1$, which corresponds to the case of positive impulses, the jump $\Delta X_3(\tau)$ can assume any value in the interval

$$\left[(\exp\{\mathscr{V}_2\} - 1)x_3(\tau-) + \mathscr{V}_1\{\tau\},\ (\exp\{\mathscr{V}_2\} - 1)x_3(\tau-) + \exp\{\mathscr{V}_2\{\tau\}\}\mathscr{V}_1\{\tau\}\right].$$

However, in the case of sign-variable impulses, i.e., when $-1 \leq e_1, e_2 \leq 1$, the third component may undergo a nonzero jump even if $\mathscr{V}_1\{\tau\} = \mathscr{V}_2\{\tau\} = 0$. Suppose that

$$\begin{aligned} &\int_0^{|\mathscr{V}|\{\tau\}} e_1(s)ds = \int_0^{|\mathscr{V}|\{\tau\}} e_2(s)ds = 0, \\ &\int_0^{|\mathscr{V}|\{\tau\}} |e_1(s)|ds = |\mathscr{V}_1|\{\tau\}, \quad \int_0^{|\mathscr{V}|\{\tau\}} |e_2(s)|ds = |\mathscr{V}_2|\{\tau\}, \\ &|\mathscr{V}_1|\{\tau\} + |\mathscr{V}_2|\{\tau\} = |\mathscr{V}|\{\tau\}. \end{aligned} \tag{4.108}$$

Then

$$\Delta X_1(\tau) = 0, \quad \Delta X_2(\tau) = 0,$$

but

$$\Delta X_3(\tau) = \int_0^{|\mathscr{V}|\{\tau\}} \exp\{-\xi_2(s)\}e_1(s)ds.$$

Since $e_1(s) = e_1^+(s) + e_1^-(s)$, where

$$e_1^+(s) = \max\{0, e_1(s)\} \geq 0 \quad \text{and} \quad e_1^-(s) = \min\{0, e_1(s)\} \leq 0,$$

we have $|e_1(s)| = e_1^+(s) - e_1^-(s)$ and

$$\int_0^{|\mathscr{V}|\{\tau\}} e_1^+(s)ds = -\int_0^{|\mathscr{V}|\{\tau\}} e_1^-(s)ds = \frac{|\mathscr{V}_1|\{\tau\}}{2}.$$

Therefore,

$$\Delta X_3(\tau) \leq \left(\max_{s\in[0,|\mathscr{V}|\{\tau\}]} \exp\{-\xi_2(s)\} - \min_{s\in[0,|\mathscr{V}|\{\tau\}]} \exp\{-\xi_2(s)\}\right) \frac{|\mathscr{V}_1|\{\tau\}}{2}.$$

Using the constraint (4.108) and the equation $\dot{\xi}_2(s) = e_2(s)$ one can derive the inequalities

$$\max_{[0,|\mathscr{V}|\{\tau\}]} \xi_2(s) - \min_{[0,|\mathscr{V}|\{\tau\}]} \xi_2(s) \leq \frac{|\mathscr{V}_2|\{\tau\}}{2}, \tag{4.109}$$

and

$$\max_{[0,|\mathscr{V}|\{\tau\}]} \exp\{-\xi_2(s)\} - \min_{[0,|\mathscr{V}|\{\tau\}]} \exp\{-\xi_2(s)\} \leq \exp\left\{\frac{|\mathscr{V}_2|\{\tau\}}{2}\right\} - 1. \tag{4.110}$$

Indeed, for any two points τ_1 and τ_2 with $\tau_1 < \tau_2$,

$$|\xi_2(\tau_1) - \xi_2(0)| + |\xi_2(\tau_2) - \xi_2(\tau_1)| + |\xi_2(T) - \xi_2(\tau_2)| \leq \operatorname*{Var}_{[0,|\mathscr{V}|\{\tau\}]} \xi_2(s) = |\mathscr{V}_2|.$$

However, $\xi_2(0) = \xi_2(T) = 0$ and if τ_1 and τ_2 are the points of maximum and minimum of the function ξ_2, where $\xi_2(\tau_1) \geq 0$ and $\xi_2(\tau_2) \leq 0$, respectively, we obtain

$$\begin{gathered} \xi_2(\tau_1) + |\xi_2(\tau_2) - \xi_2(\tau_1)| - \xi_2(\tau_2) = 2|\xi_2(\tau_2) - \xi_2(\tau_1)| \leq |\mathscr{V}_2|, \\ |\xi_2(\tau_2) - \xi_2(\tau_1)|,\ |\xi_2(\tau_2)|,\ |\xi_2(\tau_1)| \leq \frac{|\mathscr{V}_2|}{2}, \end{gathered} \tag{4.111}$$

and the inequality (4.109) follows. Now the inequality (4.110) is a direct consequence of (4.109) and (4.111). Hence,

$$\Delta X_3(\tau) \leq \left[\exp\left\{\frac{|\mathscr{V}_2|\{\tau\}}{2}\right\} - 1\right] \frac{|\mathscr{V}_1|\{\tau\}}{2}.$$

The same arguments applied to the minimum value of ΔX_3 give the inequality

$$\Delta X_3(\tau) \geq -\left[\exp\left\{\frac{|\mathscr{V}_2|\{\tau\}}{2}\right\} - 1\right]\frac{|\mathscr{V}_1|\{\tau\}}{2}.$$

Therefore,

$$|\Delta X_3(\tau)| = \left[\exp\left\{\frac{|\mathscr{V}_2|\{\tau\}}{2}\right\} - 1\right]\frac{|\mathscr{V}_1|\{\tau\}}{2}.$$

It readily seen that the lower and upper bounds can be achieved by suitable choices of $\{e_1(s), e_2(s)\}$. For example, to achieve the upper bound, one can choose

$$e_1(s) = \begin{cases} 1, & \text{if } s \in \left(\frac{|\mathscr{V}_2|\{\tau\}}{2}, \frac{|\mathscr{V}_2|\{\tau\}}{2} + \frac{|\mathscr{V}_1|\{\tau\}}{2}\right), \\ -1, & \text{if } s \in \left(\frac{|\mathscr{V}_1|\{\tau\}}{2} + |\mathscr{V}_1|\{\tau\} + |\mathscr{V}_2|\{\tau\}\right), \\ 0, & \text{otherwise,} \end{cases}$$

and

$$e_2(s) = \begin{cases} -1, & \text{if } s \in \left[0, \frac{|\mathscr{V}_2|\{\tau\}}{2}\right], \\ 1, & \text{if } s \in \left[\frac{|\mathscr{V}_2|\{\tau\}}{2} + \frac{|\mathscr{V}_2|\{\tau\}}{2}, \frac{|\mathscr{V}_1|\{\tau\}}{2} + |\mathscr{V}_2|\{\tau\}\right], \\ 0 & \text{otherwise.} \end{cases}$$

The fact that one can achieve the lower bound is obvious. This last example also demonstrates that the values of the jump depend on the approximation used for impulsive inputs. Moreover, it shows that the use of impulse controls extends the control possibilities. Indeed, in this example one can keep the first and second variables constant and at the same time ensure a prescribed non-vanishing evolution for the third component.

In conclusion, let us note that the same issues arise if the original system is of a mixed discrete-continuous nature. This occurs when its dynamics admit jumps, and if the instants of these jumps coincide with the instants of the impulse controls, then it is necessary to find the correct description for the response of the system. In the next section, we show that using the method of discontinuous time transformation enables one to settle this issue also for the class of systems with discrete-continuous properties.

Chapter 5
Optimal Generalized Solutions in Control Problems

5.1 Control Problem with Phase Constraints

In this chapter, we return to the general nonlinear system described by the equation

$$\dot{X}(t) = F(X(t), u(t), w(t), t), \tag{5.1}$$

which satisfies Assumption 4.1 of the previous chapter, i.e., $F(x, u, w, t)$ is a given continuous function, Lipschitz with respect to the variables $(x, w) \in \mathbb{R}^n \times \mathbb{R}^m$, and with linear growth with respect to x and w.

We consider the system (5.1) in a given interval $[0, T]$ and with the initial condition $X(0) = x_0 \in \mathbb{R}^n$.

The controls $\{u(\cdot),\ w(\cdot)\}$ are Lebesgue-measurable functions in $[0, T]$ such that $u(t) \in \mathbb{R}^k$, $w(t) \in \mathbb{R}^m$ and the following constraints are satisfied a.e. on $[0, T]$:

$$u(t) \in U \subset \mathbb{R}^k \tag{5.2}$$

where U is a compact set, and

$$w(t) \in W \subset \mathbb{R}^m \tag{5.3}$$

$$\int_0^T \|w(t)\| dt \leq M < \infty, \tag{5.4}$$

where W is an unbounded closed set and M is a given constant.

Our objective is to minimize the performance criterion

$$J[X(\cdot), u(\cdot), w(\cdot)] = \varphi_0(X(0), X(T)), \tag{5.5}$$

B. Miller, E. Rubinovich, *Optimization of Dynamical Systems with Impulse Controls and Shocks*, Systems & Control: Foundations & Applications,
https://doi.org/10.1007/978-3-031-64124-4_5

where φ_0 is continuous in all the variables. The minimization of (5.5) is done over the set of admissible controls $\{u(\cdot), w(\cdot)\}$ that satisfy the constraints (5.2)–(5.4) and such that the corresponding trajectory $X(\cdot)$ satisfies the following terminal and phase restrictions:

$$h(X(0), X(T)) = 0, \quad S(X(0), X(T)) \leq 0, \tag{5.6}$$

$$G(X(t), t) \leq 0 \quad \text{for all } t \in [0, T]. \tag{5.7}$$

Here $h = \{h_1, \dots, h_{N_1}\}$, $S = \{s_1, \dots, s_{N_2}\}$ and $G = \{g_1, \dots, g_{N_3}\}$ are $\mathbb{R}^{N_1}$-, $\mathbb{R}^{N_2}$-, and R^{N_3}-valued continuous functions, respectively, and (5.6) and (5.7) are to be understood componentwise, i.e., as

$$s_1(X(0), X(T)) \leq 0, \; \dots, \; s_{N_2}(X(0), X(T)) \leq 0$$

$$g_1(X(t), t) \leq 0, \; \dots, \; g_{N_3}(X(t), t) \leq 0, \quad \text{for all} \quad t \in [0, T].$$

Remark 5.1 Generally speaking, the set of trajectories corresponding to admissible controls is not closed, and its closure includes generalized (discontinuous) solutions. This means that the optimal control and the optimal (continuous) trajectory may not even exist. For this reason in this problem we will search for generalized solutions. It turns out that in the class of generalized solutions there exist an optimal solution and even optimal controls, though their description is more complex than just in the class of controls satisfying the constraints (5.2)–(5.4). The issue is that the control $w(\cdot)$ admits impulsive actions, and so we need to take into account the possible discontinuity of the optimal solution and accordingly search for the optimal solution within the class of generalized solutions in the sense of Definition 4.1.

Remark 5.2 For the sake of convenience, in this chapter we will assume that all trajectories, whether ordinary or generalized, are extended as $X(t) = X(0-)$ for $t < 0$ and as $X(t) = X(T)$ for $t > T$. So, they have a limit from the left at $t = 0$ and are right-continuous at $t = T$.

5.1.1 *Strong and Weak Admissible Generalized Solutions*

Following the lines of the preceding chapter, we introduce below the concepts of the admissible generalized solutions in the strong sense and in the weak sense [280].

Definition 5.1 A right-continuous function $X(\cdot)$ of bounded variation defined on the interval $[0, T]$ is called an *admissible generalized solution* of the system (5.1) under the *strong* constraints (5.6) and (5.7) if:

(i) $\{X(\cdot)\}$ satisfies the constraints (5.6) and (5.7) in the sense that

$$h(X(0-), X(T)) = 0, \qquad S(X(0-), X(T)) \le 0,$$

and

$$G(X(t), t) \le 0 \quad \text{for all } t \in [0, T];$$

(ii) there exists a sequence of admissible controls $\{u^n(\cdot), w^n(\cdot)\}$ satisfying the constraints (5.2)–(5.4), such that the corresponding sequence $\{X^n(\cdot)\}$ of solutions of the system (5.1) converges to $X(\cdot)$ at all points of continuity, and, in addition

$$\lim_n X^n(0) = X(0-), \qquad \lim_n X^n(T) = X(T),$$

$$\lim_n \sup_{[0,T]} g_k(X^n(t), t) \le 0 \quad \text{for} \quad k = 1, \ldots, N_3.$$

Remark 5.3 Since the functions h and S are continuous,

$$\lim_n h(X^n(0), X^n(T)) = h(X(0-), X(T)) = 0,$$

$$\lim_n S(X^n(0), X^n(T)) = S(X(0-), X(T)) \le 0.$$

Therefore, the terminal and phase constraints that hold for the generalized solution may be violated on the approximating sequence $X^n(\cdot)$, though this violation is uniformly small in $[0, T]$.

A specific feature of the admissible generalized solutions introduced in Definition 5.1 is the presence of the so-called "strong" constraints, when the elements of the approximating sequence almost satisfy the constraints and the violations of the constraints must be small in the uniform norm. This case is illustrated in Fig. 5.1, where $X(t)$ is the generalized discontinuous solution and $X^n(t)$ is one of the approximating functions in the sequence, and the region restricted by the constraint $G(X, t) \le 0$ is shown on the top and is not violated by the elements of the approximating sequence.

However, some typical problems arising in applications may allow constraint violations that are not uniformly small. If, for example, these violations are of impulse type, one can allow that the constraints are fulfilled only in the average.

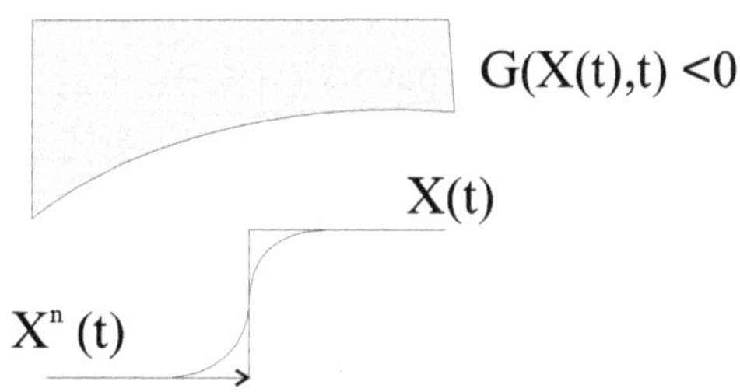

Fig. 5.1 Approximation of a generalized solution that is admissible in the strong sense

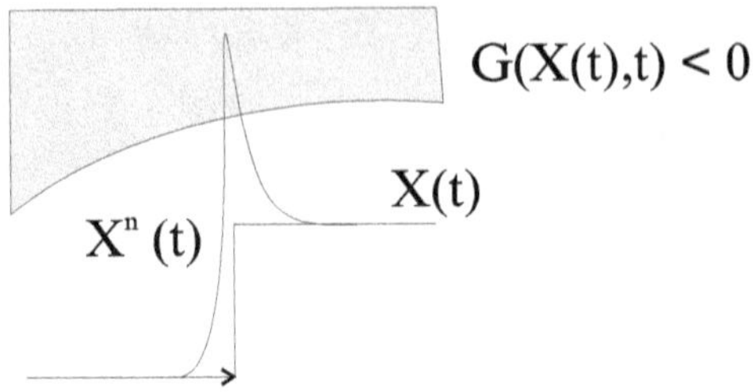

Fig. 5.2 Approximation of a generalized solution that is admissible in the weak sense

Therefore, one can consider another concept of admissible generalized solution, which could be interpreted as a solution corresponding to the case of "weak" constraints.

Definition 5.2 A right -continuous function $X(\cdot)$ of bounded variation defined on the interval $[0, T]$ is called an *admissible generalized solution* of the system (5.1) under the *weak* constraints (5.6) and (5.7) if:

(i) $\{X(\cdot)\}$ satisfies the constraints (5.6) and (5.7) in the sense that

$$h(X(0-), X(T)) = 0, \quad S(X(0-), X(T)) \le 0$$

and

$$G(X(t), t) \le 0 \quad \text{for all } t \in [0, T];$$

(ii) there exists a sequence of admissible controls $\{u^n(\cdot), w^n(\cdot)\}$, satisfying the constraints (5.2)–(5.4), such that the corresponding sequence $\{X^n(\cdot)\}$ of solutions of the system (5.1) converges to $X(\cdot)$ at all points of continuity and, in addition,

$$\lim_n X^n(0) = X(0-), \quad \lim_n X^n(T) = X(T).$$

This type of generalized solutions admissible in the weak sense is illustrated in Fig. 5.2, where in contrast to the previous case, the approximating solutions may "strongly" violate the constraint region, but only in the small vicinity of the jump point.

The convergence of $X^n(t)$ to $X(t)$ and the continuity of the function G imply that

$$\lim_n g_k(X^n(t), t) \le 0 \qquad \text{for} \quad k = 1, \ldots, N_3,$$

at all points of continuity.

Thus, we have two different types of admissible generalized solutions and we can now introduce the appropriate concept of the optimal solution.

Definition 5.3 An admissible generalized solution $\{X^0(\cdot)\}$ is said to be an *optimal generalized solution* if

$$\varphi_0(X^0(0-), X^0(T)) \leq \varphi_0(X(0-), X(T))$$

for any admissible generalized solution $\{X(\cdot)\}$.

Remark 5.4 In this definition one can use the two different types of admissible solution, and accordingly one can obtain two different optimal solutions, namely the optimal generalized solution in *the strong sense* and in *the weak sense*.

The following example demonstrates the considerable difference between these two types of solutions of the optimal control problem.

Example 5.1 Consider a system for $X(t) = (x_1(t), x_2(t))$ described by the equations

$$\dot{X}(t) = \begin{pmatrix} \dot{x}_1(t) \\ \dot{x}_2(t) \end{pmatrix} = \begin{pmatrix} x_2(t)u(t) \\ -x_1(t)u(t) \end{pmatrix}, \text{ with the initial conditions } \begin{pmatrix} x_1(0) \\ x_2(0) \end{pmatrix} = \begin{pmatrix} 0 \\ 1 \end{pmatrix},$$

and the problem of minimizing the performance criterion

$$J[x_1, x_2] = (x_2(1) + 1)^2$$

subject to the control constraints

$$u(t) \geq 0, \qquad \int_0^1 u(t)dt \leq \pi$$

and the phase constraint

$$x_1(t) \leq 0.$$

This system has the explicit solution

$$X(t) = \begin{pmatrix} x_1(t) \\ x_2(t) \end{pmatrix} = \begin{pmatrix} \sin v(t) \\ \cos v(t) \end{pmatrix}, \quad \text{where} \quad v(t) = \int_0^t u(s)ds.$$

Hence, one can easily find the optimal generalized solution which satisfies the constraints in the weak sense, as in Definition 5.2: namely, take

$$v^0(t) = \begin{cases} 0, & \text{if } t < \tau, \\ \pi, & \text{if } t \geq \tau, \end{cases}$$

where τ is an arbitrary point in the interval (0, 1). Indeed, the appropriate optimal generalized solution in the weak sense is

$$X^0(t) = \begin{pmatrix} x_1^0(t) \\ x_2^0(t) \end{pmatrix} = \begin{pmatrix} 0 \\ \begin{cases} 1, & \text{if } t < \tau, \\ -1, & \text{if } t \geq \tau \end{cases} \end{pmatrix},$$

and the corresponding value of the performance criterion is 0.

To see that $X^0(\cdot)$ is the optimal solution only in the weak sense, take any sequence of absolutely continuous functions $\{v^n(t)\}$ which approximates $v^0(t)$, so that $v^n(t) \to v^0(t)$ everywhere on [0, 1], except possibly at the point $t = \tau$. Then $X^n(t) \to X^0(t)$ everywhere on [0, 1] except possibly at the point $t = \tau$. Moreover, $\lim_n \cos v^n(t) = 1$ for $t < \tau$ and $\lim_n \cos v^n(t) = -1$ for $t > \tau$. Therefore, there exist numbers t^n such that $\cos v^n(t^n) = 0$ and $\sin v^n(t^n) = x_1(t^n) = 1$. Therefore, despite the fact that $x_1^n(t) \to 0$ everywhere on [0, 1], except possibly at the point $t = \tau$, we nevertheless have

$$\lim_n \max_{[0,1]} \; x_1^n(t) = 1.$$

If we allow only solutions admissible in the strong sense, then the constraint $\sup_{[0,1]} x_1^n(t) = \max_{[0,1]} \sin v^n(t) \to 0$ implies that

$$\max_{[0,1]} v^n(t) \to 0$$

and so there exists exactly one admissible generalized solution, namely:

$$X(t) = \begin{pmatrix} x_1(t) \\ x_2(t) \end{pmatrix} = \begin{pmatrix} 0 \\ 1 \end{pmatrix}$$

and the corresponding value of the performance criterion is equal to 4.

So, the difference between the two statements of the problem considered above is rather significant.

Remark 5.5 In ordinary optimal control problems with bounded controls, the correctness of the relaxation of constraints is very serious also, though often not so obvious, because in these problems, the violation of constraints is always uniformly small; however, when violation of constraints is forbidden, difficult issues arise, especially in deriving the optimality conditions [112]. At the same time, in problems with discontinuous trajectories this is of paramount importance [84, 85]. Every generalized solution admissible in the strong sense is also admissible in the weak sense. Therefore, the value of the performance criterion in the weak setting is in general smaller (when solving the minimization problem) and the optimality conditions for it is stronger, since the class of admissible variations is wider [128].

5.2 Existence of the Optimal Generalized Solution

5.2.1 Existence of the Weak Optimal Solution

After the statement of the optimal control problem in the class of generalized solutions, we address here the question of the existence of a solution. The standard way to prove the existence of an optimal solution is to show the compactness of the set of admissible solutions in some topology (it is natural to choose the topology in which a generalized solution is the limit of a sequence of ordinary solutions) and then establish the lower semi-continuity or the continuity of the performance criterion.

Theorem 5.1 *Let Assumption 4.1 be in force. Suppose that the set of admissible generalized solutions of the system (5.1) under the weak constraints (5.2)–(5.7) is non-empty and uniformly bounded. Then there exists an optimal generalized solution of the problem (5.1)–(5.7), admissible in the weak sense as in Definition 5.2.*

Remark 5.6 In our case it is natural to use the topology of the space of right-continuous functions of bounded variation. This topology corresponds to the weak-$*$ convergence and a necessary and sufficient condition for precompactness is given by Helly's Theorem A.5. Specifically, a subset of functions of bounded variation is precompact if and only if it is uniformly bounded and its elements have a uniformly bounded variation [204].

Proof Since the set of generalized solutions is uniformly bounded, the set of initial conditions of ordinary solutions $\{X(0)\}$ is also uniformly bounded. Therefore, all admissible generalized solutions have uniformly bounded variation. Indeed, due to Assumption 4.1, any ordinary solution of the system (5.1) satisfies the inequalities

$$\begin{aligned}
\operatorname*{Var}_{[0,T]} X(t) &\leq \|X(0)\| + \int_0^T \|F(X(t), u(t), w(t), t)\|\, dt \leq \\
&\leq \sup_{X(\cdot)} \|X(0)\| + L_2 \int_0^T (1 + \|w(t)\|)(1 + \|X(t)\|)\, dt \leq \\
&\leq \sup_{X(\cdot)} \|X(0)\| + L_2 \Big(T + M)(1 + \sup_{X(\cdot)} \|X(t)\|\Big) \leq C < \infty
\end{aligned}$$

and, therefore, all ordinary solutions have uniformly bounded variation. Hence, the set of ordinary solutions is totally bounded in the space of functions of bounded variation, and hence is precompact [204]. The set of generalized solutions is the closure of the set of ordinary solutions in the topology of weak-$*$ convergence and, consequently, it is compact.

Let $X^n(\cdot)$ be a sequence of admissible generalized solutions that minimizes the performance criterion. Since the set of generalized solutions is uniformly bounded and φ_0 is continuous, we have

$$-\infty < \inf_{X(\cdot)} J[X(\cdot)] = \lim_{n} \varphi_0(X^n(0-), X^n(T)).$$

By Helly's theorem (Theorem A.5) and following our rule of extension of a function beyond the interval $[0, T]$ (see Remark 5.2), one can extract a subsequence X^{n_k} which converges everywhere on the interval $[-\varepsilon, T+\varepsilon]$, with $\varepsilon > 0$, to some function of bounded variation that we denote by $x^0(\cdot)$ According to the extension rule, we have also

$$\begin{aligned} X^{n_k}(0-) &= X^{n_k}(-\varepsilon) \to x^0(-\varepsilon) = x^0(0-), \\ X^{n_k}(T) &= X^{n_k}(T+\varepsilon) \to x^0(T+\varepsilon) = x^0(T). \end{aligned} \tag{5.8}$$

Consider the function

$$X^0(t) = \begin{cases} x^0(t+), & \text{if } t \in [0, T], \\ x^0(0-), & \text{if } t = 0-. \end{cases}$$

Let us prove that $X^0(t)$ is the optimal generalized solution in the weak sense.

As the functions $\varphi_0, S, h,$ and g are continuous and the function $x^0(t+)$ is right-continuous, $X^0(\cdot)$ satisfies the constraints (5.6) and (5.7). Moreover,

$$J[X^0(\cdot)] = \lim_{n} \varphi_0(X^{n_k}(0-), X^{n_k}(T)) = \inf_{X(\cdot)} J[X(\cdot)].$$

It remains to establish the existence of sequence of ordinary solutions that approximates $X^0(t)$ in the sense of Definition 5.2. By the definition of admissible generalized solutions, for each of the functions from the sequence $\{X^{n_k}(\cdot)\}$ there exists a sequence of ordinary solutions $\{x^{n_k,m}(\cdot)\}$ which converges to $X^{n_k}(\cdot)$ at all points of continuity. Since $\{x^{n_k,m}(\cdot)\}$ converges to $\{X^{n_k}(\cdot)\}$ a.e. and $\{X^{n_k}(\cdot)\}$ converges to $X^0(\cdot)$ a.e. as well, one can extract a subsequence $\{x^{n_k,m_k}(\cdot)\}$ that converges to $X^0(\cdot)$ a.e. [322]. This subsequence is totally bounded in the space of functions of bounded variation; hence, by Helly's theorem, one can further extract a sub-subsequence that converges to some function $z^0(\cdot)$ of bounded variation for any $t \in [-\varepsilon, T+\varepsilon]$, $\varepsilon > 0$.

Clearly, $z^0(\cdot)$ and $x^0(\cdot)$ coincide almost everywhere on $[0, T]$.

Since $z^0(\cdot)$ and $x^0(\cdot)$ are functions of bounded variations, they admit right and left limits at any point $t \in [0, T]$, and

$$\begin{aligned} z^0(t+) &= \lim_{\tau \downarrow t} z^0(\tau) = \lim_{\tau \downarrow t} x^0(\tau) = x^0(t+), \\ z^0(t-) &= \lim_{\tau \uparrow t} z^0(\tau) = \lim_{\tau \uparrow t} x^0(\tau), = x^0(t-) \end{aligned}$$

where the sequence $\{\tau\}$ can be selected from the set of points where $z^0(\tau) = x^0(\tau)$. Hence, $z^0(t) = X^0(t)$ at all points of continuity and so $X^0(t)$ is an admissible generalized solution in the weak sense, and in fact is the weak optimal generalized solution. □

This theorem establishes the existence of a solution in the weak sense only, since we could not guarantee that the violation of constraints by the approximating sequence of ordinary solutions is uniformly small. To prove the existence of the strong solution, we use the method based on the discontinuous time transformation. Using this method one can reduce the original problem to a standard problem, which can be treated by the well-known methods of the optimal control theory. In the next subsection we formulate this auxiliary problem and prove its equivalence to the original optimization problem.

5.2.2 *Existence of the Strong Optimal Solution*

We will consider the case where the function $F(x, u, w, t)$ in (5.1) satisfies the regularity assumptions (see Definition 4.2). Then, by Theorem 4.4, there exist an $\mathbb{R}^{n+1}$-valued function $\mathscr{G}(y, \omega, \eta)$ and a closed bounded set Ω in some vector space of appropriate dimension, such that for any $(y, \eta) \in, \mathbb{R}^{n+1}$

$$
\begin{aligned}
&\overline{\operatorname{conv}}\,(\widetilde{F}(y, u_1, w_1, \eta) \mid u_1 \in U, w_1 \in W) = \\
&\quad = \overline{\operatorname{conv}}\,(\widetilde{F}(y, U, W, \eta)) = \mathscr{G}(y, \Omega, \eta) = \{\mathscr{G}(y, \omega, \eta) \mid \omega \in \Omega\},
\end{aligned}
$$

where $\mathscr{G}(y, \omega, \eta)$ is continuous in all the variables, satisfies the Lipschitz condition, and has a linear growth with respect to the variables (y, η).

Consider an optimal control problem for the system

$$
\begin{pmatrix} \dot{Y}(s) \\ \dot{\eta}(s) \end{pmatrix} = \mathscr{G}(Y(s), \omega(s), \eta(s)) \tag{5.9}
$$

with the initial conditions

$$
Y(0) = X(0), \qquad \eta(0) = 0,
$$

and with the control variable ω subject to the constraint

$$
\omega(s) \in \Omega. \tag{5.10}
$$

The purpose of this auxiliary problem is to minimize the performance criterion

$$
J'[Y(\cdot), \eta(\cdot), \omega(\cdot)] = \varphi_0(Y(0), Y(T_1)), \tag{5.11}
$$

where φ_0 is the same as in (5.5), and the admissible control $\omega(\cdot)$ obeys the constraint (5.10), so that the trajectory $\{Y(\cdot), \eta(\cdot)\}$ satisfies the following terminal and phase constraints:

$$h(Y(0), Y(T_1)) = 0, \quad S(Y(0), Y(T_1)) \le 0 \tag{5.12}$$

$$G(Y(s), \eta(s)) \le 0 \quad \text{for any } s \in [0, T_1] \tag{5.13}$$

where h, S, and G are the same as in (5.6) and (5.7), and the terminal time T_1 is such that

$$\eta(T_1) = T, \quad T_1 \le T + M. \tag{5.14}$$

The auxiliary optimal control problem (5.9)–(5.14) is equivalent to the original problem (5.1–(5.7), as established in the following result.

Theorem 5.2 *Let the assumptions of Theorem 5.1 and Assumption 4.2 be in force. Suppose the set of admissible generalized solutions in the strong sense of the system (5.1) with the constraints (5.2)–(5.7) is non-empty and uniformly bounded.*

Then there exists an optimal solution $\{Y^0(\cdot), \eta^0(\cdot)\}$ of the auxiliary problem (5.9)–(5.14), and

$$X^0(t) = Y^0(\Gamma^0(t)), \tag{5.15}$$

where

$$\Gamma^0(t) = \inf\{s : \eta^0(s) > t\}, \quad \Gamma^0(T) = T_1, \tag{5.16}$$

is the optimal generalized solution in the strong sense of the original optimal control problem.

Conversely, if $X^0(\cdot)$ is the optimal generalized solution in the strong sense of the original problem (5.1)–(5.7), then there exists a pair of functions $\{Y^0(\cdot), \eta^0(\cdot)\}$ that is the optimal solution of the auxiliary problem and such that relations (5.15) and (5.16) hold.

Proof Since the set of the admissible solutions is not empty, there exists a generalized solution $X(\cdot)$ which satisfies the conditions of Definition 5.1. Therefore, there exists a sequence of ordinary solutions $\{X^n(\cdot)\}$ such that

$$\lim_n X_n(0) = X(0-), \quad \lim_n X^n(T) = X(T),$$

$$\lim_n \sup_{[0,T]} g_k(X^n(t), t) \le 0 \qquad \text{for} \quad k = 1, \dots, N_3.$$

By Theorems 4.1 and 4.3, for each function $X^n(\cdot)$ there exists an admissible control $\omega^n(\cdot)$, defined in some interval $[0, T_1^n]$, such that corresponding solution $\{Y^n(\cdot), \eta^n(\cdot)\}$ of the system (5.9) satisfies the relations

$$
\begin{aligned}
&Y^n(0) = X^n(0), \quad \eta^n(0) = 0, \quad \eta^n(T_1^n) = T, \\
&X^n(t) = Y^n(\Gamma^n(t)), \quad \text{where} \quad \Gamma^n(t) = \inf\{s : \eta^n(s) > t\}, \\
&Y^n(s) = X^n(\eta^n(s)), \quad \Gamma^n(\eta^n(s)) = s.
\end{aligned}
\tag{5.17}
$$

Since the function $\mathscr{G}$ satisfies the linear growth condition and the set of initial conditions $\{Y^n(0), \eta^n(0)\}$ is uniformly bounded, the set of functions $\{Y^n(\cdot), \eta^n(\cdot)\}$ is uniformly bounded and equicontinuous [237]. Arguing as in the proof of Theorem 4.3, one can extract a subsequence $\{Y^n(\cdot), \eta^n(\cdot), T_1^n\}$ (we retain the same notation) that converges to some pair $\{Y(\cdot), \eta(\cdot)\}$, uniformly on the interval $[0, T_1]$, where $T_1 = \lim_n T_1^n$.

Since the set $\mathscr{G}(y, \Omega, \eta)$ is convex and closed, there exists an admissible control $\omega(\cdot)$, defined on $[0, T_1]$, such that $\{Y(\cdot), \eta(\cdot)\}$ is the corresponding solution of the system (5.9). Moreover, thanks to the uniform convergence of the sequence $\{Y^n(\cdot), \eta^n(\cdot)\}$ to $\{Y(\cdot), \eta(\cdot)\}$ and the relations (5.17), we have

$$
Y(0) = X(0-), \quad Y(T_1) = X(T)
$$

and

$$
\begin{aligned}
&g_k(Y(s), \eta(s)) = \lim_n g_k(Y^n(s), \eta^n(s)) \le \\
&\le \lim_n g_k(X^n(\eta^n(s)), \eta^n(s)) \le \lim_n \sup_{[0,T]} g_k(X^n(t), t)) \le 0, \quad k = 1, \dots, N_3.
\end{aligned}
$$

Therefore, $\{Y(\cdot), \eta(\cdot)\}$ is an admissible solution in the auxiliary control problem. Hence, the set of admissible solutions in the auxiliary problem is not empty, and thanks to the boundedness of the set of $\{Y(0)\}$, it is uniformly bounded. By the convexity of the right-hand side of (5.9) and the properties of $\mathscr{G}$, in the auxiliary problem there exists an optimal control $\omega^0(\cdot)$ such that the corresponding solution $\{Y^0(\cdot), \eta^0(\cdot)\}$ satisfies the relation

$$
J'[Y^0(\cdot), \eta^0(\cdot), \omega^0(\cdot)] = \min_{\omega(\cdot)} J'.
$$

Now, it remains to prove that the function $X^0(\cdot)$ defined by (5.15) and (5.16) is the optimal generalized solution. Since

$$
J[X^0(\cdot)] = \min_{\omega(\cdot)} J' \le \inf_{X(\cdot)} J,
$$

it suffices to establish the existence of a sequence of ordinary solutions that approximates $X^0(\cdot)$ at all points of continuity and satisfies condition (ii) of Definition 5.1.

The set of solutions of the system (5.9) is the closure of the set of solutions of the system (4.6) in the topology of uniform convergence. Therefore, there exists a sequence of controls $\{u_1^n(s), w_1^n(s)\}$ satisfying the constraints (4.7) and such that the sequence of corresponding solutions $\{Y^n(s), \eta^n(s)\}$ of (5.7) converges to $\{Y^0(s), \eta^0(s)\}$, uniformly on $[0, T_1]$. Arguing as in the proof of Lemma 2.5, we conclude that the function $\Gamma^n(t)$, that is, the inverse of $\eta^n(s)$, converges to the function $\Gamma^0(t)$ defined by (5.16) at all points of continuity.

Suppose first that $\Gamma^0(\cdot)$ is continuous at the point $t = T$. Then at all points of continuity of $\Gamma^0(\cdot)$ it holds that

$$\|Y^n(\Gamma^n(t)) - Y^0(\Gamma^0(t))\| \leq$$
$$\leq \|Y^n(\Gamma^n(t)) - Y^0(\Gamma^n(t))\| + \|Y^0(\Gamma^n(t)) - Y^0(\Gamma^0(t))\| \to 0,$$

where the first term on the right-hand side goes to zero due to the uniform convergence of $\{Y^n(\cdot)\}$ to $Y^0(\cdot)$ and the second term goes to zero due to the continuity of the function $Y^0(\cdot)$ and the convergence of $\Gamma^n(t)$ to $\Gamma^0(t)$.

Hence,

$$X^0(t) = \lim_n X^n(t) = \lim_n Y^n(\Gamma^n(t)) = Y^0(\Gamma^0(t))$$

at all points of continuity of $\Gamma^0(\cdot)$, as well as at all points of continuity of $X^0(\cdot)$. Moreover, thanks to the uniform convergence of Y^n to Y^0, we have the inequality

$$\lim_n \sup_{[0,T]} g_k(X^n(t), t) = \lim_n \sup_{[0,T]} g_k(Y^n(\Gamma^n(t)), \eta^n(\Gamma^n(t))) \leq$$
$$\leq \lim_n \sup_{[0,T_1]} g_k(Y^n(s), \eta^n(s)) \leq 0,$$

which shows that $X^0(\cdot)$ is an admissible generalized solution in the strong sense.

To complete the proof, we need to consider the case when $t = T$ is a discontinuity point of Γ^0. To define the solution, we continue $\{Y^0(s), \eta^0(s)\}$ beyond the point T_1 in a such way that $\dot{\eta}^0(s) > 0$ for $s > T_1$. Then $\Gamma^0(t)$ will be continuous for $t > T$ and

$$\lim_n Y^n(\Gamma^n(t)) = Y^0(\Gamma^0(t)) = X^0(t)$$

for $t > T$. So let us set

$$X^0(T) = X^0(T+) = \lim_{t \downarrow T} X^0(t) = \lim_{t \downarrow T} \lim_n X^n(t) = \lim_{t \downarrow T} \lim_n Y^n(\Gamma^n(t)) =$$
$$= \lim_{t \downarrow T} Y^0(\Gamma^0(t)) = \lim_{t \downarrow T} Y^0(\Gamma^0(t)) = Y^0(T_1).$$

Since $g_k(Y^0(T_1), \eta^0(T_1)) \leq 0$ and the corresponding sequence $\{Y^n, \eta^n\}$ approximates $\{Y^0, \eta^0\}$ uniformly for $\{s > T_1\}$, we conclude that the function $X^0(\cdot)$, extended beyond the point T by the value $Y^0(T_1)$, is an admissible generalized solution in the strong sense.

The last statement of the theorem is a direct consequence of Theorem 4.3. Indeed, if $X^0(\cdot)$ is an optimal generalized solution, then by Theorem 4.3 the auxiliary problem has a solution $\{Y^0(\cdot), \eta^0(\cdot)\}$ such that (5.15) and (5.16) hold and $J'[Y^0] = J[X^0]$. Suppose that Y^0 is not an optimal solution and there exists another admissible solution Y' such that $J'[Y^0] > J'[Y']$. In this case the above argument yields the existence of an admissible generalized solution X' such that $J[X'] = J'[Y'] < J[X^0]$, which contradicts the optimality of X^0 and completes the proof. □

Remark 5.7 The case where the function Γ^0 is discontinuous at $t = T$ is, generally speaking, not trivial. In this case, the situation is facilitated by the fact that, thanks to our assumptions (see Remark 5.2), the solution to the original systems may be continued for $t > T$ with preservation of the right continuity. If such a continuation is not admissible, a more complex construction is required. However, as will be seen later—for example, in the case of affine dependence on an unbounded control—this problem can be solved by the direct construction of a "correct" approximation of the control in the auxiliary problem (see Sect. 5.3, the proof of Lemma 5.2 and Remark 5.10). An analogous problem for a generalized solution in stochastic control problems was solved in [114]).

Theorem 5.2 establishes the equivalence of the original problem with unbounded controls to an ordinary problem, which can be treated by the standard methods of optimal control theory. However, for the moment this concerns only the case of strong solutions In the next subsection we consider the problem in the class of weak solution and formulate an equivalent ordinary problem.

5.2.3 Auxiliary Control Problem for a Weak Solution

By Theorem 4.3, any generalized solution, either in the weak or the strong sense, can be represented by means of the discontinuous time transformation. Hence, if X^0 is an optimal solution in the weak sense, then there exists a trajectory $\{Y^0, \eta^0\}$ of an auxiliary system such that

$$X^0(t) = Y^0(\Gamma^0(t)), \quad \text{where} \quad \Gamma^0(t) = \inf\{s : \eta^0(s) > t\}.$$

Since X^0 is the admissible solution, we have

$$g_k(X^0(t), t) = g_k(Y^0(\Gamma^0(t)), \eta^0(\Gamma^0(t)) \le 0$$

for all $t \in [0, T]$. This implies that

$$g_k(Y^0(s), \eta^0(s)) \le 0 \tag{5.18}$$

for all s for which there is a $t \in [0, T]$ such that $s = \Gamma^0(t)$. However, since $\Gamma^0(\cdot)$ may be discontinuous, inequality (5.18) is valid only if $t = \eta^0(s)$ is a point of continuity of Γ^0. Therefore, to guarantee the admissibility of $X^0(t) = Y^0(\Gamma^0(t))$ in the weak sense, it is sufficient to require that (5.18) holds only for $\dot{\eta}^0(s) > 0$.

This last constraint, which is of mixed kind (i.e., a constraint imposed simultaneously on the phase coordinates and on the control coordinates, as in [113]), can be expressed in the more convenient form

$$g_k^+(Y^0(s), \eta^0(s))\dot{\eta}^0(s) = 0, \quad \text{a. e. on} \quad [0, T_1] \tag{5.19}$$

for $k = 1, \ldots, N_3$, where $g^+ = \max\{g, 0\}$.

Remark 5.8 *Condition (5.19) is the relaxation of the stronger condition (5.18).*

Consider the auxiliary optimal control problem (5.9)–(5.14) with the constraint (5.19) instead of (5.13). Our aim is to prove that this *relaxed auxiliary problem* is equivalent to the optimization problem in the class of generalized solutions admissible in the weak sense.

Theorem 5.3 *Let the assumptions of Theorem 5.2 be in force. Let $X(\cdot)$ be a generalized solution that is admissible in the weak sense.*

Then there exists a trajectory of the auxiliary system $\{Y(\cdot), \eta(\cdot)\}$, defined on some interval $[0, T_1]$ and satisfying the constraint (5.19), such that

$$X(t) = Y(\Gamma(t)), \quad \textit{where} \quad \Gamma(t) = \inf\{s : \eta(s) > t\}.$$

If the function $X^0(\cdot)$ is an optimal generalized solution, then the corresponding trajectory $\{Y^0(\cdot), \eta^0(\cdot)\}$ is optimal in the relaxed auxiliary problem.

Proof The first assertion of the theorem follows directly from the above derivation of the constraint (5.19).

Let us prove the second assertion. As before in the proof of Theorem 5.2, suppose that the corresponding trajectory $\{Y^0(\cdot), \eta^0(\cdot)\}$ is not optimal; hence, there exists another admissible Y' such that $J'[Y^0] > J'[Y']$. If we prove that the appropriate solution $X'(\cdot)$ is admissible in the sense of Definition 5.2, then we end by contradicting the optimality of X^0, and thus complete the proof.

As follows from Theorem 4.3, $X'(t)$ is a right-continuous function and can be expressed as

$$X'(t) = Y'(\Gamma'(t)), \quad \text{where} \quad \Gamma'(t) = \inf\{s : \eta'(s) > t\}.$$

Moreover, (5.19) implies that

$$\int_0^{T_1} g_k^+(Y'(s), \eta'(s))\dot{\eta}'(s)ds = 0. \tag{5.20}$$

Performing here the change of variable $t = \eta'(s)$, we obtain

$$\int_0^{T_1} g_k^+(Y'(s), \eta'(s))\dot{\eta}'(s))ds = \int_0^{\eta'(T_1)} g_k^+(Y'(\Gamma'(t)), \eta'(\Gamma'(t)))dt =$$

$$= \int_0^T g_k^+(X'(t), t)dt = 0.$$

Therefore, $g_k^+(X^{'}(t), t) = 0$ a.e. on $[0, T]$, whence

$$g_k(X'(t), t) \leq 0 \quad \text{a.e. on} \quad [0, T].$$

But the function $X'(\cdot)$ is right-continuous, and g_k is continuous, so their composition is right-continuous and the last inequality holds for every $t \in [0, T]$. This confirms that $X'(\cdot)$ is an admissible solution in the weak sense, which, as indicated above, completes the proof of the theorem. □

Remark 5.9 The results presented above, which establish the equivalence of general problems with impulse and ordinary controls to certain problems with ordinary bounded controls are extremely important, since they are opening the way of using the available powerful apparatus of optimal control theory in a new class of problems. In particular, they play a key role in the derivation of optimality conditions (see Chap. 6).

In what follows we consider some extension of this approach to the special classes of so-called DCS with impulsive and ordinary controls and to hybrid systems with unilateral constraints.

5.3 Existence of Optimal Solutions for DCS with Ordinary and Impulse Controls

5.3.1 DCS with Ordinary and Impulse Controls

In the preceding sections, we considered the general problem of impulse control. Now we turn to a class of problems in which jumps of the solution are originally present in the system. This, on the one hand, is the case with the so-called discrete-continuous systems (DCS), in which an abrupt change in the state takes place even if there is no impulse control. An example of such problems is provided by the observation control in DCS [279, 293, 302–304]. Generally speaking, the results obtained in the previous subsection are not applicable when the state of the dynamic system under consideration undergoes jumps at some fixed time instants. Indeed, the requirement of continuity and event hat of Lipschitzianity with respect to the variable t in impulse control problems is of fundamental importance, because the coincidence of the instants of application of the impulse action with the points of singularity with respect to t does not allow an unambiguous determination of the response to an impulse control, and, as a result, leads to results that do not agree well with the physics of the problem (see Sect. 1.2.3 and Example 1.7.) Moreover, the method of discontinuous time change may itself not be applicable, since even the lack of Lipschitzianity in t results in an auxiliary problem in which time is a new phase variable, and without Lipschitzianity it is rather difficult to guarantee the existence of an optimal solution. In this section, we study a simple formulation of the problem for DCS with affine dependence on an unbounded control and a finite number of prescribed jump points. It is clear that this formulation covers the case of piecewise continuous (piecewise smooth) systems, where, for example, the right-hand side of the system experiences abrupt changes at points of discontinuity.

The problem is stated as follows: on a given finite time interval $[0, T]$ consider a process $\{X(t)\}$ described by the equations

$$\begin{cases} \dot{X}(t) = f(X(t), V(t), u(t), t) + B(X(t), V(t), u(t), t)w(t), \\ \dot{V}(t) = \|w(t)\|, \end{cases} \tag{5.21}$$

with two types of controls, $u \in U$ and $w \in K$, where $U \subset \mathbb{R}^k$ is a compact set and $K \subset \mathbb{R}^m$ is a convex closed cone. The control $w(t)$ obeys also the constraint

$$\int_0^T \|w(t)\|dt \leq M, \tag{5.22}$$

with a given constant $M < \infty$. We assume that the right-hand side of (5.21) satisfies Assumption 4.1, that is, the functions $f(X, V, u, t)$ and $B(X, V, u, t)$ are continuous with respect to all variables and satisfy the Lipschitz condition with respect to the variables (X, V).

The specific feature of our DCS is that there are given a finite set of instants $\{\tau_i,\ i = 0, \dots, N+1\}$ such that $0 = \tau_0 < \tau_1 < \cdots < \tau_i < \tau_{i+1} < \cdots < \tau_{N+1} = T$, at which the state of the DCS undergoes jumps, described by the relations

$$X(\tau_i) = \Phi(X(\tau_i-), i), \tag{5.23}$$

where the functions $\Phi(\cdot, i)$ are continuous in X.

Let the optimality criterion be

$$J[X(\cdot), u(\cdot), w(\cdot)] = \phi_0(X(0), X(T), V(T)) \to \min \tag{5.24}$$

with a continuous function ϕ_0 and subject to the terminal and phase constraints

$$h(X(0), X(T), V(T)) = 0, \quad S(X(0), X(T), V(T)) \leq 0, \tag{5.25}$$

and

$$G(X(t), V(t), t) \leq 0 \quad \text{for any} \quad t \in [0, T], \tag{5.26}$$

respectively. Here, as above, h, S, G are $\mathbb{R}^{N_1}$-, $\mathbb{R}^{N_2}$-, and $\mathbb{R}^{N_3}$-valued continuous functions, respectively, while (5.25) and (5.26) are to be understood componentwise.

5.3.2 *Auxiliary Control Problem for a DCS with Ordinary and Impulse Controls*

Following our general approach (see the preceding section) we introduce the auxiliary optimal control problem for the system

$$\begin{cases} \dot{Y}(s) = \alpha(s) f(Y(s), z(s), u_1(s), \eta(s)) + \\ \qquad + (1-\alpha(s)) B(Y(s), z(s), u_1(s), \eta(s)) e(t), \\ \dot{z}(s) = (1-\alpha(s)) \|e(s)\|, \\ \dot{\eta}(s) = \alpha(s), \end{cases} \tag{5.27}$$

with the initial conditions

$$Y(0) = X(0), \quad z(0) = 0, \quad \eta(0) = 0,$$

and controls

$$u_1(s) \in U, \quad \alpha(s) \in [0, 1], \quad e(s) \in K \cap \{\|e\| \leq 1\} \tag{5.28}$$

in some interval $[0, T_1]$, such that $T_1 \leq T + M$. The performance criterion to be minimized is

$$J'[Y(\cdot), u_1(\cdot), \alpha(\cdot), e(\cdot)] = \phi_0(Y(0), Y(T_1), z(T_1)) \to \min. \tag{5.29}$$

In addition, the problem is subject to the intermediate, terminal and phase constraints

$$Y(s_i) = \Phi(Y(s_i-), i), \quad \text{for some} \quad s_i : \eta(s_i) = \tau_i, \tag{5.30}$$

$$h(Y(0), Y(T_1), z(T_1)) = 0, \quad \eta(T_1) = T, \quad S(Y(0), Y(T_1), z(T_1)) \leq 0, \tag{5.31}$$

and

$$g(Y(s), z(s), \eta(s)) \leq 0 \quad \text{for all } s \in [0, T_1] \tag{5.32}$$

respectively. Here the set U, the cone K, the constant M, the functions f, B, ϕ_0, Φ, h, S, G and the times τ_i, $i = 0, \dots, N+1$ are the same as in the problem (5.21)–(5.26).

This auxiliary problem is the result of applying the time transformation in the original system (5.22) for the variables $\{X(t), V(t)\}$ with the "new time" $\eta(s)$. Via this transformation, the jump points τ_i go into the points s_i and the unbounded control variable $v(t)$ goes into the bounded control $\alpha(s)$.

As above, we define a concept of the generalized solution of the original system. Here and below we deal with admissible solutions in the strong sense.

Definition 5.4 A generalized solution $\{X(\cdot), V(\cdot)\}$ of the system (5.21) and (5.23) is said to be *admissible in the strong sense* if:

(i) $h(X(0-), X(T), V(T)) = 0, \quad S(X(0-), X(T), V(T)) \le 0$

(ii) $g(X(t), V(t), t) \le 0$ for all $t \in [0, T]$;

(iii) there exists a sequence of controls $\{u^n(\cdot), w^n(\cdot)\}$ satisfying the constraints (5.22), such that the corresponding sequence $\{X^n(\cdot), V^n(\cdot)\}$ of solutions of the problem (5.21) and (5.23) with the initial conditions

$$X^n(0) \to X(0-), \quad V^n(0) = 0$$

satisfies the relations:

$$\lim_n h(X^n(0), X^n(T), V^n(T)) = 0, \quad \lim_n S(X^n(0), X^n(T), V^n(T)) \le 0,$$

and

$$\limsup_n \sup_{[0,T]} g_k(X^n(t), V^n(t), t) \le 0, \qquad \text{for any } k = 1, \dots, N_3.$$

Finally, we define the concept of the optimal generalized solution.

Definition 5.5 An admissible generalized solution $X^0(\cdot)$ is called an *optimal generalized solution* if

$$\phi_0(X^0(0), X^0(T), V^0(T)) \le \phi_0(X'(0), X'(T), V'(T)).$$

for any admissible generalized solution $\{X'(\cdot), V'(\cdot)\}$.

The next theorem establishes the equivalence of the auxiliary problem and the original problem.

Theorem 5.4 *Let $\{Y^0(\cdot), z^0(\cdot), \eta^0(\cdot)\}$ be the optimal solution of the auxiliary problem (5.27)–(5.32). Then*

$$X^0(t) = Y^0(\Gamma^0(t)), \quad V^0(t) = z^0(\Gamma^0(t)), \tag{5.33}$$

where

$$\Gamma^0(t) = \inf\{s : \eta^0(s) > t\}, \quad \Gamma^0(T) = T_1, \tag{5.34}$$

is the optimal generalized solution of the original problem (5.22)–(5.26).

First we state and prove two lemmas.

Lemma 5.1 *Suppose the functions* $\{X(t), V(t), u(t), w(t)\}$ *satisfy the conditions (5.21) and (5.23) with some measurable controls* $\{u(t), w(t)\}$ *obeying (5.22).*

Then there exists a solution $\{Y(s), z(s), \eta(s)\}$ *of the system (5.27) with the intermediate conditions (5.29), defined on some interval* $[0, T_1]$, $T_1 \leq T + M$, *with some controls* $\{u_1(s), \alpha(s), e(s)\}$ *obeying (5.28), such that*

$$X(t) = Y(\Gamma(t)), \quad V(t) = z(\Gamma(t)), \quad \textit{where } \Gamma(t) = \inf\{s : \eta(s) > t\}. \tag{5.35}$$

Proof of Lemma 5.1 Let

$$\Gamma(t) = t + \int_0^t \|w(s)\|ds.$$

This function is monotonically increasing and its inverse function $\eta(s)$, given by

$$\eta(s) = \inf\{t : \Gamma(t) \geq s\},$$

is defined on the interval $[0, T_1]$, where $T_1 = \Gamma(T) \leq T + M$. The functions $\Gamma(t)$ and $\eta(s)$ are defined by (5.35) and satisfy the inverse functions relations

$$\eta(\Gamma(t)) = t, \quad \Gamma(\eta(s)) = s, \tag{5.36}$$

for all $t \in [0, T]$ and all $s \in [0, T_1]$. Moreover, the function $\eta(s)$ is increasing and absolutely continuous, and due to the properties of inverse functions [322], it is differentiable almost everywhere in $[0, T_1]$, and satisfies

$$0 < \dot{\eta}(s) = \left.\frac{1}{\dot{\Gamma}(t)}\right|_{t=\eta(s)} = \frac{1}{1 + \|w(\eta(s)\|)} \leq 1. \tag{5.37}$$

In view of the monotonicity of $\eta(s)$, the instants s_i such that $\eta(s_i) = \tau_i$ are uniquely determined.

Define the controls $\{u_1(\cdot), \alpha(\cdot), e(\cdot)\}$ by the formulas

$$u_1(s) = u(\eta(s)), \quad \alpha(s) = \dot{\eta}(s), \quad e(s) = w(\eta(s))\|w(\eta(s))\|^+, \tag{5.38}$$

where the superscript $^+$ stands for pseudoinverse (i.e., $a^+ = a^{-1}$ if $0^+ = 0$). Arguing as in the proof of Theorem 4.1, we see that these controls are Lebesgue-measurable functions and satisfy the constraint (5.28). Next, consider the functions

$$Y(s) = X(\eta(s)), \quad z(s) = V(\eta(s)). \tag{5.39}$$

Since,

$$\dot{Y}(s) = \dot{X}(\eta(s))\dot{\eta}(s), \quad \dot{z}(s) = \dot{V}(\eta(s))\dot{\eta}(s)$$

and

$$\|w(\eta(s))\| \, w(\eta(s)) \, \|w(\eta(s))\|^{+} = w(\eta(s)),$$

the functions $\{Y(s), z(s)\}$ satisfy (5.21) and (5.23) with the controls (5.38).

Substituting the relations (5.36) in (5.39), we obtain

$$Y(\Gamma(t)) = X(\eta(\Gamma(t))) = X(t), \quad z(\Gamma(t)) = V(\eta(\Gamma(t))) = V(t),$$

which gives (5.35) and completes the proof. □

Lemma 5.2 *Suppose the functions* $\{Y(s), z(s), \eta(s)\}$ *satisfy in some interval* $[0, T_1]$, *where* $T_1 \leq T + M$, *the Eq. (5.27) with the intermediate conditions (5.29) and with some control* $\{u_1(\cdot), \alpha(\cdot), e(\cdot)\}$ *satisfying (5.28). Suppose also that*

$$\eta(T_1) = T.$$

Then the pair $\{X(t), V(t)\}$ *defined by the formulas (5.35) is the generalized solution of the Eq. (5.21) with the intermediate conditions (5.23).*

Proof of Lemma 5.2 It suffices to show that there exists a sequence of controls $\{u^n(t), w^n(t)\}$ satisfying (5.22) and such that the corresponding sequence of solutions $\{X^n(t), V^n(t)\}$ of the system (5.21) and (5.23) converges to $X(t) = Y(\Gamma(t))$, $V(t) = z(\Gamma(t))$ at all points of continuity of $\Gamma(t)$, where the function $\Gamma(\cdot)$ is defined in (5.35).

First suppose that $\alpha(s) > 0$ a.e. on $[0, T_1]$. That is to say, $\eta(s)$ is a monotonically increasing function, and one can define its inverse function $\Gamma(t)$ as in (5.35). In the same manner as before, Γ and η satisfy (5.36), and therefore, the instants $s_i \in [0, T_1]$ such that $\eta(s_i) = \tau_i$ are uniquely determined.

The function $\Gamma(\cdot)$ is absolutely continuous and has the derivative

$$\dot{\Gamma}(t) = \frac{1}{\alpha(\Gamma(t))} \geq 1 \quad \text{a.e. in} \quad [0, T].$$

Let us introduce the controls

$$w(t) = \frac{1 - \alpha(\Gamma(t))}{\alpha(\Gamma(t))} e(\Gamma(t)) \quad \text{and} \quad u(t) = u_1(\Gamma(t)). \tag{5.40}$$

These controls are Lebesgue-measurable functions thanks to the absolute continuity of $\Gamma(\cdot)$ [322], satisfy the constraints $w(t) \in K$ and $u(t) \in U$, and also

$$\begin{aligned}\int_0^T \|w(t)\|dt &= \int_0^T \frac{1-\alpha(\Gamma(\tau))}{\alpha(\Gamma(\tau))}\|e(\Gamma(\tau))\|d\tau = \\ &= \int_0^T (1-\alpha(\Gamma(\tau)))\|e(\Gamma(\tau))\|\dot{\Gamma}(\tau)d\tau = \\ &= \int_0^T (1-\alpha(\Gamma(\tau)))\|e(\Gamma(\tau))\|d\Gamma(\tau) \le \\ &\le \int_0^{\Gamma(T)} (1-\alpha(s))ds = \Gamma(T) - \eta(\Gamma(T)) = T_1 - T \le M.\end{aligned}$$

Hence, the controls $w(t)$ and $u(t)$ satisfy (5.22) and the pair $\{X(t), V(t)\}$ satisfies in the intervals (τ_i, τ_{i+1}) the equations

$$\begin{cases} \dot{X}(t) = \dot{y}(\Gamma(t))\dot{\Gamma}(t) = \dfrac{\dot{y}(\Gamma(t))}{\alpha(\Gamma(t))} = f(y(\Gamma(t)), z(\Gamma(t)), u_1(\Gamma(t)), \eta(\Gamma(t))) + \\ \qquad + B(y(\Gamma(t)), z(\Gamma(t)), u_1(\Gamma(t)), \eta(\Gamma(t)))\dfrac{1-\alpha(\Gamma(t))}{\alpha(\Gamma(t))}e(\Gamma(t)), \\ \dot{V}(t) = \dfrac{1-\alpha(\Gamma(t))}{\alpha(\Gamma(t))}\|e(\Gamma(t))\| = \|w(t)\|. \end{cases}$$

Thus, the pair $\{X(t), V(t)\}$ satisfies the Eq. (5.21) with the controls (5.40). Moreover, since $Y(s_i) = \Phi(Y(s_i-), i)$ and $s_i = \Gamma(\tau_i)$, we have

$$X(\tau_i) = Y(\Gamma(\tau_i)) = Y(s_i) = \Phi(Y(s_i-), i) = \Phi(Y(\Gamma(\tau_i)-), i) = \Phi(X(\tau_i-), i).$$

Hence, $X(\cdot)$ satisfies the system (5.21) and the intermediate conditions (5.23), and so $X(\cdot)$ is an ordinary solution of the original system.

In a general case, when $\alpha(s) \ge 0$, we construct a sequence $\{\alpha^n(\cdot)\}$ such that $\alpha^n(s) > 0$ a.e. on $[0, T_1]$, $\alpha^n(s)$ converges to $\alpha(s)$ uniformly in $[0, T_1]$, and

$$\int_0^{s_i} \alpha^n(s)ds = \tau_i, \qquad \int_0^{T_1} \alpha^n(s)ds = T. \tag{5.41}$$

Namely, we consider the following Lebesgue-measurable subsets of the interval $[0, T_1]$:

$$\mathscr{N} = \{s : \alpha(s) = 0\}, \quad \text{and} \quad \mathscr{N}_i = \mathscr{N} \cap [s_{i-1}, s_i], \quad i = 1, \dots, N+1,$$

where, by definition, $s_0 = 0$ and $s_{N+1} = T_1$, by definition. For a set $\mathscr{A}$, denote by $\mu(\mathscr{A})$ its Lebesgue measure and by $\mathbf{I}\{\mathscr{A}\}(s)$ its indicator function.

Now put

$$\alpha^n(s) = \left(1 - \frac{\mu(\mathcal{N}_i)}{n(\tau_i - \tau_{i-1})}\right)\alpha(s) + \frac{1}{n}\mathbf{I}\{\mathcal{N}_i\}(s) \quad \text{for } s \in [s_{i-1}, s_i], \\ i = 1, \dots, N+1. \tag{5.42}$$

The sequence of functions $\{\alpha^n(s)\}$ converges to $\alpha(s)$ uniformly on $[0, T_1]$ and satisfies the inequality

$$0 < \alpha^n(s) \leq 1,$$

if n is sufficiently large.

Note that

$$\int_{s_{i-1}}^{s_i} \alpha(s)ds = \eta(s_i) - \eta(s_{i-1}) = \tau_i - \tau_{i-1} \quad \text{and} \quad \int_{s_{i-1}}^{s_i} \mathbf{I}\{\mathcal{N}_i\}(s)ds = \mu(\mathcal{N}_i).$$

Then using the expression (5.42) we obtain

$$\int_{s_{i-1}}^{s_i} \alpha^n(s)ds = \int_{s_{i-1}}^{s_i} \left(1 - \frac{\mu(\mathcal{N}_i)}{n(\tau_i - \tau_{i-1})}\right)\alpha(s)ds + \\ + \frac{1}{n}\int_{s_{i-1}}^{s_i} \mathbf{I}\{\mathcal{N}_i\}(s)ds = \int_{s_{i-1}}^{s_i} \alpha(s)ds = \tau_i - \tau_{i-1}, \tag{5.43}$$

which proves (5.41).

Now consider the sequence $\{Y^n(s), z^n(s), \eta^n(s)\}$ of solutions of (5.27) and (5.29) that correspond to the sequence of controls $\{u_1(s), \alpha^n(s), e(s)\}$ defined on $[0, T_1]$. As it follows from (5.41),

$$s_i^n = \inf\{s : \eta^n(s) = \tau_i\} = s_i,$$

so the jump points of any of the functions $Y^n(s)$ coincide with the jump points of the function $Y(s)$.

Let us prove that the sequence of functions $\{Y^n(s), z^n(s), \eta^n(s)\}$ converges to the functions $\{Y(s), z(s), \eta(s)\}$ uniformly on $[0, T_1]$.

Thanks to the uniform convergence of α^n to α, this assertion is obviously true for the variables z and η. Therefore, we need to prove only the uniform convergence of Y^n to Y.

First, we claim that the convergence on the half-interval $[0, s_1)$ follows from the linearity of the system (5.27) with respect to α and the uniform convergence of $\{\alpha^n(s)\}$ to $\alpha(s)$. This can be established directly by using the Lipschitz properties of f, and B and the standard argument based on the Gronwall-Bellman inequality

[237]. Alternatively, one can simply use the theorem of continuous dependence of solutions of differential equations on the right-hand side and the initial conditions ([365], Theorem 7 §1).

Further, the continuity of the function Φ in (5.28) implies the convergence of $Y^n(s_1) = \Phi(Y^n(s_i-), i)$ to $Y(s_1) = \Phi(Y(s_i-), i)$ and, as a consequence, the uniform convergence in $[s_1, s_2)$, and so on. As a result, since the number of intervals is finite, we obtain the uniform convergence on the whole interval $[0, T_1]$.

Recall that from (5.42) it follows that $\alpha^n(s) > 0$ for sufficiently large n. Therefore, for such values of n, the function $\Gamma^n(\cdot)$, defined on the interval $[0, T]$ as the inverse of $\eta^n(\cdot)$, is absolutely continuous. Hence, one can introduce the sequence of Lebesgue-measurable controls

$$w^n(t) = \frac{1-\alpha^n(\Gamma^n(t))}{\alpha^n(\Gamma^n(t))}e(\Gamma(t)), \quad u^n(t) = u_1(\Gamma^n(t)), \tag{5.44}$$

to which corresponds a sequence of solutions $\{X^n(t), V^n(t)\}$ of (5.21), namely

$$X^n(t) = Y^n(\Gamma^n(t)), \quad V^n(t) = z^n(\Gamma^n(t)),$$

such that

$$X^n(\tau_i) = Y^n(\Gamma^n(\tau_i)) = Y^n(s_i) = \Phi(Y^n(s_i-), i) = \Phi(X^n(\tau_i-), i).$$

To complete the proof we recall that, thanks to the uniform convergence of η^n to η, the sequence $\Gamma^n(t)$ converges to $\Gamma(t) = \inf\{s : \eta(s) > t\}$ at all points of continuity (see Lemma 2.5). Arguing as in the proof of Theorem 4.3 we conclude that the sequence X^n converges to the function X specified by (5.35) at all points of continuity. □

Proof of the Theorem As follows from Lemma 5.1,

$$\inf_{u,w} J[X, V] \geq \inf_{u_1,\alpha,e} J'[Y, z, T_1],$$

and if $\{Y^0, z^0, \eta^0\}$ minimizes the performance criterion and the generalized solution $\{X^0, V^0\}$ is such that

$$J'[Y^0, z^0, T_1] = J[X^0, V^0], \tag{5.45}$$

then $\{X^0, V^0\}$ provides the infimum of the performance criterion (5.24). Since $\{Y^0, z^0, \eta^0\}$ minimizes J', Lemma 5.2 shows that the pair $\{X^0, V^0\}$ defined by (5.33) and (5.34) is a generalized solution of the problem (5.21) and (5.23). It is readily verified that $\{X^0, V^0\}$ satisfies (5.45), so to complete the proof it remains only to verify that this generalized solution is admissible in the sense of Definition 5.4. By Lemmas 5.1 and 5.2, there is a sequence of admissible controls $\{u^n, w^n\}$ such that the corresponding sequence of solutions $\{X^n, V^n\}$ satisfies

$$\begin{aligned} X^n(t) &= Y^n(\Gamma^n(t)) \to X^0(t) = Y^0(\Gamma^0(t)), \\ V^n(t) &= z^n(\Gamma^n(t)) \to V^0(t) = x^0(\Gamma^0(t)), \end{aligned} \tag{5.46}$$

at all points of continuity. By virtue of (5.41),

$$\Gamma^n(0-) = 0, \quad \Gamma^n(T) = \Gamma^0(T) = T_1, \tag{5.47}$$

so we see that

$$X^0(0-) = Y^0(0), \quad X^0(T) = Y^0(T_1), \quad V^0(T) = z^0(T_1)$$

satisfy the terminal conditions **(i)** in Definition 5.4.

The representation (5.33) and (5.34) implies that the generalized solution $\{X^0, V^0\}$ satisfies also the phase constraint **(ii)**.

To verify **(iii)** one can use the relations (5.46) and the uniform convergence of $\{Y^n, z^n, \eta^n\}$ to $\{Y^0, z^0, \eta^0\}$. The convergence at the terminal points 0 and T is evident thanks to (5.47). Hence, due to the uniform convergence and continuity of $G(x, v, t)$, we have

$$\begin{aligned} \lim_n \sup_{[0,T]} g_k(X^n(t), V^n(t), t) &= \lim_n \sup_{[0,T]} g_k(Y^n(\Gamma^n(t), z^n(\Gamma^n(t), \eta^n(\Gamma^n(t)) \leq \\ &\leq \lim_n \sup_{[0,T_1]} g_k(Y^n(s), z^n(s), \eta^n(s)) \leq 0, \end{aligned}$$

for any $k = 1, \ldots, N_3$, as we needed to show.

Therefore, $\{X^0, V^0\}$ is the optimal generalized solution. □

Remark 5.10 Note that one of the essential points in the proof of equivalence is the choice of "right" approximations of the control α and correspondingly of the function η for which the relations (5.41) are satisfied. For an arbitrary choice of the approximation one runs into the issue of "terminal jumps" [162], connected to the indeterminacy of the right limit and left limit at the point $t = T$ and $t = 0$, respectively. The possibility of continuing solutions outside of the interval $[0, T]$ is generally a subject of discussion in each particular case, so it is very important that the class of problems considered here one can be dealt with without this assumption.

Remark 5.11 The proof of the theorem was done without assumptions regarding the convexity of the right-hand side of the auxiliary system, since it is only necessary to ensure the existence of optimal control, but not in the proof of the equivalence. Of course, if there is no optimal control in the auxiliary problem, then it is necessary to apply the known convexification procedures on the right-hand side, which in the present case are rather standard even in the presence of intermediate conditions [237, 446]. This is because the problem with a finite set of control-independent intermediate conditions can always be reduced to a standard control problem of a larger dimension on a fixed time interval and involving only terminal conditions. Relevant examples can be found in [251, 438, 439].

5.3.3 Description of the Optimal Generalized Solution in DCS with the Aid of a Differential Equation with Measure

We are now ready to derive a differential equation with measure for an optimal generalized solution. By Lemma 5.2, any admissible generalized solution can be represented by applying the discontinuous time transformation (5.35) to a suitable solution $\{Y(s), z(s), \eta(s)\}$ of the auxiliary system that satisfies on some interval $[0, T_1]$, $T_1 \leq T + M$, the Eq. (5.27) with the intermediate conditions (5.29) and with some controls $\{u_1(s), \alpha(s), e(s)\}$ obeying (5.28). Therefore, the functions $\{Y(s), z(s), \eta(s)\}$ satisfy the system of equations

$$\begin{cases} Y(s) = X(0) + \displaystyle\int_0^s \alpha(\tau) f(Y(\tau), z(\tau), u_1(\tau), \eta(\tau)) d\tau + \\ \quad + \displaystyle\int_0^s (1 - \alpha(\tau)) B(Y(\tau), z(\tau), u_1(\tau), \eta(\tau)) e(\tau) d\tau + \sum_{s_i \leq s} \Delta Y(s_i), \\ z(s) = \displaystyle\int_0^s (1 - \alpha(\tau)) \|e(\tau)\| d\tau, \\ \eta(s) = \displaystyle\int_0^s \alpha(\tau) d\tau, \end{cases} \tag{5.48}$$

with the intermediate conditions

$$\Delta Y(s_i) = \Phi(Y(s_i-), i) - Y(s_i-),$$

where $\eta(s_i) = \tau_i$.

As in the proof of Theorem 4.7, we introduce the regular non-negative scalar measure $\mathscr{V}(dt)$ with the distribution function given by

$$V(t) = \mathscr{V}([0, t]) = z(\Gamma(t)) = \int_0^{\Gamma(t)} (1 - \alpha(\tau)) \|e(\tau)\| d\tau,$$

which admits the Lebesgue decomposition into its continuous and discrete parts,

$$V(t) = V^c(t) + \sum_{\tau \leq t} \Delta V(\tau), \quad \text{with } \Delta V(\tau) = \mathscr{V}(\{\tau\})$$

Next, we define the controls

$$u(t) = u_1(\Gamma(t)), \quad w(t) = e(\Gamma(t))$$

and by applying the same change-of-variable argument as in the proof of Theorem 4.7, we conclude that $X(t) = Y(\Gamma(t))$ satisfies the equation

$$
\begin{aligned}
X(t) = X(0-) + \int_0^t f(X(\tau), V(\tau), u(\tau), \tau)d\tau + \\
+ \int_0^t B(X(\tau), V(\tau), u(\tau), \tau)w(\tau)dV^c(\tau) + \\
+ \sum_{\tau \le t} \int_{\Gamma(\tau-)}^{\Gamma(\tau)} B(Y(\tau), z(\tau), u_1(\tau), \tau)e(\tau)d\tau + \sum_{s_i \le \Gamma(t)} \Delta Y(s_i),
\end{aligned}
\tag{5.49}
$$

where the last two terms describe the jumps of $X(\cdot)$.

Let τ be a jump point of $X(\cdot)$ and consider three possible cases:

(i) $\tau \in \{\tau_i,\ i = 1, \dots, N\}$ and $\tau \notin D_V$;
(ii) $\tau \in D_V$ and $\tau \notin \{\tau_i,\ i = 1, \dots, N\}$;
(iii) $\tau \in D_V \cap \{\tau_i,\ i = 1, \dots, N\}$,

where $D_V = \{t : \Delta V(t) > 0\}$.

In case (i), $s_i = \Gamma(\tau_i)$ and $Y(s_i-) = Y(\Gamma(\tau_i-)) = X(\tau_i-)$, so

$$\Delta X(\tau_i) = \Phi(X(\tau_i-), i) - X(\tau_i-). \tag{5.50}$$

Case (ii) corresponds to the situation in Theorem 4.7; therefore, there exist Lebesgue-measurable controls $\{u^\tau(s), e^\tau(s)\}$, defined on the interval $[0, \Delta V(\tau)]$ and satisfying the constraints

$$u^\tau(s) \in U, \quad e^\tau(s) \in K \cap \{\|e\| \le 1\},$$

such that

$$\Delta X(\tau) = y_\tau(\Delta V(\tau)) - X(\tau-), \tag{5.51}$$

where $\{y_\tau(s), z_\tau(s)\}$ is the solution of the system the differential equations

$$
\begin{cases}
\dot{y}_\tau(s) = B(y_\tau(s), z_\tau(s), u^\tau(s), \tau)e^\tau(s), \\
\dot{z}_\tau(s) = 1,
\end{cases}
$$

with the initial conditions $y_\tau(0) = X(\tau-)$, $z_\tau(0) = V(\tau-)$.

Finally, case (iii) is specific precisely to DCS and combines the peculiarities of cases (i) and (ii).

Indeed, if $\tau \in D_V \cap \{\tau_i,\ i = 1, \dots, N\}$, then for appropriate points s_i satisfying $\eta(s_i) = \tau_i$, the fact that τ_i is a point of discontinuity of $\Gamma(\cdot)$ implies that $s_i \in [\Gamma(\tau_i-), \Gamma(\tau_i)]$. Therefore,

$$\begin{aligned}\Delta X(\tau_i) &= Y(\Gamma(\tau_i)) - Y(\Gamma(\tau_i-) = \\ &= \int_{\Gamma(\tau_i-)}^{\Gamma(\tau_i)} B(Y(\tau), z(\tau), u_1(\tau), \tau)e(\tau)d\tau + \Delta Y(s_i) = \\ &= \int_{\Gamma(\tau_i-)}^{s_i} B(Y(\tau), z(\tau), u_1(\tau), \tau)e(\tau)d\tau + \Delta Y(s_i) + \\ &+ \int_{s_i}^{\Gamma(\tau_i)} B(Y(\tau), z(\tau), u_1(\tau), \tau)e(\tau)d\tau = \\ &= \Delta^1 X(\tau_i) + \Delta^2 X(\tau_i) + \Delta^3 X(\tau_i).\end{aligned} \tag{5.52}$$

Define

$$\Delta V^1(\tau_i) = s_i - V(\tau_i-) \quad \text{and} \quad \Delta V^2(\tau_i) = V(\tau_i) - s_i.$$

Arguing as in the proof of Theorem 4.7 for each of the half-intervals $[\Gamma(\tau_i-), s_i)$ and $(s_i, \Gamma(\tau_i)]$, we deduce that the summands in the expression of the jump are as follows:

There exists a pair of Lebesgue-measurable controls

$$\{u^{\tau_i,1}(s), e^{\tau_i,1}(s)\}, \quad \{u^{\tau_i,2}(s), e^{\tau_i,2}(s)\},$$

defined on the intervals $[0, \Delta V^1(\tau_i)]$ and $[0, \Delta V^2(\tau_i)]$, respectively, and obeying the constraints

$$u^{\tau_i,j}(s) \in U, \quad e^{\tau_i,j}(s) \in K \cap \{\|e\| \leq 1\}, \quad j = 1, 2,$$

such that

$$\Delta^1 X(\tau_i) = y^1_{\tau_i}(\Delta V^1(\tau_i)) - X(\tau_i-), \tag{5.53}$$

where $y^1_{\tau_i}(s)$ is a solution of the system of differential equations

$$\begin{cases} \dot{y}^1_{\tau_i}(s) = B(y^1_{\tau_i}(s), z^1_{\tau_i}(s), u^{\tau_i,1}(s), \tau_i)e^{\tau_i,1}(s), \\ \dot{z}^1_{\tau}(s) = 1, \end{cases}$$

with the initial conditions $y^1_{\tau_i}(0) = X(\tau_i-)$, $z^1_{\tau_i}(0) = V(\tau_i-)$.

The second summand is given by

$$\Delta^2 X(\tau_i) = \Phi(X^1(\tau_i), i) - X^1(\tau_i), \tag{5.54}$$

Fig. 5.3 Jump of generalized solution at the point $t = \tau_i$

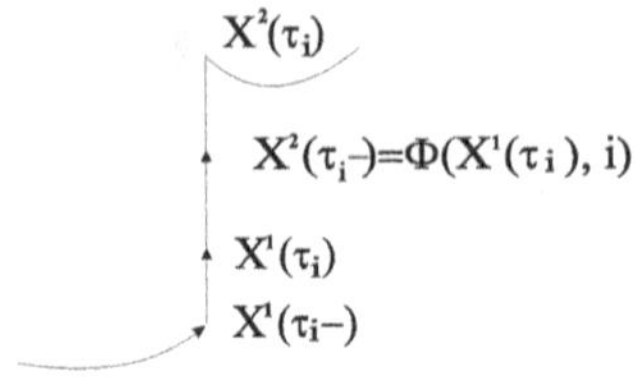

where

$$X^1(\tau_i) = X^1(\tau_i-) + \Delta^1 X(\tau_i)$$

Finally, the third summand is given by

$$\Delta^3 X(\tau_i) = y^2_{\tau_i}(\Delta V^2(\tau_i)) - X^2(\tau_i), \tag{5.55}$$

where $y^2_{\tau_i}(s)$ is a solution of the system of differential equations

$$\begin{cases} \dot{y}^2_{\tau_i}(s) = B(y^2_{\tau_i}(s), z^2_{\tau_i}(s), u^{\tau_i,2}(s), \tau_i)e^{\tau_i,2}(s), \\ \dot{z}^2_{\tau}(s) = 1, \end{cases}$$

with the initial conditions $y^2_{\tau_i}(0) = X^2(\tau_i), \quad z^2_{\tau_i}(0) = V(\tau_i-) + \Delta V^1(\tau_i)$, where $X^2(\tau_i) = X^1(\tau_i) + \Delta^2 X(\tau_i)$.

Thus, we produced a formula for generalized solutions that extends Theorem 4.7 to the discrete-continuous case. As follows from the relations (5.53)–(5.55), if the moment of application on an impulse control coincides with one of the jump points τ_i of the system itself, then possible jump of a generalized solution $X(\cdot)$ consists of three components, as illustrated in Fig. 5.3.

Each components of the jump has a specific meaning. Namely, $\Delta^1 X(\tau_i)$ is determined by the impulse control that precedes the jump itself, so that the sequence of ordinary controls that approximate this impulse control are localized to the left of the point τ_i. The second component, $\Delta^2 X(\tau_i)$, corresponds to the jump of the discrete-continuous system itself and does not depend on controls. And finally, the third component $\Delta^3 X(\tau_i)$ is determined by the impulse control that is localized just after the instant τ_i. The presence of such a complicated decomposition of a jump is typical for the so-called hybrid systems, which became a focus of interest during the recent years. In the field of observation control problems, the three types of jumps described above correspond to various possible localizations of the discrete observations relative to the location of the control instants, see [276]. For example, in the problem of observation control this situation corresponds to the possibility of observing the system's state immediately before the jump and/or immediately after the jump [276, 304, 305].

Remark 5.12 Here, as in Sect. 3.4, we see the same situation as with two different kinds of impulses. Individual impulse actions from an infinite set of impulses can coincide with impulse actions from a finite set. In this case, as above, we will have a jump generated by three neighboring impulses: the average one from a finite set of impulse actions and two adjacent impulses from an infinite set.

5.4 Optimal Generalized Solutions in Nonlinear Hybrid Systems

5.4.1 Hybrid Systems with Unilateral Constraints

Hybrid systems are frequently treated as systems characterized by a continuous and discrete behaviour. The motion of such systems can be divided into regular and singular parts, i.e., continuous and jump-like motions, respectively. These systems are typical for various mechanical applications, where discrete-continuous modes of motions may arise due to shocks and friction. There has been significant progress in this area, including the development of a rigorous mathematical framework for the description of these systems and preliminary formulations of procedures for the synthesis of control laws for them. However, their common mathematical feature is the presence of singularities, which manifest themselves as discontinuity and non-smoothness of the evolution: jumps in the dimension of the system, lack of continuous dependence on the initial conditions, and non-uniqueness of solutions of the equations of motion (see, for example, [70, 194]).

A typical approach to the description of hybrid systems is to consider that the equations of the system along with phase variables may change when the trajectory of the system hits a certain manifold. It is generally assumed that such changes occur instantly, i.e., the trajectories of such systems experience breaks (switching) of all or part of the phase variables on the manifold in question. Such models work quite well if the number of time instants when the switching manifold is reached is bounded, i.e., a "finite clearance" exists always between switching times. In that case, the trajectory is constructed by joining (*concatenating*) patches of smooth behavior. However, even in the absence of an additional control the motions of such systems often fail to satisfy regularity requirements such as uniqueness and continuous dependence of the solution on the parameters and initial conditions [365]. As shown by examples in Introduction, these difficulties only increase in the presence of controls and often result in the actual absence of optimal controls and solutions. One of the most interesting areas in the theory of dynamical systems, where this problem is most pressing, is that of mechanical systems with impact and friction (see, for example, [70, 194], Sect. 1.3.2, and Chap. 8 of this monograph).

It seems reasonable to transfer the approaches of the theory of impulse control to the above-mentioned class of systems. What both classes share is that, as with impulse-controlled systems, discontinuities of phase coordinates are nothing but a way of describing "fast" movements, which, generally speaking, is not one-to-one, as manifested in the case of non-robust DCS. However, one should emphasize that there is a significant difference between the impulsively-controlled systems where the controls are "external" actions, and fast motions in dynamical systems that are inherent in the systems's dynamics. Specifically, the theory of impulse control deals with programmed controls, whereas in the case of hybrid systems one deals with actions in the form of impulse feedback. It is precisely these aspects that introduce the main difficulties, because the theory of systems with impulse feedbacks faces the same set of problems as the theory of dynamical systems with discontinuous right-hand side.

Let us consider an example of a simple controlled hybrid system in which jumps of phase variables occur on the boundary of a certain set (at a singular point). We assume that the set of solutions is parametrized by a parameter $\mu > 0$, such that when μ tends to infinity the processes is accelerated in the vicinity of the singular point, so that in the limit as $\mu \uparrow \infty$ the trajectory becomes discontinuous. This approach is motivated by analogies with the mechanics of systems with impact, where the role of the parameter μ is played by the elasticity coefficient, so that when $\mu \to \infty$ the velocity coordinates converge to discontinuous functions [17, 70, 179–185, 215, 346–353].

So, consider the evolution of a discrete-continuous dynamical system whose behavior is described in some interval $[0, T]$ by the variable $X(t) \in \mathbb{R}^n$, which satisfies the differential equation

$$\dot{X}(t) = F(X(t), u(t)), \tag{5.56}$$

with a given initial condition $X(0) = x_0 \in \mathbb{R}^n$ and the following intermediate conditions:

$$X(\tau_i) = X(\tau_i-) + \Psi(X(\tau_i-)), \tag{5.57}$$

which are prescribed for some sequence of instants $\{\tau_i,\ i = 0, \dots, N\},\ N \le \infty$ that obey the recurrence relations

$$\begin{aligned} \tau_0 &= 0, \\ \tau_i &= \begin{cases} \inf\{\tau_{i-1} < t \le T : G(X(t-) = 0)\}, \\ \infty, \quad \text{if the above set is empty.} \end{cases} \end{aligned} \tag{5.58}$$

In (5.57), $X(\tau_i-) = \lim\limits_{t \uparrow \tau_i} X(t)$ and $\{\tau_i,\ i = 0, \dots, N\}$, $N \le \infty$, is the sequence of instants at which the state of the system changes in a discontinuous manner.

Thus, the state of the system varies continuously in the half-intervals $[0, \tau_1), \dots,$ $[\tau_{i-1}, \tau_i), \dots, [\tau_N, T]$, and undergoes a sudden change at each instant τ_i. According to the relation (5.57), the value of each jump depends on the state preceding the jump.

We assume that the control variable u in (5.56) is such that

$$u \in U \subset \mathbb{R}^m, \tag{5.59}$$

where U is a compact set, and that the function $F(X, u)$ is continuous with respect to all variables and continuously differentiable with respect to X. To ensure that the solution of (5.56) is extendable to the right, we need some additional assumptions on the functions $\Psi(X)$ and $G(X)$. Specifically, we assume that the change $X(\tau_i) = X(\tau_i-) + \Psi(X(\tau_i-))$ is the result of the action of the shift operator along the trajectories of the differential equation

$$\dot{y}(s) = B(y(s)), \quad s \in [0, \infty), \tag{5.60}$$

with the initial condition $y(0) = X(\tau_i-)$.

Therefore, if $\Phi(x, s)$ denotes the general solution of Eq. (5.60) with the initial condition $y(0) = x$, then

$$X(\tau_i) = X(\tau_i-) + \Psi(X(\tau_i-)) = \Phi(X(\tau_i-), s^*(X(\tau_i-))), \tag{5.61}$$

where

$$s^*(X(\tau_i)) = \inf\{s > 0 : G(\Phi(X(\tau_i-), s) < 0\}, \tag{5.62}$$

and in the interval $(0, s^*(X(\tau_i))$ it holds that $G(\Phi(X(\tau_i-), s) > 0$.

We assume also that $B(y)$ is continuously differentiable with respect to y.

A typical trajectory of the hybrid system is shown in Fig. 5.4, where τ is the instant at which the trajectory hits the boundary of the set $G(X) \leq 0$. The solid line between points A and B shows the visible motion after hitting the obstacle at the point A, while the real "fast" motion occurs along the dotted line. As a rationale for the proposed model of hybrid system, we introduce the following parametrization of the system (5.56)–(5.58). Consider the family $X^\varkappa(\cdot)$ of solutions of the system of equations

$$\dot{X}^\varkappa(t) = F(X^\varkappa(t), u(t)) + \varkappa\, B(X^\varkappa(t))\mathbf{I}\{t : G(X^\varkappa(t)) > 0\}, \tag{5.63}$$

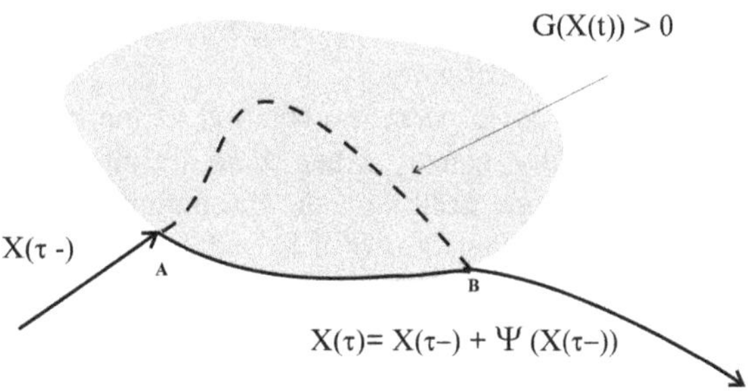

Fig. 5.4 Trajectory of a hybrid system

with the initial condition $X(0) \in \mathbb{R}^n$, $G(X(0)) < 0$ (recall that $\mathbf{I}\{\cdot\}$ stands for the indicator function).

Theorem 5.5 *Let the following assumptions be fulfilled:*

(i) The continuously differentiable functions B and G satisfy the following condition: for any initial condition $x_0 \in \mathbb{R}^n$ such that $G(x_0) > 0$, the equation $\dot{y}(s) = B(y(s))$ has a unique solution; moreover, there exists a time instant $s^*(x_0) : 0 < s^*(x_0) < \infty$ such that on the interval $(0, s^*(x_0))$ one has $G(y(s)) > 0$ and at the point $s = s^*(x_0)$ one has

$$G(y(s^*(x_0))) = 0, \quad G'_y(y)B(y)\big|_{y=(y(s^*(x_0)))} < 0.$$

(ii) Let $\tau = \inf\{t : G(X^0(t)) > 0\}$, where $X^0(t)$ is the solution of the system of equations $\dot{X}^0(t) = F(X^0(t), u(t))$ with the initial condition at $t = 0$ not belonging to the set $\{G(x) \geq 0\}$.

Then on the interval $[0, \tau + \varepsilon]$, with $\varepsilon > 0$ sufficiently small , the sequence of solutions of the system of Eq. (5.63) converges everywhere, except perhaps at the point τ, to the function

$$\bar{X}(t) = \begin{cases} X^0(t), & \text{for } t < \tau, \\ X^1(t), & \text{for } t \geq \tau. \end{cases}$$

Here $X^1(\cdot)$ is the solution of Eq. (5.56) with the following initial condition at the point $t = \tau$:

$$X^1(\tau) = X^0(\tau-) + \Psi(X^0(\tau-)) = \Phi\big(X^0(\tau-), s^*(X^0(\tau-))\big),$$

and the function Φ is the general solution of the Eq. (5.60).

The proof of this result is similar to that of Theorem 2.3 on the approximation of impulse controls and differs only in the choice of the function $\phi^{\varkappa}(t, \tau) = \varkappa\, \mathbf{I}\{G(X^{\varkappa}) > 0\}$ which, by assumption **(i)**, approximates the impulse action with intensity $s^*(X^0(\tau-))$.

Thus, if on the trajectories of the system (5.56)–(5.58), there is defined a finite set of time instants τ_i, then a sequence of solutions of the a parametrized system (5.63) converges to a solution hybrid system (5.56)–(5.58) everywhere, except perhaps at the points of discontinuity.

Difficulties arise when the solution of the hybrid system under consideration is not so regular, namely, when there are infinitely many points of discontinuity or a sliding mode occurs on the discontinuity surface $G(x) = 0$. In this case, a description in the form of a hybrid system with intermediate conditions is no longer adequate and the more general class of differential equations with measure is required. Below we look at examples in which the jumps in the hybrid system are described by a shift operator.

All the conditions stated above are not sufficient for proving the extendability of solution for an arbitrary initial condition x_0 and some measurable control $u(\cdot)$; however, if there exists a bounded solution $X(t)$ with a finite number of jumps one can establish its uniqueness and continuous dependence on the initial conditions.

In spite of all our assumptions concerning regularity of functions involved into the problem statement, this problem is extremely irregular due to the possibility of non-extendability of solution (if the set of points s in (5.62) is empty or the infimum is equal to infinity). The first case leads to a so-called "sliding mode" along the set $G(x) = 0$, as in systems with a discontinuous right-hand side [365]; the second one corresponds to the case of non-extendability of a solution beyond the jump point. However, in the case when the jump behavior is described by a shift-type operator the problem can be reduced to a more regular problem by using the discontinuous time transformation as in the case of impulse control problems [270].

The description of a jump by a shift operator may seem rather artificial; however, all discrete-continuous systems that are stable (or robust) with respect to an approximation procedure of the impulsive input admit such a representation of jumps [273, 333, 464].

Prior to further considerations it is useful to present some simple examples of hybrid systems that are rather typical and simultaneously exhibit all the specific features of the systems described above.

Example 5.2 (The Shift Operator Describing an Elastic Shock [346]) Consider a unit-mass point with generalized coordinates $\{x_1, x_2\}$ (state and velocity, respectively), which moves along a straight line towards an elastic obstacle placed at the point $x_1 = 0$. Suppose the initial state $x_1(0) < 0$ and the force F depends on the state, velocity and some control u, so that in the domain $x_1 < 0$ the equations of motion are

$$\begin{cases} \dot{x}_1(t) = x_2(t), \\ \dot{x}_2(t) = F(x_1(t), x_2(t), u(t)). \end{cases} \tag{5.64}$$

The elastic shock at the instant $\{\tau : x_1(\tau) = 0\}$ causes a sudden change in the sign of the velocity, namely,

$$x_2(\tau) = -x_2(\tau-).$$

This change can be described by a suitable shift operator along the trajectories of the system of differential equations

$$\begin{cases} \dot{y}_1(s) = y_2(s), \\ \dot{y}_2(s) = -y_1(s), \end{cases} \quad \text{with the initial conditions} \quad \begin{cases} y_1(0) = x_1(\tau-), \\ y_2(0) = x_2(\tau-). \end{cases} \tag{5.65}$$

Indeed, the solution of this system is

$$\begin{cases} y_1(s) = x_2(\tau-)\sin(s), \\ y_2(s) = x_2(\tau-)\cos(s), \end{cases}$$

so we have $y_1(s) > 0$ in the interval $(0, \pi)$ and

$$y_1(\pi) = 0, \quad y_2(\pi) = -y_2(0) = -x_2(\tau-).$$

Therefore, the shift operator along the trajectories of the system (5.65) provides a correct description of the jump behaviour.

The next example shows that one motivation of our model of the representation of jumps comes from a real problem in the mechanics of shocks as a result of the interaction with a surface with very large coefficient of elasticity.

Example 5.3 (Controllable Shock [33]) Consider a unit-mass point with phase coordinates $\{x_1, x_2\}$ (state and velocity, respectively) that moves freely along a straight line up to the collision with the obstacle at the point $x_1 = 0$. Suppose that the initial state $x_1(0)$ is < 0 and that in the region $x_1 < 0$ the force depends on the state, velocity and some control $u(\cdot)$; then in the region $x_1 < 0$ the equations of motion have the standard form (5.64). When the point hits the obstacle at the instant $\{\tau : x_1(\tau) = 0\}$, its motion is described by the equations

$$\begin{cases} \dot{x}_1(t) = x_2(t), \\ \dot{x}_2(t) = -\varkappa x_1(t) + w^{\varkappa}(t), \end{cases} \tag{5.66}$$

where $\varkappa > 0$ is a large parameter which characterizes the elasticity of the obstacle and $w^{\varkappa}(t)$ is an external impulsive force that acts only during the contact, i.e., when $x_1(t) > 0$. As an example, consider the force

$$w^{\varkappa}(t) = \mu^{1/2} w\left(\frac{t-\tau}{\varkappa^{-1/2}}\right),$$

with some integrable function $w(\cdot)$ given for $t \geq \tau$. By the space-time transformation

$$\begin{cases} y_1^{\varkappa}(s) = \varkappa^{1/2} x_1(s\varkappa^{-1/2} + \tau), \\ y_2^{\varkappa}(s) = x_2(s\varkappa^{-1/2} + \tau) \end{cases} \tag{5.67}$$

we obtain the following system of equations that describes the fast motion in the fictitious time variable $s > 0$:

$$\begin{cases} \dot{y}_1^\varkappa(s) = y_2^\varkappa(s), \\ \dot{y}_2^\varkappa(s) = -y_1^\varkappa(s) + w(s). \end{cases} \tag{5.68}$$

The initial conditions are

$$\begin{cases} y_1^\varkappa(0) = x_1(\tau), \\ y_2^\varkappa(0) = x_2(\tau-), \end{cases}$$

and the solution of the system (5.68) provides a general description of the jump of the velocity by means of the shift operator along the trajectories of that system. Take, for example

$$w(s) = \begin{cases} -2\xi y_2(s), & \text{if } s \geq 0, \ \ y_1 > 0, \\ 0, & \text{otherwise,} \end{cases}$$

where $|\xi| < 1$. This type of the contact force arises in electro-mechanical scanners, where the electromagnetic actuator operates only during the contact of a moving mass with an elastic obstacle. For this choice the solution of the system obtained above via the space-time transformation (5.67) is given by

$$\begin{cases} y_1^\varkappa(s) = \dfrac{x_2(\tau-)}{\sqrt{1-\xi^2}} \exp(-\xi s) \sin\left(s\sqrt{1-\xi^2}\right), \\ y_2^\varkappa(s) = x_2(\tau-) \exp(-\xi s) \left[\cos\left(s\sqrt{1-\xi^2}\right) - \dfrac{\xi}{\sqrt{1-\xi^2}} \sin\left(s\sqrt{1-\xi^2}\right) \right]. \end{cases}$$

So $y_1^\varkappa(s) > 0$ in the interval $\left(0, \dfrac{\pi}{\sqrt{1-\xi^2}}\right)$ and

$$\begin{cases} x_1\left(\tau + \dfrac{\varkappa^{-1/2}\pi}{\sqrt{1-\xi^2}}\right) = y_1^\varkappa\left(\dfrac{\pi}{\sqrt{1-\xi^2}}\right) = 0, \\ x_2\left(\tau + \dfrac{\varkappa^{-1/2}\pi}{\sqrt{1-\xi^2}}\right) = y_2^\varkappa\left(\dfrac{\pi}{\sqrt{1-\xi^2}}\right) = -x_2(\tau-) \exp\left(-\dfrac{\pi\xi}{\sqrt{1-\xi^2}}\right). \end{cases}$$

Letting here $\varkappa \to \infty$ we obtain a description of the jump by means of the shift operator along the trajectories of the system (5.68). Moreover, using various types of impulsive force $w^\varkappa(\cdot)$ one can describe various types of jump behavior, i.e., with damping if $\xi > 0$ and acceleration if $\xi < 0$. Generally speaking, using an impulsive force $w^{\varkappa}(t)$ subject to the condition

$$\int_{\tau}^{\tau+c\varkappa^{-1/2}} w^{\varkappa}(t)dt < \infty, \quad \text{for all } c > 0,$$

one can obtain various types of controllable jumps that admit a representation by means of a shift operator.

This example shows that describing a jump with the aid of a shift operator is rather natural in the case of surfaces with large coefficient of elasticity. Let us also mention that the motion of bouncing ball (see Example 1.11) can be also considered as an example of hybrid system in which the jump behaviour is described by a shift operator. Indeed, the Eq. (1.52) for the ball's velocity after impact can be obtained by considering the system with the state vector $X = (y, v, x, w)$ and constraint $G(X) = x - y \leq 0$, and with the motion of the ball in the "forbidden" domain described by the system of equations

$$\begin{cases} \dot{y}(t) = v(t), \\ \dot{v}(t) = -\varkappa(y(t) - x(t)) - 2\varkappa^{1/2}\xi(v(t) - w(t)), \end{cases} \tag{5.69}$$

with the initial conditions $y(\tau)$ and $v(\tau-)$, where τ is the time of impact, $x(t)$ and $w(t)$ are the position and the velocity of the racket, respectively, $\varkappa$ is the elasticity coefficient and $0 < \xi < 1$ is the damping coefficient. Applying the modified space-time transformation (5.67), namely

$$\begin{cases} y_1^{\varkappa}(s) = y(\tau) + \varkappa^{1/2}[y(s\varkappa^{-1/2} + \tau) - y(\tau)], \\ y_2^{\varkappa}(s) = v(s\varkappa^{-1/2} + \tau), \end{cases} \tag{5.70}$$

and observing that

$$x^{\varkappa}(s) = x(s\varkappa^{-1/2} + \tau) = x(\tau) + \varkappa^{-1/2}w(\tau)s + O(\varkappa^{-1}),$$

we obtain the following system (for the sake of simplicity we omit some of the calculations) that describes the motion if $\varkappa$ is sufficiently large:

$$\begin{cases} \dot{y}_1^{\varkappa}(s) = y_2^{\varkappa}(s), \\ \dot{y}_2^{\varkappa}(s) = -(y_1^{\varkappa}(s) - x(\tau)) - 2\xi(y_2^{\varkappa}(s) - w(\tau)), \end{cases} \tag{5.71}$$

This system has a solution $y_1^{\varkappa}(s)$, $y_2^{\varkappa}(s)$ such that in the interval $[0, s^*)$, where $s^* = \pi/\sqrt{1-\xi^2}$, we have $x(\tau) - y_1^{\varkappa}(s) > 0$; moreover,

$$y_1^{\varkappa}(s^*) = x(\tau), \quad y_2^{\varkappa}(s^*) > w(\tau).$$

Solving the system (5.71) and letting $\varkappa \uparrow \infty$, we obtain the expression

$$v(\tau) = y_2^{\varkappa}(s^*) = -(v(\tau-) - w(\tau))\exp\{-\xi s^*\} + w(\tau),$$

which coincides with (1.52) if

$$k = \exp\{-\xi s^*\} = \exp\left\{-\frac{\pi\xi}{\sqrt{1-\xi^2}}\right\}.$$

It is important that the representation of the jump with the aid of a shift operator enables one to reduce the optimal control problem for a (hybrid) DCS to a standard—though not smooth—optimal control problem.

5.4.2 Generalized Solution of a Hybrid System

Suppose that for a control $u(\cdot)$ there exists a solution $X(t)$ of (5.56) defined on the interval $[0, T]$ and which has a finite number of jumps at the points $\{\tau_i,\ i = 1, \dots, N\}$. This means that for every $i = 1, \dots, N$ there is defined the set of values $s_i^* = s^*(X(\tau_i-)) < \infty$ such that

$$X(\tau_i) = \Phi(X(\tau_i-), s^*(X(\tau_i-)).$$

Consider the time interval $[0, T_1]$, where

$$T_1 = T + \sum_{i=1}^{N} s_i^*, \tag{5.72}$$

and the function

$$\alpha(s) = \begin{cases} 0, & \text{if } s \in \left[\tau_i + \sum_{k<i} s_k^*,\ \tau_i + \sum_{k\le i} s_k^*\right), \\ 1, & \text{otherwise.} \end{cases} \tag{5.73}$$

Next, consider in $[0, T_1]$ the auxiliary system

$$\begin{cases} \dot{y}(s) = \alpha(s)F(y(s), u_1(s)) + (1-\alpha(s))B(y(s)), \\ \dot{\eta}(s) = \alpha(s), \end{cases} \tag{5.74}$$

with the initial conditions

$$y(0) = x_0, \quad \eta(0) = 0$$

and the control

$$u_1(s) = u(\eta(s)).$$

Then one has the following correspondence between the auxiliary and original systems, (5.74) and (5.56):

Theorem 5.6 *Let $X(t)$ be a solution of (5.56) and define the function $\eta(s)$ by the formula (5.73) and the inverse function by the formula*

$$\Gamma(t) = \inf\{s : \eta(s) > t\},$$

where by definition $\Gamma(T) = T_1$.

If $\{y(\cdot), \eta(\cdot)\}$ is the solution of the system (5.74), then

$$X(t) = y(\Gamma(t)).$$

This result is a consequence of the standard arguments based on the discontinuous time transformation.

Note that, by its definition, $\alpha(s) = 1$ if $G(y(s)) < 0$ and $\alpha(s) = 0$ if $G(y(s)) > 0$. This observation may serve as a basis for a further transformation of the original optimal control problem. Moreover, one can regard (5.74) as a system that in some sense is equivalent to (5.56), with a "new" time $\eta(s)$ and with the variable α as an additional control satisfying the constraint

$$\alpha(s) = \begin{cases} 1, & \text{if } G(y(s)) < 0, \\ 0, & \text{if } G(y(s)) > 0. \end{cases} \tag{5.75}$$

Formula (5.75) does not define α on the boundary of the constraint, that is, on the set $\{s : G(y(s) = 0\}$. However, one can relax the problem, setting $\alpha(s) \in [0, 1]$ on this set. This relaxation corresponds to a standard method of convexification of the right-hand side to guarantee the existence of a solution to an optimization problem [237].

The transformation of (5.75) gives the constraint in the integral form

$$\begin{aligned} &\int_0^{T_1} \alpha(s) G^+(y(s)) ds = 0, \\ &\int_0^{T_1} (1 - \alpha(s)) G^-(y(s)) ds = 0, \end{aligned} \tag{5.76}$$

where

$$G^+(y) = \max\{G(y), 0\}, \quad G^-(y) = \min\{G(y), 0\}.$$

Using this relaxation we obtain the following extension of Theorem 5.6.

Theorem 5.7 *Suppose that* $\{y(\cdot), \eta(\cdot)\}$ *is a solution of the system (5.74), with some Lebesgue-measurable controls* $\{\alpha(\cdot), u_1(\cdot)\}$ *that satisfy the constraints*

$$\alpha(s) \in [0, 1], \quad u_1(s) \in U \ \text{a.e. on } [0, T_1], \tag{5.77}$$

and such that (5.76) holds and $\eta(T_1) = T$. *Define the function* $\Gamma(t)$ *by*

$$\Gamma(t) = \inf\{s : \eta(s) > t\}, \quad \Gamma(T) = T_1. \tag{5.78}$$

Then for $X(t) = y(\Gamma(t))$ *there exist:*

1. *a Lebesgue-measurable control* $u(\cdot)$ *such that* $u(t) \in U$ *a.e. in* $[0, T]$,
2. *a non-negative regular measure* $\mu(dt)$, *localized on the subset* $\{t : G(X(t)) = 0\}$, *such that*

$$\mu([0, T]) = T_1 - T,$$

and its distribution function $\mu(t) = \mu([0, t])$ *has the Lebesgue decomposition*

$$\mu(t) = \mu^c(t) + \sum_{\tau \le t} \Delta\mu(\tau),$$

where $\mu^c(t)$ *is the continuous component and* $\Delta\mu(\tau) = \mu(\{\tau\})$ *is the discrete component, localized at the point* τ,

such that the functions $\{X(t), \mu(t)\}$ *are right-continuous and satisfy the nonlinear differential equation with measure*

$$\begin{aligned} dX(t) &= F(X(t), u(t))dt + B(X(t))d\mu^c(t) + \\ &+ \sum_{\tau \le t} \Psi(X(\tau-), \Delta\mu(\tau))\delta(t - \tau)dt, \end{aligned} \tag{5.79}$$

with the constraints

$$G(X(t)) \le 0, \quad \text{for all} \quad t \in [0, T], \tag{5.80}$$

and

$$\int_0^T G(X(t))d\mu(t) = 0. \tag{5.81}$$

Remark 5.13 If the boundary of the set $G(x) \geq 0$ is sufficiently regular (for example, it is a manifold of class C^3), if on the set $\{t : G(X(t)) = 0\}$ it holds that

$$\langle G'_x, B(x)\rangle\big|_{x=X(t)} \leq \delta < 0, \tag{5.82}$$

and if the number of jumps is finite, then the solution to (5.79)–(5.81) is unique. The proof of this fact is based on fairly subtle constructions of the theory of processes with reflection, which are investigated in detail in problems of stochastic impulse control. The above conditions are sufficient for the uniqueness of a continuous solution (see Theorem 1.1 [43]), and therefore can be easily transferred to the case of a finite number of jumps, because they guarantee the uniqueness of the solution in the intervals between jumps. Apparently, in control problems for deterministic systems these conditions can be weakened—for example, one could allow an infinite number of jump points with a finite set of left accumulation points, while solutions with right accumulation points may be not unique [18]; one can also try to relax the condition (5.82) by setting $\delta = 0$, but this study is beyond the scope of this book.

Proof Define $\Gamma(\cdot)$ by formula (5.78). Then the Eq. (5.79) for $X(t) = y(\Gamma(t))$ can be derived by the same arguments as in Theorem 2.14. The measure $\mu(dt)$ is given by its distribution function

$$\mu(t) = \mu([0, t]) = \int_0^{\Gamma(t)} (1 - \alpha(s))ds,$$

with

$$\mu^c(t) = \int_0^{\Gamma(t)} \mathbf{I}\{s : \eta(s) \notin D_\Gamma\}(1 - \alpha(s))ds,$$

and

$$\Delta\mu(\tau) = \Gamma(\tau) - \Gamma(\tau-), \quad \text{if } \tau \in D_\Gamma,$$

where D_Γ is the set of the jump points of Γ. Lemma 4.1 ensures that the control $u(t) = u_1(\Gamma(t))$ can be assumed to be a Lebesgue-measurable function.

Now upon applying the change of time to the constraints (5.76), we obtain

$$\begin{aligned} &\int_0^{T_1} \alpha(s)G^+(y(s))ds = \int_0^T G^+(X(s))ds = 0, \\ &\int_0^{T_1} (1 - \alpha(s))G^-(y(s))ds = \int_0^T G^-(X(s))d\mu(s) = 0, \end{aligned} \tag{5.83}$$

and consequently

$$G(X(t)) \le 0 \quad \text{a.e. on } [0, T];$$

actually, thanks to the continuity of G and the right-continuity of X, this inequality holds for all $t \in [0, T]$.

As follows from the second relation in (5.83), the non-negative measure $\mu(dt)$ is localized on the set $\{t : G^-(X(t)) = 0\}$; however, as we just pointed out above, this set coincides with $\{t : G(X(t)) = 0\}$. □

Theorem 5.7 gives a representation of a generalized solution of the DCS (5.56) in the case of infinitely many jumps, as well as in the case when a sliding mode arises along the boundary $G(x) = 0$. Both these cases are determined by the properties of the measure $\mu(dt)$; specifically, the case $\mu^c([0, T]) > 0$ corresponds to the presence of a sliding mode, while the case when $\mu(dt)$ has infinitely many point atoms corresponds to the case of infinitely many jumps.

Earlier, we discussed the fact that in hybrid systems generalized solutions can be naturally interpreted as limits, at all points continuity, of sequences of parametrized continuous solutions. The following result shows that if a generalized solution exists as a limit of solutions to a parametrized system, then it is necessarily a solution to the system (5.79)–(5.81).

Theorem 5.8 *Let the following conditions be satisfied:*

(i) *the sequence* $\int_0^T \varkappa\, \mathbf{I}\{G(X^\varkappa(t)) > 0\} dt$ *is uniformly bounded for* $\varkappa > 0$ *on the solutions* $X^\varkappa(t)$ *of the systems (5.63);*
(ii) *the set* $F(X, U)$ *is convex for any fixed* $X \in \mathbb{R}^n$*;*
(iii) *the functions* G *and* B *satisfy the conditions of Theorem 5.5.*

Then there exists a subsequence $Y^n(t) = X^{\varkappa_n}(t)$*, where* $\varkappa_n \uparrow \infty$*, such that* $Y^n(t)$ *converges at all points of continuity to a right-continuous function of bounded variation* $\bar{X}(t)$*, for which there are a non-negative measure* $\bar{\mu}(dt)$ *and a Lebesgue-measurable control* $\bar{u}(t) \in U$*, such that the pair* $\{\bar{X}(t), \bar{\mu}(dt)\}$ *satisfies the nonlinear equation with measure and constraints (5.79)–(5.81).*

For the pair $\{\bar{X}(t), \bar{\mu}(dt)\}$ *on the interval* $[0, T_1]$*, where* $T_1 = T + \mu\{[0, T]\}$*, there exists a solution* $\{y(\cdot), \eta(\cdot)\}$ *of the system(5.74), with some measurable controls* $\{\alpha(\cdot), u_1(\cdot)\}$ *satisfying the constraints (5.77), and such that the functions* $\alpha(\cdot)$ *and* $y(\cdot))$ *satisfy the relations (5.76) and* $\eta(T_1) = T$*. Moreover*

$$\bar{X}(t) = y(\Gamma(t)), \quad \textit{where} \quad \Gamma(t) = \inf\{s : \eta(s) > t\}.$$

Remark 5.14 We state this result without proof, but it is clear that the condition **(i)** allows us to introduce the sequence non-negative integrable controls

$$w^\varkappa(t) = \varkappa\, \mathbf{I}\{G(X^\varkappa(t)) > 0\}$$

and perform the change of time, which leads to the auxiliary system (5.74) with some measurable controls $\{\alpha^{\varkappa}(\cdot), u_1^{\varkappa}(\cdot)\}$ that satisfy the constraints (5.77), on the interval $[0, T_1^{\varkappa}]$, where

$$T_1^{\varkappa} = T + \int_0^T w^{\varkappa}(t)dt, \quad \text{with} \quad \sup_{\mu>0} T_1^{\varkappa} < \infty.$$

The corresponding sequence of solutions $\{y^{\varkappa}(\cdot), \eta^{\varkappa}(\cdot)\}$ is equicontinuous on the interval $[0, T_1^{\varkappa}]$. Therefore, one can extract a subsequence that converges uniformly to $\{y(\cdot), \eta(\cdot)\}$ on some common interval $[0, T_1]$ so that for $s = T_1$ the terminal constraints (5.76) and $\eta(T_1) = T$ are met. This limit function defines a generalized solution via of the reverse change of time.

Remark 5.15 Clearly, any generalized solution, if it exists, is necessarily described by the system of Eqs. (5.79)–(5.81); however, it is possible that this system admits also solutions that are not generalized ones. A direct answer to this question would be provided if one could prove a theorem establishing the uniqueness of solutions of the system (5.79)–(5.81). This system belongs to the class of so-called complementary systems [73], which are very actively studied in recent years, especially in the field of control of electronic switching circuits and of mechanical and electro-mechanical systems. However, as demonstrated by some examples, the uniqueness of the solution of a complementary system may not hold even in very simple cases [18].

5.4.3 *Existence of the Generalized Solution in the Optimal Control Problem for Hybrid Systems*

Consider the optimal control system problem (5.56)–(5.58) with the performance criterion

$$J[X(\cdot), u(\cdot)] = \phi_0(X(T)), \tag{5.84}$$

for some continuously differentiable function ϕ_0. We consider here this problem as an optimization problem in the class generalized solutions, which enables us to reduce it to an optimal control problem for an ordinary (not hybrid) dynamic system.

Auxiliary Problem Consider the optimal control system (5.74) with controls $\{\alpha, u_1\}$ satisfying the constraints (5.77) and the integral constraints (5.76). More precisely, this system is considered on some, not a priori fixed, interval $[0, T_1]$ such that $T_1 < \infty$ and $\eta(T_1) = T$, and with the performance criterion

$$J'\{y(\cdot), \alpha(\cdot), u_1(\cdot), T_1\} = \phi_0(y(T_1)) \to \min,$$

where ϕ_0 is the same function as in (5.84).

Theorem 5.9 *Suppose that the set $F(X, U)$ is convex for any fixed $X \in \mathbb{R}^n$, the set of admissible controls in the auxiliary problem is not empty, and the set of admissible values T_1 such that $\eta(T_1) = T$ is uniformly bounded. Then the auxiliary problem admits an optimal solution and the corresponding optimal generalized solution of the original problem satisfies the Eq. (5.79) with some control $u(\cdot)$ and measure $\mu(dt)$.*

Proof For any control $u(\cdot)$ and any measure $\mu(dt)$ that correspond to some admissible generalized solution, one can associate suitable controls $\{\alpha(s), u_1(s)\}$ that are admissible in the auxiliary problem. Indeed, if

$$\Gamma(t) = t + \mu((0, t]) \quad \text{and} \quad \eta(s) = \inf\{t : \Gamma(t) \geq s\},$$

then the appropriate admissible controls $\{\alpha(s), u_1(s)\}$ can be defined by the formulas

$$\alpha(s) = \dot{\eta}(s), \quad u_1(s) = u(\eta(s))$$

and then

$$J[X(\cdot), u(\cdot)] = J'\{y(\cdot), \alpha(\cdot), u_1(\cdot), T_1\},$$

since $X(T) = y(T_1)$. Therefore,

$$\inf_{u(\cdot)} J \geq \inf_{\{\alpha(\cdot), u_1(\cdot)\}} J'.$$

Thanks to the convexity assumptions and the boundedness of T_1, the set of admissible trajectories of the auxiliary problems is compact (see [237]), therefore the optimal control exists and the infimum in the right-hand side of the above inequality can be achieved on some controls $\{\alpha^0, u_1^0\}$, defined on some interval $[0, T_1]$.

To prove the existence of the optimal solution in the original problem it suffices to apply Theorem 5.7. Indeed, if $\{y^0, \alpha^0, u_1^0, T_1\}$ is the optimal solution of the auxiliary problem, then one can define $\{X^0, u^0, \mu^0\}$ by the formulas

$$X^0(t) = y^0(\Gamma(t)), \quad u^0(t) = u^1(\Gamma(t)), \quad \mu^0([0, t]) = \int_0^{\Gamma(t)} (1 - \alpha(s))ds.$$

Conditions (5.76) imply that

$$\int_0^T G^+(X^0(t))dt = \int_0^{T_1} \alpha^0(s)G^+(y^0(s))ds = 0,$$

$$\int_0^T G^-(X^0(t))d\mu^0(t) = \int_0^{T_1} (1 - \alpha^0(s))G^-(y^0(s))ds = 0.$$

Therefore, $G(X^0(t)) \leq 0$ a.e. in $[0, T]$ and the measure $\mu^0(dt)$ is localized on the set $\{t : G(X^0(t)) = 0\}$. Since $X^0(\cdot)$ is a right-continuous function, the constraint $G(X^0(t)) \leq 0$ is met for all $t \in [0, T]$.

Finally, we have $X^0(T) = y^0(T_1)$,

$$J[X^0(\cdot), u^0(\cdot)] = J'\{y^0(\cdot), \alpha^0(\cdot), u_1^0(\cdot), T_1\} = \inf J',$$

and so the triple $\{X^0, u^0, \mu^0\}$ is an optimal solution of the original problem in the class of generalized solutions.

Remark 5.16 In view of the non-differentiability of the functions G^+ and G^-, the above auxiliary problem belongs to a class of non-smooth optimization problems. However, by applying methods recently developed for this class of non-smooth problems (see, e.g., [93]), it becomes possible to derive the necessary optimality conditions in the form of a maximum principle and to design relevant computational algorithms. In the next chapter we derive the optimality conditions for various optimal control problems, in particular, for the optimal control problem described above (see Sect. 6.2).

Concluding this chapter, it should be noted that the use of the method singular transformations in problems with generalized solutions performs well in other tasks than the ones discussed above. On the one hand, numerous control problems can be reduced to the tasks that were considered here: for example, problems with monotone controls [23–25, 111, 443], or with constraints on the control variation [90, 91]. On the other hand, the discussed methods can be successfully applied in control problems for systems with distributed parameters [44], in stochastic observation control problems [265, 267, 276, 279, 300], as well as in stochastic control problems with unbounded and singular controls [114, 115, 162].

Chapter 6
The Maximum Principle in Problems of Generalized Optimal Control

6.1 Introduction

The problem of optimality conditions plays a central role in optimal control problems. Well-known optimality conditions in the form of the maximum principle were obtained in the classical control problems in the 1960s, and since then were widely used in solving practical problems and creating numerical optimization methods. The maximum principle is convenient in that the original infinite-dimensional optimization problem is reduced to solving some boundary value problem for a system of ordinary differential equations with initial and terminal conditions, in which the optimal control at every time instant is calculated as the solution to a finite-dimensional optimization problem. Optimality conditions in discrete-continuous systems have always aroused great interest, but it also immediately became clear that a simple transfer of methods introduced for classical optimal control problems, for example, the Dubovitskii-Milyutin schemes [113, 146], is not possible. Indeed, systems with impulse controls are described correctly by a special class of equations, namely, differential equations with measure, and for systems described by such equations, general schemes for the derivation of optimality conditions are applicable only under very strong restrictions, which narrows considerable the classes of treatable problems. Although there is a plethora of approaches to the derivation of optimality conditions, a general principle always remains in force: the trajectory to be tested for optimality must allow variations such that small changes in the control result in small changes in the trajectory and in the quality criterion, and then the choice of a specific class of admissible variations leads to optimality conditions of different strengths. In the impulse control problems, especially in the nonlinear case, the class of such admissible variations is rather narrow. As demonstrated by examples in the Introduction, a simple decomposition

B. Miller, E. Rubinovich, *Optimization of Dynamical Systems with Impulse Controls and Shocks*, Systems & Control: Foundations & Applications,
https://doi.org/10.1007/978-3-031-64124-4_6

of one impulse action into two such actions with preservation of the total amplitude (that is, a "small variation of the impulsive measure-control"), results in a "strong" change of the trajectory, since if the time difference between these two impulses tends to zero, the corresponding sequence of trajectories does not converge to the original trajectory. If we interpret the problem in strict terms, then the derivation of optimality conditions by classical schemes requires at least the continuity of the mapping from the set of controls to the set of trajectories, but in impulse control problems such continuity does not hold in general.

Therefore, if one does not want to restrict the class of variations, then one need to distinguish the class of "continuous" systems, or, if there is no continuity, propose a method for choosing a nontrivial class of variations. The first optimality conditions for nonlinear systems with impulse controls were obtained as soon as the role of such continuity (robustness) became clear [64, 205, 263, 333]. However, the robustness requirement is very restrictive, since it leads to the Frobenius condition, and therefore for many very important classes of control problems these optimality conditions were not applicable.

Incidentally, it is precisely in the process of deriving optimality conditions for robust systems that a universal approach was proposed and then successfully implemented in problems where the robustness condition is not satisfied [66, 269, 270]. The first example of using this approach can be found in Chap. 3, where we considered an impulse control problem with a restricted number of impulses. This general approach, based on the method of discontinuous time change, consists of the following steps:

1. transform the initial optimization problem posed in the class of generalized solutions into an equivalent auxiliary problem with bounded controls;
2. derive optimality conditions in the auxiliary problem;
3. use the inverse discontinuous time transformation to turn these optimality conditions into optimality conditions for the original problem.

The present chapter demonstrates the implementation of this approach for various classes of optimal control problems. We start with a general problem with terminal and phase constraints and derive for it the necessary optimality condition in the form of a generalized maximum principle. Then we obtain a simpler form of these conditions for systems satisfying the robustness conditions, and for systems with a linear dependence on an unbounded control. As an example, optimality conditions in a linear-convex problem are derived. It turns out that for this class of problems the generalized maximum principle is also a sufficient optimality condition. This result is important for systems described by the Riccati equation that arise in problems of observation control.

Finally, as an illustration of the general approach, we present the optimality condition for control problems of a hybrid system with phase constraints.

6.2 Necessary Optimality Condition: Generalized Maximum Principle

In this section, a necessary optimality condition will be derived for generalized solutions in the form of the maximum principle in a problem with phase and terminal constraints. The problem is considered only for systems with affine dependence on an unbounded control since in this case, the results take on the most explicit form. Indeed, in this case (see Sect. 4.2.1) the equation for the generalized solution is derived directly in terms of the original problem (4.63) (see Theorem 4.7), while in the general case the dynamics of a generalized solution is described implicitly and, in addition, the representation of the equation describing a generalized solution is not unique (see Remark 4.3). Accordingly, the optimality conditions may depend on the representation of the auxiliary system in control form, so one needs to present them in terms of the original problem, a task that remains open to this day.

6.2.1 Problem with Phase Constraints

We consider the following optimal control problem: on a given interval $[0, T]$, the controlled process $\{X(t), V(t)\}$ is described by the system of equations

$$\begin{cases} \dot{X}(t) = f(X(t), V(t), u(t), t) + B(X(t), V(t), u(t), t)w(t), \\ \dot{V}(t) = \|w(t)\|, \end{cases} \tag{6.1}$$

with two types of controls, $u \in U \subset \mathbb{R}^k$ and $w \in K \subset \mathbb{R}^m$, where U is a compact set and K is a convex closed cone ($nm < \infty$). The control $w(t)$ satisfies the constraint

$$\int_0^T \|w(t)\|dt \leq M < \infty \tag{6.2}$$

with some constant M. We assume that the right-hand side of (6.1) satisfies Assumption 4.1, i.e., the functions $f(X, V, u, t)$ and $B(X, V, u, t)$ are continuous in all the variables and satisfy the Lipschitz condition with respect to the variables (X, V).

We consider the problem with terminal and phase constraints

$$h(X(0), X(T), V(T)) = 0, \quad S(X(0), X(T), V(T)) \leq 0, \tag{6.3}$$

$$g(X(t), V(t), t) \leq 0 \quad \text{for any} \quad t \in [0, T]. \tag{6.4}$$

The criterion to be minimized is

$$J[X(\cdot), u(\cdot), w(\cdot)] = \phi_0(X(0), X(T), V(T)) \to \min \tag{6.5}$$

with a continuous function ϕ_0. Here, as above, h, S, g are $\mathbb{R}^{N_1}$-, $\mathbb{R}^{N_2}$-, and $\mathbb{R}^{N_3}$-valued continuous functions, respectively, and (6.3) and (6.4) can be understood component-wise.

Following the approach presented earlier, we consider the auxiliary control problem for the system

$$\begin{cases} \dot{Y}(s) = \alpha(s) f(Y(s), z(s), u_1(s), \eta(s)) + \\ \qquad + (1 - \alpha(s)) B(Y(s), z(s), u_1(s), \eta(s)) e(s), \\ \dot{z}(s) = (1 - \alpha(s)) \|e(s)\|, \\ \dot{\eta}(s) = \alpha(s), \end{cases} \tag{6.6}$$

with the initial conditions

$$Y(0) = X(0), \quad z(0) = 0, \quad \eta(0) = 0,$$

and the controls

$$u_1(s) \in U, \quad \alpha(s) \in [0, 1], \quad e(s) \in K \cap \{\|e\| \leq 1\}. \tag{6.7}$$

This problem is considered in some interval $[0, T_1]$ such that $T_1 \leq T + M$, with performance criterion to be minimized given by

$$J'[Y(\cdot), u_1(\cdot), \alpha(\cdot), e(\cdot)] = \phi_0(Y(0), Y(T_1), z(T_1)) \to \min, \tag{6.8}$$

and with the terminal and phase constraints

$$\begin{cases} h(Y(0), Y(T_1), z(T_1)) = 0, \quad \eta(T_1) = T, \\ S(Y(0), Y(T_1), z(T_1)) \leq 0, \quad z(T_1) \leq M, \end{cases} \tag{6.9}$$

$$g(Y(s), z(s), \eta(s)) \leq 0 \quad \text{for any} \quad s \in [0, T_1]. \tag{6.10}$$

Here the set U, the cone K, the constant M, and the functions f, B, ϕ_0, φ, h, S and g are the same as in the original problem (6.1)–(6.5).

This auxiliary problem is the result of replacing the time variable in the original system (6.1) for the variables $\{X(t), V(t)\}$ by the "new time" $\eta(s)$. As a consequence, the unbounded control $w(t)$ is transformed into the pair of bounded controls $\{\alpha(s), e(s)\}$.

The optimal control problem is considered in the class of strong generalized solutions (see Definition 5.1), which provides a relaxation of the original problem; according to Theorems 5.2 and 5.4, in this way the original problem (6.1)–(6.5) with unbounded control $w(t)$ is reduced to the standard problem (6.6)–(6.10) with bounded controls. Using Theorem 4.7 one can prove that any optimal control in the original problem corresponds to the optimal control in the auxiliary problem.

To avoid additional complications, we assume that the right-hand side of (6.1) satisfies the convexity conditions (otherwise, we have first to apply the convexification procedure, as described in the proof of Theorem 4.4). If the convexity conditions are satisfied, we obtain the following representation of the optimal generalized solution.

Theorem 6.1 *Let the sets*

$$\mathscr{L}_1(x, v, t) = f(x, v, U, t) = \{l \in \mathbb{R}^n : l = f(x, v, u, t),\ u \in U\},$$

$$\mathscr{L}_2(x, v, t) = \left\{ l = \begin{pmatrix} l_1 \in \mathbb{R}^n \\ l_2 \in \mathbb{R}^1 \end{pmatrix} \mid \begin{array}{l} l_1 = B(x, v, u, t)w, \\ l_2 = \|w\|, \end{array} \quad u \in U,\ w \in K \right\} \tag{6.11}$$

be convex for any fixed (x, v, t).

Then for any optimal generalized solution in the strong sense $\{X^0(\cdot), V^0(t)\}$ *of problem (6.1)–(6.5) there exist*

1. *a non-negative scalar measure* $\mathscr{V}(dt)$*, defined on the Borel subsets of the interval* $[0, T]$*, with distribution function*

$$V^0(t) = \mathscr{V}([0, t]),$$

such that

$$\mathscr{V}([0, T]) = V^0(T) \leq M;$$

2. *a pair of Lebesgue- and* $\mathscr{V}$*-measurable controls* $\{u^0(t), w^0(t)\}$*, satisfying on* $[0, T]$ *the constraints*

$$u^0(t) \in U, \quad w^0(t) \in K, \quad \text{a.e. with respect to the Lebesgue and } \mathscr{V} \text{ measures;}$$

3. *a set of Lebesgue-measurable controls* $\{u^0_\tau(\cdot), e^0_\tau(\cdot)\}$*, defined on the segments* $[0, \Delta V^0(\tau)]$ *for each* $\tau \in D^{\mathrm{d}}_{\mathscr{V}}$*, and satisfying on* $[0, \Delta V^0(\tau)]$ *the constraints*

$$u^0_\tau(s) \in U, \quad e^0_\tau(s) \in K \cap \{\|e\| = 1\} \quad \text{a.e.,}$$

such that the following holds.

The optimal generalized solution $\{X^0(t), V^0(t)\}$ *satisfies the equations*

$$
\begin{aligned}
X^0(t) = X(0-) + \int_0^t f(X^0(\tau), V^0(\tau), u^0(\tau), \tau)d\tau + \\
+ \int_0^t B(X^0(\tau), V^0(\tau), u^0(\tau), \tau)w^0(\tau)dV^c(\tau) + \sum_{\tau \le t} \Delta X^0(\tau),
\end{aligned}
\tag{6.12}
$$

where $V^c(\cdot)$ *is the continuous component in the Lebesgue decomposition of the function* $V^0(\cdot)$ *of bounded variation [322],*

$$
\Delta X^0(\tau) = y_\tau^0(\Delta V^0(\tau)) - X^0(\tau-),
$$

and $y_\tau^0(s)$ *is the solution of the system*

$$
\begin{cases}
\dot{y}_\tau^0(s) = B(y_\tau^0(s), z_\tau^0(s), u_\tau^0(s), \tau)e_\tau^0(s), \\
\dot{z}_\tau^0(s) = 1
\end{cases}
$$

with the initial conditions

$$
y_\tau^0(0) = X^0(\tau-), \quad z_\tau^0(0) = V^0(\tau-).
$$

Moreover, for the optimal generalized solution $\{X^0(\cdot), V^0(\cdot)\}$ *there exists the optimal control* $\{\alpha^0(\cdot), u_1^0(\cdot), e^0(\cdot)\}$ *in the auxiliary problem (6.6)–(6.10), satisfying the constraints*

$$
u_1^0(s) \in U, \quad \alpha^0(s) \in [0, 1], \quad e^0(s) \in K \cap \{\|e\| = 1, \}
$$

and such that the corresponding optimal solution $\{Y^0(\cdot), z^0(\cdot), \eta^0(\cdot)\}$ *of the auxiliary problem satisfies the relations (5.15) and (5.16).*

Proof The first statement is a direct consequence of Theorem 4.7. Therefore, there exist functions

$$
\{\mathscr{V}(dt), u^0(t), w^0(t)\} \quad \text{and} \quad \{u_\tau^0(s), e_\tau^0(s)\} \quad \text{for all} \quad \tau \in D_{\mathscr{V}}^{\mathrm{d}},
\tag{6.13}
$$

which can be regarded as the optimal generalized control in the optimization problem posed in the class of strong generalized solutions, and which generate the optimal strong generalized solution $\{X^0(\cdot), V^0(\cdot)\}$. Moreover, since the magnitude of the jump is defined via the solution of a differential equation with Lipschitz right-hand side, it is readily seen that $\Delta X(\tau) = \Psi(X(\tau-), \tau, V(\tau-), e_\tau(\cdot))$ satisfies the Lipschitz condition with respect to $(X(\tau-), V(\tau-))$ (this follows from the continuous dependence of the solutions on initial conditions [365]). Therefore, (6.13) defines the unique generalized solution (see Theorem A.24).

Remark 6.1 The second statement of Theorem 6.1 means that the optimal control can be always chosen to satisfy $\|e^0(s)\| = 1$.

This statement follows from Theorem 5.2 if one sets

$$\alpha^0(s) = \dot{\eta}^0(s), \quad \text{where } \eta^0(s) = \inf\{t : t + V^0(t) = \Gamma^0(t) > s\}, \tag{6.14}$$

$$u_1^0(s) = u^0(\eta^0(s)), \quad e^0(s) = w^0(\eta^0(s))\|w^0(\eta^0(s))\|^{\oplus}, \quad \text{for } \eta^0(s) \notin D_V, \tag{6.15}$$

and

$$u_1^0(s) = u_\tau^0(s - \Gamma(\tau-)), \quad e^0(s) = e_\tau^0(s - \Gamma(\tau-)), \quad \text{for } \eta^0(s) = \tau \in D_V. \tag{6.16}$$

Remark 6.2 The superscript $\oplus$ stands for pseudoinversion, that is $a^{\oplus} = a^{-1}$ if $a \neq 0$, and $0^{\oplus} = 0$.

In view of Remark 4.5, the control $e^0(\cdot)$ can be chosen to satisfy

$$\|e^0(s)\| = 1 \quad \text{on the set } \{s : \eta^0(s) = \tau \in D_V\}. \tag{6.17}$$

If the Lebesgue measure of the set D_V equals T_1, then the function $e^0(\cdot)$ defined in (6.15), (6.16), satisfies the condition (6.17) a.e. on the segment $[0, T_1]$, where $T_1 = T + V^0(T)$.

However, if the equality (6.17) does not hold a.e. on the whole interval $[0, T_1]$, then, by (6.15) $\|e^0(s)\| \in \{0, 1\}$. On the set $\{s : \|e^0(s)\| = 0\}$ the system (6.6) has the form

$$\begin{cases} \dot{Y}^0(s) = \alpha^0(s) f(Y^0(s), z^0(s), u_1^0(s), \eta^0(s)), \\ \dot{z}^0(s) = 0, \\ \dot{\eta}^0(s) = \alpha^0(s); \end{cases}$$

therefore, one can shrink the subset $\{s : \|e^0(s)\| = 0\}$ to ensure that on it $\alpha^0 = 1$ and then the equation for z^0 becomes

$$\dot{z}^0(s) = 0 = (1 - \alpha^0(s))\|e^0(s)\|,$$

with an arbitrary function $e^0(s)$ such that $\|e^0(s)\| = 1$.

We implement this idea by analogy with the proof of Theorem 4.7. We take the non-decreasing function

$$\rho(s) = z^0(s) + \eta^0(s);$$

then $\rho(T_1) = z^0(T_1) + \eta^0(T_1) \leq M + T$, and by the standard arguments of change of variables in the Lebesgue integral we obtain the collection of functions

$$\{\tilde{Y}^0(s'), \tilde{\eta}^0(s'), \tilde{z}^0(s')\} = \{Y^0(\rho^{-1}(s')), \eta^0(\rho^{-1}(s')), z^0(\rho^{-1}(s'))\},$$

$$\{\tilde{\alpha}^0(s'), \tilde{u}_1^0(s'), \tilde{e}^0(s')\} = \{\alpha^0(\rho^{-1}(s')), u_1^0(\rho^{-1}(s')), e^0(\rho^{-1}(s'))\},$$

where $\rho^{-1}(s') = \inf\{\tau : \rho(\tau) > s\}$. These functions satisfy the system of equations

$$\begin{cases} \dfrac{\tilde{Y}^0(s')}{ds'} = \left(\dfrac{d\rho^{-1}(s')}{ds'}\right)^{\oplus} \tilde{\alpha}^0(s') f(\tilde{Y}^0(s')), \tilde{z}^0(s'), \tilde{u}_1^0(s'), \tilde{\eta}^0(s')) + \\ \qquad + \left(\dfrac{d\rho^{-1}(s')}{ds'}\right)^{\oplus} (1 - \tilde{\alpha}^0(s'))B(\tilde{Y}^0(s'), \tilde{z}^0(s'), \tilde{u}_1^0(s'), \tilde{\eta}^0(s'))\tilde{e}^0(s'), \\ \dfrac{d\tilde{z}^0(s')}{ds'} = \left(\dfrac{d\rho^{-1}(s')}{ds'}\right)^{\oplus} (1 - \tilde{\alpha}^0(s'))\|\tilde{e}^0(s')\|, \\ \dfrac{d\tilde{\eta}^0(s')}{ds'} = \left(\dfrac{d\rho^{-1}(s')}{ds'}\right)^{\oplus} \tilde{\alpha}^0(s') \end{cases} \tag{6.18}$$

on the interval $[0, \mu(T_1)] = [0, \tilde{T}_1]$ where

$$\left(\frac{d\rho^{-1}(s')}{ds'}\right)^{\oplus} = \left(\tilde{\alpha}^0(s') + (1 - \tilde{\alpha}^0(s'))\|\tilde{e}^0(s')\|\right)^{\oplus}.$$

If we put in (6.18)

$$\bar{\alpha}^0(s') = \left(\frac{d\rho^{-1}(s')}{ds'}\right)^{\oplus} \tilde{\alpha}^0(s') = \begin{cases} \tilde{\alpha}^0(s'), & \text{if } \|\tilde{e}^0(s')\| = 1, \\ 1, & \text{if } \|\tilde{e}^0(s')\| = 0, \end{cases}$$

$$\bar{e}^0(s') = \begin{cases} \tilde{e}^0(s'), & \text{if } \|\tilde{e}^0(s')\| = 1, \\ \text{arbitrary}, & \text{but of unit norm}, \end{cases}$$

and

$$\bar{u}_1^0(s') = \tilde{u}_1^0(s')$$

we obtain a set of admissible controls $\{\bar{\alpha}^0(s'), \bar{e}^0(s'), \bar{u}_1^0(s')\}$, and on an interval $[0, \tilde{T}_1]$, with $\tilde{T}_1 \leq T + M$, we have that

$$\frac{d\tilde{z}^0(s')}{ds'} = 1 - \bar{\alpha}(s').$$

Theorem 6.1 establishes a complete correspondence between optimal controls in the original generalized optimization problem and the auxiliary problem with bounded controls. The auxiliary problem belongs to a class of optimal control problems with phase constraints, for which optimality conditions in the form of the maximum principle are well known. Hence, we can formulate these conditions in the auxiliary problem and then transfer them to the original problem with generalized controls. In this way we avoid the rather complex issue of dealing with variations (i.e., small changes) of discontinuous functions of bounded variation [90, 91]. However, in this approach, the type of extremum we are studying is quite difficult to define exactly. We can say that it belongs to a class of strong extrema only because "needle-shaped" variations of controls in the auxiliary problem lead to shifts and decomposition of impulses, which undoubtedly are "strong" variations of the control in the original problem. However, since the representation of a generalized control is not unique, especially at jump points, the characterization of the type of extremum, and more precisely, of the type of the optimality conditions is a nontrivial task. Nevertheless, there are some cases in which defining the type of extremum is possible; such results can be found in [110, 118, 119, 121–124, 128, 206].

6.2.2 *Generalized Maximum Principle in a Problem with Phase Constraints*

We introduce the following notations:

$$\begin{aligned}
&\mathscr{H}_0(p_x, p_t, x, v, u, t) = \langle p_x, f(x, v, u, t)\rangle + p_t,\\
&\mathscr{H}_1(p_x, p_v, x, v, u, t, e) = \langle p_x, B(x, v, u, t)e\rangle + p_v,\\
&\mathscr{L}(x_0, x_1, v) = \lambda_0\phi_0(x_0, x_1, v) + \langle l_1, h(x_0, x_1, v)\rangle +\\
&\qquad + \langle l_2, S(x_0, x_1, v)\rangle + \gamma_0(v - M),
\end{aligned} \tag{6.19}$$

where $\langle\cdot,\cdot\rangle$ denotes the scalar product, $p_x \in \mathbb{R}^n$, $p_t, p_v \in \mathbb{R}^1$, $u \in \mathbb{R}^k$, $e \in \mathbb{R}^m$, $x \in \mathbb{R}^n$, $v \in \mathbb{R}^1$, $\lambda_0 \in \mathbb{R}^1$, $l_1 \in \mathbb{R}^1$, and $l_2 \in \mathbb{R}^2$.

Denote by $D_{\mathscr{V}}$ the support of the measure $\mathscr{V}$ on $[0, T]$, where $\mathscr{V}([0, T]) \le M < \infty$. The set $D_{\mathscr{V}}$ admits the decomposition

$$D_{\mathscr{V}} = D_{\mathscr{V}}^{\mathrm{c}} \cup D_{\mathscr{V}}^{\mathrm{d}},$$

where the disjoint subsets $D_{\mathscr{V}}^{\mathrm{c}}$ and $D_{\mathscr{V}}^{\mathrm{d}}$ are the supports of the continuous and discrete components of measure $\mathscr{V}(dt)$, respectively. The subset $D_{\mathscr{V}}^{\mathrm{d}}$ is countable;

$$D^{\mathrm{d}}_{\mathscr{V}} = \{\tau \in [0,T] : \mathscr{V}(\{\tau\}) > 0\} = \{\tau \in [0,T] : \Delta V^0(\tau) > 0\}.$$

Assumption 6.1 The functions f and B are continuous with respect to all the variables and continuously differentiable with respect to (x, v, t). The functions ϕ_0, S, h, and g are continuously differentiable with respect to all the variables.

The partial derivative of a function F with respect to the variable z will be denoted by F'_z. If F is vector valued function and z is a vector variable, then F'_z is the matrix-valued function with the components $(F'_z)_{ij} = \dfrac{\partial F_i}{\partial z_j}$.

The sought-for necessary optimality condition is given by the following result.

Theorem 6.2 *Suppose Assumption 6.1 is in force. Let $\{X^0(\cdot), V^0(\cdot)\}$ be the optimal generalized solution of the problem (6.1)–(6.5), with corresponding optimal control*

$$\{\mathscr{V}(dt), u^0(t), w^0(t);\ u^0_\tau(s), e^0_\tau(s) \textit{ for } \tau \in D^{\mathrm{d}}_{\mathscr{V}}\}.$$

Then there exists a collection of elements

$$\Lambda = \{\lambda_0, \beta_0, \gamma_0, l_1, l_2, p^0_x(\cdot), p^0_v(\cdot), p^0_t(\cdot), \lambda_k(dt),$$
$$p^0_{x,\tau}(\cdot), p^0_{v,\tau}(\cdot), p^0_{t,\tau}(\cdot), \nu^k_\tau(ds) \quad \textit{for } k = 1, \dots, N_3,\ \tau \in D^{\mathrm{d}}_{\mathscr{V}}\},$$

not simultaneously equal to zero, where

(1) $\lambda_0 \geq 0$, $\gamma_0 \geq 0$, *and* β_0 *are real numbers;*
(2) $l_1 \in \mathbb{R}^{N_1}$ *and* $l_2 \in \mathbb{R}^{N_2}$ *are vectors such that the components of* l_2 *are non-negative and the functions* $\{X^0(\cdot), V^0(\cdot)\}$ *satisfy the condition*

$$\left\langle l_2, S(X^0(0-), X^0(T), V^0(T))\right\rangle + \gamma_0(V^0(T) - M) = 0, \tag{6.20}$$

such that the following holds.
(3) *The vector-valued functions* $\{p^0_x(\cdot)$, $p^0_v(\cdot)$ *and* $p^0_t(\cdot)\}$ *satisfy the equations*

$$\begin{aligned} p^0_x(t) = &-\mathscr{L}'_{x(T)}(X^0(0-), X^0(T), V^0(T)) + \\ &+ \int_t^T (\mathscr{H}_0)'_x\big(p^0_x(\tau), p^0_t(\tau), X^0(\tau), V^0(\tau), u^0(\tau), \tau\big) d\tau + \\ &+ \int_t^T (\mathscr{H}_1)'_x\big(p^0_x(\tau), p^0_v(\tau), X^0(\tau), V^0(\tau), u^0(\tau), \tau, w^0(\tau)\big) dV^{\mathrm{c}}(\tau) + \\ &+ \sum_{\{\tau > t, \tau \in D^{\mathrm{d}}_{\mathscr{V}}\}} \Delta p^0_x(\tau) - \int_t^T \sum_{k=1}^{N_3} g'_{k_x}(X^0(\tau), V^0(\tau), \tau) d\lambda_k(\tau), \end{aligned} \tag{6.21}$$

with p_x^0 subject at $t = 0$ to the boundary condition

$$p_x^0(0-) = \mathscr{L}'_{x(0)}(X^0(0-), X^0(T), V^0(T)), \tag{6.22}$$

$$\begin{aligned} p_v^0(t) = &-\mathscr{L}'_v(X^0(0-), X^0(T), V^0(T)) + \\ &+ \int_t^T (\mathscr{H}_0)'_v\big(p_x^0(\tau), p_t^0(\tau), X^0(\tau), V^0(\tau), u^0(\tau), \tau\big)d\tau + \\ &+ \int_t^T (\mathscr{H}_1)'_v\big(p_x^0(\tau), p_v^0(\tau), X^0(\tau), V^0(\tau), u^0(\tau), \tau, w^0(\tau)\big)dV^c(\tau) + \\ &+ \sum_{\{\tau>t, \tau\in D^{\mathrm{d}}_{\mathscr{V}}\}} \Delta p_v^0(\tau) - \int_t^T \sum_{k=1}^{N_3} g'_{k_v}(X^0(\tau), V^0(\tau), \tau)d\lambda_k(\tau), \end{aligned} \tag{6.23}$$

and

$$\begin{aligned} p_t^0(t) = \beta_0 &+ \int_t^T (\mathscr{H}_0)'_t(p_x^0(\tau), p_t^0(\tau), X^0(\tau), V^0(\tau), u^0(\tau), \tau)d\tau + \\ &+ \int_t^T (\mathscr{H}_1)'_t(p_x^0(\tau), p_v^0(\tau), X^0(\tau), V^0(\tau), u^0(\tau), \tau, w^0(\tau))dV^c(\tau) + \\ &+ \sum_{\{\tau>t, \tau\in D^{d}_{\mathscr{V}}\}} \Delta p_t^0(\tau) - \int_t^T \sum_{k=1}^{N_3} g'_{k_t}(X^0(\tau), V^0(\tau), \tau)d\lambda_k(\tau); \end{aligned} \tag{6.24}$$

(4) *The non-negative measures $\lambda_k(dt)$ in Eq. (6.21) are localized on the subsets*

$$T_k^0 = \{t \in [0, T] \setminus D^{\mathrm{d}}_{\mathscr{V}} : g_k(X^0(t), V^0(t), t) = 0\}. \tag{6.25}$$

(5) *The jumps of the functions $p_x^0(\cdot)$, $p_v^0(\cdot)$ and $p_t^0(\cdot)\}$ at the points $\tau \in D^{\mathrm{d}}_{\mathscr{V}}$ are defined by the relations*

$$\begin{aligned} \Delta p_x^0(\tau) &= p_{x,\tau}^0(\Delta V^0(\tau)) - p_{x,\tau}^0(0), \\ \Delta p_v^0(\tau) &= p_{v,\tau}^0(\Delta V^0(\tau)) - p_{v,\tau}^0(0), \\ \Delta p_t^0(\tau) &= p_{t,\tau}^0(\Delta V^0(\tau)) - p_{t,\tau}^0(0), \end{aligned} \tag{6.26}$$

where $p_{x,\tau}^0(\cdot)$, $p_{v,\tau}^0(\cdot)$ and $p_{t,\tau}^0(\cdot)$ satisfy on $[0, \Delta V^0(\tau)]$ the system of equations

$$p^0_{x,\tau}(s) = p^0_x(\tau) - \int_s^{\Delta V^0(\tau)} \sum_{k=1}^{N_3} g'_{k_x}\big(y^0_\tau(\omega), z^0_\tau(\omega), \tau\big) dv^k_\tau(\omega) +$$

$$+ \int_s^{\Delta V^0(\tau)} (\mathscr{H}_1)'_x\big(p^0_{x,\tau}(\omega), p^0_v(\omega), y^0_\tau(\omega), z^0_\tau(\omega), u^0_\tau(\omega), \tau, e^0_\tau(\omega)\big) d\omega,$$

$$p^0_{v,\tau}(s) = p^0_v(\tau) - \int_s^{\Delta V^0(\tau)} \sum_{k=1}^{N_3} g'_{k_v}\big(y^0_\tau(\omega), z^0_\tau(\omega), \tau\big) dv^k_\tau(\omega) +$$

$$+ \int_s^{\Delta V^0(\tau)} (\mathscr{H}_1)'_v\big(p^0_{x,\tau}(\omega), p^0_v(\omega), y^0_\tau(\omega), z^0_\tau(\omega), u^0_\tau(\omega), \tau, e^0_\tau(\omega)\big) d\omega,$$

$$p^0_{t,\tau}(s) = p^0_t(\tau) - \int_s^{\Delta V^0(\tau)} \sum_{k=1}^{N_3} g'_{k_x}\big(y^0_\tau(\omega), z^0_\tau(\omega), \tau\big) dv^k_\tau(\omega)+$$

$$+ \int_s^{\Delta V^0(\tau)} (\mathscr{H}_1)'_t\big(p^0_{x,\tau}(\omega), p^0_v(\omega), y^0_\tau(\omega), z^0_\tau(\omega), u^0_\tau(\omega), \tau, e^0_\tau(\omega)\big) d\omega, \tag{6.27}$$

where the functions $y^0_\tau(s)$ *and* $z^0_\tau(s)\}$ *are the solutions of the system*

$$\begin{cases} \dot{y}^0_\tau(s) = B\big(y^0_\tau(s), z^0_\tau(s), u^0_\tau(s), \tau\big) e^0_\tau(s), \\ \dot{z}^0_\tau(s) = 1 \end{cases} \tag{6.28}$$

with the initial conditions

$$y^0_\tau(0) = X^0(\tau-), \quad z^0_\tau(0) = V^0(\tau-).$$

(6) *The non-negative measures* $v^k_\tau(ds)$ *in (6.27) are localized on the subsets*

$$T^k_\tau = \{s \in [0, \Delta V^0(\tau)] : g_k\big(y^0_\tau(s), z^0_\tau(s), \tau\big) = 0\}. \tag{6.29}$$

For this collection of elements the optimal generalized solution $\{X^0(\cdot), V^0(\cdot)\}$ *and the corresponding optimal generalized control*

$$\{\mathscr{V}(dt), u^0(t), w^0(t); \quad u^0_\tau(s), e^0_\tau(s) \quad \text{for } \tau \in D^{\mathrm{d}}_{\mathscr{V}}\}$$

satisfy the generalized maximum principle, *which is expressed by the following system of inequalities on the subsets* $[0, T] \setminus D_{\mathscr{V}}$, $D^{\mathrm{c}}_{\mathscr{V}}$ *and* $D^{\mathrm{d}}_{\mathscr{V}}$, *respectively:*

- *For a.e.* $t \in [0, T] \setminus D_{\mathscr{V}}$,

$$\max_{u\in U,\, e\in K\cap\{\|e\|=1\}} \mathscr{H}_1\big(p_x^0(t), p_v^0(t), X^0(t), V^0(t), u, t, e\big) \leq$$
$$\leq \mathscr{H}_0\big(p_x^0(t), p_t^0(t), X^0(t), V^0(t), u^0(t), t\big) = \tag{6.30}$$
$$= \max_{u\in U} \mathscr{H}_0\big(p_x^0(t), p_t^0(t), X^0(t), V^0(t), u, t\big) = 0.$$

- *For $\mathscr{V}(dt)$-a.e. $t \in D^{\mathrm{c}}_{\mathscr{V}}$,*

$$\max_{u\in U,\, e\in K\cap\{\|e\|=1\}} \mathscr{H}_1\big(p_x^0(t), p_v^0(t), X^0(t), V^0(t), u, t, e\big) =$$
$$= \mathscr{H}_1\big(p_x^0(t), p_v^0(t), X^0(t), V^0(t), u^0(t), t, w^0(t)\big) =$$
$$= \mathscr{H}_0\big(p_x^0(t), p_t^0(t), X^0(t), V^0(t), u^0(t), t\big) = \tag{6.31}$$
$$= \max_{u\in U} \mathscr{H}_0\big(p_x^0(t), p_t^0(t), X^0(t), V^0(t), u, t\big) = 0.$$

- *For $t \in D^{\mathrm{d}}_{\mathscr{V}}$,*

$$0 = \max_{u\in U,\, e\in K\cap\{\|e\|=1\}} \mathscr{H}_1\big(p_x^0(t), p_v^0(t), X^0(t), V^0(t), u, t, e\big) \geq$$
$$\geq \max_{u\in U} \mathscr{H}_0\big(p_x^0(t), p_t^0(t), X^0(t), V^0(t), u, t\big). \tag{6.32}$$

Moreover, for a.e. $\tau \in D^{\mathrm{d}}_{\mathscr{V}}$, the following inequality holds on the corresponding segment $[0, \Delta V^0(\tau)]$*:*

$$\max_{u\in U} \mathscr{H}_0\big(p_{x,\tau}^0(s), p_{t,\tau}^0(s), y_\tau^0(s), z_\tau^0(s), u, \tau\big) \leq$$
$$\leq \max_{u\in U,\, e\in K\cap\{\|e\|=1\}} \mathscr{H}_1\big(p_{x,\tau}^0(s), p_{v,\tau}^0(s), y_\tau^0(s), z_\tau^0(s), u, t, e\big) = \tag{6.33}$$
$$= \mathscr{H}_1\big(p_{x,\tau}^0(s), p_{v,\tau}^0(s), y_\tau^0(s), z_\tau^0(s), u_\tau^0(s), \tau, e_\tau^0(s)\big) = 0.$$

Proof By Theorem 6.1, in the problem (6.6)–(6.10) there exist the optimal control and optimal solution

$$\{\alpha^0(\cdot), u_1^0(\cdot), e_1^0(\cdot), Y^0(\cdot), z^0(\cdot), \eta^0(\cdot)\},$$

defined on some interval $[0, T_1]$. The auxiliary problem is supplemented by the terminal constraints $z(T_1) \leq M$ and $\eta(T_1) = T$, which ensure that the constraint $T_1 \leq T + M$ is satisfied. Therefore, the auxiliary problem belongs to the class of free terminal problems and the right-hand side of the system (6.6) does not depend on s. All these properties enable us to formulate the strong necessary optimality condition in the following form [113, 146, 173].

There exists a collection of elements,

$$\tilde{\Lambda} = \{\lambda_0, \beta_0, \gamma_0, l_1, l_2, \psi_y^0(\cdot), \psi_z^0(\cdot), \psi_\eta^0(\cdot), \tilde{\lambda}_k(ds), \ k = 1, \dots, N_3\},$$

not simultaneously equal to zero, where

1. $\lambda_0 \geq 0$, $\gamma_0 \geq 0$ and β_0 are the real numbers;
2. $l_1 \in \mathbb{R}^{N_1}$ and $l_2 \in \mathbb{R}^{N_2}$ are vectors such that the components of l_2 are non-negative, and the solution $\{Y^0(\cdot), z^0(\cdot), \eta^0(\cdot)\}$ satisfies the condition

$$\big\langle l_2, S(Y^0(0), Y^0(T_1), z^0(T_1))\big\rangle + \gamma_0\big(z^0(T_1) - M\big) = 0; \tag{6.34}$$

3. the vector-functions $\{\psi_y^0(\cdot), \psi_z^0(\cdot), \psi_\eta^0(\cdot)\}$ satisfy the system of equations

$$\begin{aligned}
&\psi_y^0(s) = -\mathscr{L}'_{x(T)}\big(y^0(0), y^0(T_1), z^0(T_1)\big) + \\
&+ \int_s^{T_1} \alpha^0(\tau)(\mathscr{H}_0)'_x\big(\psi_y^0(\tau), \psi_\eta^0(\tau), Y^0(\tau), z^0(\tau), u_1^0(\tau), \eta^0(\tau)\big)d\tau + \\
&+ \int_s^{T_1} (1-\alpha^0(\tau))(\mathscr{H}_1)'_x\big(\psi_y^0(\tau), \psi_z^0(\tau), Y^0(\tau), \\
&z^0(\tau), u_1^0(\tau), \eta^0(\tau), e^0(\tau)\big)d\tau - \\
&- \int_s^{T_1} \sum_{k=1}^{N_3} g'_{k_x}\big(Y^0(\tau), z^0(\tau), \eta^0(\tau)\big)d\tilde{\lambda}_k(\tau),
\end{aligned} \tag{6.35}$$

$$\begin{aligned}
&\psi_z^0(s) = -\mathscr{L}'_v\big(y^0(0), y^0(T_1), z^0(T_1)\big) + \\
&+ \int_s^{T_1} \alpha^0(\tau)(\mathscr{H}_0)'_v\big(\psi_y^0(\tau), \psi_\eta^0(\tau), Y^0(\tau), z^0(\tau), u_1^0(\tau), \eta^0(\tau)\big)d\tau + \\
&+ \int_s^{T_1} (1-\alpha^0(\tau))(\mathscr{H}_1)'_v\big(\psi_y^0(\tau), \psi_z^0(\tau), Y^0(\tau), \\
&z^0(\tau), u_1^0(\tau), \eta^0(\tau), e^0(\tau)\big)d\tau - \\
&- \int_s^{T_1} \sum_{k=1}^{N_3} g'_{k_v}\big(Y^0(\tau), z^0(\tau), \eta^0(\tau)\big)d\tilde{\lambda}_k(\tau),
\end{aligned} \tag{6.36}$$

$$\psi_\eta^0(s) = \beta_0 + \int_s^{T_1} \alpha^0(\tau)(\mathscr{H}_0)'_t\big(\psi_y^0(\tau), \psi_\eta^0(\tau), Y^0(\tau), z^0(\tau), u_1^0(\tau), \eta^0(\tau)\big)d\tau +$$

$$+ \int_s^{T_1} (1-\alpha^0(\tau))(\mathscr{H}_1)'_t\big(\psi_y^0(\tau), \psi_z^0(\tau), Y^0(\tau),$$

$$z^0(\tau), u_1^0(\tau), \eta^0(\tau), e^0(\tau)\big)d\tau -$$

$$- \int_s^{T_1} \sum_{k=1}^{N_3} g'_{k_t}(Y^0(\tau), z^0(\tau), \eta^0(\tau))d\tilde{\lambda}_k(\tau), \tag{6.37}$$

with terminal conditions for ψ_y^0 at $s = 0$:

$$\psi_y^0(0) = \mathscr{L}'_{x(0)}\big(Y^0(0), Y^0(T_1), z^0(T_1)\big); \tag{6.38}$$

4. the non-negative measures $\tilde{\lambda}_k(ds)$ in (6.35) are localized on the sets

$$\tilde{T}_k = \{s \in [0, T_1] : g_k\big(Y^0(s), z^0(s), \eta^0(s)\big) = 0\}, \tag{6.39}$$

and the elements $\tilde{\Lambda}$ satisfy the *maximum principle* a.e. on $[0, T_1]$, i.e.,

$$\mathscr{H}\big(\psi_y^0(s), \psi_z^0(s), \psi_\eta^0(s), Y^0(s), z^0(s), \eta^0(s), \alpha, u_1, e\big) \le$$

$$\le \mathscr{H}\big(\psi_y^0(s), \psi_z^0(s), \psi_\eta^0(s), Y^0(s), z^0(s), \eta^0(s), \alpha^0(s), u_1^0(s), e^0(s)\big) = 0, \tag{6.40}$$

for all $\{\alpha, u_1, e\}$ satisfying the constraints (6.7). Here

$$\begin{aligned} &\mathscr{H}\big(\psi_y, \psi_z, \psi_\eta, y, z, \eta, \alpha, u_1, e\big) = \\ &\quad = \alpha\mathscr{H}_0\big(\psi_y, \psi_\eta, y, z, u_1, \eta\big) + (1-\alpha)\mathscr{H}_1\big(\psi_y, \psi_z, y, z, u_1, \eta, e\big), \end{aligned} \tag{6.41}$$

where $\mathscr{H}_0$ and $\mathscr{H}_1$ are defined in (6.19).

Since $\mathscr{H}$ is linear with respect to $\alpha \in [0, 1]$, we have

$$\max_{\alpha, u_1, e} \mathscr{H} = \max\big\{\max_{u_1} \mathscr{H}_0, \max_{u_1, e} \mathscr{H}_1\big\}$$

and

- if $\alpha^0(s) = 1$, then $0 = \max\limits_{u_1} \mathscr{H}_0 \ge \max\limits_{u_1, e} \mathscr{H}_1$;
- if $\alpha^0(s) = 0$, then $\max\limits_{u_1} \mathscr{H}_0 \le \max\limits_{u_1, e} \mathscr{H}_1 = 0$;
- if $\alpha^0(s) \in (0, 1)$, then $\max\limits_{u_1} \mathscr{H}_0 = \max\limits_{u_1, e} \mathscr{H}_1 = 0$.

Also,

$$\alpha^0(s) = \begin{cases} 1, & \text{if} \quad \max\limits_{u_1} \mathscr{H}_0 \geq \max\limits_{u_1,e} \mathscr{H}_1, \\ 0, & \text{if} \quad \max\limits_{u_1} \mathscr{H}_0 \leq \max\limits_{u_1,e} \mathscr{H}_1, \\ \in [0,1], & \text{if} \quad \max\limits_{u_1} \mathscr{H}_0 = \max\limits_{u_1,e} \mathscr{H}_1. \end{cases} \tag{6.42}$$

Consider the set $[0, T] \setminus D_{\mathscr{V}}$. Then $\dot{\Gamma}^0(t) = 1$ a.e. on this set. Therefore, since α^0 is the derivative of he function η^0, which is the inverse of the function Γ^0, we have $\alpha^0(\Gamma^0(t)) = 1$ a.e. on $[0, T] \setminus D_{\mathscr{V}}$. Hence, for almost every s such that $\eta^0(s) \in [0, T] \setminus D_{\mathscr{V}}$ we have $\alpha^0(s) = 1$. In view of (6.42),

$$\begin{aligned} &\max_{u_1,e} \mathscr{H}_1\big(\psi_y^0(s), \psi_z^0(s), Y^0(s), z^0(s), u_1, e, \eta^0(s)\big) \leq \\ &\qquad \leq \max_{u_1} \mathscr{H}_0\big(\psi_y^0(s), \psi_\eta^0(s), Y^0(s), z^0(s), u_1, \eta^0(s)\big) = \\ &\qquad = \max_{u_1} \mathscr{H}_0\big(\psi_y^0(s), \psi_\eta^0(s), Y^0(s), z^0(s), u_1^0(s), \eta^0(s)) = 0, \end{aligned} \tag{6.43}$$

a.e. on the set $\{s : \eta^0(s) \in [0, T] \setminus D_{\mathscr{V}}\}$. Since

$$Y^0(\Gamma^0(t)) = X^0(t), \quad z^0(\Gamma^0(t)) = V^0(t),$$

$$\eta^0(\Gamma^0(t)) = t, \quad u_1^0(\Gamma^0(t)) = u^0(t),$$

upon setting

$$p_x^0(t) = \psi_y^0(\Gamma^0(t)), \quad p_v^0(t) = \psi_z^0(\Gamma^0(t)), \quad p_t^0(t) = \psi_\eta^0(\Gamma^0(t)). \tag{6.44}$$

we obtain the assertion (6.30) of the theorem.

Next, if $t = \tau \in D_{\mathscr{V}}^{\mathrm{d}}$, we have $\eta^0(s) = \tau$ on the segment $[\Gamma^0(\tau-), \Gamma^0(\tau)]$, and so $\alpha^0(s) = 0$. In view of (6.42), we get

$$\begin{aligned} &\max_{u_1} \mathscr{H}_0\big(\psi_y^0(s), \psi_\eta^0(s), Y^0(s), z^0(s), u_1, \eta^0(s)\big) \leq \\ &\qquad \leq \max_{u_1,e} \mathscr{H}_1\big(\psi_y^0(s), \psi_z^0(s), Y^0(s), z^0(s), u_1, \eta^0(s), e\big) = \\ &\qquad = \mathscr{H}_1\big(\psi_y^0(s), \psi_z^0(s), Y^0(s), z^0(s), u_1^0(s), \eta^0(s), e^0(s)\big) = 0, \end{aligned} \tag{6.45}$$

for $s \in [\Gamma^0(\tau-), \Gamma^0(\tau)]$. Making the time change $s' = s - \Gamma(\tau-)$, setting

$$
\begin{aligned}
&y_\tau^0(s') = Y^0(s' + \Gamma(\tau-)), \qquad z_\tau^0(s') = z^0(s' + \Gamma(\tau-)),\\
&u_\tau^0(s') = u_1^0(s' + \Gamma(\tau-)), \qquad e_\tau^0(s') = e^0(s' + \Gamma(\tau-)),\\
&p_{x,\tau}^0(s') = \psi_y^0(s' + \Gamma(\tau-)), \quad p_{v,\tau}^0(s') = \psi_z^0(s' + \Gamma(\tau-)),\\
&p_{t,\tau}^0(s') = \psi_\eta^0(s' + \Gamma(\tau-)),
\end{aligned}
\tag{6.46}
$$

and using the relations (6.44), we obtain assertions (6.32) and (6.33) of the theorem.

The set $D_{\mathcal{V}}^{\mathrm{c}}$ is the disjoint union of the support $D_{\mathcal{V}}^{\mathrm{a}}$ of the absolutely continuous Lebesgue component of the measure $\mathcal{V}(dt)$ and the support $D_{\mathcal{V}}^{\mathrm{s}}$ of the singular component of $\mathcal{V}(dt)$. Moreover, the Lebesgue measure of the set $D_{\mathcal{V}}^{\mathrm{s}}$ equals zero (see [322]) and we have

$$
\{s : \eta^0(s) \in D_{\mathcal{V}}^{\mathrm{c}}\} = \{s : \eta^0(s) \in D_{\mathcal{V}}^{\mathrm{a}}\} \cup \{s : \eta^0(s) \in D_{\mathcal{V}}^{\mathrm{s}}\}.
$$

The derivative of $V^0(t)$ exists for Lebesgue-almost all $t \in D_{\mathcal{V}}^{a}$; therefore, for s such that $\eta^0(s) \in D_{\mathcal{V}}^{\mathrm{a}}$, the value of $\alpha^0(s)$, which is equal

$$
\alpha^0(s) = \left.\frac{1}{1 + \dot{V}^0(t)}\right|_{t=\eta^0(s)}
$$

satisfies the inequality $0 < \alpha^0(s) < 1$. On the set $\{s : \eta^0(s) \in D_{\mathcal{V}}^{\mathrm{s}}\}$ we have $\alpha^0(s) = \dot{\eta}^0(s) = 0$, provided the corresponding derivative exists. Therefore, $0 \le \alpha^0(s) < 1$ a.e. on the set $\{s : \eta^0(s) \in D_{\mathcal{V}}^{\mathrm{c}}\}$. As above, we then conclude that a.e. on the set $\{s : \eta^0(s) \in D_{\mathcal{V}}^{\mathrm{c}}\}$ it holds that

$$
\begin{aligned}
&\max_{u_1, e} \mathcal{H}_1\big(\psi_y^0(s), \psi_z^0(s), Y^0(s), z^0(s), u_1, e, \eta^0(s)\big) =\\
&\quad = \mathcal{H}_1\big(\psi_y^0(s), \psi_z^0(s), Y^0(s), z^0(s), u_1^0(s), e^0(s), \eta^0(s)\big) =\\
&\quad = \mathcal{H}_0\big(\psi_y^0(s), \psi_\eta^0(s), Y^0(s), z^0(s), u_1^0(s), \eta^0(s)\big) =\\
&\quad = \max_{u_1} \mathcal{H}_0\big(\psi_y^0(s), \psi_\eta^0(s), Y^0(s), z^0(s), u_1^0(s), \eta^0(s)\big) = 0.
\end{aligned}
\tag{6.47}
$$

So, as above, taking into account the relations

$$
\begin{aligned}
&Y^0(\Gamma^0(t)) = X^0(t), \quad z^0(\Gamma^0(t)) = V^0(t),\\
&\eta^0(\Gamma^0(t)) = t, \quad u_1^0(\Gamma^0(t)) = u^0(t), \quad w^0(t) = e^0(\Gamma^0(t))
\end{aligned}
$$

we obtain assertion (6.30) of the theorem.

To complete the proof it remains to derive equations for the set of conjugate variables (6.44) and (6.46). Consider the variable $p_x^0(t) = \psi_y^0(\Gamma^0(t))$. It follows from Eq. (6.35) and relation $\Gamma^0(T) = T_1$ that

$$p_x^0(t) = \psi_y^0(\Gamma^0(t)) = -\mathscr{L}'_{x(T)}\big(Y^0(0), Y^0(T_1), z^0(T_1)\big) +$$

$$+ \int_{\Gamma^0(t)}^{\Gamma^0(T)} \alpha^0(\tau)(\mathscr{H}_0)'_x\big(\psi_y^0(\tau), \psi_\eta^0(\tau), Y^0(\tau), z^0(\tau), u_1^0(\tau), \eta^0(\tau)\big)d\tau +$$

$$+ \int_{\Gamma^0(t)}^{\Gamma^0(T)} (1-\alpha^0(\tau))(\mathscr{H}_1)'_x\big(\psi_y^0(\tau), \psi_z^0(\tau), Y^0(\tau),$$

$$z^0(\tau), u_1^0(\tau), \eta^0(\tau), e^0(\tau)\big)d\tau -$$

$$- \int_{\Gamma^0(t)}^{\Gamma^0(T)} \sum_{k=1}^{N_3} g'_{k_x}\big(Y^0(\tau), z^0(\tau), \eta^0(\tau)\big)d\tilde{\lambda}_k(\tau) = p_x^0(T) + I_1(t) + I_2(t) + I_3(t). \tag{6.48}$$

Since

$$\begin{aligned} &Y^0(\Gamma^0(t)) = X^0(t), \quad z^0(\Gamma^0(t)) = V^0(t), \\ &\eta^0(\Gamma^0(t)) = t, \quad u_1(\Gamma^0(t) = u^0(t), \end{aligned} \tag{6.49}$$

applying the time transformation $\tau = \Gamma(t)$ we obtain

$$p_x^0(T) = -\mathscr{L}'_{x(T)}\big(Y^0(0), Y^0(T_1), z^0(T_1)\big) = \mathscr{L}'_{x(T)}\big(X^0(0-), X^0(T), V(T)\big). \tag{6.50}$$

Further, applying the same transformation in the integral $I_1(t)$ we obtain

$$I_1(t) = \int_{\Gamma^0(t)}^{\Gamma^0(T)} \alpha^0(\tau)(\mathscr{H}_0)'_x\big(\psi_y^0(\tau), \psi_\eta^0(\tau), Y^0(\tau), z^0(\tau), u_1^0(\tau), \eta^0(\tau)\big)d\tau =$$

$$= \int_{\eta^0(\Gamma^0(t))}^{\eta^0(\Gamma^0(T))} (\mathscr{H}_0)'_x\big(\psi_y^0(\Gamma^0(\tau)), \psi_\eta^0(\Gamma^0(\tau)), Y^0(\Gamma^0(\tau)), z^0(\Gamma^0(\tau)),$$

$$u_1^0(\Gamma^0(\tau)), \eta^0(\Gamma^0(\tau)))\alpha^0(\Gamma^0(\tau)\big)d\tau =$$

$$= \int_t^T (\mathscr{H}_0)'_x\big(p_x^0(\tau), p_t^0(\tau), X^0(\tau), V^0(\tau), u^0(\tau), \tau\big)d\tau. \tag{6.51}$$

Next, to calculate the term $I_2(t)$, we use (6.49) and (6.46) and the relation

$$\int_0^{\Gamma^0(t)} \mathbf{I}\{\tau : \eta^0(\tau) \notin D^{\mathrm{d}}_{\mathscr{V}}\}(1-\alpha^0(\tau))e^0(\tau)d\tau = \int_0^t w^0(\tau)dV^{\mathrm{c}}(\tau), \tag{6.52}$$

which defines $w^0(\cdot)$. Since $\alpha^0(s) = 0$ for $\eta^0(s) = \tau \in D^{\mathrm{d}}_{\mathscr{V}}$, we get the expression

$$I_2(t) =$$

$$= \int_{\Gamma^0(t)}^{\Gamma^0(T)} (1-\alpha^0(\tau))(\mathscr{H}_1)'_x\big(\psi^0_y(\tau), \psi^0_z(\tau), Y^0(\tau),$$

$$z^0(\tau), u^0_1(\tau), \eta^0(\tau), e^0(\tau)\big)d\tau =$$

$$= \int_{\Gamma^0(t)}^{\Gamma^0(T)} \mathbf{I}\{\tau : \eta^0(\tau) \notin D^{d}_{\mathscr{V}}\}(1-\alpha^0(\tau))\times$$

$$\times (\mathscr{H}_1)'_x\big(\psi^0_y(\tau), \psi^0_z(\tau), Y^0(\tau), z^0(\tau), u^0_1(\tau), \eta^0(\tau), e^0(\tau)\big)d\tau +$$

$$+ \sum_{\{\tau>t, \tau\in D^{\mathrm{d}}_{\mathscr{V}}\}} \int_{\Gamma^0(\tau-)}^{\Gamma^0(\tau)} (\mathscr{H}_1)'_x\big(\psi^0_y(\omega), \psi^0_z(\omega), Y^0(\omega), z^0(\omega), u^0_1(\omega), \tau, e^0(\omega)\big)d\omega =$$

$$= \int_t^T (\mathscr{H}_1)'_x\big(p^0_x(\tau), p^0_v(\tau), X^0(\tau), V^0(\tau), u^0(\tau), \tau, w^0(\tau)\big)dV^{\mathrm{c}}(\tau) +$$

$$+ \sum_{\{\tau>t, \tau\in D^{\mathrm{d}}_{\mathscr{V}}\}} \Delta I_2(\tau), \tag{6.53}$$

where

$$\Delta I_2(\tau) =$$

$$= \int_{\Gamma^0(\tau-)}^{\Gamma^0(\tau)} (\mathscr{H}_1)'_x\big(\psi^0_y(\omega), \psi^0_z(\omega), Y^0(\omega), z^0(\omega), u^0_1(\omega), \tau, e^0(\omega)\big)d\omega =$$

$$= \int_0^{\Delta V^0(\tau)} (\mathscr{H}_1)'_x\big(\psi^0_{x,\tau}(\omega), \psi^0_{v,\tau}(\omega), y^0_\tau(\omega), z^0_\tau(\omega), u^0_\tau(\omega), \tau, e^0_\tau(\omega)\big)d\omega. \tag{6.54}$$

Finally, the term $I_3(t)$ is transformed as follows:

$$
\begin{aligned}
I_3(t) &= \int_{\Gamma^0(t)}^{\Gamma^0(T)} \sum_{k=1}^{N_3} g'_{k_x}(Y^0(\tau), z^0(\tau), \eta^0(\tau)) d\tilde{\lambda}_k(\tau) = \\
&= \int_{\Gamma^0(t)}^{\Gamma^0(T)} \sum_{k=1}^{N_3} \mathbf{I}\{\tau : \eta^0(\tau) \bar{\in} D^{\mathrm{d}}_{\mathcal{V}}\} g'_{k_x}(Y^0(\tau), z^0(\tau), \eta^0(\tau)) d\tilde{\lambda}_k(\tau) + \\
&\quad + \sum_{\{\tau > t, \tau \in D^{\mathrm{d}}_{\mathcal{V}}\}} \int_{\Gamma^0(\tau-)}^{\Gamma^0(\tau)} \sum_{k=1}^{N_3} g'_{k_x}(Y^0(\tau), z^0(\tau), \eta^0(\tau)) d\tilde{\lambda}_k(\tau) = \\
&= \int_t^T \sum_{k=1}^{N_3} g'_{k_x}(X^0(\tau), V^0(\tau), \tau) d\lambda_k(\tau) + \\
&\quad + \sum_{\{\tau > t, \tau \in D^{\mathrm{d}}_{\mathcal{V}}\}} \int_0^{\Delta V^0(\tau)} \sum_{k=1}^{N_3} g'_{k_x}(y^0_\tau(\omega), z^0_\tau(\omega), \tau) d\nu^k_\tau(\omega),
\end{aligned}
\tag{6.55}
$$

where the measures $\lambda_k(dt)$ and $\nu^k_\tau(ds)$ are defined by the formulas

$$
\lambda_k\{(a, b]\} = \int_{\Gamma^0(a)}^{\Gamma^0(b)} \mathbf{I}\{\tau : \eta^0(\tau) \bar{\in} D^{\mathrm{d}}_{\mathcal{V}}\} d\tilde{\lambda}_k(\tau)
$$

and

$$
\nu^k_\tau\{(c, d]\} = \tilde{\lambda}_k\{(\Gamma^0(\tau-) + c, \Gamma^0(\tau-) + d]\}.
$$

In view of the definition of the measure $\tilde{\lambda}_k(ds)$ (see (6.39)), the measure $\lambda_k(dt)$ is non-negative and localized on the set

$$
T^k = \{t \in [0, T] \setminus D^d_{\mathcal{V}} : g_k(X^0(t),\ V^0(t), t) = 0\}.
$$

Similarly, the measures $\nu^k_\tau(ds)$, which are defined for all $\tau \in D^{\mathrm{d}}_{\mathcal{V}}$, are non-negative and localized on the sets

$$
T^k_\tau = \{s \in [0, \Delta V^0(\tau)] : g_k(y^0_\tau(s), z^0_\tau(s), \tau) = 0\}.
$$

The relations $p^0_x(t) = \psi^0_y(\Gamma^0(t))$ and (6.46) for $t \in D^{\mathrm{d}}_{\mathcal{V}}$ imply that

$$
\begin{aligned}
\Delta p^0_x(\tau) &= p^0_x(\tau) - p^0_x(\tau-) = \psi^0_y(\Gamma^0(\tau)) - \psi^0_y(\Gamma^0(\tau-)) = \\
&= p^0_{x,\tau}(\Delta V^0(\tau)) - p^0_{x,\tau}(0).
\end{aligned}
\tag{6.56}
$$

The function $p^0_{x,\tau}(s)$ satisfies on the segment $[0, \Delta V^0(\tau)]$ the equation

$$
\begin{aligned}
p^0_{x,\tau}(s) &= \psi^0_y(s+\Gamma^0(\tau-)) = \psi^0_y(\Gamma^0(\tau)) + \\
&+ \int_s^{\Delta V^0(\tau)} (\mathscr{H}_1)'_x\Big(\psi^0_y(\omega+\Gamma^0(\tau-)), \psi^0_z(\omega+\Gamma^0(\tau-)), \\
&Y^0(\omega+\Gamma^0(\tau-)), z^0(\omega+\Gamma^0(\tau-)), u^0_1(\omega+\Gamma^0(\tau-)), \tau, e^0(\omega+\Gamma^0(\tau-))\Big)d\omega - \\
&\int_s^{\Delta V^0(\tau)} \sum_{k=1}^{N_3} g'_{k_x}(Y^0(\omega+\Gamma^0(\tau-)), z^0(\omega+\Gamma^0(\tau-)), \tau)d\tilde{\lambda}_k(\omega+\Gamma^0(\tau-)).
\end{aligned}
\tag{6.57}
$$

Therefore, upon substituting the expressions (6.44) and (6.46), and using the definition of the measure $\nu^k_\tau(ds)$ and relation (6.47), we finally conclude that

$$
\Delta p^0_x(\tau) = \Delta I_2(\tau) + \int_0^{\Delta V^0(\tau)} \sum_{k=1}^{N_3} g'_{k_x}\big(y^0_\tau(\omega), z^0_\tau(\omega), \tau\big)d\nu^k_\tau(\omega),
$$

which yields (6.27).

Thus, we have shown that the conjugate variable $p^0_x(t) = \psi^0_y(\Gamma^0(t))$ satisfies the Eq. (6.21) with the boundary conditions (6.22) and the intermediate conditions (6.26) and (6.27) at the jump points caused by the application of the impulse control. The equations for $p^0_v(\cdot)$, $p^0_t(\cdot)$ can be can be derived by the same arguments. This completes the proof. □

The next corollary states the optimality condition in a more familiar form.

Corollary 6.1 *Suppose the conditions of Theorem 6.2 are satisfied. Then the optimal solution* $\{X^0(\cdot), V^0(\cdot)\}$ *and the set of conjugate variables* Λ *obey the following inequalities:*

$$
\begin{aligned}
\max_{u\in U} \mathscr{H}_0\big(p^0_x(t), p^0_t(t), X^0(t), V^0(t), u, t\big) &= \\
= \mathscr{H}_0\big(p^0_x(t), p^0_t(t), X^0(t), V^0(t), u^0(t), t\big) &= 0
\end{aligned}
\tag{6.58}
$$

a.e. on $[0, T]$;

$$
\begin{aligned}
\max_{u\in U,\, e\in K\cap\{\|e\|=1\}} \mathscr{H}_1\big(p^0_x(t), p^0_v(t), X^0(t), V^0(t), u, t, e\big) &= \\
= \mathscr{H}_1\big(p^0_x(t), p^0_v(t), X^0(t), V^0(t), u^0(t), t, w^0(t)\big) &= 0
\end{aligned}
\tag{6.59}
$$

a.e. with respect to the measure $(V^0)^c(dt)$, *and*

$$\max_{u\in U,\, e\in K\cap\{\|e\|=1\}} \mathscr{H}_1(p_x^0(t), p_v^0(t), X^0(t), V^0(t), u, t, e) = 0 \tag{6.60}$$

a.e. with respect to the measure $V^0(dt)$.

Proof The proof follows from the relations (6.30)–(6.33) if we note that, by (6.52),

$$w^0(t) = \frac{d(\mathscr{W}^0)^c(t)}{dV^0(t)} \tag{6.61}$$

V^0-a.e. on $[0, T]$, where

$$\mathscr{W}^0([0,t]) = \int_0^{\Gamma^0(t)} (1-\alpha^0(s))e^0(s)ds,$$

and the Lebesgue measure of the set $D_{\mathscr{V}}^{d}$ is zero. Therefore, the relations (6.30) and (6.31) hold on the set $[0, T] \setminus D_{\mathscr{V}}$ and on the set where $w^0(\cdot)$ exists as the Radon-Nikodym derivative. However, this derivative exists a.e. with respect to the Lebesgue measure, which completes the proof. □

Further, we note that the measure $V^0(dt)$ is localized on the set where the function $\max_{u,e} \mathscr{H}_1(p_x^0(t), p_v^0(t), X^0(t), V^0(t), u, t, e)$ achieves the maximum with respect to t, which is equal to zero. This allows us to obtain the following more familiar integral form of the maximum principle , which is a straightforward consequence of the relations (6.31), (6.32), and (6.61).

Corollary 6.2 *Suppose the conditions of Theorem 6.2 hold. Then for any admissible control* $u(\cdot)$ *and for any vector-valued measure* $\mathscr{A}(dt)$ *taking values in the cone* K, *the optimal generalized solution* $\{X^0(\cdot), V^0(\cdot)\}$ *and the set of conjugate variables* Λ *obey the inequality*

$$\int_0^T \mathscr{H}_1\Big(p_x^0(t), p_v^0(t), X^0(t), V^0(t), u(t), t, \frac{d\mathscr{A}(t)}{d|\mathscr{A}|(t)}\Big)\, d|\mathscr{A}|(t) \le$$

$$\le \int_0^T \mathscr{H}_1\Big(p_x^0(t), p_v^0(t), X^0(t), V^0(t), u^0(t), t, w^0(t)\Big)\, d(V^0)^c(t) = 0, \tag{6.62}$$

where $|\mathscr{A}|(t) - \operatorname*{Var}_{[0,t]} \mathscr{A}(s)$.

The necessary optimality condition established in Theorem 6.2 has the form of a (strong) pointwise maximum principle. Since it is a consequence of the strong maximum principle in the auxiliary problem, it should be interpreted as a *strong optimality condition*, as illustrated by Corollary 6.1. Indeed, the class of control variations in the auxiliary problem is a class of strong variations that are not small in the $C([0, T])$ norm. The corresponding variations of the control and measures in the original problem are also not small in the $C([0, T])$ norm. Until now attempts to obtain optimality conditions by direct simultaneous variation of ordinary and impulsive controls were not successful, since in this case it is difficult to control the fulfillment of the phase constraints, especially at the instants of access of the constraint boundary at the points of impulse control implementation. The discontinuous time change method allows to do this automatically. It is interesting that this approach applies also to nonsmooth problems and remains the most convenient method for deriving optimality conditions in problems with impulsive actions [8–10, 78, 126, 128, 129, 315, 317, 356–358, 440–442]. The following sections details various classes of systems for which the discontinuous time change method allows us to obtain necessary optimality condition in the form of a generalized maximum principle.

6.3 Necessary Optimality Conditions in Generalized Optimization Problems

6.3.1 Optimality Conditions for Robust DCS

In this section we consider optimality conditions for strong generalized solutions in DCS that satisfy stability conditions (see Sect. 2.2). The dynamics of such a system is governed by equations of the form

$$\begin{cases} \dot{X}(t) = f(X(t), V(t), u(t), t) + B(X(t), t)w(t), \\ \dot{V}(t) = \|w(t)\|, \end{cases} \tag{6.63}$$

with two types of controls: $u \in U \subset \mathbb{R}^k$ and $w \in K \subseteq \mathbb{R}^m$, where U is a compact set and K is a closed convex cone. It is assumed that the control $w(t)$ is unbounded and satisfies the constraint

$$\int_0^T \|w(t)\| dt \leq M < \infty. \tag{6.64}$$

The matrix-valued function $B(X, t)$ satisfies the Frobenius condition (2.2). The generalized solution of the system (6.63) is described by a differential equation with measure.

Remark 6.3 Theorems 2.6 and 4.7 yield the following representation of the generalized solution.

Theorem 6.3 *Suppose the function B satisfies Assumption 2.2 and the sets of vectors*

$$\mathscr{L}_1(x, v, t) = f(x, v, U, t) = \{l_1 \in \mathbb{R}^n : l_1 = f(x, v, u, t), \ \ u \in U\},$$
$$\mathscr{L}_2(x, t) = \left\{ l_2 = \begin{pmatrix} l_2' \in \mathbb{R}^n \\ l_2'' \in \mathbb{R}^1 \end{pmatrix} : l_2' = B(x, t)w, \ l_2'' = \|w\|, \ \ w \in K \right\} \tag{6.65}$$

are convex for any fixed (x, v, t).

Then for any generalized solution $\{X(\cdot), V(\cdot)\}$ of the system (6.63) there exist

1. *a non-negative scalar measure $\mathscr{V}(dt)$, defined on the Borel subsets of the interval $[0, T]$, such that*

$$\mathscr{V}([0, T]) \leq M,$$

and therefore its distribution function $V(t) = \mathscr{V}([0, t])$ satisfies the inequality

$$V(T) = \mathscr{V}([0, T]) \leq M;$$

2. *a pair of Lebesgue- and $\mathscr{V}$-measurable controls $\{u_0(\cdot), w_0(\cdot)\}$, satisfying on $[0, T]$ the constraints*

$$u_0(t) \in U, \quad w_0(t) \in K \cap \{\|w_0\| \leq 1\}$$

a.e. with respect to both the Lebesgue and the $\mathscr{V}$ measures;

such that

$$X(t) = X(0-) + \int_0^t f(X(\tau), V(\tau), u_0(\tau), \tau)d\tau +$$
$$+ \int_0^t B(X(\tau), \tau)w_0(\tau)dV^c(\tau) + \sum_{\tau \leq t} \Psi(X(\tau-), \tau, w_0(\tau)\Delta V(\tau)), \tag{6.66}$$

where $V^c(\cdot)$ is the continuous component in the Lebesgue decomposition of the function of bounded variation $V(\cdot)$ (see [322]), and

$$\Delta V(\tau) = V(\tau) - V(\tau-) = \mathscr{V}(\{\tau\}).$$

Proof The representation of the continuous part of $X(\cdot)$ is standard, so we focus on the discontinuous part. By Theorem 4.7, the jump of the generalized solution at a point $\tau \in D^{\mathrm{d}}_{\mathscr{V}}$ is represented with the aid of the auxiliary system

$$\dot{y}_\tau(s) = B(y_\tau(s), \tau)e^\tau(s), \quad y_\tau(0) = X(\tau-), \tag{6.67}$$

where $e^\tau(s)$ is some control defined on the interval $[0, \Delta V(\tau)]$ such that $e^\tau(s) \in K \cap \{\|e\| \leq 1\}$, and

$$\Delta X(\tau) = y_\tau(\Delta V(\tau)) - X(\tau-).$$

Since $B(X, t)$ satisfies the Frobenius condition, Lemma 2.7 shows that the solution of Eq. (6.67) admits the representation

$$y_\tau(s) = X(\tau-) + \Psi\big(X(\tau-), \tau, \int_0^s e^\tau(s)ds\big).$$

Therefore,

$$X(\tau) = X(\tau-) + \Psi\big(X(\tau-), \tau, \int_0^{\Delta V(\tau)} e^\tau(s)\,ds\big) = \Psi\big(X(\tau-), \tau, w_0(\tau)\Delta V(\tau)\big),$$

where w_0 is defined by

$$w_0(\tau) = \frac{1}{\Delta V(\tau)} \int_0^{\Delta V(\tau)} e^\tau(s)\,ds \in K \cap \{\|w_0\| \le 1\}. \tag{6.68}$$

□

Remark 6.4 To investigate the optimality condition the general case one needs the jump representation with the aid of the auxiliary control $e_\tau(\cdot) \in \{K \cap \|e\| = 1\}$, because one cannot set this control to be constant on the interval $[0, \Delta V(\tau)]$, especially under phase constraints. For this reason we derived the optimality condition for the general case.

Thus, let us consider the optimal control problem with the performance criterion

$$J[X(\cdot), u(\cdot), w(\cdot)] = \phi_0(X(0), X(T), V(T)) \to \min, \tag{6.69}$$

with a continuous function ϕ_0 and with the terminal and phase constraints

$$h(X(0), X(T), V(T)) = 0, \quad S(X(0), X(T), V(T)) \le 0, \tag{6.70}$$

$$g(X(t), V(t), t) \le 0 \quad \text{for} \quad t \in [0, T]. \tag{6.71}$$

Here, as above, h, S and g are continuous vector functions with values in $\mathbb{R}^{N_1}$, $\mathbb{R}^{N_2}$ and $\mathbb{R}^{N_3}$, respectively, and the inequalities (6.70) and (6.71) understood componentwise.

Consider the auxiliary control problem for the system

$$\begin{cases} \dot{Y}(s) = \alpha(s) f(Y(s), z(s), u_1(s), \eta(s)) + (1 - \alpha(s)) B(Y(s), \eta(s)) e(s), \\ \dot{z}(s) = 1 - \alpha(s), \\ \dot{\eta}(s) = \alpha(s), \end{cases} \tag{6.72}$$

with the initial conditions

$$Y(0) = X(0), \quad z(0) = 0, \quad \eta(0) = 0,$$

with the controls

$$u_1(s) \in U, \quad \alpha(s) \in [0, 1], \quad e(s) \in K \cap \{\|e\| = 1\} \tag{6.73}$$

on some interval $[0, T_1]$ such that $T_1 \leq T + M$, with the performance criterion

$$J'[Y(\cdot), u_1(\cdot), \alpha(\cdot), e(\cdot)] = \phi_0(Y(0), Y(T_1), z(T_1)) \to \min, \tag{6.74}$$

and with the terminal and phase constraints

$$\begin{aligned} h(Y(0), Y(T_1), z(T_1)) = 0, &\quad \eta(T_1) = T \\ S(Y(0), Y(T_1), z(T_1)) \leq 0, &\quad T_1 \leq T + M \end{aligned} \tag{6.75}$$

$$g(Y(s), z(s), \eta(s)) \leq 0 \quad \text{for all } s \in [0, T_1]. \tag{6.76}$$

Remark 6.5 By Theorems 6.1 and 6.2, the auxiliary optimal control problem (6.72)–(6.76) is equivalent to the original problem (6.63), (6.64), (6.69)–(6.71). Moreover, if for given generalized controls

$$\{u^0(\cdot), w^0(\cdot), \mathcal{V}(dt); \ e^0_\tau(\cdot), \ \text{for } \tau \in D^d_{\mathcal{V}}\}$$

in the original problem the corresponding generalized solution $\{X^0(\cdot), V^0(\cdot)\}$ is optimal in the strong sense, then the set of functions

$$\{\alpha^0(\cdot), u^0_1(\cdot), e^0_1(\cdot)\},$$

defined on the interval $[0, T_1]$, where $T_1 = T + V^0(T)$, by the formulas

$$\begin{cases} \alpha^0(s) = \dot{\eta}^0(s) = \begin{cases} \left.\dfrac{1}{1 + \dot{V}^0(t)}\right|_{t=\eta^0(s)}, & \text{for } \eta^0(s) \notin \mathrm{D}^{\mathrm{d}}_{\mathcal{V}}, \\ 0, & \text{for } \eta^0(s) \in \mathrm{D}^{\mathrm{d}}_{\mathcal{V}}, \end{cases} \\ u^0_1(s) = u^0(\eta^0(s)), \\ e^0(s) = \begin{cases} w^0(\eta^0(s)), & \text{for } \eta^0(s) \bar{\in} D^d_{\mathcal{V}}, \\ e^0_\tau(s - V^0(\tau-)), & \text{for } \eta^0(s) \in \mathrm{D}^{\mathrm{d}}_{\mathcal{V}}, \end{cases} \end{cases} \tag{6.77}$$

where

$$\eta^0(s) = \inf\{t : t + V^0(t) > s\}$$

is the optimal control in the auxiliary problem.

The necessary optimality condition in the maximum principle is given by the following theorem (where * stands for transposition).

Theorem 6.4 *Suppose Assumption 6.1 is in force and* $\{X^0(\cdot), V^0(\cdot)\}$ *is the optimal generalized solution of the problem (6.63), (6.64), (6.69)–(6.71) in the strong sense, with the corresponding optimal control*

$$\{u^0(t), w^0(t), \mathscr{V}(dt);\ e^0_\tau(\cdot),\ for\ \tau \in D^{\mathrm{d}}_{\mathscr{V}}\}.$$

Then there exists a collection of elements

$$\Lambda = \Big\{\lambda_0, \beta_0, l_1, l_2, p^0_x(\cdot), p^0_v(\cdot), p^0_t(\cdot), \lambda_k(dt),\ \nu^k_\tau(ds)\\ for\ k = 1, \ldots, N_3,\ \tau \in D^{\mathrm{d}}_{\mathscr{V}}\Big\},$$

not simultaneously equal to zero, where

1. $\lambda_0 \geq 0$ *and* β_0 *are real numbers;*
2. $l_1 \in \mathbb{R}^{N_1}$ *and* $l_2 \in \mathbb{R}^{N_2}$ *are vectors such that the components of* l_2 *are non-negative and the trajectory* $\{X^0(\cdot), V^0(\cdot)\}$ *satisfies the condition*

$$\big\langle l_2, S\big(X^0(0-), X^0(T), V^0(T)\big)\big\rangle + \gamma_0(V^0(T) - M) = 0; \tag{6.78}$$

3. *the vector-valued functions* $\{p^0_x(\cdot), p^0_v(\cdot), p^0_t(\cdot)\}$ *satisfy the system of equations*

$$\begin{aligned} p^0_x(t) = &-\mathscr{L}'_{x(T)}\big(X^0(0-), X^0(T), V^0(T)\big) + \\ &+ \int_t^T (\mathscr{H}_0)'_x\big(p^0_x(\tau), p^0_t(\tau), X^0(\tau), V^0(\tau), u^0(\tau), \tau\big)d\tau + \\ &+ \int_t^T (\mathscr{H}_1)'_x\big(p^0_x(\tau), p^0_v(\tau), X^0(\tau), \tau, w^0(\tau)\big)dV^c(\tau) + \\ &+ \sum_{\tau>t,\, \tau\in D^{\mathrm{d}}_{\mathscr{V}}} \Delta p^0_x(\tau) - \int_t^T \sum_{k=1}^{N_3} g'_{k_x}\big(X^0(\tau), V^0(\tau), \tau\big)d\lambda_k(\tau), \end{aligned} \tag{6.79}$$

$$
\begin{aligned}
p_v^0(t) &= -\mathscr{L}_v'\big(X^0(0-), X^0(T), V^0(T)\big) + \\
&+ \int_t^T (\mathscr{H}_0)_v'\big(p_x^0(\tau), p_t^0(\tau), X^0(\tau), V^0(\tau), u^0(\tau), \tau\big)d\tau + \\
&+ \sum_{\tau>t,\, \tau\in D_{\mathscr{V}}^{\mathrm{d}}} \Delta p_v^0(\tau) - \int_t^T \sum_{k=1}^{N_3} g_{k_v}'\big(X^0(\tau), V^0(\tau), \tau\big)d\lambda_k(\tau),
\end{aligned}
\tag{6.80}
$$

$$
\begin{aligned}
p_t^0(t) &= \beta_0 + \int_t^T (\mathscr{H}_0)_t'\big(p_x^0(\tau), p_t^0(\tau), X^0(\tau), V^0(\tau), u^0(\tau), \tau\big)d\tau + \\
&+ \int_t^T (\mathscr{H}_1)_t'\big(p_x^0(\tau), p_v^0(\tau), X^0(\tau), \tau, w^0(\tau)\big)dV^{\mathrm{c}}(\tau) + \\
&+ \sum_{\tau>t\, \tau\in D_{\mathscr{V}}^{\mathrm{d}}} \Delta p_t^0(\tau) - \int_t^T \sum_{k=1}^{N_3} g_{k_t}'\big(X^0(\tau), V^0(\tau), \tau\big)d\lambda_k(\tau),
\end{aligned}
\tag{6.81}
$$

with the terminal condition for $p_x^0(\cdot)$ at point $t = 0$:

$$
p_x^0(0-) = \mathscr{L}_{x(0)}'\big(X^0(0-), X^0(T), V^0(T)\big); \tag{6.82}
$$

4. *the non-negative measures $\lambda_k(dt)$ in the Eqs. (6.79)–(6.81) are concentrated on the sets*

$$
T_k^0 = \big\{t \in [0, T] \setminus D_{\mathscr{V}}^{\mathrm{d}} : \ g_k(X^0(t), V^0(t), t) = 0\big\}; \tag{6.83}
$$

5. *the jumps of the functions $\{p_x^0(\cdot), p_v^0(\cdot), p_t^0(\cdot)\}$ at the points $\tau \in D_{\mathscr{V}}^{\mathrm{d}}$ are defined by the formulas*

$$
\begin{aligned}
\Delta p_x^0(\tau) &= p_x^0(\tau) - p_x^0(\tau-) = p_{x,\tau}^0(\Delta V^0(\tau)) - p_{x,\tau}^0(0), \\
\Delta p_v^0(\tau) &= p_v^0(\tau) - p_v^0(\tau-) = p_{v,\tau}^0(\Delta V^0(\tau)) - p_{v,\tau}^0(0), \\
\Delta p_t^0(\tau) &= p_t^0(\tau) - p_t^0(\tau-) = p_{t,\tau}^0(\Delta V^0(\tau)) - p_{t,\tau}^0(0),
\end{aligned}
$$

where

$$
\begin{aligned}
p_x^0(\tau-) &= \Big[I_{(n\times n)} + \Psi_x'\big(X^0(\tau-), \tau, w^0(\tau)\Delta V^0(\tau)\big)\Big]^* p_x^0(\tau) - \\
&- \int_0^{\Delta V^0(\tau)} \left[I_{(n\times n)} + \Psi_x'\Big(X^0(\tau-), \tau, \int_0^u e_\tau^0(\omega)d\omega\Big)\right]^* \times \\
&\times \sum_{k=1}^{N_3} g_{k_x}'\big(y_\tau^0(u), V^0(\tau-) + u, \tau\big)\, dv_k^\tau(u),
\end{aligned}
\tag{6.84}
$$

$$p_v^0(\tau-) = p_v^0(\tau) - \int_0^{\Delta V^0(\tau)} \sum_{k=1}^{N_3} g'_{k_v}\left(y_\tau^0(u), V^0(\tau-)+u, \tau\right) d\nu_k^\tau(u), \tag{6.85}$$

$$\begin{aligned} p_t^0(\tau-) = {} & p_t^0(\tau) + \left\langle \Psi_t'\left(X^0(\tau), \tau, w^0(\tau)\Delta V^0(\tau)\right), \ p_x^0(\tau)\right\rangle - \\ & - \int_0^{\Delta V^0(\tau)} \left\langle \Psi_t'\left(X^0(\tau-), \tau, \int_0^u e_\tau^0(\omega)d\omega\right), \right. \\ & \left. \sum_{k=1}^{N_3} g'_{k_x}\left(y_\tau^0(u), V^0(\tau-)+u, \tau\right) d\nu_k^\tau(u) \right\rangle - \\ & - \int_0^{\Delta V^0(\tau)} \sum_{k=1}^{N_3} g'_{k_t}\left(y_\tau^0(u), V^0(\tau-)+u, \tau\right) d\nu_k^\tau(u); \end{aligned} \tag{6.86}$$

6. *the non-negative measures* $\nu_\tau^k(dt)$ *in the Eqs. (6.84)–(6.86) are concentrated on the sets*

$$T_\tau^k = \left\{ s \in [0, \Delta V^0(\tau)] |\ g_k\left(y_\tau^0(s), V^0(\tau-)+s, \tau\right) = 0 \right\}; \tag{6.87}$$

7. *the functions* $\{p_{x,\tau}(s), p_{v,\tau}(s), p_{t,\tau}(s)\}$ *are defined on the segment* $[0, \Delta V^0(\tau)]$ *for* $\tau \in D_{\mathcal{V}}^d$ *by the formulas*

$$\begin{aligned} p_{x,\tau}^0(s) = {} & \left[I_{(n\times n)} + \Psi_x'\left(y_\tau^0(s), \tau, \int_s^{\Delta V^0(\tau)} e^0(\omega)d\omega\right) \right]^* p_x^0(\tau) - \\ & - \int_s^{\Delta V^0(\tau)} \left[I_{(n\times n)} + \Psi_x'\left(y_\tau^0(s), \tau, \int_s^u e_\tau^0(\omega)d\omega\right) \right]^* \times \\ & \times \sum_{k=1}^{N_3} g'_{k_x}\left(y_\tau^0(u), V^0(\tau-)+u, \tau\right) d\nu_k^\tau(u); \end{aligned} \tag{6.88}$$

where

$$y_\tau^0(s) = X^0(\tau-) + \Psi\left(X^0(\tau-), \tau, \int_0^s e_\tau^0(\omega)d\omega\right);$$

$$p_{v,\tau}^0(s) = p_v^0(\tau) - \int_s^{\Delta V^0(\tau)} \sum_{k=1}^{N_3} g'_{k_v}\left(y_\tau^0(u), V^0(\tau-)+u, \tau\right) d\nu_k^\tau(u); \tag{6.89}$$

$$p^0_{t,\tau}(s) = p^0_t(\tau) + \langle \Phi'_t(y^0_\tau(s), \tau, s, \Delta V^0(\tau)), p^0_x(\tau)\rangle - \\ - \int_s^{\Delta V^0(\tau)} \langle \Phi'_t(y^0_\tau(s), \tau, s, \omega), \sum_{k=1}^{N_3} g'_{k_x}(y^0_\tau(\omega), z^0_\tau(\omega), \tau)\, dv^\tau_k(\omega)\rangle - \\ - \int_s^{\Delta V^0(\tau)} \sum_{k=1}^{N_3} g'_{k_t}(y^0_\tau(u), z^0_\tau(u), \tau)\, dv^k_\tau(u). \tag{6.90}$$

For this collection of elements, the optimal solution $\{X^0(\cdot), V^0(\cdot)\}$ *and the optimal control*

$$\{\mathscr{V}(dt), u^0(t), w^0(t);\ e^0_\tau(s) \text{ for } \tau \in D^{\mathrm{d}}_{\mathscr{V}}\}$$

*satisfy the following inequalities (*Maximum Principle*):*

For a.e. $t \in [0, T] \setminus D_{\mathscr{V}}$,

$$\max_{e\in K\cap\{\|e\|=1\}} \mathscr{H}_1(p^0_x(t), p^0_v(t), X^0(t), t, e) \le \\ \le \mathscr{H}_0(p^0_x(t), p^0_t(t), X^0(t), V^0(t), u^0(t), t) = \\ = \max_{u\in U} \mathscr{H}_0(p^0_x(t), p^0_t(t), X^0(t), V^0(t), u, t) = 0. \tag{6.91}$$

For $\mathscr{V}(dt)$*-a.e.* $t \in D^{\mathrm{c}}_{\mathscr{V}}$,

$$\max_{e\in K\cap\{\|e\|=1\}} \mathscr{H}_1(p^0_x(t), p^0_v(t), X^0(t), t, e) = \\ = \mathscr{H}_1(p^0_x(t), p^0_v(t), X^0(t), t, w^0(t)) = \\ = \mathscr{H}_0(p^0_x(t), p^0_t(t), X^0(t), V^0(t), u^0(t), t) = \\ = \max_{u\in U} \mathscr{H}_0(p^0_x(t), p^0_t(t), X^0(t), V^0(t), u, t) = 0. \tag{6.92}$$

For all $t \in D^{\mathrm{d}}_{\mathscr{V}}$,

$$\max_{u\in U} \mathscr{H}_0(p^0_x(t), p^0_t(t), X^0(t), V^0(t), u, t) \le \\ \le \max_{e\in K\cap\{\|e\|=1\}} \mathscr{H}_1(p^0_x(t), p^0_v(t), X^0(t), t, e) = 0; \tag{6.93}$$

Moreover, for all $\tau \in D^{\mathrm{d}}_{\mathscr{V}}$ *on the corresponding segment* $[0, \Delta V^0(\tau)]$ *it holds that*

$$
\begin{aligned}
&\max_{u\in U} \mathcal{H}_0\big(p^0_{x,\tau}(s), p^0_{t,\tau}(s), y^0_\tau(s), z^0_\tau(s), u, \tau\big) \le \\
&\le \max_{e\in K\cap\{\|e\|=1\}} \mathcal{H}_1\big(p^0_{x,\tau}(s), p^0_{v,\tau}(s), y^0_\tau(s), t, e\big) = \\
&= \mathcal{H}_1\big(p^0_{x,\tau}(s), p^0_{v,\tau}(s), y^0_\tau(s), \tau, e^0_\tau(s)\big) = 0.
\end{aligned}
\tag{6.94}
$$

Proof In the main lines, the statement of the theorem coincides with Theorem 6.2. The main difference concerns the representation of jumps given by formulas (6.84)–(6.86) and the possibility obtaining explicit expressions for the conjugate variables in the formulas (6.88)–(6.90) on the intervals $[0, \Delta V^0(\tau)]$. Hence, for the proof it only remains to verify the relations for jumps. We need the following result describing the evolution of the variables $\{y^\tau(\cdot), z^\tau(\cdot)\}$; it is a consequence of Lemma 2.7 and known results concerning the differentiability of solutions of ordinary differential equations with respect to initial conditions and parameters [79, 96].

Lemma 6.1 *Suppose the function $B(X,t)$ is continuously differentiable with respect to both variables and satisfies Assumption 2.2. Let $y(s)$ be the solution of the equation*

$$\dot{y}(s) = B(y(s), \tau)e(s) \tag{6.95}$$

with an integrable function $e(\cdot)$. Let

$$\Phi(x, \tau, u, s) = y_{x,\tau,u}(s),$$

where $y_{x,\tau,u}(s)$ is the solution of (6.95) with the initial condition $y_{x,\tau,u}(u) = x$.

Then

$$\Phi(x, \tau, u, s) = x + \Psi\Big(x, \tau, \int_u^s e(\omega)d\omega\Big) \tag{6.96}$$

and the derivative of the solution with respect to the initial condition, that is, $\Phi'_x(x, \tau, u, s)$, is the unique solution of the matrix differential equation

$$\frac{d}{ds}\Phi'_x(x, \tau, u, s) = \big[B(y_{x,\tau,u}(s), \tau)e(s)\big]'_x \Phi'_x(x, \tau, u, s) \tag{6.97}$$

with the initial condition $\Phi'_x(x, \tau, u, u) = I_{(n\times n)}$.

The inverse matrix $\big[\Phi'_x(x, \tau, u, s)\big]^{-1}$ exists and is the unique solution of the matrix differential equation

$$\frac{d}{ds}\big[\Phi'_x(x, \tau, u, s)\big]^{-1} = -\big[\Phi'_x(x, \tau, u, s)\big]^{-1}\big[B(y_{x,\tau,u}(s), \tau)e(s)\big]'_x \tag{6.98}$$

with the initial condition $\big[\Phi'_x(x, \tau, u, u)\big]^{-1} = I_{(n\times n)}$.

For any $u \le s \le t$, one has the identities

$$\Phi'_x(x,\tau,u,t) = \Phi'_x(\Phi(x,\tau,u,s),\tau,s,t)\Phi'_x(x,\tau,u,s), \tag{6.99}$$

and

$$\Phi'_x(x,\tau,u,s) = \left[\Phi'_x(\Phi(x,\tau,u,s),\tau,s,u)\right]^{-1}, \tag{6.100}$$

where

$$\Phi(x,\tau,s,u) = x + \Psi\big(x,\tau,-\int_u^s e(\omega)d\omega\big).$$

The derivative of the solution $\Phi(x,\tau,u,s)$ with respect to the scalar parameter τ is given by the formula

$$\Phi'_\tau(x,\tau,u,s) = \int_u^s \Phi'_x(x,\tau,\mu,s)B'_\tau(y_{x,\tau,u}(\mu),\tau)e(\mu)d\mu. \tag{6.101}$$

Proof The first statement is established like Lemma 2.7, and the others follow directly from the standard results concerning differentiation of (6.96) with respect to x and τ (see [79, 96] and Theorem A.27).

By Theorem 6.2, the functions $\{p^0_{x,\tau}(s), p^0_{v,\tau}(s), p^0_{t,\tau}(s)\}$ satisfy on the interval $[0,\Delta V^0(\tau)]$ the system of equations

$$\begin{aligned} p^0_{x,\tau}(s) = p^0_{x,\tau}(\Delta V^0(\tau)) + \int_s^{\Delta V^0(\tau)} \{[B(y^0_\tau(u),\tau)e^0_\tau(u)]'_x\}^* p^0_{x,\tau}(u)du - \\ - \int_s^{\Delta V^0(\tau)} \sum_{k=1}^{N_3} g'_{k_x}\big(y^0_\tau(u),z^0_\tau(u),\tau\big)\,dv^\tau_k(u); \end{aligned} \tag{6.102}$$

$$p^0_{v,\tau}(s) = p^0_{v,\tau}(\Delta V^0(\tau)) - \int_s^{\Delta V^0(\tau)} \sum_{k=1}^{N_3} g_{k_v}\big(y^0_\tau(u),z^0_\tau(u),\tau\big)\,dv^\tau_k(u); \tag{6.103}$$

$$\begin{aligned} p^0_{t,\tau}(s) = p^0_{t,\tau}(\Delta V^0(\tau)) + \int_s^{\Delta V^0(\tau)} \langle p^0_{x,\tau}(u), B'_\tau(y^0_\tau(u),\tau)e^0_\tau(u)du\rangle - \\ - \int_s^{\Delta V^0(\tau)} \sum_{k=1}^{N_3} g'_{k_t}\big(y^0_\tau(u),z^0_\tau(u),\tau\big)\,dv^\tau_k(u), \end{aligned} \tag{6.104}$$

where $\{y^0_\tau(s), z^0_\tau(s)\}$ is the solution of the system

$$\begin{cases} \dot{y}_\tau^0(s) = B(y_\tau^0(s), \tau) e_\tau^0(s), \\ \dot{z}_\tau^0(s) = 1, \end{cases}$$

with the initial and terminal conditions

$$\begin{aligned} & y_\tau^0(0) = X^0(\tau-), \quad z_\tau^0(0) = V^0(\tau-), \\ & p_{x,\tau}^0(\Delta V^0(\tau)) = p_x^0(\tau), \quad p_{v,\tau}^0(\Delta V^0(\tau)) = p_v^0(\tau), \\ & p_{t,\tau}^0(\Delta V^0(\tau)) = p_t^0(\tau). \end{aligned} \tag{6.105}$$

Set

$$\Phi(x, \tau, s, u) = x + \Psi\big(x, \tau, \int_s^u e_\tau^0(\omega) d\omega\big).$$

Then, by Lemma 6.1,

$$y_\tau^0(s) = \Phi(X^0(\tau-), \tau, 0, s), \quad z_\tau^0(s) = V^0(\tau-) + s,$$

and the matrix-valued function

$$\{[\Phi_x'(y_\tau^0(s), \tau, s, u)]^{-1}\}^* = \left\{\left[I_{(n\times n)} + \Psi_x'\big(y_\tau^0(s), \tau, \int_s^u e_\tau^0(\omega) d\omega\big)\right]^{-1}\right\}^*$$

is the solution of the linear differential equation

$$\begin{aligned} & \frac{d}{du}\{[\Phi_x'(y_\tau^0(s), \tau, s, u)]^{-1}\}^* = \\ & \quad -\{[B(y_\tau^0(s), \tau) e^0(u)]_x'\}^* \{[\Phi_x'(y_\tau^0(s), \tau, s, u)]^{-1}\}^* \end{aligned} \tag{6.106}$$

with the initial condition

$$\Phi_x'(y_\tau^0(s), \tau, s, s) = I_{(n\times n)}.$$

The Cauchy formula the solution of the non-homogeneous linear equation (6.102) gives the relation

$$\begin{aligned} p_{x,\tau}^0(\Delta V^0(\tau)) = {} & \{[\Phi_x'(y_\tau^0(s), \tau, s, \Delta V^0(\tau))]^{-1}\}^* p_{x,\tau}^0(s) + \\ & + \int_s^{\Delta V^0(\tau)} \{[\Phi_x'(y_\tau^0(u), \tau, u, \Delta V^0(\tau))]^{-1}\}^* \sum_{k=1}^{N_3} g_{k_x}'(y_\tau^0(u), z_\tau^0(u), \tau) d v_k^\tau(u), \end{aligned}$$

which in turn provides the following representation of the solution of Eq. (6.102):

$$p^0_{x,\tau}(s) = \left[\Phi'_x(y^0_\tau(s), \tau, s, \Delta V^0(\tau))\right]^* p^0_{x,\tau}(\Delta V^0(\tau)) -$$
$$- \int_s^{\Delta V^0(\tau)} [\Phi'_x(y^0_\tau(s), \tau, s, \Delta V^0(\tau))]^* \{[\Phi'_x(y^0_\tau(u), \tau, u, \Delta V^0(\tau))]^{-1}\}^* \times$$
$$\times \sum_{k=1}^{N_3} g'_{k_x}(y^0_\tau(u), z^0_\tau(u), \tau)\, dv^\tau_k(u).$$

However, by (6.99), we have that

$$\left[\Phi'_x(y^0_\tau(s), \tau, s, \Delta V^0(\tau))\right]^* \{\left[\Phi'_x(y^0_\tau(u), \tau, u, \Delta V^0(\tau))\right]^{-1}\}^* =$$
$$= \left[\Phi'_x(y^0_\tau(s), \tau, s, u)\right]^*,$$

and finally

$$p^0_{x,\tau}(s) = \left[\Phi'_x(y^0_\tau(s), \tau, s, \Delta V^0(\tau))\right]^* p^0_{x,\tau}(\Delta V^0(\tau)) -$$
$$- \int_s^{\Delta V^0(\tau)} \left[\Phi'_x(y^0_\tau(s), \tau, s, u)\right]^* \sum_{k=1}^{N_3} g'_{k_x}(y^0_\tau(u), z^0_\tau(u), \tau) dv^\tau_k(u). \tag{6.107}$$

This expression coincides with (6.88) after the substitution

$$\Phi'_x(y^0_\tau(s), \tau, s, u) = I_{(n\times n)} + \Psi'_x\big(y^0_\tau(s), \tau, \int_s^u e^0_\tau(\omega) d\omega\big).$$

Relation (6.84) is obtained in much the same way.

Next, the expression for $p^0_{v,\tau}(s)$ follows from (6.103). We get

$$p^0_{v,\tau}(s) = p^0_{v,\tau}(\Delta V^0(\tau)) - \int_s^{\Delta V^0(\tau)} \sum_{k=1}^{N_3} g'_{k_v}(y^0_\tau(u), z^0_\tau(u), \tau)\, dv^\tau_k(u),$$

so upon substituting here the relation $p^0_{v,\tau}(\Delta V^0(\tau)) = p^0_v(\tau)$ and the expression for $y^0_\tau(s)$ given by Lemma 6.1, we obtain (6.85) and (6.89).

Finally, to derive the equation for $p^0_{t,\tau}(s)$ we substitute 6.107) into 6.104) and obtain

$$
\begin{aligned}
p^0_{t,\tau}(s) = p^0_{t,\tau}(\Delta V^0(\tau)) + \\
+ \int_s^{\Delta V^0(\tau)} \Big\langle \big[\Phi'_x(y^0_\tau(u), \tau, u, \Delta V^0(\tau))\big]^* p^0_{x,\tau}(\Delta V^0(\tau)) \\
- \int_u^{\Delta V^0(\tau)} \big[\Phi'_x(y^0_\tau(u), \tau, u, \omega)\big]^* \sum_{k=1}^{N_3} g'_{k_x}(y^0_\tau(\omega), z^0_\tau(\omega), \tau)\, d\nu^\tau_k(\omega), \\
B'_\tau(y^0_\tau(u), \tau) e^0_\tau(u) \Big\rangle du - \int_s^{\Delta V^0(\tau)} \sum_{k=1}^{N_3} g'_{k_t}(y^0_\tau(u), z^0_\tau(u), \tau)\, d\nu^k_\tau(u).
\end{aligned}
$$

If we now use (6.101), we conclude that

$$
\begin{aligned}
p^0_{t,\tau}(s) = p^0_{t,\tau}(\Delta V^0(\tau)) + \big\langle \Phi'_t(y^0_\tau(s), \tau, s, \Delta V^0(\tau)), p^0_{x,\tau}(\Delta V^0(\tau)) \big\rangle - \\
- \int_s^{\Delta V^0(\tau)} \Big\langle \Phi'_t(y^0_\tau(s), \tau, s, \omega), \sum_{k=1}^{N_3} g'_{k_x}(y^0_\tau(\omega), z^0_\tau(\omega), \tau)\, d\nu^\tau_k(\omega) \Big\rangle - \\
- \int_s^{\Delta V^0(\tau)} \sum_{k=1}^{N_3} g'_{k_t}(y^0_\tau(u), z^0_\tau(u), \tau)\, d\nu^k_\tau(u).
\end{aligned}
$$

which gives (6.86) and (6.90). □

As a consequence, we obtain expressions for the jumps of the conjugate variables in the case where at the instant the impulse control is applied the generalized solution does not touch the constraints.

Corollary 6.3 *Suppose that at the point $\tau \in D^d_{\mathcal{V}}$ it holds that*

$$
g_k(y^0_\tau(s), z^0_\tau(s), \tau) < 0 \tag{6.108}
$$

for all $s \in [0, \Delta V^0(\tau)]$ and $k = 1, \ldots, N_3$. Then

$$
\begin{aligned}
\Delta p^0_x(\tau) &= -\big[\Psi'_x(X^0(\tau-), \tau, w^0(\tau)\Delta V^0(\tau)\big]^* p^0_x(\tau), \\
\Delta p^0_v(\tau) &= 0, \\
\Delta p^0_t(\tau) &= -\Big\langle \Psi'_\tau(X^0(\tau-), \tau, w^0(\tau)\Delta V^0(\tau)), p^0_x(\tau) \Big\rangle.
\end{aligned}
\tag{6.109}
$$

The proof follows immediately from Theorem 6.4; indeed, in the present case the measures $\nu^k_\tau(ds)$ vanish on the interval $[0, \Delta V^0(\tau)]$. □

Remark 6.6 If the generalized solution does not touch the constraints at the instants of application of impulse controls, the equations for the conjugate variables $\{p^0_{x,\tau}(s), p^0_{v,\tau}(s), p^0_{t,\tau}(s)\}$ become simplify, since the integrals with respect to the measures $\nu_\tau^k(ds)$ disappear. One can show that the joint system for the pair $\{(x, p_x)\}$ satisfies the robustness conditions [333]. At the same time, the presence of phase constraints that are active at instants of application of impulse controls leads to the violation of robustness. Hence, the methods based on the robustness condition (or the equivalent so-called vibro-correctness condition introduced in [333]) may give an incorrect description of the conjugate variables.

Remark 6.7 Let us emphasize that the necessary optimality conditions obtained above concern generalized solutions that are admissible in the strong sense. One would expect that the optimality conditions for weak generalized solutions will be different. It goes without saying that the difference will concern only the conditions at the points of application of impulse controls (see Sect. 6.3.3).

The first proof of Theorem 6.4 was obtained in [263]. Note that the above optimality conditions can be used also in the case where the instants of application of impulse controls is applied are fixed. In this case it is sufficient to fix the function $\alpha(\cdot)$ in the auxiliary control problem, as was done, e.g., in Sect. 3. The resulting optimality conditions are very close to the discrete maximum principle [53, 372]. Some examples of the optimality conditions for DCS derived with the aid of the discrete maximum principle can be found in [12, 13, 149, 150]. The main difference is that for robust DCS it is not necessary to require that the attainable set for a single impulse be convex. An example of this approach for various constraints imposed on the instants of application of impulse controls is given in Chap. 3. A thorough investigation of optimality conditions for robust and non-robust DCS with intermediate conditions (of the phase-constraints type) was carried out in [128].

6.3.2 Robust DCS with Singular Controls

In robust DCS one can define in a natural manner the reaction of the system under consideration to a measure-valued control, and then the optimization problem is stated as an optimal control problem of a system described by a nonlinear differential equation with measure of the form

$$\begin{aligned} X(t) &= X(0-) + \int_0^t f(X(\tau), V(\tau), u(\tau), \tau)d\tau + \\ &+ \int_0^t B(X(\tau), \tau)dW^c(\tau) + \sum_{\tau\le t} \Psi(X(\tau-), \tau, \Delta W(\tau)), \qquad (6.110) \\ V(t) &= \operatorname*{Var}_{[0,t]} W(\tau), \end{aligned}$$

with two types of controls: $u \in U \subset \mathbb{R}^k$ and $W \in K \subseteq \mathbb{R}^m$, where U is a compact set, K is closed convex cone, and the control $W(\cdot)$ is a right-continuous function of bounded variation such that for any $t_1 \geq t_2$ the increment $W(t_1) - W(t_2) \in K$. The integral with respect to the continuous part of $W(\cdot)$ is understood as in the Lebesgue-Stieltjes sense. The control $W(\cdot)$ is subject to the constraint

$$V(T) = \operatorname*{Var}_{[0,T]} W(\tau) \leq M < \infty. \tag{6.111}$$

Clearly, the solution of Eq. (6.110) for the variable $X(t)$ is a generalized solution of the system (6.63), which is also governed by an equation with measure, namely, (6.110). Indeed, if in Eq. (6.63) one choose as the norm of the control w the same norm as in the calculation of the variation of the function W, then the generalized control

$$\{\tilde{u}(\cdot), \tilde{w}(\cdot), \tilde{e}_\tau(\cdot) \text{ for } \tau \in D_V^{\mathrm{d}}\}$$

defined by the formulas

$$\tilde{u}(t) = u(t), \quad \tilde{w}(t) = \frac{dW^{\mathrm{c}}(t)}{dV(t)}, \quad \tilde{e}_\tau(s) = \frac{\Delta W(\tau)}{\Delta V(\tau)} \quad \text{for } s \in [0, \Delta V(\tau)]$$

determines the generalized solution $\{\tilde{X}(\cdot), \tilde{V}(\cdot)\}$ of the system (6.63), which coincides with the solution $\{X(\cdot), V(\cdot)\}$ of the system (6.110). Conversely, if one takes a generalized solution and regards it as the solution of Eq. (6.110) with the measure-valued control

$$W(t) = \int_0^t w(t) dV^{\mathrm{c}}(t) + \sum_{\tau \leq t} w(\tau) \Delta V(\tau),$$

then for any $t \in [0, T]$ we have

$$\operatorname*{Var}_{[0,t]} W(\tau) \leq V(t), \tag{6.112}$$

because, by virtue of (6.68), $\|w(\tau)\| \leq 1$.

Thus, if we compare the optimization problem in the class of generalized solutions and in the class of singular controls with the same constraints, but with a different understanding of the variable $V(t)$ (as the integral of the norm of w in the case of a generalized solution, and as the variation, defined by the same norm, in the case of the singular control), we reach the following conclusions:

1. generally speaking, the value of the performance criterion in the generalized control problem is smaller (in the minimization problem) because the set of generalized solutions contains also any solution with admissible singular control;
2. in view of inequality (6.112), for the coincidence of solutions some kind of monotonicity of the constraints and of the performance criterion in the variable V is required; for example, if f and h do not depend on V and the functions S,

G and ϕ_0 are non-decreasing in V, then the optimal generalized solution is the optimal solution with singular control and vice versa;
3. if the optimal generalized solution admits a representation with functions $e_\tau(\cdot)$ that are constant on the intervals $[0, \Delta V(\tau)]$ and such that $\|e_\tau(s)\| = 1$, then this solution is also optimal in the class of solutions with singular controls.

If a solution with singular control is also optimal in the class of generalized solutions, the optimality condition coincides with the condition derived in the preceding section. An example of such a problem is provided below (see Sect. 6.3.3).

6.3.3 Maximum Principle for Weak Generalized Solutions

As we saw earlier, there are different approaches that allow to determine a feasible generalized solution in the presence of phase constraints and therefore obtain different types of optimal solutions. Above we considered optimality conditions for generalized solutions that are admissible in the strong sense, i.e., solutions that can be approximated by uniformly small ordinary solutions with uniformly small violations of the phase constraints. However, if we assume that along the approximating sequences the phase constraints admit violations that are small, for example, in the L_1-norm, or tend to zero in the weak-$*$ topology on the space of functions of bounded variations, then one can switch to another type of admissible generalized solutions. These solutions, as elements of the space of generalized functions, may be similar, since the difference in the behavior of the approximating sequences manifests only at the jump points, and the solution itself is defined as the limit at the points of continuity. In other words, the difference can be seen only after one "reveals" the singularity (see Definitions 5.1 and 5.2). The typical behavior of two types of solutions at a jump point where the solution reaches the phase constraints is shown in Fig. 6.1.

In this figure, we see two curves AwB and AsB (where "w" stands for *weak*, i.e., one has a *weak* solution, and "s" stands for *strong*, i.e., one has a *strong* solution). Assuming that the transition from the point A to the point B is very fast (ideally—instantly), it is easy to check that these solutions are indistinguishable, since the

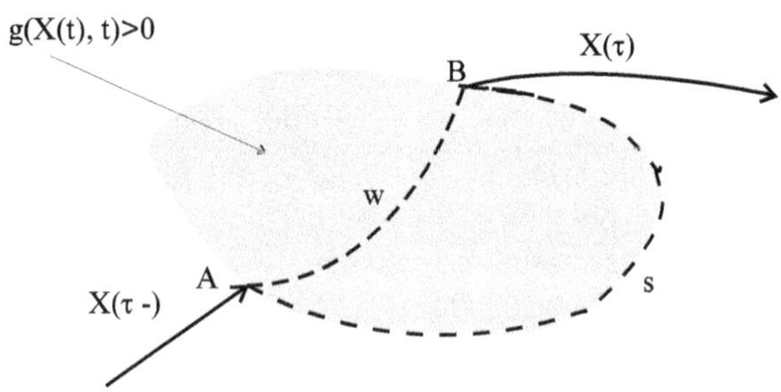

Fig. 6.1 Strong and weak generalized solutions

visible behavior of solutions is described by a solid line. However, if you look at the behavior of the approximating sequences, then on the parts shown by the dotted line and connecting the points A and B, a significant difference is observed. Roughly speaking, the strong solution goes around the forbidden region, while the weak solution can actually cross it.

The admissibility of one or another type of solution is determined, of course, by the actual physical system that one wishes to control. Moreover, the control resource needed to implement each choice and the optimality conditions corresponding to weak and strong solutions should be different.

In Sect. 5.2.3 we saw that the auxiliary optimal control problem corresponding to the weak generalized control problem has an additional non-smooth constraint that allows one to "remove" the phase constraint in a case of impulsive exit to the boundary; however, at the remaining points, where the solution is continuous, the phase constraint remains in force. The task, therefore, belongs to the class of problems with non-smooth "mixed constraints" [113]; however, the possibility of reducing the constraint to an integral form somewhat simplifies the situation.

As above, we consider the system described by the system of equations

$$\begin{cases} \dot{X}(t) = f(X(t), u(t), t) + B(X(t), t)w(t), \\ \dot{V}(t) = \|w(t)\|, \end{cases} \tag{6.113}$$

with two types of controls,

$$u \in U \subset \mathbb{R}^k \quad \text{and} \quad w \in K \subset \mathbb{R}^m, \tag{6.114}$$

where U is a compact set, K is a closed convex cone, and $w(t)$ satisfies the constraint

$$\int_0^T \|w(t)\| dt < \infty. \tag{6.115}$$

It is assumed that B satisfies the Frobenius condition (2.2).

We consider the problem in a simplified setting, assuming that the performance criterion has the form

$$J[X(\cdot), u(\cdot), w(\cdot)] = \phi_0(X(T), V(T)) \to \min \tag{6.116}$$

with a continuous function ϕ_0 that is monotonically increasing in V and satisfies the condition

$$\lim_{v \to +\infty} \inf_{x \in \mathbb{R}^n} \phi_0(x, v) = +\infty. \tag{6.117}$$

The problem is considered under the phase constraint

$$g(X(t), t) \le 0 \quad \text{for any } t \in [0, T], \tag{6.118}$$

where g is a scalar function.

Remark 6.8 Condition (6.117) means that the set

$$\{(x, v) : \phi_0(x, v) \le c\}$$

is either uniformly bounded or empty. Therefore, if the set of admissible solutions is not empty, then one can choose a uniformly bounded minimizing sequence $\{X^n, V^n\}$, which according to Helly's theorem will converge to the optimal generalized solution. To complete the proof, one resorts to Theorem 5.1.

In view of Theorem 5.3, one can formulate the following auxiliary problem that is equivalent to problem (6.113)–(6.118) on the class of weak generalized solutions. The auxiliary system is described by the equations

$$\begin{cases} \dot{Y}(s) = \alpha(s) f(Y(s), u_1(s), \eta(s)) + (1 - \alpha(s)) B(Y(s), \eta(s)) e(s), \\ \dot{z}(s) = 1 - \alpha(s), \\ \dot{\eta}(s) = \alpha(s), \\ \dot{\mu}(s) = \alpha(s) g^+(Y(s), \eta(s)), \end{cases} \tag{6.119}$$

where $g^+(x, s) = \max\{g^+(x, s), 0\}$, with the initial conditions

$$Y(0) = X(0), \quad z(0) = 0, \quad \eta(0) = 0, \quad \mu(0) = 0$$

and controls

$$u_1(s) \in U, \quad \alpha(s) \in [0, 1], \quad e(s) \in K \cap \{\|e\| = 1\}. \tag{6.120}$$

The problem is considered on an interval $[0, T_1]$ such that $z(T_1) < \infty$, and with the performance criterion

$$J'[Y(\cdot), u_1(\cdot), \alpha(\cdot), e(\cdot)] = \phi_0(Y(T_1), z(T_1)) \to \min \tag{6.121}$$

and terminal conditions

$$\eta(T_1) = T, \quad \mu(T_1) = 0. \tag{6.122}$$

The optimal generalized (weak) solution $\{X^0(\cdot), V^0(\cdot)\}$ is realized by the generalized control

$$\mathscr{U}^0 = \{u^0(\cdot), w^0(\cdot), \mathscr{V}(dt), e^0_\tau(\cdot) \text{for } \tau \in D^{\mathrm{d}}_{\mathscr{V}}\}.$$

Note also that in this problem (see Sect. 6.3.2) a weak solution admits a representation by a singular control measure because the functions $\{e^0_\tau(\cdot) : \tau \in D^{\mathrm{d}}_{\mathscr{V}}\}$ can be taken to be constant on $[0, \Delta V(\tau)]$. This control measure is defined by the formula

$$\mathscr{W}\{[0,t]\} = \int_0^t w^0(s)d\mathscr{V}^{\mathrm{c}}(s) + \sum_{\tau \le t,\, \tau \in D^{\mathrm{d}}_{\mathscr{V}}} \int_0^{\Delta V^0(\tau)} e^0_\tau(s)ds. \tag{6.123}$$

Indeed, for the control $\mathscr{U}^0$ we have

$$\operatorname*{Var}_{[0,t]} \mathscr{W}^{\mathrm{c}}\{[0,s]\} = \int_0^t \|w^0(s)\| d\mathscr{V}^{\mathrm{c}}(s) \le V^{\mathrm{c}}(t),$$

since $\|w^0(s)\| \le 1$, and, for any $\tau \in D^{\mathrm{d}}_{\mathscr{V}}$,

$$\|\mathscr{W}\{\tau\}\| = \left\| \int_0^{\Delta V^0(\tau)} e^0_\tau(s)ds \right\| \le \Delta V^0(\tau).$$

Consequently,

$$\operatorname*{Var}_{[0,t]} \mathscr{W}\{[0,s]\} \le V^{\mathrm{c}}(t) + \sum_{\tau \le t,\, \tau \in D^{\mathrm{d}}_{\mathscr{V}}} \Delta V^0(\tau) = V^0(t).$$

Since the function B is assumed to satisfy the Frobenius condition, the solution $X'(\cdot)$ corresponding to the control $u^0(\cdot)$ and the measure $\mathscr{W}(dt)$ satisfies the equation

$$X'(t) = X(0) + \int_0^t f(X'(s), u^0(s), s)ds +$$

$$+\ B(X'(s), s)d\mathscr{W}^{\mathrm{c}}(s) + \sum_{\tau \le t,\, \tau \in D^{\mathrm{d}}_{\mathscr{V}}} \Psi(X'(\tau-), \tau, \mathscr{W}\{\tau\}),$$

and coincides with $X^0(\cdot)$, and consequently satisfies the phase constraint for all $t \in [0, T]$. Moreover,

$$V'(t) = \operatorname*{Var}_{[0,t]} \mathscr{W}\{[0,s]\} \le V^0(t)$$

and, due to the monotonicity of ϕ_0,

$$\phi_0(X'(\cdot), V'(\cdot)) \le \phi_0(X^0(\cdot), V^0(\cdot)).$$

However, due to the optimality of $\{X^0, V^0\}$ we have equality, and so $\{X'(\cdot), V'(\cdot)\}$ is the optimal generalized solution. Thus, the optimal solution $\{X^0(\cdot), V^0(\cdot)\}$

is defined by the optimal control $\{u^0(\cdot), \mathscr{W}^0(\cdot)\}$, where $u^0(t)$ is a Lebesgue-measurable control satisfying (6.114), and $\mathscr{W}^0(dt)$ is a vector measure taking values in the cone K.

Now let us define the optimal control in the auxiliary problem (6.119)–(6.122). To this end we introduce the monotonically increasing function

$$\Gamma^0(t) = t + \operatorname*{Var}_{[0,t]} \mathscr{W}^0\{[0, s]\} = t + V^0(t),$$

together with the inverse function on the interval $[0, T_1]$, where $T_1 = \Gamma^0(T)$, given by the formula

$$\eta^0(s) = \inf\{t \in [0, T] \,:\, \Gamma^0(t) > s\}.$$

The optimal control in the auxiliary problem is defined as follows:

$$\begin{cases} \alpha^0(s) = \dot{\eta}^0(s), \\ u_1^0(s) = \begin{cases} u^0(\eta^0(s)), & \text{for } \eta^0(s) \bar{\in} D_{\mathscr{V}}^{\mathrm{d}}, \\ u_1^0(s) \in U \text{ arbitrary}, & \text{otherwise}, \end{cases} \\ e^0(s) = \begin{cases} w^0(\eta^0(s)), & \text{for } \eta^0(s) \bar{\in} D_{\mathscr{V}}^{\mathrm{d}}, \\ \dfrac{\mathscr{W}\{\tau\}}{\|\mathscr{W}\{\tau\}\|}, & \text{for } \eta^0(s) = \tau \in D_{\mathscr{V}}^{\mathrm{d}}. \end{cases} \end{cases} \tag{6.124}$$

A specific aspect of the auxiliary problem is the non-smoothness of the function $g^+(x, t)$, which, generally speaking, satisfies the Lipschitz condition, but is not differentiable even if the function $g(x, t)$ itself is differentiable. To derive the necessary optimality conditions in the auxiliary problem, we use the weak maximum principle obtained in [93]. Thus, let us set

$$\begin{aligned} &\mathscr{H}\big(\psi_y, \psi_z, \psi_\eta, \psi_\mu, y, \eta, \alpha, u_1, e\big) = \alpha\mathscr{H}_0 + (1 - \alpha)\mathscr{H}_1, \\ &\mathscr{H}_0\big(\psi_y, \psi_\eta, \psi_\mu, y, \eta, u_1\big) = \langle\psi_y, f(y, u_1, \eta)\rangle + \psi_\eta + \psi_\mu g^+(y, \eta), \\ &\mathscr{H}_1\big(\psi_y, \psi_z, y, \eta, e\big) = \langle\psi_y, B(y, \eta)e\rangle + \psi_z, \\ &\mathscr{L}(y, z, \eta, \mu) = \lambda_0\phi_0(y, z) + l_1(\eta - T) + l_2\mu. \end{aligned} \tag{6.125}$$

The next formulation of the optimality conditions follows directly from Clarke's general result (see [93], Chapter 5, Theorem 5.2.1).

Proposition 6.1 *Suppose that all the functions in the right-hand side of the auxiliary system satisfy Assumption 6.1 and the control* $\{\alpha^0(s), u_1^0(s), e^0(s)\}$*, defined on some interval* $[0, T_1]$*, is the optimal control in the problem (6.119)–(6.122).*

Let $\{Y^0(\cdot), z^0(\cdot), \eta^0(\cdot), \mu^0(\cdot)\}$ *be the optimal solution. Then there exists a collection of constants and functions, not simultaneously equal to zero, as follows:*

1. *constants*

$$\{\lambda_0 \geq 0,\ l_1,\ \psi_\mu^0 = l_2,\ \psi_z^0 = -\lambda_0(\phi_0)'_z(Y^0(T_1), z^0(T_1))\};$$

2. *functions* $\{\psi_y^0(s), \psi_\eta^0(s)\}$, *which satisfy the system of differential inclusions*

$$-\begin{pmatrix} \dot{\psi}_y^0(s) \\ \dot{\psi}_\eta^0(s) \end{pmatrix} \in$$

$$\in \begin{pmatrix} \partial_y \\ \partial_\eta \end{pmatrix} \mathscr{H}\big(\psi_y^0(s), \psi_z^0, \psi_\eta^0(s), \psi_\mu^0, Y^0(s), \eta^0(s), \alpha^0(s), u_1^0(s), e^0(s)\big), \tag{6.126}$$

with terminal conditions

$$\psi_y^0(T_1) = -\mathscr{L}'_y(Y^0(T_1), z^0(T_1), T, 0) = -\lambda_0(\phi_0)'_y(Y^0(T_1), z^0(T_1)),$$

$$\psi_\eta^0(T_1) = l_1,$$

such that

$$\begin{aligned} &\mathscr{H}\big(\psi_y^0(s), \psi_z^0, \psi_\eta^0(s), \psi_\mu^0, Y^0(s), \eta^0(s), \alpha^0(s), u_1^0(s), e^0(s)\big) = \\ &= \max \mathscr{H}\big(\psi_y^0(s), \psi_z^0, \psi_\eta^0(s), \psi_\mu^0, Y^0(s), \eta^0(s), \alpha, u_1, e\big), \end{aligned} \tag{6.127}$$

for almost all $s \in [0, T_1]$, *where the maximum is taken over* $\alpha \in [0, 1]$, $u_1 \in U$, *and* $e \in K \cap \{\|e\| = 1\}$.

Remark 6.9 In the differential inclusion (6.126) the symbols $\partial_y \mathscr{H}$, $\partial_\eta \mathscr{H}$ stand for the *subdifferential* of the function $\mathscr{H}$, that is, the closure of all partial limits that can be obtained by calculating the partial derivative $\partial \mathscr{H} / \partial y$ [93].

Before we analyze the optimality condition, let us bring Eq. (6.126) to a standard form. Since all the functions involved, except for g^+ in $\mathscr{H}$, are differentiable, they admit derivatives in the ordinary sense, while the subdifferential $\partial_{x,t} g^+(x,t)$ has the form

$$\partial_{x,t} g^+(x,t) = \begin{cases} \{0\}, & \text{if } g(x,t) < 0, \\ \{g'_{x,t}(x,t)\}, & \text{if } g(x,t) > 0, \\ \{\lambda g'_{x,t}(x,t) |\ \lambda \in [0,1]\}, & \text{if } g(x,t) = 0. \end{cases}$$

Thus, there exist measurable functions

$$\lambda_i(s) = \begin{cases} 0, & \text{if } g(Y^0(s), \eta^0(s)) < 0, \\ 1, & \text{if } g(Y^0(s), \eta^0(s)) > 0, \quad i = 1, 2, \\ \in [0, 1], & \text{if } g(Y^0(s), \eta^0(s)) = 0, \end{cases}$$

such that the differential inclusion (6.126) may be recast as the systems of ordinary differential equations

$$\begin{cases} -\dot{\psi}_y^0 = \\ \quad = \alpha^0(s)\left\{\{f_y'(Y^0(s), u_1^0(s), \eta^0(s))\}^* \psi_y^0(s) + \lambda_1(s)\psi_\mu^0 g_y'(Y^0(s), \eta^0(s))\right\} + \\ \qquad + (1-\alpha^0(s))\left\{[B(Y^0(s), \eta^0(s))e^0(s)]_y'\right\}^* \psi_y^0(s), \\ -\dot{\psi}_\eta^0(s) = \\ \quad = \alpha^0(s)\left\{\langle\psi_y^0(s), f_\eta'(Y^0(s), u_1^0(s), \eta^0(s))\rangle + \lambda_2(s)\psi_\mu^0 g_t'(Y^0(s), \eta^0(s))\right\} + \\ \qquad + (1-\alpha^0(s))\big\langle\psi_\eta^0(s), B_\eta'(Y^0(s), \eta^0(s))e^0(s)\big\rangle. \end{cases} \tag{6.128}$$

By analogy with the proof of Theorem 6.4, we introduce the set of conjugate variables

$$\begin{aligned} &p_x^0(t) = \psi_y^0(\Gamma^0(t)), \qquad p_t^0(t) = \psi_\eta^0(\Gamma^0(t)) \\ &p_v^0 = \psi_z^0, \qquad p_g^0 = \psi_\mu^0. \end{aligned} \tag{6.129}$$

Since

$$\begin{aligned} &X^0(t) = Y^0(\Gamma^0(t)), \quad V^0(t) = z^0(\Gamma^0(t)), \quad t = \eta^0(\Gamma^0(t)), \\ &u^0(t) = u_1^0(\Gamma^0(t)), \quad \mathscr{W}^0\{[0,t]\} = \int_0^{\Gamma^0(t)} (1-\alpha^0(\tau))e^0(\tau)d\tau, \end{aligned} \tag{6.130}$$

the conjugate variables satisfy the system of equations

$$\begin{aligned} p_x^0(t) = &-\lambda_0(\phi_0)_x'(X^0(T), V^0(T)) + \sum_{\tau>t,\, \tau\in D_{\mathscr{W}}^{\mathrm{d}}} \Delta p_x^0(\tau) + \\ &+ \int_t^T [f_x'(X^0(\tau), u^0(\tau), \tau)]^* p_x^0(\tau)d\tau + \\ &+ \int_t^T \{[B(X^0(\tau), \tau)d\mathscr{W}^{\mathrm{c}}(\tau)]_x'\}^* p_x^0(\tau) - \int_t^T p_g^0 g_x'(X^0(\tau), \tau)d\widetilde{\lambda}_1(\tau), \end{aligned} \tag{6.131}$$

$$p_t^0(t) = l_1 + \int_t^T \langle p_x^0(\tau), f_t'(X^0(\tau), u^0(\tau), \tau)\rangle d\tau + \sum_{\tau > t\, \tau \in D_{\mathscr{W}}^{\mathrm{d}}} \Delta p_t^0(\tau) +$$

$$+ \int_t^T \langle p_x^0(\tau), B_t'(X^0(\tau), \tau) d\mathscr{W}^{\mathrm{c}}(\tau)\rangle - \int_t^T p_g^0 g_t'(X^0(\tau), \tau) d\widetilde{\lambda}_2(\tau), \tag{6.132}$$

where the non-negative measures $\widetilde{\lambda}_{1,2}(dt)$ are defined by their respective distribution functions,

$$\widetilde{\lambda}_{1,2}\{[0, T]\} = \int_0^{\Gamma^0(t)} \alpha^0(s)\lambda_{1,2}(s)ds. \tag{6.133}$$

These measures are continuous: indeed, their possible atoms may be concentrated only at points $\tau \in D_{\mathscr{W}}^{\mathrm{d}}$ and are equal to

$$\widetilde{\lambda}_{1,2}\{\tau\} = \int_{\Gamma^0(\tau-)}^{\Gamma^0(\tau)} \alpha^0(s)\lambda(s)ds = 0, \tag{6.134}$$

because $\alpha^0(s) = 0$ almost surely on the interval $[\Gamma^0(\tau-), \Gamma^0(\tau)]$. Moreover, the measures $\widetilde{\lambda}_{1,2}(dt)$ are concentrated on the set $\{t \in [0, T] \setminus D_{\mathscr{W}}^d : g(X^0(t), t) = 0.\}$ In view of (6.134), it is sufficient to show that

$$\int_0^T \mathbf{I}\{t : g(X^0(t), t) \neq 0\} d\widetilde{\lambda}_{1,2}(t) = 0.$$

Let us do it for the measure $\widetilde{\lambda}_1(dt)$. Using the change of variable $t = \eta^0(s)$ and definition (6.133) we get

$$\int_0^T \mathbf{I}\{t : g(X^0(t), t) \neq 0\} d\widetilde{\lambda}_1(t) = \int_0^{T_1} \mathbf{I}\{s : g(Y^0(s), \eta^0(s)) < 0\}\alpha^0(s)\lambda_1(s)ds +$$

$$+ \int_0^{T_1} \mathbf{I}\{s : g(Y^0(s), \eta^0(s)) > 0\}\alpha^0(s)\lambda_1(s)ds = \mathscr{I}_1 + \mathscr{I}_2.$$

The integral $\mathscr{I}_1$ is equal to zero, since by the definition of the subdifferential, $\lambda_1(s) = 0$ for $g(Y^0(s), \eta^0(s)) < 0$. In view of the inequalities $\lambda_1(s) \leq 1$ and $\alpha^0(s) \leq 1$, we have

$$\mathscr{I}_2 \leq \int_0^T \mathbf{I}\{t : g(X^0(t), t) > 0\} dt = 0.$$

This completes the proof of the localization property of the measure $\widetilde{\lambda}_1$ (the proof for $\widetilde{\lambda}_2$ is similar).

The jumps of the conjugate variables are calculated in a same way as in the case of a strong solution provided the solution does not touch the constraints during the application of an impulse (see (6.109)), namely:

$$\begin{aligned}\Delta p_x^0(\tau) &= p_x^0(\tau) - p_x^0(\tau-) = -\left[\Psi_x'(X^0(\tau-), \tau, \mathscr{W}^0\{\tau\})\right]^* p_x^0(\tau),\\ \Delta p_t^0(\tau) &= p_t^0(\tau) - p_t^0(\tau-) = -\left\langle\Psi_\tau'(X^0(\tau-), \tau, \mathscr{W}^0\{\tau\}),\ p_x^0(\tau)\right\rangle.\end{aligned} \tag{6.135}$$

Finally, since in the auxiliary problem the equality

$$\alpha^0(s)g^+(Y^0(s), \eta^0(s)) = 0$$

holds a.e., the inequality $\alpha^0(s) > 0$ implies that $g^+(Y^0(s), \eta^0(s)) = 0$. We are now ready to formulate the necessary optimality condition for a weak generalized solution.

Theorem 6.5 *Let Assumption 6.1 be in force. Let $\{X^0(\cdot), V^0(\cdot)\}$ be the optimal generalized solution of the problem (6.113)–(6.118) with the optimal control*

$$\{u^0(t), \mathscr{W}^0(dt)\}.$$

Then there exists a collection of elements,

$$\Lambda = \left\{\lambda_0, p_v^0, p_g^0, p_x^0(\cdot), p_t^0(\cdot), \widetilde{\lambda}_{1,2}(dt)\right\},$$

not simultaneously equal to zero, where

1. *$\lambda_0 \geq 0$ and p_v^0, p_g^0 are real numbers, and*

$$p_v^0 = -\lambda_0(\phi_0)_v'(X^0(T), V^0(T));$$

2. *the vector-valued functions $\{p_x^0(\cdot)$ and $p_t^0(\cdot)\}$ satisfy the equations (6.131) and (6.132);*
3. *the non-negative measures $\widetilde{\lambda}_i(dt)$, $i = 1, 2$, in the equations (6.131) and (6.132) are concentrated on the set*

$$T_g^0 = \left\{t \in [0, T] \setminus D_{\mathscr{W}}^{\mathrm{d}} : g(X^0(t), t) = 0\right\}; \tag{6.136}$$

4. *the jumps of the functions $p_x^0(\cdot)$ and $p_t^0(\cdot)$ at the points $\tau \in D_{\mathscr{W}}^{\mathrm{d}}$ are given by the formulas (6.135).*

For this collection Λ, the optimal solution satisfies following inequalities (weak maximum principle):

- *For almost every $t \in [0, T] \setminus D_{\mathscr{W}}$ we have $g^+(X^0(t), t) = 0$ and*

$$\max_{e \in K \cap \{\|e\|=1\}} \mathscr{H}_1\big(p_x^0(t), p_v^0, X^0(t), t, e\big) \leq$$

$$\leq \mathscr{H}_0\big(p_x^0(t), p_t^0(t), 0, X^0(t), t, u^0(t)\big) = \tag{6.137}$$

$$= \max_{u \in U} \mathscr{H}_0\big(p_x^0(t), p_t^0(t), 0, X^0(t), t, u\big) = 0.$$

- *For almost every $t \in D^{\mathrm{c}}_{\mathscr{W}}$ with respect to the measure $V^0(dt)$ we have $g^+(X^0(t), t) = 0$ and*

$$\max_{e \in K \cap \{\|e\|=1\}} \mathscr{H}_1(p_x^0(t), p_v^0, X^0(t), t, e) =$$

$$= \mathscr{H}_1\big(p_x^0(t), p_v^0, X^0(t), t, w^0(t)\big) =$$

$$= \mathscr{H}_0\big(p_x^0(t), p_t^0(t), 0, X^0(t), t, u^0(t)\big) = \tag{6.138}$$

$$= \max_{u \in U} \mathscr{H}_0(p_x^0(t), p_t^0(t), 0, X^0(t), t, u) = 0,$$

where $w^0(t) = d\mathscr{W}^{\mathrm{c}}(t)/dV^0(t)$.
- *For $t \in D^{\mathrm{d}}_{\mathscr{W}}$*

$$\max_{u \in U} \mathscr{H}_0(p_x^0(t), p_t^0(t), p_g^0, X^0(t), t, u) \leq$$

$$\leq \max_{e \in K \cap \{\|e\|=1\}} \mathscr{H}_1(p_x^0(t), p_v^0, X^0(t), t, e) = \tag{6.139}$$

$$= \mathscr{H}_1\left(p_x^0(t), p_v^0, X^0(t), \mathscr{W}^0\{t\}/\|\mathscr{W}^0\{t\}\|\right) = 0.$$

Moreover, for any $\tau \in D^d_{\mathscr{W}}$ on the corresponding interval $[0, \Delta V^0(\tau)]$ one has that $g^+(y_\tau^0(s), \tau) \geq 0$ and

$$\max_{u \in U} \mathscr{H}_0\big(p_{x,\tau}^0(s), p_{t,\tau}^0(s), p_g^0, y_\tau^0(s), \tau, u\big) \leq$$

$$\leq \max_{e \in K \cap \{\|e\|=1\}} \mathscr{H}_1\left(p_{x,\tau}^0(s), p_v^0, y_\tau^0(s), t, e\right) = \tag{6.140}$$

$$= \mathscr{H}_1\left(p_{x,\tau}^0(s), p_v^0, y_\tau^0(s), \tau, \mathscr{W}^0\{t\}/\|\mathscr{W}^0\{t\}\|\right) = 0,$$

where the functions $\{y^0_\tau(s), p^0_{x,\tau}(s), p^0_{t,\tau}(s)\}$ *are given by the formulas*

$$y^0_\tau(s) = X^0(\tau-) + \Psi\left(X^0(\tau-), \tau, s\mathscr{W}^0\{t\}/\|\mathscr{W}^0\{t\}\|\right),$$

$$p^0_{x,\tau}(s) = \left[I_{(n\times n)} + \Psi'_x\left(y^0_\tau(s), \tau, s\mathscr{W}^0\{t\}/\|\mathscr{W}^0\{t\}\|\right)\right]^* p^0_x(\tau),$$

$$p^0_{t,\tau}(s) = p^0_t(\tau) + \left\langle p^0_x(\tau), \Psi'_t\left(y^0_\tau(s), \tau, s\mathscr{W}^0\{t\}/\|\mathscr{W}^0\{t\}\|\right)\right\rangle.$$

6.3.4 Maximum Principle for Optimal Generalized Solutions of Hybrid Systems

In this section we consider the necessary optimality condition for hybrid systems with phase constraints. This problem was formulated in Sect. 5.4, where we introduced the concept of the optimal generalized solution and proved the existence of such a solution (see Theorem 5.9). The generalized solution $X^0(\cdot)$ of problem (5.56)–(5.58) with the performance criterion (5.84) is defined by some measurable control $u^0(t) \in U$ and a non-negative measure $\mu(dt) \geq 0$ concentrated on the boundary of phase constraint, that is, on the set $\{t : G(X^0(t)) = 0\}$. By Theorems 5.7 and 5.9, the optimal trajectory is the solution of the equation

$$\begin{aligned} dX^0(t) &= F(X^0(t), u^0(t))dt + B(X^0(t))d\mu^c(dt) + \\ &+ \sum_{\tau\leq t}[\Phi(X^0(\tau-), \mu(\{\tau\})) - X^0(\tau-)]\delta(t-\tau)dt. \end{aligned} \tag{6.141}$$

In the present case the auxiliary control problem is formulated for a system given on some interval $[0, T_1]$, where $T_1 < \infty$, and described by the following system of equations for the variables $\{y(\cdot), \eta(\cdot)\}$:

$$\begin{cases} \dot{y}(s) = \alpha(s)F(y(s), u_1(s)) + (1-\alpha(s))B(y(s)), \\ \dot{\eta}(s) = \alpha(s), \end{cases} \tag{6.142}$$

with initial condition

$$y(0) = x_0, \quad \eta(0) = 0$$

and controls

$$\alpha(s) \in [0, 1], \quad u_1(s) \in U. \tag{6.143}$$

The controls $\{\alpha(\cdot), u_1(\cdot)\}$ are admissible if for $s = T_1$ the following constraints are satisfied:

$$\eta(T_1) = T, \quad \int_0^{T_1} \alpha(s)G^+(y(s))ds = 0, \quad \int_0^{T_1} (1-\alpha(s))G^-(y(s))ds = 0, \tag{6.144}$$

where, as above, $G^+(y) = \max\{G(y), 0\}$ and $G^-(y) = \min\{G(y), 0\}$. The performance criterion in the auxiliary problem is

$$J'[y(\cdot), \alpha(\cdot), u_1(\cdot)] = \phi_0(y(T_1)) \to \min. \tag{6.145}$$

By Theorem 5.9, the auxiliary control problem (6.142)–(6.145) is equivalent to the original problem: if $\{y^0(\cdot), \eta^0(\cdot)\alpha^0(\cdot), u_1^0(\cdot)\}$ is the optimal solution of the auxiliary problem, then the functions

$$X^0(t) = y^0(\Gamma^0(t)), \quad u^0(t) = u_1^0(\Gamma^0(t)), \quad \mu([0,t]) = \int_0^{\Gamma^0(t)} (1-\alpha^0(s)ds, \tag{6.146}$$

where

$$\Gamma^0(t) = \inf\{s : \eta^0(s) > t\}, \tag{6.147}$$

constitute the optimal solution of the original problem.

A necessary optimality condition in the auxiliary problem can be formulated as follows.

Proposition 6.2 *Let $\{y^0(\cdot), \eta^0(\cdot), \alpha^0(\cdot), u_1^0(\cdot)\}$ be the optimal solution of the auxiliary problem. Then there exists a collection of conjugate variables and constants*

$$\Lambda = \{\lambda_0, \psi_y^0, \psi_\eta^0, \psi_{\mu,1}^0, \psi_{\mu,2}^0\},$$

not simultaneously equal to zero, such that the following holds:

1. $\lambda_0 \geq 0$;
2. $\psi_y^0(\cdot)$ *is an absolutely continuous function satisfying the differential inclusion*

$$\dot{\psi}_y^0(s) \in -\partial_y \mathscr{H}(\psi_y^0(s), \psi_\eta^0(s), \psi_{\mu,1}^0, \psi_{\mu,2}^0, y^0(s), \alpha^0(s), u_1^0(s)), \tag{6.148}$$

where

$$\begin{aligned} &\mathscr{H}(\psi_y, \psi_\eta, \psi_{\mu,1}, \psi_{\mu,2}, y, \alpha, u_1) = \\ &\quad = \alpha\{\langle\psi_y, F(y,u_1)\rangle + \psi_\eta + \psi_{\mu,1}G^+(y)\} + \\ &\quad + (1-\alpha)\{\langle\psi_y, B(y)\rangle + \psi_{\mu,2}G^-(y)\} = \\ &\quad = \alpha\mathscr{H}_0(\psi_y, \psi_\eta, \psi_{\mu,1}, y, u_1) + (1-\alpha)\mathscr{H}_1(\psi_y, \psi_{\mu,2}, y), \end{aligned} \tag{6.149}$$

and $\psi_y^0(T_1) = -\lambda_0(\varphi_0)'_y(y(T_1)$;

3. ψ_η^0, $\psi_{\mu,1}^0$ *and* $\psi_{\mu,2}^0$ *are constants;*
4. *the function* $\mathscr{H}$ *achieves its maximum on the optimal control, that is*

$$0 = \mathscr{H}\big(\psi_y^0(s), \psi_\eta^0, \psi_{\mu,1}^0, \psi_{\mu,2}^0, y^0(s), \alpha^0(s), u_1^0(s)\big) = \\ = \max_{u_1 \in U, \alpha \in [0,1]} \mathscr{H}\big(\psi_y^0(s), \psi_\eta^0, \psi_{\mu,1}^0, \psi_{\mu,2}^0, y^0(s), \alpha, u_1\big). \tag{6.150}$$

Let us write the interval $[0, T_1]$ as the union of three disjoint sets:

$$[0, T_1] = \mathscr{A}_1 \cup \mathscr{A}_2 \cup \mathscr{A}_3,$$

where $\mathscr{A}_1 = \{s : \eta^0(s) \in [0, T] \setminus D_\mu\}$, $\mathscr{A}_2 = \{s : \eta^0(s) \in D_\mu^{\mathrm{d}}\}$ and $\mathscr{A}_3 = \{s : \eta^0(s) \in D_\mu^{\mathrm{c}}\}$. Here D_μ, D_μ^{d} and D_μ^{c} are the supports of the measures $\mu(dt)$, $\mu^{\mathrm{d}}(dt)$ and $\mu^{\mathrm{c}}(dt)$, respectively.

On the set $\mathscr{A}_1 = \{s : \eta^0(s) \in [0, T] \setminus D_\mu\}$ we have $\alpha^0(s) = 1$ almost surely and $G^+(y^0(s)) = 0$, so inequality (6.150) gives

$$0 = \mathscr{H}_0\big(\psi_y^0(s), \psi_\eta^0, 0, y^0(s), u_1^0(s)\big) = \\ = \max_{u_1 \in U} \mathscr{H}_0\big(\psi_y^0(s), \psi_\eta^0, 0, y^0(s), u_1\big) \geq \mathscr{H}_1\big(\psi_y^0(s), \psi_{\mu,2}, y^0(s)\big). \tag{6.151}$$

Via the change of variable $s = \Gamma^0(t)$ we deduce that the variables

$$X^0(t) = y^0(\Gamma^0(t)), \quad p_x^0(t) = \psi_y^0(\Gamma^0(t)), \quad u^0(t) = u_1^0(\Gamma^0(t))$$

satisfy the inequality

$$0 = \mathscr{H}_0\big(p_x^0(t), \psi_\eta^0, 0, X^0(t), u^0(t)\big) = \\ = \max_{u \in U} \mathscr{H}_0\big(p_x^0(t), \psi_\eta^0, 0, X^0(t), u\big) \geq \mathscr{H}_1\big(p_x^0(t), \psi_{\mu,2}, X^0(t)\big), \tag{6.152}$$

which holds a.e. on the set $\{t \in [0, T] \setminus D_\mu\}$.

Similarly, on the set $\mathscr{A}_2$, where $\alpha^0(s) = 0$ and $G^-(y^0(s)) = 0$, observing that

$$\{s : \eta^0(s) \in D_\mu^d\} = \bigcup_{\tau \in D_\mu^d} [\Gamma^0(\tau-), \Gamma^0(\tau)],$$

we deduce that

$$\max_{u \in U} \mathscr{H}_0\big(\psi_y^0(s), \psi_\eta^0, \psi_{\mu,1}, y^0(s), u\big) \leq \mathscr{H}_1\big(\psi_y^0(s), 0, y^0(s)\big) = 0, \tag{6.153}$$

which holds on each segment $s \in [\Gamma^0(\tau-), \Gamma^0(\tau)]$ with $\tau \in D_\mu^{\mathrm{d}}$.

And finally, on the set $\mathscr{A}_3$, where $\alpha^0(s) \in [0, 1]$ and $G^-(y^0(s)) = G^+(y^0(s)) = 0$, we get that

$$\max_{u \in U} \mathscr{H}_0\big(\psi_y^0(s), \psi_\eta^0, 0, y^0(s), u\big) \leq \mathscr{H}_1\big(\psi_y^0(s), 0, y^0(s)\big) = 0. \tag{6.154}$$

Now using the same change of variables in (6.154), we obtain the optimality condition in the original problem in the form

$$\max_{u \in U} \mathscr{H}_0\big(p_y^0(t), \psi_\eta^0, 0, X^0(t), u\big) \leq \mathscr{H}_1\big(p_x^0(t), 0, X^0(t)\big) = 0, \tag{6.155}$$

a.e. with respect to the measure $\mu(dt)$ on the set $\{t \in D_\mu^{\mathrm{c}}\}$.

It remains to derive the equations for the conjugate variables $p_x^0(t) = \psi_y^0(\Gamma^0(t))$. By using the expression of the subdifferential (see [93]),

$$\partial_x G^+(x) = \begin{cases} \{0\}, & \text{if } G(x) < 0, \\ \{G_x'(x)\}, & \text{if } G(x) > 0, \\ \{\lambda G_x'(x) \mid \lambda \in [0, 1]\}, & \text{if } G(x) = 0 \end{cases}$$

and

$$\partial_x G^-(x) = \begin{cases} \{0\}, & \text{if } G(x) > 0, \\ \{G_x'(x)\}, & \text{if } G(x) < 0, \\ \{\lambda G_x'(x) \mid \lambda \in [0, 1]\}, & \text{if } G(x) = 0, \end{cases}$$

we conclude that there exists two measurable functions $\lambda_1(s)$ and $\lambda_2(s)$ taking values in $[0, 1]$ such that $\psi_y^0(s)$ satisfies the ordinary differential equation

$$\begin{aligned} \dot{\psi}_y^0(s) = &- \left\{\alpha^0(s) F_y'(y^0(s), u_1^0(s)) + (1 - \alpha^0(s)) B_y'(y^0(s))\right\}^* \psi_y^0(s) - \\ &- \psi_{\mu,1}^0 \lambda_1(s) \alpha^0(s) G_y'(y^0(s)) \mathbf{I}\{s : G(y^0(s)) \geq 0\} - \\ &- \psi_{\mu,2}^0 \lambda_2(s)(1 - \alpha^0(s)) G_y'(y^0(s)) \mathbf{I}\{s : G(y^0(s)) \leq 0\}. \end{aligned} \tag{6.156}$$

If we now write this equation in integral form and use the change of variables $s = \Gamma^0(t)$ we obtain

$$
\begin{aligned}
p_x^0(t) = \psi_y^0(\Gamma^0(t)) = \psi_y^0(0) - \\
- \int_0^{\Gamma^0(t)} \left\{\alpha^0(s)F_y'(y^0(s), u_1^0(s)) + (1-\alpha^0(s))B_y'(y^0(s))\right\}^* \psi_y^0(s)ds - \\
- \int_0^{\Gamma^0(t)} \psi_{\mu,1}^0\lambda_1(s)\alpha^0(s)G_y'(y^0(s))\mathbf{I}\{s : G(y^0(s)) \ge 0\}ds - \\
- \int_0^{\Gamma^0(t)} \psi_{\mu,2}^0\lambda_2(s)(1-\alpha^0(s))G_y'(y^0(s))\mathbf{I}\{s : G(y^0(s)) \le 0\}ds.
\end{aligned}
\tag{6.157}
$$

Further, we have

$$
\begin{aligned}
\int_0^{\Gamma^0(t)} \left\{\alpha^0(s)F_y'(y^0(s), u_1^0(s))\right\}^* \psi_y^0(s)ds = \\
= \int_0^t \left\{F_y'(y^0(\Gamma^0(\tau)), u_1^0(\Gamma^0(\tau)))\right\}^* \psi_y^0(\Gamma^0(\tau))d\tau = \\
= \int_0^t \left\{F_x'(X^0(\tau), u^0(\tau))\right\}^* p_x^0(\tau)d\tau,
\end{aligned}
\tag{6.158}
$$

and

$$
\begin{aligned}
\int_0^{\Gamma^0(t)} \left\{(1-\alpha^0(s))B_y'(y^0(s))\right\}^* \psi_y^0(s)ds = \\
= \int_0^t \left\{B_y'(y^0(\Gamma^0(\tau)))\right\}^* \psi_y^0(\Gamma^0(\tau))d\mu^c(\tau) + \\
+ \sum_{\tau\le t,\, \tau\in D_\mu^{\mathrm{d}}} \int_{\Gamma^0(\tau-)}^{\Gamma^0(\tau)} \left\{B_y'(y^0(s))\right\}^* \psi_y^0(s)ds = \\
= \int_0^t \left\{B_x'(X^0(\tau))\right\}^* p_x^0(\tau))d\mu^c(\tau) + \\
+ \sum_{\tau\le t,\, \tau\in D_\mu^{\mathrm{d}}} \int_{\Gamma^0(\tau-)}^{\Gamma^0(\tau)} \left\{B_y'(y^0(s))\right\}^* \psi_y^0(s)ds.
\end{aligned}
\tag{6.159}
$$

To transform the two remaining integrals in (6.157), note that $\alpha^0(s) = 0$ a.e. on the set $\{s : G(y^0(s) > 0\}$ and $\alpha^0(s) = 1$ a.e. on the set $\{s : G(y^0(s) < 0\}$.

Consequently,

$$\begin{aligned}
&\int_0^{\Gamma^0(t)} \psi^0_{\mu,1}\lambda_1(s)\alpha^0(s)G'_y(y^0(s))I_{\{s:\,G(y^0(s))\geq 0\}}ds = \\
&= \int_0^{\Gamma^0(t)} \psi^0_{\mu,1}\lambda_1(s)\alpha^0(s)G'_y(y^0(s))\mathbf{I}\{s:\,G(y^0(s))=0\}ds = \\
&= \int_0^{t} \psi^0_{\mu,1}\lambda_1(\Gamma^0(\tau))G'_x(X^0(\tau))\mathbf{I}\{\tau:\,G(X^0(\tau))=0\}d\tau = \\
&= \int_0^{t} \psi^0_{\mu,1}G'_x(X^0(\tau))d\widetilde{\lambda}_1(\tau),
\end{aligned} \tag{6.160}$$

and

$$\begin{aligned}
&\int_0^{\Gamma^0(t)} \psi^0_{\mu,2}\lambda_2(s)(1-\alpha^0(s))G'_y(y^0(s))\mathbf{I}\{s:\,G(y^0(s))\leq 0\}ds = \\
&= \int_0^{\Gamma^0(t)} \psi^0_{\mu,2}\lambda_2(s)(1-\alpha^0(s))G'_y(y^0(s))\mathbf{I}\{s:\,G(y^0(s))=0\}ds = \\
&= \int_0^{t} \psi^0_{\mu,2}\lambda_2(\Gamma^0(\tau))G'_x(X^0(\tau))\mathbf{I}\{\tau:\,G(X^0(\tau))=0\}d\mu^{\mathrm{c}}(\tau) + \\
&+ \sum_{\tau\leq t,\,\tau\in D^{\mathrm{d}}_\mu} \int_{\Gamma^0(\tau-)}^{\Gamma^0(\tau)} \psi^0_{\mu,2}\lambda_2(s)G'_y(y^0(s))\mathbf{I}\{s:\,G(y^0(s))=0\}ds = \\
&= \int_0^{t} \psi^0_{\mu,2}G'_x(X^0(\tau))d\widetilde{\lambda}_2(\tau) + \sum_{\tau\leq t,\,\tau\in D^{\mathrm{d}}_\mu} \int_{\Gamma^0(\tau-)}^{\Gamma^0(\tau)} \psi^0_{\mu,2}G'_y(y^0(s))d\widetilde{\lambda}^\tau_2(s),
\end{aligned} \tag{6.161}$$

where the non-negative measures $\widetilde{\lambda}_1(dt)$, $\widetilde{\lambda}_2(dt)$ and the family of non-negative measures $\widetilde{\lambda}^\tau_2(ds)$, defined for each $\tau \in D^{\mathrm{d}}_\mu$ on the segments $[\Gamma^0(\tau-), \Gamma^0(\tau)]$, are defined by

$$\begin{aligned}
&\widetilde{\lambda}_1(dt) = \lambda_1(\Gamma^0(t))\mathbf{I}\{t:\,G(X^0(t))=0\}dt, \\
&\widetilde{\lambda}_2(dt) = \lambda_2(\Gamma^0(t))I_{\{t:\,G(X^0(t))=0\}}d\mu^{\mathrm{c}}(t), \\
&\widetilde{\lambda}^\tau_2(ds) = \lambda_2(s)\mathbf{I}\{s\in[\Gamma^0(\tau-),\Gamma^0(\tau)]:\,G(y^0(s))=0\}ds.
\end{aligned} \tag{6.162}$$

The measures $\widetilde{\lambda}_1(dt)$ and $\widetilde{\lambda}_2(dt)$ are concentrated on the set $\{t|\,G(X^0(t))=0\}$, and the measures $\widetilde{\lambda}^\tau_2$ are concentrated on the set $\{s \in [\Gamma^0(\tau-),\Gamma^0(\tau)]\,|\, G(y^0(s)) = 0\}$. Moreover, $\widetilde{\lambda}_1(dt) =$ and $\widetilde{\lambda}^\tau_2(ds)$ are absolutely continuous with respect to the Lebesgue measure, and $\widetilde{\lambda}_2(dt)$ is absolutely continuous with respect to the measure $\mu^c(dt)$.

Thus, if we use the relations (6.157)–(6.162), the equation for the adjoint variable $p_x^0(t)$ will have the form

$$
\begin{aligned}
p_x^0(t) &= -\lambda_0(\varphi_0)'_x(X(T)) + \int_t^T \left\{F'_x(X^0(\tau), u^0(\tau))\right\}^* p_x^0(\tau)d\tau + \\
&+ \int_t^T \left\{B'_x(X^0(\tau))\right\}^* p_x^0(\tau)d\mu^c(\tau) + \int_t^T \psi^0_{\mu,1} G'_x(X^0(\tau))d\widetilde{\lambda}_1(\tau) + \\
&+ \int_t^T \psi^0_{\mu,2} G'_x(X^0(\tau))d\widetilde{\lambda}_2(\tau) + \sum_{\tau>t,\, \tau\in D^d_\mu} \Delta p_x^0(\tau),
\end{aligned}
\tag{6.163}
$$

where $\Delta p_x^0(\tau)$ is defined by

$$
\Delta p_x^0(\tau) = p_x^0(\tau) - p_x^0(\tau-) = p_{\tau,x}(\mu\{\tau\}) - p_{\tau,x}(0),
\tag{6.164}
$$

and $p_{\tau,x}(s)$ satisfies on the segment $[0, \mu\{\tau\}]$ the equation

$$
\begin{aligned}
p_{\tau,x}(s) &= p_x^0(\tau) + \int_s^{\mu\{\tau\}} \left\{B'_x(\Phi(X^0(\tau-), \omega))\right\}^* p_{\tau,x}(\omega)d\omega + \\
&+ \int_s^{\mu\{\tau\}} \psi_{\mu,2} G'_x(\Phi(X^0(\tau-), \omega)d\widetilde{\lambda}_2^\tau(\omega).
\end{aligned}
\tag{6.165}
$$

This equation follows from (6.157) upon observing that $\mu\{\tau\} = \Gamma^0(\tau) - \Gamma^0(\tau-)$ and on the segment $[\Gamma^0(\tau-), \Gamma^0(\tau)]$ one has $y^0(s) = \Phi(X^0(\tau-), s - \Gamma^0(\tau-))$. Indeed, after the time change $\omega = s - \Gamma^0(\tau-)$ the segment $[\Gamma^0(\tau-), \Gamma^0(\tau)]$ transforms into $[0, \mu\{\tau\}]$, and using the same notation for the transformed measure $\widetilde{\lambda}_2^\tau$ we obtain (6.165).

Therefore, condition (6.153) can be recast as follows: for any $\tau \in D^d_\mu$,

$$
\begin{aligned}
&\max_{u\in U} \mathcal{H}_0\big(p_{\tau,x}(\omega), \psi^0_\eta, \psi_{\mu,1}, \Phi(X^0(\tau-), \omega), u\big) \le \\
&\le \mathcal{H}_1\big(p_{\tau,x}(\omega), 0, \Phi(X^0(\tau-), \omega)\big) = 0
\end{aligned}
\tag{6.166}
$$

almost everywhere on the corresponding segment $[0, \mu\{\tau\}]$. Thus, we established the following optimality condition in the maximum principle form.

Theorem 6.6 *Let $\{X^0(\cdot), u^0(\cdot), \mu(\cdot)\}$ be the optimal generalized solution in the problem (5.56)–(5.58),(5.84). Then there exists a collection of elements, not simultaneously equal to zero, consisting of:*

1. *constants* $\lambda_0 \geq 0$, ψ_η^0, $\psi_{\mu,1}^0$, $\psi_{\mu,2}^0$;
2. *non-negative measures* $\widetilde{\lambda}_1(dt)$ *and* $\widetilde{\lambda}_2(dt)$ *that are concentrated on the set* $\{t \,|\, G(X^0(t)) = 0\}$ *and absolutely continuous with respect to the Lebesgue measure and the measure* $\mu^c(dt)$;
3. *a family of non-negative measures* $\lambda_2^\tau(ds)$, *parametrized by* $\tau \in D_\mu^d$, *such that each* $\lambda_2^\tau(ds)$ *is defined on the corresponding segment* $[0, \mu\{\tau\}]$ *and concentrated on the set* $\{\omega \in [0, \mu\{\tau\}] \,|\, G(\Phi(X^0(\tau-), \omega)) = 0)\}$;
4. *the adjoint variable* $p_x^0(\cdot)$, *which satisfies the system of Eqs. (6.163)–(6.165);*

such that the generalized maximum principle is valid, that is

- *the inequality (6.152) holds on the set* $t \in [0, T] \setminus D_\mu$;
- *the inequality (6.155) holds on the set* $t \in D_\mu^c$;
- *the inequality (6.166) holds for all* $\tau \in D_\mu^d$.

To complete our discussion of the optimality condition for hybrid systems with phase constraints, we note that the the proof method used allows us to extend it to more general cases. A series of interesting and complex problems are connected with the control of mechanical systems with impacts and controlled obstacles, for which very substantial results have been obtained recently (see [33, 67, 70, 283–285]). However, the consideration of this very wide circle of problems is beyond the scope of this monograph, so in this chapter we present them only as an example of how to use the method of singular space-time transformations (discontinuous time change).

6.4 Linear and Convex Generalized Optimization Problems

6.4.1 Introduction

The necessary optimality conditions obtained above are a very effective "tool" for solving practical problems that plays a two-fold role. On the one hand, they allow one to distinguish among the whole class of admissible solutions only those that satisfy the maximum principle, thereby significantly narrowing down the set of solutions "suspected" to be optimal. On the other hand, the maximum principle is the basis of numerical optimization algorithms, especially for those problems where this principle is not only necessary, but also a sufficient optimality condition. This class of problems has convexity properties, and results concerned with problems with bounded controls are well known [133, 237]. There are also cases when the generalized maximum principle is both a necessary and sufficient condition for generalized optimization problems in systems described by linear differential equations with measure [5, 218, 328, 401, 466]. However, in all the listed works the derivation of the optimality conditions is based on the linearity of system under study, which allows to apply methods of convex optimization and functional analysis, methods that are not really suitable for nonlinear problems with general constraints. The first more or less general formulation of the problem using the time change method was suggested by R. Rishel [377], who obtained a generalized maximum principle. However, the applicability of his approach was limited by some "constancy" condition, i.e., the possibility of choosing a special representation of the generalized control in the form of a measure. This condition is a special case of equivalence condition of the generalized and singular control problems (see Sect. 6.3.2) and, needless to say, narrows significantly the class of admissible systems. The necessity and sufficiency of the generalized maximum principle for linear systems with a convex performance criterion and constraints was established for the first time in [264], and subsequently in [265] for a special class of nonlinear systems arising in the observation control problem. This section presents general results first published in [269].

6.4.2 Maximum Principle in Problems with Convexity

We consider a system described by the equations

$$\begin{cases} \dot{X}(t) = f(X(t), t) + B(u(t), t)w(t), \\ \dot{V}(t) = \|w(t)\|, \quad V(0) = 0, \end{cases} \tag{6.167}$$

where $X \in \mathbb{R}^n$ and $V \in \mathbb{R}^1$ are phase variables and

$$u \in U, \quad w \in K \tag{6.168}$$

are controls. As above, $U \subseteq \mathbb{R}^k$ is a compact set and $K \subset \mathbb{R}^m$ is a closed convex cone.

The optimal control problem is considered on the set of admissible generalized solutions of system (6.167) satisfying the following terminal and phase constraints:

$$\begin{aligned} &S(X(0-), X(T), V(T)) \le 0, \quad h(X(0-), X(T)) = 0, \\ &g(X(t), V(t), t) \le 0, \quad \text{for all } t \in [0, T] \end{aligned} \tag{6.169}$$

where the vector-valued functions S, h and g have the respective dimensions N_1, N_2 and N_3, and the inequalities (6.169) are understood componentwise.

The performance criterion is

$$J[X(\cdot), V(\cdot), u(\cdot), w(\cdot)] = \phi_0(X(0-), X(T), V(T)) \to \min. \tag{6.170}$$

Assumption 6.2 The functions S, g_k and ϕ_0 are non-decreasing in the scalar variable v.

By Corollary 4.1, if the Assumption 6.2 holds and the set of vectors

$$\mathscr{L}(t) = \left\{ l = \begin{pmatrix} l_1 \in \mathbb{R}^n \\ l_2 \in \mathbb{R}^1 \end{pmatrix} : \begin{array}{l} l_1 = B(u,t)w, \\ l_2 = \|w\|, \end{array} \quad u \in U, \ w \in K \right\} \tag{6.171}$$

is convex for any $t \in [0, T]$, then for the optimal generalized solution $\{X(\cdot), V(t)\}$ of the system (6.167) there exist:

1. a non-negative scalar measure $\mathscr{V}(dt)$ with the distribution function

$$V(t) = \mathscr{V}([0, t]),$$

defined on the Borel subsets of the segment $[0, T]$, such that

$$\mathscr{V}([0, T]) = V(T) \leq M;$$

2. a pair of controls

$$\{u_0(t), w_0(t)\},$$

which are both Lebesgue- and $\mathscr{V}$-measurable and obey on the segment $[0, T]$ the constraints

$$u_0(t) \in U, \quad w_0(t) \in K \cap \{\|w\| \leq 1\} \tag{6.172}$$

a.e. with respect to both the Lebesgue and the $\mathscr{V}$ measure,

such that $X(t)$ satisfies the equation

$$X(t) = X(0-) + \int_0^t f(X(\tau), \tau)d\tau + \int_0^t B(u_0(\tau), \tau)w_0(\tau)dV(\tau). \tag{6.173}$$

Moreover, if one defines the vector-valued measure $\mathscr{W}(dt)$ by the formula

$$\mathscr{W}([0, t]) = \int_0^t w_0(\tau)\, dV(\tau), \tag{6.174}$$

then Eq. (6.173) can be recast as the equation with measure

$$\begin{cases} X(t) = X(0-) + \int_0^t f(X(\tau), \tau)d\tau + \int_0^t B(u_0(\tau), \tau)d\mathscr{W}(\tau), \\ V(t) = \operatorname*{Var}_{[0,T]} \mathscr{W}(\tau). \end{cases} \tag{6.175}$$

Indeed, by definition (6.174) and the constraints (6.172) we have

$$\operatorname*{Var}_{[0,T]} \mathscr{W}(\tau) \le V(t), \tag{6.176}$$

and consequently the generalized control $\{u_0(\cdot), \mathscr{W}(\cdot)\}$ defines the same trajectory $X(t)$ and, thanks to Assumption 6.2, is admissible. Moreover, the corresponding value of the performance criterion is not larger than $\phi_0(X(0-), X(T), V(T))$, which readily shows that the control is optimal, that is, inequality (6.176) must be replaced by a strict equality.

Since the solution is sought in the class of strong generalized solutions of the problem (6.167)–(6.170), its behavior at the points where the boundary is reached requires special consideration. However, if the functions $g_k(x, v, t)$ are convex in the variables (x, v), then any generalized solution $\{X, V\}$ satisfying the constraints (6.169) is admissible in the strong sense.

This is a consequence of the following result:

Proposition 6.3 *Suppose the functions $g_k(x, v, t)$ are convex in the variables (x, v). Then for any jump point $\tau \in D^{\mathrm{d}}_{\mathscr{V}}$ of a solution $\{X, V\}$ satisfying the constraint $g_k(X(t), V(t), t) \le 0$, and for any pair*

$$\{u_0(\tau) \in U, w^0(\tau) \in K \cap \{\|w\| \le 1\}\}$$

such that

$$\Delta X(\tau) = B(u_0(\tau), \tau) w_0(\tau) \Delta V(\tau) = B(u_0(\tau), \tau) \Delta \mathscr{W}(\tau), \tag{6.177}$$

the variables $\{y_\tau(s), z_\tau(s)\}$, which determine the jump by means of the auxiliary system

$$\begin{cases} \dot{y}_\tau(s) = B(u_0(\tau), \tau) w_0(\tau), \\ \dot{z}_\tau(s) = 1, \end{cases} \tag{6.178}$$

with the initial conditions $y_\tau(0) = X(\tau-), z_\tau(0) = V(\tau-)$, satisfy on the corresponding segment $[0, \Delta V(\tau)]$ the inequality

$$g_k(y_\tau(s), z_\tau(s), \tau) \le 0 \quad \textit{for all } s \in [0, \Delta V(\tau)]. \tag{6.179}$$

Remark 6.10 Inequality (6.179) means that the generalized solution $\{X(\cdot), V(\cdot)\}$ satisfies the phase constraints in the strong sense.

To prove Proposition 6.3, we use the convexity of the functions g_k and the inequalities

$$g_k(X(\tau-), V(\tau-), \tau) \le 0, \quad g_k(X(\tau), V(\tau), \tau) \le 0,$$

which hold for any generalized solution. It follows from (6.178) that

$$y_\tau(s)=X(\tau-)+[X(\tau)-X(\tau-)]\frac{s}{\Delta V(\tau)}=\frac{\Delta V(\tau)-s}{\Delta V(\tau)}X(\tau-)+\frac{s}{\Delta V(\tau)}X(\tau),$$

$$z_\tau(s)=V(\tau-)+s=\frac{\Delta V(\tau)-s}{\Delta V(\tau)}V(\tau-)+\frac{s}{\Delta V(\tau)}V(\tau).$$

Consequently, thanks to the convexity of g_k, setting $\lambda=(\Delta V(\tau)-s)/\Delta V(\tau)\in[0,1]$, we have that

$$\begin{aligned} &g_k(y_\tau(s),z_\tau(s),\tau)=\\ &\qquad=g_k(\lambda X(\tau-)+(1-\lambda)X(\tau),\lambda V(\tau-)+(1-\lambda)V(\tau),\tau)\le\\ &\qquad\le\lambda g_k(X(\tau-),V(\tau-),\tau)+(1-\lambda)g_k(X(\tau),V(\tau),\tau)\le 0. \end{aligned}$$

Remark 6.11 Thus, if the functions g_k are convex, then there exists a strong optimal generalized solution. This solution is given by the 4-tuple

$$\{X^0(\cdot),V^0(\cdot),u^0(\cdot),\mathscr{W}^0(\cdot)\},$$

satisfying the Eqs. (6.174) and (6.175), where $\{X^0,V^0\}$ are the phase variables and $\{u^0(t),\mathscr{W}^0(dt)\}$ are the controls, subject to the constraints

$$u^0(t)\in U,\quad \mathscr{W}^0(dt)\in K,\quad \operatorname*{Var}_{[0,T]}\mathscr{W}^0([0,t])\le M. \tag{6.180}$$

Next, using the general optimality condition (see Theorem 6.2 and Corollary 6.2), we formulate the analogous result for the problem (6.167)–(6.170). We use Assumptions 6.1 and 6.2 and the notations introduced in Sect. 6.2.2:

$$\begin{aligned} &\mathscr{H}_0\big(p_x,x,t\big)=\langle p_x,f(x,t)\rangle\\ &\mathscr{H}_1\big(p_x,p_v,u,t,e\big)=\langle p_x,B(u,t)e\rangle+p_v,\\ &\mathscr{L}(x_0,x_1,v)=\lambda_0\phi_0(x_0,x_1,v)+\langle l_1,h(x_0,x_1)\rangle+\\ &\qquad\qquad+\langle l_2,S(x_0,x_1,v)\rangle+\gamma_0(v-M),\\ &D^{\mathrm{d}}_{\mathscr{V}}=\{\tau:\mathscr{V}(\{\tau\})>0\}=\{\tau:\Delta V^0(\tau)>0\}. \end{aligned} \tag{6.181}$$

Theorem 6.7 *Let the Assumptions 6.1 and 6.2 be in force, and let*

$$\{X^0(\cdot),V^0(\cdot),u^0(t),\mathscr{W}^0(dt)\}$$

be the optimal solution of problem (6.167)–(6.170). Then there exist a collection of elements

$$\begin{aligned}\Lambda = \{\lambda_0, \gamma_0, l_1, l_2, p_x^0(\cdot), p_v^0(\cdot), \\ \lambda_k(dt), \nu_\tau^k(ds) \text{ for } k = 1, \ldots, N_3, \quad \tau \in D_{\mathcal{V}}^{\mathrm{d}}\},\end{aligned} \tag{6.182}$$

not simultaneously equal to zero, where

1. $\lambda_0 \geq 0$ *and* $\gamma_0 \geq 0$ *are real numbers;*
2. $l_1 \in \mathbb{R}^{N_1}$ *and* $l_2 \in \mathbb{R}^{N_2}$ *are vectors, such that the component of* l_2 *are non-negative, and the trajectory* $\{X^0(\cdot), V^0(\cdot)\}$ *satisfies the relation*

$$\langle l_2, S(X^0(0-), X^0(T), V^0(T))\rangle + \gamma_0(V^0(T) - M) = 0; \tag{6.183}$$

3. *the vector-valued functions* $\{p_x^0(\cdot), p_v^0(\cdot)\}$ *satisfy the equations*

$$\begin{aligned}p_x^0(t) = &-\mathscr{L}'_{x(T)}\big(X^0(0-), X^0(T), V^0(T)\big) \\ &+ \int_t^T (\mathscr{H}_0)'_x\big(p_x^0(\tau), X^0(\tau), \tau\big)d\tau + \sum_{\tau>t,\, \tau\in D_{\mathcal{V}}^{\mathrm{d}}} \Delta p_x^0(\tau) - \\ &- \int_t^T \sum_{k=1}^{N_3} g'_{k_x}\big(X^0(\tau), V^0(\tau), \tau\big)d\lambda_k(\tau),\end{aligned} \tag{6.184}$$

$$\begin{aligned}p_v^0(t) = &-\mathscr{L}'_v\big(X^0(0-), X^0(T), V^0(T)\big) + \\ &+ \sum_{\tau>t,\, \tau\in D_{\mathcal{V}}^{d}} \Delta p_v^0(\tau) - \int_t^T \sum_{k=1}^{N_3} g'_{k_v}\big(X^0(\tau), V^0(\tau), \tau\big)d\lambda_k(\tau),\end{aligned} \tag{6.185}$$

with the following boundary condition for p_x^0 *at* $t = 0$:

$$p_x^0(0-) = \mathscr{L}'_{x(0)}(X^0(0-), X^0(T), V^0(T)); \tag{6.186}$$

4. *the non-negative measures* $\lambda_k(dt)$ *in Eqs. (6.184) and (6.185) are concentrated on the sets*

$$T_k^0 = \{t \in [0, T] \setminus D_{\mathcal{V}}^{\mathrm{d}} \,:\, g_k(X^0(t), V^0(t), t) = 0\}; \tag{6.187}$$

5. *the jumps of the functions* $\{p_x^0(\cdot), p_v^0(\cdot)\}$ *at the points* $\tau \in D_{\mathcal{V}}^{\mathrm{d}}$ *satisfy the relations*

$$
\begin{cases}
\Delta p_x^0(\tau) = \displaystyle\int_0^{\Delta V^0(\tau)} \sum_{k=1}^{N_3} g'_{k_x}\Big(X^0(\tau-)+ \\
\qquad + \omega B(u^0(\tau),\tau)\dfrac{\Delta \mathscr{W}^0(\{\tau\})}{\Delta V^0(\tau)}, V^0(\tau-)+\omega,\tau\Big)\, d\nu_\tau^k(\omega), \\
\Delta p_v^0(\tau) = \displaystyle\int_0^{\Delta V^0(\tau)} \sum_{k=1}^{N_3} g'_{k_v}\Big(X^0(\tau-)+ \\
\qquad + \omega B(u^0(\tau),\tau)\dfrac{\Delta \mathscr{W}^0(\{\tau\})}{\Delta V^0(\tau)}, V^0(\tau-)+\omega,\tau\Big)\, d\nu_\tau^k(\omega);
\end{cases}
\tag{6.188}
$$

6. *the collection of non-negative measures* $\nu_\tau^k(dt)$ *in Eq. (6.188), defined for all* $\tau \in \mathrm{D}^{\mathrm{d}}_{\mathscr{V}}$ *and concentrated on the sets*

$$
\begin{aligned}
T_\tau^k = \Big\{ & s \in [0, \Delta V^0(\tau)] : \\
& g_k\Big(X^0(\tau-) + sB(u^0(\tau),\tau)\frac{\Delta \mathscr{W}^0(\{\tau\})}{\Delta V^0(\tau)}, V^0(\tau-)+s,\tau\Big) = 0\Big\}.
\end{aligned}
\tag{6.189}
$$

For the collection Λ *described above, the optimal path* $\{X^0(\cdot), V^0(\cdot)\}$ *and the optimal control* $\{u^0(t), \mathscr{W}^0(dt)\}$ *satisfy the generalized maximum principle: For any admissible generalized control* $\{u(\cdot), \mathscr{A}(\cdot)\}$ *such that* $u(t) \in U$ *and* $\mathscr{A}(dt) \in K$ *the following inequality holds:*

$$
\begin{aligned}
&\int_0^T \mathscr{H}_1\Big(p_x^0(t), p_v^0(t), u(t), t, \frac{d\mathscr{A}(t)}{d|\mathscr{A}|(t)}\Big) d|\mathscr{A}|(t) \le \\
&\qquad \le \int_0^T \mathscr{H}_1\Big(p_x^0(t), p_v^0(t), u^0(t), t, \frac{d\mathscr{W}^0(t)}{dV^0(t)}\Big) dV^0(t) = 0,
\end{aligned}
\tag{6.190}
$$

where $|\mathscr{A}|(t) = \operatorname*{Var}_{[0,t]} \mathscr{A}(s)$.

Remark 6.12 Theorem 6.7 is a direct consequence of the general maximum principle; however, we need the following strengthening of this result for the case of convex functions g_k.

Proposition 6.4 *If the functions* $g_k(x, v, t)$ *are convex in* (x, v)*, then in inequality (6.190) one can replace* $\{p_x^0(t), p_v^0(t)\}$ *with* $\{p_x^0(t-), p_v^0(t-)\}$.

Proof As it follows from the proof of Theorem 6.2, for any $\{u, e\}$ such that $u \in U$ and $e \in \{K \cap \|e\| \le 1\}$ we have

$$
\mathscr{H}_1\big(p_x^0(t), p_v^0(t).u, t, e\big) \le 0.
$$

Since function $\mathscr{H}_1$ is continuous and the functions $\{p_x^0(t), p_v^0(t)\}$ admit left limits, for the left-hand side of inequality (6.190) we have that

$$\int_0^T \mathscr{H}_1\left(p_x^0(t-), p_v^0(t-), u(t), t, \frac{d\mathscr{A}(t)}{d|\mathscr{A}|(t)}\right) d|\mathscr{A}|(t) \leq 0. \tag{6.191}$$

At the same time,

$$\int_0^T \mathscr{H}_1\left(p_x^0(t-), p_v^0(t-), u^0(t), t, \frac{d\mathscr{W}^0(t)}{dV^0(t)}\right) d(V^0)^{\mathrm{c}}(t) = 0, \tag{6.192}$$

since on the set where the measure $V^0(dt)$ is concentrated the integrand vanishes.

Let $\tau \in D^{\mathrm{d}}_{\mathscr{V}}$. Then for any $s \in [0, \Delta V^0(\tau)]$ one has the inequality (see proof of Theorem 6.2)

$$\mathscr{H}_1\left(p_{x,\tau}^0(s), p_{v,\tau}^0(s), u^0(\tau), \tau, \frac{\Delta\mathscr{W}^0(\{\tau\})}{\Delta V^0(\tau)}\right) = 0,$$

and consequently

$$\int_0^{\Delta V^0(\tau)} \mathscr{H}_1\left(p_{x,\tau}^0(s), p_{v,\tau}^0(s), u^0(\tau), \tau, \frac{\Delta\mathscr{W}^0(\{\tau\})}{\Delta V^0(\tau)}\right) ds = 0, \tag{6.193}$$

where

$$p_{x,\tau}^0(s) = p_x^0(\tau-) + \int_0^s \sum_{k=1}^{N_3} g'_{k_x}\left(y_\tau^0(\omega), z_\tau^0(\omega), \tau\right) dv_\tau^k(\omega),$$

$$p_{v,\tau}^0(s) = p_v^0(\tau-) + \int_0^s \sum_{k=1}^{N_3} g'_{k_v}\left(y_\tau^0(\omega), z_\tau^0(\omega), \tau\right) dv_\tau^k(\omega),$$

$$y_\tau^0(s) = X^0(\tau-) + sB(u^0(\tau), \tau)\frac{\Delta\mathscr{W}^0(\{\tau\})}{\Delta V^0(\tau)},$$

$$z_\tau^0(s) = V^0(\tau-) + s.$$

Substituting these expressions in (6.193), we obtain

$$
\begin{aligned}
0 = & \int_0^{\Delta V^0(\tau)} \left[\left\langle p_x^0(\tau-), B(u^0(\tau), \tau) \frac{\Delta \mathscr{W}^0(\{\tau\})}{\Delta V^0(\tau)} \right\rangle + p_v^0(\tau-) \right] ds + \\
& + \int_0^{\Delta V^0(\tau)} \left[\int_0^s \sum_{k=1}^{N_3} \left\langle g'_{k_x}(y_\tau^0(\omega), z_\tau^0(\omega), \tau), \dot{y}_\tau^0(\omega) \right\rangle + \right. \\
& \left. + g_{k_v}(y_\tau^0(\omega), z_\tau^0(\omega), \tau)\, \dot{z}_\tau^0(\omega) \right] dv_\tau^k(\omega) d\omega = \mathscr{I}_1 + \mathscr{I}_2.
\end{aligned}
\tag{6.194}
$$

The first integral $\mathscr{I}_1$ is equal to

$$
\mathscr{I}_1 = \mathscr{H}_1\left(p_x^0(\tau-), p_v^0(\tau-), u^0(\tau), \tau, \frac{\Delta \mathscr{W}^0(\{t\})}{\Delta V^0(\tau)} \right) \Delta V^0(\tau),
$$

and the second one, $\mathscr{I}_2$, can be expressed as

$$
\mathscr{I}_2 = \sum_{k=1}^{N_3} \int_0^{\Delta V^0(\tau)} \int_0^s G_k(\omega) dv_\tau^k(\omega) ds = \sum_{k=1}^{N_3} \int_0^{\Delta V^0(\tau)} \left[\Delta V^0(\tau) - \omega \right] G_k(\omega) dv_\tau^k(\omega).
$$

Here $G_k(s) = \frac{d}{ds} \widetilde{g}_k(s)$, where

$$
\widetilde{g}_k(s) = g_k\left(X^0(\tau-) + sB(u^0(\tau), \tau) \frac{\Delta \mathscr{W}^0(\{t\})}{\Delta V^0(\tau)}, V^0(\tau-) + s, \tau \right).
$$

Note that $\widetilde{g}_k(s)$ is a smooth convex function and the measure $v_\tau^k(ds)$ is concentrated on the set where $\widetilde{g}_k(s)$ achieves its maximum, which is equal to zero. Thanks to its convexity, the function $\widetilde{g}_k(s)$ may achieve its maximum either on the whole segment, or at the endpoints. In the first case its derivative is equal to zero for all $s \in [0, \Delta V^0(\tau)]$, while in the second case the integral is equal to

$$
\int_0^{\Delta V^0(\tau)} \left[\Delta V^0(\tau) - \omega \right] G_k(\omega)\, dv_\tau^k(\omega) = \Delta V^0(\tau) G_k(0) v_\tau^k(\{0\}).
$$

However, the equality $\widetilde{g}_k(0) = 0$ implies $G_k(0) \le 0$, and therefore, all these integrals in the sum $\mathscr{I}_2$ are non-positive, which implies that $I_2 \le 0$.

Now from (6.194) it follows that

$$
\int_0^T \mathscr{H}_1\left(p_x^0(t-), p_{v-}^0(t), u^0(t), t, \frac{d\mathscr{W}^0(t)}{dV^0(t)} \right) dV^0(t) \ge 0,
$$

which in conjunction with the inequalities (6.191) and (6.192) leads to the following statement:

If the functions $g_k(x, v, t)$ *are convex in* (x, v), *then in addition to* (6.190) *we have the inequality*

$$\int_0^T \mathscr{H}_1\left(p_x^0(t-), p_v^0(t-), u(t), t, \frac{d\mathscr{A}(t)}{d|\mathscr{A}|(t)}\right) d|\mathscr{A}|(t) \leq$$

$$\leq \int_0^T \mathscr{H}_1\left(p_x^0(t-), p_v^0(t-), u^0(t), t, \frac{d\mathscr{W}^0(t)}{dV^0(t)}\right) dV^0(t) = 0. \tag{6.195}$$

The next result shows that if the constraints and the Hamiltonian enjoy convexity properties, then the maximum principle is a sufficient optimality condition.

Theorem 6.8 *Suppose the generalized solution* $\{X^0(\cdot), V^0(\cdot), u^0(\cdot), \mathscr{W}^0(\cdot)\}$ *satisfies the terminal and phase constraints, as well as the maximum principle (in the form of relation (6.195)) with a collection of elements (6.182) such that*

1. $\lambda_0 > 0$;
2. *the function* $\mathscr{L}\big(x_0, x_1, v\big)$ *is convex in the variables* $\{x_0, x_1, v\}$;
3. *the functions* $g_k(x, v, t)$, $k = 1, \dots, N_3$, *are convex in the variables* (x, v) *for any fixed* $t \in [0, T]$;
4. *the function* $\mathscr{H}_0\big(p_x^0(t), x, t\big)$ *is concave in the variable* x *for any fixed* $t \in [0, T]$.

Then $\{X^0(\cdot), V^0(\cdot), u^0(\cdot), \mathscr{W}^0(\cdot)\}$ *is the optimal generalized solution of problem (6.167)–(6.170).*

□

Proof Let $\{X(\cdot), V(\cdot), u(\cdot), \mathscr{W}(\cdot)\}$, i.e., $\{X(\cdot), V(\cdot)\}$ be an arbitrary admissible generalized solution of the system (6.167), satisfying the constraints (6.169), and let $\{u(\cdot), \mathscr{W}(\cdot)\}$ be controls such that the constraints (6.180) are satisfied.

Consider the function

$$\mathscr{M}_{X,V}(t) = \langle p_x^0(t), X(t)\rangle + p_v^0(t)V(t), \tag{6.196}$$

where the subscript X, V corresponds to the selected generalized solution.

Since the functions $\{X(\cdot), V(\cdot)\}$ and $\{p_x^0(\cdot), p_v^0(\cdot)\}$ are right-continuous, the integration-by-parts formula for functions of bounded variation [257] (see also Sect. A.14) implies that

$$\Delta \mathscr{M}_{X,V} = \mathscr{M}_{X,V}(T) - \mathscr{M}_{X,V}(0-) =$$

$$= \int_0^T \langle dp_x^0(t), X(t)\rangle + \langle p_x^0(t-), dX(t)\rangle + dp_v^0(t)V(t) + p_v^0(t-)dV(t) =$$

$$= -\int_0^T \Big[\langle (\mathscr{H}_0)'_x(p_x^0(t), X^0(t), t), X(t)\rangle - \mathscr{H}_0(p_x^0(t-), X(t), t)\Big]dt +$$

$$+\sum_{k=1}^{N_3}\int_0^T \Big[\langle g'_{k_x}(X^0(t), V^0(t), t), X(t)\rangle + g'_{k_v}(X^0(t), V^0(t), t)V(t)\Big]\, d\lambda_k(t) +$$

$$+\sum_{k=1}^{N_3}\sum_{\tau\in D^{\mathrm{d}}_{\mathscr{V}}}\int_0^{\Delta V^0(\tau)} \Big\{\Big\langle g'_{k_x}(X^0(\tau-) + sB(u^0(\tau), \tau)\frac{\Delta \mathscr{W}^0(\{\tau\})}{\Delta V^0(\tau)},$$

$$V^0(\tau-) + s, \tau), X(\tau)\Big\rangle +$$

$$+ g'_{k_v}(X^0(\tau-) + sB(u^0(\tau), \tau)\frac{\Delta \mathscr{W}^0(\{\tau\})}{\Delta V^0(\tau)}, V^0(\tau-) + s, \tau)V(\tau)\Big\}dv_\tau^k(s)+$$

$$+\int_0^T \mathscr{H}_1\Big(p_x^0(t-), p_v^0(t-), u(t), t, \frac{d\mathscr{W}(t)}{dV(t)}\Big)dV(t). \tag{6.197}$$

In view of the terminal and initial conditions for $\{p_x^0, p_v^0\}$, and of (6.184), (6.185), and (6.186), we obtain

$$\begin{aligned}\Delta \mathscr{M}_{X,V} &= \mathscr{M}_{X,V}(T) - \mathscr{M}_{X,V}(0-) =\\ &= -\langle \mathscr{L}'_{x_0}(X^0(0-), X^0(T), V^0(T)), X(0-)\rangle -\\ &- \langle \mathscr{L}'_{x_1}(X^0(0-), X^0(T), V^0(T)), X(T)\rangle -\\ &- \mathscr{L}'_v(X^0(0-), X^0(T), V^0(T))V(T).\end{aligned} \tag{6.198}$$

Now let us take the trajectory $\{X^0, V^0\}$ satisfying the maximum principle and calculate the difference $\Delta\mathscr{M}_{X,V} - \Delta\mathscr{M}_{X^0,V^0}$. We get

$$\begin{aligned}
&\Delta\mathscr{M}_{X,V} - \Delta\mathscr{M}_{X^0,V^0} = \\
&= -\langle \mathscr{L}'_{x_0}(X^0(0-), X^0(T), V^0(T)), X(0-) - X^0(0-)\rangle - \\
&- \langle \mathscr{L}'_{x_1}(X^0(0-), X^0(T), V^0(T)), X(T) - X^0(T)\rangle - \\
&- \mathscr{L}'_{v}(X^0(0-), X^0(T), V^0(T))(V(T) - V^0(T)).
\end{aligned} \tag{6.199}$$

Since for any smooth convex function $F(z)$ one has

$$F(z) - F(z_0) \geq \langle \text{grad}_z F(z_0), \ z - z_0\rangle,$$

relation (6.199) shows that

$$\begin{aligned}
&\Delta\mathscr{M}_{X,V} - \Delta\mathscr{M}_{X^0,V^0} \geq \\
&\geq \mathscr{L}(X^0(0-), X^0(T), V^0(T)) - \mathscr{L}(X(0-), X(T), V(T)) = \\
&= \lambda_0[\phi_0(X^0(0-), X^0(T), V^0(T)) - \phi_0(X(0-), X(T), V(T))] + \\
&+ \langle l_1, h(X^0(0-), X^0(T))) - h(X(0-), X(T))\rangle + \\
&+ \langle l_2, S(X^0(0-), X^0(T), V^0(T)) - S(X(0-), X(T), V(T))\rangle + \\
&+ \gamma_0[(V^0(T) - M) - (V(T) - M)].
\end{aligned} \tag{6.200}$$

Since the trajectories $\{X, V\}$ and $\{X^0, V^0\}$ satisfy the constraints (6.169), and the trajectory $\{X^0, V^0\}$ satisfies the condition (6.183), we obtain the relations

$$\begin{aligned}
&\langle l_1, h(X^0(0-), X^0(T)) - h(X(0-), X(T))\rangle = 0, \\
&\langle l_2, S(X^0(0-), X^0(T), V^0(T))\rangle + \gamma_0(V^0(T) - M) = 0, \\
&\langle l_2, S(X(0-), X(T), V(T))\rangle + \gamma_0(V^0(T) - M) \leq 0,
\end{aligned}$$

which upon substitution in (6.198) yield

$$\begin{aligned}
&\lambda_0[\phi_0(X^0(0-), X^0(T), V^0(T)) - \phi_0(X(0-), X(T), V(T))] \leq \\
&\leq \Delta\mathscr{M}_{X,V} - \Delta\mathscr{M}_{X^0,V^0} + \\
&+ \langle l_2, S(X(0-), X(T), V(T))\rangle + \gamma_0(V(T) - M).
\end{aligned} \tag{6.201}$$

The last term in (6.201) is non-positive and $\lambda_0 > 0$ by the assumptions of the theorem, therefore if we are able to show that

$$\Delta\mathscr{M}_{X,V} - \Delta\mathscr{M}_{X^0,V^0} \le 0, \tag{6.202}$$

then it will follow that

$$\begin{aligned}\phi_0(X^0(0-), X^0(T), V^0(T)) - \phi_0(X(0-), X(T), V(T)) =\\ = J(X^0(\cdot), V^0(\cdot)) - J(X(\cdot), V(\cdot)) \le 0,\end{aligned}$$

thereby completing the proof.

So it remains to check that inequality (6.202) holds. Using (6.197), we express the left-hand side of (6.202) as

$$\Delta\mathscr{M}_{X,V} - \Delta\mathscr{M}_{X^0,V^0} = \mathscr{I}_1 + \mathscr{I}_2 + \mathscr{I}_3 + \mathscr{I}_4, \tag{6.203}$$

where

$$\begin{aligned}\mathscr{I}_1 = &-\int_0^T \Big[\Big\langle (\mathscr{H}_0)'_x(p_x^0(t), X^0(t), t), X(t) - X^0(t)\Big\rangle -\\ &- \mathscr{H}_0(p_x^0(t-), X(t), t) + \mathscr{H}_0(p_x^0(t-), X^0(t), t)\Big]\, dt,\end{aligned}$$

$$\begin{aligned}\mathscr{I}_2 = &\sum_{k=1}^{N_3}\int_0^T \Big[\Big\langle g'_{k_x}(X^0(t), V^0(t), t), X(t) - X^0\Big\rangle +\\ &+ g'_{k_v}(X^0(t), V^0(t), t)(V(t) - V^0)\Big]\, d\lambda_k(t),\end{aligned}$$

$$\begin{aligned}\mathscr{I}_3 = &\int_0^T \mathscr{H}_1\Big(p_x^0(t-), p_v^0(t-), u(t), t, \frac{d\mathscr{W}(t)}{dV(t)}\Big)\, dV(t) -\\ &- \int_0^T \mathscr{H}_1\Big(p_x^0(t-), p_v^0(t-), u(t), t, \frac{d\mathscr{W}^0(t)}{dV^0(t)}\Big)\, dV^0(t),\end{aligned}$$

$$\begin{aligned}\mathscr{I}_4 = &\sum_{k=1}^{N_3}\sum_{\tau\in D^d_{\mathscr{V}}}\int_0^{\Delta V^0(\tau)} \Big[\Big\langle g'_{k_x}(y_\tau^0(s), z_\tau^0(s), \tau), X(\tau) - X^0(\tau)\Big\rangle +\\ &+ g'_{k_v}(y_\tau^0(s), z_\tau^0(s), \tau)(V(\tau) - V^0(\tau))\Big]\, dv_\tau^k(s),\end{aligned}$$

and

$$y_\tau^0(s) = X^0(\tau-) + sB(u^0(\tau), \tau)\frac{\Delta\mathscr{W}^0(\{\tau\})}{\Delta V^0(\tau)}, \quad z_\tau^0(s) = V^0(\tau-) + s.$$

By the assumptions of the theorem, the function $\mathscr{H}_0(p_x^0(t), x, t)$ is concave in the variable x for any fixed $t \in [0, T]$. Moreover, since the variation of the function $p_x^0(\cdot)$ is bounded, $\mathscr{H}_0\big(p_x^0(t), x, t\big)$ coincides with $\mathscr{H}_0\big(p_x^0(t-), x, t\big)$ a.e. on $[0, T]$. Therefore, the integrand in $\mathscr{I}_1$ is non-positive, which implies that $\mathscr{I}_1 \le 0$.

Next, thanks to the convexity of the functions g_k, the integral $\mathscr{I}_2$ is estimated as

$$\mathscr{I}_2 \le \sum_{k=1}^{N_3} \int_0^T \Big[g_k(X(t), V(t), t) - g_k(X^0(t), V^0(t), t)\Big] d\lambda_k(t) \le 0;$$

indeed, since $g_k(X(t), V(t), t) \le 0$, $\lambda_k(dt) \ge 0$, and since each measure λ_k is concentrated on the set where g_k equals zero, we have

$$\sum_{k=1}^{N_3} \int_0^T g_k(X^0(t), V^0(t), t)\, d\lambda_k(t) = 0.$$

Next, the term $\mathscr{I}_3$ is smaller that 0 due to the maximum principle (6.195).

Finally, the term $\mathscr{I}_4$ can be written in the form

$$\mathscr{I}_4 = \sum_{k=1}^{N_3} \sum_{\tau \in D^{\mathrm{d}}_{\mathscr{V}}} \Bigg\{ \int_0^{\Delta V^0(\tau)} \Big[\Big\langle g'_{k_x}(y^0_\tau(s), z^0_\tau(s), \tau), X(\tau) - y^0_\tau(s)\Big\rangle +$$

$$+ g'_{k_v}(y^0_\tau(s), z^0_\tau(s), \tau)(V(\tau) - z^0_\tau(s))\Big] dv^k_\tau(s) +$$

$$+ \int_0^{\Delta V^0(\tau)} (\Delta V^0(\tau) - s)\Big[\Big\langle g'_{k_x}(y^0_\tau(s), z^0_\tau(s), \tau), B(u^0(\tau), \tau)\frac{\Delta \mathscr{W}^0(\{\tau\})}{\Delta V^0(\tau)}\Big\rangle +$$

$$+ g'_{k_v}(y^0_\tau(s), z^0_\tau(s), \tau)\Big] dv^k_\tau(s)\Bigg\} =$$

$$= \sum_{k=1}^{N_3} \sum_{\tau \in D^{\mathrm{d}}_{\mathscr{V}}} (\mathscr{I}'^k_\tau + \mathscr{I}''^k_\tau).$$

That $\mathscr{I}'^k_\tau \le 0$ can be proved as in the case of the term $\mathscr{I}_2$, because g_k is convex. And finally, we already know that $\mathscr{I}''^k_\tau \le 0$ (see the proof of Proposition 6.4 and the estimation of the same term in formula (6.194)).

Therefore, inequality (6.202), and along with it the optimality of $\{X^0, V^0\}$, are proved. □

Remark 6.13 The condition $\lambda_0 > 0$ is satisfied if the so-called Slater's condition is satisfied [173]. For example, if there are no equality constraints and there exists an admissible solution $\{X(\cdot), V(\cdot)\}$ such that

$$S(X(0), X(T), V(T)) < 0, \quad g_k(X(t), V(t), t) < 0$$

for all $t \in [0, T]$ and $k = 1, \ldots, N_3$, then $\lambda_0 > 0$ and one can put $\lambda_0 = 1$.

Linear systems with convex constraints constitute a natural area of application of this result. But, rather unexpectedly, there is a whole class of optimal control problems of nonlinear systems that also have a convex structure. These problems arise in the observation control, where the maximum principle is also a necessary and sufficient optimality condition. A detailed discussion of such problems is found in Chap. 7.

6.4.3 Generalized Optimization of Linear Systems

In this subsection we consider the control problem for a linear system described by equations of the form

$$\begin{cases} \dot{X}(t) = A(t)X(t) + B(t, u(t))w(t), \\ \dot{V}(t) = \|w(t)\|, \end{cases} \tag{6.204}$$

on a given segment $t \in [0, T]$, with a continuous matrix-function $A(t)$ and with a function $B(t, u)$, which assumed to be continuous in (t, u). The controls (u, w) satisfy the constraints

$$u \in U, \quad w \in K, \quad \int_0^T \|w(t)\| dt \leq M < \infty, \tag{6.205}$$

where $U \subset \mathbb{R}^k$ is a compact set and $K \subset \mathbb{R}^m$ is a closed convex cone.

Let Assumption 6.1 be in force and assume that the set of vectors

$$\mathscr{L}(t) = \left\{ l = \begin{pmatrix} l_1 \in \mathbb{R}^n \\ l_2 \in \mathbb{R}^1 \end{pmatrix} : \begin{matrix} l_1 = B(u, t)w, \\ l_2 = \|w\|, \end{matrix} \quad u \in U, \ w \in K \right\} \tag{6.206}$$

is convex for any fixed $t \in [0, T]$. Then, by Corollary 4.1, for any generalized solution $\{X(\cdot), V(t)\}$ of the system (6.204) there exist:

1. a scalar non-negative measure $\mathscr{V}(dt)$ with distribution function

$$V(t) = \mathscr{V}([0, t]),$$

defined on the Borel subsets of the segment $[0, T]$, and such that

$$\mathscr{V}([0,T]) = V(T) \le M;$$

2. a pair of both Lebesgues- and $\mathscr{V}$-measurable controls

$$\{u_0(t), w_0(t)\},$$

satisfying the constraints

$$u_0(t) \in U, \quad w_0(t) \in K \cap \{\|w\| \le 1\}$$

on the segment $[0, T]$, a.e. with respect to the Lebesgue and $\mathscr{V}$ measures

such that $X(t)$ satisfies the equation

$$X(t) = X(0-) + \int_0^t A(\tau)X(\tau)d\tau + \int_0^t B(u_0(\tau), \tau)w_0(\tau)\,dV(\tau).$$

Moreover, if the vector measure $\mathscr{W}(dt)$ is given by means of a distribution function,

$$\mathscr{W}([0,t]) = \int_0^t w_0(\tau)dV(\tau),$$

then the above equation becomes

$$X(t) = X(0-) + \int_0^t A(\tau)X(\tau)d\tau + \int_0^t B(u_0(\tau), \tau)d\mathscr{W}(\tau). \tag{6.207}$$

Consider the optimal control problem for the system (6.204) with the constraints (6.169) and the performance criterion (6.170). If the optimal generalized solution exists, it satisfies the necessary optimality condition that follows directly from Theorem 6.7. With the standard notations

$$\begin{aligned}
&\mathscr{H}_0(p_x, x, t) = \langle p_x, A(t)x\rangle,\\
&\mathscr{H}_1(p_x, p_v, u, t, e) = \langle p_x, B(u,t)e\rangle + p_v,\\
&\mathscr{L}(x_0, x_1, v) = \lambda_0\phi_0(x_0, x_1, v) + \langle l_1, h(x_0, x_1)\rangle + \\
&\qquad + \langle l_2, S(x_0, x_1, v)\rangle + \gamma_0(v - M),\\
&D^{\mathrm{d}}_{\mathscr{V}} = \{\tau : \mathscr{V}(\tau) > 0\} = \{\tau |\, \Delta V^0(\tau) > 0\},
\end{aligned} \tag{6.208}$$

we obtain the following statement.

Theorem 6.9 *Let the assumptions of Theorem 6.1 be in force and let*

$$\{X^0(\cdot), V^0(\cdot), u^0(t), \mathscr{W}^0(dt)\}$$

be the optimal generalized solution of the problem (6.204), (6.168)–(6.170). Then there exists a collection of elements

$$\begin{aligned}\Lambda = \big\{\lambda_0, \gamma_0, l_1, l_2, p_x^0(\cdot), p_v^0(\cdot),\ \ \lambda_k(dt)\ \textit{for}\ k = 1, \dots, N_3,\\ \nu_\tau^k(ds)\ \textit{for}\ k = 1, \dots, N_3,\ \tau \in D_{\mathscr{V}}^{\mathrm{d}}\big\},\end{aligned} \tag{6.209}$$

not simultaneously equal to zero, where

1. $\lambda_0 \geq 0$, $\gamma_0 \geq 0$, *and* β_0 *are real numbers;*
2. $l_1 \in \mathbb{R}^{N_1}$ *and* $l_2 \in \mathbb{R}^{N_2}$ *are vectors such that the components of* l_2 *are non-negative and the trajectory* $\{X^0(\cdot), V^0(\cdot)\}$ *satisfies the condition*

$$\big\langle l_2, S(X^0(0-), X^0(T), V^0(T))\big\rangle + \gamma_0(V^0(T) - M) = 0; \tag{6.210}$$

3. *the vector-valued functions* $\{p_x^0(\cdot), p_v^0(\cdot)\}$ *satisfy the equations*

$$p_x^0(t) = -\mathscr{L}'_{x(T)}\big(X^0(0-), X^0(T), V^0(T)\big) + \\ + \int_t^T A^*(\tau)p_x^0(\tau)d\tau - \int_t^T \sum_{k=1}^{N_3} g'_{k_x}\big(X^0(\tau), V^0(\tau), \tau\big)\, d\lambda_k(\tau) + \\ + \sum_{\tau>t,\, \tau\in D_{\mathscr{V}}^{\mathrm{d}}} \Delta p_x^0(\tau), \tag{6.211}$$

$$p_v^0(t) = -\mathscr{L}'_v\big(X^0(0-), X^0(T), V^0(T)\big) + \\ + \sum_{\tau>t,\, \tau\in D_{\mathscr{V}}^{\mathrm{d}}} \Delta p_v^0(\tau) - \int_t^T \sum_{k=1}^{N_3} g'_{k_v}\big(X^0(\tau), V^0(\tau), \tau\big)\, d\lambda_k(\tau), \tag{6.212}$$

with the following boundary conditions for p_x^0 *at* $t = 0$*:*

$$p_x^0(0-) = \mathscr{L}'_{x(0)}\big(X^0(0-), X^0(T), V^0(T)\big); \tag{6.213}$$

4. *the non-negative measures* $\lambda_k(dt)$ *in Eqs. (6.211) and (6.212) are concentrated on the sets*

$$T_k^0 = \big\{t \in [0, T] \setminus D_{\mathscr{V}}^{\mathrm{d}} \,|\, g_k(X^0(t), V^0(t), t) = 0\big\} \tag{6.214}$$

5. *the jumps of the functions* $\{p_x^0(\cdot), p_v^0(\cdot)\}$ *at the points* $\tau \in D_{\mathscr{V}}^{\mathrm{d}}$ *are given by the formulas*

$$\Delta p_x^0(\tau) =$$

$$= -\int_s^{\Delta V^0(\tau)} \sum_{k=1}^{N_3} g'_{k_x}\left(X^0(\tau-) + \omega\frac{\Delta \mathscr{W}^0(\{\tau\})}{\Delta V^0(\tau)},\, V^0(\tau-) + \omega, \tau\right) d\nu_\tau^k(\omega),$$

$$\Delta p_v^0(\tau) =$$

$$= -\int_s^{\Delta V^0(\tau)} \sum_{k=1}^{N_3} g'_{k_v}\left(X^0(\tau-) + \omega\frac{\Delta \mathscr{W}^0(\{\tau\})}{\Delta V^0(\tau)},\, V^0(\tau-) + \omega, \tau\right) d\nu_\tau^k(\omega); \tag{6.215}$$

6. *the non-negative measures* $\nu_\tau^k(dt)$ *in (6.215) are concentrated on the sets*

$$T_\tau^k = \left\{s \in [0, \Delta V^0(\tau)] \,:\, g_k\left(X^0(\tau-) + s\frac{\Delta \mathscr{W}^0(\{\tau\})}{\Delta V^0(\tau)},\, V^0(\tau-) + s, \tau\right)\right\} = 0. \tag{6.216}$$

For this collection Λ*, the optimal solution* $\{X^0(\cdot), V^0(\cdot), u^0(t), \mathscr{W}^0(dt)\}$ *satisfies the generalized maximum principle, as follows:*

For any control $\{u(\cdot), \mathscr{A}(dt)\}$ *admitted by the constraints it holds that*

$$\begin{aligned}
&\int_0^T \mathscr{H}_1\left(p_x^0(t), p_v^0(t), u(t), t, \frac{d\mathscr{A}(t)}{d|\mathscr{A}|(t)}\right) d|\mathscr{A}|(t) = \\
&\quad = \int_0^T \langle p_x^0(t), B(u(t), t)d\mathscr{A}(t)\rangle + p_v^0(t)d|\mathscr{A}|(t) \le \\
&\quad \le \int_0^T \mathscr{H}_1\left(p_x^0(t), p_v^0(t), u^0(t), t, \frac{d\mathscr{W}^0(t)}{dV^0(t)}\right) dV^0(t) = \\
&\quad = \int_0^T \langle p_x^0(t), B(u^0(t), t)d\mathscr{W}^0(t)\rangle + p_v^0(t)dV^0(t) = 0,
\end{aligned} \tag{6.217}$$

where $|\mathscr{A}|(t) = \operatorname*{Var}_{[0,t]} \mathscr{A}(s)$.

It is obvious that Proposition 6.4 is valid in this case as well, and the following sufficient optimality condition holds.

Theorem 6.10 *Suppose that the admissible generalized solution*

$$\{X^0(\cdot), V^0(\cdot), u^0(\cdot), \mathscr{W}^0(\cdot)\}$$

satisfies the maximum principle with the collection Λ *(6.182) such that*

1. $\lambda_0 > 0$;
2. *the function* $\mathscr{L}(x_0, x_1, v)$ *is convex in* $\{x_0, x_1, v\}$;
3. *the functions* $g_k(x, v, t)$ *are convex in the variables* (x, v).

Then $\{X^0(\cdot), V^0(\cdot), u^0(\cdot), \mathscr{W}^0(\cdot)\}$ *is the optimal generalized solution of the problem (6.204), (6.205) with the constraints (6.169) and the performance criterion (6.170).*

This theorem is a consequence of Theorem 6.9 because $\mathscr{H}_0$ in (6.208) is linear in x. The next, useful result follows from the maximum principle (see relation (6.217)). Assume that the phase constraints in (6.169) are absent, i.e., $g_k(x, v, t) \equiv 0$. Then the conjugate variables $\{p_x^0, p_v^0\}$ satisfy the equations

$$\begin{aligned} p_x^0(t) &= p_x^0(T) + \int_t^T A^*(\tau) p_x^0(\tau) d\tau, \\ p_v^0(t) &= p_v^0(T), \end{aligned} \tag{6.218}$$

and so

$$p_x^0(t) = \Phi^*(T, t) p_x^0(T),$$

where $\Phi(t, s)$ is the fundamental solution of the linear system $\dot{X}(t) = A(t)X(t)$. Substituting this expression in (6.217), we obtain

$$\begin{aligned} &\int_0^T \langle p_x^0(t), B(u(t), t) d\mathscr{A}(t)\rangle + p_v^0(t)\, d|\mathscr{A}|(t) = \\ &= \int_0^T \langle \Phi^*(T, t) p_x^0(T), B(u(t), t) d\mathscr{A}(t)\rangle + p_v^0(T)\, d|\mathscr{A}|(t) = \\ &= \langle p_x^0(T), \int_0^T \Phi(T, t) B(u(t), t) d\mathscr{A}(t)\rangle + p_v^0(T) |\mathscr{A}|(T) \le \\ &\le \langle p_x^0(T), \int_0^T \Phi(T, t) B(u^0(t), t) d\mathscr{W}^0(t)\rangle + p_v^0(T) V^0(T) = \\ &= \langle p_x^0(T), X^0(T) - \Phi(T, 0) X^0(0-)\rangle + p_v^0(T) V^0(T) = \\ &= \langle p_x^0(T), X^0(T)\rangle - \langle p_x^0(0), X^0(0-)\rangle + p_v^0(T) V^0(T) = 0. \end{aligned} \tag{6.219}$$

However, for any admissible solution $\{X(\cdot), V(\cdot)\}$,

$$X(t) = \Phi(t, 0) X(0-) + \int_0^t \Phi(t, s) B(s) d\mathscr{W}(s).$$

Thus, for any admissible solution of the system (6.207) we have the inequality

$$
\begin{aligned}
&\langle p_x^0(T), X^0(T) - X(T)\rangle - \langle p_x^0(0), X^0(0-) - X(0-)\rangle + \\
&\quad + p_v^0(T)(V^0(T) - V(T)) \geq 0,
\end{aligned}
\tag{6.220}
$$

which means that

$$
\{X^0(T), X^0(0-), V^0(T)\} = \underset{(x_1, x_0, v)\in C}{\operatorname{argmax}} \left\{\langle p_x^0(T), x_1\rangle - \langle p_x^0(0), x_0\rangle + p_v^0(T)v\right\},
$$

where C is the set of admissble triples $\{X(T), X(0), V(T)\}$ for the system of Eq. (6.204) with the constraints (6.168) and (6.169).

This allows us to obtain a more detailed description of the optimal solution.

Assumption 6.3 The initial condition are fixed and the functions ϕ_0, S depend only on the variables (x_1, v) and are non-decreasing in the variable v, and there are no inequality constraints.

Theorem 6.11 *Suppose Assumption 6.3 is in force. Then there exists the optimal generalized solution with control* $\{u(\cdot), \mathscr{W}(\cdot)\}$ *such that the measure* $\mathscr{W}(\cdot)$ *is concentrated in at most* $n+2$ *points of the segment* $[0, T]$*, where* n *is the dimension of the variable* x.

Remark 6.14 Results of this kind are already known for linear systems with generalized controls (see, e.g., [5, 218, 237, 328],) where a representation of the optimal impulse control with a finite number of impulses was obtained with the aid of *moment problem* approach. The application of the generalized maximum principle allows one to extend these results to a wider class of problems [264]. The original result of V. I. Geiman, which concerns the optimal impulse control with a finite number of impulses in a bilinear (nonlinear) system is presented in [465].

The proof of Theorem 6.11 is employs some important concepts of convex analysis [379]. First we observe that the attainability set $\mathscr{X}(T)$ for the variables $\{X(\cdot), V(\cdot)\}$, satisfying (6.207) at the time instant $t = T$ under the constraints (6.168), (6.169) and for a given initial condition $X(0-)$, is a compact convex set in $\mathbb{R}^{n+1}$. Compactness follows from Corollary 4.1, while convexity is a consequence of the linearity of Eq. (6.207), the convexity of the set of constraints on the controls (6.168), and the convexity of the set bounded by the inequalities $S(x_1, v) \leq 0$. So the set of admissible pairs $\{X(T), V(T)\}$ is compact convex subset in $\mathbb{R}^{n+1}$. By the Carathéodory theorem, any point of a convex compact set C in M-dimensional space can be expressed as a convex combination of no more than $M + 1$ *extreme points* of the set C [379].

Definition 6.1 A point x^* of a convex set C is called *extreme point of* C, if it cannot be written as a convex combination of other points of C.

Thus, the theorem will be proved if we show that any extreme point of the set $\mathscr{X}(T)$ can be obtained by means of a control with a measure that is concentrated a just one point of the segment $[0, T]$.

Thanks to the compactness of the set $\mathscr{X}(T)$, for any extreme point (x^*, v^*) there is a linear functional $l(x, v)$ in $\mathbb{R}^{n+1}$, which achieves its maximum on the set $\mathscr{X}(T)$ at (x^*, v^*). This functional defines the *supporting hyperplane* [379] to the set $\mathscr{X}(T)$ at the point (x^*, v^*), namely, the hyperplane defined by the equation

$$l(x, v) = l(x^*, v^*),$$

and for any $(x, v) \in \mathscr{X}(T)$ one has $l(x, v) \leq l(x^*, v^*)$. However, the functional $l(x, v)$ may achieve a maximum in more than one point of $\mathscr{X}(T)$, and so it does not define the extreme point unambiguously. At the same time, the set of extreme points contains the subset of points that are unambiguously defined as the maximum points of some linear functional. This is the set of *exposed points* [379].

Definition 6.2 A point x^* of a convex set C is said to be an *exposed point of* C if there is a linear functional $l(\cdot)$ such that

$$\max_C l(x) = l(x^*) \quad \text{and} \quad l(x) < l(x^*), \qquad \text{for all } x \in C \setminus \{x^*\}.$$

One replace the set of extreme points by the set of exposed points [379].

Theorem 6.12 (Straszewicz Theorem [419]) . *For any closed convex set C the set of its exposed points is dense in the set of its extreme points. In other words, any extreme point is the limit of a sequence of exposed points.*

Next we obtain the representation of an arbitrary outstanding point of $\mathscr{X}(T)$ with the aid of "single-impulse" control.

Lemma 6.2 *Let $\{X^*, V^*\}$ be an exposed point of $\mathscr{X}(T)$. Then there exists a control $\{u(\cdot), \mathscr{W}(\cdot)\}$ which ensures that the point $\{X^*, V^*\}$ is attained, and such that the measure $\mathscr{W}(\cdot)$ is concentrated in a single point of the segment $[0, T]$. That is, the measure $\mathscr{W}(\cdot)$ admits the representation*

$$\mathscr{W}(dt) = w_0 \delta(t - t_0) dt$$

and

$$X^* = X(T) = \Phi(T, 0) X(0-) + \Phi(T, t_0) B(u(t_0), t_0) w_0,$$

$$V^* = V(T) = \|w_0\|.$$

In other words, any exposed point of the set $\mathscr{X}(T)$ can be attained by means of a "single-impulse" control.

Proof By the definition of exposed points, there exists a linear functional

$$l(x, v) = \langle l_x, x \rangle + l_v v,$$

where $l_x \in \mathbb{R}^n$ and $l_v \in \mathbb{R}^1$, such that $l(\cdot)$ achieves its maximum on $\mathscr{X}(T)$ at exactly one point $\{X^*, V^*\}$. Let us take an admissible control $\{u(\cdot), \mathscr{W}(\cdot)\}$ and calculate the value $l(X(T), V(T))$. Since $X(t)$ satisfies Eq. (6.207), we obtain

$$
\begin{aligned}
&l(X(T), V(T)) = \\
&= \Big\langle l_x, \Phi(T,0)X(0-) + \int_0^T \Phi(T,s)B(u(s),s)\frac{d\mathscr{W}(s)}{dV(s)}\Big\rangle dV(s) + l_v V(T) = \\
&= \big\langle l_x, \Phi(T,0)X(0-)\big\rangle + \int_0^T \Big\langle l_x, \Phi(T,s)B(u(s),s)\frac{d\mathscr{W}(s)}{dV(s)}\Big\rangle dV(s) + l_v V(T) \le \\
&\le \big\langle l_x, \Phi(T,0)X(0-)\big\rangle + \\
&\qquad + \left[\max_{s\in[0,T]} \max_{u\in U,\, e\in K\cap\{\|e\|\le 1\}} \big\langle l_x, \Phi(T,s)B(u,s)e\big\rangle + l_v\right] V(T).
\end{aligned}
\tag{6.221}
$$

The function under the maximum symbol is continuous and since the set of admissible values $\{s, u, e\}$ is compact, it follows that there a exists triple $\{t_0, u_0, e_0\}$, such that

$$
\max_{s\in[0,T]} \max_{u\in U,\, e\in K\cap\{\|e\|\le 1\}} \big\langle l_x, \Phi(T,s)B(u,s)e\big\rangle = \big\langle l_x, \Phi(T,t_0)B(u_0,t_0)e_0\big\rangle.
$$

Hence, one can take the admissible control

$$
\begin{aligned}
u_0(t) &= \begin{cases} u_0, & \text{for } t = t_0, \\ \text{arbitrary, but continuous on } [0, T], \end{cases} \\
\mathscr{W}(dt) &= \begin{cases} e_0 M\delta(t-t_0)dt, & \text{for } \big\langle l_x, \Phi(T,t_0)B(u_0,t_0)e_0\big\rangle + l_v > 0, \\ 0, & \text{otherwise.} \end{cases}
\end{aligned}
\tag{6.222}
$$

$$
V(T) = \begin{cases} M\|e_0\| = \operatorname*{Var}_{[0,T]} \mathscr{W}([0,s]), & \text{for } \big\langle l_x, \Phi(T,t_0)B(u_0,t_0)e_0\big\rangle + l_v > 0, \\ 0, & \text{otherwise.} \end{cases}
$$

By relation (6.221), we have

$$
\begin{aligned}
&l(X(T), V(T)) \le \\
&\le \Big\langle l_x, \Phi(T,0)X(0-) + \int_0^T \Phi(T,s)B(u_0(s),s)\, d\mathscr{W}(s)\Big\rangle + l_v \operatorname*{Var}_{[0,T]} \mathscr{W}([0,s]) = \\
&= l(X^*(T), V^*(T)),
\end{aligned}
$$

where $\{X^*(T), V^*(T)\}$ denotes the terminal point corresponding to the control $\{u_0(\cdot), \mathscr{W}(\cdot)\}$. This terminal point admits a representation with "single-impulse" control. At the same time, the exposed point $\{X^*, V^*\}$ is the unique maximum of the functional $l(\cdot)$. Therefore, $\{X^*(T), V^*(T)\}$ coincides with $\{X^*, V^*\}$, which completes the proof. □

As the corollary of Straszewicz's theorem one can obtain the "single-impulse" representation for any extreme point of the set $\mathscr{X}(T)$.

Lemma 6.3 *Any extreme point of the set $\mathscr{X}(T)$ admits a "single-impulse" representation.*

Proof Let $\{X^*, V^*\}$ be an extreme point of $\mathscr{X}(T)$, By the Straszewicz's theorem, there exists a sequence $\{X^n, V^n\}$ of exposed points such that

$$\lim_n \{X^n, V^n\} = \{X^*, V^*\}.$$

By Lemma 6.2, for any n there exists a triple $\{t^n, u^n, e^n\}$, such that the "single-impulse" control (6.222) provides the "single-impulse" representation of the point $\{X^n, V^n\}$. Since all the members of the triplet $\{t^n, u^n, e^n\}$ are uniformly bounded, there exists a subsequence (we keep the same notation for it) that converges to some triplet $\{\bar{t}, \bar{u}, \bar{e}\}$. However, thanks to the continuity of all functions involved, we have

$$\begin{aligned}\lim_n \{X^n, V^n\} &= \lim_n \left\{[\Phi(T,0)X(0-) + \Phi(T,t^n)B(u^n,t^n)e^n V^n],\ V^n\right\} = \\ &= \left\{\Phi(T,0)X(0-) + \Phi(T,\bar{t})B(\bar{u},\bar{t})\bar{e}V^*,\ V^*\right\} = \{X^*,\ V^*\},\end{aligned}$$

where $\{X^*, V^*\}$ is the terminal point corresponding to the impulse control

$$\bar{u}(t) = \begin{cases} \bar{u}, & \text{for } t = \bar{t}, \\ \text{arbitrary}, & \text{but continuous on the segment } [0, T], \end{cases}$$

$$\mathscr{W}(dt) = \bar{e}V^*\delta(t - \bar{t})dt.$$

This complete the proof. □

Now we are ready to establish the main result.

Proof of Theorem 6.11 Let $\{X^0(\cdot), V^0(\cdot)\}$ be the optimal solution. Then the point $\{X^0(T), V^0(T)\} \in \mathscr{X}(T)$ and, by the Carathéodory theorem [379], it can be written as a convex combination of no more than $n+2$ extreme points of the set $\mathscr{X}(T)$. By Lemma 6.3, each extreme point admits a "single-impulse" representation.

Assume first that these impulse controls are concentrated at different points of the segment $[0, T]$, i.e., they are of the form

$$\{u_i(t), \mathscr{W}_i(dt)\}, \quad i \le n+2, \quad \text{where} \quad \mathscr{W}_i(dt) = V_i(T)e_i\delta(t - t_i)dt.$$

Since $\{X^0(T), V^0(T)\}$ can be expressed as a convex combination, we obtain

$$\{X^0(T), V^0(T)\} = \sum_i \lambda_i \{X_i(T), V_i(T)\}, \tag{6.223}$$

where $\lambda_i > 0$ and $\sum_i \lambda_i = 1$. But then one can choose the control

$$u^0(t) = \begin{cases} u_i(t_i), & \text{for } t = t_i, \\ \text{arbitrary}, & \text{but continuous on the segment} \quad [0, T], \end{cases}$$

$$\mathscr{W}(dt) = \sum_i \lambda_i \mathscr{W}_i(dt).$$

Now substituting this control in the Eq. (6.207) for $\{X(t), V(t)\}$, we obtain

$$X(T) = \Phi(T, 0)X(0-) + \sum_i \lambda_i \Phi(T, t_i)B(u_i, t_i)e_i V_i(T) =$$

$$= \sum_i \lambda_i X_i(T) = X^0(T),$$

$$V(T) \leq \sum_i \lambda_i V_i(T) = V^0(T).$$

The last inequality holds because the *variation of a sum of functions is not larger than the sum of the individual variations* [322]. However, the functions ϕ_0 and S are non-decreasing in v and since $\{X^0, V^0\}$ is the optimal solution, the solution $\{X(\cdot), V(\cdot)\}$ is also optimal. This completes the proof of the claimed representation in the case when the measures are concentrated at different points.

In the case when some points of where the measures are concentrated coincide, one can merge them as follows. For the sake of simplicity, we consider the case of two points (the general case is then obvious). So, assume that there are two extreme points, $\{X_1(T), V_1(T)\}$ and $\{X_2(T), V_2(T)\}$, with "single-impulse" representations and with measures concentrated at a common point t^*. Then

$$X_1(T) = \Phi(T, 0)X(0-) + \Phi(T, t^*)B(u_1, t^*)e_1 V_1(T)$$

$$X_2(T) = \Phi(T, 0)X(0-) + \Phi(T, t^*)B(u_2, t^*)e_2 V_2(T).$$

A linear combination of $X_1(T)$ and $X_2(T)$ with positive coefficients λ_1 and λ_2 can be written in the form

$$\lambda_1 X_1(T) + \lambda_2 X_2(T) = (\lambda_1 + \lambda_2)\Phi(T,0)X(0-) +$$
$$+ \Phi(T,t^*)(\lambda_1 V_1(T) + \lambda_2 V_2(T))\times$$
$$\times \left[\frac{\lambda_1 V_1(T)}{\lambda_1 V_1(T) + \lambda_2 V_2(T)} B(u_1, t^*)e_1 + \frac{\lambda_2 V_2(T)}{\lambda_1 V_1(T) + \lambda_2 V_2(T)} B(u_2, t^*)e_2 \right].$$

The expression in the square brackets, together with $\lambda_1 V_1(T)+\lambda_2 V_2(T)$, is a convex combination of two vectors from the set $\mathscr{L}(t^*)$ (see (6.206)), which is convex. Therefore, there exists an admissible pair $\{u^*, e^*\}$ such that the expression in the brackets equals $B(u^*, t^*)e^*$. We conclude that

$$\lambda_1 X_1(T) + \lambda_2 X_2(T) = (\lambda_1 + \lambda_2)\times$$
$$\times \left[\Phi(T,0)X(0-) + \Phi(T,t^*)\frac{\lambda_1 V_1(T) + \lambda_2 V_2(T)}{\lambda_1 + \lambda_2} B(u^*, t^*)e^* \right],$$

and the admissible control with the measure

$$\mathscr{W}(dt) = \frac{\lambda_1 V_1(T) + \lambda_2 V_2(T)}{\lambda_1 + \lambda_2} e^* \delta(t - t^*)dt,$$

and continuous control $u(\cdot)$, satisfying the condition $u(t^*) = u^*$, gives the terminal point $\{\bar{X}(T), \bar{V}(T)\}$ such that

$$\bar{X}(T) = \frac{\lambda_1 X_1(T) + \lambda_2 X_2(T)}{\lambda_1 + \lambda_2},$$

$$\bar{V}(T) \leq \frac{\lambda_1 V_1(T) + \lambda_2 V_2(T)}{\lambda_1 + \lambda_2}.$$

Thus, the point $\{\bar{X}(T), \bar{V}(T)\}$, which admits a "single-impulse" representation, can be included in the convex combination that gives a representation of the point $\{X^0(T), V^0(T)\}$ (6.223) with the coefficient $\lambda_1+\lambda_2$. In this way the issue is reduced to the previous case with different points of concentration of the measures, and the theorem is proved. □

This result completes our discussion of problems with convex structure. Some results that relate to the application of the maximum principle for optimal generalized solutions in linear system can be found in [264, 333]. In the next section systems with convex structure arise in observation control problems. It will be shown that systems described by the matrix Riccati equation in the optimal observation control problem also have the property of convexity, and for them the necessary optimality condition given by the maximum principle is also sufficient.

Chapter 7
Observation Control in Discrete-Continuous Stochastic Systems

7.1 Problem Statement

As a rule, in problems of stochastic optimization one studies issues of optimization of procedures for processing information coming from measuring devices (filtration problems) and/or issues of optimal control of the evolution of phase coordinates in stochastic systems based on such information. A procedure for collecting such information, referred to as observation or measurement, is assumed to be given. However, in many practical tasks, it is possible to control the information retrieval processes, i.e., apply observation control. The essence of observation control is that it allows to improve the "quality" of the received information, i.e., ultimately, improve the accuracy of the estimates of unobserved parameters and/or processes. The observation control itself consists of the control of the *composition* of measurements, i.e., of the choice of measurement channels and sensors of measuring devices, and the allocation of resources (time or energy), which are available at a given moment, as well as of the control of the *moments* of observations in the case they are made as discrete time instants. From the side of a mobile observer, it may sometimes turn out that the quality of the information and the estimation accuracy depend in essential manner on the form of the observer's trajectory. In this event, there arises the so-called problem of *trajectory-controlled* observation, i.e., the problem of controlling the trajectory of the observer.

Observation control problems were considered by various authors (see, for example, [26, 27, 76, 87, 91, 153, 231, 246, 249, 265, 267, 276, 302]). For nonlinear systems, solving such problems reduces to rather complicated stochastic optimization procedures that can be investigated only by numerical methods. In the linear case, constructive results can be obtained analytically. In particular, the application of the "separation principle" [87, 153] makes it possible to reduce the observation control problem to a deterministic optimal control problem [153] that is solvable by standard methods.

B. Miller, E. Rubinovich, *Optimization of Dynamical Systems with Impulse Controls and Shocks*, Systems & Control: Foundations & Applications,
https://doi.org/10.1007/978-3-031-64124-4_7

The general statement of the observation control problem is as follows. On a given time interval $[0, T]$, consider a partially observable process $\{x(t), y(t), z(t)\}$ satisfying the system of stochastic differential equations

$$\begin{cases} dx(t) = A(t)x(t)dt + B(t)dw_t^1, \\ \qquad x(t) \in \mathbb{R}^n, \ \ x(0-) = x_0, \\ dy(t) = \alpha(t, u(t), z(t))x(t)v(t)dt + \beta(t, u(t), z(t))v^{1/2}(t)dw_t^2, \\ \qquad y(t) \in \mathbb{R}^m, \ \ y(0) = 0, \\ dz(t) = f(t, u(t), z(t))dt + G(t, u(t), z(t))v(t)dt, \\ \qquad z(t) \in \mathbb{R}^k, \ z(0) = z_0, \end{cases} \tag{7.1}$$

where $x(t)$ is an unobservable process describing the evolution of the stochastic system, and $\{y(t), z(t)\}$ are observation processes. The vector variable $z(t)$ may be regarded as the state of the moving observer.

The observation controls $\{u(t), v(t)\}$ satisfy the constraints

$$u(t) \in U \subset \mathbb{R}^r, \quad v(t) \geq 0, \quad \int_0^T v(t)dt \leq M < \infty, \tag{7.2}$$

where U is a compact set and M is a given positive constant.

The initial condition x_0 is assumed to be a Gaussian random vector with the parameters

$$\mathbf{E}\, x_0 = m_0, \quad \operatorname{cov}(x_0, x_0^*) = \gamma_0,$$

$w_t^1 \in \mathbb{R}^l$ and $w_t^2 \in \mathbb{R}^p$ are Wiener processes, and x_0, w_t^1 and w_t^2 are assumed to be independent. Here $\mathbf{E}$ stands for mathematical expectation.

In the system (7.1), $A(t)$ and $B(t)$ are the integrable matrix-valued functions, with $B(t)$ also square integrable, and the functions $\alpha(t, u, z)$ and $\beta(t, u, z)$ are continuous and bounded. The matrix-valued functions A, B, α, β, f and G have dimensions $n \times n$, $n \times l$, $m \times n$, $m \times p$, $k \times 1$, and $k \times 1$, respectively. Moreover, the function β is assumed to be *uniformly non-degenerate* in the following sense:

there exists a positive constant C such that, for any vector $h \in \mathbb{R}^p$,

$$h^* \beta^*(t, u, z)\beta(t, u, z)h \geq C\|h\|^2,$$

for all $t \in [0, T]$, $u \in U$ and $z \in \mathbb{R}^k$.

We assume also that the matrix-valued functions α, f and G are Lipschitz in the variable z.

The assumptions listed above ensure that the system (7.1) has a unique solution for any measurable deterministic controls $\{u(t), v(t)\}$ satisfying the constraints (7.2).

Let us assume that the controls $u(\cdot)$, $v(\cdot)$ are deterministic. Then the mean-square optimal estimate of $x(t)$ from the observations $\{y(s), z(s) | 0 \leq s \leq t\}$, i.e., the conditional mean

$$m(t) = \mathbf{E}\{x(t) | y_0^t, z_0^t\},$$

is given by the Kalman filter equations [244]

$$\begin{cases} dm(t) = A(t)m(t)dt + \\ \quad + \gamma(t)\alpha^*(t, u(t), z(t))(\beta^*\beta)^{-1}(t, u(t), z(t))(dy(t) - \\ \quad - \alpha(t, u(t), z(t))m(t)v(t)dt), \\ \dot{\gamma}(t) = A(t)\gamma(t) + \gamma(t)A^*(t) + B(t)B^*(t) - \\ \quad - \gamma(t)\alpha^*(t, u(t), z(t))(\beta^*\beta)^{-1}(t, u(t), z(t))\alpha(t, u(t), z(t))\gamma(t)v(t)dt, \\ dz(t) = f(t, u(t), z(t))dt + G(t, u(t), z(t))v(t)dt, \end{cases} \tag{7.3}$$

with given initial conditions

$$m(0) = m_0 = \mathbf{E}x_0, \quad \gamma(0) = \gamma_0 = \text{cov}(x_0, x_0^*) \text{ and } z(0-) = z_0.$$

The system (7.3) consists of two types of equations. The equation describing the evolution of the optimal estimate $m(t)$ is a stochastic differential equation of Itô type. The other equations, for the matrix function $\gamma(t)$ and for the variable $z(t)$, are ordinary differential equations, since it is assumed that the functions $\{u(\cdot), v(\cdot)\}$ are deterministic.

Moreover, note that since

$$\gamma(t) = \text{cov}\{x(t), x^*(t) | y_0^t, z_0^t\} = \mathbf{E}\{(x(t) - m(t))(x(t) - m(t))^* | y_0^t, z_0^t\},$$

and the vector $x(t) - m(t)$ follows a Gaussian distribution (with zero mean and covariance $\gamma(t)$), the equation for $\gamma(t)$ governs the evolution of the accuracy of the estimation of the state $x(t)$ of the stochastic system. The observation control problem can be formulated as follows: minimize the terminal functional

$$J[u(\cdot), v(\cdot)] = \phi_0(z(T), \gamma(T)) \to \min \tag{7.4}$$

over the class of admissible controls $\{u(\cdot), v(\cdot)\}$ satisfying the constraints (7.2) and such that $\gamma(t)$ and $z(t)$ satisfy the terminal and phase constraints

$$\begin{aligned} &h(z(T), \gamma(T)) = 0, \quad S(z(T), \gamma(T)) \le 0, \\ &g(z(t), \gamma(t), t) \le 0 \quad \text{for all } t \in [0, T]. \end{aligned} \tag{7.5}$$

The functions ϕ_0, h, S and g in (7.4) and (7.5) are continuous, and the vector-valued functions h, S and g have dimensions N_1, N_2 and N_3, respectively.

The independence of the performance criterion (7.4) and the constraints (7.5) from the random process $m(t)$ allows one to regard the observation control problem as a standard optimal control problem for a deterministic system governed by the Eq. (7.3), except, of course, for the equation for $m(t)$. This optimal control problem belongs to the class of problems treated in the previous chapters.

Remark 7.1 We will seek the solution of the observation control problem in the class of programmed controls, i.e., controls that depend only on the time t, and not on the current and/or past states of the observation process. As shown in [153, 233, 385], in typical observation control problems, widening the class of programmed controls to the class controls with feedback does not improve the quality of the control process. This important conclusion follows from the following reasoning.

Consider the stochastic control problem with controls

$$u(t) = u(t, y_0^t), \quad v(t) = v(t, y_0^t),$$

satisfying the constraints (7.2), and with the performance criterion

$$J[u(\cdot), v(\cdot)] = \mathbf{E}\big\{\phi_0(\gamma^{(u,v)}(T), z^{(u,v)}(T))\big\},$$

where $\{\gamma^{(u,v)}(T), z^{(u,v)}(T)\}$ is a solution of Eq. (7.3) for this choice of stochastic controls.

If there exists the optimal generalized solution $\{\gamma^0(\cdot), z^0(\cdot)\}$, such that

$$\inf_{(u,v)} J[u(\cdot), v(\cdot)] = \phi_0(\gamma^0(T), z^0(T)),$$

then for any admissible (even stochastic) control (u, v) we have

$$\phi_0(\gamma^{(u,v)}(T), z^{(u,v)}(T)) \ge \phi_0(\gamma^0(T), z^0(T)),$$

and therefore,

$$\mathbf{E}\big\{\phi_0(\gamma^{(u,v)}(T), z^{(u,v)}(T))\big\} \ge \phi_0(\gamma^0(T), z^0(T)).$$

Thus, a deterministic generalized solution realizes the absolute minimum to the performance criterion, which allows to narrow the class of admissible controls to the deterministic controls.

A detailed proof of this fact is based on the application of the method dynamic programming to stochastic optimization problems and is beyond the scope of this book. Relevant results can be found in [153, 233, 385].

Remark 7.2 Let us emphasize once again that the two types of observation controls under consideration provide control over both the composition of the observations (function $u(t)$) and the distribution over time of the observation resource (function $v(t)$). The impulsive character of $v(t)$ corresponds to discrete observations localized at the points where the impulse control is applied. Moreover, in this model the observation process $y(t)$ depends on the controlled process $z(t)$. As noted earlier, this model is typical for problems in which there is a so-called moving observer [385], and the accuracy of the estimation depends on its position and speed. Such problems arise in optical observation and/or radiolocation [265, 267, 301, 304].

7.2 Generalized Solutions in Observation Control Problems

Following the ideas developed in Chap. 5, we define next the notion of a strong generalized solution of an observation control problem.

Definition 7.1 A pair of right-continuous functions $\{\gamma(\cdot), z(\cdot)\}$ of bounded variation is called a *generalized solution* of the system (7.3), if there exists a sequence $\{u^n(\cdot), v^n(\cdot)\}$ satisfying the constraints (7.2) and such that the corresponding sequence of solutions $\{\gamma^n(\cdot), z^n(\cdot)\}$ of the system (7.3) with initial conditions $\gamma^n(0-) = \gamma(0-)$ and $z^n(0) = z(0-)$ converges to $\{\gamma(\cdot), z(\cdot)\}$ at all points of continuity.

A generalized solution $\{\gamma(\cdot), z(\cdot)\}$ of the system (7.3) is called an *admissible generalized solution* if

1. it satisfies the constraints (7.5);
2. there exists a sequence of controls $\{u^n(\cdot), v^n(\cdot)\}$, satisfying the constraints (7.2), such that the corresponding sequence of solutions $\{\gamma^n(\cdot), z^n(\cdot)\}$ of the system (7.3) with initial conditions $\gamma^n(0-) = \gamma(0-)$ and $z^n(0) = z(0-)$ converges to $\{\gamma(t), z(t)\}$ at all points of continuity and the following conditions are satisfied:

$$\begin{gathered} \lim_n h(z^n(T), \gamma^n(T)) = 0, \quad \lim_n S(z^n(T), \gamma^n(T)) \leq 0, \\ \lim_n \sup_{[0,T]} g_k(z^n(t), \gamma^n(t), t) \leq 0 \end{gathered} \tag{7.6}$$

for all $k = 1, \ldots, N_3$.

And finally, we define the notion of the optimal generalized solution.

Definition 7.2 The admissible generalized solution $\{\gamma^0(\cdot), z^0(\cdot)\}$ is called an *optimal generalized solution* if

$$\phi_0(z^0(T), \gamma^0(T)) \le \phi_0(z'(T), \gamma'(T))$$

for any admissible generalized solution $\{\gamma'(\cdot), z'(\cdot)\}$.

7.2.1 Representation of Generalized Solutions in Observation Control Problems

First let us produce a representation of generalized solutions in the following auxiliary control problem with variables $\{\gamma_1(s), z_1(s), \eta(s)\}$:

$$\begin{cases} \dot{\gamma}_1(s) = \kappa(s)\big[A(\eta(s))\gamma_1(s) + \gamma_1(s)A^*(\eta(s)) + B(\eta(s))B^*(\eta(s))\big] - \\ \qquad - (1-\kappa(s))\gamma_1(s)\alpha^*(\eta(s), u_1(s), z_1(s)) \times \\ \qquad \times (\beta^*\beta)^{-1}(\eta(s), u_1(s), z_1(s))\alpha(\eta(s), u_1(s), z_1(s))\gamma_1(s), \\ \dot{z}_1(s) = \kappa(s)f(\eta(s), u_1(s), z_1(s)) + (1-\kappa(s))G(\eta(s), u_1(s), z_1(s)), \\ \dot{\eta}(s) = \kappa(s) \end{cases} \tag{7.7}$$

with the initial conditions

$$\gamma_1(0) = \gamma_0, \quad z_1(0) = z_0, \quad \eta(0) = 0.$$

This system is considered on some interval $[0, T_1]$, with $T_1 \le T + M$ such that $\eta(T_1) = T$. The controls $\{\kappa(s), u_1(s)\}$ are assumed to satisfy the constraints

$$0 \le \kappa(s) \le 1, \quad u_1(s) \in U. \tag{7.8}$$

The correspondence between the solutions of the system (7.2) and (7.3) and those of the system (7.7) and (7.8) is established by the following theorem, which is a consequence of the more general results obtained in Chap. 5.

Theorem 7.1 *Let $\{\gamma(t), z(t)\}$ be a generalized solution of the system (7.3). Then there exists a sequence of controls $\{\kappa^n(s), u_1^n(s)\}$, defined on some interval $[0, T_1]$, with $T_1 \le T + M$, and satisfying (7.8), such that the corresponding sequence of solutions $\{\gamma_1^n(s), z_1^n(s), \eta^n(s)\}$ of the auxiliary system (7.7) converges to the triplet $\{\gamma_1(s), z_1(s), \eta(s)\}$ uniformly on $[0, T_1]$, and*

$$\gamma(t) = \gamma_1(\Gamma(t)), \quad z(t) = z_1(\Gamma(t)), \tag{7.9}$$

where

$$\Gamma(t) = \inf\{s : \eta(s) > t\}. \tag{7.10}$$

Let $\mathscr{L} = \mathscr{L}(\gamma, z, t, U) = \{(m, l) \in \mathbb{R}^{n\times n} \times \mathbb{R}^k\}$ be the set of pairs (matrix,vector) of the form

$$\begin{pmatrix} m \\ l \end{pmatrix} = \begin{pmatrix} (1-\kappa)\gamma\alpha^*(t,u,z)(\beta^*\beta)^{-1}(t,u,z)\alpha(t,u,z)\gamma \\ \kappa f(t,u,z) + (1-\kappa)G(t,u,z) \end{pmatrix}, \tag{7.11}$$

with $\kappa \in [0, 1]$ and $u \in U$.

Theorem 7.2 *Assume that the set $\mathscr{L}$ is convex for any fixed $\{\gamma, z, t\}$, where $\gamma \in \mathbb{R}^{n\times n}$ is a symmetric non-negative definite matrix, $z \in \mathbb{R}^k$, and $t \in [0, T]$.*

Then for any generalized solution $\{\gamma(\cdot), z(\cdot)\}$ there exist controls $\{\kappa(\cdot), u_1(\cdot)\}$, satisfying the constraints (7.8) on some interval $[0, T_1]$ with $T_1 \le T + M$, and such that the corresponding solution $\{\gamma_1(s), z_1(s), \eta(s)\}$ of the system (7.7) satisfies the relations (7.9) and (7.10).

Moreover, the generalized solution satisfies the following differential equation with measure.

Theorem 7.3 *Suppose the set $\mathscr{L}$ satisfies the assumptions of Theorem 7.2.*

Then for any generalized solution $\{\gamma(t), z(t)\}$ there exist

1. *a non-negative measure $V(dt)$, concentrated on the interval $[0, T]$, such that*

$$V([0, T]) \le M;$$

2. *a Lebesgue- and $V(dt)$-measurable function $u(\cdot)$ such that $u(t) \in U$ a.e. with respect to both the Lebesgue and the $V(dt)$ measures;*
3. *a collection of functions $u_\tau(s) \in U$, defined on segments $[0, V(\{t\})]$ for each $\tau \in D_V^d$, where $D_V^d = \{t : V(\{t\}) > 0\}$,*

such that the generalized solution $\{\gamma(\cdot), z(\cdot)\}$ satisfies the following differential equations with measure

$$\begin{cases} d\gamma(t) = \big[A(t)\gamma(t) + \gamma(t)A^*(t) + B(t)B^*(t)\big]dt - \\ \quad - \gamma(t)\alpha^*(t, u(t), z(t))(\beta^*\beta)^{-1}(t, u(t), z(t))\alpha(t, u(t), z(t))\gamma(t)dV^c + \\ \quad + \sum_{\tau\le t} \Delta\gamma(\tau)\delta(t-\tau)dt, \\ dz(t) = f(t, u(t), z(t))dt + G(t, u(t), z(t))dV^c(t) + \\ \quad + \sum_{\tau\le t} \Delta z(\tau)\delta(t-\tau)dt, \end{cases} \tag{7.12}$$

where $V^{c}(dt)$ is the continuous component of the measure $V(dt)$. The jumps of the functions $\{\gamma(t), z(t)\}$ at the points $\tau \in D_V^{d}$ satisfy the relations

$$\Delta\gamma(\tau) = \gamma_\tau(V(\{t\})) - \gamma_\tau(0), \quad \Delta z(\tau) = z_\tau(V(\{t\})) - z_\tau(0),$$

where $\{\gamma_\tau(s), z_\tau(s)\}$ is the solution of the system

$$\begin{cases} \dot{\gamma}_\tau(s) = -\gamma_\tau(s)\alpha^*(\tau, u_\tau(s), z_\tau(s))(\beta^*\beta)^{-1}(\tau, u_\tau(s), z_\tau(s))\times \\ \quad \times\, \alpha(\tau, u_\tau(s), z_\tau(s))\gamma_\tau(s), \\ \dot{z}_\tau(s) = G(\tau, u_\tau(s), z_\tau(s)), \end{cases}$$

with the initial conditions

$$\gamma_\tau(0) = \gamma(\tau-), \quad z_\tau(0) = z(\tau-).$$

If $G(t, u, z) \equiv 0$, then the expressions for jumps can be derived explicitly. In this case the jumps of the function z at the points of application of the impulse actions vanish, while the jumps of γ at the points $\tau \in D_V^{d}$ can be expressed as

$$\begin{aligned} &\Delta\gamma(\tau) = \gamma(\tau-)\alpha^*(\tau, u(\tau), z(\tau))\times \\ &\times \{(\beta^*\beta)(\tau, u(\tau), z(\tau)) + \alpha(\tau, u(\tau), z(\tau))\gamma(\tau-)\alpha^*(\tau, u(\tau), z(\tau))V(\{\tau\})\}^{-1}\times \\ &\times \alpha(\tau, u(\tau), z(\tau))\gamma(\tau-)V(\{\tau\}), \end{aligned} \tag{7.13}$$

with some control $u(\tau) \in U$.

To prove this theorem we need the following auxiliary lemma.

Lemma 7.1 *Suppose the symmetric $n \times n$ matrix-valued function $X(t)$ satisfies on the interval $[0, T]$ the differential equation*

$$\dot{X}(t) = -X(t)N^*(t)N(t)X(t), \tag{7.14}$$

and the initial condition $X(0)$, where $X(0)$ is a symmetric non-negative definite matrix, and $N(t)$ is a square-integrable $m \times n$ matrix-valued function. Assume also that the $m \times n$ matrix-valued function $M(t)$ is such that

$$\int_0^t N^*(s)N(s)ds = M^*(t)M(t). \tag{7.15}$$

Then $X(t)$ admits the representation

$$X(t) = X(0) - X(0)M^*(t)\big[I_{(m\times m)} + M(t)X(0)M^*(t)\big]^{-1}M(t)X(0). \tag{7.16}$$

Proof Assume first that $X(0)$ is positive definite. Thanks to its continuity, $X(t)$ remains positive definite in some small vicinity of $t = 0$, and so it admits an inverse $X^{-1}(t)$. This function satisfies the differential equation

$$\dot{X}^{-1}(t) = N^*(t)N(t),$$

and the initial condition $X^{-1}(0)$, where $X^{-1}(0)$ is also positive definite. As the function N^*N is integrable, the solution of last equation exists on the whole interval $[0, T]$ and can be written in the form

$$X^{-1}(t) = X^{-1}(0) + \int_0^t N^*(s)N(s)ds = X^{-1}(0) + M^*(t)M(t).$$

Therefore, $X^{-1}(t)$ is positive definite on $[0, T]$ and the difference $X(t) - X(0)$ satisfies the relation

$$X(t) - X(0) = -X(t)(X^{-1}(t) - X^{-1}(0))X(0) = -X(t)M^*(t)M(t)X(0). \tag{7.17}$$

Multiplying this relation on the right on $M^*(t)$ we obtain

$$X(t)M^*(t)\big[I_{(m\times m)} + M(t)X(0)M^*(t)\big] = X(0)M^*(t),$$

or

$$X(t)M^*(t) = X(0)M^*(t)\big[I_{(m\times m)} + M(t)X(0)M^*(t)\big]^{-1},$$

and then

$$X(t)M^*(t)M(t)X(0) =$$
$$= X(0)M^*(t)\big[I_{(m\times m)} + M(t)X(0)M^*(t)\big]^{-1}M(t)X(0).$$

Substituting this last expression in (7.17) gives (7.16), which completes the proof in the case $X(0) > 0$.

In the general case when $X(0) \geq 0$ one can choose a sequence of positive definite matrices $X^n(0)$ such that $X^n(0) \downarrow X(0)$. Then for any solution of Eq. (7.14) with initial condition $X^n(0)$ we have

$$X^n(t) = X^n(0) - X^n(0)M^*(t)\big[I_{(m\times m)} + M(t)X^n(0)M^*(t)\big]^{-1}M(t)X^n(t),$$

where the right-hand side is continuous in $X^n(0)$, and consequently $X^n(t)$ converges to the function $X(t)$ defined by formula (7.16), for any $t \in [0, T]$. To complete the proof it remains to show that $X(t) = \lim_n X^n(t)$ is a solution of Eq. (7.14).

The family of solutions of Eq. (7.14) corresponding to some bounded set of non-negative definite matrices as initial conditions is uniformly bounded. This follows immediately from (7.14), because

$$X(t) = X(0) - \int_0^t X(s)N^*(s)N(s)X(s)ds \leq X(0),$$

where the inequality means that the matrix $X(0) - X(t)$ is non-negative definite.

Further, the right-hand side of (7.14) is locally Lipschitz, so the solution of Eq. (7.14) is unique and depends continuously on the initial conditions [96]. Thus, $X(t)$, which is the solution of (7.14), has the form (7.16). □

In order to apply Lemma 7.1 to the representation of $\Delta\gamma$ we note that $\gamma_\tau(s)$ satisfies on the interval $[0, V(\{\tau\})]$ an equation of the form (7.14) with $X(0) = \gamma_\tau(0) = \gamma(\tau-)$ and the matrix

$$N(s) = (\beta^*\beta)^{-1/2}(\tau, u_\tau(s), z(\tau))\alpha(\tau, u_\tau(s), z(\tau)).$$

By formula (7.15), the corresponding matrix $M(s)$ satisfies at $s = V(\{\tau\})$ the relation

$$M^*(V(\{\tau\}))M(V(\{\tau\})) = \int_0^{V(\{\tau\})} N^*(s)N(s)ds =$$

$$= \int_0^{V(\{\tau\})} \alpha^*(\tau, u_\tau(s), z(\tau))(\beta^*\beta)^{-1}(\tau, u_\tau(s), z(\tau))\alpha(\tau, u_\tau(s), z(\tau))ds.$$

Next, the convexity of the set $\mathscr{L}(t, U, z, \gamma)$ ensures that there exists a control $u(\tau) \in U$ such that

$$\int_0^{V(\{\tau\})} \alpha^*(\tau, u_\tau(s), z(\tau))(\beta^*\beta)^{-1}(\tau, u_\tau(s), z(\tau))\alpha(\tau, u_\tau(s), z(\tau))ds =$$

$$= \alpha^*(\tau, u(\tau), z(\tau))(\beta^*\beta)^{-1}(\tau, u(\tau), z(\tau))\alpha(\tau, u(\tau), z(\tau))V(\{\tau\}). \tag{7.18}$$

Consequently, the matrix $M(V(\{\tau\}))$ admits the representation

$$M(V(\{\tau\})) = (\beta^*\beta)^{-1/2}(\tau, u(\tau), z(\tau))\alpha(\tau, u(\tau), z(\tau))(V(\{\tau\}))^{1/2}.$$

Now applying formula (7.16) to $\gamma(\tau) = \gamma_\tau(V(\{\tau\})$, we obtain

$$\gamma(\tau) - \gamma(\tau-) = \gamma_\tau(V(\{\tau\}) - \gamma_\tau(0) =$$

$$= -\gamma_\tau(0)M^*(V(\{\tau\}))\big[I_{(m\times m)} + M(V(\{\tau\}))\gamma_\tau(0)M^*(V(\{\tau\}))\big]^{-1}\times$$

$$\times M(V(\{\tau\}))\gamma_\tau(0) =$$

$$= \gamma(\tau-)M^*(V(\{\tau\}))\big[I_{(m\times m)} + M(V(\{\tau\}))\gamma(\tau-)M^*(V(\{\tau\}))\big]^{-1}\times$$

$$\times M(V(\{\tau\}))\gamma(\tau-) = \gamma(\tau-)\alpha^*(\beta^*\beta)^{-1/2}\times$$

$$\times \left\{I_{(m\times m)} + (\beta^*\beta)^{-1/2}\alpha\gamma(\tau-)\alpha^*(\beta^*\beta)^{-1/2}V(\{\tau\})\right\}^{-1}\times$$

$$\times (\beta^*\beta)^{-1/2}\alpha\gamma(\tau-)V(\{\tau\}) =$$

$$= \gamma(\tau-)\alpha^*\left\{(\beta^*\beta) + \alpha\gamma(\tau-)\alpha^*V(\{\tau\})\right\}^{-1}\alpha\gamma(\tau-)V(\{\tau\}),$$

where for simplicity the arguments of the functions α and β are omitted. The last relation coincides with (7.13).

Remark 7.3 The generalized solution and the original filtering problem are connected as follows. (We consider the case when $G(t, u, z) \equiv 0$.) Consider an observation program defined by a control $u(\cdot)$ and a non-negative measure $V(dt)$, such that the observation process is described by the equation

$$dy(t) = \alpha(t, u(t), z(t))x(t)dV(t) + \beta(t, u(t), z(t))d\xi_t,$$

where ξ_t is a Gaussian martingale with a quadratic characteristic

$$\langle\xi\rangle_t = \operatorname{cov}\left\{\xi_t, \xi_t^*\right\} = I_{(p\times p)}V(t).$$

Here the process ξ does not depend on x_0 and w^1. This equation describes the continuous and discrete observations in a unified manner. Discrete observations are localized on the set D_V^{d} and have the representation

$$\Delta y(\tau) = \alpha(\tau, u(\tau), z(\tau))x(\tau)\Delta V(\tau) + \beta(\tau, u(\tau), z(\tau))\Delta\xi_\tau,$$

where $\Delta\xi_\tau$ is a Gaussian vector with zero mean and the covariance matrix

$$\operatorname{cov}\left\{\Delta\xi_t, \Delta\xi_t^*\right\} = I_{(p\times p)}\Delta V(t).$$

The process governing the continuous observations is given by the equation

$$dy^{\mathrm{c}}(t) = \alpha(t, u(t), z(t))x(t)dV^{\mathrm{c}}(t) + \beta(t, u(t), z(t))d\xi_t^{\mathrm{c}},$$

and we have

$$y(t) = y^{c}(t) + \sum_{\tau \le t} \Delta y(\tau).$$

In the case under consideration the optimal mean-square estimate of the unobservable process satisfies the equations of a generalized Kalman filter [291],

$$\begin{aligned}
dm(t) &= A(t)m(t)dt + \gamma(t)\alpha^{*}(t, u(t), z(t))(\beta^{*}\beta)^{-1}(t, u(t), z(t)) \times \\
&\quad \times \left[dy^{c}(t) - \alpha(t, u(t), z(t))m(t)dV^{c}(t)\right] + \sum_{\tau \le t} \Delta m(\tau)\delta(t-\tau)dt, \\
d\gamma(t) &= \left[A(t)\gamma(t) + \gamma(t)A^{*}(t) + B(t)B^{*}(t)\right]dt - \\
&\quad - \gamma(t)\alpha^{*}(t, u(t), z(t))(\beta^{*}\beta)^{-1}(t, u(t), z(t))\alpha(t, u(t), z(t))\gamma(t)dV^{c} + \\
&\quad + \sum_{\tau \le t} \Delta\gamma(\tau)\delta(t-\tau)dt, \\
dz(t) &= f(t, u(t), z(t))dt,
\end{aligned} \tag{7.19}$$

where $V^{c}(dt)$ is the continuous component of the measure $V(dt)$ and $y^{c}(t)$ is the continuous component of observations. The jumps of the variables m and γ are localized on the set D_{V}^{d}. More precisely, the jumps $\Delta m(\tau)$ are expressed as

$$\Delta m(\tau) = \alpha^{*}\left\{(\beta^{*}\beta) + \alpha\gamma(\tau-)\alpha^{*}\Delta V(\tau)\right\}^{-1}\left[\Delta y(\tau) - \alpha m(\tau-)\Delta V(\tau)\right],$$

where $\Delta V(\tau) = V(\{\tau\})$ (for simplicity, the arguments of the functions α and β are omitted), and the jumps of $\Delta\gamma(\tau)$ are calculated by formula (7.13).

7.2.2 *Discontinuous Time Change in Observation Control Problems*

The observation control problem considered above belongs to the class of impulse control problems and may be transformed into a classical optimal control problem with the aid of a discontinuous time change.

Let us introduce the auxiliary control problem for a system whose state is described by a triplet of variables $\{\gamma_1(s), z_1(s), \eta(s)\}$ satisfying the Eq. (7.7) with the initial conditions

$$\gamma_1(0) = \gamma_0, \quad z_1(0) = z_0, \quad \eta(0) = 0.$$

The problem is considered on an interval $[0, T_1]$, where $T_1 \leq T + M$. The controls $\{\kappa(s), u_1(s)\}$ obey the constraints (7.8), and admissible trajectories satisfy the conditions

$$\begin{aligned} &h(z_1(T_1), \gamma_1(T_1)) = 0, \quad \eta(T_1) = T, \\ &S(z_1(T_1), \gamma_1(T_1)) \leq 0, \\ &g(z_1(s), \gamma_1(s), \eta(s)) \leq 0, \quad \text{for all} \quad s \in [0, T_1]. \end{aligned} \tag{7.20}$$

The aim of the control is to minimize the terminal cost function

$$J'[u_1(\cdot), \kappa(\cdot)] = \phi_0(z_1(T_1), \gamma_1(T_1)). \tag{7.21}$$

The equivalence of this auxiliary problem with the original observation control problem follows directly from Theorems 5.1 and 5.2.

Theorem 7.4 *Assume that the set of admissible solutions of the problem (7.2)–(7.5) is non-empty and bounded. Then there exists the optimal generalized solution* $\{\gamma^0(\cdot), z^0(\cdot)\}$.

If the conditions of Theorem 7.2 are fulfilled, then there exists the optimal solution

$$\{\gamma_1^0(s), z_1^0(s), \eta^0(s)\},$$

of the auxiliary problem (7.7), (7.8), (7.20), (7.21) such that

$$\gamma^0(t) = \gamma_1^0(\Gamma^0(t)), \quad z^0(t) = z_1^0(\Gamma^0(t)), \quad \textit{where} \quad \Gamma^0(t) = \inf\{s : \eta^0(s) > t\}.$$

As in the cases considered above, using this theorem we can reduce the original problem with generalized controls to the auxiliary regular problem. In its turn, the auxiliary problem can be solved by standard methods of optimal control theory. The next section will explore in detail such problems and will demonstrate the effectiveness of the proposed methodology in solving problems of practical interest.

Remark 7.4 In this section we omitted details concerning the stochastic nature of the problems under consideration. More details on this topic can be found in the works [279, 291, 304, 305].

7.3 Convexity in Observation Control Problems

Let the matrix γ_0 be positive definite. Since for any admissible controls $\{u(\cdot), v(\cdot)\}$ we have

$$\int_0^T \alpha^*(t,u(t),z(t))(\beta^*\beta)^{-1}(t,u(t),z(t))\alpha(t,u(t),z(t))v(t)dt < \infty,$$

the matrix-function $\gamma(t)$, which is the solution of the system (7.3), remains positive definite for all $t \in [0, T]$. Then the relation

$$\frac{d}{dt}\gamma^{-1}(t) = -\gamma^{-1}(t)\left[\frac{d}{dt}\gamma(t)\right]\gamma^{-1}(t)$$

and (7.3) show that the inverse matrix-function $\gamma^{-1}(t) = D(t)$ (if it exists) satisfies the differential equation

$$\begin{aligned}\dot{D}(t) &= -\big[A(t)D(t) + D(t)A^*(t) + D(t)B(t)B^*(t)D(t)\big] + \\ &\quad + \alpha^*(t,u(t),z(t))(\beta^*\beta)^{-1}(t,u(t),z(t))\alpha(t,u(t),z(t))v(t)\end{aligned} \tag{7.22}$$

with the initial condition $D(0) = \gamma^{-1}(0)$. This equation has a unique bounded solution, which implies the existence of $\gamma^{-1}(t)$ for all $t \in [0, T]$.

Thus, the optimal control problem can be reformulated in terms of $D(t)$ instead of $\gamma(t)$. Then for the generalized solution of Eq. (7.22), in accordance with (7.7), we obtain the differential equation with measure

$$\begin{aligned}dD(t) &= -\big[A(t)D(t) + D(t)A^*(t) + D(t)B(t)B^*(t)D(t)\big]dt + \\ &\quad + \alpha^*(t,u(t),z(t))(\beta^*\beta)^{-1}(t,u(t),z(t))\alpha(t,u(t),z(t))dV^{\mathrm{c}}(t) + \\ &\quad + \sum_{\tau \le t} \Delta D(\tau)\delta(t-\tau)dt,\end{aligned} \tag{7.23}$$

where the jumps of the variables $\{D(t), z(t)\}$ at the points $\tau \in D_v^{\mathrm{d}}$ satisfy the relations

$$\Delta D(\tau) = D_\tau(V(\{t\})) - D_\tau(0), \quad \Delta z(\tau) = z_\tau(V(\{t\})) - z_\tau(0).$$

Here $\{D_\tau(s), z_\tau(s)\}$ is the solution of the system

$$\begin{cases}\dot{D}_\tau(s) = \alpha^*(\tau,u_\tau(s),z_\tau(s))(\beta^*\beta)^{-1}(\tau,u_\tau(s),z_\tau(s))\alpha(\tau,u_\tau(s),z_\tau(s)),\\ \dot{z}_\tau(s) = G(\tau,u_\tau(s),z_\tau(s)),\end{cases}$$

with the initial conditions

$$D_\tau(0) = D(\tau-), \quad z_\tau(0) = z(\tau-).$$

Note that the right-hand side of the equation for $D_\tau(s)$ does not depend on D_τ. If the jumps of $z(\cdot)$ are equal zero, then, as in the case when $G(t,u,z) \equiv 0$, we obtain for $\Delta D(\tau)$ the expression

$$\Delta D(\tau) = \alpha^*(\tau, u(\tau), z(\tau))(\beta^*\beta)^{-1}(\tau, u(\tau), z(\tau))\alpha(\tau, u(\tau), z(\tau))\Delta V(\tau),$$

with some $u(\tau) \in U$ satisfying (7.18). Then the equation for the generalized solution has the form

$$dD(t) = -[A(t)D(t) + D(t)A^*(t) + D(t)B(t)B^*(t)D(t)]dt + \\ + \alpha^*(t, u(t), z(t))(\beta^*\beta)^{-1}(t, u(t), z(t))\alpha(t, u(t), z(t))dV(t). \tag{7.24}$$

The jumps of the variables $\{D, z\}$ admit an explicit representation also in the case when $\alpha(t,u,z) \equiv \alpha(t,u)$, $\beta(t,u,z) \equiv \beta(t,u)$, $f(t,u,z) \equiv f(t,z)$ and $G(t,u,z) \equiv G(t,u)$.

Hence, the auxiliary system for the jumps decouples into two independent equations, for $D_\tau(s)$ and $z_\tau(s)$. Therefore, if the set $\mathscr{L}$ defined in (7.11) satisfies the convexity conditions, we obtain the following representation for the generalized solution $\{D(\cdot), z(\cdot)\}$:

$$\begin{cases} dD(t) = -\big[A(t)D(t) + D(t)A^*(t) + D(t)B(t)B^*(t)D(t)\big]dt + \\ \qquad + \alpha^*(t, u(t))(\beta^*\beta)^{-1}(t, u(t))\alpha(t, u(t))dV(t), \\ dz(t) = f(t, z(t))dt + G(t, u(t))dV(t). \end{cases} \tag{7.25}$$

This shows that, in terms of $D(t)$, the optimal control problem belongs to the class of linear convex problems considered above in Sect. 6.4.

In the present problem, the first variable in $\{D(\cdot), z(\cdot)\}$ is a matrix-function. Therefore, one needs to reformulate the necessary optimality conditions in the from of a *matrix maximum principle* [15]. To this end we introduce the following notations:

$$\mathscr{H}_0\big(P_D, p_z, D, z, t\big) = \mathrm{tr}\big\{ - \big[A(t)D + DA^*(t) + DB(t)B^*(t)D\big]P_D^*\big\} + \\ + \langle p_z, f(t,z)\rangle,$$

$$\mathscr{H}_1\big(P_D, p_z, p_v, u, t, e\big) = \mathrm{tr}\big\{\big[\alpha^*(t,u)(\beta^*\beta)^{-1}(t,u)\alpha(t,u))e\big]P_D^*\big\} + \\ + \langle p_z, G(t,u)e\rangle + p_v,$$

$$\mathscr{L}(z, D) = \lambda_0\widetilde{\phi}_0(z, D) + \langle l_1, \widetilde{h}(z, D)\rangle + \langle l_2, \widetilde{S}(z, D)\rangle + \gamma_0(v - M), \tag{7.26}$$

where, by the definition of the trace,

$$\operatorname{tr}\{AB^*\} = \sum_{i=1}^{n} \{AB^*\}_{i\,i} = \sum_{i=1}^{n}\sum_{k=1}^{n} A_{i\,k} B_{i\,k},$$

$\langle \cdot, \cdot \rangle$ stands for the scalar product, and

$$\widetilde{\phi}_0(z, D) = \phi_0(z, D^{-1}), \quad \widetilde{h}(z, D) = h(z, D^{-1}),$$

$$\widetilde{S}(z, D) = S(z, D^{-1}), \quad \widetilde{g}(z, D, t) = g(z, D^{-1}, t).$$

In order to obtain the necessary optimality conditions, we need to apply Theorem 6.7 directly. Suppose $\{D^0(\cdot), z^0(\cdot)\}$ is an optimal trajectory. Then the equation for the conjugate variable $P_D^0(\cdot)$ has the form

$$\begin{aligned} P_D^0(t) = &-\mathscr{L}'_D\big(z^0(T), D^0(T)\big) + \\ &+ \int_t^T (\mathscr{H}_0)'_D\big(P_D^0(\tau), p_z^0(\tau), D^0(\tau), z^0(\tau), \tau\big) d\tau + \\ &+ \sum_{\tau > t,\, \tau \in D^{\mathrm{d}}_{\mathscr{V}}} \Delta P_D^0(\tau) - \int_t^T \sum_{k=1}^{N_3} \widetilde{g}'_{kD}\big(z^0(\tau), D^0(\tau), \tau\big) d\lambda_k(\tau). \end{aligned} \tag{7.27}$$

This equation contains terms that involve derivatives with respect to matrix arguments. Following [15], the derivative $F'_D(D)$ will be understood as the matrix-function, such that

$$F(D + \Delta D) - F(D) = \operatorname{tr}\{F'_D(D) \Delta D^*\} + \mathscr{O}(\Delta D),$$

where

$$\frac{\mathscr{O}(\Delta D)}{\|\Delta D\|} \to 0 \quad \text{as } \|\Delta D\| \to 0.$$

According to this definition, the derivatives of functions $F(X)$ of a matrix argument X have the following list of properties (see [15] for details):

$$\begin{aligned} &\frac{d}{dX}\operatorname{tr}\{AX\} = \frac{d}{dX}\operatorname{tr}\{XA\} = A^*, \\ &\frac{d}{dX}\operatorname{tr}\{AXB\} = A^* B^*, \\ &\frac{d}{dX}\operatorname{tr}\{XAXB\} = (AXB)^* + (XA)^* B^* = B^* X^* A^* + A^* X^* B^*, \\ &\frac{d}{dX}\operatorname{tr}\{AX^{-1}\} = \frac{d}{dX}\operatorname{tr}\{X^{-1}A\} = -(X^{-1})^* A^* (X^{-1})^*. \end{aligned} \tag{7.28}$$

Further, in view of the definition of the function $\mathscr{H}_0$ (7.26) and formulas (7.28), we can rewrite Eq. (7.27) in the form

$$\begin{aligned}P_D^0(t) = &-\mathscr{L}'_D\big(z^0(T), D^0(T)\big) - \int_t^T \Big[A^*(\tau)P_D^0(\tau) + P_D^0(\tau)A^*(\tau) + \\ &+ P_D^0(\tau)D^0(\tau)B(\tau)B^*(\tau) + B(\tau)B^*(\tau)D^0(\tau)P_D^0(\tau)\Big]d\tau + \\ &+ \sum_{\tau>t,\, \tau\in D^{\mathrm{d}}_{\mathscr{V}}} \Delta P_D^0(\tau) - \int_t^T \sum_{k=1}^{N_3} \widetilde{g}'_{kD}(z^0(\tau), D^0(\tau), \tau)d\lambda_k(\tau).\end{aligned} \tag{7.29}$$

So, the general theory of necessary optimality conditions is straightforwardly applicable to observation control problems. A specific feature of Eq. (7.29) is that the matrix-function P_D^0 preserves its sign. In other words, one can show that in the typical observation control problem, the matrix $P_D^0(\cdot)$ is non-negative definite on the segment $[0, T]$, and consequently, the term in the function $\mathscr{H}_0$ depending on D is concave. In conjunction with Theorem 6.8, this implies that the maximum principle is a sufficient optimality condition .

Lemma 7.2 *[265] Let $F(D) = \mathrm{tr}\{DMDP\}$, where M and P are non-negative definite matrices. Then F is a convex function in D on the set of symmetric matrices D.*

Proof Since F is a quadratic form in the entries of D, it suffices to show that the condition $D = D^*$ implies the inequality $F(D) \geq 0$. Indeed, this last inequality expresses the positive definiteness of the matrix $\left\| \dfrac{\partial^2}{\partial D_{ij}\partial D_{lk}} F(D) \right\|_{i,j=1:n;\, l,k=1:n}$, and hence the convexity of the function F.

Assume first that $M > 0$. Then there exists the inverse matrix $C = M^{-1}$, which is positive definite and symmetric. For the pair of matrices D and C there exists a non-singular matrix N that diagonalizes both D and C [137], i.e., the matrices

$$D' = NDN^* \quad \text{and} \quad C' = NCN^*$$

are diagonal.

Consequently, in view of the properties of the trace $\mathrm{tr}(\cdot)$, we have

$$\begin{aligned}F(D) &= \mathrm{tr}\{DMDP\} = \mathrm{tr}\{DC^{-1}DP\} = \\ &= \mathrm{tr}\{NDC^{-1}DPN^{-1}\} = \\ &= \mathrm{tr}\{NDN^*(N^*)^{-1}C^{-1}N^{-1}NDN^*(N^*)^{-1}PN^{-1}\} = \\ &= \mathrm{tr}\{D'(C')^{-1}D'P'\},\end{aligned} \tag{7.30}$$

where we used the notation $P' = (N^*)^{-1}PN^{-1}$.

Notice that the matrices D' and $(C')^{-1}$ are diagonal and that $(C')^{-1}$ and P' is non-negative definite. Continuing the calculations in (7.30), we conclude that

$$F(D) = \sum_i (D'_{ii})^2 (C'_{ii})^{-1} P'_{ii} \geq 0.$$

To complete the proof in the general case, i.e., when the matrix M is non-negative definite, we note that the function F is continuous in the variable M. Therefore, for any non-negative definite matrix M we have

$$F(D) = \operatorname{tr}\{DMDP\} = \lim_n \operatorname{tr}\{DM^{(n)}DP\} \geq 0,$$

where $M^{(n)}$ is some sequence of positive definite matrices such that $M^{(n)} \downarrow M$. □

In various observation control problems, one encounters penalty functions of the form $F(\gamma) = f[\operatorname{tr}\{L\gamma\}]$, where the function f is monotonically increasing and L is a positive definite matrix. This type of penalty function of a matrix argument arises when it is necessary to restrict or minimize the accuracy of observations, which is a linear function of the covariance matrix γ. If we reformulate the problem in terms of $D = \gamma^{-1}$, then it is necessary to investigate properties of functions of the form

$$\widetilde{F}(D) = f(\operatorname{tr}\{LD^{-1}\}). \tag{7.31}$$

Lemma 7.3 *The function* $\operatorname{tr}\{LD^{-1}\}$ *is convex on the set of symmetric positive definite matrices D.*

Proof To prove this assertions we need to check that the inequality

$$\operatorname{tr}\{L(\lambda D_1 + (1-\lambda)D_2)^{-1}\} \leq \lambda \operatorname{tr}\{LD_1^{-1}\} + (1-\lambda)\operatorname{tr}\{LD_2^{-1}\}$$

holds for all matrices $D_1 > 0$ and $D_2 > 0$ and all $\lambda \in [0, 1]$, If D_1 and D_2 are positive definite and symmetric, then there exists a nonsingular matrix N such that both ND_1N^* and ND_2N^* are diagonal. Therefore,

$$\begin{aligned}
&\operatorname{tr}\{L(\lambda D_1 + (1-\lambda)D_2)^{-1}\} = \\
&\qquad = \operatorname{tr}\{NL(\lambda D_1 + (1-\lambda)D_2)^{-1}N^{-1}\} = \\
&\qquad = \operatorname{tr}\{NLN^*(N^*)^{-1}(\lambda D_1 + (1-\lambda)D_2)^{-1}N^{-1}\}.
\end{aligned}$$

Moreover,

$$\begin{aligned}
&(N^*)^{-1}\big(\lambda D_1 + (1-\lambda)D_2)^{-1}N^{-1}\big) = \\
&\qquad = \big(N(\lambda D_1 + (1-\lambda)D_2\big)N^*)^{-1} = \operatorname{diag}\left\{\frac{1}{\lambda(D_1)_{ii} + (1-\lambda)(D_2)_{ii}}\right\},
\end{aligned}$$

where

$$\frac{1}{\lambda(D_1)_{ii} + (1-\lambda)(D_2)_{ii}} \leq \frac{\lambda}{(D_1)_{ii}} + \frac{1-\lambda}{(D_2)_{ii}}$$

thanks to the convexity of the function $f(x) = 1/x$ for positive x. We conclude that

$$\operatorname{tr}\{L(\lambda D_1 + (1-\lambda)D_2)^{-1}\} = \sum_i \frac{(NLN^*)_{ii}}{\lambda(D_1)_{ii} + (1-\lambda)(D_2)_{ii}} \leq$$

$$\leq \sum_i (NLN^*)_{ii} \left\{\frac{\lambda}{(D_1)_{ii}} + \frac{1-\lambda}{(D_2)_{ii}}\right\} = \lambda \operatorname{tr}\{LD_1^{-1}\} + (1-\lambda)\operatorname{tr}\{LD_2^{-1}\},$$

which completes the proof. □

Lemma 7.4 *Suppose the non-decreasing function f satisfies the relations*

$$f(x+y) \leq f(x) + f(y) \quad \text{and} \quad f(\alpha x) \leq \alpha f(x)$$

for all $x, y > 0$ and $\alpha > 0$. Then the function $\widetilde{F}(D) = f(\operatorname{tr}\{LD^{-1}\})$ is convex on the set of symmetric positive definite matrices D. If in addition f is differentiable, then the derivative $F'_D(D)$ is a non-positive definite matrix on the set of positive definite matrices D.

Proof The first statement follows immediately from Lemma 7.3 thanks to the sequence of inequalities

$$f(\operatorname{tr}\{L(\lambda D_1 + (1-\lambda)D_2)^{-1}\}) \leq f\left[\lambda \operatorname{tr}\{LD_1^{-1}\} + (1-\lambda)\operatorname{tr}\{LD_2^{-1}\}\right] \leq$$

$$\leq f\left[\lambda \operatorname{tr}\{LD_1^{-1}\}\right] + f\left[(1-\lambda)\operatorname{tr}\{LD_1^{-1}\}\right] \leq$$

$$\leq \lambda f\left[\operatorname{tr}\{LD_1^{-1}\}\right] + (1-\lambda) f\left[\operatorname{tr}\{LD_1^{-1}\}\right].$$

The second statement follows from properties (7.28), because

$$\frac{d}{dD} f[\operatorname{tr}\{LD^{-1}\}] = f'_D\left[\operatorname{tr}\{LD^{-1}\}\right]\left[-(D^{-1})^* L^* (D^{-1})^*\right],$$

where $f'_x\left[\operatorname{tr}\{LD^{-1}\}\right] \geq 0$ and $-(D^{-1})^* L^* (D^{-1})^*$ is a non-positive definite matrix. □

Definition 7.3 The function $F(D)$ is *non-decreasing (resp., non-increasing)* on the set of positive definite symmetric matrices D if $D' - D \geq 0$ implies $F(D') \geq F(D)$ (resp., $F(D') \leq F(D)$).

From Lemma 7.4 it follows that functions of the form (7.31) are non-increasing. Indeed, the difference $F(D') - F(D)$ can be expressed as

$$F(D') - F(D) = \mathrm{tr}\{F'_D(\widetilde{D})(D' - D)\},$$

where $\widetilde{D} = D + \xi(D' - D)$ for some $\xi \in [0, 1]$. Moreover,

$$\mathrm{tr}\{F'_D(\widetilde{D})(D' - D)\} = \mathrm{tr}\{((D' - D)^{1/2})^* F'_D(\widetilde{D})(D' - D)^{1/2}\},$$

where $(D' - D)^{1/2}$ exists when $D' - D \geq 0$. If the matrix F'_D is non-positive definite, then the matrix under the tr symbol in the right-hand side of the last relation is positive definite, and so its trace is non-positive as well.

The next lemma describes properties of the matrix conjugate variable $P^0_D(\cdot)$, which satisfies the Eq. (7.29).

Lemma 7.5 *Assume that the functions $\widetilde{g}_k(z, D)$ are non-increasing in D and the matrix $P^0_D(T)$ is positive definite. Then the matrix $P^0_D(t)$ is positive definite for all $t \in [0, T]$.*

Proof The matrix-function $P^0_D(t)$ satisfies the Eq. (7.29), which can be transformed into a linear differential equations with measure of the form

$$dP^0_D(t) = \left[A^*_1(t)P^0_D(t) + P^0_D A_1(t)\right]dt + d\mathscr{M}(t), \tag{7.32}$$

where

$$A_1(t) = A(t) + D^0(t)B(t)B^*(t)$$

and the matrix measure $\mathscr{M}(dt)$ is defined by its distribution function,

$$\mathscr{M}([0, t]) = -\sum_{\tau \leq t,\, \tau \in D^{\mathrm{d}}_{\mathscr{V}}} \Delta P^0_D(\tau) + \int_t^T \sum_{k=1}^{N_3} \widetilde{g}'_{kD}\left(z^0(\tau), D^0(\tau), \tau\right) d\lambda_k(\tau). \tag{7.33}$$

□

Remark 7.5 Further, using Theorem 6.7, we obtain the following expression for the jumps of the function P^0_D:

$$\Delta P^0_D(\tau) = -\int_0^{\Delta V^0(\tau)} \sum_{k=1}^{N_3} \widetilde{g}'_{kD}\Big(z^0(\tau-) + \omega G(\tau, u(\tau)),$$
$$D^0(\tau-) + \omega\alpha^*(\tau, u(\tau))(\beta^*\beta)^{-1}(\tau, u(\tau))\alpha(\tau, u(\tau)), \tau\Big)\, d\nu^k_\tau(\omega). \tag{7.34}$$

Since the functions g_k are non-increasing in the sense of Definition 7.3, their derivatives with respect to D are non-positive definite matrices. Consequently, the integrals in (7.33) and (7.34) with respect to the non-negative measures λ_k and ν^k are also non-positive definite matrices. Moreover, the matrix measure $\mathscr{M}(dt)$ is non-positive definite. Therefore, for any Borel subset A of the segment $[0, T]$, the matrix $\mathscr{M}(A)$ is non-positive definite.

By Theorem A.23 in Appendix, the solution of Eq. (7.32) may be written in the form

$$P_D^0(t) = \Phi_1(t, 0) P_D^0(0) \Phi_1^*(t, 0) + \int_0^t \Phi_1(t, s) d\mathscr{M}(s) \Phi_1^*(t, s) \tag{7.35}$$

where $\Phi_1(t, s)$ is the matrix solution of the linear equation

$$\dot{\Phi}_1(t, s) = A_1(t)\Phi_1(t, s), \quad \Phi_1(s, s) = I_{(n\times n)}. \tag{7.36}$$

Using the representation (7.35), we obtain the formula

$$P_D^0(t) = \Phi_1(T, t) P_D^0(T) \Phi_1^*(T, t) - \int_t^T \Phi_1(t, s) d\mathscr{M}(s) \Phi_1^*(t, s).$$

Hence, due to the non-positive definiteness of the matrix $P_D^0(T)$ and of the matrix measure $\mathscr{M}(dt)$, the matrix $P_D^0(t)$ is non-negative definite for all $t \in [0, T]$.

Now we are in the position to formulate sufficient optimality conditions in the observation control problem in terms of the variable $D(t) = \gamma^{-1}(t)$.

Consider the optimal control problem for the system described by the Eq. (7.25) under the constraints

$$\begin{aligned} &\widetilde{h}(z(T), D(T)) = 0, \quad \widetilde{S}(z(T), D(T)) \leq 0, \\ &\widetilde{g}(z(t), D(t), t) \leq 0 \quad \text{for all} \quad t \in [0, T] \end{aligned} \tag{7.37}$$

and with the performance criterion

$$J'[u(\cdot), v(\cdot)] = \widetilde{\phi}_0(z(T), D(T)) \tag{7.38}$$

and the admissble controls $\{u(\cdot), v(\cdot)\}$ obeying the constraints (7.2) .

Theorem 7.5 *Suppose the generalized solution $\{D^0(\cdot), z^0(\cdot)\}$ corresponding to the controls $\{u^0(\cdot), V^0(\cdot)\}$ is admissible for the constraints (7.37) and satisfies the maximum principle for some collection Λ as in* (6.182) *and such that*

1. $\lambda_0 > 0$;
2. *the function $\mathscr{L}(z, D)$ is convex in (z, D) and non-decreasing in D;*
3. *the function $\mathscr{H}_0\big(P_D^0(t), p_0^z(t), D, z, t\big)$ is concave in z for all $t \in [0, T]$;*
4. *the functions $g_k(z, D, t)$ are convex in (z, D) and non-decreasing in D.*

Then $\{D^0(\cdot), z^0(\cdot), u^0(\cdot), V^0(\cdot)\}$ *is the optimal generalized solution of the problem (7.25), (7.37), and (7.38).*

Remark 7.6 We do not provide here the necessary optimality conditions in the form of the maximum principle for the observation control problems under consideration, since these conditions were formulated in the most general form in Theorem 6.7. In the sequel, in specific examples we will provide particular formulations of the maximum principle, that take into consideration the specifics of the problem at hand.

Proof To prove the theorem, it suffices to apply Theorem 6.8. Indeed by Lemma 7.5, the conjugate variable $P_D^0(t)$ is a positive definite matrix for all $t \in [0, T]$. Therefore, the component of the function $\mathscr{H}_0$ depending on the variable D from (7.26) is concave. In addition, the z-dependent part of $\mathscr{H}_0$ is concave by the assumptions of the theorem. Thus, $\mathscr{H}_0$ is concave with respect to the pair $\{z, D\}$ and, therefore, Theorem 6.8 is applicable. The proof is complete. □

Corollary 7.1 *Suppose that each of the functions* $\widetilde{\phi}_0$, $\widetilde{S}$, $\widetilde{h}$ *and* $\widetilde{g}$ *is of the form* $f(\mathrm{tr}\{L(t)D^{-1}\})$ *with a non-negative definite matrix* $L(t)$ *and a function* $f(\cdot)$ *satisfying the conditions of Lemma* 7.4. *Suppose the following conditions are satisfied:*

1. $\lambda_0 > 0$;
2. *the function* $\mathscr{L}(z, D)$ *is convex in* (z, D);
3. *the function* $\mathscr{H}_0(P_D^0(t), p_0^z(t), D, z, t)$ *is concave in* z *for all* $t \in [0, T]$.

Then the solution $\{D^0(\cdot), z^0(\cdot), u^0(\cdot), V^0(\cdot)\}$ *is optimal.*

In the next section we will show how to apply these results to typical observation control problems.

7.4 Examples of Observation Control Problems

There exist sufficiently many practical problems where observation control is applicable. Here are some examples:

- selection of the observation channel in multichannel observation systems [87, 151, 153, 249];
- choice of the type of sensing pulses in location systems [290];
- optimal coding of random signals (see [271] and many other literature items).

In all these cases, the efficiency of the observation system is highly dependent on both the actual moments of observation and the composition of observations. Accordingly, the optimization of the observation process is a highly relevant task. This section discusses some formulations of observation control problems. The class of problems that will be considered is limited to those where the methodology developed earlier in the book is the most effective. These are problems that admit generalized solutions and have a convex structure, for which, as it was shown above, the necessary optimality conditions in the form of a generalized maximum principle are also sufficient.

7.4.1 Stochastic Systems with Deterministic Dynamics

Consider the following partially observed systems described by equations of the type (7.1):

$$\begin{cases} dx(t) = A(t)x(t)dt, \quad x(t) \in \mathbb{R}^n, \ x(0-) = x_0, \\ dy(t) = \alpha(t, u(t))x(t)v(t)dt + \beta(t, u(t))v^{1/2}(t)dw_t, \\ \qquad y(t) \in \mathbb{R}^m, \ y(0) = 0, \\ dz(t) = f(t, z(t))dt + G(t, u(t))v(t)dt, \quad z(t) \in \mathbb{R}^k, \ z(0) = 0. \end{cases} \tag{7.39}$$

The observation controls $\{u(t), v(t)\}$ satisfy the constraints

$$u(t) \in U \subset \mathbb{R}^r, \qquad v(t) \geq 0, \qquad \int_0^T v(t)dt \leq M < \infty \tag{7.40}$$

where U is a compact subset of $\mathbb{R}^r$ and M is a given positive number.

The uncertainty in the evolution and estimation of $x(\cdot)$ is due to the random initial condition x_0, which is a Gaussian vector with the parameters

$$\mathbf{E}x_0 = m_0, \quad \operatorname{cov}(x_0, x_0^*) = \gamma_0,$$

as well as to the Wiener process $w_t \in \mathbb{R}^p$ figuring in the observation Eq. (7.39); moreover, x_0 and w_t are assumed to be independent.

The matrix-functions in the right-hand sides of the Eq. (7.39) satisfy the conditions stated at the beginning of the chapter for the system (7.1). The aim of the observation control is the minimization of the performance criterion

$$J[u(\cdot), v(\cdot)] = \phi_0(z(T), \gamma(T)), \tag{7.41}$$

subject to the constraints

$$\begin{aligned} &h(z(T), \gamma(T)) = 0, \quad S(z(T), \gamma(T)) \leq 0, \\ &g(z(t), \gamma(t), t) \leq 0, \quad \text{for all } t \in [0, T]. \end{aligned} \tag{7.42}$$

We will assume also that the conditions of the Theorem 7.4 are fulfilled, i.e., the set of admissible solutions of the problem (7.42) is nonempty and bounded.

Consider the set $\mathscr{L}(t, U, \gamma)$ of $n \times n$ matrices m and vectors $l \in \mathbb{R}^k$ of the form

$$\begin{pmatrix} m \\ l \end{pmatrix} = \begin{pmatrix} (1-\kappa)\gamma\alpha^*(t,u)(\beta^*\beta)^{-1}(t,u)\alpha(t,u)\gamma \\ \kappa f(t,z) + (1-\kappa)G(t,u), \end{pmatrix} \tag{7.43}$$

with $\kappa \in [0, 1]$ and $u \in U$ and assume that $\mathscr{L}(t, U, \gamma)$ is convex for any fixed $t \in [0, T]$, any non-negative definite symmetric matrix γ and any vector $z \in \mathbb{R}^k$. Under these assumptions, the problem has an optimal generalized solution satisfying a differential equation with measure of the form (7.12). After transition to the new variable $D(t) = \gamma^{-1}(t)$ (we assume that $\gamma_0 > 0$), we obtain a new problem for the system described by the equations

$$\begin{cases} dD(t) = -\big[A(t)D(t) + D(t)A^*(t)\big]dt + \\ \qquad + \alpha^*(t, u(t))(\beta^*\beta)^{-1}(t, u(t))\alpha(t, u(t))dV(t), \\ dz(t) = f(t, z(t))dt + G(t, u(t))dV(t) \end{cases} \tag{7.44}$$

with the constraints

$$\begin{aligned} &\widetilde{h}(z(T), D(T)) = 0, \quad \widetilde{S}(z(T), D(T)) \le 0, \\ &\widetilde{g}(z(t), D(t), t) \le 0 \quad \text{for all } t \in [0, T] \end{aligned} \tag{7.45}$$

and the performance criterion

$$J'[u(\cdot), V(\cdot)] = \widetilde{\phi}_0(z(T), D(T)) \to \min_{u,V}. \tag{7.46}$$

This problem belongs to the class of linear convex problems considered in Sect. 6.4.3. Hence, all the results obtained therein can be applied straightforwardly provided the functions in the constraints (7.45) and the performance criterion (7.46) satisfy the conditions of Theorem 7.5. In this case the maximum principle gives necessary and sufficient optimality conditions. Moreover, if there are no phase constraints (i.e., $g(t, z, \gamma) \equiv 0$) and $f(t, z) = H(t)z$, then the optimal solution may be obtained such that the number of impulses involved is no larger than $\frac{N(N+1)}{2} + k + 2$. This conclusion follows from Theorem 6.11, because the dimension of the conjugate system for the variables $\{z, D\}$ is $\frac{N(N+1)}{2} + k$, due to the symmetry of the matrix D.

From the point of view of possible applications, the above problem is typical for cases when the observation (measurement) program obeys some restrictions imposed on the composition of observations and/or on the total observation time by each measuring channel [152, 153, 265]. Some examples of explicit solutions to such problems can be found in [87, 152, 264].

7.4.2 Observation Control Under Time and Energy Constraints

This section discusses a class of problems with constraints imposed on the total observation time and/or on the "energy resources" used for observations. The problem is formulated as follows:

We consider a partially observable system described by equations of the form (7.1)

$$\begin{cases} dx(t) = A(t)x(t)dt + B(t)dw_t, \quad x(t) \in \mathbb{R}^n, \quad x(0-) = x_0, \\ dy(t) = \alpha(t, u(t))x(t)v(t)dt + \beta(t, u(t))v^{1/2}(t)dw_t^1, \\ \qquad y(t) \in \mathbb{R}^m, \quad y(0) = 0, \\ dz(t) = f(t, u(t))dt + G(t, u(t))v(t)dt, \\ z(t) \in \mathbb{R}^k, \quad z(0) = 0. \end{cases} \tag{7.47}$$

The observation control process $\{u(t), v(t)\}$ satisfies the constraints (7.40). The initial condition x_0 is a Gaussian random vector with the parameters

$$\mathbf{E}x_0 = m_0, \quad \mathrm{cov}(x_0, x_0^*) = \gamma_0.$$

The Wiener processes $w_t \in \mathbb{R}^l$ and $w_t^1 \in \mathbb{R}^p$ and the vector x_0 are assumed to be independent.

The matrix-functions in the right-hand side of the Eq. (7.47) satisfy the conditions formulated at the beginning of this chapter for the system (7.1). The observation control problem we are considering is to minimize the performance criterion (7.41) under the terminal and phase constraints (7.42).

The additional equation for $z(t)$ is introduced in order to account for the total observation time and the observation energy resource. The example below is typical for multichannel observation systems.

Example 7.1 ([152, 153]) Suppose that in a multichannel observation system at any moment of time the observation can be carried out in only one of M available channels. The vector observation processes $y^i(t)$ for channel i $(i = 1, \ldots, M)$ satisfies the equations

$$dy^i(t) = \alpha_i(t))x(t)v(t)dt + \beta_i(t))v^{1/2}(t)dw_t^i,$$

where w^i are Wiener processes independent on x_0 and w. In other words, at each moment of time the observation vector $y(t)$ in Eq. (7.47) is selected from the set $\{y^i(t),\ i = 1, \ldots, M\}$, so one can write

$$dy(t) = \begin{pmatrix} u_1(t)dy^1(t) \\ u_2(t)dy^2(t) \\ \vdots \\ u_M(t)dy^M(t) \end{pmatrix},$$

where

$$u_i(t) \in \{0, 1\}, \quad \sum_{i=1}^{M} u_i(t) = 1.$$

The corresponding equation for the matrix-function $D(t)$ has the form

$$\dot{D}(t) = -\big[A(t)D(t) + D(t)A^*(t) + D(t)B(t)B^*(t)D(t)\big] + \dot{\mathscr{M}}(t),$$

where

$$\dot{\mathscr{M}}(t) = \sum_{i=1}^{M} u_i^2(t)\alpha_i^*(t)(\beta_i^*\beta_i)^{-1}(t)\alpha_i(t)v(t).$$

Since the admissible control $u_i(t)$ satisfies the equality $u_i(t) = u_i^2(t)$, the corresponding set of matrices $\mathscr{L}$ (7.43), described here as

$$\mathscr{L}(t, U, \gamma) =$$

$$= \overline{\text{conv}}\left\{\gamma\left[\sum_{i=1}^{M} u_i^2\alpha_i^*(t)(\beta_i^*\beta_i)^{-1}(t)\alpha_i(t)\right]\gamma\ \ u_i \in \{0, 1\},\ \sum_i u_i = 1\right\} =$$

$$= \left\{\gamma\left[\sum_{i=1}^{M} u_i\alpha_i^*(t)(\beta_i^*\beta_i)^{-1}(t)\alpha_i(t)\right]\gamma\ :\ u_i \in \{0, 1\},\ \sum_i u_i = 1\right\}$$

is convex. The equation for the variable $z(t)$ has either the form

$$\dot{z}_i(t) = u_i(t),$$

if it describes the total observation time in the i-th channel, or

$$\dot{z}_i(t) = u_i(t)v(t),$$

if it describes the expenditure of "observation resources" in the i-th channel, $i = 1, \dots, M$. Indeed, in the first case, the quantity $z_i(T) = \int_0^T u_i(t)dt$ is equal to the total observation time in the i-th channel, while in the second case $z_i(T) = \int_0^T u_i(t)v(t)dt$, which is equal to total "energy resource of observations" spent on observations using the i-th channel. As for the constraints mentioned in the beginning of the subsection, in the present example it makes sense to consider constraints of the form (7.42). Note that the imposition of phase constraints on $\{z(t), D(t)\}$ allows one to take into account the current physical constraints on the observations accuracy and on the expenditure of "observation resources".

In the case when the performance criterion as well as the constraints satisfy convexity conditions (see the conditions in Theorem 7.5 and in Corollary 7.1), this example belongs to the class of convex problems, and therefore the maximum principle is a necessary and sufficient optimality condition.

Moreover, one can show that the solution of Eq. (7.25) enjoys the following monotonicity property.

Lemma 7.6 *Let $D(t)$ be a solution of Eq. (7.25), i.e.,*

$$dD(t) = -\big[A(t)D(t) + D(t)A^*(t) + D(t)B(t)B^*(t)D(t)\big]dt + d\mathscr{M}(t),$$

where

$$\mathscr{M}([0,t]) = \int_0^t \alpha^*(\tau, u(\tau))(\beta^*\beta)^{-1}(\tau, u(\tau))\alpha(\tau, u(\tau))dV(\tau). \tag{7.48}$$

Let $D'(t)$ be another solution of the same equation, with the same initial condition $D'(0-) = D(0-)$, but with another matrix measure $\mathscr{M}'(dt)$, such that the difference $\mathscr{M}'(dt) - \mathscr{M}(dt)$ is non-negative definite.

Then the difference $D'(t) - D(t)$ is non-negative definite. Moroever, if the difference $\mathscr{M}'([0,t]) - \mathscr{M}([0,t]) > 0$ for all $t \in [0,T]$, then $D'(t) - D(t) > 0$. (Recall that the notation $M > 0$ means that M is positive definite).

Proof Let $Z(t) = D'(t) - D(t)$. Then $Z(t)$ satisfies the equation

$$dZ(t) = -\big[A_1(t)Z(t) + Z(t)A_1^*(t) - Z(t)B(t)B^*(t)Z(t)\big]dt + d\big[\mathscr{M}' - \mathscr{M}\big](t),$$

where

$$A_1(t) = A(t) + D'(t)B(t)B^*(t),$$

and the initial condition is $Z(0) = 0$.

The solution of this equation is given by the formula (see Sect. A.2.5)

$$Z(t) = \int_0^t \Phi_1(t,s)d\widetilde{M}(s)\Phi_1^*(t,s)$$

with the matrix measure

$$\widetilde{\mathscr{M}}(dt) = Z(t)B(t)B^*(t)Z(t)dt + \left[\mathscr{M}' - \mathscr{M}\right](dt) \geq 0.$$

Further, the matrix $\Phi_1(t, s)$, being the fundamental solution of the Eq. (7.36), is non-singular. Therefore, $\mathscr{M}'([0, t]) - \mathscr{M}([0, t]) \geq 0$ implies $Z(t) \geq 0$ for all $t \in [0, t]$. Moreover, if $\mathscr{M}'([0, t]) - \mathscr{M}([0, t]) > 0$ for all $t \in [0, t]$, then $Z(t)$ is non-singular and hence positive definite. □

Corollary 7.2 *Assume the set of all admissible measures (7.48) is non-empty (needless to say, under the terminal and phase constraints (7.42)) and has a* maximal element, *that is, a measure $\mathscr{M}^0(dt)$ such that $\mathscr{M}^0(dt) - \mathscr{M}(dt) \geq 0$ for any admissible measure $\mathscr{M}(dt)$. If, in addition, the function $\phi_0(z, \gamma) \equiv \phi_0(\gamma)$ is non-decreasing (see Definition* 7.3*), then the controls $\{u_0(\cdot), V_0(\cdot)\}$ corresponding to the measure $\mathscr{M}^0(dt)$ generate the optimal generalized solution of the observation control problem.*

The existence of the maximal element is usually easily established for a given location of the observations, that is for a given measure $V(dt) \geq 0$, as illustrated by the following example.

Example 7.2 Consider the observation control problem where a scalar unobservable process $x(t)$ is described by the equation

$$dx(t) = a(t)x(t)dt + b(t)dw_t, \quad x(0) = x_0,$$

and the observation process $y(t) \in \mathbb{R}^2$ is governed by the equation

$$dy(t) = \begin{pmatrix} dy^1(t) \\ dy^2(t) \end{pmatrix} = \begin{pmatrix} \alpha_1(t)u_1(t) \\ \alpha_2(t)u_2(t) \end{pmatrix} x(t)dV(t) + \begin{pmatrix} d\xi_t^1 \\ d\xi_t^2 \end{pmatrix}.$$

Here x_0 is Gaussian random variable, w is a scalar Wiener process, and ξ^1 and ξ^2 are Gaussian martingales with quadratic characteristics (see [244]):

$$\langle \xi^1 \rangle_t = \int_0^t u_1(s)dV(s), \qquad \langle \xi^2 \rangle_t = \int_0^t u_2(s)dV(s).$$

It is assumed that x_0, w, ξ^1 and ξ^2 are mutually independent.

Remark 7.7 The presence of discrete observations requires an extension of the class of noise processes affecting the observations up to class of Gaussian martingales, in comparison with Wiener processes in (7.47), because in this case the measure $V(dt)$ may have a discrete component corresponding to the discrete observations. For details concerning the Kalman filter in the case of discrete-continuous observations see [291] and Remark 7.3. In the present case only the matrix $D(t) = \gamma^{-1}(t)$ matters.

It is assumed that the observation controls $\{u_1, u_2\}$ satisfy the constraints

$$u_1(t) \in \{0, 1\}, \quad u_2(t) \in \{0, 1\}, \quad u_1(t) + u_2(t) = 1$$

and the measure $V(dt)$ is given fixed.

Using the expression for the generalized solution (see Sect. 7.2), and the representation obtained in Example 7.1, we derive the following equations for the variables $\gamma(t)$ and $D(t) = \gamma^{-1}(t)$, respectively:

$$d\gamma(t) = \big[2a(t)\gamma(t) + b^2(t)\big]dt - \gamma^2(t)\big(\alpha_1^2(t)u_1(t) + \alpha_2^2 u_2(t)\big)dV(t),$$

and

$$dD(t) = -\big[2a(t)D(t) + b^2(t)D^2(t)\big]dt + \big(\alpha_1^2(t)u_1(t) + \alpha_2^2(t)u_2(t)\big)dV(t). \tag{7.49}$$

The set of admissible controls leading to the convexification of the right-hand sides of these equations is subject to the constraints

$$u_1(t) \in [0, 1], \quad u_2(t) \in [0, 1], \quad u_1(t) + u_2(t) = 1.$$

Suppose the task in the original problem is to minimize the terminal cost function

$$\gamma(T) \to \min_{\{u_1(\cdot), u_2(\cdot)\}} .$$

Then in the auxiliary problem, in terms of the variable $D(t)$, one needs to minimize the criterion

$$-D(T) \to \min_{\{u_1(\cdot), u_2(\cdot)\}} .$$

A necessary optimality condition for the auxiliary problem reads: if $\{u_1^0(t), u_2^0(t)\}$ is the optimal controls, then for any admissible control $\{u_1(t), u_2(t)\}$ it holds that

$$\begin{aligned} &\int_0^T p_{\mathrm{d}}^0(t)\big[\alpha_1^2(t)u_1^0(t) + \alpha_2^2(t)u_2^0(t)\big]dV(t) \geq \\ &\geq \int_0^T p_{\mathrm{d}}^0(t)\big[\alpha_1^2(t)u_1(t) + \alpha_2^2(t)u_2(t)\big]dV(t). \end{aligned} \tag{7.50}$$

The conjugate variable $p_{\mathrm{d}}^0(t)$ satisfies the linear equation

$$\dot{p}_{\mathrm{d}}^0(t) = 2\big[a(t) + b^2(t)D^0(t)\big]p_{\mathrm{d}}^0(t), \qquad p_{\mathrm{d}}^0(T) = 1, \tag{7.51}$$

and so $p_{\mathrm{d}}^0(t) > 0$ for all $t \in [0, T]$. Therefore, the optimal control is given by

$$\{u_1^0(t), u_2^0(t)\} = \operatorname*{argmax}_{u_1,u_2}\big(\alpha_1^2(t)u_1 + \alpha_2^2(t)u_2\big), \tag{7.52}$$

and

$$u_1^0(t) = \begin{cases} 1, & \text{if } \alpha_1^2(t) \geq \alpha_2^2(t), \\ 0, & \text{otherwise}, \end{cases} \qquad u_2^0(t) = 1 - u_1^0(t). \tag{7.53}$$

This result is quite obvious, since in order to have highest-quality observations one needs to use the channel with maximal signal-to-noise ratio, which is equal to $\alpha_1^2(t)$ and $\alpha_2^2(t)$ for the first and the second channel, respectively. Moreover, the measure

$$\mathscr{M}^0(dt) = \big[\alpha_1^2(t)u_1^0(t) + \alpha_2^2(t)u_2^0(t)\big]dV(t)$$

is indeed the maximal element of the set of admissible measures, defined by the formula

$$\mathscr{M}(dt) = \big[\alpha_1^2(t)u_1(t) + \alpha_2^2(t)u_2(t)\big]dV(t).$$

Thus, in the case when the measure $V(dt)$ is given, the optimal observation control can be obtained by the formula (7.53).

However, the solution becomes considerably more complicated when the measure $V(dt)$ is not fixed a priori, but must be determined by means an optimization procedure. In this case it is necessary to solve a two-point boundary value problem for the pair of variables $\{D(t), p_d(t)\}$, which satisfy the Eqs. (7.49) and (7.51) with the measure $V(dt)$, ensuring the satisfaction of the maximum condition (7.50). Therefore, upon substituting the expression (7.53) of the optimal control, we obtain the following equation for optimal variable $D^0(t)$:

$$dD^0(t) = -\big[2a(t)D^0(t) + b^2(t)(D^0)^2(t)\big]dt + \alpha^0(t)dV(t), \tag{7.54}$$

where

$$\alpha^0(t) = \max_{\{u_1,u_2\}}\big[(\alpha_1^2(t)u_1(t) + \alpha_2^2(t)u_2(t)\big] = \max\big\{\alpha_1^2(t), \alpha_2^2(t)\big\}.$$

As it follows from the maximum principle, the optimal measure $V(dt)$ is concentrated on the set where the function $p_d(t)\alpha^0(t)$ achieves its maximum. But the evolution of the variable p_d depends on this measure. So we are coming to a rather complicated problem which may be solved, as a rule, only numerically. Nevertheless, sometimes the solution can be obtained analytically, as the next example shows.

Example 7.3 Consider the preceding observation control problem with a scalar unobservable process $x(t)$ described by the equation

$$dx(t) = a(t)x(t)dt + bdw_t, \quad x(0) = x_0.$$

The observation process $y(t) \in \mathbb{R}^1$ is governed by the equation

$$dy(t) = x(t)dV(t) + d\xi_t.$$

Here x_0 is a Gaussian random variable, w is a scalar Wiener process, and ξ is a Gaussian martingale with quadratic characteristic (see [244]) $\langle \xi \rangle_t = V(t)$. The problem asks to minimize the performance criterion

$$\gamma(T) \to \min_{\{V(\cdot)\}}$$

under the constraint $V(T) \leq M < \infty$.

As above, we reformulate the problem in terms of the variable $D(t) = \gamma^{-1}(t)$, with performance criterion $-D(T) \to \min$. It is readily verified that the optimal control satisfies $V(T) = M$, hence the maximum principle takes on the form

$$\int_0^T p_{\mathrm{d}}^0(t)dV^0(t) = \max_{\{V(\cdot)\}}. \tag{7.55}$$

Here the measure $V^0(dt)$ is concentrated on the set where the adjoint variable $p_{\mathrm{d}}^0(t)$ achieves its maximum. The function $p_{\mathrm{d}}^0(t)$ satisfies the equation

$$\dot{p}_{\mathrm{d}}^0(t) = 2\big[a(t) + D^0(t)b^2\big]p_d^0(t), \tag{7.56}$$

with the terminal constraint $p_{\mathrm{d}}^0(T) = 1$, and thanks to the linearity of this equation it remains positive for all $t \in [0, T]$.

Assume that the function $a(t)$ is differentiable. Then the function $p_{\mathrm{d}}^0(t)$ is also differentiable on $[0, T]$ and achieves its maximum either at points where $\dot{p}_{\mathrm{d}}^0(t) = 0$, or at the endpoints $t = 0$, $t = T$.

First, let us show that the discrete components of the measure $V^0(dt)$ cannot be concentrated on the set where $\dot{p}_{\mathrm{d}}^0(t) = 0$. Indeed, assume, by contradiction, that $\Delta V^0(\tau) > 0$ and $\dot{p}_{\mathrm{d}}^0(\tau) = 0$. Then in view of (7.56), $a(\tau) + D^0(\tau)b^2 = 0$ and

$$\Delta\dot{p}_{\mathrm{d}}^0(\tau) = 2p_{\mathrm{d}}^0(\tau)b^2\Delta D^0(\tau) = 2p_{\mathrm{d}}^0(\tau)b^2\Delta V^0(\tau) > 0.$$

However, at the maximum points of the differentiable function $p_{\mathrm{d}}^0(t)$ we must have $\Delta\dot{p}_{\mathrm{d}}^0(\tau) \leq 0$, so we reached a contradiction.

On the other hand, the continuous component of the optimal measure $V^0(dt)$ may be concentrated only on the set

$$\mathcal{T}_1 = \{t : \dot{p}_{\rm d}^0(t) = 0\} = \{t : a(t) + D^0(t)b^2 = 0\}.$$

Indeed, as it follows from the equations

$$\begin{aligned} dD^0(t) &= -\big[2a(t)D^0(t) + b^2(D^0)^2(t)\big]dt + dV^0(t), \\ a(t) &+ D^0(t)b^2 = 0, \end{aligned} \tag{7.57}$$

the measure $V^0(dt)$ must be of the form

$$V^0(dt) = -\frac{\dot{a}(t) + a^2(t)}{b^2}\,dt. \tag{7.58}$$

Thus, the continuous part of the measure $V^0(dt)$ can be positive only on the set

$$\mathcal{T}_2 = \{t : \dot{a}(t) + a^2(t) < 0\}, \tag{7.59}$$

and consequently the continuous part of the measure $V^0(dt)$ may be concentrated only on the intersection of sets $\mathcal{T}_1$ and $\mathcal{T}_2$. Further, let $[\tau_1, \tau_2]$ be a segment of concentration of the continuous part of the measure $V^0(dt)$. Since in this segment the function $p_{\rm d}^0$ achieves its maximum value and keeps this value on the whole segment, at the endpoints τ_1 and τ_2 we necessarily have

$$\frac{d\dot{p}_{\rm d}^0(\tau_1-)}{dt} \geq 0 \quad \text{and} \quad \frac{d\dot{p}_{\rm d}^0(\tau_2+)}{dt} \geq 0. \tag{7.60}$$

Moreover, Eq. (7.56) imply that

$$d\dot{p}_{\rm d}^0(t) = p_d^0(t)\big[\{\dot{a}(t) + a^2(t)\}dt + b^2 dV^0(t)\big],$$

and hence

$$\frac{d\dot{p}_{\rm d}^0(\tau_1-)}{dt} = \dot{a}(\tau_1) + a^2(\tau_1) \quad \text{and} \quad \frac{d\dot{p}_d^0(\tau_2+)}{dt} = \dot{a}(\tau_2) + a^2(\tau_2).$$

This means that if $\dot{a}(t)$ is continuous, then the continuous components of the measure $V^0(t)$, starting from the point τ_1, remains continuous until the point τ_2, where condition (7.60) will hold again.

Thus, the optimal control has the following form: the discrete component of the measure may be concentrated at the points $t = 0$ or $t = T$, while the continuous component, defined by relation (7.58), is concentrated on intervals from the set $\mathcal{T}_2$ defined by (7.59).

The example just considered illustrates a general property of the observation control problem. Namely, the discrete components of the optimal control can be concentrated only at the endpoints of the total observation interval and the values these components can be determined by solving appropriate mathematical programming problems. The continuous component is defined directly in the feedback control form and is concentrated on finite subintervals of the total observation interval. Moreover, once a solution to the two-point boundary value problem is found, this solution turns will obviously be optimal thanks to the convexity of the optimization problem.

It should be noted that the absence of discrete components inside the entire observation interval is a characteristic property of a whole class of observation control problems with continuously incoming data.

Example 7.4 Let us consider the observation control problem for the system (7.47) with a continuously differentiable matrix-valued function $A(t)$ and $B(t) \equiv B$ in the observation equation, and with constant matrices $\alpha(t, u) \equiv \alpha$ and $\beta(t, u) \equiv \beta$ such that $\beta\beta^* > 0$ in the observation equation. In other words, we consider the problem of "optimal allocation of observation resources". This means that we are required to find a measure $V(dt) \geq 0$ that satisfies the constraint $V([0, T]) \leq M$ and minimizes the cost function

$$\phi_0(\gamma(T)) = \mathrm{tr}[G_0\gamma(T)] \to \min_{V(\cdot)},$$

with some positive definite matrix G_0.

Reformulating the problem in terms of the function $D(t) = \gamma^{-1}(t)$, we find that the optimal control $V^0(dt)$ satisfies the maximum principle

$$\int_0^T \mathrm{tr}\big[P_D^0(t)\Lambda\big]dV^0(t) \geq \int_0^T \mathrm{tr}\big[P_D^0(t)\Lambda\big]dV(t), \tag{7.61}$$

where

$$\Lambda = \alpha^*(\beta\beta^*)^{-1}\alpha$$

for all non-negative scalar measures $V(dt)$ such that $V([0, T]) \leq M$. In (7.61) the matrix-function $P_D^0(t)$ is the solution of equation

$$\dot{P}_D^0(t) = \big[A(t) + BB^*D^0(t)\big]P_D^0(t) + P_D^0(t)\big[A(t) + BB^*D^0(t)\big]^* \tag{7.62}$$

with terminal constraint $P_D^0(T) = (D^0)^{-1}G_0(D^0)^{-1}$. As we saw above, the maximum condition (7.61) means that the measure $V^0(dt)$ is concentrated on the set, where the function $\mathrm{tr}\big[P_D^0(t)\Lambda\big]$ achieves the maximum in the variable t. Therefore, on the concentration set of the measure $V^0(dt)$ inside the interval $(0, T)$ we have that

$$\frac{d\mathrm{tr}\big[P_D^0(t)\Lambda\big]}{dt} = \mathrm{tr}\big[\dot{P}_D^0(t)\Lambda\big] = 0.$$

Since the matrix-function $P_D^0(t)$ satisfies the Eq. (7.62), we have the equality

$$\mathrm{tr}\big\{\big[A(t) + BB^*D^0(t)\big]P_D^0(t)\Lambda\big\} = 0,$$

which determines the concentration set of the measure $V^0(dt)$.

However, the discrete components of the measure $V^0(dt)$ cannot be concentrated on this set. Indeed, if $\Delta V^0(\tau) > 0$, then

$$\Delta\,\mathrm{tr}\big[\dot{P}_D^0(t)\Lambda\big] = 2\mathrm{tr}\big[BB^*\Lambda P_D^0(\tau)\Lambda\big] \geq 0,$$

which contradicts the maximum condition for the function $\mathrm{tr}\big[\dot{P}_D^0(t)\Lambda\big]\Delta V^0(\tau)$ at the point τ.

Thus, the discrete components of the measure $V^0(dt)$ may be concentrated only at the endpoints of the segment $[0, T]$.

7.4.3 Observation Control with Phase Constraints

In this subsection we consider a somewhat different formulation of the observation control problem, namely, a formulation in which the energy constraints appear in implicit form. For the sake of simplicity, we consider a problem with an additional equation for a variable $z(t)$:

$$\dot{z}(t) = \phi(t, z(t)) + v(t), \qquad z(0) = 0, \tag{7.63}$$

subject to the phase constraint

$$z(t) \leq z_0. \tag{7.64}$$

We assume that $\phi(t, z) < 0$ for all $t \in [0, T]$ and all $z > 0$, that $\phi(t, 0) = 0$, and that ϕ is convex in z. Note that the constraint (7.64) is equivalent to a constraint imposed on the value of the impulse at a given point. Moreover, this constraint depends on the "prehistory" of the control. After transition to a generalized control, Eq. (7.63) becomes

$$dz(t) = \phi(t, z(t))dt + dV(t), \qquad z(0) = 0. \tag{7.65}$$

Assume an impulse control is used. Then at the moment τ_1 of the first impulse action the magnitude of the impulse w_1 obeys the constraint

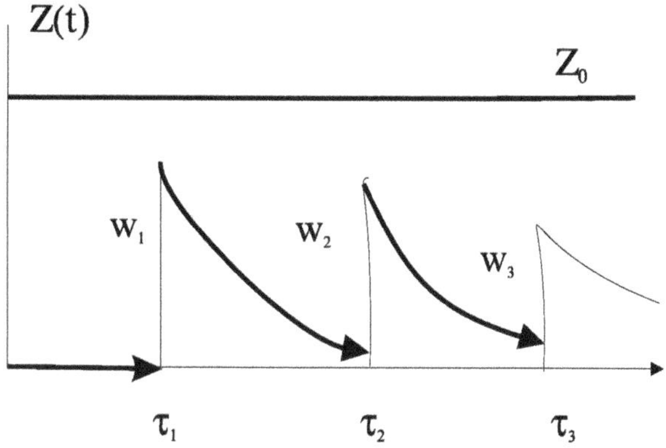

Fig. 7.1 Evolution of the variable $z(t)$

$$z(\tau_1) = z(\tau_1-) + w_1 \le z_0, \tag{7.66}$$

or, equivalently, $w_1 \le z_0 - z(\tau_1-)$. Analogously, for the subsequent impulses w_i at the points τ_i we have

$$w_i \le z_0 - z(\tau_i-).$$

Between impulses the variable $z(t)$ decreases, because

$$\dot{z}(t) = \phi(t, z(t)) < 0;$$

the corresponding admissible magnitude of the impulse increases. A typical evolution of the variable $z(t)$ is displayed in Fig. 7.1.

Thus, this model allows us to describe a situation where resource recovery for further impulse control is possible, because the value of the admissible impulse action at an arbitrary moment of time $t \in [0, T]$ is equal to the difference $z_0 - z(t)$, which increases in the intervals between impulses. The presence of phase constraints leads to a modification of the maximum principle. Indeed, the additional variable $z(t)$ satisfying Eq. (7.65) generates an additional conjugate variable $p_z^0(t)$ that obeys the equation

$$dp_z^0(t) = -\phi_z'(t, z^0(t))p_z^0(t)dt + d\lambda(t), \quad p_z^0(T) = 0, \tag{7.67}$$

where $\lambda(dt)$ is a non-negative regular measure concentrated on the set $\{t : z^0(t) = z_0\}$. The maximum principle for the standard observation control problem gives a necessary and sufficient optimality condition provided that, in accordance to Theorem 7.5, the part of function $\mathscr{H}_0$ depending on z is concave. To establishing the concavity property, we note that the function $p^0(z)(t)$ is non-positive due to the non-negativity of $\lambda(dt)$ and the null terminal condition. Hence, the part of $\mathscr{H}_0$ depending on z is equal to $p_z^0(t)\phi(t, z)$, which is obviously concave. Thus, if we are able to find the trajectory satisfying the maximum principle, then this trajectory is the optimal one.

7.4.4 Optimal Switching of Observation Channels

In this subsection, we give an example concerned with the optimal switching of observation channels [279]. Specifically, we consider the problem of optimizing the structure of observations of a stochastic system described by the equation

$$dx(t) = Ax(t)dt + Cdw_t,$$

where $x \in \mathbb{R}^3$, $w_t \in \mathbb{R}^1$ is a standard Wiener process, and

$$A = \begin{pmatrix} 0 & 1 & 0 \\ 0 & 0 & 1 \\ 0 & 0 & -a \end{pmatrix}, \quad C = \begin{pmatrix} 0 \\ 0 \\ (2a)^{1/2}\sigma_s \end{pmatrix},$$

with $a > 0$ and $\sigma_s > 0$.

This model describes the motion of a dynamical object with random acceleration. The object can be observed via one of two channels. The first (resp., second) channel is described by an equation of the form

$$dy^1(t) = \frac{x^1(t)}{\sigma_x^1} v(t)dt + d\zeta_t^1,$$

(resp.,

$$dy^2(t) = \frac{x^1(t)}{\sigma_x^2 f(\gamma(t))} v(t)dt + d\zeta_t^2.,)$$

In these equations the Gaussian martingales ζ_t^1 and ζ_t^2, the Wiener process w_t, and the Gaussian initial condition $x(0)$ are mutually independent. It is assumed that

$$\langle \zeta^1 \rangle_t = \langle \zeta^2 \rangle_t = \int_0^t v(s)ds.$$

The function $f(\gamma)$ depends on the moments of a random process $\widehat{x}(t) - x(t)$ and satisfies the condition

$$f(\gamma) = f_0(\gamma_{11}),$$

where $f_0(x) > 0$ and $f_0'(x) > 0$ for all $x \geq 0$.

This type of dependence is typical for correlation-extremal tracking, in which the accuracy of the measurement of the position depends on the accuracy of the matching between the current image to the reference one [267].

We consider the case $\sigma_x^1 \gg \sigma_x^2$ and assume that using only the first channel provides a stable estimation, that is, $\gamma(t) \to \gamma^1$ for $t \to \infty$ for any initial condition

$\gamma(0) > 0$. When only the second channel is used, the quality of the estimation depends on the initial condition $\gamma(0)$. If $\gamma(0)$ is large enough, then due to the large a priory uncertainty in $x^1(0)$ (large value of $\gamma_{11}(0)$) it may happen that $\gamma(t) \to \infty$ as $t \to \infty$. However, for small values of $\gamma(0)$ we have $\gamma(t) \to \gamma^2$, where $\gamma^1 \gg \gamma^2$ due to the condition $\sigma_x^1 \gg \sigma_x^2$.

The task at hand is to determine the optimal switching of the observation channels so as to minimize the performance criterion

$$J = \operatorname{tr}\{D\gamma(T)\},$$

where D is a positive definite matrix.

We consider the case of discrete observations at times $\{t_i,\ i = 1, \ldots, M\}$. The observation process will have a form

$$dy(t) = \alpha(u^1(t), u^2(t), \gamma(t-))x(t)dV(t) + d\zeta_t,$$

where

$$\alpha(u^1, u^2, \gamma) = \begin{pmatrix} \dfrac{u^1}{\sigma_x^1} & 0\ 0 \\ \dfrac{u^2}{\sigma_x^2 f(\gamma_{11})} & 0\ 0 \end{pmatrix},$$

$$y(t) = \begin{pmatrix} y^1(t) \\ y^2(t) \end{pmatrix}, \qquad \zeta(t) = \begin{pmatrix} \zeta^1(t) \\ \zeta^2(t) \end{pmatrix},$$

$$V(t) = \sum_{t_i \le t} I(t - t_i),$$

and the controls u^1, u^2 satisfy the constraints

$$u^1 \ge 0, \quad u^2 \ge 0, \quad u^1 + u^2 \le 1. \tag{7.68}$$

Let us introduce the new variable $g(t) = \gamma^{-1}(t)$. The governing equation for $g(t)$ reads

$$\begin{aligned} dg(t) = &-\big[g(t)A + A^* g(t) + g(t)CC^* g(t)\big]dt + \\ &+ \widetilde{\alpha}^*(u^1(t), u^2(t), g(t-))\widetilde{\alpha}(u^1(t), u^2(t), g(t-))dV(t), \end{aligned} \tag{7.69}$$

where

$$\widetilde{\alpha}(u^1, u^2, g) = \alpha(u^1, u^2, g^{-1}) \tag{7.70}$$

and

$$g(0) = \gamma^{-1}(0). \tag{7.71}$$

In terms of $g(t)$, the task is to minimize the criterion

$$J = \text{tr}\{Dg^{-1}(T)\} \tag{7.72}$$

on the set of controls obeying the constraints (7.68). Now let us consider the associated auxiliary problem. Following the procedure of discontinuous time change, we introduce the variables

$$\eta(s) = \inf\{t : t + V(t) > s\}, \tag{7.73}$$

and

$$\kappa(s) = \dot{\eta}(s) = \begin{cases} 1, & \text{if } s \in [t_i + V(t_i),\, t_{i+1} + V(t_i)], \\ 0, & \text{otherwise.} \end{cases} \tag{7.74}$$

On the segment $[0, T + M]$, where $M = V(T)$, the auxiliary controlled system is governed by the equations

$$\dot{g}_1(s) = -\kappa(s)\big[g_1(s)A + A^* g_1(s) + g_1(s)CC^* g_1(s)\big] +$$
$$+ \sum_{i=1}^{N} (1 - \kappa(s))\widetilde{\alpha}^*(u_1^1(s), u_1^2(s), g_1^i(s))\alpha(u_1^1(s), u_1^2(s), g_1^i(s))\mathbf{I}\{s \in [s_i, s_{i+1}]\}, \tag{7.75}$$

where

$$\dot{g}_1^i(s) = \mathbf{I}\{s \le s_i\}\, \dot{g}_1(s), \quad g_1^i(0) = g_1(0), \tag{7.76}$$

and

$$s_i = t_{i+1} + V(t_i) = t_{i+1} + i. \tag{7.77}$$

The controls $\{u_1^1(s), u_1^2(s)\}$ satisfy the constraints (7.68), and the performance criterion has the form

$$J = \text{tr}\{Dg_1^{-1}(T + M)\} \to \min, \tag{7.78}$$

By Lemma 7.1, for any controls $\{u_1^1(s), u_1^2(s)\}$ satisfying the constraints (7.68), the application of the controls

$$u^1(t_i) = \int_{s_i}^{s_i+1} u_1^1(s)ds \quad \text{and} \quad u^2(t_i) = \int_{s_i}^{s_i+1} u_1^2(s)ds$$

in (7.69) leads to the following relations between $g(t)$ from (7.69) and $g_1(s)$, $g_1^i(s)$:

$$g(t) = g_1(\Gamma(t)), \quad g(t_i-) = g_1^i(s), \quad \text{for } s > s_i. \tag{7.79}$$

Here $\Gamma(t)$ and $\eta(s)$ are related by the usual formulas of discontinuous time change.

To solve the auxiliary problem one can resort to the known methods. In particular, using the standard maximum principle in the problem (7.73)–(7.78) one finds that on the optimal controls $\{u_1^1(s), u_1^2(s)\}$ the function

$$\begin{aligned} &\mathscr{H}_1(u^1, u^2, s) = \\ &= \mathrm{tr}\Big\{\Big(\psi_g(s) + \sum_{k=1}^{M} \psi_g^k(s)\mathbf{I}\{s : s \le s_k\}\Big)\widetilde{\alpha}^*(u^1, u^2, g_1^i(s))\widetilde{\alpha}(u^1, u^2, g_1^i(s))\Big\} \end{aligned} \tag{7.80}$$

achieves its maximum with respect to the variables $\{u^1, u^2\}$ subject to the constraints (7.68) at the point $\{u_1^1(s), u_1^2(s)\}$ on the segment $[s_i, s_i + 1]$.

In formula (7.80) the matrix-functions $\psi_g(s)$ and $\psi_g^k(s)$ play the role of conjugate variables for $g_1(s)$ and $g_1^k(s)$, respectively, and satisfy the corresponding system of differential equations with terminal conditions at the point $T_1 = T + M$.

By ordinary differentiation, with the properties of $f_0(x)$ and the form of the performance criterion accounted for, we deduce that the matrix-function

$$\widetilde{\psi}_g(s) = \psi_g(s) + \sum_{k=1}^{M} \psi_g^k(s)\mathbf{I}\{s : s \le s_k\}$$

remains non-negative definite on the segment $[0, T_1]$, and that on each segments $[s_i, s_i + 1]$ we have

$$\mathscr{H}_1(u^1, u^2, s) = \widetilde{\psi}_{11}(s)\left[\left(\frac{u^1}{\sigma_x^1}\right)^2 + \left(\frac{u^2}{\sigma_x^2 f(\mathrm{tr}\{d\, g_1^i(s)\})}\right)^2\right], \tag{7.81}$$

where

$$d = \begin{pmatrix} 1 & 0 & 0 \\ 0 & 0 & 0 \\ 0 & 0 & 0 \end{pmatrix},$$

$\widetilde{\psi}_{11}(s) \ge 0$, and the function $g_1^i(s)$ is constant on the segment $[s_i, s_i + 1]$.

The fact that the optimal control provides the maximum of (7.81) implies that the control law

$$u_1^1(s) = \begin{cases} 1, & \text{if } \dfrac{1}{\sigma_x^1} \geq \dfrac{1}{\sigma_x^2 f(\mathrm{tr}\{dg_1^i(s)\})}, \\ 0, & \text{otherwise} \end{cases} \qquad \text{and} \quad u_1^2(s) = 1 - u_1^1(s),$$

is optimal on $[s_i, s_i + 1]$.

Hence, the optimal control in the original problem has the form

$$u^1(t_i) = \begin{cases} 1, & \text{if } \dfrac{1}{\sigma_x^1} \geq \dfrac{1}{\sigma_x^2 f(\gamma_{11}(t_i-))}, \\ 0, & \text{otherwise} \end{cases} \qquad \text{and} \quad u^2(t_i) = 1 - u^1(t_i).$$

This follows directly from (7.79) by performing the inverse time change. The resulting control law not only provides the optimal switching of the observation channels, it also guarantees the stability of the estimation procedure. Similar problems arise in observation control for systems where the accuracy of estimation depends on the unobservable coordinates of the object as well as on the current estimation error [305].

7.5 Observation Control with a Finite Number of Observation Times

In this section, we consider another example of observation control, the solution of which is found by the method described in Chap. 3 for a special class of impulse control problems, namely, for problems with a finitely number of impulse actions. Note that, as in previous examples, the approach to the solution of this example is to use the method of discontinuous time change.

Let us assume that the unobservable n-dimensional vector-valued process $X = (X(t))_{0 \leq t \leq T}$ satisfies the Itô stochastic differential equation

$$dX(t) = a(t)X(t)\,dt + c(t)\,dt + b(t)\,dW(t), \qquad X(0) = X_0, \tag{7.82}$$

where $W = (W(t))_{0 \leq t \leq T}$ is an m-dimensional Wiener process with independent components, and $a(t)$, $c(t)$ and $b(t)$ are $n \times n$, $n \times 1$ and $n \times m$ matrix-valued functions with piecewise smooth components.

We further assume that there are r different k-dimensional observation channels, exactly one of which at a time may be used to observe the process X at given discrete time instants τ_i, $i = 1, \ldots, N$ $(0 < \tau_1 < \cdots < \tau_N < T)$. The collection of admissible observations times must satisfy the constraints

$$\tau_{i-1} + T_i \leq \tau_i \leq T - \sum_{k=i+1}^{N+1} T_k, \quad i = 1, \dots, N, \tag{7.83}$$

which specify the minimal time intervals between successive observations. Here $\tau_0 = 0$ and T_i are given non-negative constants.

The observation process via the j-th channel is described by a sequence of k-dimensional vectors $\xi_j = (\xi_j(\tau_i))_{i=1,\dots,N}$, $j = 1, \dots, r$, defined by the equation

$$\xi_j(\tau_i) = A_j(\tau_i)X(\tau_i) + B_j(\tau_i)\epsilon_j(\tau_i), \tag{7.84}$$

where $\epsilon_j = (\epsilon_j(\tau_i))_{i=1,\dots,N}$ is a sequence of d-dimensional Gaussian vectors with independent components, with zero mean and unit covariance matrix; $A_j(\tau_i)$ and $B_j(\tau_i)$ are $k \times n$ and $k \times d$ matrices, such that $B_j(\tau_i)B_j^*(\tau_i)$ is non-singular for any τ_i, $i = 1, \dots, N$. The processes W and ϵ_j, $j = 1, \dots, r$ and the Gaussian initial condition X_0 are assumed to be independent.

To solve the problem we introduce the (time-dependent) rk-dimensional vector ξ, rd-dimensional vector ϵ, and the $(rk \times n)$- and $(rk \times rd)$-dimensional block matrices A and B as follows:

$$\xi(t) = \begin{pmatrix} \xi_1(t) \\ \xi_2(t) \\ \vdots \\ \xi_r(t) \end{pmatrix}, \quad \epsilon(t) = \begin{pmatrix} \epsilon_1(t) \\ \epsilon_2(t) \\ \vdots \\ \epsilon_r(t) \end{pmatrix},$$
$$A(\alpha(t)) = \begin{pmatrix} \alpha_1(t)A_1(t) \\ \alpha_2(t)A_2(t) \\ \vdots \\ \alpha_r(t)A_r(t) \end{pmatrix}, \quad B(t) = \operatorname{diag}\big(B_1(t)\ B_2(t)\ \dots\ B_r(t)\big), \tag{7.85}$$

where $\alpha(t) = (\alpha_1(t), \alpha_2(t), \dots, \alpha_r(t))^*$, and the functions $\alpha_j(t)$, $j = 1, \dots, r$, with values in $\{0, 1\}$, control the switching of the observation channel (see the example in Sect. 7.4.2). From the definition of the function $\alpha(t)$ it follows that

$$\begin{cases} \alpha_1(\tau_i) + \alpha_2(\tau_i) + \dots + \alpha_r(\tau_i) = 1, \quad i = 1, \dots, N, \\ \alpha_j(\tau_i)\alpha_k(\tau_i) = 0, \quad j \neq k. \end{cases} \tag{7.86}$$

In the notations (7.85), the observation equations read

$$\xi(\tau_i) = A(\alpha(\tau_i))X(\tau_i) + B(\tau_i)\epsilon(\tau_i).$$

Thus, the observation control problem is stated as an ordinary impulse control problem, namely: it is required to determine time instants τ_i, $i = 1, \dots, N$, and functions $\alpha_j(\tau_i)$, $j = 1, \dots, r$ which minimize the performance criterion

$$J = \mathrm{tr}\{h\gamma(T)\}, \tag{7.87}$$

where h is a given $n \times n$ weight matrix and γ is the $n \times n$ covariance matrix of the estimation errors for the process X from the observations ξ at the terminal time T (i.e., at the end of the observation process).

7.5.1 *Solution of the Problem with a Finite Number of Observations*

One can show (see [153, 385]) that in this formulation of the observation control problem the optimal control is deterministic. In other words, one can use just programmed controls. For this reason, the processes described by the system (7.82)–(7.85) are Gaussian and the solution of the estimation problem can be found by means of Kalman filtering. Consequently, the $n \times n$ covariance matrix of estimation errors $\gamma(t)$ satisfies the following differential equation with intermediate conditions (hereafter we omit the time argument in the matrices a, b and B):

$$\begin{aligned} \gamma(t) = \gamma_0 + \int_0^t \left[bb^* + a\gamma(s) + \gamma(s)a^*\right] ds - \sum_{\tau_i \le t} \gamma(\tau_i-)A^*(\alpha(\tau_i))\times \\ \times \left[BB^* + A(\alpha(\tau_i))\gamma(\tau_i-)A^*(\alpha(\tau_i))\right]^{-1} A(\alpha(\tau_i))\gamma(\tau_i-), \end{aligned} \tag{7.88}$$

where

$$\gamma_0 = \mathrm{cov}(X_0,\ X_0).$$

Thus, the problem reduces to the impulse control problem for the Eq. (7.88), performance criterion (7.87), and constraints (7.86) and (7.83).

As above, we reformulate the problem in terms of the variable $D(t) = \gamma^{-1}(t)$, with $D(0) = \gamma_0^{-1}$. Then the Eq. (7.88) and the performance criterion (7.87) take on the form

$$\begin{aligned} D(t) = D_0 - \int_0^t \left[D(s)a + a^*D(s) + D(s)bb^*D(s)\right] ds + \\ + \sum_{\tau_i \le t} \left[\alpha_1(\tau_i)\overline{A}_1 + \alpha_2(\tau_i)\overline{A}_2 + \cdots + \alpha_r(\tau_i)\overline{A}_r\right], \end{aligned} \tag{7.89}$$

where $\overline{A}_j = \overline{A}_j(\tau_i) = A_j^*(\tau_i)[B_j(\tau_i)B_j^*(\tau_i)]^{-1}A_j(\tau_i)$, $j = 1, \dots, r$, $i = 1, \dots, N$, and

$$J = \text{tr}\{hD^{-1}(T)\}. \tag{7.90}$$

Remark 7.8 In the derivation of Eq. (7.89) we took into account the following properties of the control, which are valid for all τ_i, $i = 1, \ldots, N$:

$$\alpha_j^2(\tau_i) = \alpha_j(\tau_i) \quad \text{and} \quad \alpha_j(\tau_i)\alpha_k(\tau_i) = 0, \quad j \neq k, \quad j, k = 1, \ldots, r.$$

These relations mean that at each given time τ_i, $i = 1, \ldots, N$, only one observation channel can be used.

To take into consideration the specifics of the constraints on the observation time instants (see (7.83)), we use the special form of the discontinuous time change that was described in Sect. 3.

As prescribed in Sect. 3.2.1, we perform the change of time $t = t(\theta)$, $\theta \in [0,\ 2N + 1]$, and we introduce the equivalent auxiliary control problem with the variables

$$z(\theta) = D(t(\theta)),\ t_j(\theta),\ \rho_j(\theta), \quad i = 1, \ldots, N,\ j = 1, \ldots, N+1,$$

which satisfy the equations

$$\begin{cases} \overset{\circ}{z}(\theta) = -\big[z(\theta)a + a^* z(\theta) + z(\theta)bb^* z(\theta)\big]\beta(\theta)v(\rho(\theta), u(\theta)) + \\ \qquad + (1 - \beta(\theta))\big[\ell_1(\theta)\overline{A}_1 + \ell_2(\theta)\overline{A}_2 + \cdots + \ell_r(\theta)\overline{A}_r\big], \\ \overset{\circ}{t}(\theta) = \beta(\theta)v(\rho(\theta), u(\theta)), \\ \overset{\circ}{t}_i(\theta) = \beta_i(\theta)\beta(\theta)v(\rho(\theta), u(\theta)), \quad i = 1, \ldots, N, \\ \overset{\circ}{\rho}_j(\theta) = 0, \quad j = 1, \ldots, N+1, \end{cases} \tag{7.91}$$

with the terminal constraints

$$\begin{cases} z(0) = D_0, \quad t(0) = 0, \quad t(2N+1) = T, \quad t_i(0) = 0, \\ \rho_i(2N+1) = T - t_{i-1}(2N+1) - \sum\limits_{k=i+1}^{N+1} T_k, \quad i = 1, \ldots, N, \\ \rho_{N+1}(2N+1) = T - t_N(2N+1), \quad t_0(2N+1) = 0 \end{cases} \tag{7.92}$$

where $a = a(t(\theta))$, $b = b(t(\theta))$, $\overline{A}_j = \overline{A}_j(t(\theta))$, $j = 1, \ldots, r$, and the functions $\beta(\theta)$ and $\beta_i(\theta)$ are defined by formulas (3.9) and (3.20); the controls $\ell_j(\theta)$, $j = 1, \ldots, r$ are piecewise constant on the "even" intervals (where $\beta(\theta) = 0$), but are arbitrary on the "odd" intervals (where $\beta(\theta) = 1$) of the segment $[0,\ 2N+1]$, and obey the constraints

$$\begin{cases} \ell_j(\theta) \in [0, 1], \quad j = 1, \ldots, r, \\ \sum_{j=1}^{r} \ell_j(\theta) = 1. \end{cases} \tag{7.93}$$

In turn, the function $v(\rho, u)$ is piecewise constant on the "odd" intervals (where $\beta(\theta) = 1$), and arbitrary on the "even" intervals (where $\beta(\theta) = 0$) of the segment $[0, 2N + 1]$:

$$v(\rho(\theta), u(\theta)) = \begin{cases} u_i, \ \text{for } \theta \in [2i - 2, 2i - 1], \ i = 1, \ldots, N + 1, \\ \text{arbitrary function, for } \theta \in]\,2i - 2, 2i\,[, \ \ i = 1, \ldots, N, \end{cases} \tag{7.94}$$

where the constants u_i satisfy the relations (3.64), for $u_i \in [-1, 1]$, and define according (3.65) the step control function $u(\theta) \in [-1, 1]$.

To derive the optimality conditions in the auxiliary problem we use (as instructed by Theorem 3.2) the classical maximum principle. Thus, we introduce the Hamiltonian H for this problem on "odd" ($\beta(\theta) = 1$) and "even" ($\beta(\theta) = 0$) intervals of the segment $[0, 2N + 1]$:

$$H = \beta H_{\text{odd}} (z, t, \rho_i, p_z, p_t, p_{t_i}, u) + (1 - \beta) H_{\text{even}}(t, p_z, \ell), \tag{7.95}$$

where

$$\begin{aligned} &H_{\text{odd}} (z, t, \rho_i, p_z, p_t, p_{t_i}, u) = \left[p_z \cdot f(z, t) + p_t + \sum_{i=1}^{N} p_{t_i} \beta_i \right] v(\rho, u), \\ &H_{\text{even}}(t, p_z, l) = \sum_{j=1}^{r} \ell_j p_z \cdot \overline{A}_j(t). \end{aligned} \tag{7.96}$$

Here the symbol "$\cdot$" stands for the matrix scalar product $A \cdot B = \text{tr}\{AB^*\}$,

$$f(z, t) = -(za(t) + a^*(t)z + zb(t)b^*(t)z), \tag{7.97}$$

and the conjugate variables $p_z, p_t, p_{t_i}, p_{\rho_i}, p_{\rho N+1}, i = 1, \ldots, N$, obey the following equations (the argument θ is omitted):

$$\begin{aligned} \overset{\circ}{p}_z = \beta [&a(t^0) p_z a^*(t^0) + b(t^0) b^*(t^0) z^0 p_z + \\ &+ p_z z^0 b(t^0) b^*(t^0)] v(\rho^0, u^0), \end{aligned} \tag{7.98}$$

$$\overset{\circ}{p}_t = -\beta p_z \cdot \frac{\partial f(z^0, t^0)}{\partial t} v(\rho^0, u^0) - (1-\beta) \sum_{j=1}^{r} l_j^0 p_z \cdot \frac{\partial \overline{A}_j(t^0)}{\partial t}, \tag{7.99}$$

$$\overset{\circ}{p}_{t_i} = 0, \quad \text{i.e.,} \ p_{t_i}(\theta) = \text{const.}, \tag{7.100}$$

$$\overset{\circ}{p}_{\rho_i} = \left[p_z \cdot f(z^0, t^0) + p_t + \sum_{j=i}^{N} p_{t_j} \right] \frac{1}{2}(u_i^0 + 1)\mathbf{I}\{[2i-2,\ 2i-1]\}, \tag{7.101}$$

$$\overset{\circ}{p}_{\rho_{N+1}} = [p_z \cdot f(z^0, t^0) + p_t]\mathbf{I}\{[2N,\ 2N+1]\}. \tag{7.102}$$

To determine the terminal conditions for the conjugate variables at the time instant $\theta = 2N + 1$, put

$$g_1(t(\theta)) = (t(\theta) - T)^2.$$

Then, by (3.75),

$$\lambda_1 g_1(t(2N+1)) = \lambda_1 (t(2N+1) - T)^2$$

and

$$\lambda_1 \frac{\partial g_1(t^0(2N+1))}{\partial t} = 2\lambda_1(t^0(2N+1) - T) = 0,$$

since $t^0(2N+1) = T$. Further, it follows from (3.75) that for $\theta = 2N + 1$

$$p_z(2N+1) = \lambda_0 h[z^0(2N+1)]^{-2}, \quad \lambda_0 \geq 0,$$

$$\begin{aligned} p_t(2N+1) &= \sum_{k=1}^{N+1} \nu_k, \\ p_{t_i}(2N+1) &= -\nu_{i+1}, \quad i = 1, \ldots, N, \\ p_{\rho_i}(2N+1) &= -\nu_i, \quad i = 1, \ldots, N+1. \end{aligned} \tag{7.103}$$

By the maximum principle, the following inequalities must hold on the optimal control and the corresponding trajectories:

$$\begin{aligned} H_{\text{odd}}\,(z^0, t^0, \rho_i^0, p_z, p_t, p_{t_i}, u^0) &\geq H_{\text{even}}\,(z^0, t^0, \rho_i^0, p_z, p_t, p_{t_i}, u), \\ H_{\text{odd}}(t^0, p_z, \ell^0) &\geq H_{\text{even}}(t^0, p_z, \ell). \end{aligned} \tag{7.104}$$

Remark 7.9 The linearity of Eq. (7.98) in the variables $p_z(\theta)$ and the positive definiteness of the terminal conditions for $p_z(2N+1)$ (see (7.103)) imply the positive definiteness of $p_z(\theta)$ for all $\theta \in [0,\ 2N+1]$. From this fact and the quadratic dependence of $f(z,t)$ on z (see (7.97)), it follows that the Hamiltonian (7.95) is concave in the matrix variable z (see [265], Theorem 3). Therefore, the requirement that the matrix-functions $a(t)$, $b(t)$ and $\overline{A}_j(t)$, $j = 1,\dots,r$ be such that the Hamiltonian (7.95) is concave in the variable t provides a sufficient optimality conditions for the controls u^0, l^0 satisfying (7.104). In particular, this takes place when the said functions do not depend on t at all.

Remark 7.10 As it follows from (7.96) and (7.100)–(7.102), the constants ν_i, $i = 1,\dots,N+1$, in (7.103) may be set equal to zero.

7.5.2 *Example of Observation Control in a Location System*

Consider a location system which, by sending probing pulses over a given time interval $[0,\ T]$, can measure the current distance R_t and the radial velocity $v_t = \dot{R}_t$ of a moving object. Let N be the admissible number of pulses and let $\tau_i, i = 1,\dots,N$ be the time instants when the pulses are radiated, which satisfy the constraints (7.83).

Assume that the accuracy of measurement of the quantities R_t and v_t depends on the type of pulse modulation, namely: for one type of modulation the distance R_t is measured with better precision, but the velocity v_t is measured with worse precision, while for the second type of pulses the situation is reversed.

One is required to find times τ_i and type of pulses under which the estimation error of the values R_t and v_t at time T is minimal.

Let $x(t) = (x_1(t), x_2(t), x_3(t))^*$, $t \in [0,T]$ denote the vector of phase coordinates of the moving object, where $x_1(t) = R_t$, $x_2(t) = v_t$ and $x_3(t) = \dot{v}_t$. Then estimation process of x_1 and x_2 by means of two types of probing pulses can be described in terms of a two-channel linear observation system of the following form:

The first channel provides the measurements:

$$\begin{cases} \xi_{11}(\tau_i) = x_1(\tau_i) + \sigma_{11}\epsilon_{11}(\tau_i), \\ \xi_{12}(\tau_i) = x_2(\tau_i) + \sigma_{12}\epsilon_{12}(\tau_i), \end{cases} \tag{7.105}$$

The second channel provides the measurements:

$$\begin{cases} \xi_{21}(\tau_i) = x_1(\tau_i) + \sigma_{21}\epsilon_{21}(\tau_i), \\ \xi_{22}(\tau_i) = x_2(\tau_i) + \sigma_{22}\epsilon_{22}(\tau_i), \end{cases} \tag{7.106}$$

where $\sigma_{kj} = \text{const} > 0$, $k, j = 1, 2$, is the mean squared errors of measurement of the j-th coordinate in the k-th channel, i.e., by using a probing pulse of k-th type; and $\epsilon_{kj}(\tau_i)$ is a sequence of $(i = 1, \ldots, N)$ of independent Gaussian random variables with zero mean and unit covariance.

The problem is to find the time instants $\tau_i, i = 1, \ldots, N$ and the type of channels that minimize the performance criterion

$$J = \gamma_{11}(T) + \gamma_{22}(T), \tag{7.107}$$

where $\gamma_{11}(T)$ and $\gamma_{22}(T)$ are the covariances of the estimation errors for the components $x_1(T)$ and $x_2(T)$, respectively, at time T.

To find the channel-switching control rule in explicit form, we change the variables in Eqs. (7.105) and (7.106) as follows. Denote $\varkappa_{kj} = \sigma_{kj}^{-1}$, $k, j = 1, 2$, and put

$$z(t) = (z_1(t), z_2(t), z_3(t), z_4(t))^*,$$

where

$$z_1(t) = \xi_{11}(t)\sigma_{11}^{-1}, \qquad z_2(t) = \xi_{12}(t)\sigma_{12}^{-1},$$

$$z_3(t) = \xi_{21}(t)\sigma_{21}^{-1}, \qquad z_4(t) = \xi_{22}(t)\sigma_{22}^{-1},$$

Next, we introduce the matrices

$$A_1 = \begin{pmatrix} \varkappa_{11} & 0 & 0 \\ 0 & \varkappa_{12} & 0 \end{pmatrix}, \quad A_2 = \begin{pmatrix} \varkappa_{21} & 0 & 0 \\ 0 & \varkappa_{22} & 0 \end{pmatrix}, \quad \overline{A}_1 = A_1^* A_1, \quad \overline{A}_2 = A_2^* A_2,$$

as well as the block matrix

$$A(\alpha) = \begin{pmatrix} \alpha_1 A_1 \\ \alpha_2 A_2 \end{pmatrix},$$

where $\alpha = (\alpha_1, \ \alpha_2)$ is a two-dimensional vector playing the role of the channel-switching control, with components

$$\alpha_1 \in \{0, \ 1\}, \quad \alpha_2 \in \{0, \ 1\},$$

such that $\alpha_1(\tau_i) + \alpha_2(\tau_i) = 1$ for each $t = \tau_i, i = 1, \ldots, N$.

In these notations, Eqs. (7.105) and (7.106) take on the vector form

$$z(\tau_i) = A(\alpha(\tau_i))x(\tau_i) + \epsilon(\tau_i), \tag{7.108}$$

where $\epsilon(\tau_i) = (\epsilon_{11}(\tau_i), \epsilon_{12}(\tau_i), \epsilon_{21}(\tau_i), \epsilon_{22}(\tau_i))^*, i = 1, \ldots, N$.

Now let us assume that the evolution of the coordinates of the moving object are described by an Itô-type stochastic vector differential equation

$$dx(t) = ax(t)\,dt + dW(t), \qquad t \in [0,\ T],$$

where $W(t)$ is the scalar Wiener process,

$$a = \begin{pmatrix} 0 & 1 & 0 \\ 0 & -\kappa & 1 \\ 0 & 0 & -\nu \end{pmatrix}, \qquad b = \begin{pmatrix} 0 \\ 0 \\ \sigma \end{pmatrix},$$

and κ, ν, σ are given positive constants characterizing the inertial properties of the object. Finally, the initial conditions $x_1(0) = R_0$, $x_2(0) = v_0$, $x_3(0) = x_{30}$ are Gaussian random variables with known means and covariances.

The matrix $D(t) = \gamma^{-1}(t)$, where $\gamma(t)$ is the 3×3 covariance matrix of the estimation errors of $x(t)$ by the observations (7.108), obeys Eq. (7.89) with $\overline{A}_j = A_j^* A_j$, $j = 1, 2$. In terms of the variable $D(t)$, the performance criterion (7.107) has the form (7.90) with the weight matrix h =diag(1, 1, 0).

The equivalent auxiliary problem is described by the Eqs. (7.91) and (7.92) with the constraints (7.93) and (7.94). In view of Remark 7.9, the maximum principle in the auxiliary problem represents a necessary and sufficient optimality condition.

We report here results of numerical modeling of the problem performed for the following values of the parameters:

$$T = 10, \quad N = 3, \quad T_1 = 0, \quad T_2 = T_3 = 0.5, \quad T_4 = 0$$

(the values $T_1 = T_4 = 0$ mean that the first and the last probing pulses are permitted at $t = 0$ and $t = T$, that is at the very beginning and at the very end, respectively, of the observation interval). Moreover,

$$\kappa = 0.128, \quad \nu = 3\lambda_r, \quad \lambda_r = 0.03, \quad \sigma = 2N_r\sqrt{\lambda_r},$$

$$N_r = 0.33 \cdot 10^{-4}, \quad \gamma_{11}(0) = (\sigma_{11} \vee \sigma_{21})^2, \quad \gamma_{22}(0) = (\sigma_{12} \vee \sigma_{22})^2,$$

$$\gamma_{33}(0) = N_r^2, \quad \gamma(0) = \mathrm{diag}(\gamma_{11}(0),\ \gamma_{22}(0),\ \gamma_{33}(0)),$$

where $a \vee b$:=max$\{a, b\}$.

Thanks to the linearity in controls of the auxiliary problem, the optimal control turned out to be of relay type. The modeling results are presented in Table 7.1.

We considered four different variants of channel parameters σ_{ij}, $i, j = 1, 2$, indicated in the table by upper indices 0,1,2,3. Index "0" corresponds to the optimal observation control, and the other index, i.e., "1,2,3," to other non-optimal controls.

The control corresponding to a particular case is indicated in the corresponding column u_i, $i = 1, 2, 3$, and $\alpha_1^i = \alpha_1(\tau_i)$, $i = 1, 2, 3$ (with $\alpha_2^i = \alpha_2(\tau_i) = 1 - \alpha_1^i$). For example, in case 1^0 we have $u_1 = -1$ and $u_2 = u_3 = 1$. This means that the

Table 7.1

Variant		1st channel		2nd channel		Performance criterion	u_i for $i = 1, 2, 3$	α_1^i for $i = 1, 2, 3$
		σ_{11}	σ_{12}	σ_{21}	σ_{22}			
1st	1^0	0.25	0.03	0.05	0.15	0.00149	−1, 1, 1	0, 0, 0
	1^1					0.00308	1, 1, 1	0, 1, 1
	1^2					0.01688	1, 1, 1	1, 1, 1
	1^3					0.01202	−1, 1, 1	0, 1, 1
2nd	2^0	0.25	0.03	0.15	0.15	0.00983	1, 1, 1	1, 0, 0
	2^1					0.03473	−1, −1, −1	1, 1, 0
	2^2					0.05175	−1, −1, −1	0, 1, 0
	2^3					0.01688	1, 1, 1	1, 1, 1
3rd	3^0	0.25	0.05	0.15	0.30	0.00149	−1, 1, 1	0, 0, 0
	3^1					0.00308	−1, 1, 1	1, 0, 1
	3^2					0.02062	−1, −1, 1	0, 1, 1
	3^3					34.89119	−1, −1, −1	0, 0, 0
4th	4^0	0.25	0.03	0.15	0.30	0.00991	1, 1, 1	1, 0, 0
	4^1					1.41645	−1, −1, −1	0, 0, 0
	4^2					0.01930	−1, 1, 1	1, 1, 1
	4^3					0.02220	−1, −1, 1	1, 0, 0

first of the probing pulses is radiated at $t = 0$, and two other pulses are radiated at the end of the interval $[0, T]$, with minimal time difference equal to 0.5. In this case $\alpha_1^1 = \alpha_1^2 = \alpha_1^3 = 0$, i.e., $\alpha_2^1 = \alpha_2^2 = \alpha_2^3 = 1$, which corresponds to the operation of the second probing channel.

A comparison of the variants presented in Table 7.1 leads to following conclusions.

1. The choice of a rational (optimal) way of observation control enables one to increase the accuracy of the estimation of coordinates by several orders. (Compare the cases 4^0 and 4^1, 3^0 and 3^3, or 1^0 and 1^1 in the column "performance criterion").
2. When the accuracy of the distance measurement is sufficiently high, it make sense to use the first probing pulse at the beginning of the observations interval, so as to estimate first the initial position of the object (see 1^0 and 3^0).
3. When, on the other hand, the accuracy of the distance measurement is relatively low, making the initial position of the object more precise will have little influence at the end of the observation interval, and so it is advisable to radiate all sounding pulses at the end of the observation interval (see 1^0 and 4^0).
4. In the case of equal requirements on the accuracy in the estimation of the distance $(\gamma_{11}(0))$ and of the radial velocity $(\gamma_{22}(0))$, the most essential influence on the decrease of the performance criterion $J = \gamma_{11} + \gamma_{22}$ is exerted by increasing the accuracy of the distance measurement (compare 3^0 and 3^1 or 1^0 and 1^3).

Chapter 8
Impulsive Control in Collision Mechanics

8.1 Introduction

8.1.1 Collisions in Mechanical Systems

The dynamics of mechanical systems with collisions has attracted the interest of researchers for a very long time. On the one hand, the classical mechanics of systems with collisions is known to us from the time we attended the physics course in school, where its application is limited to simple examples of colliding elastic and/or inelastic bodies, and each high-school graduate probably has extensive experience in solving problems that require to determine velocities of billiard balls before and after impact. On the other hand, impact phenomena arise in many fields of engineering applications, which have only expanded in recent years due to the development of micro- and nano-technologies. We list here just a few examples: vibrotechnique and vibroimpact mechanisms [17], robotics and legged robots [7, 134, 155, 253, 254, 289], mechanics of vibromotion [374, 433], micromechanics[166, 238, 471], ultra sound transducers [14], atomic force microscopy [202, 203].

The real process of collision of elastic and/or inelastic bodies is a complex phenomenon which depends on the surfaces and shapes of the colliding bodies, external conditions, and other factors. Many outstanding scientists: Huygens, Newton, Rayleigh, Kelvin, Hertz have devoted attention, to various degrees, to the problem of describing mechanical collisions. Their works laid down the foundations of the classical approaches to impact theory. In recent years, the dynamics of systems with impacts is treated in many monographs, where various impact models and numerical methods of analysis and control are considered, with extended references [17, 54, 70, 74, 183, 215, 308, 362]. However, most authors note that despite the relative clarity that can be achieved when one describes a single impact, the general dynamics of controlled systems with impacts remains a very confusing area, where ambiguity in determining the solution and extreme sensitivity of the behavior of

B. Miller, E. Rubinovich, *Optimization of Dynamical Systems with Impulse Controls and Shocks*, Systems & Control: Foundations & Applications,
https://doi.org/10.1007/978-3-031-64124-4_8

a system's behavior to changes in parameters and initial conditions are typical features. This is why the stream of new publications, concerning the phenomenon of impact in controlled dynamic systems is not running out, which can be illustrated by the following far from complete list of works published just in recent years [18, 77, 81, 100, 143, 179–185, 194, 199, 213, 239, 240, 310, 311, 336, 346–354, 363, 370, 384, 389, 391–393, 449, 453].

Let us mention here that is the sequel the terms "shock", "collision" and "impact" will be frequently interchanged, and assumed to have the same meaning.

8.1.2 Shocks from the Point of View of Control Theory

From the point of view of control theory, the presence of active unilateral constraints, such as an elastic obstacle, leads to a special class of control problems for systems of discrete-continuous type, in which when the trajectory of the system hits some surface discontinuities of part of the phase coordinates arise. However, in recent years, an increasing number of authors studying control problems in systems with unilateral constraints come to the interpretation of constraints as some additional control that allows to ensure stability or modify the attainability subset [69, 71, 72, 86, 256, 413, 429]. One can go even further if one regards the collision phase as a controlled process. We do not discuss here the control of the position and speed of the surface against which collision occurs. This part of the problem does not refer to the collision itself, it only determines its external conditions preceding the contact phase, although during the impact, the velocity of the surface may change. When we are talking about control in the shock phase, we mean the effects that occur at the pace of the contact, i.e., fast enough. One of the most exotic examples of this kind can be observed in ping-pong, when an experienced player manages, during the contact of the ball with the surface of the racket, to quickly turn it around, giving the ball an additional, very strong rotation (a so-called *top spin* hit). Another example is selective capture or, on the contrary, repulsion of a metal object, for example, by an electromagnet. It is clear that the electromagnet needs to be turned on to create a force of attraction or one of repulsion only when the object touches its surface. In this interpretation, control problems in mechanical systems with unilateral constraints are close to impulse control problems, since, obviously, the description of impacts and jump-like changes in velocities in terms of some fast motion is extremely close to the concept of impulsive control. There are many works devoted to control problems in mechanical systems with discontinuous velocities approached from the point of view of impulse control [58, 59].

A series of works use a transformation of some of the phase coordinates of a mechanical system into controls [60–62]. This approach allows the description of velocity discontinuities in terms of discontinuities of some control variables and is very close to the technique proposed by Gurman and Dykhta [118, 158, 160] for singular control problems. However, a basic limitation of this approach in the case of the dynamics of systems with shocks is that, despite a qualitatively similar behavior, the intensity and localization of impulse action are exogenous for the system under consideration and it can be applied, generally speaking, arbitrarily as a certain programmed control. At the same time, the impulse action caused by collision with an obstacle is, firstly, localized in time and space and, secondly, its intensity depends on the properties of the obstacle and the values of the phase coordinates (or of part of them) immediately preceding the impact. Therefore, the impulse action in a system with shocks is an impulse action with feedback, and the description of the corresponding behavior of the system is more complicated. It is all the more tempting to produce a description of the impact phenomenon in sufficiently transparent terms, for example, in the form of a connection between the velocities of bodies before and after impact. Some publications use the term *law of restitution* [70]. This law is sometimes derived using various hypotheses characterizing the collision process, such as, for example, the hypothesis of elastic or inelastic impact or, for example, in terms of a restitution coefficient of the normal velocity components (*Newton's hypothesis* [215]). However, if one studies not only a single shock, but also the possibility that after collision the bodies involved can move together for some time and then disengage, a more detailed description is needed, requiring methods of the theory of differential inclusions [308] or using the so-called *complementarity* approach [73].

8.1.3 Systems with Controllable Shocks

The analysis of different impact models in systems with many degrees freedom has been a research subject over the past 150 years, starting with the works of Poisson and Routh [183]. These systems exhibit rather unusual properties, such as discontinuities and non-smoothness of phase trajectories, jumps of dimension, lack of continuous dependence of solutions on the initial conditions, and lack of uniqueness of solutions. At the same time, this class of systems become increasingly important in connection with the development of micro- and nano-technology, where microimpact is one of the main means of providing a given movement in microelectromechanical systems (MEMS) with massive microactuators and thin-film shock microactuators [48, 49, 166, 238]. Microshocks and controlled vibration actions allow to implement movements with a significant reduction in friction and adhesion, thus overcoming an overpresent serious obstacle for the creation of microelectromechanical actuators [382, 383].

Numerous examples of shock effects arise in the design and construction of robotic manipulators [289, 362, 429], vibro-impact mechanisms [17], walking or jumping robots [7, 134, 155] and many others applications described, in particular, in the monographs [70, 72]. In mechanical systems arising in games and sports [69, 413, 460] and microelectromechanical systems [48, 49, 166], and [238] movement can be realized only by means shocks. Moreover, it is known that in electrical grids the propagation of shock overloads can be stopped with the aid of shock-type actions, which are carried out by reconfigurations of the network's topology

[89, 225, 428]. Hence, the singularities arising from shocks of this type are not only important from a theoretical point of view, but also play an important practical role in the implementation of control laws in such systems.

Usually, control actions are carried out either during a nonsingular phase of motion (motions in a constraint-free region) or during the interaction with an obstacle, but then the position of surfaces is rigidly fixed and cannot be changed during the singular phase of the movement. Thus, for example, in the presence of singularities caused by collisions with an obstacle [70, 155, 429], these collisions must be regarded as *passive constraints*, and determine only the law of restitution. In this sense, switching systems preventing the occurrence of short circuits in power grids also belong to the class of passive constraints, since they do not allow corrections during the switching phase.

In the same manner one can consider the control in some impact games akin to air hockey [413], where the control action consists only of setting the position and speed of the club at the moment before the strike, but cannot be changed in the very short period of the strike phase. However, in some games, such as table tennis, with a sufficiently soft racket, a well-trained player can turn the racket during the shock to give the ball extra spin (top spin hit). In this situation, the player can be regarded as the source of an *active controllable constraint* that can radically modify the reachability set. Accordingly, the phase of interaction at which the additional control action takes place is called the phase of motion with *active singularity*.

To make effective use of the new opportunities introduced by controlled singularities, it is necessary to:

- provide a complete and adequate description of the managed singular phase of motion (controlled shock);
- find a way to describe the motion of the system as a whole, including phases of regular and singular motiont and their junction;
- develop a methodology for managing such systems by using two types of controls, namely, controls in regular and in singular phases of motion.

Analysis of existing methods for describing shock effects shows that these methodologies (namely, detailed methods based on the establishment of the law of restitution) cannot be directly transferred to this class of systems. The main tasks in the implementation of such models belong to the field of kinematics, i.e., determination of speeds and positions at the moment immediately preceding the impact, which is a separate rather complex problem for systems with many degrees of freedom [17, 70, 71, 81, 143, 336, 362, 453]. Although these methods have undergone very major development in recent years, especially for problems with moving boundaries, they are not applicable for problems with active constraints.

A complete description of the trajectories of systems with controlled impacts should take into account the "fast" transient process that occurs in the moment of contact and the control action that can be implemented in the phase of controlled singularity. One way to describe the controlled singular phase is to use the *penalization* methodology popular in shock mechanics, in which during the singular phase of motion a large but finite force arises, depending on some *generalized rigidity* parameter, which then let go infinity. In the limit, the contact time tends to zero, the motion satisfies unilateral phase constraints, and the generalized velocities experience discontinuities at the moment of contact with the constraint boundary [346, 351–353, 393].

The concept of the controllable mechanical shock, which is a development of the penalty method, was first proposed by Bentsman and Miller [33, 283] and developed further in the cycle of works [34–42, 284–288]. It should be noted that such controls are the natural generalization of the impulse control concept [292] and lead to the consideration of a new class of dynamical systems *with active or controlled singularities,* i.e., systems that allow control during the singular phase of motion.

8.1.4 Controllable Systems with Shocks

Control in systems with distributed impacts, i.e., without accumulation points and without sliding modes along the constraint surface, can be approached from the point of view of traditional control problems in discrete-continuous systems. More complex problems arise when describing systems where such modes can arise. Known methods for describing solutions use *quasi-differential equations* [337], differential inclusions [365], complementarity approaches [73], and the penalty method [346, 393]. The first two approaches give a description in terms of solely a bundle of trajectories (integral funnel). To select an isolated path in shock mechanics usually requires additional assumptions regarding the nature of the impact (i.e., energy and momentum conservation laws), which ultimately lead to differential equations with measures [194, 308, 310, 311, 450, 451]. It should be noted that the assumptions regarding the nature of the impact are just needed for the derivation of the laws of restitution, although they insufficient to determine the law of restitution in the case controlled shock. Using a complementarity approach also does not solve the problem, because:

- the complementarity description of a system with unilateral constraints leads to an equation with measure in implicit form, where the measure, concentrated on the boundary of the constrained region, must be determined together with trajectory and distinguished among the set of admissible pairs of trajectories and measures based on some additional assumptions, for example, by means of variational principles;
- as a rule, the solution of the system of complementarity equations system is not unique, the uniqueness of the solution being established only for when the right-hand side of the system is analytic [18, 20, 21, 391, 392], which is an excessive requirement for the control tasks;
- even the existence of a solution to the system of complementarity equations is a very difficult problem, especially in the presence of discontinuous forces of *dry friction* type (see the well-known Painlevé paradoxes [143, 144, 394]).

In this book, we analyze models that describe impacts, which from the point of view of control theory can be considered as some disturbance or control action modifying the motion of the unperturbed system (without constraints) and which allow to achieve certain goals, such as a significant modification of the attainability set and ensuring stability, properties that the system may or may not have in the absence of shock interaction.

Thus, we consider next various models for describing impact from the viewpoint of control theory. First we consider single impact models based on the principles of classical mechanics; these are followed by methods for describing trajectories globally, where the main problems are the existence and the uniqueness of solutions to systems of the differential equations with measures arising from the description of a single impact. In some examples, we consider the feedback control problems and programmed control. This is an area where many nontrivial features of problems with impacts arise, since the feedback control provides significant advantages in terms of robustness and values of quality criteria. We assign an important place to problem modeling and numerical methods used in models of systems with shock interactions, as well as various methods based on space-time transformations of systems of differential equations that allow to reveal the dynamics of "fast" motion in the vicinity of points of impact interaction. Here it is possible to establish analogies between systems with impacts and systems with impulse control. Finally, a brief discussion of problems related to the description of systems that allow multiple collisions arising in the case of the so-called "constrained" impacts, or more precisely, in the presence of nonsmooth unilateral constraints.

The next section is devoted to control problems that can be successfully solved in the framework of various simplified impact models. We consider problems in which the impacts are described in terms of restitution conditions, as well as some simple problems with control in the contact phase under various conditions for its description. Necessary optimality conditions are given in the form of the maximum principle for control problems with impacts. A special subsection is devoted to some control problems in systems with multiple impacts.

8.2 Models of Shock Action from the Point of View of Control Theory

8.2.1 Classical Mechanics: Elastic and Inelastic Impacts

In classical mechanics, an impact (collision) is considered as a short-time interaction of bodies during which the bodies do not change their positions, while their velocities receive finite increments. Therefore one of the central objectives of impact theory is finding relationship between the velocities before and after impact [215]. In the highschool course on mechanics, the definition this relationship made on the basis of conservation laws: energy, momentum, angular momentum, which hold under certain assumptions concerning the geometry of the impact and the properties of the colliding bodies. To be more precise, let ($\mathbf{V}_1^-$ and $\mathbf{V}_2^-$) and ($\mathbf{V}_1^+, \mathbf{V}_2^+$) be the velocities of two bodies with masses m_1 and m_2 before and after collision, respectively. In the case of a perfectly inelastic central collision, the interaction forces arising during the collision are point in opposite directions, so the total momentum of the system of bodies is conserved; moreover, the assumption inelasticity assumption means that after collision the bodies move as unified whole with a common velocity $\mathbf{V}^+ = \mathbf{V}_1^+ = \mathbf{V}_2^+$. That is, in a perfectly inelastic collision, the only relation that follows from the momentum conservation law,

$$m_1\mathbf{V}_1^- + m_2\mathbf{V}_2^- = m_1\mathbf{V}_1^+ + m_2\mathbf{V}_2^+ = (m_1 + m_2)\mathbf{V}^+, \tag{8.1}$$

is sufficient for determining the post-collision velocity $\mathbf{V}^+$. In a perfectly elastic collision, it is assumed that the kinetic energy is also conserved, which allows one to supplement the relations for velocities by the relation

$$m_1|\mathbf{V}_1^-|^2 + m_2|\mathbf{V}_2^-|^2 = m_1|\mathbf{V}_1^+|^2 + m_2|\mathbf{V}_2^+|^2. \tag{8.2}$$

In the case of a central collision, where the velocities $\mathbf{V}_1^-$ and $\mathbf{V}_2^-$ are collinear, this yields two supplementary equations, which enable one to determine the post-collision velocities. However if we abandon the hypothesis that the collision is perfectly elastic, the problem becomes indetermined if no additional assumptions on the nature of the collision are made [183]. The first step forward in solving this problem was *Newton's hypothesis*; ; based on a set of experiments (see [183]), Newton proposed that the velocities of the moving bodies before and after collision, ($\mathbf{V}_1^-, \mathbf{V}_2^-$) and ($\mathbf{V}_1^+, \mathbf{V}_2^+$), respectively, satisfy the law (*restitution equation*)

$$\mathbf{V}_1^+ - \mathbf{V}_2^+ = -æ(\mathbf{V}_1^- - \mathbf{V}_2^-), \quad æ \in (0, 1), \tag{8.3}$$

where the parameter æ is called the *restitution coefficient*, and cannot be determined directly from the basic theorems of dynamics. However, the supplementation of the momentum conservation law (8.1) by relation (8.3) alows one to determine the post-collision velocities and express them in terms of ($\mathbf{V}_1^-$ and $\mathbf{V}_2^-$).

Without touching here upon the wave theory of collisions, which takes into account nonhomogeneous and time-varying deformations of bodies during impact, described by partial differential equations, we indicate a general approach to collision modeling based on so-called *discrete models*. The use of such models makes it possible to analyze the collision process in terms of a finite number of variables described by a system of ordinary differential equations, and based on this to evaluate various characteristics of a collision, such as its duration and the energy loss. A detailed overview of discrete models is contained in the monograph [183]. Here we provide only a brief description of this approach needed for the subsequent exposition.

The basic concept of the theory of discrete models is that of *deformable element*, which specifies the relationship between load and deformation in the form

$$F = F(\varepsilon, \dot{\varepsilon}), \tag{8.4}$$

where ε is the deformation and F is the acting force. Here are some examples of deformable elements:

- elastic element

$$F = c\,\varepsilon,$$

  where c is the rigidity coefficient;
- viscous element (damper)

$$F = k\,\dot{\varepsilon},$$

  where k is the coefficient of viscous friction;
- Saint-Venant element, used for the description of systems with dry, friction

$$F = k_0 \operatorname{sign} \dot{\varepsilon},$$

  where k_0 is a constant sometimes referred to as the *coefficient of dry friction*.

We note that a deformable element is an inertialess link, so the forces applied to its boundary points are equal. A deformable element can be composed of several elements of simplest type connected in parallel or series.

Various combinations of such simplest elements lead to models that are widely used in the analysis of collision phenomena [338]. For example, the parallel connection of a rigid element and a damper yields the well-known Kelvin-Voigt model [183, 215]. In the analysis of two bodies with an intermediate element of Kelvin-Voigt type, the acting force is the linear combination of the deformation and its derivative,

$$F = c\,\varepsilon + k\,\dot{\varepsilon},$$

and the equation of motion reduces to a second-order differential equation for the deformation, namely

$$\begin{aligned} &\ddot{\varepsilon} + 2\xi\omega_0\dot{\varepsilon} + \omega_0^2\varepsilon = 0, \\ &2\xi\omega_0 = (m_1^{-1} + m_2^{-1})k, \quad \omega_0^2 = (m_1^{-1} + m_2^{-1})c, \end{aligned} \tag{8.5}$$

with the initial conditions

$$\varepsilon(0) = 0, \quad \dot{\varepsilon}(0) = -|\mathbf{V}_1^- - \mathbf{V}_2^-|.$$

This equation must be solved until the end of the collision, that is, until the moment when the second derivative of the deformation, which is proportional to the load, vanishes:

$$\ddot{\varepsilon}(\tau) = 0.$$

The relative rebound velocity is equal to

$$|\mathbf{V}_1^+ - \mathbf{V}_2^+| = \dot{\varepsilon}(\tau) = æ|\mathbf{V}_1^- - \mathbf{V}_2^-|.$$

The solution of Eq. (8.5) is given by the expression

$$\varepsilon(t) = \dot{\varepsilon}(0)\,\frac{e^{\lambda_1 t} - e^{\lambda_2 t}}{\lambda_1 - \lambda_2},$$

where λ_1, λ_2 are the roots of the characteristic equation

$$\lambda^2 + 2\xi\omega_0\lambda + \omega_0^2 = 0, \quad \lambda_{1,2} = \omega_0\Big[\xi \pm \sqrt{\xi^2 - 1}\,\Big].$$

This solution is valid until the time $\tau > 0$, which is defined as the smallest positive solution of the equation

$$\ddot{\varepsilon}(\tau) = \frac{\lambda_1^2 e^{\lambda_1 \tau} - \lambda_2^2 e^{\lambda_2 \tau}}{\lambda_1 - \lambda_2} = 0,$$

which gives

$$\tau = \frac{2}{\lambda_1 - \lambda_2}\ln\left(\frac{\lambda_2}{\lambda_1}\right), \quad æ = \exp\left\{-\frac{k\tau}{2}\right\}.$$

After this instant, the bodies move freely, although the deformation remains for some time. An example of the evolution of deformation in the Kelvin-Voigt model is shown in Fig. 8.1

Since the coefficients in Eq. (8.5) are non-negative, the characteristic numbers $\lambda_{1,2}$ have non-positive real parts, and consequently for as long as the contact of the bodies is maintained, the solution of Eq. (8.5) has different forms for $0 \le \xi < 1$ and for $\xi \ge 1$, namely:

$$\varepsilon(t) = \begin{cases} \dot{\varepsilon}(0)\,\dfrac{e^{-\xi\omega_0 t}\sin\left(\omega_0 t\sqrt{1-\xi^2}\right)}{\omega_0\sqrt{1-\xi^2}}, & \text{for } 0 \le \xi < 1, \\ \dot{\varepsilon}(0)\,\dfrac{e^{-\xi\omega_0 t}\,\mathrm{sh}\left(\omega_0 t\sqrt{\xi^2-1}\right)}{\omega_0\sqrt{\xi^2-1}}, & \text{for } \xi > 1, \\ \dot{\varepsilon}(0) t e^{-\omega_0 t}, & \text{for } \xi = 1. \end{cases}$$

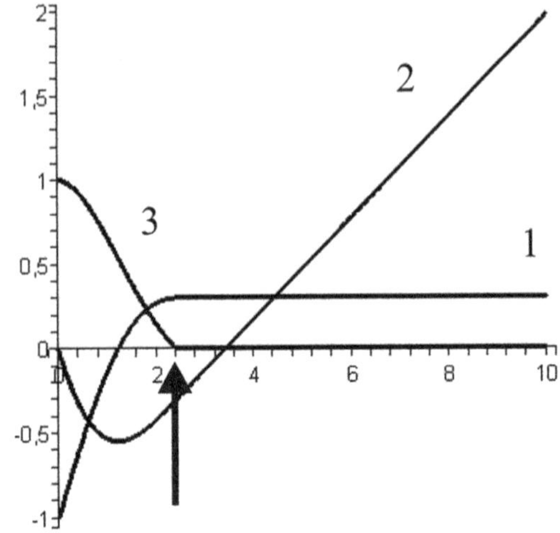

Fig. 8.1 Evolution of the deformation (curve 2), its derivative (curve 1) and the tension in the deformable element (curve 3) in the Kelvin-Voigt model. The discharge moment is indicated by an arrow, the relaxation coefficient is $\xi = 0.5$, and the initial velocity of deformation is -1. One can see that the interaction force becomes zero when the deformation remains, but after that instant the bodies move independendly

The corresponding values of the restitution coefficient are

$$æ = \begin{cases} e^{-2\xi}, & \text{for } 0 \le \xi < 1, \\ \exp\left\{ \dfrac{-2\xi \ln\left|\xi + \sqrt{\xi^2 - 1}\right|}{\sqrt{\xi^2 - 1}} \right\}, & \text{for } \xi > 1, \\ e^{-2}, & \text{for } \quad \xi = 1. \end{cases}$$

For $\xi = 0$ the restitution coefficient is equal to 1, and this case of corresponds to an elastic collision; in all the other cases the restitution coefficient is positive and goes to zero as $\xi \to \infty$. Hence, this model does not describe a perfectly inelastic collision (Fig. 8.2).

A model which describes perfectly inelastic collisions is the *Maxwell model* [183]. In this model, the elastic element and the damper are connected in series, and, in contrast to the preceding model, deformations are summed instead of forces. (Note the analogy with the inclusion of passive electrical elements). If the deformation is expressed in the form

$$\varepsilon = \varepsilon_{\mathrm{e}} + \varepsilon_{\mathrm{d}},$$

where the subscripts refer to the elastic element (e) and to the damper (d), the equilibrium condition gives

$$F = c\,\varepsilon_{\mathrm{e}} = k\,\dot{\varepsilon}_{\mathrm{d}}.$$

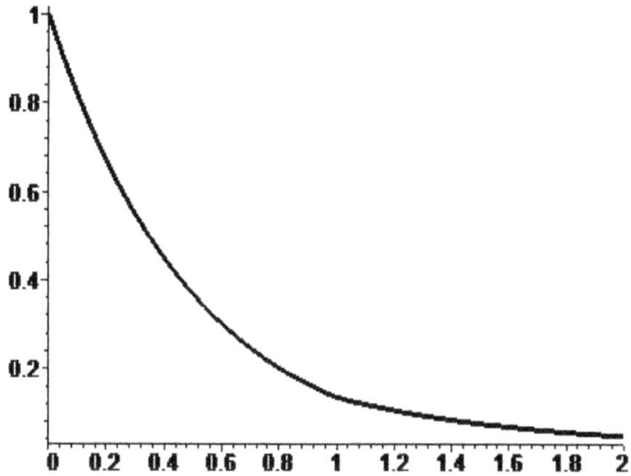

Fig. 8.2 Restitution coefficient for $0 \leq \xi < \infty$ (Kelvin-Voigt model)

Differentiating the first equality, multiplying it by c, and substituting the result in the expression for the deformation, we obtain the equation satisfied by the force:

$$\frac{k}{c}\dot{F} + F = k\,\dot{\varepsilon}.$$

The solution of this linear equation with the null initial condition has the form

$$F(t) = \exp(-ct/k)\int_0^t \exp(cs/k)\dot{\varepsilon}(s)ds. \tag{8.6}$$

Thus, in the Maxwell model, the simple functional relation (8.4) is replaced by an operator representation of the relationship between force and deformation, in which the force depends on the history of the system.

Remark 8.1 This property is also inherent in other, more complex viscoelastic elements, including series-connected elements. Moroever, many models with serial connection of element lead to incorrect mathematical models. For example, [28] considers models with two elastic elements and two Saint-Venant elements, all connected in series. It is shown that these models are characterized by jamming, i.e., the corresponding system of equations of dynamics does not admit a solution.

In this case, one can obtain the equations of dynamics in differential form using the relation

$$\ddot{\varepsilon} = (m_1^{-1} + m_2^{-1})F(\varepsilon, \dot{\varepsilon}),$$

where the expression for the force is understood as the operator (8.6). If we write the derivative of the force in terms of $d^3\varepsilon/dt^3$, we arrive at the following differential equation for the deformation:

$$\frac{d^3\varepsilon}{dt^3} + \frac{c}{k}\ddot{\varepsilon} + c(m_1^{-1} + m_2^{-1})\,\dot{\varepsilon} = 0, \tag{8.7}$$

with the initial conditions

$$\varepsilon(0) = 0, \quad \dot{\varepsilon}(0) = -|\mathbf{V}_1^- - \mathbf{V}_2^-|, \quad \ddot{\varepsilon}(0) = 0; \tag{8.8}$$

this equation must be solved until the collision ends, that is, until the time when the second derivative of the deformation, which is proportional to the load, vanishes:

$$\ddot{\varepsilon}(\tau) = 0.$$

Let us rewrite Eq. (8.7) in the form

$$\frac{d^3\varepsilon}{dt^3} + 2\alpha\,\omega_0\,\ddot{\varepsilon} + \omega_0^2\,\dot{\varepsilon} = 0, \tag{8.9}$$

where

$$\omega_0^2 = c\left(\frac{1}{m_1} + \frac{1}{m_2}\right) = \frac{c}{m^*}, \quad \alpha = \frac{\sqrt{cm^*}}{2k}, \quad \text{and} \quad \frac{1}{m^*} = \frac{1}{m_1} + \frac{1}{m_2}.$$

The duration of the contact and the restitution coefficient are determined by the parameters (α, ω_0). Indeed, the roots of characteristic equation are

$$\lambda_{1,2} = \omega_0\big[-\alpha \pm \sqrt{\alpha^2 - 1}\big], \quad \lambda_3 = 0,$$

and if $\alpha \geq 1$, then the solution of Eq. (8.9) remains stable for $\varepsilon \leq 0$ and the final time of contact is not reached (see Fig. 8.3. If, however, $0 \leq \alpha < 1$, then the solution of Eq. (8.9) with the corresponding initial conditions must be sought in the form

$$\varepsilon(t) = e^{-\alpha\omega_0 t}\left[C_1 \sin\left(\omega_0 t\sqrt{1-\alpha^2}\right) + C_2 \cos\left(\omega_0 t\sqrt{1-\alpha^2}\right)\right] + C_3,$$

and taking into account the initial conditions (8.8) we obtain the expression

$$\ddot{\varepsilon}(t) = -\frac{\dot{\varepsilon}(0)\,\omega_0}{\sqrt{1-\alpha^2}}\, e^{-\alpha\omega_0 t} \sin\left(\omega_0 t\sqrt{1-\alpha^2}\right).$$

We see that the final time of contact does not depend on $\dot{\varepsilon}(0)$ and equals

$$\tau = \frac{\pi}{\omega_0\sqrt{1-\alpha^2}}.$$

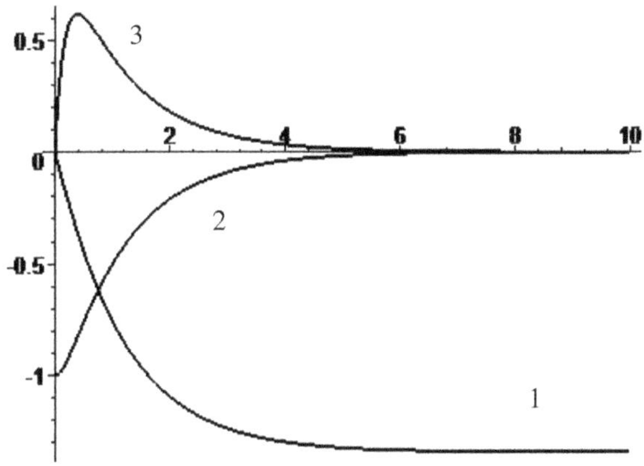

Fig. 8.3 Evolution of the deformation in the Maxwell model for $\alpha > 1$. In this example, $\omega_0^2 = 5$, $\alpha = 1.5$, $\dot{\varepsilon}(0) = -1$. The graphs represent the defomation (curve 1), rate of deformation (curve 2), and the second derivative of deformation (curve 3), which is proportional to the load on the deformable element

Moreover, the derivative of the deformation derivative at the time of contact termination is equal to

$$\dot{\varepsilon}(\tau) = -\dot{\varepsilon}(0)\, e^{-\alpha\omega_0\tau}.$$

We conclude that the restitution coefficient does not depend on the initial velocity and is equal to

$$æ = \begin{cases} \exp\left\{\dfrac{-\pi\alpha}{\sqrt{1-\alpha^2}}\right\}, & \text{for } 0 \le \alpha < 1, \\ 0, & \text{for } \alpha \ge 1. \end{cases} \tag{8.10}$$

For $\alpha = 0$ we obtain a model of perfectly elastic collision, for $0 < \alpha < 1$ a model of collision with energy absorption, and for $\alpha \ge 1$ a model of perfectly inelastic collision. A detailed analysis of Eq. (8.9) [338] shows that at the time of contact termination the deformation is different from zero.

Let us formulate some general properties of the models considered above:

- the contact dynamics is determined by two parameters: the generalized rigidity ω_0 , which is relatively large for real mechanical systems, and the parameter ξ or α, which characterizes the rate of oscillation relaxation;
- the maximal deformation and the contact duration are of the order of ω_0^{-1}, so they tend to zero as $\omega_0 \to \infty$;
- the restitution coefficient does not depend on the initial relative velocity of bodies and is completely determined by their rheological properties.

One can say that these models confirm the approaches adopted in classical mechanics when the generalized rigidity parameter tends to infinity. From the point of view of the description of trajectories of the dynamical system for infinitely large values of ω_0, at the collision instant the position of the bodies does not change, while their relative velocity $\mathbf{V}$ changes sign and magnitude according to the rule

$$\mathbf{V}^{+} = -e\mathbf{V}^{-} \quad \text{or} \quad \Delta\mathbf{V} = \mathbf{V}^{+} - \mathbf{V}^{-} = -(1 + æ)\mathbf{V}^{-}. \tag{8.11}$$

However, such simple relations are valid only in the case of central collision. Already in the case of an "oblique" collision, the law of velocity change must be different for the normal and the tangential components of the relative velocity, and the presence of friction (dry or viscous) makes the task even more difficult.

Remark 8.2 In both models considered above, the deformation at the end of the contact is not zero. If we assume that in the Kelvin-Voigt model the interaction occurs until the deformation vanishes, then the restoration law takes on the form (8.10), with the parameter α replaced by ξ. This observation is often used in phenomenological models of collisions [215], but in fact cannot claim to be physically justified. There is a physically absurd picture in these models, namely, that an obstacle keeps a body in contact, i.e., the force is directed opposite the direction of travel. On the other hand, the presence such forces does not contradict the setting of controlled contact, for example, when an electromagnet is used to hold a body. Such models are interesting from the point of view of control theory; however, in mechanical systems with relatively massive bodies similar phenomena do not occur. It is interesting to note that for contacting bodies of small size, for example, when the probe head of an atomic force microscope is in contact with the surface being examined, adhesive forces arise that are proportional to the probe's velocity [202, 203].

8.2.2 Controlled Impact in Systems with Many Degrees of Freedom: General Approach

In recent years, the concept of a controlled impact has undergone a major development based of the widely known in classical impact mechanics approach based on the penalization methodology. As in the preceding examples describing the motion in the impact phase, in the penalization framework one introduces in the differential equations governing the dynamics of a system with singular behavior a large parameter $\mu \to \infty$, and one examines the corresponding family of trajectories that converge to discontinuous functions. The limit discontinuous trajectories are called generalized solutions of the given system with controlled singularities. The description of the dynamics of such a system is obtained by resorting to a special space-time transformation, which allows to get a description of jumps in terms of a controlled shift operator along the trajectorics of a certain auxiliary system of differential equations of fast motion. Key ideas of this approach were developed in the works [33–36, 284], and finalized in the work [37], which also provides a number of examples demonstrating the limitations of the complementarity approach in problems with controlled singularities. Those examples, such as the shock-impulse action on an elastic body lying on an elastic surface (Example 1.13), or the fast rotation of a tennis racket at the moment of contact with the ball, demonstrate the need for a thorough analysis of the fast (singular) phase of motion. In fact, these examples served as a basis for a general approach that can be formulated as follows:

- the interaction that occurs upon contact with an obstacle (or a contact force, in the case of a mechanical system), is a result of a small violation of the constraint that occurs at the time of contact and continues until the system leaves the domain forbidden by the constraint;
- this interaction prevents entrance to the constrained domain and, in the case of a mechanical system, leads to the change of the sign of the normal component of the velocity;
- the controlled motion in the constrained domain is characterized by a differential equation whose right-hand side increases unboundedly (as $\mu \to \infty$), so that in the limit the trajectory of the system converges to a function that is discontinuous in the generalized velocity;
- to simplify calculations, the parameter μ is assumed to be dimensionless, and taking this into account in problems of mechanics of systems with impacts, the time, length and mass scales are selected appropriately.

Note that from the point of view of the theory of control systems, the action arising as a contact force can be considered as some specially designed impulse control. In the next subsection, we describe systems with a controlled singular phase of motion.

Natural and Singular Phases of Motion

Suppose the evolution of a controlled dynamical system with an elastic constraint is described by the pair of variables $x_p(t) \in \mathbb{R}^n$, $x_v(t) \in \mathbb{R}^n$, with $t \in [0, T]$, where vectors x_p and x_v are the generalized coordinates and generalized velocities, respectively, and the time T is large enough.

Assume that the set constraining the motion is given in the form

$$\{(x_p, t) : G(x_p, t) \leq 0\}, \tag{8.12}$$

where $G : \mathbb{R}^n \times [0, T] \to \mathbb{R}$ is a sufficiently smooth function. The motion in the constraint-free domain $\{(x_p, t) : G(x_p, t) > 0\}$ is regarded as the natural (regular) phase of motion, while the motion in the domain $\{(x_p, t) : G(x_p, t) \leq 0\}$ is regarded as singular motion.

Natural Phase of Motion

The motion is described by the equations

$$\begin{cases} \dot{x}_p(t) = F_p^{\mathrm{r}}\big(x_p(t), x_v(t), t\big), \\ \dot{x}_v(t) = F_v^{\mathrm{r}}\big(x_p(t), x_v(t), u(t), t\big) + \\ \qquad + \mu F_v^{\mathrm{s}}\big(x_p(t), x_v(t), w_1^{\mu}(\xi, t), t, \mu\big)\, \mathbf{I}\{t : G(x_p(t), t) \le 0\} + \\ \qquad + \mu F_v^{\mathrm{rs}}\big(x_p(t), x_v(t), w_2^{\mu}(\xi, t), t, \mu\big)\, \mathbf{I}\{t : G(x_p(t), t) > 0\}, \end{cases} \tag{8.13}$$

where $t \mapsto u(t) \in U \subset \mathbb{R}^m$ (a measurable function) is the control in the constraint-free domain and U is a compact set (the superscripts "r", "s" and "rs" stand for "regular," "singular," and "regular-singular".) Also, $F_p^{\mathrm{r}}(x_p, x_v, t)$ and $F_v^{\mathrm{r}}(x_p, x_v, u, t)$ are generalized forces in the constraint-free domain, and $\mu F_v^{\mathrm{s}}\big(x_p, x_v, w_1^{\mu}(\xi, t), t, \mu\big)$ is a generalized controllable (with control $w_1^{\mu}(\xi, t)$) contact force, which arises due to the violation of the constraint (8.12). Finally, $\mu F_v^{\mathrm{rs}}\big(x_p, x_v, w_2^{\mu}(\xi, t), t, \mu\big)$ is an additional generalized force acting under the control $w_2^{\mu}(\xi, t)$ in the constraint-free domain, but on a time interval of duration of order $\mu^{-1/2}$. Here ξ is the output signal of a sensor that measures the state of the system during its motion in the constrained area, that is, under violation of the constraint.

The force $\mu F_v^{\mathrm{s}}\big(x_p, x_v, w_1^{\mu}(\xi, t), t, \mu\big)$ satisfies the following conditions:

$$\begin{aligned} & F_v^{\mathrm{s}}\big(x_p, x_v, w_1^{\mu}, t, \mu\big) = 0, \text{ if} \\ & \qquad (1)\ G\big(x_p, t\big) > 0, \text{ or} \\ & \qquad (2)\ G\big(x_p, t\big) = 0 \text{ and} \\ & \qquad \left.\frac{d}{dt}\right|_{F_p^{\mathrm{r}}} G(x_p, t) = G'_{x_p}(x_p, t) F_p^{\mathrm{r}}(x_p, x_v, t) + G'_t(x_p, t) = 0, \end{aligned} \tag{8.14}$$

where G'_{x_p} and G'_t are the partial derivatives with respect to x_p and t, respectively, and $\left.\frac{d}{dt}\right|_{F_p^{\mathrm{r}}} G(x_p, t)$ stands for the partial derivative of the function $G(x_p, t)$ by virtue of the system

$$\dot{x}_p(t) = F_p^{\mathrm{r}}\big(x_p(t), x_v(t), t\big).$$

Note that $G(x_p, t)$ does not depend on x_v, and the second term in (8.14) defines the partial derivative of the function $G(x_p, t)$ with respect to time along the trajectories of the system (8.13).

The force $\mu F_v^{\mathrm{rs}}\big(x_p, x_v, w_2^{\mu}(t), t, \mu\big)$ in (8.13) characterizes the external impulsive action in the so-called *intersingular* phase of motion; an example of such a motion is given later in Sect. 8.7.3. In this phase of motion the constraints are not violated, however its duration is of an order of $\mu^{-1/2}$, i.e., it tends to zero if $\mu \to \infty$. Over the duration of the intersingular phase, the system seems to "emerge" from the forbidden domain, only to "dive" there again under the action of an impulsive external force. For this reason

$$F_v^{rs}(x_p, x_v, w_2^\mu, t, \mu) = 0 \quad \text{for } G(x_p, t) < 0. \tag{8.15}$$

We assume that in the singular phase, that is, when $G\big(x_p(t), t\big) \le 0$, the components of the state vector $\big(x_p(t),\ x_v(t)\big)$ are not measurable directly, but the output signal $\xi(t) \in \mathbb{R}^k$ of the sensor is available for measurement; this signal satisfies the observation equation

$$\dot{\xi}(t) = \mu H\big(x_p(t), x_v(t), \alpha^\mu(\xi, t), t, \mu\big)\, \mathbf{I}\{t : G(x_p(t), t) \le 0\}, \tag{8.16}$$

where $\alpha^\mu(\xi, t)$ is the observation control. Here

$$H(x_p, x_v, \alpha^\mu, t, \mu) = 0 \quad \text{for } G(x_p, t) > 0. \tag{8.17}$$

Let τ be the time instant the singular phase of motion begins, that is, τ is the first time instant when

$$G\big(x_p(\tau), \tau\big) = 0 \quad \text{and} \quad \frac{d}{dt}\Big|_{F_p^r} G\big(x_p(\tau), \tau\big) < 0. \tag{8.18}$$

Let γ be any of the controls w_1, w_2, or α. We define the dependence of these controls on t and μ in the singular (or the intersingular) phase, coordinated with the starting moment τ, as follows.

$$\gamma^\mu(\xi, t) = \begin{cases} \gamma\big(\xi, \sqrt{\mu}(t-\tau)\big), & \text{for } t \ge \tau, \\ 0, & \text{otherwise.} \end{cases} \tag{8.19}$$

Moreover, assume that $\gamma^\mu(\xi, t)$ is measurable in t and satisfies the Lipschitz condition with respect to ξ:

$$|\gamma(\xi', t) - \gamma(\xi'', t)| \le L\|\xi' - \xi''\|_t, \quad L = \text{const}, \tag{8.20}$$

where, for example,

$$\|\xi\|_t = \operatorname*{ess\,sup}_{\tau \le s \le t} |\xi(s)| := \min\big\{\lambda : |\xi(s)| \le \lambda, \quad \text{a.e. } s \in [\tau,\ t]\big\},$$

or

$$\|\xi\|_t = \left(\int_\tau^t |\xi(s)|^2\, ds\right)^{1/2}.$$

Definition 8.1 A control $w_1^\mu(\xi, t)$ in the singular phase of motion is said to be *admissible* if $w_1^\mu(\xi, t)$ is functional that is bounded and measurable in t, and moreover the dependence on τ, t, μ and ξ is given by the relations (8.19) and (8.20), and the constraint has the form $w_1^\mu(\xi, t) \in W_1 \subset \mathbb{R}^{r_1}$, where W_1 is a compact set containing zero. Admissible controls $w_2^\mu(\xi, t)$ and $\alpha^\mu(\xi, t)$ are defined analogously.

Omitting further details, we assume that the right-hand sides of the Eqs. (8.13) and (8.16) are smooth enough to ensure existence and uniqueness of their solutions in the corresponding domains defined by the function $G(x_p, t)$. In order to determine the limit motion for $\mu \to \infty$ we introduce a special space-time transformation in the vicinity of a singular point.

Space-Time Transformation in the Neighborhood of a Singular Point

Suppose the motion starts at some initial point $x_p(0)$, $x_v(0)$, such that $G(x_p(0), 0) > 0$. Let τ be the first time instant the trajectory of the system reaches the constraint boundary, so the conditions (8.18) and (8.19) are fulfilled.

As is known, in the case of mechanical systems with viscoelastic constraints the typical size of the space and time intervals of constraint violation is of the order of $\sim \mu^{-1/2}$ [215, 346], as shown in Fig. 8.4, and in the limit, this motion is continuous in the state variable and discontinuous in the velocity, as shown in Fig. 8.5.

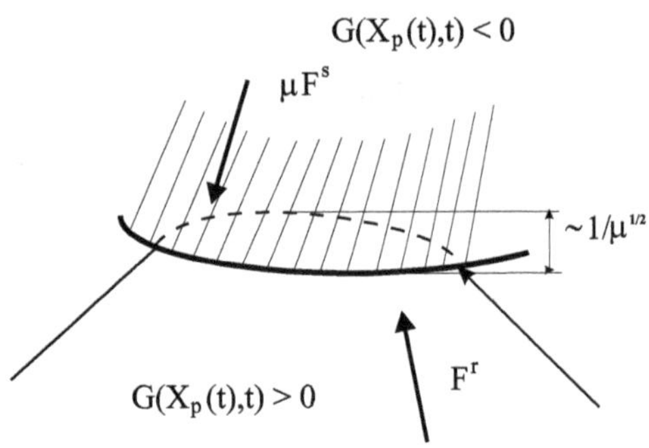

Fig. 8.4 Motion of a dynamical system with a constraint for $\mu < \infty$

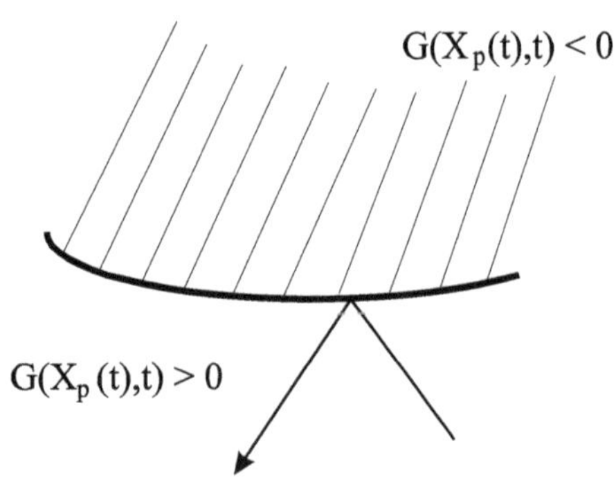

Fig. 8.5 Motion of a dynamical system with a constraint in the limit $\mu \to \infty$

For finite values of μ, there exists a non-zero time interval in which the constraint is violated, therefore one can use the following space-time transformation of coordinates in a neighborhood of the impact point for $s > 0$:

$$\begin{cases} y_p^\mu(s) = x_p(\tau) + \mu^{1/2}\big[x_p(\tau + \mu^{-1/2}s) - x_p(\tau)\big], \\ y_v^\mu(s) = x_v(\tau + \mu^{-1/2}s), \\ \eta^\mu(s) = \xi(\tau + \mu^{-1/2}s), \\ t^\mu(s) = \tau + \mu^{-1/2}s. \end{cases} \tag{8.21}$$

The new set of variables $\{y_p^\mu(s), y_v^\mu(s), \eta^\mu(s)\}$ satisfies on the interval

$$\{s > 0 : G\big(y_p^\mu(s), t^\mu(s)\big) < 0\}$$

the system of equations

$$\begin{cases} \dot{y}_p^\mu(s) = F_p^{\mathrm{r}}\Big(\dfrac{y_p^\mu(s) - x_p(\tau)}{\mu^{1/2}} + x_p(\tau), y_v^\mu(s), \tau + \dfrac{s}{\mu^{1/2}}\Big), \\ \dot{y}_v^\mu(s) = \mu^{1/2} F_v^{\mathrm{s}}\Big(\dfrac{y_p^\mu(s) - x_p(\tau)}{\mu^{1/2}} + x_p(\tau), y_v^\mu(s), w_1(\eta^\mu, s), \tau + \dfrac{s}{\mu^{1/2}}, \mu\Big) + \\ + \mu^{1/2} F_v^{\mathrm{rs}}\Big(\dfrac{y_p^\mu(s) - x_p(\tau)}{\mu^{1/2}} + x_p(\tau), y_v^\mu(s), w_2(\eta^\mu, s), \tau + \dfrac{s}{\mu^{1/2}}, \mu\Big) + \\ + \mu^{-1/2} F_v^{\mathrm{r}}\Big(\dfrac{y_p^\mu(s) - x_p(\tau)}{\mu^{1/2}} + x_p(\tau), y_v^\mu(s), u(\tau + \mu^{-1/2}s), \tau + \dfrac{s}{\mu^{1/2}}\Big), \\ \dot{\eta}^\mu(s) = \dfrac{1}{\mu^{1/2}} H\Big(\dfrac{y_p^\mu(s) - x_p(\tau)}{\mu^{1/2}} + x_p(\tau), y_v^\mu(s), \alpha(\eta^\mu, s), \tau + \dfrac{s}{\mu^{1/2}}, \mu\Big). \end{cases} \tag{8.22}$$

with the initial conditions

$$y_p^\mu(0) = x_p(\tau), \quad y_v^\mu(0) = x_v(\tau-), \quad \eta^\mu(0) = \xi(\tau).$$

The systems of Eq. (8.13) for the natural phase of motion, the transformation (8.21), and the system of Eq. (8.22) for the singular phase of motion are regarded as a *multiscale representation of the motion* of the original system, since they combine the *slow* and the *fast* motions in the natural and singular phases of motion.

8.2.3 The Limit Motion in the Synthesis of Impulse Control Rules

In what follows we present several technical results (see [37]) which serve as the basis for the synthesis of impulse control laws. Theorem 8.1 describes the limit behavior of a singularly perturbed system (8.22) as $\mu \uparrow \infty$. This theorem and its corollary show that the jump in the generalized velocity at the time of the impact is described by means of the shift operator along the trajectories of the limiting system of equations for the fast motion. This theorem also shows how the control in the singular (fast) phase of motion acts and thus *builds a bridge* between impact mechanics and the theory of impulsive control. A sufficient condition for the existence of a representation in terms of the shift operator is given by the 8.1 assumption (see Assumption 8.1). Further, Theorems 8.2 and 8.4 give a description of the limit motion in terms of differential equations with measures, where the jumps of generalized velocities are given in terms of the shift operator. And finally, Theorem 8.5 shows how the control synthesized for the limiting system can be used for the initial systems (8.12)–(8.13). Theorems 8.1–8.5 enable us to reduce the ill-conditioned problem of synthesis of control signals $\gamma^{\mu}(\xi, t)$ from (8.19) in the Eqs. (8.13) and (8.16) (the exact value of μ is usually unknown) to the following regular two-step procedure:

(a) synthesis of bounded controls $w_{\tau_i}(\cdot)$, $i = 1, \dots, N$, which define the weight of the δ-function in the equations of the limit motion given by the Theorems 8.2 and 8.4, and
(b) finding control signals $\gamma^{\mu}(\xi, t)$ for the original system by simple scaling (in accordance with Theorem 8.5) of the control signals $w_{\tau_i}(\cdot)$, $i = 1, \dots, N$, found in stage (a).

Discontinuous Limit Motion of the Dynamical System

Assumption 8.1 The functions F_v^{s}, F_v^{rs} and H satisfy the Lipshitz condition in the following form: There are constants $L > 0$ and $\mu_0 > 0$, such that for any $(x_p', x_p'', x_v', x_v'')$, $t \in [0, T]$, $T < \infty$, $w_1^{\mu} \in W_1$ and $\mu \geq \mu_0$ it holds that

$$\begin{aligned}\left\| F_v^{\mathrm{s}}(x_p', x_v', w_1^{\mu}, t, \mu) - F_v^{\mathrm{s}}(x_p'', x_v'', w_1^{\mu}, t, \mu)\right\| &\leq \\ \leq L\big(\|x_p' - x_p''\| + \mu^{-1/2}\|x_v' - x_v''\|\big).\end{aligned} \tag{8.23}$$

Condition similar to (8.23) are satisfied for known mechanical systems with visco-elastic properties (for example, in the Kelvin-Voight and Maxwell models) and are usually related with the so-called Rayleigh dissipation potential (function) in the domain $\{(x_p, t) : G(x_p, t) \leq 0\}$ [215].

Consider first the case of a single violation of the constraint. In this case

$$F_v^{\mathrm{rs}}\big(x_p, x_v, w_2^\mu(\xi, t), t, \mu\big) \equiv 0.$$

Theorem 8.1 *Suppose that in addition to Assumption 8.1 the following conditions are fulfilled:*

(1) *For any admissible controls w_1 and α, and any (x_p, τ), such that $G(x_p, \tau) = 0$ and $\left.\frac{d}{dt}\right|_{F_p^{\mathrm{r}}} G\big(x_p(\tau), \tau\big) < 0$, there exist the limits*

$$\lim_{\mu\uparrow\infty} F_p^{\mathrm{r}}\left(\frac{y_p - x_p}{\mu^{1/2}} + x_p, y_v, \tau + \mu^{-1/2}s)\right) = F_p^{\mathrm{r}}(x_p, y_v, \tau),$$

$$\begin{aligned}&\lim_{\mu\uparrow\infty} \mu^{1/2} F_v^{\mathrm{s}}\left(\frac{y_p - x_p}{\mu^{1/2}} + x_p, y_v, w_1(\eta^\mu, s), \tau + \mu^{-1/2}s, \mu\right) = \\ &\quad = \bar{F}_v^{\mathrm{s}}\big(y_p, y_v, w_1(\eta, s), x_p, \tau\big),\end{aligned} \tag{8.24}$$

$$\begin{aligned}&\lim_{\mu\uparrow\infty} \mu^{1/2} H\left(\frac{y_p - x_p}{\mu^{1/2}} + x_p, y_v, \alpha(\eta^\mu, s), \tau + \mu^{-1/2}s, \mu\right) = \\ &\quad = \bar{H}\big(y_p, y_v, \alpha(\eta, s), x_p, \tau\big),\end{aligned}$$

where the convergence is uniform in any bounded neighborhood of (y_p, y_v, η, s);

(2) *the system of differential equations*

$$\begin{cases} \dot{y}_p(s) = F_p^{\mathrm{r}}\big(x_p(\tau), y_v(s), \tau\big), \\ \dot{y}_v(s) = \bar{F}_v^{\mathrm{s}}\big(y_p(s), y_v(s), w_1(\eta, s), x_p(\tau), \tau\big), \\ \dot{\eta}(s) = \bar{H}\big(y_p(s), y_v(s), \alpha(\eta, s), x_p(\tau), \tau\big) \end{cases} \tag{8.25}$$

with the initial conditions $y_p(0) = x_p(\tau)$, $y_v(0) = x_v(\tau-)$, $\eta(0) = \xi(\tau)$ has exactly one solution on some segment $[0, s^*(\tau) + \varepsilon]$, where $\varepsilon > 0$ and

$$s^*(\tau) = \inf_{s>0} \left\{ \begin{aligned} &G_t'\Big|_{(x_p(\tau),\tau)} s + G_{x_p}'\Big|_{(x_p(\tau),\tau)} \big(y_p(s) - x_p(\tau)\big) = 0, \\ &G_t'\Big|_{(x_p(\tau),\tau)} + G_{x_p}'\Big|_{(x_p(\tau),\tau)} F_p^{\mathrm{r}}\big(x_p(\tau), y_v(s), \tau\big) > 0. \end{aligned} \right\} \tag{8.26}$$

Then as $\mu \to \infty$,

$$\big(y_p^\mu(s), y_v^\mu(s), \eta^\mu(s)\big) \to \big(y_p(s), y_v(s), \eta(s)\big) \tag{8.27}$$

uniformly on $[0, s^*(\tau) + \varepsilon]$, and for all sufficiently large μ there exists the time

$$s_\mu^*(\tau) = \inf_{s>0} \left\{ \begin{array}{l} G\big(x_p(\tau+\mu^{-1/2}s), \tau+\mu^{-1/2}s\big) = 0, \\ G_t'\Big|_{(x_p(\tau+\mu^{-1/2}s),\tau+\mu^{-1/2}s)} + G_{x_p}'\Big|_{(x_p(\tau+\mu^{-1/2}s),\tau+\mu^{-1/2}s)} \times \\ \times F_p^{\mathrm{r}}\big(x_p(\tau+\mu^{-1/2}s), x_v(\tau+\mu^{-1/2}s), \tau+\mu^{-1/2}s\big) > 0 \end{array} \right\}, \tag{8.28}$$

such that

$$s_\mu^*(\tau) \to s^*(\tau). \tag{8.29}$$

Proof Due to the continuity of the function F_p^{r}, it is evident that

$$\lim_{\mu\uparrow\infty} F_p^{\mathrm{r}}\left(\frac{y_p - x_p}{\mu^{1/2}} + x_p, y_v, \tau + \mu^{-1/2}s)\right) = F_p^{\mathrm{r}}(x_p, y_v, \tau).$$

Next, it follows from conditions (8.23) and (8.24) that the functions

$$\bar{F}_v^{\mathrm{s}}\big(y_p, y_v, s, w_1(s), x_p, \tau\big) \quad \text{and} \quad \bar{H}\big(y_p, y_v, s, \alpha(s), x_p, \tau\big)$$

satisfy the Lipshitz condition in the variables (y_p, y_v).

Thanks to the continuity of solutions of a differential equation with respect to parameters (see [96], p. 71 or [365], Theorem 7), $\big(y_p^\mu(s), y_v^\mu(s), \eta^\mu(s)\big)$ converges to $\big(y_p(s), y_v(s), \eta(s)\big)$ uniformly on $[0, s^* + \varepsilon]$.

To establish the convergence (8.27), we introduce the functions

$$f^0(s) = G_{x_p}'\big(x_p(\tau), \tau\big)\big(y_p(s) - x_p(\tau)\big) + G_t'\big(x_p(\tau), \tau\big)s,$$

and

$$f^\mu(s) = G_x'\big(x_p(\tau+\mu^{-1/2}s), \tau+\mu^{-1/2}s\big)\big(y_p^\mu(s) - x_p(\tau)\big) + \\ + G_t'\big(x_p(\tau+\mu^{-1/2}s), \tau+\mu^{-1/2}s\big)s.$$

In view of (8.26), the derivative

$$\dot{f}^0(s) = G_{x_p}'\big(x_p(\tau), \tau\big)\dot{y}_p(s) + G_t'\big(x_p(\tau), \tau\big)$$

is positive at the point $s^*(\tau)$ and $f^\mu(s) \to f^0(s)$ uniformly on $[0, s^*(\tau) + \varepsilon]$. The claim is now a consequence of the uniform convergence of $\big(y_p^\mu(s), y_v^\mu(s), \eta^\mu(s)\big)$ to $\big(y_p(s), y_v(s), \eta(s)\big)$ and the following auxiliary result. □

Lemma 8.1 *For any continuous and smooth real function $f(\cdot) \in C^1$ defined for some $a > 0$, the exit time*

$$\tau^{\delta}(f) = \begin{cases} \inf\{s > a : f(s) \geq \delta\}, \\ \infty, \ if \ \{s > a : f(s) \geq \delta\} = \emptyset. \end{cases}$$

Let $f^0(\cdot) \in C^1$ be a real-valued function such that $f^0(a) < 0$, the exit time $\tau^0(f^0) < \infty$, and the derivative of $f^0(\cdot)$ is positive at $t = \tau^0(f^0)$. Then the functional $f \mapsto \tau^0(f)$ is continuous at the point $f^0(\cdot)$ in the topology of uniform convergence. That is, if $f^n(s)$ converges uniformly to $f^0(s)$ on some interval $[a, \tau^0(f^0) + \varepsilon]$, with $\varepsilon > 0$, then

$$\tau^0(f^n) \to \tau^0(f^0), \quad n \to \infty.$$

Proof of Lemma 8.1 Since the derivative of the function f^0 at the point $\tau^0(f^0)$ is positive, there exists $\delta_0 > 0$ such that $\tau^{-\delta}(f^0) < \tau^0(f^0) < \tau^{\delta}(f^0)$ for any $0 < \delta < \delta_0$. Moreover, $\tau^{\delta}(f^0) \downarrow \tau^0(f^0)$ and $\tau^{-\delta}(f^0) \uparrow \tau^0(f^0)$ as $\delta \to 0$. Thanks to the uniform convergence of $\{f^n(\cdot)\}$ to $f^0(\cdot)$, for any $\delta < \delta_0$ one can find $n(\delta)$ such that for $n \geq n(\delta)$

$$f^n\big(\tau^{\delta}(f^0)\big) > f^0\big(\tau^{\delta}(f^0)\big) - \frac{\delta}{2} = \frac{\delta}{2} > 0,$$

and

$$f^n(s) < f^0\big(\tau^{-\delta}(f^0)\big) + \frac{\delta}{2} = -\frac{\delta}{2} < 0 \quad \text{for all } s \leq \tau^{-\delta}(f^0).$$

Hence, due to the continuity of f^n for $n \geq n(\delta)$, we conclude that

$$\tau^0(f^n) \in \big(\tau^{-\delta}(f^0), \tau^{\delta}(f^0)\big),$$

which establishes the claimed convergence because of the arbitrariness of δ. The proof of the lemma is complete. □

Continuing, the assumptions of the theorem ensure that there exists $a > 0$ such that $f^0(s) < 0$ for $s \in (0, a]$. By Lemma 8.1, there exists the limit $\tau^0(f^{\mu}) \to \tau^0(f^0) = s^*$. For the sake of convenience, denote

$$x_p\big(\tau + \mu^{-1/2}s\big) = x_p^{\mu}(s).$$

Then

$$G\big(x_p^{\mu}(s), \tau + \mu^{-1/2}s\big) - G\big(x_p(\tau), \tau\big) = G\big(x_p^{\mu}(s), \tau + \mu^{-1/2}s\big) =$$
$$= \mu^{-1/2}\big[G'_{x_p}\big(x_p^{\mu}(s), \tau + \mu^{-1/2}s\big)\big(y_p^{\mu}(s) - x_p(\tau)\big) + G'_t\big(x_p^{\mu}(s), \tau + \mu^{-1/2}s\big)s\big] +$$
$$+ O(\mu^{-1}),$$

where $O(\mu^{-1})\mu^{1/2} \to 0$ (as $\mu \to \infty$), uniformly on the interval $s \in [0, s^* + \varepsilon]$. Since $G\big(x_p(\tau), \tau\big) = 0$, we obtain

$$\mu^{1/2} G\big(x_p^\mu(s), \tau + \mu^{-1/2} s\big) = f^\mu(s) + \frac{O(\mu^{-1})}{\mu^{-1/2}}. \tag{8.30}$$

Next, since the derivative of f^0 at the point $\tau^0(f^0)$ is positive, there exists δ_0 such that for any $\delta \in (0, \delta_0)$,

$$\tau^{-\delta}(f^0) < \tau^{\delta}(f^0) \le s^* + \varepsilon.$$

Then the uniform convergence of f^μ to f^0 for the chosen δ and sufficiently large $\mu > \mu_1(\delta)$ implies that

$$f^\mu\big(\tau^\delta(f^0)\big) \ge \frac{\delta}{2} \quad \text{and} \quad f^\mu\big(\tau^{-\delta}(f^0)\big) \le -\frac{\delta}{2}.$$

At the same time, we can choose $\mu_2(\delta)$ such that for $\mu > \mu_2(\delta)$ we will have $\big|O(\mu^{-1})\mu^{1/2}\big| \le \delta/4$. Consequently, (8.30) implies that, for sufficiently large μ,

$$\mu^{1/2} G\big(x_p(\tau + \mu^{-1/2}\tau^\delta(f^0)), \tau + \mu^{-1/2}\tau^\delta(f^0)\big) > \frac{\delta}{4}$$

and

$$\mu^{1/2} G\big(x_p(\tau + \mu^{-1/2}\tau^{-\delta}(f^0)), \tau + \mu^{-1/2}\tau^{-\delta}(f^0)\big) < -\frac{\delta}{4}.$$

This means that the function $G\big(x_p^\mu(s), \tau + \mu^{-1/2}s\big)$ changes sign on the interval $\big(\tau^{-\delta}(f^0), \tau^\delta(f^0)\big)$. Due to the uniform convergence and the expression (8.30), there are no other sign changes for $s < \tau^{-\delta}(f^0)$ and therefore there exists

$$s_\mu^* \in \big(\tau^{-\delta}(f^0), \tau^\delta(f^0)\big), \tag{8.31}$$

such that the first equality in (8.28) is satisfied. The inequality for the derivative follows from the expression

$$\dot f^\mu(s) = G'_{x_p}\big(x_p^\mu(s), \tau + \mu^{-1/2}s\big)\, \dot y_p^\mu(s) + G'_t\big(x_p^\mu(s), \tau + \mu^{-1/2}s\big) + O(\mu^{-1/2}),$$

and the uniform convergence of $\big(y_p^\mu(s), \dot y_p^\mu(s)\big)$ to $\big(y_p(s), \dot y_p(s)\big)$. Finally, the convergence of s_μ^* to s^* follows from (8.31) and the arbitrariness of δ. □

Remark 8.3 What Theorem 8.1 says is that the limit system (8.25) behaves in the same manner as the original system for large values of μ. However, the main result is the representation of a single jump by means of the solution of the limit system and the possibility of representing the jump using the shift operator along the trajectories of the limit system (8.25).

Corollary 8.1 *For sufficiently small $\varepsilon > 0$ the solution of the system (8.12)–(8.13) converges on the interval $[0, \tau + \varepsilon)$ to a discontinuous functions $\big(\bar{x}_p(t), \bar{x}_v(t)\big)$ such that*

$$\bar{x}_p(t) = x_p(t), \quad \bar{x}_v(t) = x_v(t), \quad t < \tau,$$

and

$$\bar{x}_p(\tau) = \lim_{\mu \uparrow \infty} x_p\big(\tau + \mu^{-1/2} s_\mu^*(\tau)\big) = x_p(\tau),$$

$$\bar{x}_v(\tau) = \lim_{\mu \uparrow \infty} x_v\big(\tau + \mu^{-1/2} s_\mu^*(\tau)\big) = y_v\big(s^*(\tau)\big).$$

The functions $\bar{x}_p(t), \bar{x}_v(t)$ may be regarded as a *generalized solution* of the original system (8.13). The variables $\bar{x}_p(t), \bar{x}_v(t)$ satisfy a specific system of differential equations exhibited below in Theorem 8.2.

In the case of a single shock in mechanical systems with elastic constraints, the above result was obtained in the works [346, 350, 393] with the aid of penalty function methods. Theorem 8.1 is, however, more general; it is not limited to special cases of elastic or viscoelastic obstacles and is valid for general dynamical systems with constraints.

"Repulsive" Constraint

The following results provide a representation of the solution to the limit system, where we use an explicit expression for the jump of the generalized velocity in terms of the shift operator along trajectories of the system (8.25). A similar representation is known for discrete-continuous systems that are robust with respect to impulsive actions [273, 275, 292, 294, 464, 465]. However, generally speaking, for systems with constraints the existence of such a representation for any initial conditions $x_p(\tau), x_v(\tau-)$ and arbitrary impulse controls $w_1(\cdot)$ in (8.25) cannot be guaranteed, because one cannot guarantee the existence of the exit instant $s^* < \infty$ from the constrained domain in (8.26). Assumption (2) of Theorem 8.1 implies that for the existence of such an instant $s^* < \infty$ it is necessary that the constraint possess the repulsiveness property. This property can be expressed in terms of the so-called *restitution force*.

In the domain $\big\{(y_p, s) : G'_{x_p}\big(x_p(\tau), \tau\big)\big[y_p - x_p(\tau)\big] + G'_t\big(x_p(\tau), \tau\big)s < 0\big\}$, along the trajectories of the limit system (8.25), the term

$$Z\big(y_p(s), s\big) = G'_x\big(x_p(\tau), \tau\big)\big[y_p(s) - x_p(\tau)\big] + G'_t\big(x_p(\tau), \tau\big)s \tag{8.32}$$

characterizes the principal part (of the order $\sim \mu^{-1/2}$) of the constraint violation for finite values of μ, and the restitution force on the trajectories of the limit system is usually expressed in terms of Z as $F(s) = \frac{d^2}{ds^2} Z(s)$.

Let us assume that the restitution force satisfies the inequality

$$\frac{d^2}{ds^2} Z(s) \geq -k_1 Z(s) - k_2 \dot{Z}(s). \tag{8.33}$$

This case is typical for mechanical systems with impacts and the Kelvin-Voigt and Maxwell models considered above. The next result is useful for estimation of properties of the constraint [37].

Proposition 8.1

(1) *Suppose the restitution force satisfies the inequality (8.33) in the domain $Z < 0$. Suppose also that $k_1, k_2 \geq 0$ and satisfy the inequality*

$$\left(\frac{k_2}{2}\right)^2 < k_1. \tag{8.34}$$

Let $Z(0) = 0$ and $\dot{Z}(0) < 0$ Then there exists the exit time from the domain $Z < 0$,

$$s^* = \inf\{s > 0 : Z(s) = 0,\ \dot{Z}(s) > 0\} < \infty,$$

with

$$s^* \leq \frac{\pi}{\omega}, \quad where \quad \omega^2 = k_1 - \left(\frac{k_2}{2}\right)^2.$$

(2) *If the restoring force satisfies in the domain $Z < 0$ the inequality*

$$\frac{d^2}{ds^2} Z(s) \leq -k_1 Z(s) - k_2 \dot{Z}(s), \tag{8.35}$$

then the exit time s^*, if it exists, satisfies the opposite inequality

$$s^* \geq \frac{\pi}{\omega}.$$

Remark 8.4 This result means that a restitution force of the type (8.33) guarantees the repulsion from the constrained domain in a finite time. Condition 1) is fulfilled for typical models of mechanical systems with shocks [70, 215].

Proof of Proposition 8.1 (1) The function $Z(s)$ satisfies the equation

$$\frac{d^2}{ds^2}Z(s) = f(s) - k_1 Z(s) - k_2 \dot{Z}(s),$$

with some function $f(s) \geq 0$. This equation has the explicit solution

$$Z(s) = e^{-\lambda s}\left[\dot{Z}(0)\frac{\sin \omega s}{\omega} + \int_0^s e^{\lambda\tau}\frac{\sin \omega(s-\tau)}{\omega} f(\tau)\,d\tau\right], \tag{8.36}$$

where

$$\omega^2 = k_1 - \left(\frac{k_2}{2}\right)^2, \quad \lambda = \frac{k_2}{2}.$$

It is readily verified that $Z\left(\dfrac{\pi}{\omega}\right) \geq 0$. Consider first the case when the Lebesgue measure of the set $\{s : f(s) > 0\}$ is positive. Then

$$\int_0^{\pi/\omega} e^{\lambda\tau}\frac{\sin \omega(\pi/\omega - \tau)}{\omega} f(\tau)\,d\tau > 0,$$

due to the positiveness of $\sin \omega(\pi/\omega - \tau)$ and the non-negativity of f. Since $\dot{Z}(0) < 0$, there is an interval $[0, \varepsilon)$ where $Z(s) < 0$. Therefore, one can find a time instant $s^* \in (0, \pi/\omega)$ such that $Z(s^*) = 0$ and $Z(s) < 0$ for all $s \in [0, s^*)$. At the same time,

$$\dot{Z}(s^*) = e^{-\lambda s^*}\left[\dot{Z}(0)\cos \omega s^* + \int_0^{s^*} e^{\lambda\tau}\cos \omega(s^* - \tau) f(\tau)\,d\tau\right]. \tag{8.37}$$

Then taking relations (8.36) and (8.37) into account, we obtain

$$\dot{Z}(s^*)\sin \omega s^* = e^{-\lambda s^*}\int_0^{s^*} e^{\lambda\tau}\sin \omega\tau\, f(\tau)\,d\tau > 0.$$

This implies that $\dot{Z}(s^*) > 0$ if $\sin \omega s^* \neq 0$. If, however, $\sin \omega s^* = 0$, we have that $s^* = \pi/\omega$ and $\displaystyle\int_0^{s^*} e^{\lambda\tau}\sin \omega\tau\, f(\tau)\,d\tau = 0$. If $f(s) = 0$ almost surely on $[0, s^*]$, then as it follows from (8.37),

$$\dot{Z}(s^*) = -e^{-\lambda s^*}\dot{Z}(0) > 0.$$

The second statement is proved similarly.

Representation of Jumps of the Generalized Velocity

Here we provide a description of the jump in terms of the shift operator along the trajectories of the “fast motion” system.

Suppose the system starts with the initial condition

$$x_p(0), x_v(0) \quad \text{such that } G\big(x_p(0), 0\big) > 0.$$

Let τ is the first instant of time the path hits the constraint, that is, (8.18) holds. To describe the jump of the velocity at the point τ, we introduce the system of equations of “fast motion”

$$\begin{cases} \dot{y}_p(s) = F_p^{\mathrm{r}}\big(x_p(\tau), y_v(s), \tau\big), \\ \dot{y}_v(s) = \bar{F}_v\big(y_p(s), y_v(s), s, w_\tau(s), x_p(\tau), \tau\big), \end{cases} \qquad s \in [0, \infty), \tag{8.38}$$

with the initial condition $y_p(0) = x_p(\tau)$, $y_v(0) = x_v(\tau-)$ and the admissible (in sense of Definition 8.1) control

$$w_\tau(s) \in W \subset \mathbb{R}^k. \tag{8.39}$$

Let

$$\Psi\big(x_p(\tau), x_v(\tau-), w_\tau(\cdot), \tau\big) = \begin{pmatrix} \Psi_p\big(x_p(\tau), x_v(\tau-), w_\tau(\cdot), \tau\big) \\ \Psi_v\big(x_p(\tau), x_v(\tau-), w_\tau(\cdot), \tau\big) \end{pmatrix} \tag{8.40}$$

be the shift operator along the trajectories of the system (8.38) over the time the trajectories lie in the domain $\{Z\big(y_p(s), s\big) < 0\}$. Then the discontinuity of the generalized velocity $x_v(\cdot)$ at the instant $t = \tau$ is expressed as

$$x_v(\tau) = x_v(\tau-) + \Psi_v\big(x_p(\tau), x_v(\tau-), w_\tau(\cdot), \tau\big). \tag{8.41}$$

Relation (8.41) expresses the jump at the point $t = \tau$ in terms of the solution of Eq. (8.38) (equations of “fast dynamics ”). If

$$\Phi\big(\cdot, s, w(\cdot), \tau\big) = \begin{pmatrix} \Phi_p\big(\cdot, s, w_\tau(\cdot), \tau\big) \\ \Phi_v\big(\cdot, s, w_\tau(\cdot), \tau\big) \end{pmatrix} \tag{8.42}$$

is the general solution of (8.38) with the initial condition $y_p(0) = x_p(\tau)$, $y_v(0) = x_v(\tau-)$, then

$$x_v(\tau) = \Phi_v\big(x_p(\tau), x_v(\tau-), s^*(x_p(\tau), x_v(\tau-)), w_\tau(\cdot), \tau\big), \tag{8.43}$$

where

$$s^*\big(x_p(\tau), x_v(\tau-)\big) =$$

$$= \inf_{s>0} \left\{ \begin{array}{l} G_t'\Big|_{(x_p(\tau),\tau)} s + \\ + G_{x_p}'\Big|_{(x_p(\tau),\tau)} \Big(\Phi_p\big(x_p(\tau), x_v(\tau-), s, w_\tau(\cdot), \tau\big) - x_p(\tau)\Big) = 0, \\ G_t'\Big|_{(x_p(\tau),\tau)} + \\ + G_{x_p}'\Big|_{(x_p(\tau),\tau)} F_p^{\mathrm{r}}\Big(x_p(\tau), \Phi_v\big(x_p(\tau), x_v(\tau-), s, w_\tau(\cdot), \tau\big), \tau\Big) > 0 \end{array} \right\},$$

the inequality

$$G_t'\Big|_{(x_p(\tau),\tau)} s + G_x'\Big|_{(x_p(\tau),\tau)} \Big(\Phi_p\big(x_p(\tau), x_v(\tau-), s, w_\tau(\cdot), \tau\big) - x_p(\tau)\Big) < 0$$

holds on the entire interval $\big(0, s^*(x_p(\tau), x_v(\tau-))\big)$, and

$$\Psi_v\big(x_p(\tau), x_v(\tau-), w_\tau(\cdot), \tau\big) =$$
$$= \Phi_v\big(x_p(\tau), x_v(\tau-), s^*(x_p(\tau), x_v(\tau-)), w_\tau(\cdot), \tau\big) - x_v(\tau-).$$

We define a generalized solution of the system (8.12)–(8.13) as the pointwise limit of a sequence of ordinary solutions for $\mu \to \infty$. Then, by Corollary 8.1 and the jump representation with the aid of the shift operator, we have the following result.

Theorem 8.2 *Let $(x_p^\mu(t), x_v^\mu(t))$ be the solution of the system (8.13) in the domain (8.12) for finite values of the parameter μ (the superscript indicates that the solution depends on μ). Then the generalized solution $(\bar{x}_p(t), \bar{x}_v(t))$ of the original system (8.13) is the pointwise limit of the ordinary solution $(x_p^\mu(t), x_v^\mu(t))$ as $\mu \to \infty$ and satisfies on the segment $[0, \tau + \varepsilon]$ the system of differential equations*

$$\begin{cases} \dot{\bar{x}}_p(t) = F_p^r\big(\bar{x}_p(t), \bar{x}_v(t), t\big), \\ \dot{\bar{x}}_v(t) = F_v^r\big(\bar{x}_p(t), \bar{x}_v(t), u(t), t\big) + \Psi_v\big(\bar{x}_p(\tau), \bar{x}_v(\tau-), w_\tau(\cdot), \tau\big)\,\delta(t-\tau), \end{cases} \tag{8.44}$$

with the boundary conditions $\bar{x}_p(0) = x_p(0)$, $\bar{x}_v(0) = x_v(0)$, $\bar{x}_p(\tau) = x_p(\tau)$, $\bar{x}_v(\tau-) = x_v(\tau-)$, where the arbitrary ε is small enough.

Here $\Psi_v(\cdot)$ is the v-component of the shift operator along the trajectories of the system (3.5), so that

$$y_v\big(s^*(\tau)\big) = y_v(0) + \Psi_v\big(y_p(0), y_v(0), w_\tau(\cdot), \tau\big),$$

where $w_\tau(\cdot) = \{w(\eta, s) : 0 \le s \le s^*(\tau)\}$ (we will use the same notation for the control α). The time $s^*(\tau)$ is given by (8.26); moreover, we used the fact that $\big(\bar{x}_p(\tau), \bar{x}_v(\tau-)\big) = \big(x_p(\tau), x_v(\tau-)\big) = \big(y_p(0), y_v(0)\big)$.

Since (8.44) describes both the regular and the singular phases of motion, we will call this system the *complete limit system.*

Implementation of Impulse Controls in Systems with Collisions

The next statement is, in some sense, a converse to Theorem 8.2. It shows how, given controls $u(t)$, $w_\tau(\cdot)$, $\alpha_\tau(\cdot)$ that generate a generalized solution $(\bar{x}_p(t), \bar{x}_v(t))$ of system (8.44), to obtain controls generating a sequence $(x_p^\mu(t), x_v^\mu(t))$ of solutions of the original system (8.13) that converge pointwise to $(\bar{x}_p(t), \bar{x}_v(t))$ as $\mu \to \infty$.

First we consider a single impulse action at the shock point.

Theorem 8.3 *For any solution $(\bar{x}_p(t), \bar{x}_v(t))$ of the system (8.44), corresponding to some admissible controls $u(t)$, $w_\tau(\cdot)$, $\alpha_\tau(\cdot)$, there exists a sequence of solutions $(x_p^\mu(t), x_v^\mu(t))$ of the original system (8.13) that converges to the solution $(\bar{x}_p(t), \bar{x}_v(t))$ as $\mu \to \infty$ everywhere on $[0, T]$ except at the point τ, provided the μ-parametrized controls in the original system are chosen as follows:*

$$
\begin{aligned}
u^\mu(t) &= \begin{cases} u(t), & if\ t \in [0, \tau], \\ any\ admissible\ value, & if\ t > \tau, \end{cases} \\
\gamma^\mu(\xi, t) &= \begin{cases} \gamma(\xi, \sqrt{\mu}(t-\tau)), & if\ t \in [\tau, \tau + t_\mu^*], \\ any\ admissible\ value,\ otherwise, \end{cases}
\end{aligned}
\tag{8.45}
$$

where γ stands for w or α, and $t_\mu^* = \min\left\{\dfrac{s^*(\tau)}{\sqrt{\mu}}, \dfrac{s_\mu^*(\tau)}{\sqrt{\mu}}\right\}$.

Proof To obtain the solution $(\bar{x}_p(t), \bar{x}_v(t))$ under the controls $u(t)$, $w_\tau(\cdot)$, $\alpha_\tau(\cdot)$, we find the solution of the system (8.13) with the control $u(t)$, $t \in [0, \tau]$, and then calculate the shift operator $\Psi_v\big(x_p(\tau), x_v(\tau-), w_\tau(\cdot), \tau\big)$ by integrating the "fast motion" system (8.25), (8.26) for given controls $w(\eta, s)$, $\alpha(\eta, s)$, $s \in [0, s^*(\tau)]$, where the functional dependence on η is also given. Further, we consider the system (8.13) with sufficiently large $\mu > \mu_0$ and with controls of the type (8.45) equal to those applied in (8.44), where $w_\tau(\cdot)$ and $\alpha_\tau(\cdot)$ can be continued, if necessary, beyond the point $s^*(\tau)$ in any admissible way. Denote the solution of this system by $(x_p^\mu(t), x_v^\mu(t))$. Then, by (8.45), for $t \in [0, \tau)$ we have

$$(x_p^\mu(t), x_v^\mu(t)) = (\bar{x}_p(t), \bar{x}_v(t));$$

moreover, by Theorem 8.1 and Corollary 8.1 there exists a time instant $\tau^{\mu*} = \tau + \mu^{-1/2} s_\mu^*(\tau)$ such that $\tau^{\mu*} \to \tau$ and $(x_p^\mu(\tau^{\mu*}), x_v^\mu(\tau^{\mu*})) \to (\bar{x}_p(\tau), \bar{x}_v(\tau))$

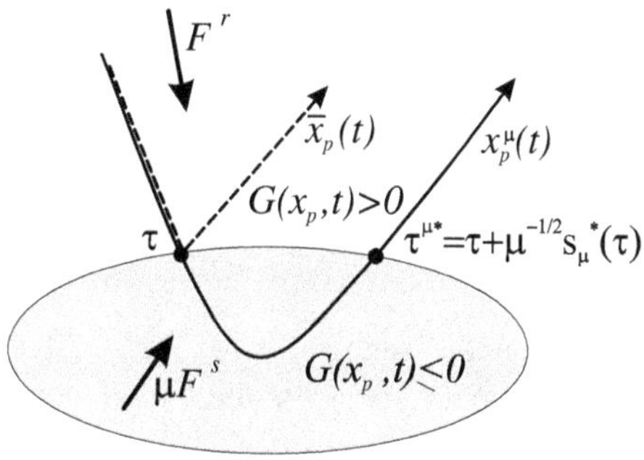

Fig. 8.6 Pre-limit and limit (dash-dot) trajectories of the original system and of the limit system

as $\mu \to \infty$. Now the convergence of the solution on any segment $[\tau+\varepsilon, T]$, $\varepsilon > 0$, follows from the continuous dependence of the solution on the initial condition at the point $\tau+\varepsilon$.

A schematic comparison of the trajectories of the original system (8.13) and those of the limit system (8.44) is shown in Fig. 8.6.

Remark 8.5 Thus, the conclusion of Theorem 8.3 is a consequence of the uniform convergence with respect to the parameter μ of $(x_p^\mu(t), x_v^\mu(t))$ to $(\bar{x}_p(t), \bar{x}_v(t))$ on $[0, T]$, except possibly, at the point of discontinuity. This means that for sufficiently large μ in the Eq. (8.13), the distance between $(x_p^\mu(t), x_v^\mu(t))$ and $(\bar{x}_p(t), \bar{x}_v(t))$ in the topology of pointwise convergence is small everywhere on $[0, T]$, except possibly in a small neighborhood of the discontinuity point τ.

The construction described above can be extended to the case of a finite number of shocks.

In order to determine the solution in the case of subsequent jumps of the velocity, suppose that the motion starts in the domain $G\big(x_p(0), 0\big) > 0$, and define the sequence of instants

$$0 < \tau_1 < \cdots < \tau_i < \cdots < T \tag{8.46}$$

by the rule

$$\tau_1 = \begin{cases} \inf\limits_{0<t\leq T} \left\{ t : \begin{array}{l} G\big(x_p(t), t\big) < 0, \\ G_t'\big(x_p(t), t\big) + G_{x_p}'\big(x_p(t), t\big) F_p^{\mathrm{r}}\big(x_p(t), x_v(t), t\big) < 0 \end{array} \right\}, \\ T, \quad \text{if the above set is empty,} \end{cases} \tag{8.47}$$

and

$$
\tau_i = \begin{cases} \inf\limits_{\tau_{i-1} < t \le T} \left\{ t : \begin{array}{l} G\big(x_p(t), t\big) < 0, \\ G_t'\big(x_p(t), t\big) + G_{x_p}'\big(x_p(t), t\big) F_p^r\big(x_p(t), x_v(t), t\big) < 0 \end{array} \right\}, \\ \\ T, \quad \text{if the above set is empty,} \end{cases} \tag{8.48}
$$

where on the segment $t \in [\tau_{i-1}, T]$ the pair $(x_p(t), x_v(t))$ is the solution of the system (8.13) with the initial conditions $(x_p(\tau_{i-1}), x_v(\tau_{i-1}))$ at the point $t = \tau_{i-1}$. Note that τ_i, $i = 1, 2, \ldots$ is the sequence of exit times from the domain $\{(x_p, t) : G(x_p, t) \geq 0\}$, and since the trajectory $x_p(\cdot)$ is absolutely continuous on the intervals (τ_{i-1}, τ_i), we have that

$$
\begin{aligned}
&G\big(x_p(\tau_i), \tau_i\big) = 0, \\
&G_t'\big(x_p(\tau_i), \tau_i\big) + G_{x_p}'\big(x_p(\tau_i), \tau_i\big) F_p^{\mathrm{r}}\big(x_p(\tau_i), x_v(\tau_i-), \tau_i\big) < 0.
\end{aligned} \tag{8.49}
$$

Generally speaking, the sequence of instants (8.46) is not determined in advance and, as in the example of the bouncing ball ([294], Section 1.2.3), can have an accumulation point after which the motion has to be defined in a special way. However, if the number of jumps on $[0, T]$ is finite, the motion is uniquely determined.

Theorem 8.4 *Suppose that on the segment* $[0, T]$ *there is given the control* $u(t) \in U$ *and a set of time instants*

$$
0 < \tau_1 < \cdots < \tau_i < \cdots < T, \quad i \leq N < \infty, \tag{8.50}
$$

satisfying (8.48). Let

$$
\Psi_v\big(x_p(\tau_i), x_v(\tau_i-), w_{\tau_i}(\cdot), \tau_i\big)
$$

be the corresponding shift operators, defined with the aid of the impulse controls $w_{\tau_i}(\cdot)$*, for all* τ_i*.*

Then as $\mu \to \infty$ *the sequence of ordinary solutions of the system (8.13) converges everywhere on* $[0, T]$*, except possibly at the points* $\{\tau_i\}$*, to the generalized solution* $\big(\bar{x}_p(t), \bar{x}_v(t)\big)$*, which satisfies on* $[0, T]$ *the system of generalized differential equations*

$$
\begin{cases}
\dot{\bar{x}}_p(t) = F_p^{\mathrm{r}}\big(\bar{x}_p(t), \bar{x}_v(t), t\big), \\
\dot{\bar{x}}_v(t) = F_v^{\mathrm{r}}\big(\bar{x}_p(t), \bar{x}_v(t), u(t), t\big) + \\
\qquad + \sum\limits_{\tau_i \le t} \Psi_v\big(\bar{x}_p(\tau_i), \bar{x}_v(\tau_i-), w_{\tau_i}(\cdot), \tau_i\big)\, \delta(t - \tau_i).
\end{cases} \tag{8.51}
$$

with $\bar{x}_p(0) = x_p(0)$, $\bar{x}_v(0) = x_v(0)$.

Proof The argument is divided into a number of steps.

Step 1. For an arbitrary admissible control $u(t)$, $t \in [0, T]$, take $\tau_1 = \tau$ as in (8.47), that is

$$\tau_1 = \begin{cases} \inf\limits_{0<t\le T} \left\{t : G(\bar{x}_p(t), t) = 0, \ \dfrac{d}{dt}\Big|_{F_p^{\mathrm{r}}} G(\bar{x}_p(t), t) < 0\right\}, \\ T, \quad \text{if the above set is empty.} \end{cases} \tag{8.52}$$

Here we can write $\bar{x}_p(t)$ instead of $x_p(t)$ because these two functions coincide on the segment $[0, \tau_1]$.

Step 2. For arbitrary admissible controls $w_{\tau_1}(\cdot)$ and $\alpha_{\tau_1}(\cdot)$ find the solution of the limit system (8.25). This gives $y_v(s^*(\tau_1))$, and consequently the shift operator

$$\Psi_v\big(y_p(0), y_v(0), w_{1\tau_1}(\cdot), \tau_1\big) = y_v(s^*(\tau_1)) - y_v(0).$$

Step 3. Take

$$\bar{x}_p(\tau_1) \quad \text{and} \quad \bar{x}_v(\tau_1) = y_v(s^*(\tau_1)) = \bar{x}_v(\tau_1-) + \Psi_v\big(\bar{x}_p(\tau_1), \bar{x}_v(\tau_1-), w_{1\tau_1}(\cdot), \tau_1\big)$$

as initial values, define τ_2 from (8.48) by analogy with τ_1, namely

$$\tau_2 = \begin{cases} \inf\limits_{\tau_1<t\le T} \left\{t : G(\bar{x}_p(t), t) = 0, \ \dfrac{d}{dt}\Big|_{F_p^{\mathrm{r}}} G(\bar{x}_p(t), t) < 0\right\}, \\ T, \quad \text{if the above set is empty,} \end{cases} \tag{8.53}$$

and so on. By induction, we have

$$\bar{x}_v(\tau_i) = \bar{x}_v(\tau_i-) + \Psi_v\big(\bar{x}_p(\tau_i), \bar{x}_v(\tau_i-), w_{\tau_i}(\cdot), \tau_i\big). \tag{8.54}$$

The functions $(\bar{x}_p(t), \bar{x}_v(t))$ generated by this recipe satisfy on $[0, T]$ the system of differential equations (8.51) with $\bar{x}_p(0) = x_p(0)$, $\bar{x}_v(0) = x_v(0)$. This completes the proof. □

As one can see from above description, the state of the limit system (8.51) changes continuously on the half-intervals $[0, \tau_1), \ldots, [\tau_{i-1}, \tau_i), \ldots$ and jumps at the instants τ_i. By (8.54), the magnitudes of such a jump depends on the state preceding the jump and the impulse control $w_{\tau_i}(\cdot)$ applied during the singular phase corresponding to τ_i. The Lipschitz property of $\Psi_v(\cdot)$ guarantees the existence and uniqueness of the solution to the system (8.51) [294].

The implementation of such an impulse control law is provided by the utilization of μ-parametrized controls, , while the convergence to the limit motion is guaranteed by Theorem 8.4. In the framework of this approach, the generalized solution "corresponding to $\mu = \infty$" and optimal in the sense of the planning motion, is the supporting solution for the implementation of an approximate control law that is robust with respect to real value of μ. The next result plays a fundamental role in the implementation of the impulse control.

Consider the system (8.13) with some fixed $\mu > \mu_0$ and with the same controls as in the system (8.51), where $w_{\tau_i}(\cdot), \alpha_{\tau_i}(\cdot)$, $i = 1, \ldots, N$, can be defined in any admissible manner beyond the point $s^*(\tau_i)$ (according to (8.45)).

Let $(x_p^\mu(t), x_v^\mu(t))$ be the solution of this system. The sequence of time instants

$$0 < \tau_1^\mu < \tau_1^{\mu *} < \cdots < \tau_i^\mu < \tau_i^{\mu *} < \cdots \leq T, \quad i \leq N < \infty$$

at which the trajectory $(x_p^\mu(t), x_v^\mu(t))$ hits the constraint boundary surface is defined recursively as follows:

Step 1. Put $\tau_1^\mu = \tau_1 = \tau$, that is,

$$\tau_1^\mu = \begin{cases} \inf\limits_{0<t\leq T} \left\{t : G(x_p^\mu(t), t) = 0, \ \left.\dfrac{d}{dt}\right|_{F_p^\Gamma} G(x_p^\mu(t), t) < 0\right\}, \\ T, \quad \text{if the above set is empty.} \end{cases} \tag{8.55}$$

Step 2. As in (8.19), let γ denote any of the aforementioned controls w, α. Under the controls $\gamma\big(\xi, \sqrt{\mu}(t - \tau_1^\mu)\big)$, $t \geq \tau_1^\mu$, define the exit time from the constrained domain by

$$\tau_1^{\mu *} = \begin{cases} \inf\limits_{\tau_1^\mu<t\leq T} \left\{t : G(x_p^\mu(t), t) = 0, \ \left.\dfrac{d}{dt}\right|_{F_p^\Gamma} G(x_p^\mu(t), t) > 0\right\}, \\ T, \quad \text{if the above set is empty.} \end{cases} \tag{8.56}$$

Step 3. Define the second instant τ_2^μ when the path hits the constraint by analogy with (8.48), that is,

$$\tau_2^\mu = \begin{cases} \inf\limits_{\tau_1^\mu<t\leq T} \left\{t : G(x_p^\mu(t), t) = 0, \ \left.\dfrac{d}{dt}\right|_{F_p^\Gamma} G(x_p^\mu(t), t) < 0\right\}, \\ T, \quad \text{if the above set is empty.} \end{cases} \tag{8.57}$$

Step 4. Continuing in this manner, define the second instant $\tau_2^{\mu *}$ of exit from the constrained domain by analogy with $\tau_1^{\mu *}$, as shown in Fig. 8.7, and so on.

Theorem 8.5 *Let $(x_p^\mu(t), x_v^\mu(t))$ be the solutions of the original system (8.13) with admissible controls $u(t)$, $t \in [0, T]$ and $\gamma^\mu(\xi, t) = \gamma\big(\xi, \sqrt{\mu}(t - \tau_i^\mu)\big)$, $t \in [\tau_i^\mu, \tau_i^{\mu *}]$, defined above. Then when $\mu \to \infty$ the sequence $(x_p^\mu(t), x_v^\mu(t))$ converges everywhere on $[0, T]$, except perhaps at the points $\{\tau_i\}$, to the appropriate generalized solution $(\bar{x}_p(t), \bar{x}_v(t))$ of the system (8.51).*

Proof By the assumptions of the theorem, $\tau_1^\mu = \tau_1$ and, for $t \in \big[0, \tau_1\big)$,

$$\big(x_p^\mu(t), x_v^\mu(t)\big) = \big(\bar{x}_p(t), \bar{x}_v(t)\big).$$

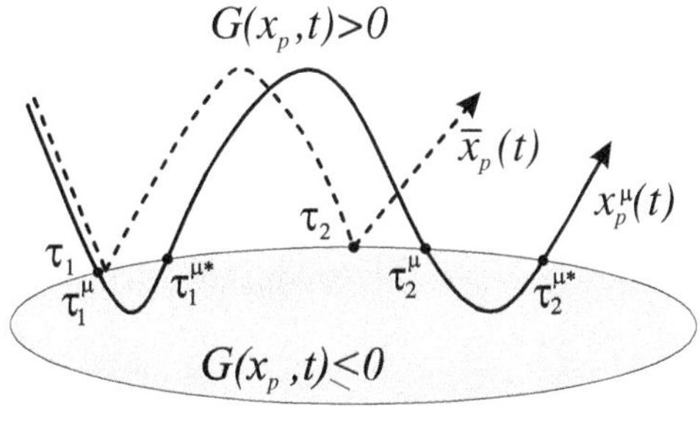

Fig. 8.7 Sequence of collision interactions with an obstacle. Pre-limit and limit (dash-dot) paths

By Theorem 8.1, there exists the time instant $\tau_1^{\mu*} = \tau_1^{\mu} + \mu^{-1/2} s_{\mu}^{*}(\tau_1^{\mu})$, such that

$$\tau_1^{\mu*} \to \tau_1 \quad \text{and} \quad \left(x_p^{\mu}(\tau_1^{\mu*}), x_v^{\mu}(\tau_1^{\mu*})\right) \to \left(\bar{x}_p(\tau_1), \bar{x}_v(\tau_1)\right) \quad \text{as } \mu \to \infty.$$

Moreover, in view of the definition of τ_2^{μ} and τ_2 and the continuous dependence of solutions on initial conditions, when $\mu \to \infty$

$$\tau_2^{\mu} \to \tau_2, \; (x_p^{\mu}(\tau_2^{\mu*}), x_v^{\mu}(\tau_2^{\mu*})) \to (\bar{x}_p(\tau_2), \bar{x}_v(\tau_2-))$$

and

$$(x_p^{\mu}(t), x_v^{\mu}(t)) \to (\bar{x}_p(t), \bar{x}_v(t))$$

for all $t \in (\tau_1, \tau_2)$. Next, use the same arguments for τ_2^{μ}, and so on. Note that the uniformity of the convergence $(x_p^{\mu}(t), x_v^{\mu}(t))$ to $(\bar{x}_p(t), \bar{x}_v(t))$ in μ on $[0, T]$, except perhaps at the points $\{\tau_i\}$, presumes that if μ in (8.13) is sufficiently large, the distance between $(x_p^{\mu}(t), x_v^{\mu}(t))$ and $(\bar{x}_p(t), \bar{x}_v(t))$ on $[0, T]$ is very small, except possibly in small neighborhoods of the points τ_i. This completes the proof. □

Multi-Impulse Control in Systems with Collisions

To treat the case where $F_v^{\text{rs}}\left(x_p, x_v, w_2^{\mu}(\xi, t), t, \mu\right) \not\equiv 0$, we introduce several definitions.

Definition 8.2 The motion in the constraint-free domain free that arises between two singular phases will be called *intersingular* if its duration tend to zero as $\mu \to \infty$.

Definition 8.3 An arbitrary sequence of adjacent singular and intersingular phases of motion will be referred to as *bi-alternative singular phase*.

Definition 8.4 A pair of admissible controls $w_1^{\mu}(\xi, t)$ and $w_2^{\mu}(\xi, t)$ in the system (8.13) such that

$$w_1^\mu(\xi, t) = 0 \text{ for } G(x_p(t), t) > 0 \quad \text{and} \quad w_2^\mu(\xi, t) = 0 \text{ for } G(x_p(t), t) < 0$$

will be referred to as an *extended multi-impulse control* if the components $w_1^\mu(\xi, t)$ and $w_2^\mu(\xi, t)$ are defined on disjoint adjacent intervals that include intersingular phases of motion.

The presence of intersingular phases requires an extension of the theorems proved above.

Assumption 8.2 In condition (1) of Theorem 8.1, in addition to the system (8.24), the relation

$$\begin{aligned} &\lim_{\mu\uparrow\infty} \mu^{1/2} F_v^{\mathrm{rs}}\Big(\frac{y_p - x_p}{\mu^{1/2}} + x_p, y_v, w_2(\eta^\mu, s), \tau + \mu^{-1/2}s, \mu\Big) = \\ &\quad = \bar{F}_v^{\mathrm{rs}}\big(y_p, y_v, w_2(\eta, s), x_p, \tau\big), \end{aligned} \tag{8.58}$$

holds for any admissible control w_2.

Now fix some admissible controls u, w_1, w_2 and α on $[0, T]$ and define the finite sequence of time instants

$$0 = s_1(\tau) < s_1^*(\tau) < \cdots < s_j(\tau) < s_j^*(\tau) < \cdots,$$

$j = 1, \ldots, N_1$, where $s_1^*(\tau)$ is taken from (8.26), by the following recurrent procedure:

Step 1. Along with the limit system (8.25), consider its analogue for $F_v^{\mathrm{rs}} \not\equiv 0$,

$$\begin{cases} \dot{y}_p(s) = F_p^{\mathrm{r}}\big(x_p(\tau), y_v(s), \tau\big), \\ \dot{y}_v(s) = \bar{F}_v^{\mathrm{rs}}\big(y_p(s), y_v(s), w_2(\eta, s), x_p(\tau), \tau\big), \\ \dot{\eta}(s) = 0. \end{cases} \tag{8.59}$$

We assume that the system (8.59) has exactly one solution on the closed interval $[s_1^*(\tau), s_2(\tau) + \varepsilon_1]$, where $\varepsilon_1 > 0$, and that $y_p\big(s_1^*(\tau)\big)$, $y_v\big(s_1^*(\tau)\big)$, $\eta\big(s_1^*(\tau)\big)$ coincide with the terminal values of the solution of system (8.25) at the time instant $s^*(\tau)$; in other words, the solution of the system (8.59) is a continuous extension of solutions of (8.25).

Define

$$s_2(\tau) = \inf_{s > s_1^*(\tau)} \left\{ \begin{aligned} &G_t'\Big|_{(x_p(\tau),\tau)} s + G_{x_p}'\Big|_{(x_p(\tau),\tau)} \big(y_p(s) - x_p(\tau)\big) = 0, \\ &G_t'\Big|_{(x_p(\tau),\tau)} + G_{x_p}'\Big|_{(x_p(\tau),\tau)} F_p^{\mathrm{r}}\big(x_p(\tau), y_v(s), \tau\big) < 0 \end{aligned} \right\}. \tag{8.60}$$

Step 2. Next, take the terminal values $y_p(s_2(\tau))$, $y_v(s_2(\tau))$, $\eta(s_2(\tau))$ of the solutions of system (8.59) as initial values for the system (8.25), and assume that the latter has a unique solution on the segment $[s_2(\tau), s_2^*(\tau)+\varepsilon_2]$, where $\varepsilon_2 > 0$ and

$$s_2^*(\tau) = \inf_{s>s_2(\tau)} \left\{ \begin{array}{l} G_t'\Big|_{(x_p(\tau),\tau)} s + G_{x_p}'\Big|_{(x_p(\tau),\tau)} \big(y_p(s) - x_p(\tau)\big) = 0, \\ G_t'\Big|_{(x_p(\tau),\tau)} + G_{x_p}'\Big|_{(x_p(\tau),\tau)} F_p^r\big(x_p(\tau), y_v(s), \tau\big) > 0 \end{array} \right\}, \tag{8.61}$$

and so on.

Now put $\varepsilon_0 = \min_j\{\varepsilon, \varepsilon_j\}$ and denote by $(y_p(s), y_v(s), \eta(s))$ the solution defined above of the systems (8.25) and (8.59) for $s \in [0, s_{N_1}^*(\tau)+\varepsilon_0]$. Further, for sufficiently large $\mu > \mu_0$, consider the systems (8.13) and (8.4) under the fixed controls defined above. Denote the solutions of these systems by $(x_p^\mu(t), x_v^\mu(t))$ and $(y_p^\mu(s), y_v^\mu(s), \eta^\mu(s))$, respectively, and define recursively the finite sequence of time instants when the trajectory intersects the boundary of the constraint,

$$0 = s_{\mu 1}(\tau) < s_{\mu 1}^*(\tau) < \cdots < s_{\mu j}(\tau) < s_{\mu j}^*(\tau) < \cdots, \qquad j = 1, \ldots, N_1,$$

starting with $s_{\mu 1}^*(\tau)$ from (8.28), as follows.

Step 1. Define $s_{\mu 2}(\tau)$ as

$$\inf_{s>s_{\mu 1}^*(\tau)} \left\{ \begin{array}{l} G\big(x_p^\mu(\tau+\mu^{-1/2}s), \tau+\mu^{-1/2}s\big) = 0, \\ G_t'\Big|_{(x_p^\mu(\tau+\mu^{-1/2}s),\tau+\mu^{-1/2}s)} + G_{x_p}'\Big|_{(x_p^\mu(\tau+\mu^{-1/2}s),\tau+\mu^{-1/2}s)} \times \\ \times F_p^r\big(x_p^\mu(\tau+\mu^{-1/2}s), x_v^\mu(\tau+\mu^{-1/2}s), \tau+\mu^{-1/2}s\big) < 0 \end{array} \right\}. \tag{8.62}$$

Step 2. Define $s_{\mu 2}^*(\tau)$ as

$$\inf_{s>s_{\mu 2}(\tau)} \left\{ \begin{array}{l} G\big(x_p^\mu(\tau+\mu^{-1/2}s), \tau+\mu^{-1/2}s\big) = 0, \\ G_t'\Big|_{(x_p^\mu(\tau+\mu^{-1/2}s),\tau+\mu^{-1/2}s)} + G_{x_p}'\Big|_{(x_p^\mu(\tau+\mu^{-1/2}s),\tau+\mu^{-1/2}s)} \times \\ \times F_p^r\big(x_p^\mu(\tau+\mu^{-1/2}s), x_v^\mu(\tau+\mu^{-1/2}s), \tau+\mu^{-1/2}s\big) > 0, \end{array} \right\}, \tag{8.63}$$

and continue in this manner.

Here the intervals $[s_{\mu j}(\tau), s_{\mu j}^*(\tau)]$, $j = 1, \ldots, N_1$ (respectively, $(s_{\mu j}^*(\tau), s_{\mu, j+1}(\tau))$, $j = 1, \ldots, N_1 - 1$) correspond to the singular (respectively, intersingular) phase of motion.

Theorem 8.6 *Suppose the conditions of Theorem 8.1 and Assumption 8.2 hold. Then when* $\mu \to \infty$,

$$\big(y_p^{\mu}(s), y_v^{\mu}(s), \eta^{\mu}(s)\big) \to \big(y_p(s), y_v(s), \eta(s)\big) \tag{8.64}$$

uniformly on $[0, s^*_{N_1}(\tau) + \varepsilon_0]$ *and*

$$s_{\mu j}(\tau) \to s_j(\tau), \tag{8.65}$$

$$s^*_{\mu j}(\tau) \to s^*_j(\tau) \tag{8.66}$$

for $j = 1, \ldots, N_1$.

Proof As in the case of Theorem 8.1, the convergence (8.64) follows from the continuous dependence of solutions of ordinary differential equations on parameters [96].

Statements (8.65) and (8.66) are proved by induction.

Step 1. For $j = 1$, (8.65) and (8.66) are consequences of Theorem 8.1.

Step 2. For $j = 2$, consider first the trajectories $(y_p^{\mu}(s), y_v^{\mu}(s), \eta^{\mu}(s))$ and $(y_p(s), y_v(s), \eta(s))$ for the pre-limit and the limit system, respectively, for $s \in [0, s_2(\tau) + \varepsilon_0]$ and sufficiently large μ, ignoring the intermediate time instants $s^*_{\mu 1}(\tau)$ and $s^*_1(\tau)$. Now, by virtue of their definition, the time instants $s_{\mu 2}(\tau)$ and $s_2(\tau)$ play the same role as $s^*_{\mu}(\tau)$ and $s^*(\tau)$ in Theorem 8.1, so by that theorem (8.65) is true.

Step 3. As in the previous step for $j = 2$, consider $(y_p^{\mu}(s), y_v^{\mu}(s), \eta^{\mu}(s))$ and $(y_p(s), y_v(s), \eta(s))$ for $s \in [0, s^*_2(\tau) + \varepsilon_0]$ and sufficiently large μ, ignoring the time instants $s^*_{\mu 1}(\tau)$, $s^*_1(\tau)$, $s_{\mu 2}(\tau)$, $s_2(\tau)$. Arguing exactly as above we conclude that (8.66) is true, and so on. The proof is complete. □

Remark 8.6 Clearly, with small modifications Corollary 8.1 and Theorems 8.2 and 8.3 remain valid in the case of multi-impulse control. Indeed, it suffices to put $s^*(\tau) = s^*_{N_1}(\tau)$ and $s^*_{\mu}(\tau) = s^*_{\mu N_1}(\tau)$, and then integrate the system (8.25), (8.59) for $s \in [0, s^*(\tau)]$ to calculate the v-component of the shift operator $\Psi_v(\cdot)$, which is then used in (8.44). Therefore,

$$y_v(s^*(\tau)) = y_v(0) + \Psi_v\big(y_p(0), y_v(0), w_{1\tau}(\cdot), w_{2\tau}(\cdot), \tau\big). \tag{8.67}$$

Theorem 8.5 also admits a natural generalization to the case of multi-impulse control. The sequence $\{\tau_i\}$, $i = 1, \ldots, N$, is defined analogously. However, the double sequence $\{\tau_i^{\mu}, \tau_i^{\mu *}\}$, $i = 1, \ldots, N$ splits into N finite series

$$\tau_i^{\mu} = \tau_{i1}^{\mu} < \tau_{i1}^{\mu *} < \tau_{i2}^{\mu} < \tau_{i2}^{\mu *} < \cdots < \tau_{iN_i}^{\mu} < \tau_{iN_i}^{\mu *} = \tau_i^{\mu *},$$

where

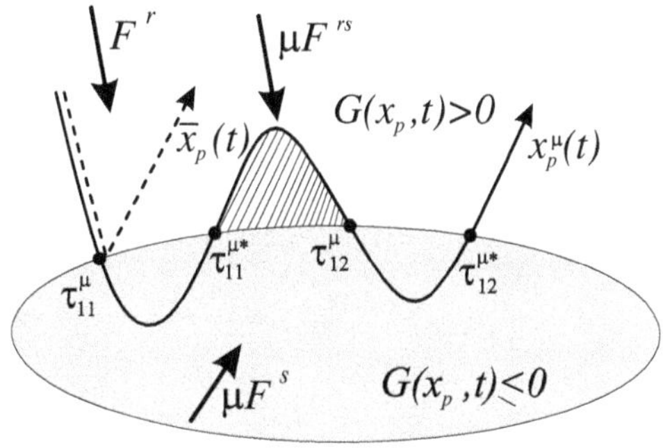

Fig. 8.8 Single extended multi-impulse. Pre-limit and limit (dash-dot) trajectories. The hatched area corresponds to the intersingular motion

$$\tau_{ij}^{\mu} = \tau_i^{\mu} + \mu^{-1/2} s_{\mu j}(\tau_i^{\mu}), \quad \tau_{ij}^{\mu *} = \tau_i^{\mu} + \mu^{-1/2} s_{\mu j}^{*}(\tau_i^{\mu}), \quad j = 1, \ldots, N_i.$$

Here each series corresponds to a single multi-impulse (see Fig. 8.8).

Theorem 8.7 *Let $(\bar{x}_p(t), \bar{x}_v(t))$, $t \in [0, T]$ be the generalized solution of the system (8.51) for some admissible controls u, w_1, w_2 and α, where the shift operators are defined by (8.67). Let $(x_p^{\mu}(t), x_v^{\mu}(t))$ be the corresponding ordinary solutions of the system (8.13) under some controls $u(t)$ for $t \in [0, T]$ and $\gamma^{\mu}(\xi, t) = \gamma(\xi, \sqrt{\mu}(t - \tau_i^{\mu}))$ for $t \in [\tau_i^{\mu}, \tau_i^{\mu *}]$. Here γ stands for any of the controls w_1, w_2 and α, which can be extended in the neighborhood of the points $\tau_i^{\mu *}$ as in (8.45).*

Then as $\mu \to \infty$, the sequence $(x_p^{\mu}(t), x_v^{\mu}(t))$ converges to $(\bar{x}_p(t), \bar{x}_v(t))$ everywhere on $[0, T]$, except possibly at the points $\{\tau_i\}$.

Proof The argument is analogous to that in the proof of Theorem 8.5. □

8.2.4 The "Rotating Racket-Ball" Model

In the paper [37] the methodology described above is considered for the simple example of the motion of a ball colliding with an elastic surface (a racket). It is assumed that the rotation of the surface is coordinated with the generalized rigidity, so that during the contact time the surface manages to turn by a certain fixed angle.

We consider the case of a perfectly elastic surface (the damping coefficient is equal to zero), so if the racket is stationary, then the speeds of the ball before and after hitting are equal, and the angle of incidence is equal to the angle of reflection. A similar result should be expected if the racket rotates at a given finite speed; however, if the speed of rotation is coordinated with the duration of the contact so that during this time the player manages to turn the racket by a finite angle, then both the rebound angle and the rebound speed change. Let us consider this example in detail (Fig. 8.9).

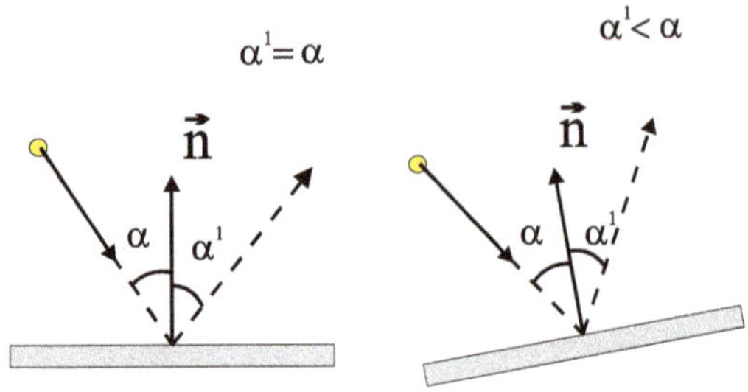

Fig. 8.9 Additional rotation of the ball's velocity vector during the collision, when the normal vector to the racket also turns by a certain angle

Natural Phase of Free (Uncontrolled) Motion

Let the motion of the ball be described by a pair of vectors (x_p, x_v), $x_p, x_v \in \mathbb{R}^3$ corresponding to the position and velocity of the ball, and related by the simple kinematic equations

$$\dot{x}_p(t) = x_v(t), \quad \dot{x}_v(t) = 0.$$

The surface of the racket is defined by the equation $\langle y, n(t)\rangle = 0$, where $\langle \cdot, \cdot \rangle$ denotes the scalar product, and $n(t)$ is the unit normal vector to the surface. For simplicity, we assume that the center of rotation coincides with the contact point, which is taken as the origin of coordinates, and the axis of rotation of the racket is orthogonal to the plane spanned by the vectors x_v and $n(t)$.

Singular Phase of Controlled Motion

In the singular phase, which starts when ball hits the surface of the racket, the contact force is orthogonal to the surface, so that

$$G(x_p(t), t) = \langle x_p(t), \ n(t)\rangle < 0,$$

and the equations of motion have the form

$$\dot{x}_p(t) = x_v(t), \quad \dot{x}_v(t) = -\mu n(t)\langle x_p(t), n(t)\rangle,$$

where μ is the elasticity coefficient. The controlled singularity is introduced by the rotation of the racket, which is described by the following equation for $n(t)$:

$$\dot{n}(t) = \omega(t, \mu) \times n(t)$$

where $a \times b$ denotes the vector product and $\omega(t, \mu)$ is the angular velocity of rotation of the racket's surface. This velocity can be considered as the impulse control, which makes the angle of inclination of the surface jump during the collision. Thus, the velocity $\omega(t, \mu)$ is described by the formula

$$\omega(t,\mu) \equiv w^{\mu}(t) = \omega_0 \mu^{1/2} w_{\tau}\big(\mu^{1/2}(t-\tau)\big) \tag{8.68}$$

where ω_0 is the unit vector directed along the axis of rotation, $w_{\tau}(\cdot)$ is the impulse control, and τ is the collision time.

Singular Transformation of Coordinates and the Equations of "Fast" Motion

Applying the transformation (8.3), we obtain the system of equations

$$\begin{cases} y_p^{\mu}(s) = \mu^{1/2} x_p(\tau + \mu^{-1/2} s) \\ y_v^{\mu}(s) = x_v(\tau + \mu^{-1/2} s), \quad n^{\mu}(s) = n(\tau + \mu^{-1/2} s), \\ t = \tau + \mu^{-1/2} s, \end{cases} \tag{8.69}$$

which in the limit as $\mu \to \infty$ gives the following system of equations of "fast motion" for the variables $(y_p(s), y_v(s), \bar{n}(s))$ defined for $s > 0$:

$$\begin{pmatrix} \dot{y}_p(s) \\ \dot{y}_v(s) \\ \dot{\bar{n}}(s) \end{pmatrix} = \begin{pmatrix} y_v(s) \\ -\bar{n}(s)\langle y_p(s), \bar{n}(s)\rangle \\ (\omega_0 \times \bar{n}(s)) w_{\tau}(s) \end{pmatrix}. \tag{8.70}$$

This system is solved for the initial conditions

$$y_p(0) = 0, \quad y_v(0) = x_v(\tau-), \quad \bar{n}(0) = n(\tau-),$$

such that

$$\langle x_v(\tau-), n(\tau-)\rangle < 0,$$

until the time

$$s^* = \inf\left\{ s > 0 : \langle y_p(s), \bar{n}(s)\rangle = 0, \ \frac{d}{ds}\langle y_p(s), \bar{n}(s)\rangle > 0 \right\}. \tag{8.71}$$

Put

$$\bar{n}(0) = n(\tau-) = \begin{pmatrix} n_1(\tau-) \\ n_2(\tau-) \\ 0 \end{pmatrix}.$$

Then the solution of the third equation of the system (8.70) is given by the formula

$$\bar{n}(s) = \begin{pmatrix} n_1(\tau-)\cos\phi(s) + n_2(\tau-)\sin\phi(s) \\ -n_1(\tau-)\sin\phi(s) + n_2(\tau-)\cos\phi(s) \\ 0 \end{pmatrix},$$

where

$$\phi(s) = \int_0^s w_\tau(u)du \tag{8.72}$$

and $w_\tau(\cdot)$ are the rotation angle and the angular velocity of the racket in "fast" motion. Thus, for an arbitrary control law of $w(\cdot)$ we obtain for (y_p, y_v) the linear system of differential equations

$$\begin{pmatrix} \dot{y}_p(s) \\ \dot{y}_v(s) \end{pmatrix} = \begin{pmatrix} y_v(s) \\ -\bar{n}(s)\langle y_p(s), \bar{n}(s)\rangle \end{pmatrix} = \begin{pmatrix} y_v(s) \\ A(\phi(s), \bar{n}(\tau-))y_p(s) \end{pmatrix}, \tag{8.73}$$

where $A(\phi, n)$ is a matrix function that depends on the current rotation angle $\phi(s)$ and the original orientation $n(\tau-)$ of the racket surface. Therefore, (8.73) represents the equation of motion in the controlled singular phase, with explicit dependence on the original orientation of the racket.

The general solution of the system (8.73) has a form

$$\big(y_p(s) y_v(s)\big) = \Phi(s, 0, \phi(\cdot)) \begin{pmatrix} y_p(0) \\ y_v(0) \end{pmatrix}.$$

The matrix $\Phi(s, 0, \phi(\cdot))$ admits the block decomposition

$$\Phi(s, 0, \phi(\cdot)) = \begin{pmatrix} \Phi_{pp} & \Phi_{pv} \\ \Phi_{vp} & \Phi_{vv} \end{pmatrix}(s, 0, \phi(\cdot)).$$

If there exists the exit time s^* defined by relation (8.71), then the restitution law can be defined as

$$\Delta x_v(\tau) = y_v(s^*) - y_v(0) = \big[\Phi_{vv}(s^*, 0, \phi(\cdot)) - I_{(3\times3)}\big]x_v(\tau-),$$

and the equations of motion for the limit system have the form

$$\begin{cases} \dot{\bar{x}}_p(t) = \bar{x}_v(t), \\ \dot{\bar{x}}_v(t) = \big[\Phi_{vv}(s^*, 0, \phi(\cdot)) - I_{(3\times3)}\big]\bar{x}_v(\tau-)\delta(t-\tau), \\ \bar{x}_p(0) = x_p(0),\ \bar{x}_v(0) = x_v(0),\ \bar{x}_v(\tau-) = x_v(\tau-). \end{cases} \tag{8.74}$$

Implementation of the Control Law

First we provide results of computations using the limit motion model for the initial conditions

$$x_{v1}(\tau-) = -\sqrt{3}/2, \quad x_{v2}(\tau-) = 1/2, \; x_{v3}(\tau-) = 0,$$

$$n_1(\tau) = 1, \quad n_2(\tau) = n_3(\tau) = 0.$$

That is, the absolute value of the initial velocity is equal to 1 and the incidence angle is $-\pi/6$. We consider two cases: clockwise rotation and counterclockwise rotation, that is, $w_\tau(s) = \bar{w} = 0.1$ and $w_\tau(s) = \bar{w} = -0.1$, respectively (see Figs. 8.10 and 8.11). The variables $\bar{y}_v^{1,2}$ correspond to the velocity components in the x and y directions, and $Z = \langle y_p, \bar{n}(s)\rangle$ measures the violation of the constraint.

With the aid of these curves, we define s^* as the first instant of time at which $Z(s) = \langle y_p(s), \bar{n}(s)\rangle = 0$ and $dZ(s)/ds > 0$. The rotation angle of the racket may be taken as $\Delta\phi(s^*) = \bar{w}s^*$, and then one defines the velocity components after collision as $\bar{y}_v^1(s^*)$, $\bar{y}_v^2(s^*)$ (Fig. 8.11).

The increment of the magnitude of the velocity ΔV and the change of bouncing angle $\Delta\psi$ with respect to the nominal value $\pi/6$ are given by

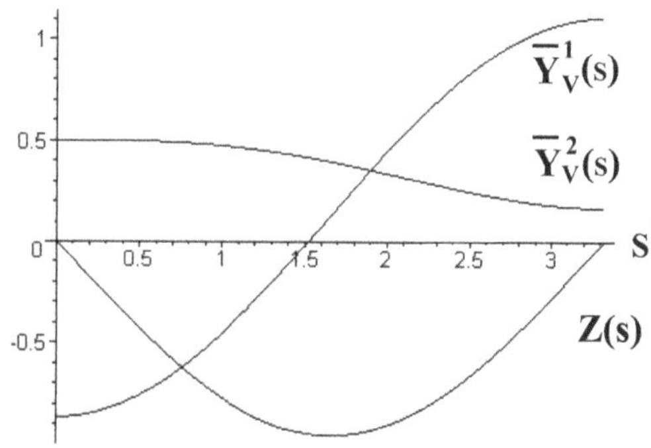

Fig. 8.10 Solution of the equations of "fast" motion for clockwise rotation of the racket

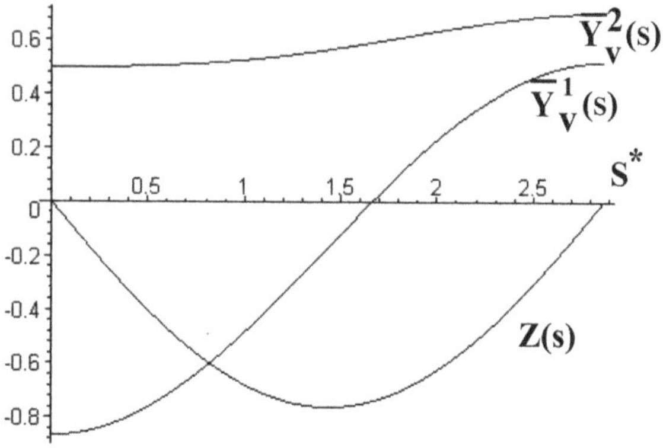

Fig. 8.11 Solution of the equations of "fast" motion for counterclockwise rotation of the racket

$$\Delta V = \sqrt{(y_{v1}(s^*))^2 + (y_{v2}(s^*))^2 - 1} \quad \text{and} \quad \Delta\psi = \arctan(y_{v2}(s^*)/y_{v1}(s^*)) - \pi/6,$$

respectively. For clockwise and counterclockwise rotation of the racket we have

$$s^* = 3.32, \quad \Delta\phi(s^*) = -0.332\ \text{rad}, \quad \Delta V = 0.1 > 0, \quad \Delta\psi = -0.343\ \text{rad},$$

and

$$s^* = 2.87, \quad \Delta\phi(s^*) = 0.287\ \text{rad}, \quad \Delta V = -0.12 < 0, \quad \Delta\psi = 0, 404\ \text{rad},$$

respectively.

The implementation of this control depends on the elasticity of the racket's surface. Assume that the duration of the contact phase is equal to Δt_μ and provides the necessary rotation velocity of the racket,

$$\|\omega(t, \mu)\| = \Delta\phi(s^*)/\Delta t_\mu \quad \text{for } t \in [\tau, \ \tau + \Delta t_\mu].$$

Then, for example, an increment of $\Delta\phi(s^*) = -0.332$ rad can be obtained over the time $\Delta t_\mu = 0.1$ sec with angular velocity $\|\omega(t, \mu)\|l \approx 3$ rad/s, with a racket of rigidity $\mu = \left(s^*/\Delta + t_\mu\right)^2 \approx 10^3$, and over the time $\Delta t_\mu = 0.3$ s with angular velocity $\|\omega(t, \mu)\| \approx 1$ rad/s with a softer racket of rigidity $\mu \approx 10^2$, respectively. Thus, in full agreement with the physical picture, a softer racket allows you to achieve the same effect, but for a lower angular velocity of rotation of the racket and with a more extended contact phase.

8.3 Description of the Singular Phase of Motion in Classical Mechanics

8.3.1 Termination Condition of the Singular Phase

In Sect. 8.2.1, we discussed various models describing the singular phase of motion in mechanical systems with impacts. Then in Sect. 8.2.3 a generic model for the description of the "fast" motion in the singular phase in the form of the auxiliary system of differential equations (8.25) was proposed. The termination condition of the singular phase is defined in terms of the exit time (8.26), which provides a geometric condition for the breaking of contact at the moment when the "deformation" (or, more precisely, the function $G(x_p, t)$) becomes again positive. This geometric condition is rather natural and easily verifiable, because the function $G(x_p(t), t)$ is continuous and is continuously differentiable along the trajectories of the system. However, as we already discussed (see Sect. 8.2.1 (Remark 8.2)), in mechanical models the condition for contact termination corresponds to the vanishing of the contact force, namely, the vanishing of the second derivative of the "deformation" (see, e.g., the Maxwell and Kelvin-Voigt models in Sect. 8.2.1). Such a "physical" condition for the termination of the singular phase can be easily derived if the function $\ddot{G}(x_p(t), t)$ is continuous (for example, if $w_1(\cdot)$ is continuous). Under the assumptions of Theorem 8.1, part 2), this means that the instant of time s^{**} when the singular phase terminates is given by

$$s^{**} = \inf \left\{ s > 0 : \begin{array}{l} G'_t|_{(x_p(\tau),\tau)} s + G'_x\Big|_{(x_p(\tau),\tau)} (y_p(s) - x_p(\tau)) = 0, \\ G'_t\Big|_{(x_p(\tau),\tau)} + G'_x\Big|_{(x_p(\tau),\tau)} F^{\mathrm{r}}_p(x_p(\tau), y_v(s), \tau) > 0, \\ \text{or} \\ G'_x Big|_{(x_p(\tau),\tau)} (F^{\mathrm{r}}_p)'_{x_v}(x_p(\tau), y_v(s), \tau)\times \\ \times \bar{F}_v(y_p(s), y_v(s), s, w_1(s), x_p(\tau), \tau) < 0. \end{array} \right\} \tag{8.75}$$

The expression in the second condition of (8.75) is the second derivative of the principal term of the expansion of the constraint violation (of the order of $\mu^{-1/2}$) (see (8.32)) along the trajectories of the auxiliary system (8.25). One can give a condition under which the moment of termination of the singular phase can always be determined from the physical condition. Indeed, suppose that

$$F^{\mathrm{s}}_v(x_p, x_v, w, t, \mu) = A(x_p, x_v, w, t)G(x_p, t) + \mu^{-1/2} B(x_p, x_v, w, t)\dot{G}(x_p, x_v, t),$$

with continuous and smooth functions A and B. Then

$$\begin{aligned} &\bar{F}_v(y_p(s), y_v(s), s, w_1(s), x_p(\tau), \tau) = \\ &\qquad = A(x_p(\tau), y_v(s), w_1(s), \tau)Z(s) + B(x_p(\tau), y_v(s), w_1(s), \tau)\dot{Z}(s), \end{aligned}$$

where

$$Z(s) = G'_t\Big|_{(x_p(\tau),\tau)} s + G'_x\Big|_{(x_p(\tau),\tau)} (y_p(s) - x_p(\tau))$$

and

$$\dot{Z}(s) = G'_t\Big|_{(x_p(\tau),\tau)} + G'_x\Big|_{(x_p(\tau),\tau)} F^{\mathrm{r}}_p(x_p(\tau), y_v(s), \tau).$$

The second derivative of the "deformation" is expressed as

$$\ddot{Z}(s) = G'_x\Big|_{(x_p(\tau),\tau)} (F^{\mathrm{r}}_p)'_{x_v}(x_p(\tau), y_v(s), \tau)[AZ(s) + B\dot{Z}(s)],$$

(below, for simplicity, we omit the arguments of the functions G, F_p, A and B). Then if the inequalities

$$G'_x(F_p)'_{x_v}A \leq 0, \quad G'_x(F_p)'_{x_v}B \leq 0,$$

which express a natural physical condition ("the violation of the constraint generates a reaction force"), are fulfilled, we see that for $s = 0$ we have $\ddot{Z}(0) \geq 0$, and for $s = s^*$ we have $\ddot{Z}(s^*) \leq 0$. Thanks to the continuity of the function $\ddot{Z}(s)$ for a continuous control $w_1(\cdot)$, we get

$$s^{**} \leq s^*,$$

and for purely elastic interaction ($B = 0$), we get $s^{**} = s^*$. Note that all the ensuing results related to the representation of jumps of the generalized velocity remain valid when s^* is replaced by s^{**}.

8.3.2 Painlevé's Problem

Many classical mechanical systems, especially those with dry friction, often exhibit such a complex behavior in the singular phase that the use of the auxiliary system is simply unavoidable. As such an example it is worth considering the mathematical model of one of the systems in which the Painlevé paradox occurs [335]. Specifically, the model describes the impact of an absolutely rigid free-falling rotating rod on a rough surface. The essence of the paradox is that for some relations involving the speeds and parameters in systems with dry friction that forbid interpenetration of bodies, jamming or the impossibility of continuing the solution occurs. Attempts at solving the paradox and problems related to it are presented in numerous recent studies [144, 154, 324, 355, 362, 389, 415]. Many works try to solve the problem in terms of the restitution law, using the methodology of hybrid systems via *linear complementarity problems*. This approach originates in Painlevé's classical book [335], where it was observed that for a small friction coefficient ($k < 4/3$), no ambiguities arise in the description of the dynamics. Several recent works using the complementarity approach [144] show that the range of applicability and the unambiguous definition of the solution can be expanded to systems with the friction coefficient equal to $k = 8/3\sqrt{3} \approx 1.53$, but serious difficulties arise for larger values of k. Apparently these difficulties are not related to the physics of the process, but rather to the limitations of the complementarity approach in application to mechanical systems with collisions, since for such systems the mechanism of the interaction is important. The complementarity approach ignores the law of interaction with the obstacle in the impact process and attempts to derive the post-impact velocities based only on general mechanical principles expressed in the form of conservation laws. Note that the approach we are proposing, based on singular space-time transformations in the vicinity of the impact point, is free of these shortcomings, and allows us not only to analyze and optimize the controlled impact [38, 40, 41], but also, as it will be shown below, to obtain meaningful results for uncontrollable dynamical systems with complex (hybrid) dynamics. This methodology is connected with the so-called *penalization methods* developed by M. Schatzman and L. Paoli [346, 350], in which violation of constraints is permitted, which results in the singular phase of motion, and in which the velocities of the bodies after impact are determined by the velocities at its completion [394]. However, this last cited reference focuses on obtaining solutions in analytical form, which greatly restricts the domain of applicability of its results; for example, the case of "sliding" in the "fast" phase of motion is not treated. The utilization of the space-time transformation method and the application of numerical methods in the auxiliary system for the description of the "fast" motion phase leads to a solution in all possible situations.

We will illustrate the application of our approach to a modified example from [394].

Since the method of singular space-time transformations is oriented towards numerical methods, we can treat a more complex and realistic case, which assumes that the interaction with the obstacle is described by a viscoelastic model and the time the contact terminate is determined from physical conditions like (8.75). The even more complex model of a rod that falls at an angle and has an initial transverse velocity is described in [297, 299, 472, 473].

The geometry of a falling rod of unit mass is pictured in Fig. 8.12.

We work with the generalized coordinates

$$x_p = (x, y, \theta) = (x_p^1, x_p^2, x_p^3), \quad x_v = (\dot{x}, \dot{y}, \dot{\theta}) = (\dot{x}_p^1, \dot{x}_p^2, \dot{x}_p^3) = (x_v^1, x_v^2, x_v^3),$$

that is

$$\dot{x}_p(t) = x_v(t). \tag{8.76}$$

In the constraint-free domain (regular phase of motion) the equations of motion have the form

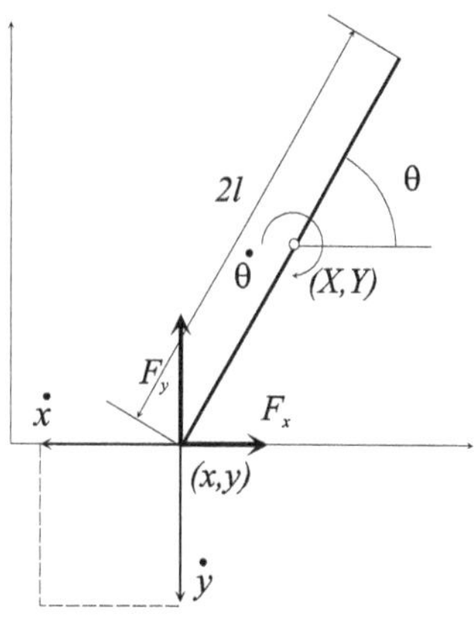

Fig. 8.12 Geometry of a falling rod of length $2l$; (X, Y) are the coordinates of the center of mass, (x, y) are the coordinates of the rod's bottom tip, θ is the angle the rod makes with the x-axis; the force of gravity is directed along the y-axis

$$\dot{x}_v(t) = F_v^{\mathrm{r}}(x_p(t), x_v(t)) = \begin{pmatrix} l\dot{\theta}(t)^2 \cos\theta(t) \\ l\dot{\theta}(t)^2 \sin\theta(t) - g \\ 0 \end{pmatrix} = \begin{pmatrix} l(x_v^3(t))^2 \cos x_p^3(t) \\ l(x_v^3(t))^2 \sin x_p^3(t) - g \\ 0 \end{pmatrix} \tag{8.77}$$

The constraint is described as

$$G(x_p) = y = x_p^2 \leq 0, \tag{8.78}$$

where $F_v^{\mathrm{r}}(x_p(t), x_v(t))$ is the generalized force, acting on the rod in the constraint-free domain (the superscript "r" stands for "regular" motion), and g is the acceleration of gravity.

During the penetration into the obstacle (singular phase of motion), the rod is acted upon by a supplementary force—the reaction of the support. The equation of motion in the singular phase has the form

$$\dot{x}_v(t) = F_v^{\mathrm{r}}(x_p(t), x_v(t)) + \mu F_v^{\mathrm{s}}(x_p(t), x_v(t), \mu), \tag{8.79}$$

and the viscoelastic force $\mu F_v^s(x_p, x_v, \mu)$ that arises when the rod penetrates the obstacle is given by (see (8.13)) the formula

$$\mu F_v^s(x_p, x_v, \mu) =$$

$$= -\mu\left[y + 2\xi\mu^{-1/2}\dot{y}\right] \begin{pmatrix} -k \operatorname{Sign}\dot{x} - 3\sin\theta[\cos\theta + k\sin\theta \operatorname{Sign}\dot{x}] \\ 1 + 3\cos\theta[\cos\theta + k\sin\theta \operatorname{Sign}\dot{x}] \\ -\frac{3}{l}[\cos\theta + k\sin\theta \operatorname{Sign}\dot{x}] \end{pmatrix} =$$

$$= -\mu[x_p^2 + 2\xi\mu^{-1/2}x_v^2] \begin{pmatrix} -k \operatorname{Sign} x_v^1 - 3\sin x_p^3[\cos x_p^3 + k\sin x_p^3 \operatorname{Sign} x_v^1] \\ 1 + 3\cos x_p^3[\cos x_p^3 + k\sin x_p^3 \operatorname{Sign} x_v^1] \\ -\frac{3}{l}[\cos x_p^3 + k\sin x_p^3 \operatorname{sgn} x_v^1] \end{pmatrix}, \tag{8.80}$$

where μ is the stiffness coefficient (which in the analysis of the shock interaction tends to infinity), ξ is the viscosity coefficient, k is the coefficient of dry friction, and the multivalued function $\operatorname{Sign} x$ is defined as s

$$\operatorname{Sign} x = \begin{cases} -1, & \text{if } x < 0, \\ [-1, 1], & \text{if } x = 0, \\ 1, & \text{if } x > 0. \end{cases}$$

For the sake of brevity, we omit the dependence on t.

Behavior of the Solution for Finite μ

Let us investigate the solution of the pre-limit system for finite μ on some finite segment $[0, T_1]$, choosing the scale $l = 1$. Our aim is to prove that the set of solutions corresponding to values of $\mu < \infty$ in the direct product of spaces $C \times BV$ of generalized three-dimensional coordinates and velocities is precompact. This will allow us to establish the existence of the limit of some subsequence of solutions corresponding to finite μ, which converges uniformly with respect to the generalized coordinates and pointwise at each point of continuity of the generalized velocities, and thereby establish the existence of the limit solution for $\mu \to \infty$ [204].

As the next step, we prove the uniqueness of the solution to the limit system and thus obtain a method for analyzing the behavior of the system when $\mu \to \infty$, that is, for the limit case of the absolutely rigid body.

Let K_0 be the kinetic energy of the rod at the moment τ when the contact begins, from which we start measuring the new time t (i.e., in the new time t we have $\tau = 0$); then

$$K_0 = \frac{1}{2}\Big[(\dot{X}(0))^2 + (\dot{Y}(0))^2 + \frac{1}{3}(\dot{\theta}(0))^2\Big],$$

where (X, Y) are the coordinates of the center of mass of the rod. From the energy conservation law we have

$$K_0 = K_t + g\big[Y(t) - Y(0)\big] + \int_0^{-y(t)} -\mu\big[y(s) + 2\xi\mu^{-1/2}\dot{y}(s)\big]dy(s) -$$
$$- k\int_0^t \mu\big[y(s) + 2\xi\mu^{-1/2}\dot{y}(s)\big]\dot{x}(s)ds \geq K_t + g\big[Y(t) - Y(0)\big],$$

where K_t is the kinetic energy at time t:

$$K_t = \frac{1}{2}\Big[(\dot{X}(t))^2 + (\dot{Y}(t))^2 + \frac{1}{3}(\dot{\theta}(t))^2\Big].$$

This implies, via the Cauchy-Bunyakovsky-Schwarz inequality, that

$$\begin{aligned} K_t &\leq K_0 - g\big[Y(t) - Y(0)\big] \leq K_0 + g\Big|\int_0^t \dot{Y}(s)ds\Big| \leq \\ &\leq K_0 + g\int_0^t 1^2 ds \int_0^t (\dot{Y}(s))^2 ds. \end{aligned} \tag{8.81}$$

Consider the three-dimensional vector-function $Z(t) = \big(\dot{X}(t), \dot{Y}(t), 3^{-1/2}\dot{\theta}(t)\big)$. Then, by the last inequality,

$$|Z(t)|^2 \leq C_1 + C_2 \int_0^t |Z(s)|^2 ds, \tag{8.82}$$

where C_1 and C_2 are constants. Hence, by the Gronwall-Bellman inequality, we deduce that $Z(t)$ is uniformly bounded:

$$|Z(t)|^2 \leq C_1 e^{C_2 T} = C = \text{const.} \tag{8.83}$$

Further, since

$$\dot{x}(t) = \dot{X}(t) + \dot{\theta}(t) \sin \theta(t),$$
$$\dot{y}(t) = \dot{Y}(t) - \dot{\theta}(t) \cos \theta(t),$$

the derivatives $\dot{x}(t) = \dot{x}_p^1(t)$ and $\dot{y}(t) = \dot{x}_p^2(t)$ are uniformly bounded.

Now from the momentum conservation law we deduce that

$$\dot{Y}(T) - Y(0) + gT = \int_0^T -\mu\big[y(s) + 2\xi\mu^{-1/2}\dot{y}(s)\big]ds \leq \text{const},$$

because the left-hand side is bounded. Here and above

$$T = \inf\big\{t : -\mu\big[y(s) + 2\xi\mu^{-1/2}\dot{y}(s)\big] = 0\big\}.$$

Hence, the right-hand sides of the differential equations of the pre-limit (with respect to μ) system are uniformly bounded. Moreover, the impulse of the y-components of the force (8.80) is uniformly bounded, which implies the uniform boundedness of variation of generalized velocities. Therefore, the solutions of the pre-limit system are equicontinuous functions with respect to the generalized coordinates. By the Arzelà-Ascoli theorem, there exists a subsequence of solutions that converges uniformly (with respect to μ) on $[0, T]$ in the generalized coordinates. Since the μ-parametrized set of generalized velocities is uniformly bounded and has uniformly bounded variations, Helly's second theorem (see Appendix) ensures that this set is precompact in the topology of pointwise convergence. In other words, there exists a subsequence (in μ) which converges at all points of continuity with respect to the generalized velocities.

The Limit System for $\mu \to \infty$

As mentioned earlier, for finite values of μ there exists a non-trivial time interval where the constraint is violated. Therefore, one can use the following space-time transformation around the impact point for $s > 0$:

$$
\begin{cases} y_p^\mu(s) = x_p(\tau) + \mu^{1/2}\big[x_p(\tau + \mu^{-1/2}s) - x_p(\tau)\big] \\ y_v^\mu(s) = x_v(\tau + \mu^{-1/2}s) \\ t^\mu(s) = \tau + \mu^{-1/2}s. \end{cases} \tag{8.84}
$$

Here τ is the instant the rod hits the obstacle (the starting instant of the contact with the constraint (8.78)) in the real time scale, and $\big(x_p(\tau), x_v(\tau)\big)$ are the values of the generalized coordinates at this instant. The new set of variables $\big(y_p^\mu(s), y_v^\mu(s)\big)$ satisfies on the interval $\{s > 0 : G\big(y_p^\mu(s)\big) \le 0\}$ the system of equations

$$
\begin{cases} \dot y_p^\mu(s) = y_v^\mu(s), \\ \dot y_v^\mu(s) = \mu^{1/2} F_v^{\mathrm{s}}\left(\dfrac{y_p^\mu(s) - x_p(\tau)}{\mu^{1/2}} + x_p(\tau), y_v^\mu(s), \mu\right) + \\ \qquad + \mu^{-1/2} F_v^{\mathrm{r}}\left(\dfrac{y_p^\mu(s) - x_p(\tau)}{\mu^{1/2}} + x_p(\tau), y_v^\mu(s)\right), \\ y_p^\mu(0) = x_p(\tau), \quad y_v^\mu(0) = x_v(\tau-), \end{cases} \tag{8.85}
$$

which is be obtained directly by differentiating with respect to s the system (8.84), with the Eqs. (8.76), (8.77), and (8.85) accounted for.

Letting $\mu \to \infty$ in (8.85) and using (8.80), we obtain the limit system of equations which described the "fast" motion in the new fast time and stretched spatial scale [282]. This system describes the impact that begins at the time instant τ (i.e., at $s = 0$), i.e., at the moment of initiation of the contact with the constraint, and reads

$$
\begin{cases} \dot y_p(s) = \begin{pmatrix} y_v^1(s) \\ y_v^2(s) \\ y_v^3(s) \end{pmatrix}, \\ \dot y_v(s) = -[y_p^2(s) + 2\xi y_v^2(s)] \times \\ \times \begin{pmatrix} -k \,\mathrm{Sign}\, y_v^1(s) - 3\sin\theta(\tau)\big[\cos\theta(\tau) + k \sin\theta(\tau)\,\mathrm{Sign}\, y_v^1(s)\big] \\ 1 + 3\cos\theta(\tau)\big[\cos\theta(\tau) + k\sin\theta(\tau)\,\mathrm{Sign}\, y_v^1(s)\big] \\ -\dfrac{3}{l}\Big[\cos\theta(\tau) + k\sin\theta(\tau)\,\mathrm{Sign}\, y_v^1(s)\Big] \end{pmatrix}. \end{cases} \tag{8.86}
$$

Here for $s = 0$ one has

$$
y_p(0) = (x(\tau), y(\tau), \theta(\tau)), \quad y_v(0) = (\dot x(\tau-), \dot y(\tau-), \dot\theta(\tau-)).
$$

This system must be solved for the initial condition

$$y_p^2(0) = 0, \quad y_v^2(0) = \dot{y}(\tau-) < 0$$

until the time s^{**} when the contact force vanishes, that is, until the time s^{**} when

$$y_p^2(s^{**}) + 2\xi y_v^2(s^{**}) = 0$$

(physical condition of contact termination), or until the time when $y_p^2(s^*) = 0$ (geometrical condition of contact termination); clearly, $s^* \geq s^{**}$ [215]. In what follows we use only the physical condition of termination of contact. The justification of the limit system (8.86) will be given below, after we establish the uniqueness of its solution.

Uniqueness of the Solution of the Limit System

Let us prove the uniqueness of the solution for the limit system (8.86). Take some solution $(\bar{y}_p(s), \bar{y}_v(s))$ and perform the time change $s = s(\vartheta)$ by the rule

$$\frac{ds(\vartheta)}{d\vartheta} = -\frac{1}{Y_p^2(\vartheta) + 2\xi Y_v^2(\vartheta)}, \qquad s(0) = 0,$$

where

$$Y_p(\vartheta) \stackrel{\text{def}}{=} \bar{y}_p(s(\vartheta)), \quad Y_v(\vartheta) \stackrel{\text{def}}{=} \bar{y}_v(s(\vartheta)),$$

and Y_p^2 and Y_v^2 are the second components of the two-dimensional vectors Y_p and Y_v, respectively. This time substitution is made on the interval of interaction between the rod and obstacle, that is for

$$\left\{s : \quad -\left[\bar{y}_p^2(s) + 2\xi \bar{y}_v^2(s)\right] > 0\right\}$$

and respectively

$$\left\{\vartheta : \quad -\left[Y_p^2(\vartheta) + 2\xi Y_v^2(\vartheta)\right] > 0\right\}.$$

Denote the derivative with respect to the new time ϑ by "$\circ$". Then the first vector equation in the system governing the motion of the rod (in the scale $l = 1$) reads

$$\overset{\circ}{Y}_p(\vartheta) = Y_v(\vartheta)\frac{ds(\vartheta)}{d\vartheta}, \qquad Y_p(0) = \bar{y}_p(0); \tag{8.87}$$

in particular,

$$\overset{\circ}{Y_p^2}(\vartheta) = -\frac{Y_v^2(\vartheta)}{Y_p^2(\vartheta) + 2\xi Y_v^2(\vartheta)}, \qquad Y_p^2(0) = 0,\ Y_v^2(0) = \bar{y}_v^2(0) < 0. \tag{8.88}$$

The second vector equation in the system reads

$$\overset{\circ}{Y_v}(\vartheta) = \frac{d}{d\vartheta}\begin{pmatrix} Y_v^1(\vartheta) \\ Y_v^2(\vartheta) \\ Y_v^3(\vartheta) \end{pmatrix} = $$
$$= \begin{pmatrix} -\frac{3}{2}\sin\big(2\theta(\tau)\big) - k\big(1 + 3\sin^2\theta(\tau)\big)\operatorname{Sign} Y_v^1(\vartheta) \\ 1 + 3\cos^2\theta(\tau) + \frac{3}{2}k\sin\big(2\theta(\tau)\big)\operatorname{Sign} Y_v^1(\vartheta) \\ -3\cos\theta(\tau) - 3k\sin\theta(\tau)\operatorname{Sign} Y_v^1(\vartheta) \end{pmatrix}. \tag{8.89}$$

Let us consider the scalar equation (8.88) and the first two scalar equations in (8.89). The resulting system of equations is complete in the variables $Y_p^2(\vartheta)$, $Y_v^1(\vartheta)$, $Y_v^2(\vartheta)$. Let us prove that it has exactly one solution. To this end we integrate the first equation in(8.89) with respect to $Y_v^1(\vartheta)$, rewriting it in the form

$$\overset{\circ}{Y_v^1}(\vartheta) = -b - ka\operatorname{Sign} Y_v^1(\vartheta), \qquad Y_v^1(0) = x_v^1(\tau-), \tag{8.90}$$

where the constants a and b are given by

$$a = 1 + 3\sin^2\theta(\tau) > 0, \qquad b = \frac{3}{2}\sin\big(2\theta(\tau)\big) > 0.$$

Suppose Eq. (8.90) admits a second solution $\widetilde{Y}_v^1(\vartheta)$. Then

$$\overset{\circ}{\widetilde{Y}_v^1}(\vartheta) = -b - ka\operatorname{Sign}\widetilde{Y}_v^1(\vartheta), \qquad \widetilde{Y}_v^1(0) = x_v^1(\tau-).$$

Consider the squared difference $\big[Y_v^1(\vartheta) - \widetilde{Y}_v^1(\vartheta)\big]^2$. We have

$$\frac{d}{d\vartheta}\big[Y_v^1(\vartheta) - \widetilde{Y}_v^1(\vartheta)\big]^2 = 2\big[Y_v^1(\vartheta) - \widetilde{Y}_v^1(\vartheta)\big](-ka)\big[\operatorname{Sign} Y_v^1(\vartheta) - \operatorname{Sign}\widetilde{Y}_v^1(\vartheta)\big].$$

Since

$$(x - \widetilde{x})(\operatorname{Sign} x - \operatorname{Sign}\widetilde{x}) \geq 0 \quad \text{for any real } x,\ \widetilde{x}, \tag{8.91}$$

we have that

$$\frac{d}{d\vartheta}\left[Y_v^1(\vartheta) - \widetilde{Y}_v^1(\vartheta)\right]^2 \leq 0.$$

But $Y_v^1(0) - \widetilde{Y}_v^1(0) = 0$, hence $Y_v^1(\vartheta) - \widetilde{Y}_v^1(\vartheta) \equiv 0$. Therefore, the solution of the first equation in (8.89) is unique. Further, by substituting this solution in (8.90) we find $\operatorname{Sign} \widetilde{Y}_v^1(\vartheta)$. Finally, substituting this function $\operatorname{Sign} \widetilde{Y}_v^1(\vartheta)$ in the second and third equations of (8.89), we find their unique solutions. Clearly, all these solutions are bounded in any finite time interval.

Remark 8.7 For a given a solution $Y_v^2(\vartheta)$ that is bounded in any finite interval, the solution of Eq. (8.88) in the interval $[0, \bar{\vartheta}]$, where

$$\bar{\vartheta} = \inf\{\vartheta > 0 : V_p^2(\vartheta) + 2\xi V_v^2(\vartheta) < 0\},$$

exists and is unique. Indeed, (8.88) is an equation of the form

$$\overset{\circ}{y}(\vartheta) = \frac{\alpha(\vartheta)}{y(\vartheta) + c\alpha(\vartheta)}, \qquad y(0) = 0,\ \alpha(0) < 0,\quad c = \text{const},$$

where $\alpha(\vartheta)$ is a continuous piecewise-linear function.

Now let us prove the uniqueness of the solution of the system (8.86). Suppose that there exists another solution $\left(\widetilde{y}_p(s), \widetilde{y}_v(s)\right)$. Arguing in much the same way as above, we perform the time change $s = s(\vartheta)$ with $s(0) = 0$ and

$$\frac{ds(\vartheta)}{d\vartheta} = -\frac{1}{\widetilde{y}_p^2\big(s(\vartheta)\big) + 2\xi\widetilde{y}_v^2\big(s(\vartheta)\big)} \overset{\text{def}}{=} -\frac{1}{\widetilde{Y}_p^2(\vartheta) + 2\xi\widetilde{Y}_v^2(\vartheta)},$$

where $\widetilde{Y}_p(\vartheta) \overset{\text{def}}{=} \widetilde{y}_p\big(s(\vartheta)\big)$ and $\widetilde{Y}_v(\vartheta) \overset{\text{def}}{=} \widetilde{y}_v\big(s(\vartheta)\big)$. In the new time variable we get a system of equations with "tilde", which differs only in notation from the system considered above for the variables without "tilde". Hence, the unique solutions of these systems coincide. It follows that the inverse time transformation yields the unique solution of the system (8.86). Therefore, the generalized solution corresponding to $\mu \to \infty$ is unique and the jump of the velocity is defined unambiguously with the aid of the solution of the equations of "fast" motion.

Equations of the Limit System

Denote $\theta^\mu(s) \overset{\text{def}}{=} \theta\big(\tau + \mu^{-1/2}s\big)$. Then by (8.84) for the third coordinate we have

$$\frac{y_p^{\mu 3}(s) - \theta(\tau)}{\sqrt{\mu}} + \theta(\tau) = \theta^\mu(s) \to \theta(\tau), \quad \mu \to \infty,$$

thanks to the obvious uniform boundedness of $y_p^{\mu 3}(s)$. Next,

$$\frac{d}{ds}\|y_v(s) - y_v^{\mu}(s)\|^2 = 2\langle y_v(s) - y_v^{\mu}(s), \dot{y}_v(s) - \dot{y}_v^{\mu}(s)\rangle, \tag{8.92}$$

where $\langle \cdot, \cdot \rangle$ denotes the scalar product of three-dimensional vectors. Here, in view of (8.85) and (8.86),

$$\dot{y}_v^{\mu}(s) = F_1^{\mu}(s) + F_2^{\mu}(s)\sigma^{\mu}(s), \qquad \dot{y}_v(s) = F_1(s) + F_2(s)\sigma(s),$$

where we use the notation

$$F_1^{\mu}(s) \stackrel{\text{def}}{=} -\big[y_p^{\mu 2} + 2\xi y_v^{\mu 2}\big]\varphi\big(\theta^{\mu}(s)\big), \qquad F_1(s) \stackrel{\text{def}}{=} -\big[y_p^2 + 2\xi y_v^2\big]\varphi\big(\theta(\tau)\big),$$

$$F_2^{\mu}(s) \stackrel{\text{def}}{=} k\big[y_p^{\mu 2} + 2\xi y_v^{\mu 2}\big]\psi\big(\theta^{\mu}(s)\big), \qquad F_2(s) \stackrel{\text{def}}{=} k\big[y_p^2 + 2\xi y_v^2\big]\psi\big(\theta(\tau)\big),$$

$$\sigma^{\mu}(s) \stackrel{\text{def}}{=} \operatorname{Sign}\big(y_v^{\mu 1}\big), \qquad \sigma(s) \stackrel{\text{def}}{=} \operatorname{Sign}\big(y_v^1\big),$$

and

$$\varphi(\theta) \stackrel{\text{def}}{=} \begin{pmatrix} -(3/2)\sin 2\theta \\ 1 + 3\cos^2\theta \\ -3\cos\theta \end{pmatrix}, \qquad \psi(\theta) \stackrel{\text{def}}{=} \begin{pmatrix} 1 + 3\sin^2\theta \\ -(3/2)\sin 2\theta \\ 3\sin\theta \end{pmatrix}.$$

In these notations and omitting the argument s, we have

$$F_1 - F_1^{\mu} = -\big[y_p^2 + 2\xi y_v^2\big]\varphi\big(\theta(\tau)\big) + \big[y_p^{\mu 2} + 2\xi y_v^{\mu 2}\big]\varphi\big(\theta^{\mu}(s)\big).$$

where y_p^2 and y_v^2 are the limit values (by respective subsequences) of $y_p^{\mu 2}$ and $y_v^{\mu 2}$.
Analogously

$$F_2\sigma - F_2^{\mu}\sigma^{\mu} = \big(F_2 - F_2^{\mu}\big)\sigma + F_2^{\mu}(\sigma - \sigma^{\mu}),$$

where

$$\big(F_2 - F_2^{\mu}\big)\sigma = k\big[y_p^2 + 2\xi y_v^2\big]\psi\big(\theta(\tau)\big) - k\big[y_p^{\mu 2} + 2\xi y_v^{\mu 2}\big]\psi\big(\theta^{\mu}(s)\big).$$

Next, (8.92) implies that

$$
\begin{aligned}
&\langle y_v - y_v^\mu, F_2^\mu(\sigma - \sigma^\mu)\rangle = \langle (y_v - y_v^\mu)(\sigma - \sigma^\mu), F_2^\mu\rangle = \\
&\quad = k\big[y_p^{\mu 2} + 2\xi y_v^{\mu 2}\big]\langle (y_v - y_v^\mu)(\sigma - \sigma^\mu), \psi(\theta^\mu)\rangle = \\
&\quad = k\big[y_p^{\mu 2} + 2\xi y_v^{\mu 2}\big]\Big\{\big(\bar{y}_v^1 - y_v^{\mu 1}\big)(\sigma - \sigma^\mu)\psi_1(\theta^\mu) + \\
&\quad + \big(y_v^2 - y_v^{\mu 2}\big)(\sigma - \sigma^\mu)\psi_2(\theta^\mu)\big(\theta(\tau) - \theta^\mu\big)(\sigma - \sigma^\mu)\psi_3(\theta^\mu)\Big\}.
\end{aligned}
\tag{8.93}
$$

Here, ψ_i, $i = 1, 2, 3$, are the components of the vector ψ. Note that in (8.93) the factor in square brackets is negative, and in the braces we have three terms, the first of which is negative thanks to the inequality (8.91), and the second and third terms involve the uniformly bounded factors $(\sigma - \sigma^\mu)\psi_1(\theta^\mu)$ and $(\sigma - \sigma^\mu)\psi_3(\theta^\mu)$, respectively.

Then integrating (8.92), we find

$$
\|y_v(s) - y_v^\mu(s)\|^2 = 2\int_0^s \big\langle y_v(\vartheta) - y_v^\mu(\vartheta), \dot{y}_v(\vartheta) - \dot{y}_v^\mu(\vartheta)\big\rangle d\vartheta. \tag{8.94}
$$

Denoting $\delta^\mu(s) \overset{\text{def}}{=} y_v(s) - y_v^\mu(s)$ and using the fact that $\delta^\mu(0) = 0$ and the preceding remark, we deduce, by analogy with the inequalities (8.81), (8.82) and (8.83) , that

$$
\|\delta^\mu(s)\|^2 \le C_1^\mu + C_2 \int_0^s \|\delta^\mu(\vartheta)\|^2 d\vartheta, \tag{8.95}
$$

where $C_2 = \text{const}$, and $C_1^\mu \to 0$ because $\theta^\mu(s) - \theta(\tau) \to 0$ as $\mu \to \infty$. This in turn implies that

$$
\|\delta^\mu(s)\|^2 \le C_1^\mu e^{C_2 T} = C = \text{const}. \tag{8.96}
$$

Letting here $\mu \to \infty$, we see that $y_v^\mu(s) \to y_v(s)$ (uniformly). From the first equation in the system (8.85) it follows that also $y_p^\mu(s) \to y_p(s)$ uniformly.

Numerical Analysis

The theoretical analysis of solutions of the system (8.86) is quite complex (see [394]), even for the simpler case of purely elastic collision, in which $\xi = 0$. However, if we want to calculate solutions numerically, we note that the general behavior can be described as follows: for as long as the variable $y_v^1(s)$ preserves its sign, the system is integrated in the usual way. After that, two options are possible:

(a) the speed of the rod's bottom extremity changes sign and the system is integrated further,

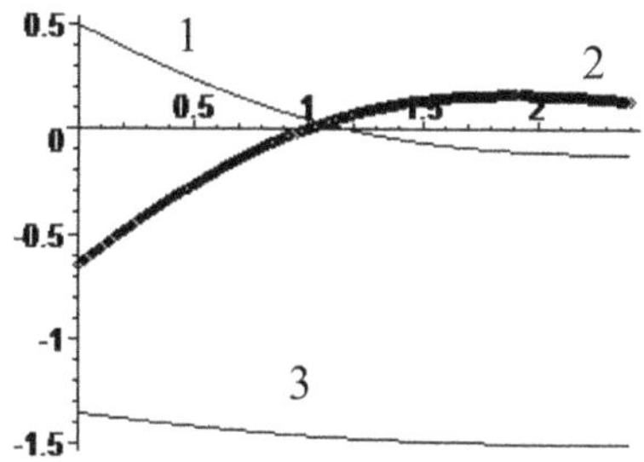

Fig. 8.13 Case (a). The sign of the variable $y_v^1(s)$ changes at $s \approx 1.0$. The friction force is not enough ($k = 0.5$) to fix the null velocity, so the horizontal velocity changes sign and the bottom tip of the rod starts sliding in the opposite direction. A "sliding mode" does not occur. Notation: 1—horizontal velocity of the contact point, 2—vertical velocity of the contact point, 3—angular velocity of rotation of the rod. The post-shock velocities are determined from the physical condition of termination of contact, i.e., at the moment of time when the variable 2 achieves a local maximum, for $s \approx 2.3$

(b) a "sliding mode" arises, in which $\dot{y}_v^1(s) = 0$, which gives the value of the function

$$\text{Sign } y_v^1(s) = -\frac{3\sin\theta(\tau)\cos\theta(\tau)}{k(1+3\sin^2\theta(\tau))}, \tag{8.97}$$

provided this value lies inside the interval $[-1, 1]$. In other words, if the maximum value of the friction force is sufficient to fix the zero speed, then a " sliding mode" occurs; in the opposite case, the speed changes sign. You can see that for $k > 3/4$ the value of the expression (8.97) is always less than 1 in modulus, therefore, for large friction coefficients, the lower extremity of the bar stops and the ensuing motion is described by a "sliding mode". This terminology is borrowed from the theory of automatic control systems [365], where it describes modes with infinitely frequent switchings. In the presents case, of course, no switching occurs in the real motion; however, in the numerical solution, the variable corresponding to the friction force changes sign as one moves along the sliding surface with a high frequency, which justifies using this terminology.

Examples of the two types of solutions discussed are shown in Figs. 8.13 and 8.14. The calculations were carried out for the following values of the variables:

$$l = 1,\ \xi = 0.4,\ \mu = 10^5,\ \dot{X}(\tau) = -1,\ \dot{Y}(\tau) = -1,\ \dot{\theta}(\tau) = 0.5,$$

that is, the initial velocities were prescribed for the rod's center mass (Fig. 8.14).

We note that the limit solution, corresponding to $\mu = \infty$, behaves completely similarly (see Fig. 8.15).

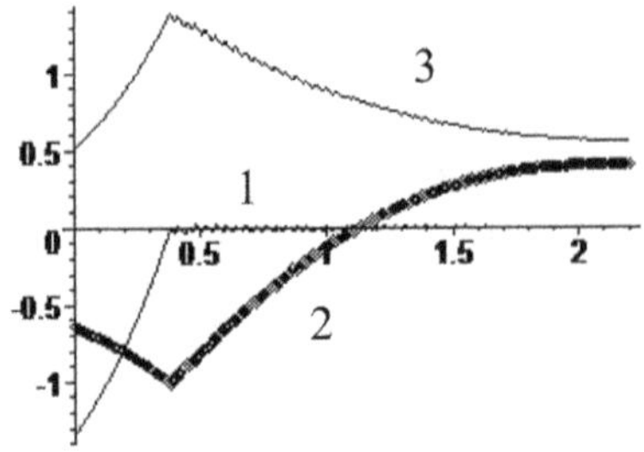

Fig. 8.14 Case (b). The variable $y_v^1(s)$ vanishes at $s \approx 0.4$. The friction force is large enough ($k = 2.4$) and the "sliding mode" starts, i.e., the velocity of the bottom extremity of the rod becomes zero. Notation: 1—horizontal speed of the contact point, 2—vertical velocity of the contact point, 3—angular velocity of rotation of the rod. The post-shock velocities are determined from the physical condition of the termination of the contact, i.e., at the moment of time when variable 2 achieves the local maximum, for $s \approx 2.2$

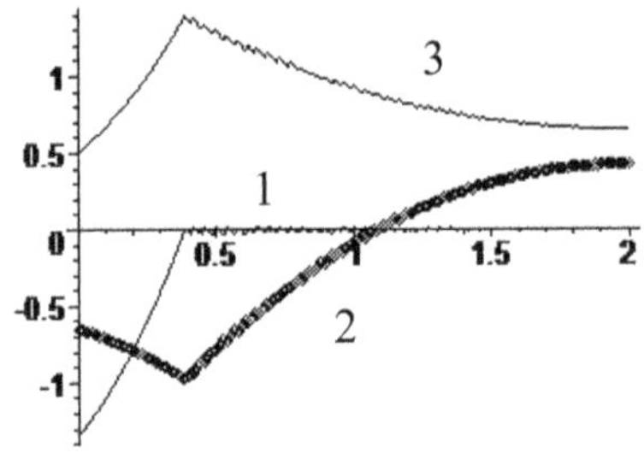

Fig. 8.15 Limit motion for $\mu = \infty$. The graph is almost identical to that in Fig. 8.14

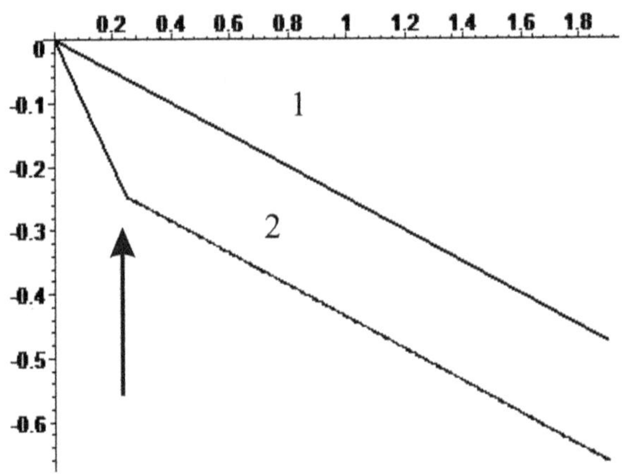

Fig. 8.16 The start of the"sliding mode" at $s \approx 0.4$. Notation: 1—the value of the integral of the friction force obtained via numerical integration of the system, 2—the value of the integral of the friction force for the "sliding mode", calculated by the formula (8.97). Upon increasing of the picture resolution, one can see that the value of the horizontal velocity changes the sign with increasing frequency

The next figure, Fig. 8.16, compares the real friction force found by numerical integration of the system (8.86) and the same integral, except that the value of the function Sign $y_v^1(s)$ for the "sliding mode" is calculated by formula (8.97). One can see that after the "sliding mode" starts, the two graphs are almost parallel, which means that the calculated and experimental values of the friction force coincide.

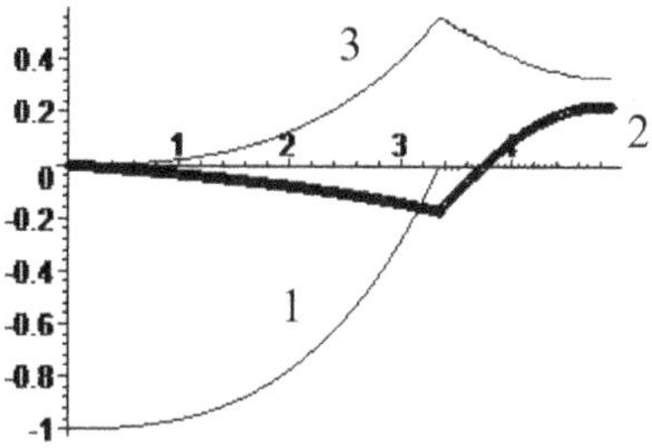

Fig. 8.17 Shockless rebound. The initial vertical velocity is zero, however, due to the horizontal velocity and the large coefficient of friction, the bottom tip of the rod embeds in the deformable medium and is then pushed out at a nonzero velocity at the end of the compression-relaxation phase. Notation: 1—horizontal velocity, 2—vertical velocity, 3—angular velocity. One sees that all the velocities change abruptly

Remark 8.8 In fact, in a real physical system, there is no sliding mode, i.e., there are no changes in the direction of the friction force; here the sliding mode is only a convenient abstraction, allowing one to calculate the value of the friction force in the range when $|\,\mathrm{Sign}\; y_v^1(s)| < 1$. Nevertheless, for the numerical solution, due to the impossibility of setting a so-called "equivalent" control [365], the variable corresponding to the friction force changes sign frequently.

Remark 8.9 These results were first published in [307], and relate to the simplest case of so-called flat shock. However, the proposed approach can be successfully applied also to the case of "oblique" shock, when during the sliding phase the bottom tip of the rod moves along some complicated curve [355, 472, 473], and as a result, the rod gains a velocity increment in the transverse direction [297, 299]

Shockless Rebound

Another effect that is frequently associated to the so-called the Painlevé paradoxes [335] is the shockless rebound. For a large coefficient of friction, a rod with no vertical velocity at the beginning of the interaction gains, due to the friction force, an additional angular velocity (as if the rod stumbles), resulting in a vertical component of velocity which leads to compression. The bottom tip of the rod embeds in the deformable surface, and then is pushed out at a nonzero velocity at the end of the compression-relaxation phase (see Fig. 8.17).

Resonance Nature of Energy Absorption

In the literature devoted to the analysis of the impact of a rod on a rough surface, a resonance in the energy absorption during impact was observed. In [171] and [414], the rod is described in terms of the theory of elastic media and it is shown that energy

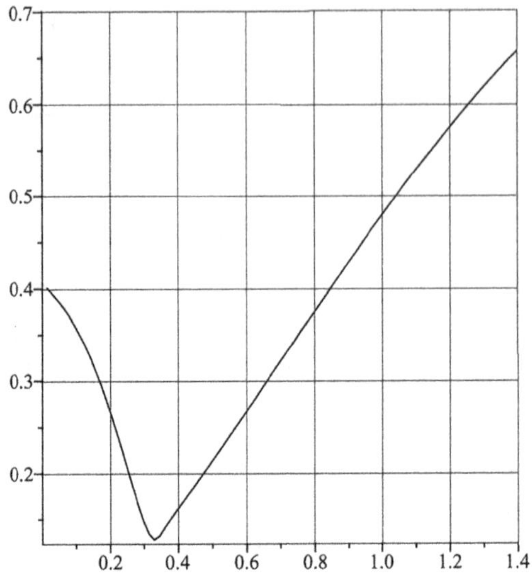

Fig. 8.18 Resonance in the energy absorption, $k = 2.4$

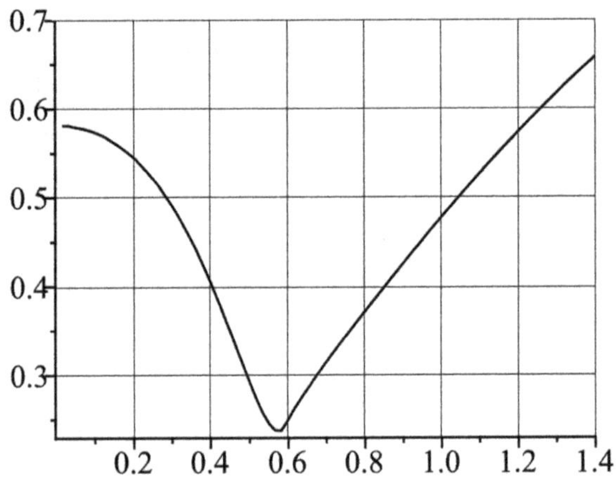

Fig. 8.19 Resonance in the energy absorption, $k = 1.5$

absorption has indeed a resonant nature; this was then confirmed experimentally. However, in subsequent works [354], it was noted that numerical solutions often lead to instability and exhibit a chaotic behavior. In our model, irregularities in the numerical solution corresponding to different angles of incidence of the rod were not noticed, but the resonant nature of energy absorption is indeed observed. Figures 8.18 and 8.19 show results corresponding to the value of the "energy" recovery factor as a function of the angle of incidence,

$$k(\theta(\tau)) = \frac{E_{\text{init}}}{E_{\text{fin}}},$$

where $0 \leq \theta(\tau) \leq \pi/2$ is the angle of incidence, E_{init} and E_{fin} are the initial and final kinetic energy of the rod before and after impact, respectively. These figures correspond to different values of the coefficient of friction, $k = 2.4$ in Fig. 8.18 and $k = 1.5$ in Fig. 8.19. It is interesting that the angle corresponding to the maximum absorption is less than $\theta_{\text{min}} \approx 0.3$ in the first case and less than $\theta_{\text{min}} \approx 0.6$ in the second cases.

Analysis of the Painlevé Paradox and Attempts of its Resolution

The clearest formulation of the limitations of the complementarity approach to the falling rod model is provided in the monograph [362], pp. 59–63. Combining the Eq. (8.86), which describe the complete motion of the rod at the moment of impact, including the regular and singular phases, leads to the following equation for a variable $x_v^2(t)$ that corresponds to the velocity of the rod's bottom tip in the vertical direction:

$$\begin{aligned} \dot{x}_v^2(t) &= -\mu\left[x_p^2(t) + 2\xi\mu^{-1/2}x_v^2(t)\right] \times \\ &\times \left[1 + 3\cos x_p^3(t)[\cos x_p^3(t) + k\sin x_p^3(t)\,\mathrm{Sign}\,x_v^1(t)]\right] + \\ &+ l(x_v^2(t))^2 \sin x_p^3(t) - g = F_Y(t)A(t) + B(t), \end{aligned} \tag{8.98}$$

which is valid during the singular phase of motion until the time when the support reaction $F_Y(t)$ vanishes. That is, during the singular phase of motion one has

$$F_Y(t) = -\mu\left[x_p^2(t) + 2\xi\mu^{-1/2}x_v^2(t)\right] > 0.$$

At the same time one allows violation of the constraint $x_p^2 \geq 0$, and it is precisely due to this violation that a positive support reaction arises which pushes the rod out of the "forbidden domain" $x_p^2 < 0$. Note that with such an approach, there is no lack of uniqueness in the equations of motion, which as shown above can be integrated in a unique manner till the singular phase ends.

When the complementarity approach is used, the equation for $x_v^2(t)$ is written in the same way as (8.98) (see [362] p. 61), but in view of the complementarity conditions or of the condition of non penetration of the rod in the forbidden domain $x_p^2 < 0$, the following relations must be satisfied:

$$\dot{x}_v^2(t) = F_Y(t)A(t) + B(t) \geq 0, \quad \dot{x}_v^2(t)F_Y(t) = 0, \tag{8.99}$$

where the variables $\dot{x}_v^2(t)$ and $F_Y(t)$ are determined simultaneously and satisfy the constraints

$$\dot{x}_v^2(t) \geq 0 \quad \text{and} \quad F_Y(t) \geq 0.$$

As first shown by Painlevé in his classical work (see [335], p. 102), it is at this point that one runs into the paradoxical situation that the determination of these variables from Eq. (8.99) lacks uniqueness or is even impossible.

Indeed, the functions $A(t)$ and $B(t)$ depend on the initial conditions at the moment of collision and change slowly; moreover, they can assume values of both signs, so the system of Eq. (8.99) admits a unique solution only if $A(t) > 0$. If $A(t)$ is negative, then depending on the sign of $B(t)$, two cases arise: if $B(t) > 0$, two solutions are possible, namely sliding or detachment from the surface, whereas if $B(t) < 0$ there is no solution at all [362]. In order that the function $A(t)$ be positive for all values of the angle $x_p^3 \in [0, \pi/2]$ and $\operatorname{Sign} x_v^1 \in [-1, 1]$, it is necessary that the coefficient of friction k be smaller than 4/3. Indeed, the condition that $A(t) > 0$ implies the inequality

$$1 + 3\cos\theta[\cos\theta + k\cos\theta \operatorname{Sign} x] > 0,$$

which must hold for all $\theta \in [0, \pi/2]$ and $x \in \mathbb{R}$. Consequently,

$$\frac{5}{2} + \frac{3}{2}(\cos 2\theta - k\sin 2\theta) > 0$$

for all $2\theta \in [0, \pi]$, and so $k < 4/3$.

In the monograph [362], it is noted that the domain of applicability of the complementarity approach may be expanded by choosing the limit value of the coefficient of friction depending on the angle of incidence θ, and in the article [144] this is done successfully. Perhaps the most successful attempt of solving the dynamical problem of rigid bodies in the presence of friction is the approach of D. E. Stewart [415], who used the implicit scheme of the differential equation solution for systems with Coulomb friction. The principal result of this work is the proof of the solution existence, while the question of its uniqueness remains open. The problem that all authors have tried to get around is that the contact force at impact is the impulse one, and the dynamics of the system should be described by a differential equation (or inclusion) with measures. As noted by Stewart in his review article [416], defining this force by using only the complementarity approach is impossible; moreover, if one uses the restitution law describing the velocity after impact in Newton's model, it is even possible to obtain a positive increment of the kinetic energy [420]. It is believed that to determine the impulse force one needs the law of interaction and knowledge of the stiffness and viscosity coefficients. From this point of view, our approach, which guarantees the existence of exactly one solution for sufficiently large values of the stiffness, as well as the convergence to the unique solution when the stiffness tends to infinity, is apparently a reasonable alternative implicit scheme for the numerical solution; indeed, it allows to unambiguously determine the velocity after impact and agrees with experimental results demonstrating the dependence of the restitution law on the angle of incidence of the rod [171, 414]. This discussion would not be complete without mentioning other approaches to the analysis of other mechanical models where the Painlevé paradox is inherent [324], or other approaches to analysis, such as, for example, from the point of view of the theory of contrast structures or singularly perturbed systems [323, 325–327]. However, these topics would take the presentation quite far from the general approach developed in the present book.

8.4 Controllable Systems with Shocks

In this section, we consider the problem of describing the trajectories of systems with shocks, first as a general problem of dynamics, and then as a problem of the dynamics of controlled motion.

8.4.1 Systems with Shocks in Classical Mechanics: Lagrange Equations

In classical mechanics, the presence of unilateral constraints generates special contact forces, which, by virtue of the d'Alembert-Lagrange principle, result in equations of motion in the form [3]

$$\frac{d}{dt}\left(\frac{\partial T}{\partial \dot{x}}\right) - \frac{\partial T}{\partial x} = F + \lambda \frac{\partial f}{\partial x}, \qquad \lambda \geq 0, \quad \lambda f(x) = 0, \tag{8.100}$$

where x is the vector of generalized coordinates, T is the kinetic energy, which is quadratic in the generalized velocities, F is the vector of generalized forces, and the unilateral constraint is given by a scalar (or vector) function $f(x) \geq 0$, such that on the boundary $\{x : f(x) = 0\}$ the gradient of f does not vanish (in the vector case the inequalities are understood componentwise, and λ is a vector-valued measure). A very large number of works are devoted to the discussion of the Eq. (8.100) and its justification, but from the point of view of the dynamics of systems with impacts, one is interested in the behavior of the Lagrange multiplier λ. In the case of collision with a unilateral constraint, this multiplier can no longer be a bounded function of coordinates and time, but must take infinite values, which corresponds to the occurrence of impulsive actions upon impact on the constraint; this in turn leads to discontinuities in velocities at the moment of impact. The Eq. (8.100) themselves together with the constraint linking λ and f express the so-called *complementarity principle*. A detailed analysis of the equations of motion in the case of impact interaction can be found, e.g., in the monograph [183].

Generally speaking, writing the equations of motion in the form (8.100) and other forms of complementarity equations does not allow to integrate them, since the Lagrange multiplier is undetermined. Considerable efforts were made in the attempt to solve this problem and substantiate the d'Alembert-Lagrange principle. Recently, a lot of attention is devoted to the approach based on the assumptions that the interacting bodies have a finite stiffness, a contact phase arises at the initial moment of the interaction, and the motion corresponding to an absolutely rigid body is the limit of a sequence of solutions when the contact stiffness coefficient tends to infinity [215] (the so-called genetic approach to the theory of motion of systems with shocks).

Further, we consider the following issues, without the resolution of which it is impossible to construct a meaningful theory of controlled systems with shock interactions.

1. When can one guarantee the existence and uniqueness of solutions to the complementarity equations of dynamics for systems with unilateral constraints?
2. Can we justify the complementary equations by the method of passage to the limit in a sequence of solutions when the contact stiffness coefficient tends to infinity?
3. How to describe the motion of controlled systems with shocks?

8.4.2 Complementarity Equations in the Dynamics of Systems: Existence and Uniqueness of Solutions

The answer to the question about existence of a solution depends on the initial conditions, and if these conditions are prescribed in a constraint-free domain, then the solution exists and is unambiguously extendable until it reaches the constraints for the first time. Moreover, to guarantee existence and uniqueness it is sufficient that the functions on the right-hand sides of the system of differential equation that describes the dynamics of the system under consideration satisfy the usual Lipschitz and linear growth conditions [237]. However, already when the constraint is reached for the first time it becomes necessary to determine how the motion should be continued. The classical theory of the Lagrange equation (8.100) does not provide an answer and to resolve the issue successfully it is necessary to introduce some additional assumptions on the nature of the interaction at the moment of contact with the constraint. The classification of possible motions can be found in [183]; it agrees very well with the results of mechanical experiments as well as with the physical intuition. However, many motions are highly nontrivial, for example, the motion along the constraint belongs to the class of so-called "sliding modes," well-known in the theory of dynamical systems [365]. Even when the existence of such a regime is allowed, the conditions for its occurrence and stability require a detailed study. An example of such a study can be found in the work [143], devoted to the study of the Painlevé paradox. The most serious difficulties arise in the analysis of the mode of exit from the constraint. It is here that the phenomenon of non-uniqueness of the solution arises, as was discovered for the first time by Aldo Bressan [56]. It should be noted that nonuniqueness occurs even in the simplest example of the motion of a point of unit mass (bouncing ball) in which the constraint domain is specified by the relation $x_p(t) \leq 0$. Since nonuniqueness of solutions is closely related to the controllability properties of a system, let us consider it in more detail.

8.4.3 Nonuniqueness of Solutions: External Control Force

Suppose the external force $f(t)$ depends only on time and that $f(t) \leq 0$. Then the system of complementarity equations has the form

$$\dot{x}_p(t) = x_v(t), \quad dx_v(t) = f(t)dt + \lambda(dt),$$
$$\lambda(dt) \geq 0, \quad x_p(t) \geq 0, \quad \int_0^t x_p(s)\lambda(ds) = 0. \tag{8.101}$$

If the initial conditions $x_p(0) = 0$, $x_v(0) = 0$ are prescribed, then the set of functions

$$x_p(t) = 0, \quad x_v(t) = 0, \quad \lambda(dt) = -f(t)dt$$

will always satisfy the system (8.101). This result is quite plausible from the point of view of physical intuition, since when a point mass at rest is pressed against an absolutely rigid surface by some force, it is difficult to imagine that it can be made to move. However, if one admit the possibility of elastic impact, in which

$$x_p(\tau) = 0, \quad x_v(\tau) = x_v(\tau-),$$

then another solution may arise, which starts from the point $(0, 0)$ and leaves it through an infinite sequence of impacts. Similar examples, starting with [56], were also given in the [18, 20, 391]. Our presentation is closest to [391], although it does not repeat it. First of all, we should note that the above examples illustrate the possibility of nonuniqueness of the solution in the case of an infinitely smooth external action; however, from our point of view, the smoothness or the irregularity of the external action have absolutely nothing to do with it, rather, nonuniqueness is a consequence of the possibility of stabilizing a systems with impacts in a finite time by means of a bounded external action. To maintain continuity with the aforementioned studies, we will construct examples with an infinitely smooth external action, although for the emergence of multiple solutions it is enough that the external action satisfies certain integral relations. So, let us consider the system (8.101) supplemented by the restitution law (8.102)

$$x_v(\tau) = -\gamma x_v(\tau-), \quad 0 \leq \gamma \leq 1, \quad \text{for} \quad x_p(\tau) = 0. \tag{8.102}$$

Fig. 8.20 The function $f_0(s)$ for $A = 1$ and $B = 3$

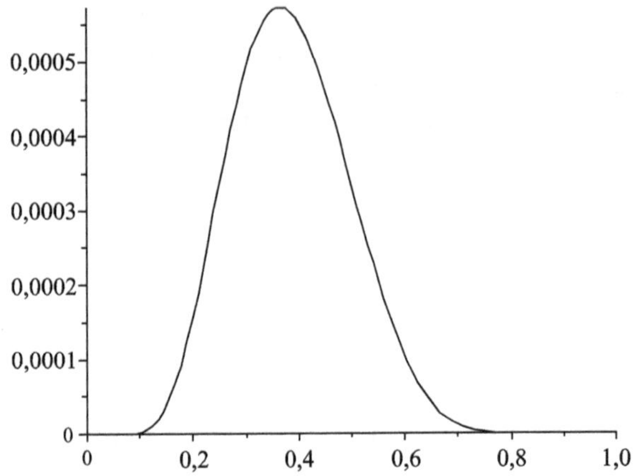

Note that the system (8.101) with the initial conditions $x_p(0) = 0,\ x_v(0) = 0$ does always admit the identically null solution for any non-positive external force $f(t) \leq 0$, independently on the constant γ in the restitution law (8.102).

Consider the infinitely smooth function

$$f_0(s) = \begin{cases} \exp\left\{\dfrac{A}{-s} + \dfrac{B}{1-s}\right\}, & A, B > 0, \text{ for } s \in (0,1), \\ 0, & \text{for the remaining values of } s. \end{cases}$$

The profile of this force is shown in Fig. 8.20. For an arbitrary interval $t_1 < t_2$, we put $x_p(t_1) = 0,\ x_v(t_1) > 0$, and chose the force $f(t)$ in the form

$$f(t) = C f_0\left(\frac{t - t_1}{t_2 - t_1}\right),$$

so that we will have $x_p(t_2) = 0$ and $x_v(t_2) = -\gamma_1 x_v(t_1)$, with some constant $0 < \gamma_1 < 1$. It is readily checked that the solutions of the system (8.101) satisfy the relations

$$\begin{aligned} x_v(t_2) &= x_v(t_1) + C(t_2 - t_1)\int_0^1 f_0(s)ds = x_v(t_1) + C(t_2 - t_1)I_1, \\ 0 = x_p(t_2) &= x_v(t_1)(t_2 - t_1) + C(t_2 - t_1)^2 \int_0^1 (1 - s) f_0(s)ds = \\ &= x_v(t_1)(t_2 - t_1) + C(t_2 - t_1)^2 I_2, \end{aligned} \tag{8.103}$$

where obviously

$$I_1 > I_2 > 0.$$

From here we define

$$C = -\frac{x_v(t_1)}{(t_2 - t_1)I_2} \quad \text{and} \quad \gamma_1 = \frac{I_1}{I_2} - 1 > 0;$$

for example, for $A = 1$, $B = 1$, we have $\gamma_1 = 1$; for $A = 1$, $B = 3$, we have $\gamma_1 \approx 0.628$, and for $A = 1$, $B = 0.3$ we have $\gamma_1 \approx 1.501$. Generally, by choosing the values of the parameters $A, B > 0$ one can obtain any value $\gamma_1 \in (0, \infty)$. Note that there exists a number of other two-parametric families of functions $f_0(s)$, for which, by a change of scale, one can obtain relations analogous to (8.103).

Let us assume that at the moment t_2 a shock occurs with the restitution law $x_v(t_2+) = -\gamma x_v(t_2)$. Choose the function f_0 (more precisely, the parameters $A, B > 0$) in such a way that $\gamma_1\gamma < 1$ and the desired time instant of the next "flight" is such that $t_3 - t_2 = \gamma_1 k(t_2 - t_1)$, where $k < 1$, which ensures that the amplitude of the force decreases. Then

$$|x_v(t_3)| = \gamma_1 |x_v(t_2+)| = \gamma_1 \gamma k |x_v(t_2)|.$$

Continuing in the similar manner the construction of the force on the interval $[t_n, t_{n+1}]$ so that

$$t_{n+1} - t_n = \gamma_1 k(t_n - t_{n-1}), \quad n = 2, 3, \ldots,$$

we obtain

$$t_{n+1} - t_n = (\gamma_1 k)^{n-1}(t_n - t_{n-1}), \qquad x_v(t_{n+1}+) = (\gamma_1 \gamma)^{n-1} x_v(t_1).$$

Thus, the sequence of values of the velocity decreases like a geometric progression with common ration $\gamma_1\gamma k$, while the sequence of intervals between impacts decreases like a geometric progression with common ratio $\gamma_1 k < 1$. Consequently, the sequence of impact times converges, and the motion stabilizes in the finite time $T^* = (t_2 - t_1)/(1 - \gamma_1 k)$ (see Fig. 8.21). The profile of the force $f(t)$ is shown in Fig. 8.22.

Since for a mechanical system with perfectly elastic impacts time is reversible, one can easily construct for such the system (8.101), (8.102) an example of a solution that starts from the initial point (0,0). The corresponding example of solution is shown in Figs. 8.23, 8.24, and 8.25. In the reverse time, the sequence of intervals between impacts is a geometrical progression with common ratio $1/(\gamma_1 k) \approx 0.567$, and the absolute values of post-impact velocities decrease as a geometrical progression with common ratio $1/\gamma_1 \approx 0.510$. Hence, the sum of the intervals between impacts converges, which means that after a time approximately equal to $1/[1 - 1/(\gamma_1 k)] \approx 2.3$, the inter-impact duration and the velocity reach values of the order of 1.

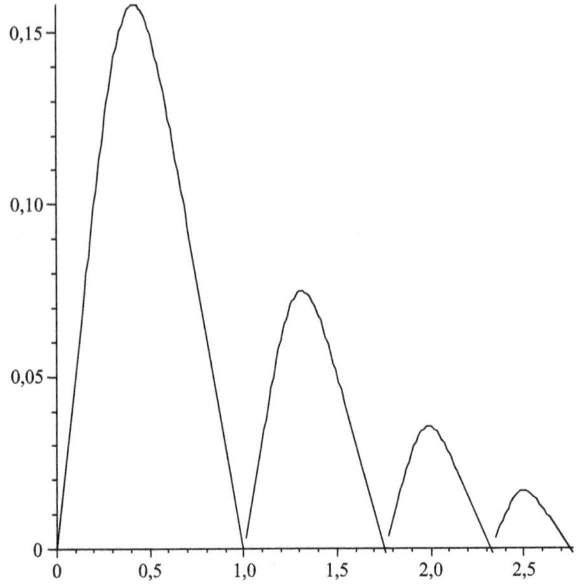

Fig. 8.21 Finite-time stabilization of the bouncing ball. Initial velocity $x_v(0) = 0.5$; the impact is perfectly elastic, i.e., $\gamma = 1$; the decrease of the velocity between two successive flights occurs with the common ratio $\gamma_1 k \approx 0.628 * 1.2 \approx 0.7536$; the time of complete stabilization is $T^* = 1/(1 - \gamma_1 k) \approx 4$

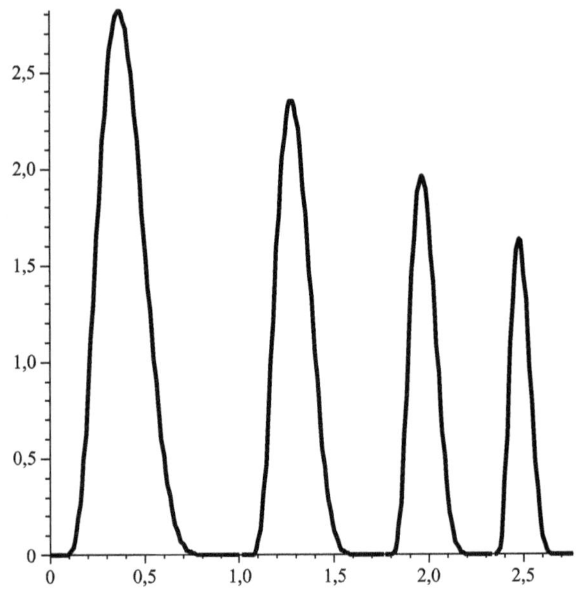

Fig. 8.22 Profile of the force change. Note that the force amplitude decreases as a geometric progression with common ratio $k < 1$, which guarantees the infinite differentiability of the force everywhere on the interval $[0, T^*)$ up to complete stabilization

It may seem that the point here is that, realistically, there is no perfectly elastic impact; however, by choosing a suitable force profile $f(t)$ one can construct an example of a non-zero solution also for an inelastic impact with $\gamma < 1$; it only suffices to ensure that $\gamma_1 \gamma > 1$ and choose the sequence of moments between impacts according to the rule $t_{n+1} - t_n = \gamma_1 k(t_n - t_{n-1})$ so that $\gamma_1 k > 1$. This

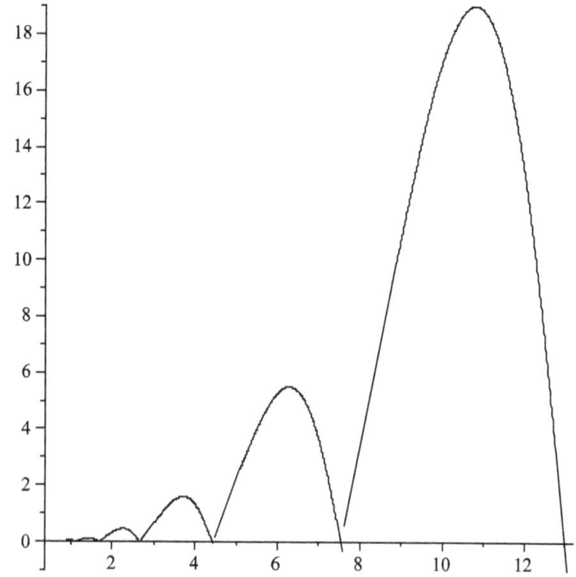

Fig. 8.23 Solution emerging from the point (0,0) through an infinite sequence of impacts. The impact is perfectly elastic, i.e., $\gamma = 1$

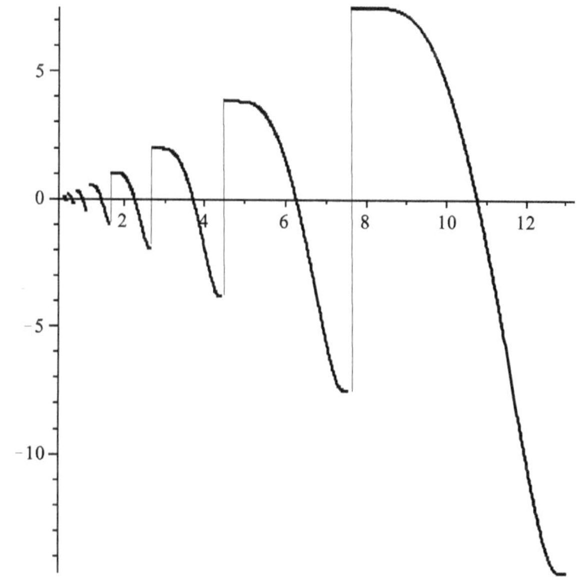

Fig. 8.24 Profile of the velocity. Since the impact is perfectly elastic, the absolute values of the velocities before and after impact coincide. The profile of the force is shown in Fig. 8.25

ensures the divergence of the series of impact times in the positive time direction and its convergence in the negative direction (see Figs. 8.26 and 8.27). In this example $\gamma_1 \approx 1.959$, $\gamma = 0.85$, $k = 0.65$, i.e., $(\gamma_1 k)^{-1} \approx 0.785$ and $(\gamma_1 \gamma)^{-1} \approx 0.6$. Thus, in a time of order $T^* = 1/(1 - (\gamma_1 k)^{-1}) \approx 4.65$, the interval between collisions reaches a value of the order of 1.

Fig. 8.25 Profile of the force corresponding to a solution that emerges from point (0,0) through an infinite sequence of impacts

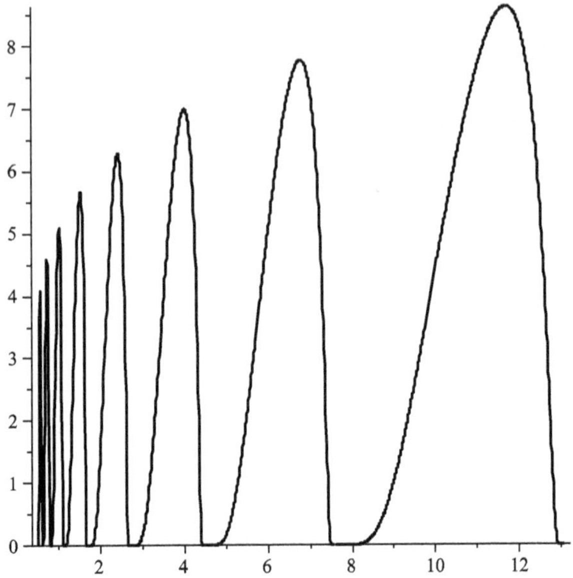

Fig. 8.26 Solution emerging from the point (0,0) through an infinite sequence of impacts. Note that the impact is viscoelastic and the velocity after impact decreases ($\gamma = 0.85$). Notation: 1—profile of $x_p(t)$, 2—profile of $x_v(t)$

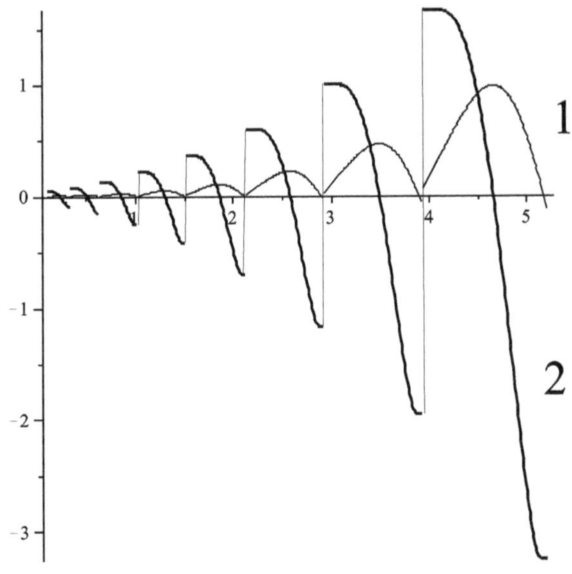

8.4.4 *Nonuniqueness of the Solution: Discussion*

Nonuniqueness of the solution may look like some artificial mathematical exercise that leads to results that do not hold in reality. However, it is worth looking at this phenomenon from a physics viewpoint and remember that since the rigidity of the obstacle is generally finite, the system (8.101), (8.102) should be written in the form

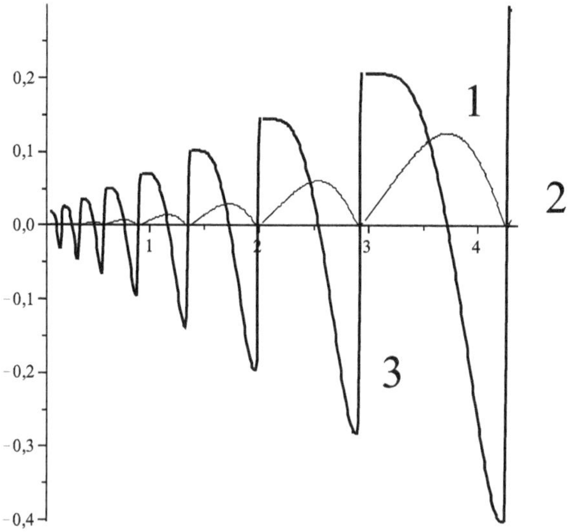

Fig. 8.27 Solution emerging from the point (0,0) through an infinite sequence of shocks for real (not ideally rigid surface) with $\mu = 10^4$. Notation: 1—profile of $x_p(t)$, 2—profile of $x_v(t)$, 3—the segment corresponding to "fast" motion in the forbidden area

$$\begin{cases} \dot{x}_p(t) = x_v(t), \\ dx_v(t) = f(t)dt + \mu(-x_p(t) + 2\xi\mu^{-1/2}x_v(t))\mathbf{I}\{x_p(t) < 0\}dt. \end{cases} \tag{8.104}$$

Let us consider the following example, in which at the initial moment of time the point mass is at rest. If we act on this point mass, which is at rest at the initial moment of time, with some constant force, pressing it inward on obstacle, then due to finite rigidity μ of the obstacle the point mass will penetrate into the domain $\{x_p < 0\}$, and after force is removed will exit this domain in a time of the order of $\mu^{-1/2}$, with a velocity of order $\mu^{-1/2}$. We will use the force profile described above, and suppress the force when the point mass is in the domain $\{x_p < 0\}$ (we intentionally simplify the conditions in order to get an explicit answer); then the velocity that the point mass will acquire during the flight on segment $[t_n, t_{n+1}]$ will be equal to $x_v(t_{n+1}) = -x_v(t_n)\gamma_1$. If while moving in the domain $\{x_p < 0\}$ the external force is equal to zero, then starting the motion with the initial velocity $x_v(t_{n+1})$, the point mass will exit this domain in time

$$\tau_\mu = \frac{\pi}{\mu^{1/2}\sqrt{1-\xi^2}},$$

with the velocity

$$x_v(t_{n+1} + \tau_\mu) = -x(t_{n+1}) \exp\left\{-\frac{\pi\xi}{\sqrt{1-\xi^2}}\right\} = \gamma_1\gamma x_v(t_n).$$

If we impose the same relations as in the previous example, then $(\gamma_1\gamma)^{-1} = \alpha \approx 0.6$, and the duration of the next flight will be equal to $(\gamma_1 * k)(t_{n+1} - t_n) = \beta(t_{n+1} - t_n)$, where $\beta = (\gamma_1 * k)^{-1} \approx 0.785$. Suppose that the velocity at time $t = 0$ is equal to $1/\mu^{1/2}$. Let us denote the number of flights by N^μ. We want to estimate the time T^μ necessary for the velocity to reach a value of the order of 1. More precisely, we will solve problem in the reverse time. We start with the velocity $x_v(T^\mu) = 1$ and choose a force profile such that during a flight time of 1 the velocity on the segment $[T^\mu - 1, T^\mu]$ increases by a factor of $\gamma_1\gamma > 1$. Then a direct calculation gives the following estimate for the number of flights needed to reduce the velocity from to value 1 to a value smaller than $1/\mu^{1/2}$:

$$\alpha^{N^\mu} \leq 1/\mu^{1/2} \leq \alpha^{N^\mu - 1},$$

or

$$N^\mu = \left[-\frac{\ln \mu}{2 \ln \alpha} \right] + 1.$$

The time T^μ is the sum of flight times that form a geometric progression with common ration β and the sum of time durations during which the point is in domain $\{x_p(t) < 0\}$. Therefore,

$$T^\mu = \sum_{n=1}^{N^\mu} \beta^n + N^\mu \tau_\mu = \frac{1 - \beta^{N^\mu}}{1 - \beta} + \left(\left[-\frac{\ln \mu}{2 \ln \alpha} \right] + 1 \right) \frac{\pi}{\mu^{1/2}\sqrt{1 - \xi^2}}.$$

It readily seen that as $\mu \to \infty$, the "build-up" time T^μ tends to the limit $T^* = (1 - \beta)^{-1}$, which is equal to the "build-up" time of the oscillations in the case of a viscoelastic obstacle (see Fig. 8.27). Note that in this case $N^\mu \to \infty$, i.e., if the stiffness tends to infinity, the sequence of impacts that push the point out of equilibrium becomes infinite.

Thus, in the class of generalized solutions, the motions emerging from equilibrium positions are physically realizable within the class of external actions in feedback form. However, if we choose only programmed controls, this phenomenon is impossible. In the article [288] we have shown that, for any strictly negative external force that depends only on time, and having a certain degree of regularity, the solution of the system (8.104) tends to zero as $\mu \to \infty$. Hence, in the class of generalized solutions generated by programmed actions, only the zero solution is physically realizable. Here we observe a phenomenon that is known in control theory for strongly nonlinear systems: *controls in feedback form can provide a much wider class motions than programmed controls*. Thus, when solving control problems, we must understand exactly which class of controls is admissible for the approximation of the optimal generalized solution as $\mu \to \infty$. This issue will be addressed in more detail in the following sections.

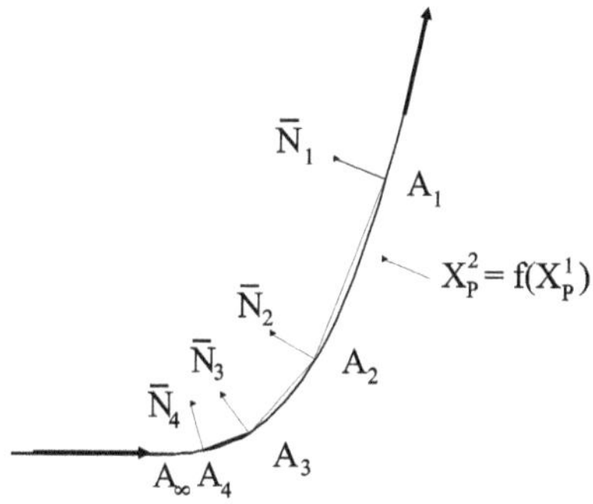

Fig. 8.28 A solution tangent to the slip line. Two possible solutions: one slides along the line $x_p^2 = f(x_p^1)$ and the other leaves the point A_∞ through an infinite sequence of reflections in this line

Lack of Uniqueness and Regularity of Solutions in Free Motions

The following example of lack of uniqueness of the solution was constructed for mechanical systems in [391], though an earlier example related to the propagation of light rays along the interface of two media (*grazing rays*) appears in [427]. Consider the free motion of a point mass in $\mathbb{R}^2$, described by the variables $x_p = (x_p^1, x_p^2)$, in a convex domain $\{x_p^2 \geq f(x_p^1)\}$ bounded by a smooth curve (see Fig. 8.28). It is assumed that upon collision with the boundary $C = \{x_p^2 = f(x_p^1)\}$, Snell's law holds: *the angle of incidence is equal to the angle of reflection and the velocity does not change*; this corresponds to elastic impact with an ideally rigid surface and is completely analogous to propagation of light. In [427] it is shown that such a curve can be constructed by taking as an initial approximation a convex polygon $(A_1, A_2, \ldots, A_\infty)$, which represents the trajectory of a ray with infinitely many reflections, and then finding a smooth convex curve C which passes through the vertices of the polygon and such that the normal to the curve at each vertex of the polygon halves the angle between the edges; denote the sequence of corresponding normals by $(\bar{N}_1, \bar{N}_2, \ldots)$. It is shown that such a curve exists and its contact with the final line at the point A_∞ is infinitely smooth. Hence, if one considers solutions that leave the final point of A_∞ tangentially to the curve C, then there are at least two of them: one solution that slides along the curve C, and a second one that is reflected in C infinitely many times. A simple algorithm for constructing such a convex curve is outlined in [183] pp. 178-179. For this phenomenon it is rather difficult to come up with a simple mechanical analogue, although one can assume the existence of a "convex billiard" with similar properties. However, the propagation of light in optical fibers is quite similar to sliding along the boundary in the example just described. As for the mechanical system, the reason for lack of uniqueness of the solution is the same here as in the previous example, namely, the possibility of stabilization in a finite time by means of an infinite sequence of impacts. Other examples of occurrence of branching solutions (*C-bifurcations*) are given in [181].

More complex examples of motions with infinitely smooth boundaries that change the sign of their curvature in the vicinity of points of contact with the motion infinitely many times and are free of constraints were constructed in [212] and [215].

Lack of regularity or *ill conditioning* is usually understood as the lack of continuous dependence of solutions on the initial conditions and parameters. In problems with reflections, examples such as reflections from the interior of an obtuse or acute angle are well known already for a very long time. If a ray or point mass experiences a sequences of reflections from boundaries and the initial condition is that the ray hits the vertex of the angle, then, on the one hand, uniqueness of solution does not hold, and on the other hand small changes in the initial condition lead to large (non-microscopic) changes in the solutions. Therefore, contrary examples, when the presence of many constraints makes it possible to build a stable solution (examples: corner reflector in optics, or absolutely elastic reflection, without loss of energy, from surfaces making a right angle with), are very important. Some conditions of this type are given in the next section.

8.5 Existence and Uniqueness of a Solution for the System of Complementarity Equations

8.5.1 System of Complementarity Equations

The first results on the existence of a solution were obtained G. Rubin and P. Ungar in [384]. They actually outlined the main direction concerning families of smooth problems, where constraints are in the form of some "penalty potentials" parameterized by a set parameters tending to infinity. The key step is their proofs is establishing the compactness of the set of solutions corresponding to a set of parameters, and selecting a converging subsequence. Later, this approach, known as the "penalty" method, was systematically developed by M. Schatzman [391–394], M. Schatzman and L. Paoli [346, 348, 350, 353], L. Paoli, [340, 342, 343], F. Bornemann [54]. Here one should mention also the work [346], where the penalty method was first presented, with a complete proof, for the existence of solutions to a classical second-order multidimensional system (we write this system in our setting)

$$\dot{x}_p(t) = x_v(t), \quad \dot{x}_v(t) = f(t, x_p(t), x_v(t)), \quad x_p(t) \in K, \tag{8.105}$$

with a convex bounding set $K \subset \mathbb{R}^n$, and the work [393], where this result was generalized to the class of dynamical systems with shocks in generalized coordinates. The result of [346] deserves a complete presentation, since after it first appearance it has been cited and repeated multiple times by various authors.

Let us introduce some definitions and notations. We assume that the set K has a regular boundary ∂K of class C^2. Let $P_K(\cdot)$ be the projection map to K, that is,

$$P_K(x) = \begin{cases} x, & \text{if } x \in K, \\ \operatorname*{argmin}_{y \in K} \|x - y\|, & \text{if } x \in \mathbb{R}^n \setminus K. \end{cases}$$

Define the "penalty potential" as

$$\phi_\mu(x_p) = \mu\|x - P_K(x)\|^2,$$

and the bilinear form $T(u, w)$, related to Rayleigh's dissipation function (see [215], pp. 42–43), by

$$T(u, w) = \begin{cases} \dfrac{\langle v, w\rangle v}{\|v\|^2}, & \text{if } v \neq 0, \\ 0, & \text{if } v = 0. \end{cases}$$

Next, define the restitution coefficient γ under visco-elastic impact by choosing the parameter $\xi \in [0, 1]$ in the system of Eq. (8.104) so that

$$\gamma = \exp\left\{-\frac{\pi\xi}{\sqrt{1-\xi^2}}\right\}$$

Now consider the system of equations with the penalty potential ϕ_μ and dissipation function T:

$$\begin{aligned} &\dot{x}_p^\mu(t) = x_v^\mu(t), \\ &\dot{x}_v^\mu(t) = f(t, x_p^\mu(t), x_v^\mu(t)) - \\ &\quad - \mu\big(x_p^\mu(t) - P_K(x_p^\mu(t)) + 2\xi - \mu^{-1/2}T(x_p^\mu(t) - P_K(x_p^\mu(t))x_v^\mu(t))\big), \end{aligned} \tag{8.106}$$

with the initial conditions $x_p^\mu(0) = x_p(0)$, $x_v^\mu(0) = x_v(0)$.

Definition 8.5 The solution of the system (8.105) is sought in the class of functions

$$(x_p(\cdot), x_v(\cdot)) \in AC[0, T] \times BV^+[0, T]$$

with the following properties: at the points of continuity of x_v they satisfy the system of Eq. (8.105), and at the points of discontinuity of x_v such that $x_p(\tau) \in \partial K$ they obey the restitution law

$$x_v(\tau) = -\gamma x_v^{\mathrm{N}}(\tau-) + x_v^{\mathrm{T}}(\tau-), \tag{8.107}$$

where x_v^{N} and x_v^{T} are the normal and the tangential component, respectively, of the vector x_v with respect to the set K at the corresponding point of the boundary ∂K. Here $AC[0, T] \times BV^+[0, T]$ is the direct product of the spaces of absolutely continuous and of right-continuous functions of bounded variation on the segment $[0, T]$.

Theorem 8.8 *[346] Let the function $f(t, x_p, x_v)$ be continuous in all the variables and Lipschitz in the variables x_p, x_v, and let the set K be convex, closed and with nonempty interior. Suppose the initial conditions satisfy the constraint $x_p(0) \in K$, $x_v(0) \in \mathbb{R}^n$. Then:*

1. *the system of Eq. (8.105) has a solution $(\bar{x}_p(\cdot), \bar{x}_v(\cdot))$ in the sense of Definition 8.5;*
2. *the solution $(\bar{x}_p(\cdot), \bar{x}_v(\cdot))$ is the limit of a sequence of solutions of the system (8.106) as $\mu \to \infty$. Moreover,*

$$x_p^{\mu}(t) \to \bar{x}_p(t) \quad \textit{uniformly on} \quad [0, T],$$

and

$$x_v^{\mu}(t) \to \bar{x}_v(t) \quad \textit{at all the points of continuity in } [0, T].$$

The uniqueness of the solution was first established by M. Schatzman in [391], where it is shown that the solution exists and is unique provided the set K is strictly convex and has a boundary of class C^3 and $f \equiv 0$.

Theorem 8.9 ([391]) *Suppose the set K is strictly convex, has a boundary ∂K of class C^3 and $f \equiv 0$. Then the system of Eq. (8.105) admits exactly one solution on $[0, +\infty)$.*

Moreover, if $x_p(0) \in \partial K$ and the initial velocity $x_v(0)$ is tangent to ∂K, then the motion continues along a geodesic curve on ∂K with constant velocity $|x_v(0)|$. If the initial velocity is not tangent to ∂K, then on the continued solution the velocity never becomes tangent to ∂K and on any bounded time interval only finitely many jumps of the velocity are possible.

Remark 8.10 The examples given in the Sects. 8.6.2–8.6.5 show that uniqueness can fail in the presence of a control force $f \neq 0$ and when the bounding set is not strictly convex. This demonstrates how much the case of a real controlled system is more complicated than the case of motion in a classical mechanical system with unilateral constraints.

Nevertheless, even in classical mechanical systems, the problem of the existence and uniqueness of a solution remains very difficult and only in recent years rigorous results have been obtained in this area. Here one should mention the cycle of works by D. Percivale [359, 360], in which uniqueness of the solution is established in the case of free motion (null external forces), the impact is perfectly elastic, there is no friction, and the bounding set is given in the form $K = \{x_p : G(x_p) \leq 0\}$, where $G(\cdot)$ is a real *analytic* function. The case of inelastic impact in the presence of friction was studied in detail in the monograph [308]. This monograph continues the research initiated in the works of J. Moreau [194, 310, 311], where the equations of dynamics of systems with unilateral constraints are written in the form of differential inclusions with measure, which is natural for systems with undetermined Lagrange multipliers in the right-hand side. Monteiro-Marquez's book explores the existence of solutions, and devotes an important place to the discussion of the uniqueness (see [308], pp. 77–78, 80–81, 129–131). One of the remarks therein is rather striking. It is mentioned that traditional approaches to the proof of uniqueness are most likely not applicable, since, as a rule, the norm of the difference of solutions, i.e., the function $\phi(t) = \|x_v^1(t) - x_v^2(t)\|$, where $x_v^1(\cdot)$ and $x_v^2(\cdot)$ are two different solutions, satisfies the inequality

$$\phi^2(t) \leq \int_0^t f(s)\phi(s)ds, \quad f(s) \geq 0,$$

which, in contrast to the classical Gronwall-Bellman inequality, admits a non-zero solution.

We should also mention a number of works in which the problem of constructing solutions is reduced to a certain limited class of motions, for example, sliding along the bounding surface. Since in many systems, for a motion of this kind the reaction force of the obstacle is defined unambiguously as a function of coordinates and velocities, the problem comes down to assessing the degree of regularity of this reaction, and one is able to establish the existence and uniqueness of the solution, at least locally, till exit from the constrained domain [55, 108, 109, 214, 474]. All these works make successful use of the "penalty potential" method. In [176, 178] an approximation method using the model of a Kelvin-Voigt material is applied to establish the existence of solutions to the equations of dynamics with friction. Here the reaction force is uniquely determined and the problem reduces to the integration of a system of ordinary differential equations.

However, most of aforementioned approaches do not allow to unambiguously determine the motion of a dynamical system under an action of a more or less general type. It is all the more surprising that quite recently this problem was completely solved by P. Ballard in [18, 19], where it is shown that uniqueness of the solution holds if the dynamics of the system is described in terms of real analytic functions, i.e., in the Eq. (8.100), T, F and f are real analytic functions of coordinates and time. In this case, the solution to the Eq. (8.100) supplemented by the velocity restitution law at the impact (8.107), admits a unique continuation to the half-line $[0, +\infty)$ for any initial condition obeying the constraint. Moreover, later P. Ballard [20, 21] established a similar result for dynamical systems with unilateral constraints and in the presence of friction. Can one assert, however, that this completely solves the problem? First of all, on a finite time interval, any continuous function can be arbitrarily-well approximated by an analytic function, and so the only issue is the stability of such an approximation. However, this stability is very difficult to establish since, generally speaking, all systems involving impact actions are usually *ill-posed*, i.e., small changes in the parameters and initial conditions lead to sharp changes in the solution. Moreover, in optimal control problems one can hardly expect to have a continuous optimal control, and it is exactly at the points of discontinuity of the external action, especially during a singular phase of motion, that stability is lost.

Similar problems arise in the general theory of hybrid systems [164], where the dynamics is described by piecewise Lipschitz (or piecewise linear) systems of equations. In that theory, the problem of the correct construction of the solution is formulated in terms of rules of exit from the constraint, which is essentially the same as the definition of the concept of the solution for dynamical systems with unilateral constraints. In further works [164] it is noted that J. Moreau's method [311] for determining the velocity restitution law at impact, which is based on the choice of a physically correct solution, i.e., one that agrees with certain conservation laws, can be used for the continuation of the solution for a hybrid system. In recent years, especially after the investigations of Moreau and Monteiro-Marquez, the connection between mechanical and hybrid systems became transparent, because in both cases differential inclusions are used to describe the dynamics of nonsmooth systems. Examples of such studied include the survey [73] and a number of articles and monographs [71, 72, 74, 130, 135, 239, 240, 396, 434, 435, 457, 460, 468].

8.6 Definition of Generalized Solutions of Controllable Systems on the Whole Interval of Motion

8.6.1 Generalized Solutions of Controllable Systems

In the previous sections, we considered the motion in the vicinity of points of discontinuity, and in special cases on the whole interval $[0, T]$. Since the nonuniqueness of the solution given by complementarity equations is a common property inherent to similar systems, solving optimal control problems requires a different approach, which takes into account the way in which such solutions are obtained. The penalty method allows one to build a solution as the limit of some subsequence of ordinary solutions. This approach has much in common with the construction of generalized solutions for systems that allow impulse actions. It is known that in such systems there is no unique solution; rather, each generalized solution can be constructed as the limit of some subsequences of ordinary solution, and given a generalized solution one can propose a method for its approximation by ordinary solutions. In this context, a generalized solution depends not only on the external control impulse action, but also on the method of its approximation. A general theory of representation of generalized solutions and their optimization is constructed in the works of B. Miller [258–270, 272], B. Miller and E. Rubinovich [292, 294], A. Sesekin and S. Zavalishchin [464, 466, 467], V. Dykhta [118–120, 122–125, 127], G. Kolokolnikova [127, 205, 206], and O. Samsonyuk [128], A. Bressan and M. Motta [60–62], F. Rampazzo and M. Motta [314–316], A. Bressan and F. Rampazzo [57, 63–66]. In the present section, we give results which demonstrate that a similar approach can be applied to controlled systems with impacts.

Here, based on [288], we present a theory for constructing generalized solutions. Consider, as described in Sect. 8.2.2, the controlled system with a constraint (8.12) and governed by the Eq. (8.13); to simplify the exposition, we assume that the system is with no control in the intersingular phase and with no observations $\xi(\cdot)$ (8.16). Accordingly, the pair of functions $(x_p(t), x_v(t)) \in \mathbb{R}^n \times \mathbb{R}^n$ satisfies on the interval $[0, T]$ the system of differential equations

$$\begin{cases} \dot{x}_p(t) = F_p^{\mathrm{r}}(x_p(t), x_v(t), t), \\ \dot{x}_v(t) = F_v^{\mathrm{r}}(x_p(t), x_v(t), u(t), t) + \\ \qquad + \mu F_v^{\mathrm{s}}(x_p(t), x_v(t), w(t, \mu), t, \mu)\mathbf{I}\{t : G(x_p(t), t) \le 0\}, \\ G : \mathbb{R}^n \times [0, T] \to \mathbb{R}, \end{cases} \tag{8.108}$$

with given initial conditions $(x_p(0), x_v(0)) \in \mathbb{R}^n \times \mathbb{R}^n$. By analogy with the case of mechanical systems, one can regard $(x_p(t), x_v(t))$ in (8.108) as generalized coordinates and generalized velocities, and consider that

$$u(t) \in U \subset \mathbb{R}^m, \quad w(t, \mu) \in W \subset \mathbb{R}^r,$$

where U and W are compact sets, and $u(t)$ and $w(t, \mu)$ are the controls in the regular and singular phases of motion, respectively. Recall that the control $w(\cdot)$ is coordinated with the beginning and duration of the singular phase of motion (see Sect. 8.2.2).

Assumption 8.3 The functions F_p^{r} and F_v^{r} are continuous and smooth in all the variables. The function $F_v^{\mathrm{s}}(x_p, x_v, w, t, \mu)$ satisfies the conditions (8.14), i.e., is continuous and smooth in the domain $\{G(x_p, \tau) \le 0\}$ and satisfies the Lipschitz condition (8.23) from Assumption 8.1.

Moreover, the function $G(x, t)$ has, as above, the continuity and smoothness properties. However, more precise statements require a higher degree of smoothness, so we will assume that the functions F_p^{r}, F_v^{r}, F_v^{s} and G admit continuous derivatives up to and including second order in the variables (x_p, x_v, t), for any fixed $(u, w) \in U \times W$.

Assumption 8.3 holds, for example, if

$$F_v^{\mathrm{s}} = A(x_p, w, t)G(x_p, t) + \mu^{-1/2}B(x_p, w, t)\big[G'_{x_p}F_p^{\mathrm{r}} + G'_t\big]\Big|_{(x_p, x_v, t)}, \tag{8.109}$$

where A and B are smooth functions.

The construction of a generalized solution as a limit of ordinary solutions as $\mu \to \infty$ usually requires the uniform boundedness in μ of the integral of the contact force or, equivalently, of the variation of the generalized velocity. The importance of this property was noted in [310] and [311], where the boundedness of the variation of the generalized velocity was postulated as one of the properties that allow to solve the system of complementarity equations. In [346] and [393] this fact is established for a specially constructed approximating sequence of "penalty potentials." As noted by A. P. Ivanov in [185], for a mechanical system, boundedness of the variation is a consequence of the energy conservation law, since impulsive contact forces arise as internal forces, and so the work they perform is equal to zero for an ideal contact, or negative for a contact with friction. Therefore, the change of the total energy is bounded by the amount of work of the bounded external forces and, as one can show, it bounds the variation of the generalized velocity. However, for controlled dynamical systems such considerations are no longer valid.

The uniform boundedness of the variation of the velocity can be established under special conditions, as we will see below. However, for the moment, to build a generalized solution we merely demand that such a condition be met. This will allow us, at the present stage, to propose a general procedure for constructing a generalized solution, by analogy with the case of systems with impulse controls (see Sect. 5.4). For hybrid systems with discontinuous solutions, a similar construction is described in Sect. 5.4.

Assumption 8.4 Suppose that for any μ such that $\mu_0 \le \mu < \infty$, where $\mu_0 > 0$, and for given controls $u(\cdot)$ and $w(\cdot)$, the system (8.108) has a solution $\{x_p^\mu(\cdot), x_v^\mu(\cdot)\}$, such that

$$\sup_{\mu \ge \mu_0} \int_0^T \mu \| F_v^{\mathrm{s}}(x_p^\mu(t), x_v^\mu(t), w(t, \mu), t, \mu) \| \mathbf{I}\{t : G(x_p^\mu(t), t) \le 0\} dt < \infty,$$

and

$$\lim_{\mu \to \infty} G(x_p^\mu(t), t) \ge 0, \quad \text{a.e. on } [0, T].$$

The approach described below makes it possible to determine the generalized solution over the entire time interval using a singular space-time transformation. This result generalizes the methodology developed earlier for mechanical systems with impacts in the works of L. Paoli and M. Schatzman [346] and M. Schatzman [393] and allows one to establish the uniform boundedness of the integral of the generalized forces over the whole interval of motion under the following assumption about the law of change of the force: let

$$Z_p(t) = G(x_p, t), \quad Z_v(t) = G'_x(t) F_p^{\mathrm{r}}(t) + G'_t(t)$$

and

$$F_v^{\mathrm{s}} = \left[A(x_p, x_v, w, t) Z_p + \mu^{-1/2} B(x_p, x_v, w, t) Z_v \right] \mathbf{I}\{Z_p \le 0\}. \tag{8.110}$$

Then the equations of motion in the singular phases have the form

$$\begin{cases} \dot{Z}_p = Z_v, \\ \dot{Z}_v = \widetilde{F}^{u,w}(t) + \mu\big[\alpha^w(t)Z_p + \mu^{-1/2}\beta^w(t)Z_v\big]\mathbf{I}\{Z_p \le 0\}. \end{cases} \tag{8.111}$$

Proposition 8.2 *Suppose the function $\widetilde{F}^{u,w}(t)$ and the coefficients $\alpha^w(t)$ and $\beta^w(t)$ in the system of Eq. (8.111) satisfy the constraints*

$$-\infty < c_1 \le \widetilde{F}^{u,w}(t), \alpha^w(t), \beta^w(t) \le c_2 < 0$$

and that the quantities $\int_0^T |\dot{\alpha}^w(t)|dt$ and $\int_0^T |\dot{\beta}^w(t)|dt$ are uniformly bounded (in other words, the functions $\alpha^w(t)$ and $\beta^w(t)$ have uniformly bounded variation).

Then,

$$\begin{aligned} &\sup_\mu \int_0^T \mu|Z_p(t)|\mathbf{I}\{Z_p(t) \le 0\}dt < \infty, \\ &\sup_\mu \int_0^T \mu^{1/2}|Z_v(t)|\mathbf{I}\{Z_p(t) \le 0\}dt < \infty, \\ &\sup_\mu \sup_{[0,T]} \mu^{1/2}Z_p(t)\mathbf{I}\{Z_p(t) \le 0\} < \infty. \end{aligned} \tag{8.112}$$

Remark 8.11 The conditions of Proposition 8.2 are trivially fulfilled if, for example, the functions $\widetilde{F}, \alpha^w, \beta^w$ do not depend on time and on the control [346]. The proof of Proposition 8.2 for constant functions α and β is also given in [346]. The proof of the general case for a mechanical system of general form with a linear phase constraint is carried out according to the same scheme [393].

The next results follow from the usual estimates based on the Gronwall-Bellman inequality [365] and Helly's theorem [322].

Proposition 8.3 *The sequence of solutions $\{x_p^\mu(\cdot), x_v^\mu(\cdot)\}$ lies in the space $C^1[0,T] \times AC[0,T]$ and satisfies the constraints*

$$\sup_{\mu\ge\mu_0} \max_{[0,T]} \|x_p^\mu(t)\| < \infty, \quad \sup_{\mu\ge\mu_0} \max_{[0,T]} \|x_v^\mu(t)\| < \infty, \quad \sup_{\mu\ge\mu_0} \operatorname{Var}_{[0,T]} \|x_p^\mu(t)\| < \infty.$$

Proposition 8.4 *There exists a subsequence $\mu_k \uparrow \infty$ such that corresponding sequence of solutions $\{x_p^{\mu_k}(\cdot), x_v^{\mu_k}(\cdot)\}$ converges to functions $\{\bar{x}_p(\cdot), \bar{x}_v(\cdot)\} \in C^1[0,T] \times BV[0,T]$, where $x_p^{\mu_k}$ converges to $\bar{x}_p$ uniformly on $[0,T]$ and $x_v^{\mu_k}$ converges to $\bar{x}_v$ at all the points of continuity.*

Further, when there is no danger of confusion, we will not distinguish between a subsequence and a sequence $\mu \uparrow \infty$.

Definition 8.6 The pair of functions $\{\bar{x}_p(\cdot), \bar{x}_v(\cdot)\} \in C^1[0, T] \times BV[0, T]$ whose existence is established in Proposition 8.13 is called a *generalized solution* of the system (8.108).

8.6.2 Representation of Generalized Solutions with the Aid of a Discontinuous Time Change

We present here just main steps; complete details can be found in the Subection 4.1.1 and in the article [288].

Let us introduce the function

$$\Gamma^\mu(t) = t + \int_0^t \mu \| F_v^s(x_p^\mu(\tau), x_v^\mu(\tau), w(\tau, \mu), \tau, \mu)\| \mathbf{I}\{\tau : G(x_p^\mu(\tau), \tau) \leq 0\} d\tau. \tag{8.113}$$

Since $\Gamma^\mu(\cdot)$ is absolutely continuous and monotonically increasing, it admits on the interval $[0, \Gamma^\mu(T)]$ the inverse function $\eta^\mu(\cdot)$, which has almost everywhere a derivative $\alpha^\mu(s) = \dot{\eta}^\mu(s)$, given by

$$\alpha^\mu(s) = \left. \frac{1}{1 + \mu \| F_v^s(x_p^\mu(\tau), x_v^\mu(\tau), w(\tau, \mu), \tau, \mu)\| \mathbf{I}\{\tau : G(x_p^\mu(\tau), \tau) \leq 0\}} \right|_{\tau = \eta^\mu(s)} \tag{8.114}$$

on $[0, \Gamma^\mu(T)]$. Consequently,

$$0 < \alpha^\mu(s) = \dot{\eta}^\mu(s) \leq 1.$$

Next, for every $\mu > \mu_0$, define the functions

$$y_p^\mu(s) = x_p^\mu(\eta^\mu(s)), \quad y_v^\mu(s) = x_v^\mu(\eta^\mu(s)),$$

which satisfy on $[0, \Gamma^\mu(T)]$ the system of equations

$$\begin{cases} y_p^\mu(s) = x_p(0) + \displaystyle\int_0^s \alpha^\mu(\tau) F_p^r(y_p^\mu(\tau), y_v^\mu(\tau), \eta^\mu(\tau)) d\tau, \\ y_v^\mu(s) = x_v(0) + \displaystyle\int_0^s \alpha^\mu(\tau) F_v^r(y_p^\mu(\tau), y_v^\mu(\tau), u(\eta^\mu(\tau)), \eta^\mu(\tau)) d\tau + \\ \quad + \displaystyle\int_0^s \alpha^\mu(\tau) \mu F_v^s(y_p^\mu(\tau), y_v^\mu(\tau), w(\eta^\mu(\tau), \mu), \eta^\mu(\tau), \mu) \times \\ \quad \times \mathbf{I}\{\tau : G(y_p^\mu(\tau), \eta^\mu(\tau)) \leq 0\} d\tau. \end{cases} \tag{8.115}$$

Moreover, if $G^+ = \max\{0, G\}$, then

$$\int_0^{\Gamma^\mu(T)} (1 - \alpha^\mu(\tau))\big(G(y_p^\mu(\tau), \eta^\mu(\tau))\big)^+ d\tau = 0. \tag{8.116}$$

The quantities $\Gamma^\mu(T)$ are uniformly bounded, so one can choose some T_1 such that $\sup_\mu \Gamma^\mu(T) \le T_1 < \infty$ and define the set of triples $\{y_p^\mu(\cdot), y_v^\mu(\cdot), \eta^\mu(\cdot)\}$ on the interval $[0, T_1]$, extending these functions by constants with preservation of the continuity for $\tau \ge \Gamma^\mu(T)$. It is readily seen that this yields an equicontinuous set of functions, which by the Arzelà-Ascoli theorem is precompact [204]. Consequently, there exist a subsequence of triples and corresponding times T_1^μ, such that

$$T_1^\mu \to \bar{T}_1, \quad \{y_p^\mu(\cdot), y_v^\mu(\cdot), \eta^\mu(\cdot)\} \to \{y_p(\cdot), y_v(\cdot), \bar{\eta}(\cdot)\},$$

where the convergence to the absolutely continuous functions $\{y_p(\cdot), y_v(\cdot), \bar{\eta}(\cdot)\}$ is uniform on $[0, \bar{T}_1]$.

The importance of this construction is justified by the following statement.

Proposition 8.5 *Every generalized solution of the system (8.108), in the sense of Definition 8.6 can be expressed in the form*

$$\bar{x}_p(t) = y_p(\Gamma(t)), \quad \bar{x}_v(t) = y_v(\Gamma(t)),$$

where

$$\Gamma(t) = \inf\{s : \bar{\eta}(s) > t\}.$$

Remark 8.12 The proof is standard and is omitted; details can be found in Sects. 4.1.1 and 5.4 of the present monograph. The principal step of the proof is establishing the convergence of the subsequence $\Gamma^\mu(t) = \inf\{s : \bar{\eta}^\mu(s) > t\}$ to $\Gamma(t)$ at all the points of continuity (see Lemma 2.5). Moreover, thanks to the uniform convergence of (y_p^μ, y_v^μ) to $(y_p, \bar{v}_p)$ we have

$$\bar{x}_p(t) = y_p(\Gamma(t)) = \lim_\mu y_p^\mu(\Gamma(t)), \quad \bar{x}_v(t) = y_v(\Gamma(t)) = \lim_\mu y_v^\mu(\Gamma(t)).$$

8.6.3 Representation of a Generalized Solution by Means of a Differential Equation with Measure

To get such a representation of a generalized solution one needs to pass to the limit in the Eq. (8.115) and thus define the limits

$$\mathscr{I}_p^{\mathrm{r}}(s) = \lim_{\mu} \int_0^s \alpha^{\mu}(\tau) F_p^{\mathrm{r}}(y_p^{\mu}(\tau), y_v^{\mu}(\tau), \eta^{\mu}(\tau)) d\tau, \tag{8.117}$$

$$\mathscr{I}_v^{\mathrm{r}}(s) = \lim_{\mu} \int_0^s \alpha^{\mu}(\tau) F_v^{\mathrm{r},\mu}(\tau)) d\tau, \tag{8.118}$$

and

$$\mathscr{I}_v^{\mathrm{s}}(s) = \lim_{\mu} \int_0^s \alpha^{\mu}(\tau) \mu F_v^{\mathrm{s},\mu}(\tau) I\{\tau : G^{\mu}(\tau) \leq 0\} d\tau. \tag{8.119}$$

Here and below we use the following notations

$$F_v^{\mathrm{r},\mu}(\tau) = F_v^{\mathrm{r}}(y_p^{\mu}(\tau), y_v^{\mu}(\tau), u(\eta^{\mu}(\tau)), \eta^{\mu}(\tau)),$$

$$F_v^{\mathrm{s},\mu}(\tau) = F_v^{\mathrm{s}}(y_p^{\mu}(\tau), y_v^{\mu}(\tau), w(\eta^{\mu}(\tau), \mu), \eta^{\mu}(\tau), \mu),$$

$$G^{\mu}(\tau) = G(y_p^{\mu}(\tau), \eta^{\mu}(\tau)),$$

$$\bar{G}(\tau) = G(y_p(\tau), \bar{\eta}(\tau)).$$

Properties of the Sequence $\alpha^{\mu}(\cdot)$ Since there exists a sequence $\eta^{\mu}(\cdot)$ that converges uniformly to a function $\bar{\eta}(\cdot)$, the corresponding sequence of derivatives $\alpha^{\mu}(\cdot)$ converges weakly to a Lebesgue-measurable function $\bar{\alpha}(\cdot)$ such that

$$\bar{\alpha}(s) = \dot{\bar{\eta}}(s) \in [0, 1] \quad \text{a.e.}$$

on the interval $[\bar{T}_1]$, where $\bar{T}_1 = \Gamma(T)$.

The following proposition clarifies the limit behavior of the sequence $\alpha^{\mu}(\cdot)$ as $\mu \uparrow \infty$.

Proposition 8.6 *There is a subsequence* $\alpha^{\mu}(s)$ *such that the following holds:*

$$\begin{aligned}
&\text{(a)}\ \lim_{\mu} \alpha^{\mu}(s)\mathbf{I}\{s : \bar{\alpha}(s) = 1\} = \bar{\alpha}(s)\mathbf{I}\{s : \bar{\alpha}(s) = 1\}, \quad \textit{a.e. on } [0, \bar{T}_1],\\
&\text{(b)}\ \lim_{\mu} \alpha^{\mu}(s)\mathbf{I}\{s : \bar{\alpha}(s) = 0\} = \bar{\alpha}(s)\mathbf{I}\{s : \bar{\alpha}(s) = 0\}, \quad \textit{a.e. on } [0, \bar{T}_1],\\
&\text{(c)}\ \lim_{\mu} \mathbf{I}\{s : G(y_p^{\mu}(s), \eta^{\mu}(s)) > 0\} = \mathbf{I}\{s : G(y_p(s), \bar{\eta}(s) > 0\},\\
&\text{(d)}\ \lim_{\mu} \mathbf{I}\{s : G(y_p^{\mu}(s), \eta^{\mu}(s)) \leq 0\} = \mathbf{I}\{\tau : G(y_p(s), \bar{\eta}(s) = 0\}.
\end{aligned} \tag{8.120}$$

Proof To establish item (a), note that by the C-property of measurable functions [204, 322], for any $\varepsilon > 0$, there exists a continuous function $I^{\varepsilon}(s)$ with values in $[0, 1]$, such that

$$\text{measure}\left\{s : I^{\varepsilon}(s) \neq \mathbf{I}\{s : \bar{\alpha}(s) = 0\}\right\} \leq \varepsilon.$$

Since $0 < \alpha^{\mu}(s) \leq 1$,

$$\mathbf{I}\{s : \bar{\alpha}(s) = 1\}(\alpha^{\mu}(s) - \bar{\alpha}(s)) \leq 0,$$

and

$$\left|\int_0^{T_1} \left[\mathbf{I}\{s : \bar{\alpha}(s) = 1\} - I^{\varepsilon}(s)\right](\alpha^{\mu}(s) - \bar{\alpha}(s))ds\right| \leq \varepsilon,$$

we have that

$$-\varepsilon \leq \liminf_{\mu} \int_0^{\bar{T}_1} \mathbf{I}\{s : \bar{\alpha}(s) = 1\}(\alpha^{\mu}(s) - \bar{\alpha}(s))ds \leq 0,$$

and so

$$\lim_{\mu} \int_0^{\bar{T}_1} \mathbf{I}\{s : \bar{\alpha}(s) = 1\}(\alpha^{\mu}(s) - \bar{\alpha}(s))ds = 0.$$

This means that the sequence $\mathbf{I}\{s : \bar{\alpha}(s) = 1\}(\alpha^{\mu}(s) - \bar{\alpha}(s))$ converges to 0 in the Lebesgue measure, which implies the existence of the corresponding subsequence.

Item (b) is obvious. Item (c) follows from the uniform convergence of the sequence (y_p^{μ}, η^{μ}) to $(y_p, \bar{\eta})$ and the continuity of G. Finally, item (d) follows from (c). □

8.6.4 Convergence of the Integrals in (8.117) and (8.118)

Proposition 8.7 *The following relations hold:*

$$\begin{aligned}\lim_{\mu} \int_0^{s} \alpha^{\mu}(\tau)F_p^{\mathrm{r}}(y_p^{\mu}(\tau), y_v^{\mu}(\tau), \eta^{\mu}(\tau))d\tau = \\ = \int_0^{s} \bar{\alpha}(\tau)F_p^{\mathrm{r}}(y_p(\tau), y_v(\tau), \bar{\eta}(\tau))d\tau.\end{aligned} \tag{8.121}$$

Moreover

$$\begin{aligned}\lim_{\mu}\int_0^{\Gamma^\mu(T)}(1-\alpha^\mu(\tau))(G(y_p^\mu(\tau),\eta^\mu(\tau)))^+d\tau &=\\ =\int_0^{\Gamma(T)}(1-\bar{\alpha}(\tau))(G(y_p(\tau),\bar{\eta}(\tau)))^+d\tau &= 0.\end{aligned} \tag{8.122}$$

This result is a consequence of the uniform convergence of the triple of functions $(y_p^\mu(\tau), y_v^\mu(\tau), \eta^\mu(\tau))$ to $(y_p(\tau), y_v(\tau), \bar{\eta}(\tau))$ and the continuity of the functions F_p^{r} and G.

Proposition 8.8 *Assume that the set of vectors*

$$F_v^{\mathrm{r}}(x,y,U,t)=\{l\in\mathbb{R}^n : \exists u\in U \text{such that}\, l=F_v^{\mathrm{r}}(x,y,u,t)\} \tag{8.123}$$

is compact and convex for each fixed (x, y, t). *Then there exists a measurable function* $\bar{u}(\cdot)$ *such that* $\bar{u}(t)\in U$ *a.e. on* $[0,\bar{T}_1]$ *and the function in (8.118) admits the integral representation*

$$\begin{aligned}\mathscr{I}_v^{\mathrm{r}}(s)=\lim_{\mu}\int_0^s\alpha^\mu(\tau)F_v^{\mathrm{r}}(y_p^\mu(\tau),y_v^\mu(\tau),u(\eta^\mu(\tau)),\eta^\mu(\tau))d\tau &=\\ =\int_0^s\bar{\alpha}(\tau)F_v^{\mathrm{r}}(y_p(\tau),y_v(\tau),\bar{u}(\tau),\eta(\tau))d\tau, &\end{aligned} \tag{8.124}$$

which is valid for all $s\in[0,\bar{T}_1]$.

The function $\bar{u}(\cdot)$ *can be chosen to be Borel measurable.*

If $u(\cdot)$ *is continuous, then* $\bar{u}(s)=u(\bar{\eta}(s))$.

The proof is standard and is based on Filippov's measurable selection lemma [237].

Justification of (8.119)

The function $\Gamma(\cdot)$ has bounded variation and is right-continuous [294], so it admits the decomposition [322]

$$\Gamma(t)=t+V^{\mathrm{c}}(t)+\sum_{\xi\le t}\Delta\Gamma(\xi),$$

where V^{c} is a continuous and non-decreasing function and $D_\Gamma=\{\xi : \Delta\Gamma(\xi)>0\}$ is the countable (finite or infinite) set of jumps of Γ. In turn, the interval $[0,\bar{T}_1]=[0,\Gamma(T)]$ can be written, up to a set of null Lebesgue measure, as the following union of disjoint sets:

$$[0, \bar{T}_1] = \{s : \bar{\alpha}(s) = 1\} \cup \left(\bigcup_{\xi \le T} [\Gamma(\xi-), \Gamma(\xi)] \right) \cup \{s : \eta(s) \in \operatorname{supp} V^{\mathrm{c}}\}.$$

This yields the following expression for the integral (8.119:

$$\begin{aligned} \mathscr{I}_v^{\mathrm{s}}(s) = &\int_0^s \alpha^\mu(\tau)\mu F_v^{\mathrm{s},\mu}(\tau)\mathbf{I}\{G^\mu(\tau) \le 0\}\mathbf{I}\{\bar{\alpha}(\tau) = 1\}d\tau + \\ &+ \sum_{\xi \le \bar{\eta}(s)} \int_{\Gamma(\xi-)}^{\Gamma(\xi)} \alpha^\mu(\tau)\mu F_v^{\mathrm{s},\mu}(\tau)\mathbf{I}\{G^\mu(\tau) \le 0\}d\tau + \\ &+ \int_0^s \alpha^\mu(\tau)\mu F_v^{\mathrm{s},\mu}(\tau)\mathbf{I}\{G^\mu(\tau) \le 0\}\mathbf{I}\{\tau : \eta(\tau) \in \operatorname{supp} V^{\mathrm{c}}\}d\tau. \end{aligned} \tag{8.125}$$

In each interval $[\Gamma(\xi-), \Gamma(\xi)]$ we have $\bar{\alpha}(s) = 0$ and $\bar{\eta}(s) = \xi$. By Proposition 8.5, the jumps of the generalized solution occur precisely at the points $\xi \in D_\Gamma$ and are equal to

$$\Delta\bar{x}_p(\xi) = y_p(\Gamma(\xi)) - y_p(\Gamma(\xi-)), \quad \Delta\bar{x}_v(\xi) = y_v(\Gamma(\xi)) - y_v(\Gamma(\xi-)).$$

However, as follows from Eq. (8.115) and Proposition 8.6, the function $\bar{x}_p(\cdot)$ is actually continuous at the points $\xi \in D_\Gamma$, that is, $\Delta\bar{x}_p(\xi) = 0$. At the same time

$$\Delta\bar{x}_v(\xi) = y_v(\Gamma(\xi)) - y_v(\Gamma(\xi-)) = \lim_\mu \int_{\Gamma(\xi-)}^{\Gamma(\xi)} \alpha^\mu(\tau)\mu F_v^{\mathrm{s},\mu}(\tau)\mathbf{I}\{G^\mu(\tau) \le 0\}d\tau.$$

8.6.5 Formula for a Jump of the Generalized Solution

The expression for a jump in the case of a controllable singularity was obtained in [37]. Here we use the same assumptions about the function F^{s} as in Sect. 8.2.3.

Assumption 8.5 Let the conditions of Theorem 8.1 be fulfilled. We assume also that:

(a) the system of equations

$$\begin{cases} \dot{y}_p(s) = F_p^{\mathrm{r}}(x_p(\tau), y_v(s), t), \\ \dot{y}_v(s) = \bar{F}_v(y_p(s), y_v(s), s, w_t(s), x_p, t), \end{cases} \tag{8.126}$$

with initial conditions $(x_p, x_v) = (x_p(t), x_v(t))$ such that

$$G(x_p, t) = 0, \quad G'_{x_p} F^{\mathrm{r}}_p + G'_t\big|_{(x_p, x_v, t)} < 0, \tag{8.127}$$

has a unique solution for $s \in [t, \infty)$;

(b) there exists the exit time

$$s^*(x_p, x_v, t) = \inf \left\{ s > t : \begin{array}{l} G'_{x_p}(x_p, t)(y_p(s) - x_p) + G'_t(x_p, t)(s - t) = 0, \\ G'_{x_p}(x_p, t) F^{\mathrm{r}}_p(x_p, y_v(s), t) + G'_t(x_p, t) > 0 \end{array} \right\}, \tag{8.128}$$

such that on the interval (t, s^*) one has

$$G'_{x_p}(x_p, t)(y_p(s) - x_p) + G'_t(x_p, t)(s - t) < 0; \tag{8.129}$$

(c) the set of right-hand sides $\bar{F}_v(y_p, y_v, s, W, x_p, t)$ of the system (8.126) is convex for any (y_p, y_v, s, x_p, t).

The next theorem provides a formula for the jump component of the generalized solution at a point $\xi \in D_\Gamma$. We emphasize that this formula is also valid in the case where the set D_Γ is infinite, and hence it covers the case of a trajectory of Zeno type [468] as well.

Theorem 8.10 *Let*

$$\varphi(s, w_t(\cdot), x_p, x_v, t) = \begin{pmatrix} \varphi_p(s, w_t(\cdot), x_p, x_v, t) \\ \varphi_v(s, w_t(\cdot), x_p, x_v, t) \end{pmatrix} \tag{8.130}$$

be the general solution of the system (8.126) for $s \geq t$ with initial conditions

$$y_p(t) = x_p, \quad y_v(t) = x_v, \tag{8.131}$$

that satisfy (8.127) and some measurable controls $w_t(\tau) \in W$, $\tau \geq t$. Then the jump of the variable $\bar{x}_v(\cdot)$ at any point $\xi \in D_\Gamma$ such that $(\bar{x}_p(\xi), \bar{x}_v(\xi-))$ meets the conditions (8.127) is given by the formula

$$\Delta \bar{x}_v(\xi) = \varphi_v(s^*, w_\xi(\cdot), \bar{x}_p(\xi), \bar{x}_v(\xi-), \xi) - \bar{x}_v(\xi-), \tag{8.132}$$

where $w_\xi(\cdot)$ is some admissible W-valued control and the variable $\bar{x}_p(t)$ is continuous at the point ξ.

Proof Let $\xi \in D_\Gamma$ and consider the segment $[\Gamma(\xi-), \Gamma(\xi)]$. For $s \geq \Gamma(\xi-)$ we introduce the representation

$$y^\mu_p(s) = y^\mu_p(\Gamma(\xi-)) + \mu^{-1/2}\big[y^{\mu,\xi}_p(s) - y^\mu_p(\Gamma(\xi-))\big],$$

or

$$y_p^{\mu,\xi}(s) = y_p^{\mu}(\Gamma(\xi-)) + \mu^{1/2}\big[y_p^{\mu}(s) - y_p^{\mu}(\Gamma(\xi-))\big] \tag{8.133}$$

and substitute it in the system (8.114) and (8.115). This leads to the following system of equations for the triple $(y_p^{\mu,\xi}(s), y_v^{\mu}(s), \eta^{\mu}(s))$:

$$y_p^{\mu,\xi}(s) = y_p^{\mu}(\Gamma(\xi-)) +$$
$$+ \int_{\Gamma(\xi-)}^{s} \frac{F_p^{\mathrm{r}}\Big(y_p^{\mu}(\Gamma(\xi-)) + \frac{y_p^{\mu,\xi}(\tau)-y_p^{\mu}(\Gamma(\xi-))}{\mu^{1/2}}, y_v^{\mu}(\tau), \eta^{\mu}(\tau)\Big)}{\mu^{-1/2} + \mu^{1/2}|F_v^{\mathrm{s},\mu}(\tau)|\mathbf{I}\{G^{\mu}(\tau) \le 0\}} d\tau, \tag{8.134}$$

$$y_v^{\mu}(s) = y_v^{\mu}(\Gamma(\xi-)) + \int_{\Gamma(\xi-)}^{s} \frac{\mu^{1/2} F_v^{\mathrm{s},\mu}(\tau)\mathbf{I}\{G^{\mu}(\tau) \le 0\}}{\mu^{-1/2} + \mu^{1/2}|F_v^{\mathrm{s},\mu}(\tau)|\mathbf{I}\{G^{\mu}(\tau) \le 0\}} d\tau +$$
$$+ \int_{\Gamma(\xi-)}^{s} \frac{\mu^{-1/2} F_v^{\mathrm{r}}\Big(y_p^{\mu}(\Gamma(\xi-)) + \frac{y_p^{\mu,\xi}(\tau)-y_p^{\mu}(\Gamma(\xi-))}{\mu^{1/2}}, y_v^{\mu}(\tau), u(\eta^{\mu}(\tau)), \eta^{\mu}(\tau)\Big)}{\mu^{-1/2} + \mu^{1/2}|F_v^{\mathrm{s},\mu}(\tau)|\mathbf{I}\{G^{\mu}(\tau) \le 0\}} d\tau, \tag{8.135}$$

$$\eta^{\mu}(s) = \eta^{\mu}(\Gamma(\xi-)) + \int_{\Gamma(\xi-)}^{s} \frac{\mu^{-1/2}}{\mu^{-1/2} + \mu^{1/2}|F_v^{\mathrm{s},\mu}(\tau)|\mathbf{I}\{G^{\mu}(\tau) \le 0\}} d\tau. \tag{8.136}$$

Next, we perform one more change of variables by means of the increasing function

$$\zeta^{\mu}(s) = \Gamma(\xi-) + \int_{\Gamma(\xi-)}^{s} \frac{1}{\mu^{-1/2} + \mu^{1/2}|F_v^{\mathrm{s},\mu}(\tau)|\mathbf{I}\{G^{\mu}(\tau) \le 0\}} d\tau,$$

which admits the inverse function $s^{-1}(\zeta)$, defined for $\zeta \in [\Gamma(\xi-), \Gamma(\xi)]$. Now let us rewrite the system (8.134), (8.135), and (8.6.5) for the triple

$$(\widetilde{y}_p^{\mu,\xi}(\zeta), \widetilde{y}_v^{\mu}(\zeta), \widetilde{\eta}^{\mu}(\zeta)) = \big(y_p^{\mu,\xi}(s^{-1}(\zeta)), y_v^{\mu}(s^{-1}(\zeta)), \eta^{\mu}(s^{-1}(\zeta))\big)$$

for $\zeta \ge \Gamma(\xi-)$. We have

$$\widetilde{y}_p^{\mu,\xi}(\zeta) = y_p^{\mu}(\Gamma(\xi-)) +$$
$$+ \int_{\Gamma(\xi-)}^{\zeta} F_p^{\mathrm{r}}\Big(y_p^{\mu}(\Gamma(\xi-)) + \frac{\widetilde{y}_p^{\mu,\xi}(\tau) - y_p^{\mu}(\Gamma(\xi-))}{\mu^{1/2}}, \widetilde{y}_v^{\mu}(\tau), \widetilde{\eta}^{\mu}(\tau)\Big) d\tau, \tag{8.137}$$

$$\widetilde{y}_v^{\mu}(\zeta) = y_v^{\mu}(\Gamma(\xi-)) + \int_{\Gamma(\xi-)}^{\zeta} \mu^{1/2} F_v^{\mathrm{s},\mu}(\tau)\mathbf{I}\{G^{\mu}(\tau) \le 0\} d\tau + \int_{\Gamma(\xi-)}^{\zeta} \mu^{-1/2} \times$$

$$\times F_v^{\mathrm{r}}\Big(\widetilde{y}_p^{\mu}(\Gamma(\xi-)) + \frac{\widetilde{y}_p^{\mu,\xi}(\tau) - y_p^{\mu}(\Gamma(\xi-))}{\mu^{1/2}}, \widetilde{y}_v^{\mu}(\tau), u(\widetilde{\eta}^{\mu}(\tau)), \widetilde{\eta}^{\mu}(\tau)\Big) d\tau, \tag{8.138}$$

$$\widetilde{\eta}^{\mu}(\zeta) = \eta^{\mu}(\Gamma(\xi-)) + \mu^{-1/2}[\zeta - \Gamma(\xi-)]. \tag{8.139}$$

By Assumption 8.5, the function $\mu^{1/2} F_v^{\mathrm{s},\mu}(\tau)$ in the first integral of Eq. (8.138) can be written as

$$\begin{aligned}
&\mu^{1/2} F_v^{s,\mu}(\tau) = \\
&= \bar{F}_v^{\mathrm{s}}(\widetilde{y}_p^{\mu,\xi}(\tau), \widetilde{y}_v^{\mu}(\tau), \widetilde{\eta}^{\mu}(\tau), w_{\widetilde{\eta}^{\mu}(\Gamma(\xi-))}(\widetilde{\eta}^{\mu}(\tau), \mu), y_p^{\mu}(\Gamma(\xi-)), \widetilde{\eta}^{\mu}(\Gamma(\xi-))) + \\
&\quad + O(\mu) = \\
&= \bar{F}_v^{\mathrm{s}}(\widetilde{y}_p^{\mu,\xi}(\tau), \widetilde{y}_v^{\mu}(\tau), \widetilde{\eta}^{\mu}(\tau), w_{\eta^{\mu}(\Gamma(\xi-))}(\widetilde{\eta}^{\mu}(\tau), \mu), y_p^{\mu}(\Gamma(\xi-)), \widetilde{\eta}^{\mu}(\Gamma(\xi-))) + \\
&\quad + O(\mu),
\end{aligned} \tag{8.140}$$

where $O(\mu) \to 0$ when $\mu \to \infty$. By Theorem 8.1, the function $\bar{F}_v^{\mathrm{s}}$ is Lipschitz in the variables $(\widetilde{y}_p^{\mu,\xi}, \widetilde{y}_v^{\mu})$, and the set of functions $\{\widetilde{\eta}^{\mu}(\cdot)\}$ is compact. Therefore, due to Assumption c), the subsequence of triples $(\widetilde{y}_p^{\mu,\xi}(\cdot), \widetilde{y}_v^{\mu}(\cdot), \widetilde{\eta}^{\mu}(\cdot))$ is the solution of the system (8.137)–(8.139) with some control $w^{\mu}(\zeta) \in W$. Moreover,

$$\begin{aligned}
&G^{\mu}(\tau) = G(y_p^{\mu}(\Gamma(\xi-)), \eta^{\mu}(\Gamma(\xi-))) + \\
&+ \mu^{-1/2}\big\{G'_{x_p}[\widetilde{y}_p^{\mu,\xi}(\tau) - y_p^{\mu}(\Gamma(\xi-))] + G'_t[\zeta - \Gamma(\xi-)]\big\} + o(1/\mu^{1/2}),
\end{aligned} \tag{8.141}$$

where the partial derivatives are calculated at $y_p^{\mu}(\Gamma(\xi-)), \eta^{\mu}(\Gamma(\xi-))$, and

$$\mu^{1/2} o(\mu^{-1/2}) \to 0 \quad \text{for} \quad \mu \to \infty.$$

Note also that, when $\mu \to \infty$,

$$\begin{aligned}
y_p^{\mu}(\Gamma(\xi-)) &\to y_p(\Gamma(\xi-)), \\
y_v^{\mu}(\Gamma(\xi-)) &\to y_v(\Gamma(\xi-)), \\
\eta^{\mu}(\Gamma(\xi-)) &\to \bar{\eta}(\Gamma(\xi-)) = \xi,
\end{aligned}$$

and that for fixed values of the variables $(\widetilde{y}_p^{\mu,\xi}, \widetilde{y}_v^{\mu}, \widetilde{\eta}^{\mu})$ the functions under the integral in (8.137) and under the first integral in (8.138) tend for $\mu \to \infty$ to

$$F_p\left(y_p(\Gamma(\xi-)), \widetilde{y}_v^{\mu}, \xi\right) \quad \text{and} \quad \bar{F}_v(\widetilde{y}_p^{\mu,\xi}, \widetilde{y}_v^{\mu}, \widetilde{\eta}^{\mu}, w_{\xi}, y_p(\Gamma(\xi-)), \xi),$$

respectively.

By Proposition 8.6, for the chosen subsequence, $\alpha^{\mu}(s) \to 0$ a.e. on the interval $[\Gamma(\xi-), \Gamma(\xi)]$, and therefore

$$\mu|F^{s,\mu}(\tau)|\mathbf{I}\{G^{\mu}(\tau) \leq 0\} \to \infty, \quad \text{a.e. on } [\Gamma(\xi-), \Gamma(\xi)],$$

whence

$$\lim_{\mu} \mathbf{I}\{G^{\mu}(\tau) \leq 0\} = 1, \quad \text{a.e. on } [\Gamma(\xi-), \Gamma(\xi)].$$

Moreover, as it follows from (8.141),

$$\begin{aligned} \mu^{1/2}G^{\mu}(\tau) &= G'_{x_p}[\widetilde{y}_p^{\mu,\xi}(\tau) - y_p(\Gamma(\xi-))] + G'_t[\tau - \Gamma(\xi-)] + \\ &+ \mu^{1/2}O(\mu^{-1/2}). \end{aligned} \tag{8.142}$$

Comparing the systems (8.126) and (8.137)–(8.138), taking into account (8.140) and (8.141), and using (8.141), we conclude that the right-hand side of the system (8.126) is equal to the limit of the expression (8.140). Then since the triple $(y_p(\Gamma(\xi-)), y_v(\Gamma(\xi-)), \xi)$ satisfies (8.127), the continuous dependence of the solution on the initial conditions and the right-hand side [365] implies that the sequence $(\widetilde{y}_p^{\mu,\xi}(\zeta), \widetilde{y}_v^{\mu}(\zeta), \widetilde{\eta}^{\mu}(\zeta))$ converges to the solution of the system (8.126) with some admissible control $w_{\Gamma(\xi-)}(\cdot)$, uniformly on $[\Gamma(\xi-), \Gamma(\xi-)+s^*]$. At the same time, thanks to Assumption 8.5, we have $\lim_{\mu} \mu^{1/2}G^{\mu}(\zeta) < 0$. All this yields the representation

$$y_v(\Gamma(\xi)) = \varphi_v(s^*, w_{\Gamma(\xi-)}(\cdot), y_p(\Gamma(\xi)), y_v(\Gamma(\xi-)), \bar{\eta}(\Gamma(\xi-))).$$

Finally, via the inverse time change, we get the claimed representation (8.132) [37]. □

8.6.6 *Motion Along the Constraint*

This type of motion occurs if the system falls into the constrained domain in such a way that the derivative of the deformation is zero, i.e., if $(\dot{G} = 0)$. If one considers a contact force of the type (8.109) or, generally, an arbitrary force satisfying the condition (8.14), then this derivative is formally equal to zero. However, if at this state the system nevertheless violates the constraint, then for a large values of μ a small deformation produces a significant force, which in turn makes a significant contribution to the equations of motion. In what follows, for the sake of simplicity, we consider the case when the force is described by the relation (8.109). In this case the generalized solution satisfies the conditions

$$G(\bar{x}_p(t), t) = 0, \quad \frac{d}{dt}\Big|_{F_p^{\mathrm{r}}} G(\bar{x}_p(t), t) = G'_{x_p} F_p^{\mathrm{r}} + G'_t\Big|_{(\bar{x}_p(t), \bar{x}_v(t), t)} = 0. \tag{8.143}$$

We consider the system (8.108) with an additional equation for the variable $G(x_p^\mu(t), t)$, namely

$$\begin{cases} \dot{x}_p^\mu(t) = F_p^{\mathrm{r}}(x_p^\mu(t), x_v^\mu(t), t), \\ \dot{x}_v^\mu(t) = F_v^{\mathrm{r}}(x_p^\mu(t), x_v^\mu(t), u(t), t) + \mu F_v^{s,\mu}(t)\mathbf{I}\{t : G^\mu(t) \le 0\}, \\ \ddot{G}^\mu(t) = D(x_p^\mu(t), x_v^\mu(t), u(t), t) + \\ \quad + \left[\widetilde{A}(x_p^\mu(t), x_v^\mu(t), t)\mu G^\mu(t) + \widetilde{B}(x_p^\mu(t), x_v^\mu(t), t)\mu^{1/2}\dot{G}^\mu(t)\right] \times \\ \quad \times \mathbf{I}\{t : G^\mu(t) \le 0\}, \end{cases} \tag{8.144}$$

where

$$\mu F_v^{\mathrm{s},\mu}(t) = A(x_p^\mu(t), t)\mu G^\mu(t) + B(x_p^\mu(t), t)\mu^{1/2}\dot{G}^\mu(t),$$

and

$$D(x_p^\mu(t), x_v^\mu(t), u(t), t) = \left((F_p^r)^* \vdots 1\right) \begin{pmatrix} G''_{x_p x_p} & G''_{x_p t} \\ (G''_{x_p t})^* & G''_{tt} \end{pmatrix} \begin{pmatrix} F_p^{\mathrm{r}} \\ 1 \end{pmatrix}\Bigg|_{(x_p^\mu(t), x_v^\mu(t), u(t), t)} +$$

$$+ G'_{x_p}\left[(F_p^{\mathrm{r}})'_{x_p}\dot{x}_p + (F_p^{\mathrm{r}})'_{x_v} F_v^{\mathrm{r}}(x_p, x_v, u, t) + (F_p)'_t\right]\Big|_{(x_p^\mu(t), x_v^\mu(t), u(t), t)}.$$

Here $\left((F_p^r)^* \vdots 1\right)$ denotes the block-vector with the last, $(n + 1)$-th component equal to 1. To obtain the equation for the generalized solution one needs to make unambiguous the expressions $\mu G^\mu(t)$ and $\mu^{1/2}\dot{G}(t)$ in (8.144), that is, to define the limits

$$\lim_\mu \mu\, G^\mu(t) \quad \text{and} \quad \lim_\mu \mu^{1/2}\dot{G}(t).$$

To do this, we need the following results concerning singularly perturbed second-order equations.

Proposition 8.9 *Consider the differential equation*

$$\ddot{G}^{\mu}(\tau) = U(\tau) - A\mu G^{\mu}(\tau) - B\mu^{1/2}\dot{G}^{\mu}(\tau), \quad \tau \geq t, \tag{8.145}$$

with the initial conditions

$$G^{\mu}(t) = 0, \quad \dot{G}^{\mu}(t) = 0,$$

where $U(\cdot)$ is uniformly bounded and has left limits for $\tau > t$. Suppose that

$$A > 0, \quad B > 0, \quad A - (B/2)^2 > 0, \quad U(\tau) < 0.$$

Then, for $\tau > t$,

$$\lim_{\mu\uparrow\infty} \mu\, G^{\mu}(\tau) = \frac{U(\tau-)}{A}, \qquad \lim_{\mu\uparrow\infty} \mu^{1/2}\dot{G}^{\mu}(\tau) = 0, \qquad \lim_{\mu\uparrow\infty} \ddot{G}^{\mu}(\tau) = 0.$$

Remark 8.13 Here $t \in [0, T]$ is considered as a parameter.

Proof The solution of Eq. (8.145) is given by the formula

$$G^{\mu}(\tau) = \int_t^{\tau} U(s)e^{-\lambda^{\mu}(\tau-s)}\frac{\sin\omega^{\mu}(\tau-s)}{\omega^{\mu}}ds,$$

where

$$\begin{aligned} \lambda^{\mu} &= \mu^{1/2}\lambda, \quad \omega^{\mu} = \mu^{1/2}\omega, \\ \lambda &= B/2, \quad \omega = \sqrt{A-(B/2)^2}. \end{aligned} \tag{8.146}$$

Using the transformation $u = \mu^{1/2}(\tau - s)$ we obtain the expression

$$\mu G^{\mu}(\tau) = \int_0^{\mu^{1/2}(\tau-t)} U(\tau - u\mu^{-1/2})e^{-\lambda u}\frac{\sin\omega u}{\omega}du,$$

and the estimate

$$\begin{aligned} &\left|\mu G^{\mu}(\tau) - U(\tau-)\int_0^{\infty} e^{-\lambda u}\frac{\sin\omega u}{\omega}du\right| \leq \\ &\leq \int_0^{\infty} \left|U(\tau - u\mu^{-1/2}) - U(\tau-)\right|e^{-\lambda u}\left|\frac{\sin\omega u}{\omega}\right|du + \\ &+ \int_{\mu^{1/2}(\tau-t)}^{\infty} \left|U(\tau-)\right|e^{-\lambda u}\left|\frac{\sin\omega u}{\omega}\right|du. \end{aligned} \tag{8.147}$$

Since $U(\cdot)$ is uniformly bounded and since

$$\lim_{\mu\uparrow\infty} U(\tau - u\mu^{-1/2}) = U(\tau-)$$

for any $\tau > t$ and every $u > 0$, we obtain in the limit

$$\lim_{\mu\uparrow\infty} \mu G^{\mu}(\tau) = U(\tau-)\int_0^{\infty} e^{-\lambda u}\frac{\sin\omega u}{\omega}du = \frac{U(\tau-)}{\lambda^2+\omega^2} = \frac{U(\tau-)}{A}.$$

With the same transformation we obtain

$$\lim_{\mu\uparrow\infty} \mu^{1/2}\dot{G}^{\mu}(\tau) = U(\tau-)\int_0^{\infty} e^{-\lambda u}\left[\cos\omega u - \frac{\lambda}{\omega}\sin\omega u\right]du = 0.$$

The last statement is evident. □

Proposition 8.10 *If $U(\cdot) \in C^1(t,\infty)$ and has a uniformly bounded derivative, then the convergence to the limit for $\mu \uparrow \infty$ is uniform for $\tau > t$.*

Proof The result follows immediately from (8.147), since

$$\left|U(\tau - u\mu^{-1/2}) - U(\tau)\right| \le \sup_{\tau>t}\left|U_t'(\tau)\right| |u|\mu^{-1/2}$$

and, correspondingly,

$$\left|\mu G^{\mu}(\tau) - \frac{U(\tau)}{A}\right| \le \sup_{\tau>t}|U_t'(\tau)|\mu^{-1/2}\int_0^{\infty}|u|e^{-\lambda u}\left|\frac{\sin\omega u}{\omega}\right|du + \\ + \int_{\mu^{1/2}(\tau-t)}^{\infty}|U(\tau-)|e^{-\lambda u}\left|\frac{\sin\omega u}{\omega}\right|du.$$

□

Remark 8.14 In last case one can assert that the solution of the equation

$$\ddot{G}^{\mu}(\tau) = U(\tau) - (A\mu G^{\mu}(\tau) - B\mu^{1/2}\dot{G}^{\mu}(\tau))\mathbf{I}\{G^{\mu}(\tau) \le 0\}, \quad \tau \ge t, \tag{8.148}$$

behaves analogously. Indeed, in some neighborhood $\tau > t$ the solution of Eq. (8.148) becomes negative due to the negativity of U and the initial conditions. Since $\mu G^{\mu}(\tau)$ converges uniformly to a negative function, there is a neighborhood of the point t where conditions of Proposition 8.10 are fulfilled and the equation of motion is (8.145). Below we provide examples which demonstrate that the smoothness of the input action $U(\cdot)$ does indeed ensure the convergence of $\mu G^{\mu}(\tau)$ to a negative function also on the solutions of the Eq. (8.148).

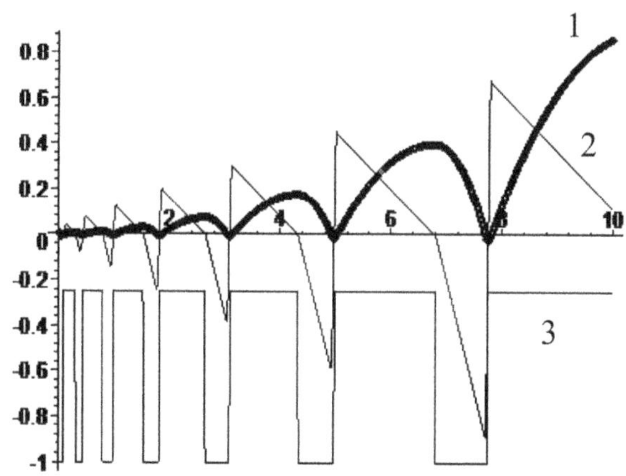

Fig. 8.29 1—displacement $G_\mu(t)$, 2—velocity $\dot{G}_\mu(t)$, and 3—profile of the external negative force $U(t)$. One can see that the external force in the feedback form pushes the system out of equilibrium. However, this does not contradict Proposition 8.9, since when μ grows the displacement converges at all points of continuity to a value proportional to the external force, as demonstrated in Fig. 8.30

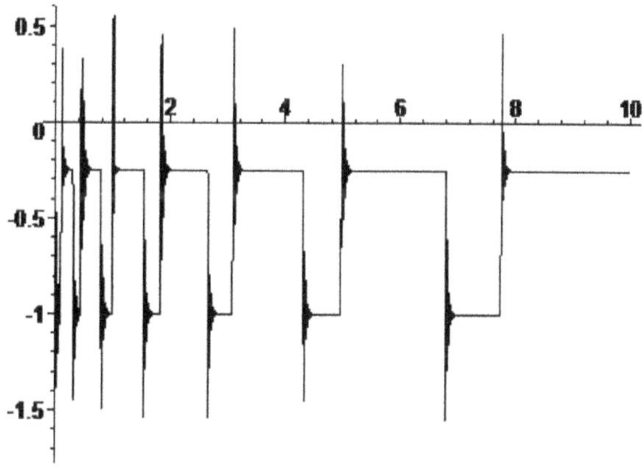

Fig. 8.30 Normalized displacement $\mu G_\mu(t)$ for $\mu = 10^5$. The example is similar to the previous one with external action calculated for $\mu = 10^3$. One can see that, at all points of continuity of the external force, the normalized displacement $\mu G_\mu(t)$ converges to $U(t)/A$ and the normalized velocity $\mu^{1/2}\dot{G}_\mu(t)$ converges to zero (see Fig. 8.31)

The following examples, illustrated in Figs. 8.29, 8.30, 8.31, and 8.32, show that our statements describe correctly the behavior of the solutions of the nonlinear system (8.148). The external perturbation $U(t)$ used is taken for a fixed value of $\mu = 10^3$ in the feedback form

$$U(t) = \begin{cases} -0.25, & \text{for } \dot{G}_\mu \geq 0, \\ -1, & \text{for } \dot{G}_\mu < 0, \end{cases} \quad \text{and} \quad A = 1,\ B = 0.1.$$

As shown in [281] and Sect. 8.6.3, this external action pushes the system out of equilibrium for given μ (Fig. 8.29). However, when one increases μ, but one keeps the same external action, the type of solution changes and the normalized solution $\mu G_\mu(t)$ stabilizes and converges at all points of continuity to the function $U(t)/A$ (Figs. 8.30 and 8.31).

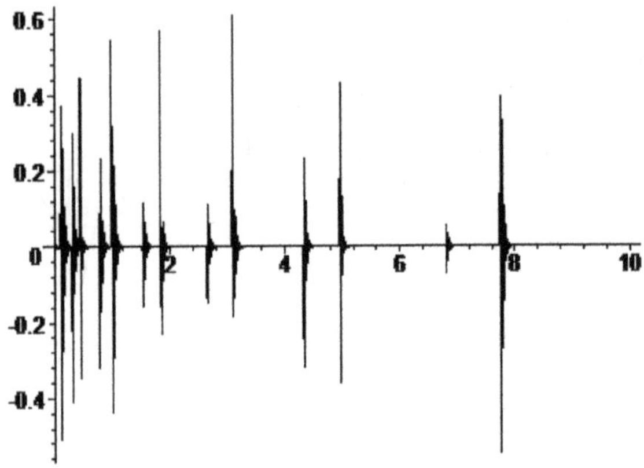

Fig. 8.31 Normalized velocity $\mu^{1/2}\dot{G}_\mu(t)$

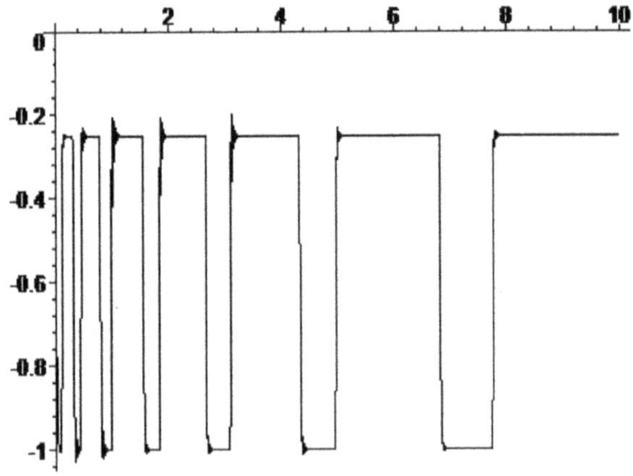

Fig. 8.32 Normalized displacement $\mu G_\mu(t)$, with $\mu = 10^4$, when the external action is smoothed (passing it through a high-frequency filter)

We remark that the discontinuities of the external action induce impulsive perturbations of states and velocities that appear when μ is increased. Figure 8.32 shows how the behavior of the system changes if the external action is made smoother. The impulsive outliers vanish or decrease, so the convergence take place on the whole time interval.

The next statements show that small perturbations of the coefficients that vanish for $\mu \to \infty$ do not change the qualitative behavior of the system.

Proposition 8.11 *Consider the differential equation*

$$\ddot{G^\mu}(\tau) = U^\mu(\tau) - A^\mu(\tau)\mu G^\mu(\tau) - B^\mu(\tau)\mu^{1/2}\dot{G}^\mu(\tau), \quad \tau \geq t, \tag{8.149}$$

with the initial conditions

$$G^\mu(t) = 0, \quad |\mu^{1/2}\dot{G}^\mu(t)| \to 0.$$

Assume that the functions $U^\mu(\cdot)$, $A^\mu(\cdot)$, $B^\mu(\cdot)$ are uniformly bounded and continuous at $\tau = t$, that

$$A(t) = \lim_{\tau\downarrow t}\lim_{\mu\uparrow\infty} A^\mu(\tau) > 0, \quad B(t) = \lim_{\tau\downarrow t}\lim_{\mu\uparrow\infty} B^\mu(\tau) > 0, \quad A(t) - (B(t)/2)^2 > 0,$$

and that

$$U(t) = \lim_{\tau\downarrow t}\lim_{\mu\uparrow\infty} U^\mu(\tau) < 0.$$

Then

$$\lim_{\tau\downarrow t}\lim_{\mu\uparrow\infty} \mu G^\mu(\tau) = \frac{U(t)}{A(t)} < 0, \quad \lim_{\tau\downarrow t}\lim_{\mu\uparrow\infty} \mu^{1/2}\dot{G}^\mu(\tau) = 0, \quad \lim_{\tau\downarrow t}\lim_{\mu\uparrow\infty} \ddot{G}^\mu(\tau) = 0.$$

Proof We change the variable to $Z^\mu(s) = \mu G^\mu(t + s\mu^{-1/2})$ and note that $Z^\mu(s)$ satisfies the equation

$$\ddot{Z}^\mu(s) = U^\mu(t + \mu^{-1/2}s) - A^\mu(t + \mu^{-1/2}s)Z^\mu(s) - B^\mu(t + \mu^{-1/2}s)\dot{Z}^\mu(s), \tag{8.150}$$

with the corresponding initial conditions

$$Z^\mu(0) = 0, \quad \dot{Z}^\mu(0) = \mu^{1/2}\dot{G}^\mu(t),$$

where t is regarded as a parameter. Comparing the solution of this equation with that of the equation

$$\ddot{\widetilde{Z}}^\mu(s) = U(t) - A(t)\widetilde{Z}^\mu(s) - B(t)\dot{\widetilde{Z}}^\mu(s), \tag{8.151}$$

with the initial conditions

$$\widetilde{Z}^\mu(0) = 0, \quad \dot{\widetilde{Z}}^\mu(0) = \mu^{1/2}\dot{G}^\mu(t),$$

we find that $Z^\mu(s)$ converges to $\widetilde{Z}^\mu(s)$ uniformly on each segment $[0, S]$ [96, 365]. Then we have

$$\widetilde{Z}^\mu(s) = \dot{\widetilde{Z}}^\mu(0)e^{-\lambda s}\frac{\sin\omega s}{\omega} + \int_0^s U(t)e^{-\lambda(s-\tau)}\frac{\sin\omega(s-\tau)}{\omega}e^{-\lambda(s-\tau)}d\tau,$$

and

$$\widetilde{Z}^\mu(s) \to \frac{U(t)}{\lambda^2 + \omega^2} = \frac{U(t)}{A(t)} < 0,$$

if $\mu, s \to \infty$. Now the desired conclusion is obtained by taking first s sufficiently large, letting $\mu \to \infty$, and performing the inverse space-time transformation. □

Remark 8.15 This result means that if the initial velocity of deformation is small, then the reaction force stabilizes and the velocity of deformation velocity is already not relevant for the motion of the system. This is valid in a more general case as well. Let us consider the equation

$$\ddot{G}^{\mu}(\tau) = U^{\mu}(\tau) - \left[A^{\mu}(\tau)\mu G^{\mu}(\tau) + B^{\mu}(\tau)\mu^{1/2}\dot{G}^{\mu}(\tau)\right]\mathbf{I}\{G^{\mu}(\tau) \leq 0\}, \tag{8.152}$$

for $\tau \geq t$. Using the space-time transformation we obtain the following equation for the variable $Z^{\mu}(s)$:

$$\begin{aligned} \ddot{Z}^{\mu}(s) &= U^{\mu}(t + \mu^{-1/2}s) - \\ &- \left[A^{\mu}(t + \mu^{-1/2}s)Z^{\mu}(s) + B^{\mu}(t + \mu^{-1/2}s)\dot{Z}^{\mu}(s)\right]\mathbf{I}\{Z^{\mu}(s) \leq 0\}, \end{aligned} \tag{8.153}$$

and for the comparison variable $\widetilde{Z}^{\mu}(s)$ the equation

$$\ddot{\widetilde{Z}}^{\mu}(s) = U(t) - [A(t)\widetilde{Z}^{\mu}(s) + B(t)\dot{\widetilde{Z}}^{\mu}(s)]\mathbf{I}\{Z^{\mu}(s) \leq 0\}. \tag{8.154}$$

Proposition 8.11 shows that for sufficiently small $Z^{\mu}(0)$ the solution remains in the domain $\{Z^{\mu} < 0\}$ for all $s > 0$ and converges to a limit. Generally speaking, we can guarantee only the estimate $|\dot{Z}^{\mu}(0)| \leq C < \infty$. It is not difficult to evaluate the solution (8.154) and observe that exits only a finite number of exits from the domain $Z^{\mu}(s) \leq 0$, and this number is bounded by a value that depends only on the constant C. Indeed, after each exit the value of $\dot{Z}^{\mu}$ is k times the initial value, where $k < 1$, and so, after a certain number of exits, the initial velocity becomes sufficiently small and satisfies the conditions of Proposition 8.11. Therefore, the corresponding of entrances and exits take some time, but the latter is uniformly bounded. Thus, after this time has elapsed, we arrive at the previous conditions. The general proof of this statement is quite tedious, so we will only demonstrate this result on an example.

Returning to the representation of the motion governed by the system (8.144), we offer the following result.

Theorem 8.11 *Suppose the generalized solution $(\bar{x}_p(t), \bar{x}_v(t))$ satisfies the conditions (8.143), and the functions D, $\widetilde{A}$ and $\widetilde{B}$ in (8.144) are continuous in all the variables and satisfy the conditions*

$$D(\bar{x}_p(t), \bar{x}_v(t), u(t), t) > 0,$$

$$\widetilde{A}(\bar{x}_p(t), \bar{x}_v(t), t) > 0, \quad \widetilde{B}(\bar{x}_p(t), \bar{x}_v(t), t) > 0,$$

and

$$\widetilde{A} - (\widetilde{B}/2)^2 > 0.$$

Then

$$\lim_{\mu \uparrow \infty} \mu G^{\mu}(t) = -\frac{D(\bar{x}_p(t), \bar{x}_v(t), u(t), t)}{\widetilde{A}(\bar{x}_p(t), \bar{x}_v(t), t)} = F^{\mathrm{sm}}(\bar{x}_p(t), \bar{x}_v(t), u(t), t),$$

and

$$\lim_{\mu \uparrow \infty} \mu G^{\mu}(t) = 0$$

at all the points of continuity of the control $u(\cdot)$.

Remark 8.16 Here the superscript sm stands for *sliding mode* and is used to indicate that we are dealing with the representation of the force acting during the sliding along the surface.

Remark 8.17 The conditions of Theorem 8.11 mean that:

- the constraint is repulsive in the sense of [36, 37],
- at the points where the system touches the constraint, the external force pushes the system inside the domain forbidden by the constraint.

Let us emphasize that the order of constraint violation in the case of smooth motion along the constraint is μ^{-1}, while in the case of impact this order is $\sim \mu^{-1/2}$ (see Eq. (8.133)).

Representation of a Generalized Solution by a Differential Equation with Measure

We can now derive the differential equation for the pair $(\bar{x}_p(t), \bar{x}_v(t))$.

Theorem 8.12

(a) *Suppose the assumptions of Proposition 8.8 and of Theorems 8.10 and 8.11 are satisfied. Let* $(\bar{x}_p(t), \bar{x}_v(t))$ *be the generalized solution of the system (8.108), subject to the unilateral constraints*

$$G(\bar{x}_p(t), t) \geq 0, \quad \text{for all } t \in [0, T].$$

Then there exist:

1. *a Borel-measurable function* $\bar{u}(\cdot)$ *such that* $\bar{u}(t) \in U$ *a.e. on* $[0, T]$;
2. *a non-negative measure* $V(dt)$ *that admits the decomposition*

$$V([0, t]) = V^{\mathrm{c}}([0, t]) + \sum_{\xi \leq t} V(\{\xi\})$$

and satisfies the slackness conditions

$$\int_0^T G(\bar{x}_p(t), t)dV(t) = 0,$$

such that the generalized solution $(\bar{x}_p(t), \bar{x}_v(t))$ *satisfies the system of equations with measure*

$$\bar{x}_p(t) = x_p(0) + \int_0^t F_p^r(\bar{x}_p(\tau), \bar{x}_v(\tau), \tau)d\tau, \tag{8.155}$$

$$\begin{aligned}\bar{x}_v(t) = x_v(0) + \int_0^t F_v^{\mathrm{r}}(\bar{x}_p(\tau), \bar{x}_v(\tau), \bar{u}(\tau), \tau)d\tau \\ + \int_0^t A(\bar{x}_p(\tau), \tau)F^{\mathrm{sm}}(\bar{x}_p(\tau), \bar{x}_v(\tau), \bar{u}(\tau), \tau)dV^{\mathrm{c}}(\tau) + \\ + \sum_{\xi \le t} \left[\varphi_v(s^*, \bar{x}_p(\xi), \bar{x}_v(\xi-), \xi) - \bar{x}_v(\xi-)\right] \times \\ \times \mathbf{I}\left\{G(\bar{x}_p(\xi), \xi) = 0,\ \dot{G}(\bar{x}_p(\xi), \bar{x}_v(\xi-), \xi) < 0\right\}.\end{aligned} \tag{8.156}$$

(b) *Suppose that the function* F^{sm} *is Lipschitz in the variables* $(\bar{x}_p, \bar{x}_v)$. *Then starting from some point* τ *such that*

$$G(\bar{x}_p(\tau), \tau) = 0, \quad \dot{G}(\bar{x}_p(\tau), \bar{x}_v(\tau), \tau) = 0,$$

the generalized solution admits a unique continuation till the time

$$t^* = \inf\left\{t > \tau :\ D(t)/\widetilde{A}(t) \ge 0\right\},$$

provided $t^* > \tau$.

Proof

(a) At any point of continuity we have

$$\bar{x}_p(t) = \lim_{\mu\uparrow\infty} x_p^\mu(t) = \lim_{\mu\uparrow\infty} y_p^\mu(\Gamma^\mu(t)) = \lim_{\mu\uparrow\infty} y_p^\mu(\Gamma(t)),$$

because at all such points $\Gamma^\mu(t) \to \Gamma(t)$ and because the family of functions $\{y^\mu(s)\}$ is equicontinuous on $[0, T_1]$. Similar relations hold for $\bar{x}_v(t)$. Using the inverse time transformation $t = \eta(\tau)$ in the first integral in (8.121), we derive the first Eq. (8.155); also, by means of the same transformation in the integral

(8.124), we derive the first integral in (8.156). To calculate other terms we pass to the limit in the second integral in (8.115). Thus, we have

$$\int_0^{\Gamma(t)} \alpha^\mu(\tau) F_v^{s,\mu}(\tau)\mathbf{I}\{G^\mu(\tau) \le 0\} d\tau =$$

$$= \int_0^{\Gamma(t)} \alpha^\mu(\tau) F_v^{s,\mu}(\tau)\mathbf{I}\{G^\mu(\tau) \le 0\}\mathbf{I}\{\bar{G}(\tau) > 0\} d\tau +$$

$$+ \int_0^{\Gamma(t)} \alpha^\mu(\tau) F_v^{s,\mu}(\tau)\mathbf{I}\{G^\mu(\tau) \le 0\}\mathbf{I}\{\bar{G}(\tau) = 0\} d\tau = I_1^\mu(t) + I_2^\mu(t).$$

By Proposition 8.6

$$I_1^\mu(t) \to 0.$$

We will use the decomposition

$$\{\bar{G}(\tau) = 0\} = \{\tau : \eta(\tau) \in \text{supp } V^c\} \bigcup_{\xi \le \Gamma(t)} [\Gamma(\xi-), \Gamma(\xi)].$$

By Theorem 8.10,

$$\lim_\mu \int_{\Gamma(\xi-)}^{\Gamma(\xi)} \alpha^\mu(\tau) \mu F_v^{s,\mu}(\tau)\mathbf{I}\{G^\mu(\tau) \le 0\} d\tau =$$

$$= \left[\varphi_v(s^*, \bar{x}_p(\xi), \bar{x}_v(\xi-), \xi) - \bar{x}_v(\xi-)\right] \times$$

$$\times\, \mathbf{I}\{G(\bar{x}_p(\xi), \xi) = 0,\ \dot{G}(\bar{x}_p(\xi), \bar{x}_v(\xi-), \xi) < 0\}.$$

Next, note that

$$\mathbf{I}\{\eta(\tau) \in \text{supp } V^c\} = \mathbf{I}\{\bar{G}(\tau) = 0,\ \dot{\bar{G}}(\tau) = 0\} =$$

$$= \mathbf{I}\left\{\{\tau : 0 \le \bar{\alpha}(\tau) < 1\} \setminus \bigcup_{\xi \le T} [\Gamma(\xi-), \Gamma(\xi)]\right\}.$$

Now applying to I_2^μ the inverse transformation $\tau = \eta^\mu(s)$ and using Theorem 8.11, we calculate the integral over the set $\{\bar{G}(\tau) = 0,\ \dot{\bar{G}}(\tau) = 0\}$. To complete the proof of item a), we define the absolutely continuous measure $V^c(dt)$ by

$$V^{c}([0,T]) = \int_0^{\Gamma(t)} \bar{\alpha}(\tau)\mathbf{I}\{\bar{G}(\tau)=0,\ \dot{\bar{G}}(\tau)=0\}d\tau =$$
$$= \int_0^{t} \mathbf{I}\{\tau : G(\bar{x}_p(\tau),\tau)=0,\ \dot{G}(\bar{x}_p(\tau),\bar{x}_v(\tau),\tau)=0\}d\tau \tag{8.157}$$

and define the control $\bar{u}(\cdot)$ by means of the measurable selection procedure).

(b) As follows from Theorem 8.11, the generalized solution satisfies on the segment $[\tau, t^*]$ the system of differential equations

$$\begin{cases} \dot{\bar{x}}_p(t) = F_p^{\mathrm{r}}(\bar{x}_p(t),\bar{x}_v(t),t), \\ \dot{\bar{x}}_v(t) = F_v^{\mathrm{r}}(\bar{x}_p(t),\bar{x}_v(t),u(t),t) + \\ \quad + A(\bar{x}_p(t),t)F^{\mathrm{sm}}(\bar{x}_p(t),\bar{x}_v(t),u(t),t)\mathbf{I}\{t : \bar{G}(t)=0,\ \dot{\bar{G}}(t)=0\}, \\ \ddot{\bar{G}}(t) = 1 - \mathbf{I}\left\{t : \dfrac{D(\bar{x}_p(t),\bar{x}_v(t),u(t),t)}{\widetilde{A}(\bar{x}_p(t),\bar{x}_v(t),t)} < 0\right\} = 0, \end{cases} \tag{8.158}$$

which is obtained by taking the limit of the right-hand sides in the system (8.144) on the set $\{t \in \operatorname{supp} V^{c}\} = \{\bar{G}(\tau)=0,\ \dot{\bar{G}}(\tau)=0\}$. Since the system of Eq. (8.158) satisfies the Lipschitz condition, its solution is unique till it exits the constrained domain. This completes the proof.

□

Stabilization Inside the Forbidden Domain: An Example

Consider the classical "bouncing ball" example, assuming the external force is constant and the collision is viscoelastic [287]. The governing system of equations has the form

$$\begin{cases} \dot{x}_p^{\mu}(t) = x_v^{\mu}(t), \\ \dot{x}_v^{\mu}(t) = -F - \mu\left[x_p^{\mu}(t) + 2\xi x_v^{\mu}(t)\mu^{-1/2}\right]\mathbf{I}\{t : x_p^{\mu}(t) \le 0\}, \end{cases} \tag{8.159}$$

where $F = \text{const} > 0$ and $0 < \xi < 1$. Let us examine the behavior of this system when $\mu \to \infty$. The system provides an approximation of the behavior of the bouncing ball, which stabilizes in finite time after an infinite number of impact. However, for any finite $\mu < \infty$, the ball stabilizes inside the domain $\{x_p^{\mu} \le 0\}$ after a **finite** number of impact. Indeed, if at the impact time τ the value of the velocity satisfies the inequality $|x_v^{\mu}(\tau)| \le C(F,\xi)/\mu^{1/2}$, then the ball remains in the domain $\{x_p^{\mu} \le 0\}$. The constant $C(F,\xi)$ does not depend on μ. This case is

Fig. 8.33 Stabilization of the "bouncing ball" for $\mu < \infty$

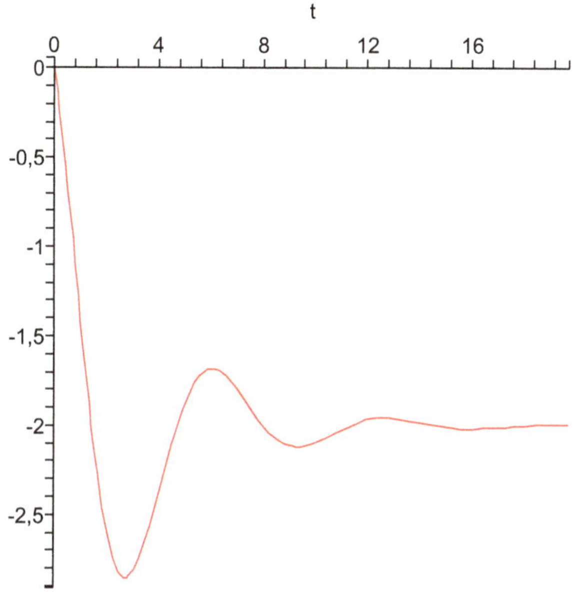

illustrated by Fig. 8.33, where the ball stabilizes around the level $x_p^\mu = -F/\mu$, exactly in accordance with Proposition 8.11. One can estimate the moment of the last impact. Indeed, if the velocity $|x_v^\mu(\tau)|$ before an impact is large enough and after the impact the ball leaves the forbidden area, we have:

1. the time Δ necessary for exit from the forbidden domain satisfies the inequality

$$\Delta \leq \frac{2\pi}{\mu^{1/2}\sqrt{1-\xi^2}}.$$

2. The velocity $x_v^\mu(\tau + \Delta)$ after exit from the forbidden domain satisfies the inequality

$$|x_v^\mu(\tau + \Delta)| < |x_v^\mu(\tau)| \exp\left\{-\frac{\pi\xi}{\sqrt{1-\xi^2}}\right\} < \alpha |x_v^\mu(\tau)|,$$

where $\alpha < 1$.

3. The next impact after exit occurs after the time

$$\frac{2|x_v^\mu(\tau + \Delta)|}{F} = \alpha \frac{2|x_v^\mu(\tau)|}{F}.$$

Consequently, the time of the last impact, T_μ^*, can be estimated as

$$T_\mu^* \le \frac{2\pi k^*}{\mu^{1/2}\sqrt{1-\xi^2}} + \frac{2|x_v^\mu(0)|}{F}\sum_{k=0}^{k^*}\alpha^k, \tag{8.160}$$

where the number k^* is such that

$$\frac{1}{\log\alpha}\log\left(\frac{C(F,\xi)}{\mu^{1/2}|x_v^\mu(0)|}\right) \le k^* \le \frac{1}{\log\alpha}\log\left(\frac{C(F,\xi)}{\mu^{1/2}|x_v^\mu(0)|}\right) + 1.$$

If $\mu \to \infty$, then $k^* \to \infty$, but the first term in (8.160) goes to zero, T_μ^* remains bounded, and

$$\lim_{\mu\to\infty} T_\mu^* = \frac{2|x_v^\mu(0)|}{F}\sum_{k=0}^{\infty}\alpha^k = \frac{2|x_v^\mu(0)|}{F(1-\alpha)},$$

which coincides with the accumulation point of the jumps of the generalized solution.

Thus, the behavior of a system with finite μ is exactly the same as described by Theorem 8.12, and in this example we can see how the solution can be continued beyond the accumulation point of the jumps. Thus, for finite μ we observe the regularization of the so-called *Zeno trajectory* with infinitely many switchings occurring in a finite time interval. Note that this method of regularization is physically stipulated, in contrast to, for example, the regularization based on the methodology of hybrid automata [468], where the switching time of the velocity is assumed to be constant and finite.

In the general case, the external force is of course not constant, but in the neighborhood of the accumulation point of the jumps, under a more or less smooth behavior of an external force, the qualitative behavior of the system will be precisely the one described.

8.6.7 Global Description of Trajectories

The approach described below makes it possible to define the generalized solution over the entire time interval using a singular space-time transformation. This result generalizes the methodology developed earlier for mechanical systems with shocks by L. Paoli and M. Schatzman [346] and M. Schatzman [393], and allows to establish the uniform boundedness of the integral of the generalized force over the entire interval of motion, under the following assumption about the force law. Let

$$Z_p(t) = G(x_p, t), \quad Z_v(t) = G'_x(t)F_p^{\mathrm{r}}(t) + G'_t(t)$$

and

$$F_v^{\mathrm{s}} = \left[A(x_p, x_v, w, t)Z_p + \mu^{-1/2} B(x_p, x_v, w, t)Z_v\right]\mathbf{I}\{Z_p \leq 0\}. \tag{8.161}$$

Then the equations of motion have the form

$$\begin{cases} \dot{Z}_p = Z_v, \\ \dot{Z}_v = \widetilde{F}^{u,w}(t) + \mu\left[\alpha^w(t)Z_p + \mu^{-1/2}\beta^w(t)Z_v\right]\mathbf{I}\{Z_p \leq 0\}. \end{cases} \tag{8.162}$$

Proposition 8.12 *Let the coefficients $\alpha^w(t)$ and $\beta^w(t)$ in Eq. (8.162) satisfy the conditions*

$$-\infty < c_1 \leq \alpha^w(t), \beta^w(t) \leq c_2 < 0$$

and let $\|\widetilde{F}^{u,w}(t)\|$, $|\dot{\alpha}^w(t)|$ and $\|\dot{\beta}^w(t)\|$ be uniformly bounded.
Then,

$$\begin{aligned} &\sup_\mu \int_0^T \mu|Z_p(t)|\mathbf{I}\{Z_p \leq 0\}dt < \infty, \\ &\sup_\mu \int_0^T \mu^{1/2}|Z_v(t)|\mathbf{I}\{Z_p \leq 0\}dt < \infty, \\ &\sup_\mu \sup_{[0,T]} \mu^{1/2}|Z_p(t)|\mathbf{I}\{Z_p(t) \leq 0\} < \infty. \end{aligned} \tag{8.163}$$

Remark 8.18 The conditions of Proposition 8.12 are trivially satisfied if, for example, the functions $\widetilde{F}^{u,w}(t)$, $\alpha^w(t)$ and $\beta^w(t)$ do not depend on time, phase coordinates and controls [346].

Proposition 8.13 *Under the assumptions of Proposition 8.12, the sequence of solutions*

$$\{x_p^\mu(\cdot), x_v^\mu(\cdot)\} \in C^1[0, T] \times AC[0, T] \tag{8.164}$$

and satisfies the constraints

$$\begin{aligned} &\sup_{\mu>0} \max_{[0,T]} \|x_p^\mu(t)\| < \infty, \\ &\sup_{\mu>0} \max_{[0,T]} \|x_v^\mu(t)\| < \infty, \\ &\sup_{\mu>0} \operatorname*{Var}_{[0,T]} \|x_p^\mu(t)\| < \infty. \end{aligned} \tag{8.165}$$

Consequently, there exist a subsequence $\mu_k \uparrow \infty$ such that the corresponding sequences $\{x_p^{\mu_k}(\cdot), x_v^{\mu_k}(\cdot)\}$ converge to some functions $\{\bar{x}_p(\cdot), \bar{x}_v(\cdot)\} \in C^1[0,T] \times BV[0,T]$, where $x_p^{\mu_k}$ converges to $\bar{x}_p$ uniformly on $[0,T]$ and $x_v^{\mu_k}$ converges to $\bar{x}_v$ at all the points of continuity.

Definition 8.7 The pair of functions $\{\bar{x}_p(\cdot), \bar{x}_v(\cdot)\} \in C^1[0,T] \times BV[0,T]$ whose existence is asserted in Proposition 8.4 is called a generalized solution of the system (8.162).

Our aim is to describe the behavior of the parametrized sequence of solutions $(x_p^\mu(t), x_v^\mu(t))$ when $\mu \to \infty$. Earlier we considered this problem in the case of a single jump or a finite number of jumps. The description of the solution in the case of possibly infinitely many jumps and sliding along the boundary of the constraint was obtained in [288].

Proof To simplify notation, we omit the dependence of Z_p^μ and Z_v^μ on μ and the dependence of $\widetilde{F}^{w,u}(t)$, $\alpha^w(t)$, and $\beta^w(t)$ on w and u. Denote $Z_p^- = Z_p\mathbf{I}\{Z_p \le 0\}$. Then inside the domain $\{Z_p \le 0\}$ it holds that

$$\begin{aligned} Z_v(t)\mathbf{I}\{Z_p(t) \le 0\}dt &= dZ_p^-(t) \\ Z_p^-(t)Z_v(t)dt &= Z_p^-(t)dZ_p^-(t). \end{aligned} \tag{8.166}$$

Let us introduce the function

$$E(t) = -\alpha(t)\frac{1}{2}\mu(Z_p^-)^2 + \frac{1}{2}(Z_v)^2.$$

Then, in view of (8.166),

$$\begin{aligned} \frac{dE(t)}{dt} &= -\dot{\alpha}(t)\frac{1}{2}\mu(Z_p^-)^2 - \alpha(t)\mu Z_p^- Z_v + \\ &+ Z_v\left(\widetilde{F}(t) + \mu\alpha(t)Z_p^- + \mu^{1/2}\beta(t)Z_v\mathbf{I}\{Z_p \le 0\}\right). \end{aligned} \tag{8.167}$$

Since $\beta(t) < 0$, integrating (8.167) one obtains

$$E(t) - E(0) \le \frac{1}{2}\int_0^t \left(-\dot{\alpha}(t)\mu(Z_p^-)^2(s) + Z_v(s)\widetilde{F}(s)\right)ds.$$

Further, under the conditions of Proposition 8.12, integrating with the inequality

$$Z_v\widetilde{F} \le \frac{1}{2}Z_v^2 + \frac{1}{2}\widetilde{F}^2$$

accounted for gives

$$E(t) - E(0) \leq c_3 \int_0^t E(s)ds + \int_0^T |\widetilde{F}(s)|ds,$$

whence, in view of the Gronwall-Bellman inequality (Lemma A.5),

$$\sup_{\mu<\infty, t\in[0,T]} E(t) \leq \infty.$$

Therefore,

$$\sup_{\mu<\infty,\ t\in[0,T]} \mu(Z_p^-)^2(t) \quad \text{and} \quad \sup_{\mu<\infty,\ t\in[0,T]} Z_v^2(t) \leq \infty. \tag{8.168}$$

As the next step, we decompose the open subset $\{t \in [0, T] : Z_p(t) < 0\} \subset [0, T]$ into a union of intervals (a_i, b_i) on each of which $Z_p(t) < 0$, in such a way that a_i is a entry point, i.e., $Z_p(a_i) = 0$ and $Z_v(a_i) \leq 0$, and b_i is a point at which the action of the contact force terminates, i.e., $Z_p(b_i) \leq 0$ and $\ddot{Z}_v(b_i) = 0$ under the physical condition of contact termination, or $Z_p(b_i) = 0$ under the geometrical condition of contact termination (see Sect. 8.3.1).

On the interval $t \in (a_i, b_i)$ we have

$$\dot{Z}_p(t) = Z_v(t), \quad Z_p(t) < 0,$$
$$\dot{Z}_v(t) = \widetilde{F}(t) + \mu\alpha(t)Z_p^-(t) + \mu^{1/2}\beta(t)Z_v(t)\mathbf{I}\{Z_p(t) \leq 0\}.$$

Further,

$$Z_p(b_i) = \int_{a_i}^{b_i} Z_v(s)ds = \int_{a_i}^{b_i} \left[Z_v(a_i) + \int_{a_i}^{s} \dot{Z}_v(\tau)d\tau\right] ds =$$
$$= \int_{a_i}^{b_i} \left[Z_v(a_i) + \int_{a_i}^{b_i} \mathbf{I}\{\tau \leq s\}\dot{Z}_v(\tau)ds\right] d\tau =$$
$$= (b_i - a_i)Z_v(a_i) + \int_{a_i}^{b_i} (b_i - \tau)\dot{Z}_v(\tau)d\tau \leq 0.$$

Continuing and substituting the expression for $\dot{Z}_v(\tau)$ we obtain

$$(b_i - a_i)Z_v(a_i) +$$
$$+ \int_{a_i}^{b_i} (b_i - \tau)\left[\widetilde{F}(\tau) + \mu\alpha(t)Z_p^-(\tau) + \mu^{1/2}\beta(\tau)Z_v(\tau)\mathbf{I}(Z_p(\tau) \leq 0)\right] d\tau \leq 0,$$

whence

$$0 \leq \int_{a_i}^{b_i} (b_i - \tau)\mu\alpha(t)Z_p^-(\tau)d\tau \leq (b_i - a_i)Z_v(a_i) + \int_{a_i}^{b_i} (b_i - \tau)\widetilde{F}(\tau)d\tau +$$

$$+ \int_{a_i}^{b_i} (b_i - \tau)\mu^{1/2}\beta(\tau)Z_v(\tau)\mathbf{I}(Z_p(\tau) \leq 0)d\tau \leq 0.$$

In view of (8.166), the last integral in this last inequality is equal to

$$\int_{a_i}^{b_i} (b_i - \tau)\mu^{1/2}\beta(\tau)\dot{Z}_p^-(\tau)d\tau,$$

so via integration by parts with (8.168) accounted for we obtain the estimate

$$\int_{a_i}^{b_i} (b_i - \tau)\mu^{1/2}\beta(\tau)\dot{Z}_p^-(\tau)d\tau =$$

$$= (b_i - b_i)\mu^{1/2}\beta(b_i)Z_p^-(b_i) - (b_i - a_i)\mu^{1/2}\beta(a_i)Z_p^-(a_i) -$$

$$- \int_{a_i}^{b_i} \mu^{1/2}\frac{d}{d\tau}\left[(b_i - \tau)\beta(\tau)\right] Z_p^-(\tau)d\tau \leq$$

$$\leq c_4(b_i - a_i),$$

with some constant $c_4 < \infty$. Since the functions $\widetilde{F}$ and Z_v are uniformly bounded, we conclude that

$$\int_{a_i}^{b_i} (b_i - \tau)\mu\alpha(\tau)Z_p^-(\tau)d\tau \leq c_5(b_i - a_i),$$

with some constant $c_5 < \infty$ that does not depend on μ and i. □

8.7 Optimal Control in the Impact Phase: Examples

8.7.1 Damping a Ball with a Racket: Geometric Constraints on the Control

Equations of Motion and Constraints

Consider a ball of unit mass that collides in free fall with a viscoelastic racket of mass M, which in turn moves towards the ball along the vertical axis at a constant speed, as shown in Fig. 8.34 [40]. The state vector of this system has the form

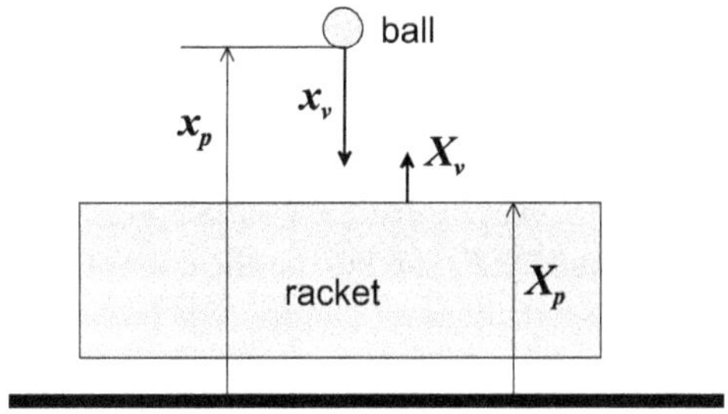

Fig. 8.34 Natural (regular) phase of motion

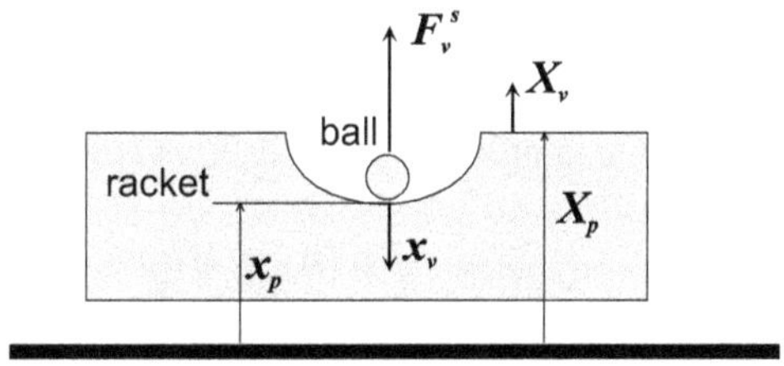

Fig. 8.35 Singular phase of motion

$X = (x_p, x_v, X_p, X_v)$, where x_p, x_v and X_p, X_v are the positions and the velocities of the ball and of the racket, respectively. The constraint-free domain is described by the inequality

$$G(X) = x_p - X_p > 0. \tag{8.169}$$

In this domain, the motion of the system is said to be in the constraint-free phase (or natural phase) and the equations of motion read

$$\begin{cases} \dot{x}_p(t) = x_v(t), & \dot{X}_p(t) = X_v(t), \\ \dot{x}_v(t) = -g, & \dot{X}_v(t) = 0, \end{cases} \tag{8.170}$$

where g denotes the acceleration of gravity. In the forbidden domain, depicted in Fig. 8.35, the motion is said to be in the singular phase and is governed by the equations

$$\begin{cases} \dot{x}_p(t) = x_v(t), \qquad \dot{X}_p(t) = X_v(t), \\ \dot{x}_v(t) = -g - \mu F_v^{\prime,\mathrm{s}}(X(t), \mu), \\ \dot{X}_v(t) = M^{-1}\big[F(t, \mu) + \mu F_v^{\prime \mathrm{s}}(X(t), \mu)\big]. \end{cases} \tag{8.171}$$

Here $\mu F_v^{\prime \mathrm{s}}(X(t), \mu)$ is the viscoelastic force acting between the ball and the racket, given by the formula

$$F_v^{\prime \mathrm{s}}(X(t), \mu) = x_p(t) - X_p(t) + 2\kappa\mu^{-1/2}\left(x_v(t) - X_v(t)\right), \tag{8.172}$$

where $\mu > 0$ is the coefficient of elasticity and $0 \le \kappa \le 1$ is the damping coefficient; $F(t, \mu)$ can be regarded as an external impulsive control force acting on the racket in the singular and the constraint-free phases as a result of introducing short-duration control signals w_1 and w_2, respectively, that may be interpreted, for example, as programmed controls. Under the control w_1 (in the singular phase), this force will be denoted by F^{s}; and under the control w_2 (in the constraint-free domain) by F^{rs}.

We assume that w_1 and w_2 depend on t and μ as

$$w_i^\mu(t) = \begin{cases} w_i(\sqrt{\mu}(t-\tau)), & \text{if } t \ge \tau, \\ 0, & \text{otherwise,} \end{cases} \qquad i = 1, 2. \tag{8.173}$$

In our example, the force $F(t, \mu)$ is given by the expression

$$F(t, \mu) = \sqrt{\mu}\, w\big(\sqrt{\mu}(t-\tau)\big), \quad t \ge \tau, \tag{8.174}$$

where w is w_1 or w_2. In (8.173) and (8.174), τ is equal to the first instant the ball meets the obstacle for w_1^μ, or the first time it exits the obstacle for w_2^μ. It is assumed that the controls w_i satisfy the constraints

$$|w_i(\cdot)| \le w_0 < \infty, \quad i = 1, 2, \tag{8.175}$$

where the constant w_0 is given. Thus, relations (8.171)–(8.175) describe motion *with active or controlled singularities.*

Suppose, in addition, that the racket includes a sensor that can measure the pressure on the racket during the contact phase, and denote by $\xi(t)$ the signal produced by the sensor, which is equal to the viscoelastic force acting on the racket:

$$\xi(t) = x_p(t) - X_p(t) + 2\kappa\mu^{-1/2}\big(x_v(t) - X_v(t)\big). \tag{8.176}$$

Equations (8.169)–(8.176), further referred to as the physically-based system, or the original system, describe the continuous motion of the ball/racket system for $\mu < \infty$.

Transformation of Variables and the Equations of Infinitesimal Dynamics

Although the original system (8.169)–(8.176) provide a physically justified description of the motion, it is not well suited for controller synthesis in the singular phase for arbitrary μ. Indeed, in practice the parameter μ varies in a wide range, depending on the material of the racket [148]. Moreover, as μ in (8.169)–(8.176) increases, the right-hand side in (8.171) becomes unacceptably large, leading to numerical ill-conditioning of the standard computational optimal control schemes.

Therefore, it is of interest to obtain a representation that is free from μ for computation purposes, but allows to derive a control law which, when reparametrized by μ, captures the entire family of optimal or near optimal control solutions for the original system. As indicated in the previous section, such a representation is, as shown in [40], provided by the so-called equations of *controlled infinitesimal dynamics*.

In the general case these equations yield a solution that is optimal only in the limit as $\mu \to \infty$ in the auxiliary time and suboptimal for a finite μ in the actual time t of the original system, where the quality of the control improves with the increase of μ. However, for a large class of systems, including the one considered below, μ-reparametrization provides a family of control laws that are optimal in the actual time, for arbitrary μ.

The equations of controlled infinitesimal dynamics also allow to obtain a system of equations, referred to as the limit system, that is satisfied by the generalized solution of the original system when $\mu \to \infty$, and which describes, in particular, the jumps of the velocity at impact. This system is turns to be useful for a numerically well-conditioned approximation of the original system for large μ.

The conceptual and analytical framework described above leads to the following three distinct objectives:

(i) *controller synthesis*: develop an analytical setting that (a) enables the computation of the optimal control law in the auxiliary time by standard controller-synthesis techniques, and (b) once this law is obtained, μ-reparametrize it so that it remains optimal for the original system;
(ii) *calculation of the control law*: calculate impulsive optimal control law that minimizes the velocity of the ball bounce after the impact in the limit system, and, finally,
(iii) *limit modeling*: derive the limit system corresponding to the original system.

These objectives are addressed below for the system (8.169)–(8.176).

To develop a setting for controller synthesis, we introduce the space-time transformation

$$
\begin{aligned}
s &= \sqrt{\mu}\,(t-\tau), \quad t \geq \tau, \\
y_p^\mu(s) &= x_p(\tau) + \sqrt{\mu}\big[x_p(\tau+\mu^{-1/2}s) - x_p(\tau)\big], \\
Y_p^\mu(s) &= X_p(\tau) + \sqrt{\mu}\big[X_p(\tau+\mu^{-1/2}s) - X_p(\tau)\big], \\
y_v^\mu(s) &= x_v(\tau+\mu^{-1/2}s), \\
Y_v^\mu(s) &= X_v(\tau+\mu^{-1/2}s), \\
\eta^\mu(s) &= \xi(\tau+\mu^{-1/2}s),
\end{aligned}
\tag{8.177}
$$

where s represents the auxiliary new time variable (stretched $\sqrt{\mu}$ times). This transformation extends the one given in [36] and [37] to accommodate sensor dynamics. Substituting the expressions (8.177) in (8.169)–(8.176) and using the variable w to denote w_1 or w_2, we arrive at the following system for the new variables $(y_p^\mu, y_v^\mu, Y_p^\mu, Y_v^\mu, \eta^\mu)$ that describes the motion in the extended space-time scale:

$$\begin{cases} \dot{y}_p^\mu(s) = y_v^\mu(s), \\ \dot{y}_v^\mu(s) = -\dfrac{g}{\sqrt{\mu}} - y_p^\mu(s) + Y_p^\mu(s) - 2\kappa y_v^\mu(s) + 2\kappa Y_v^\mu(s), \\ \dot{Y}_p^\mu(s) = Y_v^\mu(s), \\ \dot{Y}_v^\mu(s) = M^{-1}\big[w(s) + y_p^\mu(s) - Y_p^\mu(s) + 2\kappa\big(y_v^\mu(s) - Y_v^\mu(s)\big)\big], \\ \eta^\mu(s) = y_p^\mu(s) - Y_p^\mu(s) + 2\kappa\big(y_v^\mu(s) - Y_v^\mu(s)\big), \end{cases} \tag{8.178}$$

where the dot denotes the derivative d/ds, and the initial conditions are given by

$$\begin{aligned} &y_p^\mu(0) = Y_p^\mu(0) = 0, \quad y_v^\mu(0) = x_v(\tau-), \\ &Y_v^\mu(0) = X_v^\mu(\tau-), \quad \eta^\mu(0) = 2\kappa\big(x_v^\mu(\tau-) - X_v^\mu(\tau-)\big). \end{aligned} \tag{8.179}$$

In the limit as $\mu \to \infty$, (8.178) and (8.179) yield the system of controlled infinitesimal dynamics for the variables $(y_p, y_v, Y_p, Y_v, \eta)$:

$$\begin{cases} \dot{y}_p(s) = y_v(s), \\ \dot{y}_v(s) = -y_p(s) + Y_p(s) - 2\kappa y_v(s) + 2\kappa Y_v(s), \\ \dot{Y}_p(s) = Y_v(s), \\ \dot{Y}_v(s) = M^{-1}\big[w(s) + y_p(s) - Y_p(s) + 2\kappa\big(y_v(s) - Y_v(s)\big)\big], \\ \eta(s) = y_p(s) - Y_p(s) + 2\kappa\big(y_v(s) - Y_v(s)\big), \end{cases} \tag{8.180}$$

with the initial conditions

$$\begin{aligned} &y_p(0) = Y_p(0) = 0, \quad y_v(0) = x_v(\tau-), \\ &Y_v(0) = X_v(\tau-), \quad \eta(0) = 2\kappa\big(x_v(\tau-) - X_v(\tau-)\big). \end{aligned}$$

For simplicity, we introduce the relative coordinates $q(s) = \big(q_p(s), q_v(s)\big)$ as

$$q_p(s) = y_p(s) - Y_p(s), \qquad q_v(s) = y_v(s) - Y_v(s). \tag{8.181}$$

Then the system (8.180) takes on the form

$$\begin{cases} \dot{q}_p(s) = q_v(s), \\ \dot{q}_v(s) = -h(s) - aq_p(s) - 2a\kappa q_v(s), \\ \dot{y}_v(s) = -q_p(s) - 2\kappa q_v(s) = -\eta(s), \end{cases} \tag{8.182}$$

where $a \stackrel{\text{def}}{=} 1 + M^{-1}$, $q_p(0) = 0$, $q_v(0) = x_v(\tau-) - X_v(\tau-) < 0$ and $h(s) \stackrel{\text{def}}{=} M^{-1}w(s)$. The transformation (8.177) and the controlled infinitesimal dynamics equation (8.182) correspond to the objective (i). The next two objective are addressed in the text below, devoted to the calculation of the optimal law for the system (8.182) in the auxiliary time variable s.

8.7.2 Computation of the Optimal Control Law

Structure of the Problem

Suppose the viscoelastic contact force possesses the so-called restitution (repulsiveness) property [37], i.e., guarantees that the ball is repulsed by the forbidden domain in a finite time in the absence of external actions on the system (see Remark 8.4). This means that the following rebound conditions must be satisfied at the instant s^* of exit from the constraint:

$$q_p(s^*) = 0, \quad \dot{q}_p(s^*) = q_v(s^*) > 0. \tag{8.183}$$

In this case, the system (8.182) admits the explicit solution

$$q_p(s) = e^{-\lambda s}\left[q_v(0)\,\frac{\sin(\omega s)}{\omega} - \int_0^s e^{\lambda s'}\,\frac{\sin\omega(s-s')}{\omega}h(s')\,ds'\right], \tag{8.184}$$

where $\lambda \stackrel{\text{def}}{=} \kappa a$ and $\omega^2 \stackrel{\text{def}}{=} a - \lambda^2 > 0$, and so the restitution condition (8.34) takes on the form $a^{-1/2} > \kappa$ (see Proposition 8.1).

Setting, without loss of generality, $Y_v(0) = 0$, suppose first, that $h(s) \equiv 0$. Then, (8.182)–(8.184) imply that

$$y_{v0}(s_0^*) = a^{-1}y_v(0)\big(M^{-1} - \exp(-\lambda\pi\omega^{-1})\big),$$

where the subscripts 0 in y_{v0} and s_0^* correspond to the null control. Hence, since $y_v(0) < 0$, we have that

(a) $y_{v0}(s_0^*) > 0$, if $M > \exp\left(\lambda\pi\omega^{-1}\right)$,
(b) $y_{v0}(s_0^*) < 0$, if $M < \exp\left(\lambda\pi\omega^{-1}\right)$,
(c) $y_{v0}(s_0^*) = 0$, if $M = \exp\left(\lambda\pi\omega^{-1}\right)$.

In case (c), the ball stops without any external force. However, in cases (a) and (b), to reduce the velocity of the ball at the instant s_0^*, an external force must act on the racket. In these last two cases, it is natural to take

$$y_v^2(s^*) \to \min \tag{8.185}$$

as the optimality criterion. Let us denote the conjugate variables corresponding to q_p, q_v and y_v by ψ_p, ψ_v and ψ_y, respectively, and apply the Pontryagin maximum principle to the optimal control problem (8.182)–(8.185). The Hamiltonian $\mathscr{H} = \mathscr{H}(q_p, q_v, y_v, \psi_p, \psi_v, \psi_y, h)$ for this problem is given by

$$\mathscr{H} = \psi_p q_v - \psi_v h - \big(a\psi_v + \psi_y\big)\big(q_p + 2\kappa q_v\big), \tag{8.186}$$

and the resulting optimal control law (obtained by solving $\mathscr{H} \to \max\limits_h$) is given by

$$h(s) = -h_0 \operatorname{sign}\big(\psi_v(s)\big), \tag{8.187}$$

where $h_0 \overset{\text{def}}{=} M^{-1} w_0$, and the conjugate system of equations reads

$$\begin{cases} \dot{\psi}_p(s) = a\psi_v(s) + \psi_y(s), \\ \dot{\psi}_v(s) = -\psi_p(s) + 2a\kappa\psi_v(s) + 2\kappa\psi_y(s), \\ \dot{\psi}_y(s) = 0, \quad \text{i.e., } \psi_y = C_y = \text{const.} \end{cases} \tag{8.188}$$

The terminal transversality conditions at $s = s^*$ read (the time argument is omitted)

$$2y_v\,\delta y_v + \psi_p\,\delta q_p + \psi_v\,\delta q_v + \psi_y\,\delta y_v - \mathscr{H}\,\delta s = 0, \tag{8.189}$$

where $\delta q_p = 0$ due to condition (8.183). This yields

$$2y_v(s^*) + C_y = 0, \qquad \psi_v(s^*) = 0, \qquad \mathscr{H}(s^*) = 0. \tag{8.190}$$

Then $\mathscr{H}(s^*) = 0$ implies that

$$\psi_p(s^*) - 2\kappa C_y = 0, \tag{8.191}$$

and (8.188), (8.190), and (8.191) give

$$\dot{\psi}_v(s^*) = \psi_v(s^*) = 0. \tag{8.192}$$

Equation (8.188) for the conjugate variable $\psi_v(s)$, which determines the sign of the optimal control, has a null terminal condition and can be easily integrated backward in time. Indeed, suppose that in (8.190) the constant $C_y = -2y_v(s^*) \neq 0$. Then setting $\vartheta = s^* - s$ and $\varphi_v(\vartheta) = C_y^{-1}\psi_v(s^* - \vartheta)$, we deduce from (8.188) and (8.192) that

$$\ddot{\varphi}_v(\vartheta) + 2a\kappa\,\dot{\varphi}_v(\vartheta) + a\varphi_v(\vartheta) = -1, \quad \varphi_v(0) = 0. \tag{8.193}$$

This equation admits the explicit solution

$$\varphi_v(\vartheta) = a^{-1}\big[e^{-\lambda\vartheta}\left(\cos\omega\vartheta + \lambda\omega^{-1}\sin\omega\vartheta\right) - 1\big]. \tag{8.194}$$

Examination of the equation shows that $\varphi_v(\vartheta) < 0$ for $\vartheta > 0$, which means that the optimal control does not change sign. Thus, at the instant s^* two events are possible:

(i) $0 < y_v(s^*) < y_{v0}(s_0^*) < |y_v(0)|$, implying that $C_y < 0$, and consequently $\psi_v(s) = C_y\varphi_v(s^* - s) > 0$, which yields the optimal control $h(s) \equiv -h_0$. In this case the racket is subject to the external force $-w_0$, $0 \le s < s^*$, which during the contact gives the racket a real mechanical impulse $p = -w_0 s^*$ in the negative direction of the coordinate axis. However, the ball rebounds from the racket in the positive direction. The external force impulse in this case is too small to stop the ball. Since the racket remains behind the ball (the ball distances from the racket), the solution of the damping problem for the case (i) terminates with a nonzero (but decreased) velocity of the ball.

(ii) $y_v(0) < y_{v0}(s_0^*) < y_v(s^*) < 0$, which implies that $C_y > 0$, and hence $\psi_v(s) = C_y\varphi_v(s^* - s) < 0$. This yields the optimal control $h(s) \equiv h_0$. In this case the racket is subject to the external control pulse w_0, $0 \le s < s^*$, acting in the positive direction of the coordinate axis, and after the contact the ball continues to move in the negative direction. In this case, the external impulse is also too small to stop the ball, but now the racket remains in front of the ball (the ball continues to move towards the racket), hence the solution of the damping problem can be continued with new initial conditions.

Open-Loop Multi-Impact Control

In both cases considered above, the terminal time s^* is found by setting the right-hand side of (8.184) equal to zero, with $h(s') = -h_0$ or $+h_0$, respectively. This gives

$$e^{\lambda s^*} = \cos\omega s^* + \omega^{-1}\big(\lambda \mp h_0^{-1} a\, q_v(0)\big)\sin\omega s^*. \tag{8.195}$$

The increment of the ball's velocity is calculated as

$$\Delta y_v(s^*) = y_v(s^*) - y_v(0) = -\int_0^{s^*} q_p(s)\,ds. \tag{8.196}$$

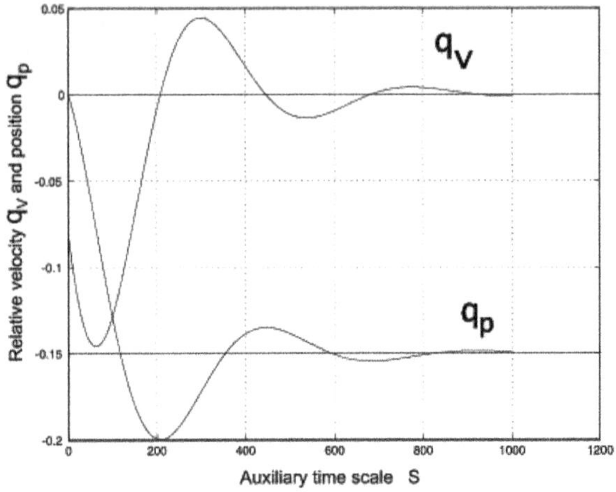

Fig. 8.36 Damped oscillations of the relative coordinates q_p and q_v in the forbidden domain

Analogously, the increment of the racket's velocity is calculated as

$$\Delta Y_v(s^*) = Y_v(s^*) - Y_v(0) = \Delta y_v(s^*) - \Delta q_v(s^*) = \\ = -M^{-1} \int_0^{s^*} q_p(s)\, ds + \int_0^{s^*} h(s)\, ds. \tag{8.197}$$

Continuing the solution in case (ii), it is easily seen that the application of a constant control $h(s) = h_0$ after the instant s^* results in the next collision after a time interval $T = 2q_v(s^*)h_0^{-1}$ with $q_v(s^*+T) = -q_v(s^*) < 0$ and $q_p(s^*+T) = 0$ under the external force w_0 applied to the racket, $s^* \leq s < s^* + T$. Of course, the values of the components of the phase vector (q_p, q_v, y_v) at $s^* + T$ can be taken as new initial conditions for the case (ii).

This process can be repeated several times for as long as the value of the initial relative velocity does not fall into the range $\hat{q}_v \leq q_v(0) \leq 0$, where $\hat{q}_v$ is some threshold velocity. After that, the ball becomes stuck to the racket; more precisely, it starts executing a damping oscillatory motion inside the forbidden domain. Physically, this means that the force of inertia acting on the ball due to the acceleration of the racket is too large to allow the ball escape the forbidden region (detach from the racket). Figure 8.36 illustrates this case for the initial data

$$a = 2, \quad \kappa = 0.25, \quad h_0 = 0.3, \quad q_p(0) = 0, \\ q_v(0) = -0.0822, \; y_v(0) = -0.3329, \; Y_v(0) = -0.2507. \tag{8.198}$$

In the original time t, as $\mu \to \infty$ we have $T \to 0$, so that the entire sequence of collisions tends to a limit point. Sequences of interactions with such characteristics are termed *multi-impulse* [41], and are considered in more detail in Sect. 8.7.3 and in the paper [42]

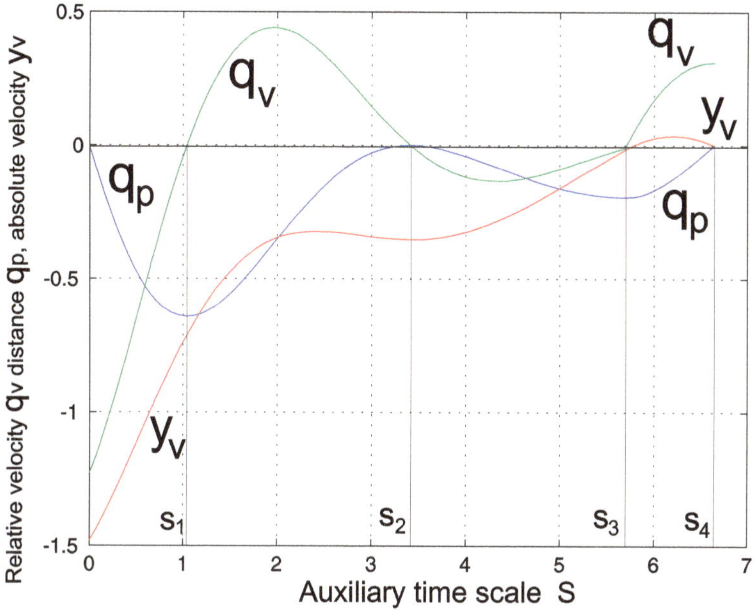

Fig. 8.37 Critical case: $q_v(0) = \hat{q}$. The evolution of the coordinates q_p, q_v and y_v in the forbidden domain

Optimal Control in the Case of a Single Impact

Increasing $|q_v(0)|$ leads to an increase in the amplitude of damped oscillations of the variable $q_p(s)$. When $|q_v(0)|$ reaches a certain critical level $q_v(0) = \hat{q}_v$, the first local maximum of the variable $q_p(s)$, at some instant $\hat{s}$ defined below, reaches zero. The conditions $q_p(\hat{s}) = 0$ and $q_v(\hat{s}) = 0$ provide the equations that determine $\hat{s}$ and $\hat{q}_v$, namely

$$\begin{aligned} e^{\lambda \hat{s}} &= \cos \omega \hat{s} + \omega^{-1}\big(\lambda + h_0^{-1} a \hat{q}_v\big) \sin \omega \hat{s}, \\ 0 &= \hat{q}_v \cos \omega \hat{s} - \omega^{-1}\big(\lambda \hat{q}_v + h_0\big) \sin \omega \hat{s}. \end{aligned} \tag{8.199}$$

This critical case with initial data (8.198) is illustrated in Fig. 8.37, where the absolute velocity of the ball $Y_v(0) = -0.2507$ has the same value as in Fig. 8.36, and

$$q_v(0) = \hat{q}_v = -1.2295, \quad y_v(0) = -1.4802, \quad \hat{s} = s_2 = 3.41.$$

In this critical case, s_1 is the first instance the relative velocity $q_v\,(s_1)$ becomes zero, while the penetration of the ball into the racket is maximal. At the instant $s_2 = 3.41$, the relative velocity and position are reset to zero, but the absolute velocity $y_v\,(s_2)$ of the ball is still significantly less than zero. For this reason, the control of $h(s) = h_0$

remains positive. At the instant $s_3 = 5.695$, the control changes sign in such a way that $h(s) = -h_0$ for $s \in [s_3, s_4]$, where $s_4 = 6.64$ is the terminal time. At the instant s_4, the ball stops, whereas the racket has a finite velocity $-q_v(s_4)$ in the negative direction.

A natural question is: what should the impact on the racket be in the case where $\hat{q}_v \leq q_v(0) \leq 0$? It seems plausible that in this case the ball can be stopped. The assumption that the ball stops gives the additional terminal condition

$$y_v(s^*) = 0, \tag{8.200}$$

which leads to the degeneration of the control law found above. In fact, as follows from (8.190) and (8.191), this condition gives $C_y = \psi_y(s^*) = 0$ and $\psi_p(s^*) = 0$, which leads to a trivial solution of the conjugate system (8.188). This circumstance requires a reformulation of the original problem, for instance, by introducing another optimality criterion. A natural choice for such a criterion is the terminal time:

$$\int_0^{s^*} C\,ds \to \min \qquad \big(\text{with } C > 0 \text{ an arbitrary constant}\big) \tag{8.201}$$

and then consider the time-optimal problem (8.182), (8.183), (8.200), and (8.201).

For this problem, the Hamiltonian, the maximum principle, and the optimal control law have the same form as in (8.186), (8.187), and (8.188). However, due to the conditions (8.200) and (8.201), the transversality conditions now reads

$$\psi_v(s^*) = 0, \quad \mathscr{H}(s^*) = C.$$

The last equation yields

$$\big(\psi_p(s^*) - 2\kappa C_y\big)q_v(s^*) = C.$$

Suppose that $C_y \neq 0$. Then, denoting $\Psi_p(s) = C_y^{-1}\psi_p(s)$ and $\Psi_v(s) = C_y^{-1}\psi_v(s)$ and assuming that $C = C_y$, it is readily verified that these variables satisfy the equations (8.188) with $\psi_y(s) = 1$ and the terminal conditions

$$\Psi_v(s^*) = 0, \quad \big(\Psi_p(s^*) - 2\kappa\big)\,q_v(s^*) = 1. \tag{8.202}$$

As in (8.193), we take the inverse time $\vartheta = s^* - s$ and put $\Phi_v(\vartheta) = \Psi_v(s^* - \vartheta)$.

Integrating the system (8.188) in inverse time under the conditions (8.202) gives

$$\Phi_v(\vartheta) = \varphi_v(\vartheta) + e^{-\lambda\vartheta}\big(\omega q_v(s^*)\big)^{-1}\sin\omega\vartheta, \tag{8.203}$$

where the function $\varphi_v(\vartheta)$ is given by formula (8.194). Due to the presence of an additional term in the right-hand side of (8.203), the function $\Phi_v(\vartheta)$ will change

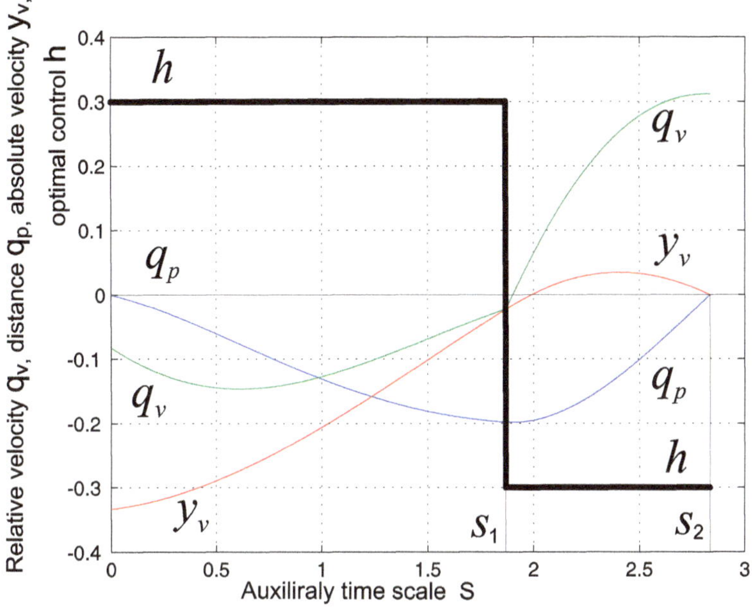

Fig. 8.38 Evolution of the coordinates q_p, q_v, and y_v in the forbidden region under a two-pulse control h (8.225)

sign at some point $\widetilde{s}$ in the segment $[0, s^*]$, which implies that the optimal control changes sign.

Figure 8.38 illustrates the behavior of the variables $q_p(s)$, $q_v(s)$, and $y_v(s)$ with the same initial data (8.198). At the time $\widetilde{s} = s_1 = 1.871$, the curve q_v has a break. This corresponds to a two-step control

$$h(s) = \begin{cases} h_0, & \text{if } 0 \leq s < \widetilde{s}, \\ -h_0, & \text{if } \widetilde{s} \leq s < s^*. \end{cases} \tag{8.204}$$

At the instant $s^* = s_2 = 2.835$ the ball stops. In real time, this control corresponds to a multi-pulse consisting of two mechanical impulses, $p_1 = w_0\widetilde{s}$ and $p_2 = -w_0(s^* - \widetilde{s})$, one immediately after the other.

Finally, let us address the case when at the instant s^* the ball and the racket have zero relative velocity, but their absolute velocity still remains negative. Taking this instant as the initial time, we consider the time-optimal ball-stopping problem with the initial conditions

$$q_p(0) = 0, \quad q_v(0) = 0, \quad y_v(0) = Y_v(0) < 0.$$

This problem is also solved by a two-pulse control. Indeed, under the control $h(s) = h_0$ and null initial conditions, system (8.182) admits the explicit solution

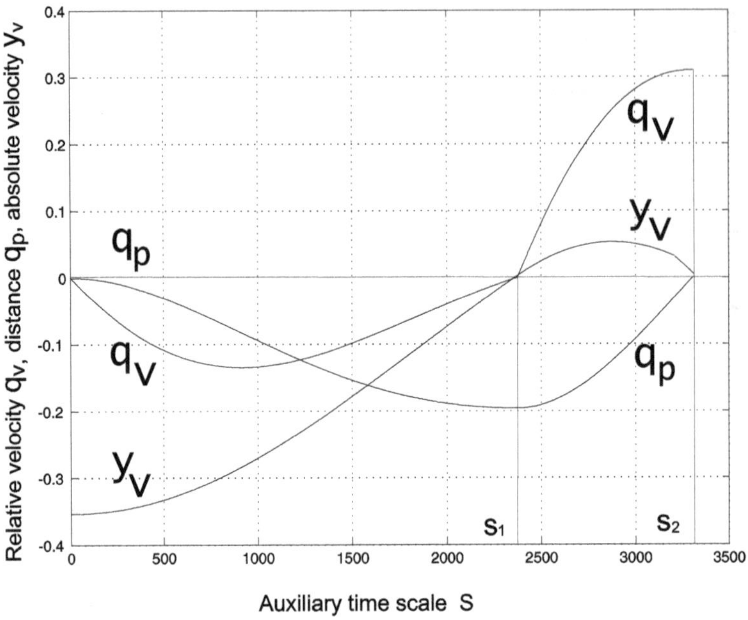

Fig. 8.39 Evolution of the coordinates q_p, q_v, and y_v in the forbidden region under a two-pulse control and zero relative initial velocity

$$q_p(s) = h_0\,\varphi_v(s), \quad q_v(s) = -h_0\,e^{-\lambda s}\omega^{-1}\sin\omega s,$$

where $\varphi_v(\vartheta)$ is given by formula (8.194), and hence, $q_p(s) < 0$ for any s. The function $q_v(s)$ changes sign at the instant $\widetilde{s} = \pi/\omega$. At this instant, the control $h(s)$ also changes sign. This case, subject to the initial conditions (8.198), is illustrated in Fig. 8.39.

8.7.3 *Optimal Multi-Impact Interaction and Control Based on Information from a Pressure Sensor*

In the above examples, the racket and the ball were assumed to have the same mass ($a = 2$). Now we consider the case of a light racket ($M = 1/3$ and, consequently, $a = 4$), moving towards the ball at a high initial relative velocity $q_v(0) = -3.2302$. The remaining parameters correspond to the example in Fig. 8.37. Figures 8.40, 8.41, and 8.42 illustrate the present case.

In the first interval $[0, s_1]$, the control is $h_1(s) = M^{-1}w_{11}(s) = h_0 = 0.3$, where the second index in w_{1j} (for the moment $j = 1$) indicates the order in which the control $w_1(s)$ is applied (the order in the constraint violation) in this interaction.

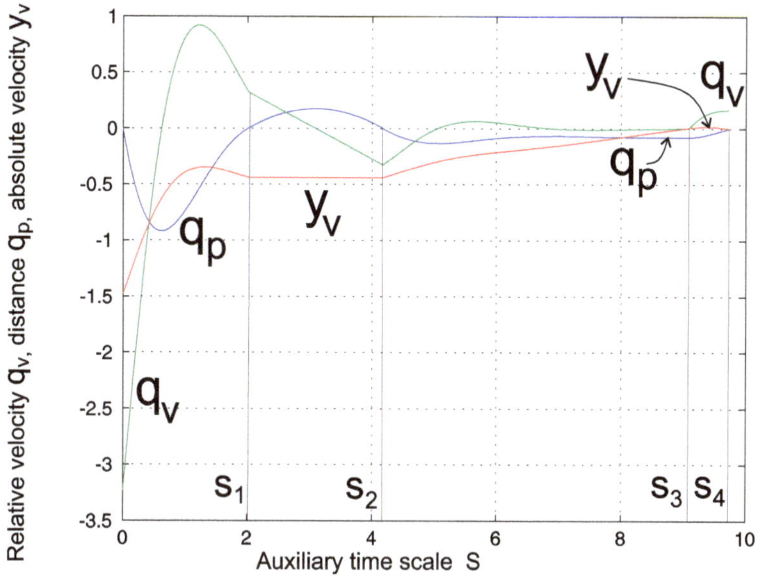

Fig. 8.40 Evolution of the coordinates q_p, q_v and y_v in the forbidden and permissible domains under a four-pulse control; case of a light racket

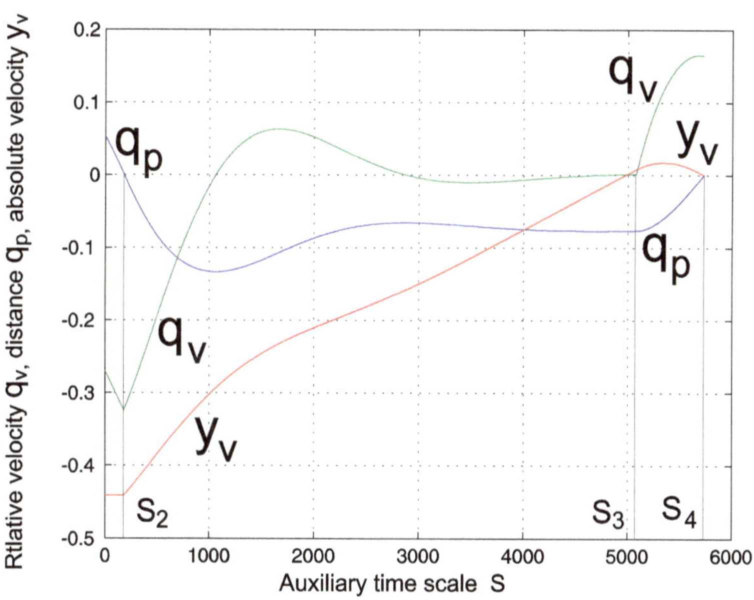

Fig. 8.41 Zoomed-in fragment of Fig. 8.40 (between the instants s_2 and s_4)

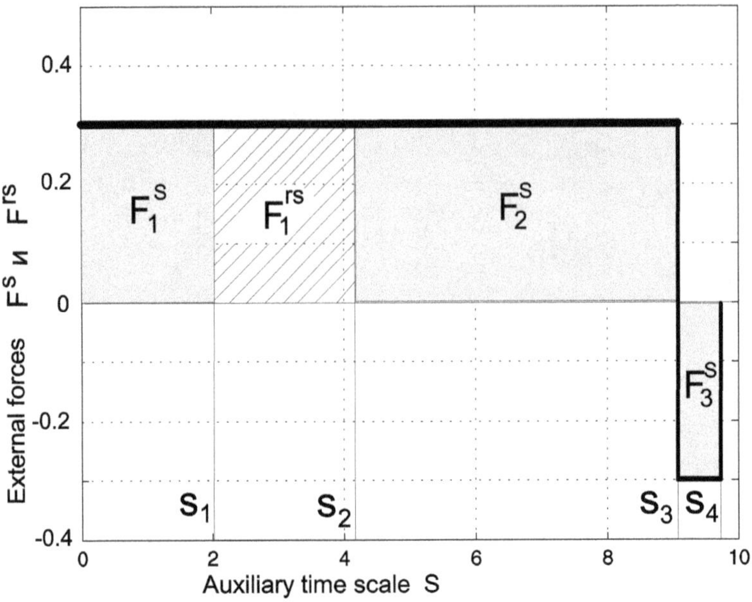

Fig. 8.42 Four-pulse control generated by F_1^{s}, F_1^{rs}, and F_2^{s}, F_3^{s}. The singular and intersingular phases are shaded and hatched, respectively

As we noted above, for finite values of μ the control is given by the external force $F_1^{\mathrm{s}} = \mu^{1/2} w_0$ applied to the racket when the ball penetrates it.

At the instant s_1, the ball-racket contact ends and during the interval $[s_1, s_2]$ they move separately: the ball (by inertia), with a constant velocity $y(s) = y(s_1) = \mathrm{const}$, in the constraint-free domain, and the racket, with a linearly increasing velocity $Y(s) = y(s_1) - q(s)$, subject to a constant external control with the same magnitude and direction, $h_2(s) = M^{-1} w_{21}(s) = h_0 = 0.3$. For finite μ this control is given by the same external force $F_1^{\mathrm{s}} = \mu^{1/2} w_0$ applied to the racket, but *outside the contact phase with the ball*, i.e., in the constraint-free domain.

Following Sect. 8.7.1, we denote this force by F_1^{rs}; the corresponding phase of motion is called *intersingular*. At the instant s_2, the racket collides again with the ball and during the interval $[s_2, s_4]$, the ball moves inside the racket: first, in the interval $[s_2, s_3]$, under the control $h_3(s) = M^{-1} w_{12}(s) = h_0 = 0.3$ associated with the force $F_2^{\mathrm{s}} = \mu^{1/2} w_0$, and then, after changing the control sign at the instant s_3, under the control $h_4(s) = M^{-1} w_{13}(s) = -h_0 = -0.3$, with the corresponding force $F_3^{\mathrm{s}} = -\mu^{1/2} w_0$. Finally, at the moment s_4, the ball stops. The control action $h_2 = M^{-1} w_{21}$ and the force F_1^{rs} are elements of the sequences $\{h_1, h_2, h_3, h_4\}$ and $\{F_1^{\mathrm{s}}, F_1^{\mathrm{rs}}, F_2^{\mathrm{s}}, F_3^{\mathrm{s}}\}$, respectively.

The sequence of force profiles is shown in Fig. 8.42 for $\mu = 9$. This sequence of force changes, which includes the contact and non-contact phases of motion over a total time interval that shrinks to zero as $\mu \to \infty$, is appropriately called a *stretched multi-pulse*.

Schematically, the interaction described is illustrated in Fig. 8.8. Here $\tau_1^\mu = \tau_{11}^\mu$ is the beginning of the first multi-pulse, and $\tau_1^{\mu *} = \tau_{12}^{\mu *}$ is the instant of its termination. Since $\tau_1^\mu \to \tau_1^{\mu *}$ as $\mu \to \infty$, the motion over the whole segment $[\tau_1^\mu, \tau_1^{\mu *}]$ (including the intersingular noncontact phase) will also be referred to as singular.

Now suppose a sensor device (such as a pressure sensor on the racket), is present, the readings of which are used by the racket controller to construct the racket control law in the damping problem [38]. It is natural to assume that the value of the initial relative velocity $q_v(0) < 0$ is not known to the controller, so it is also not known which of the cases (i) or (ii) actually takes place. For this reason, the first control action must have a negative sign, i.e., have the same direction as the initial velocity vector of the ball ($h_1(s) \equiv -h_0$). Further, during the first ball-racket impact interaction, the controller integrates (at the contact rate) the output signal of the sensor $\eta(s)$ up to the instant s^*, which is determined by the sensor. Using the Eqs. (8.182)–(8.184), one calculates the integral

$$J = \int_0^{s^*} \eta(s)\, ds = \int_0^{s^*} q_p(s)\, ds,$$

which allows us to determine the initial relative velocity $q_v(0)$ by an explicit, yet rather cumbersome formula. Using $q_v(0)$ and (8.184), the value of $q_v(s^*)$ is calculated as

$$q_v(s^*) = \frac{1}{e^{\lambda s^*}} \left[q_v(0) \cos \omega s^* - \left(\lambda q_v(0) - \zeta_0\right) \frac{\sin \omega s^*}{\omega} \right].$$

By definition, the absolute velocity of the ball is $y_v(s^*) = q_v(s^*) + Y_v(s^*)$, where the velocity $Y_v(s)$ of the racket itself is assumed to be known to the controller. If $y_v(s^*) \geq 0$, the control process ends. If $y_v(s^*) < 0$, the control process continues according to the algorithm described above, until the ball stops completely.

The above solution to the damping problem in the singular phase is illustrated in Figs. 8.40, 8.41, and 8.42 and allows us to draw the following conclusions:

(i) starting from the moment τ (respectively $s = 0$) of the first collision with the ball, the racket's controller forms a control based on the sensor output signal, consisting of a finite number of force pulses applied to the racket; the first pulse has the same direction as the initial velocity vector of the ball;
(ii) the resulting control action on the racket is *multi-pulse*;
(iii) in the general case, for a light racket the control multi-pulse implements an instantaneous (in real time) interaction of the racket with the ball, leading to its complete stop.

8.7.4 Limit System in Natural Coordinates

In natural coordinates, the limit trajectories of the ball-and-racket system are described by the equations

$$\begin{cases} \dot{\bar{x}}_p(t) = \bar{x}_v(t), \\ \dot{\bar{x}}_v(t) = -g - \Delta y_v(s^*)\,\delta(t-\tau), \\ \dot{\bar{X}}_p(t) = \bar{X}_v(t), \\ \dot{\bar{X}}_v(t) = \Delta Y_v(s^*)\,\delta(t-\tau), \end{cases} \tag{8.205}$$

where $\Delta_v(s^*)$ and $\Delta Y_v(s^*)$ are calculated by (8.196) and (8.197), respectively. Here $\delta(t)$ denotes, as above, the Dirac δ-function.

8.7.5 Applying Optimal Control in Natural Coordinates for a Finite μ

Let us consider the real situation with finite μ. Assume first that there is no gravity ($g = 0$). The space-time transformation (8.177) directly leads to the pre-limit system (8.178), which formally differs only by notations from the limit (as $\mu \to \infty$) system (8.180). This means that for any finite μ, one can perform the inverse space-time transformation and obtain an optimal solution to the original problem. The value of $\mu^{-1/2}$ plays the role of a scale factor in these transformations. In particular, the real time is $t = s\mu^{-1/2}$. The impulse of the external force acting on the racket in the limit and initial systems has the same value and the same direction. The same remark applies to the graphs of the absolute velocity of the ball, $x_v(t)$ and $y_v(s)$, in Figs. 8.37, 8.38, 8.39, 8.40, and 8.41, with the replacement of the auxiliary time s with the real time t according to the first line in (8.177).

The presence of gravity gives an additional impulse $gs_T\mu^{-1/2}$, where s_T is the calculated value of the terminal time of the ball-racket interaction in the auxiliary problem. This additional impulse will not stop the ball at the instant $s_T\mu^{-1/2}$, but the error in the deviation of the ball's velocity from zero at this instant will be infinitesimally small (about $\mu^{-1/2}$).

8.7.6 Damping with Constraints on the Control Resource

Here we introduce additional restrictions on the total mechanical impulse applied to the racket in the i-th singular phase, i.e., on the interval $[\tau_i^\mu, \tau_i^{\mu*}]$, namely

$$\int_{\tau_i^{\mu}}^{\tau_i^{\mu*}} |F(t,\mu)|\, dt \le C_i, \qquad \text{with } C_i \text{ given constants.} \tag{8.206}$$

Here τ_i^{μ} and $\tau_i^{\mu*}$ denote the beginning and the end of the i-th singular phase, respectively, and $F(t,\mu)$ is given by formula (8.174). In addition, we assume that in the regular phase, at some instants t_k, the racket can be acted on by the impulsive force $F^{\mathrm{r}}(t_k)\delta(t-t_k)$ with the bound

$$|F^{\mathrm{r}}(t_k)| \le F_0, \tag{8.207}$$

where F_0 is set. The time instants t_k, $k = 1, 2, \ldots$, of pulse actions satisfy the monotonicity condition $\tau = t_0 < t_1 < t_2 < \cdots$ and the duty-cycle constraints

$$T_k \le t_k - t_{k-1}, \quad T_k > 0. \tag{8.208}$$

The same constraints are imposed on the termination times $\tau_i^{\mu*}$ of the i-th singular phase:

$$T_i \le t_i - \tau_i^{\mu*}. \tag{8.209}$$

Due to the presence of a impulsive external force in the regular phase, the equations of motion (8.170) take on the form

$$\begin{cases} \dot{x}_p(t) = x_v(t), \\ \dot{x}_v(t) = 0, \\ \dot{X}_p(t) = X_v(t), \\ \dot{X}_v(t) = M^{-1} \sum\limits_{t_k \le t} F^{\mathrm{r}}(t_k)\delta(t-t_k). \end{cases}$$

Here, for simplicity, we do not introduce the force of gravity, i.e., we assume that the motion is one-dimensional along a horizontal line. The motion in the singular phase is subject to the previous Eq. (8.171) (for $g = 0$) with the previously adopted constraints (8.172)–(8.175)

This damping problem is solved according to the scheme presented above, from the instant τ of the first ball-racket collision, with the same space-time transformation of coordinates and transition (with complete preservation of notations) to the equations of infinitesimal dynamics in relative coordinates (8.182). In the present case the control constraints (8.175) and (8.206) in the singular phase have the form

$$|h(\cdot)| \le h_0 < \infty, \quad \text{where} \quad h_0 = M^{-1} w_0, \tag{8.210}$$

$$\int_0^{s^*} |h(s)|\, ds \le c, \quad \text{where} \quad c = M^{-1} C. \tag{8.211}$$

Here, in the notations of (8.206), $C = C_1$ for the first collision. As in the example considered above, the optimality criterion is the minimization of the absolute velocity of the ball at the instant s^* when the ball touches the surface of the racket (8.185).

Solution of the Problem

To solve the stated problem, we apply the Pontryagin maximum principle; the resulting solution depends on the initial conditions and the mass M of the racket. However, as we will see below, in the present case the additional constraints (8.208) and (8.211) modify (depending on the initial conditions) the structure of the optimal control, leading, in particular, to a sequence of multi-pulse interactions. We note that if the constant c in (8.211) is large enough, the optimal control law will obey a strict inequality in (8.211), which reduces the solution of this problem to the previous one. For this reason, we assume that the constant c satisfies the inequality

$$\int_0^{s^*} |h(s)|\, ds = c < h_0 s^*. \tag{8.212}$$

To account for (8.211), we introduce a new auxiliary variable $z(s)$ which satisfies the equation

$$\dot{z}(s) = |h(s)|, \quad z(0) = 0, \tag{8.213}$$

with the terminal condition

$$z(s^*) = c. \tag{8.214}$$

The Hamiltonian $\mathscr{H}$ for the problem with the dynamics (8.182) and (8.213), the constraints (8.210), (8.183), and (8.214), and the criterion (8.185) has the form

$$\begin{aligned} \mathscr{H} &= \mathscr{H}(q_p, q_v, y_v, z, \psi_p, \psi_v, \psi_y, \psi_z, h) = \\ &= \psi_p q_v - \psi_v h - \big(a\psi_v + \psi_y\big)\big(q_p + 2\kappa q_v\big) + \psi_z |h| \to \max, \end{aligned} \tag{8.215}$$

where the conjugate variables satisfy the equations

$$\begin{cases} \dot{\psi}_p(s) = a\psi_v(s) + \psi_y(s), \\ \dot{\psi}_v(s) = -\psi_p(s) + 2a\kappa\psi_v(s) + 2\kappa\psi_y(s), \\ \dot{\psi}_y(s) = 0, \qquad \text{i.e., } \psi_y = C_y = \text{const}, \\ \dot{\psi}_z(s) = 0, \qquad \text{i.e., } \psi_z = C_z = \text{const}. \end{cases} \tag{8.216}$$

The terminal transversality conditions at the instant $s = s^*$ read

$$2y_v\, \delta y_v + \psi_p\, \delta q_p + \psi_v\, \delta q_v + \psi_y\, \delta y_v + \psi_z\, \delta z - \mathscr{H}\, \delta s = 0, \qquad (8.217)$$

where $\delta q_p = 0$ and $\delta z = 0$ due to (8.183) and (8.214), respectively. This gives

$$2y_v(s^*) + C_y = 0, \quad \psi_v(s^*) = 0, \quad \mathscr{H}(s^*) = 0. \qquad (8.218)$$

Since $\mathscr{H}(s^*) = 0$, we have

$$\psi_p(s^*) - 2\kappa C_y = 0. \qquad (8.219)$$

Relations (8.218) and (8.219) set the terminal conditions, which allow us to integrate, in reverse time, the Eq. (8.216) for the conjugate variables $\psi_v(s)$, $\psi_p(s)$, and thus determine the optimal control. The detailed analysis of the solutions of the Eq. (8.216) is carried out in the same way as for the Eq. (8.188) in the previous example. The main difference is that in the previous example the optimal control h maximized the linear function $f(h) = -\psi_v h$, which gave $h(s) = -h_0 \operatorname{sign}\big(\psi_v(s)\big)$, whereas in the current example the optimal control h maximizes the piecewise-linear function

$$f(h) = kh - |h| \to \max, \qquad (8.220)$$

where $k = k(s) = \psi_v(s)/C_z$, with $C_z < 0$. It follows from (8.220) that

$$h(s) = \begin{cases} h_0, & \text{if } k(s) > 1, \\ 0, & \text{if } k(s) \in [-1,\ 1], \\ -h_0, & \text{if } k(s) < -1. \end{cases} \qquad (8.221)$$

Note that under certain initial conditions in the present example, a complete stop of the ball is also possible. This case gives an additional terminal condition

$$y_v(s^*) = 0. \qquad (8.222)$$

At the same time, (8.218) and (8.219) imply that $C_y = \psi_y(s^*) = 0$ and $\psi_p(s^*) = 0$, which leads to a trivial solution of the system (8.188), i.e., the problem degenerates. As in the previous example, in this case we need to set the time-optimal problem with the criterion

$$\int_0^{s^*} C_h\, ds \to \min, \quad C_h > 0. \qquad (8.223)$$

In the new time-optimal problem with the criterion (8.223) for a system with the dynamics (8.182) and the constraints (8.210), (8.183) and (8.222), the Hamiltonian, the maximum principle, the conjugate system, and the optimal control law retain the forms (8.186), (8.188), and (8.221), respectively. However, due to (8.222) and (8.223), the transversality conditions now read

$$\psi_v(s^*) = 0, \quad \mathscr{H}(s^*) = C_h. \tag{8.224}$$

As above, these equalities allow us to obtain boundary conditions for the integration of the system (8.188) in inverse time and to calculate the control switching function $k(s)$ in (8.221).

As already noted, for sufficiently small C_i the constraints (8.206) can lead to successive collisions of the ball with the racket in real time, regardless of the values of T_i in (8.208). For each successive collision, one solves the control problem in the singular phase, depending on the specific initial conditions for this collision. When the values of the quantities C_i increase to a certain critical threshold, the ball can be stopped by a single multi-pulse, as in the example above.

Let us give a numerical example of the damping problem with resource constraints (8.206) and duty cycle (8.208) with $C_i = C = 0.3$ and $T_i = T = 1$ for all i. The initial data, as in the example above, are assumed to be equal to (8.198). This implies, in particular, that $M = 1$ and $c = 0.3$ in (8.212). In this case, the maximum permissible external mechanical impulse applied to the racket is $p = w_0 s_1 = M h_0 s_1 = 0.3$, which corresponds to single-step control

$$h(s) = \begin{cases} h_0, & \text{if } s < s_1, \\ 0, & \text{if } s \geq s_1, \end{cases} \tag{8.225}$$

when $s_1 = 1$, which does not stop the ball, as shown in Fig. 8.43. In this case, the racket moves away from the ball with a relative velocity of $q_v(s^*) = 0.10414$. Here the terminal instant is $s^* = s_2 = 2.823$, and the values of the terminal velocity are $y_v(s^*) = -0.089753$ for the ball, and $Y_v(s^*) = -0.1939$ for the racket. Note that the step h_0, $s < s_1$ of the optimal control (8.225) is located at the beginning of the control interval $[0, s_2]$. Such a location maximizes the work of the external control force acting on the ball-racket system, and therefore is capable of minimizing the absolute velocity of the ball. For a finite μ, all the comments in Sect. 8.7.5 remain valid.

Successive Ball-Racket Collisions

Let in (8.207) $F_0 = 0.2$. According to (8.208), in real time $T_k = T = 1$ the racket is subject to the impulse force $F_0 = 0.2$, which will results in an increment $\Delta Y_v = 0.2$ of the velocity of the racket. As a result, the new speed of the racket will have a positive direction and the value $Y_v = 0.0061$. The relative speed of the ball and the racket will become equal $q_v = y_v - Y_v = -0.09585$. After the second collision with the racket, the ball can be stopped by a two-pulse control of the form (8.221)

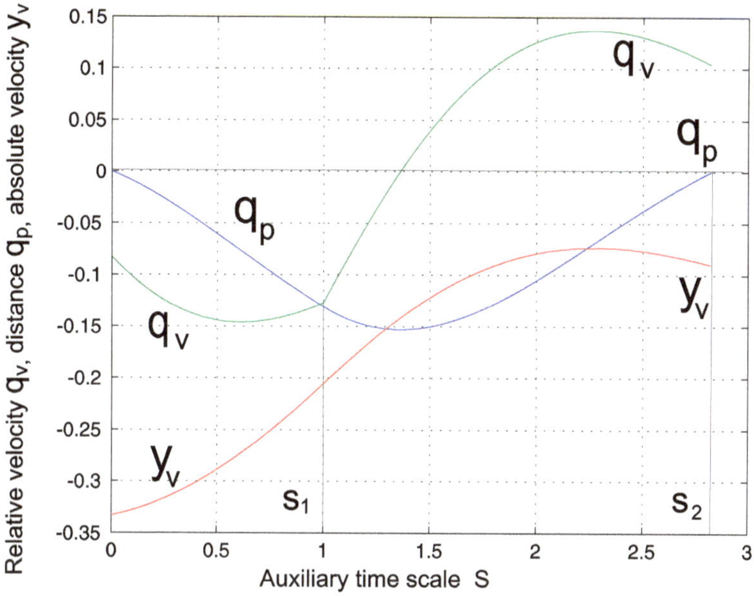

Fig. 8.43 Evolution of the coordinates q_p, q_v, and y_v in the forbidden domain under the single-step control h given by (8.225)

with pulses $p_1 = w_0 s_1$ and $p_2 = -w(s_3 - s_2)$, where $s_1 = 0.34, \quad s_2 = 0.92$, $s_3 = 1.54$ (Fig. 8.44).

As the graph of the switching function $k(s)$ shows, the optimal control $h(s)$ experiences jumps at the times s_1, s_2, s_3. The calculated stopping instant is $s^* = s_4 = 1.664$. The evolution of the phase variables under the control h given by (8.221) is shown in Fig. 8.45.

Now, as in the previous example, consider the case of a light racket ($M = 1/3$, i.e. $a = 1 + 1/M = 4$) that moves towards the ball with a higher relative initial velocity $q_v(0) = -7.5$. The other initial parameters have the following values:

$$F_0 = 1.14, \quad \kappa = 0.25, \quad h_0 = 0.3, \quad q_p(0) = 0, \quad y_v(0) = -4.1322,$$

$$Y_v(0) = 3.3678, \quad C = 0.9725, \quad \text{i.e., } c = C/M = 2.9175. \tag{8.226}$$

As we can see, here the value of the constant C, which limits the permissible impulse action on the racket in the singular phase, is almost three times greater than in the previous case.

The Figs. 8.46, 8.47, and 8.48 illustrate this case. In Fig. 8.46 on the segment $[0, s_1]$, the external influence is equal to $h(s) = h_0 = 0.3$. At this interval, the ball moves inside the racket. At the instant $s_1 = 1.892$, the ball and the racket are separated and during the segment $[s_1, s_2]$ move separately from each other: the ball is at a constant speed $y_v(s) = y_v(s_1) = const$ in a region free from constraints, and

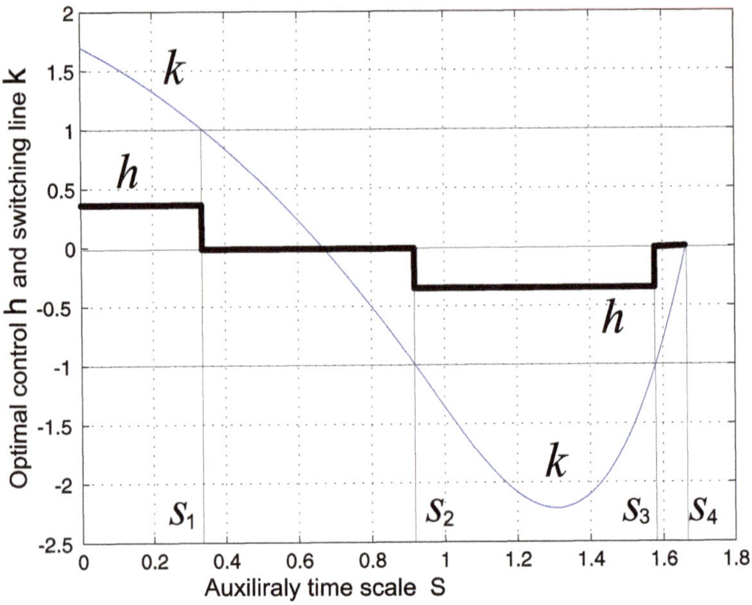

Fig. 8.44 Switching function k and two-pulse control h from (8.221), stopping the ball

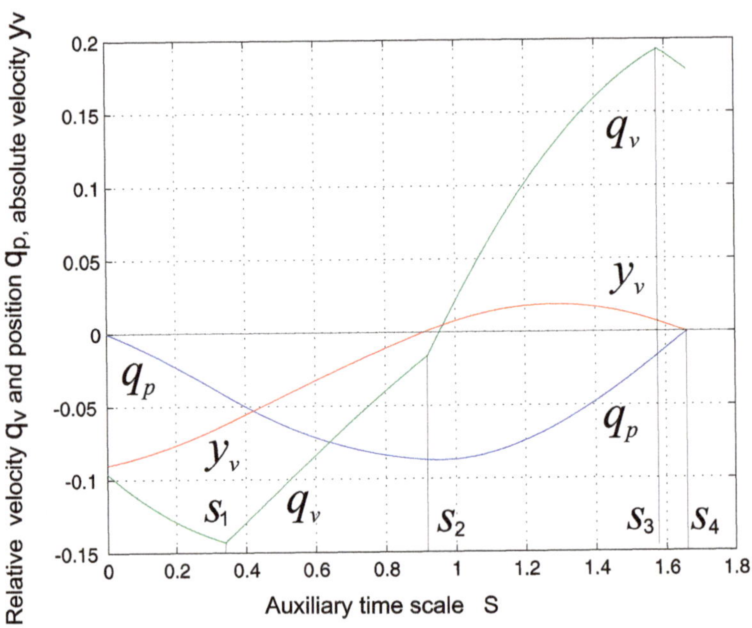

Fig. 8.45 Evolution of q_p, q_v and y_v under the two-pulse control (8.221)

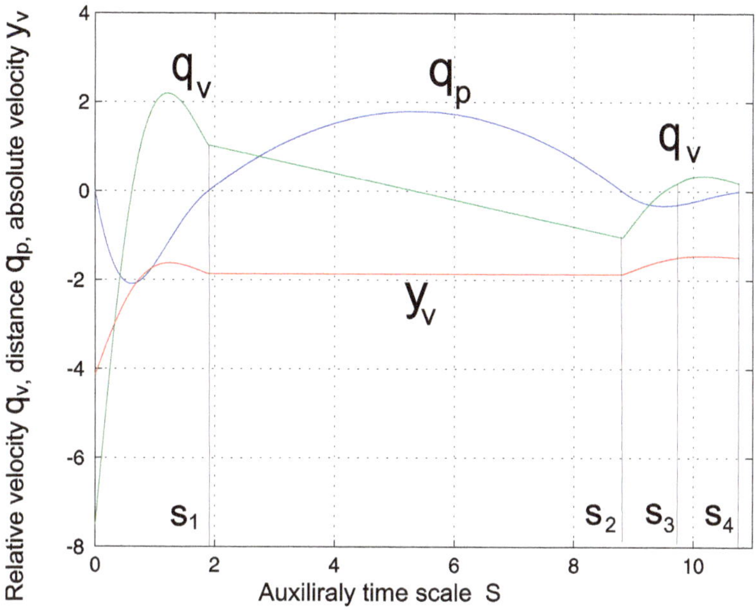

Fig. 8.46 Evolution of q_p, q_v and y_v in the first singular phase with constant control $h = h_0 = 0.3$

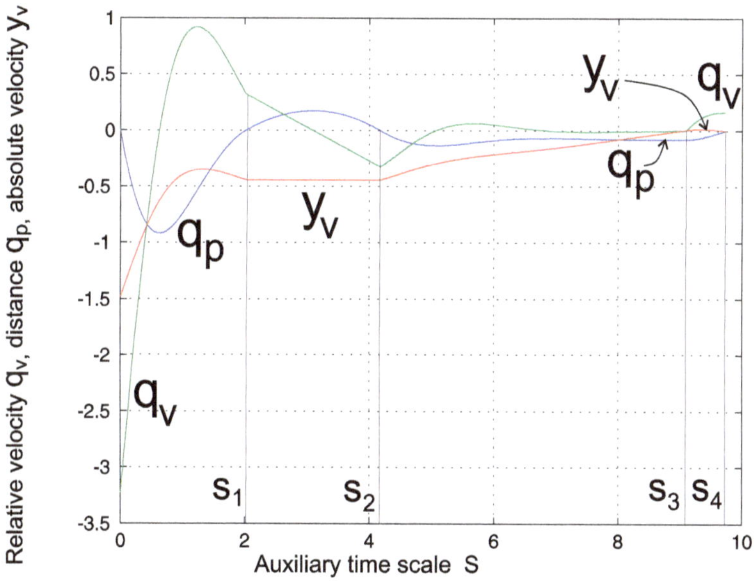

Fig. 8.47 Evolution of q_p, q_v and y_v in the second singular phase under two-stage control of the $h = \pm h_0 = \pm 0.3$ type (8.225)

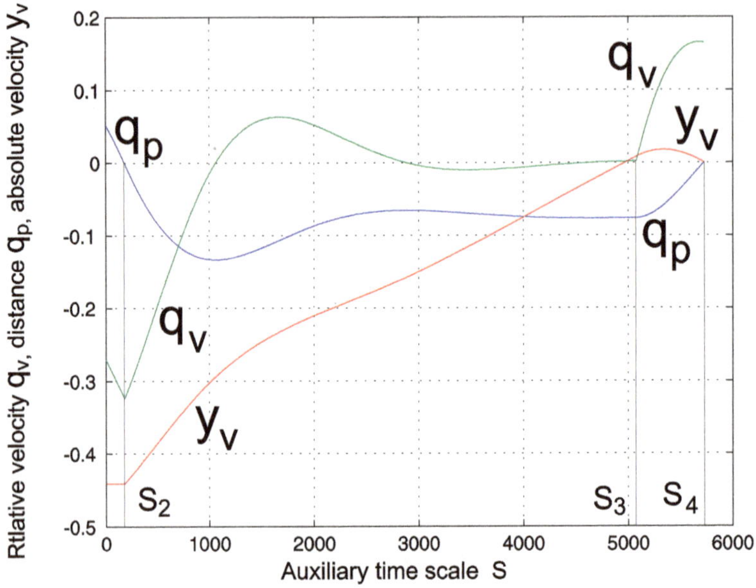

Fig. 8.48 Fragment of Fig. 8.47 between the instants s_2 and s_4

the racket—with a linearly increasing speed $Y_v(s) = y_v(s_1) - q_v(s)$ under constant external influence $h(s) = h_0 = 0.3$. As noted above, this phase of motion is called intersingular, and the external force acting on the rocket in this phase is denoted by F^{rs}. At the instant $s_2 = 8.816$ the racket collides with the ball for the second time and during the segment $[s_2, s_4]$ the ball moves inside the racket. However, at the instant $s_3 = 9.725$, the optimal control becomes zero due to the use of the resource specified by the constant C. Finally, at the instant $s_4 = s_1^* = 10.77$, the ball and the racket fly apart at a non-zero speed of the ball. The terminal values of the velocities at this instant have the following values.

$$Y_v(s_1^*) = -1.6684, \quad y_v(s_1^*) = -1.4809, \quad q_v(s_1^*) = 0.18765.$$

After the time $T_k = T = 1$, according to (8.208), a pulse action $F_0\,\delta(t - T)$ with $F_0 = 1.14$ will be applied to the racket, which will cause an increment $\Delta Y_v = 3.4177$ in the racket's velocity. As a result, the new velocity of the racket will become positive and equal to $Y_v = 1.7493$. In this case, the relative velocity will have the value $q_v = y_v - Y_v = -3.2302$. With such variable values, the ball can be stopped after another collision with the racket by the two-pulse control

$$p_1 = w_0 s_3 = 0.90766 \quad \text{and} \quad p_2 = -w_0(s_4 - s_3) = -0.06484,$$

where $s_3 = 9.0766$, $s_4 = s_2^* = 9.725$, $w_0 = M h_0 = 0.1$.

Figures 8.47 and 8.48 illustrate this case. The graphs of the variables q_p, y_m, and h_m in these figures are similar to the graphs in Fig. 8.46, where $s_1 = 2.021$, $s_2 = 4.178$, and the values s_3, s_4 are given above. Therefore, the description of the motion in this case is similar to the description of the motion after the first collision.

Appendix A
Differential Equations with Measures

A.1 Auxiliary Results

A.1.1 Functions of Bounded Variation

We begin with a description of some properties of functions of bounded variation. We do not intend to give a complete exposition of this very important area of real analysis. Rather, our aim is to present some results needed to make the present book more or less self-contained. For more details and proofs we refer to [204, 322, 388, 409].

Definition A.1 Let $f(t) = \big(f^1(t), \dots, f^m(t)\big)$ be a vector-function on the interval $[t_0, T]$ with values in $\mathbb{R}^m$. We say that $f(\cdot)$ is a *function of bounded variation* if there exists a constant C such that, for any partition $D_n = \{t_0, t_1, \dots, t_n\}$ of $[t_0, T]$ with $t_n = T$, one has

$$\sum_{i=1}^{n} \|f(t_i) - f(t_{i-1})\| = \sum_{i=1}^{n} \|\Delta f_i\| \le C, \tag{A.1}$$

where $\|\cdot\|$ denotes any norm on $\mathbb{R}^m$.

Remark A.1 Clearly, any monotone scalar function f is of bounded variation because for such f the sum in (A.1) is always equal to $\|f(T) - f(t_0)\|$. It is also clear that any vector-valued function with monotone components is a function of bounded variation. Moreover, any component of a vector-valued function of bounded variation is a scalar function of bounded variation.

Remark A.2 Since all norms on a finite-dimensional linear space are equivalent, in the sense that for any two norms $\|\cdot\|_1$ and $\|\cdot\|_2$ there are constants c_1 and c_2 such that

B. Miller, E. Rubinovich, *Optimization of Dynamical Systems with Impulse Controls and Shocks*, Systems & Control: Foundations & Applications,
https://doi.org/10.1007/978-3-031-64124-4

$$c_1\|x\|_1 \le \|x\|_2 \le c_2\|x\|_1 \quad \text{for all } x \in \mathbb{R}^m,$$

the choice of norm in (A.1) is not relevant: if the sum in (A.1) is bounded in one norm, then it will be bounded in every norm. Therefore, the property of being of bounded variation is *norm independent*.

Definition A.2 Let $f(t) = \big(f^1(t), \ldots, f^m(t)\big)$ be a function of bounded variation. The number

$$\operatorname*{Var}_{[t_0,T]} f(t) = \sup \sum_{i=1}^{n} \|\Delta f_i\|,$$

where the supremum is taken over all partitions D_n of the segment $[t_0, T]$, is called the *total variation* of $f(\cdot)$ on the segment $[t_0, T]$.

If the function $f(\cdot)$ is defined on the real line and the values $\operatorname*{Var}_{[t_0,T]} f(t)$ are uniformly bounded for any t_0 and T, then the limit

$$\lim_{\substack{T\to\infty \\ t_0\to-\infty}} \operatorname*{Var}_{[t_0,T]} f(t)$$

is called the *total variation* of the function $f(\cdot)$ on $\mathbb{R} = (-\infty, \infty)$.

Remark A.3 Of course, the actual value of the variation depends on the norm used. Usually we will consider the case of some fixed norm, so it not necessary to specify the choice of norm.

Let us list a number of obvious properties of functions of bounded variation.

1. If α is a constant, then

$$\operatorname*{Var}_{[t_0,T]} \alpha f(t) = |\alpha| \operatorname*{Var}_{[t_0,T]} f(t).$$

   This follows directly from the definition of the variation $\operatorname*{Var}_{[t_0,T]} f(t)$.
2. If f and g are functions of bounded variation, then so is their sum $f + g$, and

$$\operatorname*{Var}_{[t_0,T]} (f(t) + g(t)) \le \operatorname*{Var}_{[t_0,T]} f(t) + \operatorname*{Var}_{[t_0,T]} g(t).$$

   This follows from the properties of the norm. Consequently, the set of functions of bounded variation constitutes a linear space.
3. If $t_0 < t_1 < T$, then

$$\operatorname*{Var}_{[t_0,t_1)} f(t) + \operatorname*{Var}_{[t_1,T]} f(t) = \operatorname*{Var}_{[t_0,T]} f(t),$$

   or

$$\operatorname*{Var}_{[t_0,t_1]} f(t) + \operatorname*{Var}_{(t_1,T]} f(t) = \operatorname*{Var}_{[t_0,T]} f(t),$$

where the partition used to calculate the variation over a non-closed half-interval does not include the corresponding endpoints.

This implies that

4. The function $v(t) = \operatorname*{Var}_{[t_0,t]} f(s)$ is non-decreasing.
5. If $f(\cdot)$ is right-continuous (resp., left-continuous, continuous) at the point t^*, then $v(\cdot)$ is also right-continuous (resp., left-continuous, continuous) at t^*.

The following theorems complete this list of properties [322, 409].

Theorem A.1 *Any scalar function of bounded variation can be written as the difference of two non-decreasing functions. In other words, there exists a unique decomposition $f(t) = f(t_0)+f^+(t)-f^-(t)$, where f^+, $f^- \geq 0$ are monotonically non-decreasing functions such that*

$$\operatorname*{Var}_{[t_0,T]} f(t) = |f(t_0)| + f^+(t) + f^-(t).$$

This remarkable result admits the following consequence.

Theorem A.2 *Any function of bounded variation admits left and right limits at any point, i.e., for any $t \in (t_0, T)$ there exist the limits*

$$\lim_{\tau \uparrow t} f(\tau), \quad \lim_{\tau \downarrow t} f(\tau)$$

and

$$\lim_{\tau \uparrow T} f(\tau), \quad \lim_{\tau \downarrow t_0} f(\tau).$$

Theorem A.3 *The set of points of discontinuity of any function of bounded variation is at most countable.*

Theorem A.4 *Any function of bounded variation has a finite derivative almost everywhere with respect to the Lebesgue measure.*

And in conclusion we state a well-know result, *Helly's Second Theorem*, concerning the properties of functions of bounded variation.

Theorem A.5 (Helly's Second Theorem) *Let $\mathscr{F} = \{\Phi_\alpha(t)\}$ be an infinite family of functions on the segment $[t_0, T]$. If all the functions from $\mathscr{F}$ and their total variations are bounded by some constant $C < \infty$, i.e.,*

$$|\Phi_\alpha(t)| \leq C \quad \text{and} \quad \operatorname*{Var}_{[t_0,T]} \Phi_\alpha(t) \leq C, \quad \textit{for all } \alpha, \tag{A.2}$$

then one can find a sequence of functions $\Phi_{\alpha_n}(t) \in \mathscr{F}$ that converges at each point $t \in [t_0, T]$ to a bounded function $\Phi(t)$ of bounded variation, i.e.,

$$|\Phi(t)| \le C \quad \text{and} \quad \operatorname*{Var}_{[t_0,T]} \Phi(t) \le C.$$

Remark A.4 This property means that any set of functions satisfying (A.2) is a precompact subset in the topology of pointwise convergence on the space of functions of bounded variation [204]. Moreover, any precompact subset $\mathscr{A}$ in this topology admits the same characterization: if a set $\mathscr{A}$ of functions of bounded variation is precompact, then there exists a constant $C < \infty$ such that (A.2) holds for all functions $\Phi(\cdot) \in \mathscr{A}$.

The next result proves useful for the approximation of functions of bounded variation by a sequence of absolutely continuous functions, as well as of purely discontinuous functions.

Theorem A.6 *For any function of bounded variation $U(t)$, defined on $[0, T]$, there exist:*

1. *a sequence of absolutely continuous functions $U^n(t)$ such that*

$$\operatorname*{Var}_{[0,T]} U^n(t) \le \operatorname*{Var}_{[0,T]} U(t) \quad \text{and} \quad \lim_n U^n(t) = U(t)$$

   at all points of continuity;
2. *a sequence of piecewise-constant functions $U^n(t)$ such that*

$$\operatorname*{Var}_{[0,T]} U^n(t) \le \operatorname*{Var}_{[0,T]} U(t) \quad \text{and} \quad \lim_n U^n(t) = U(t)$$

   at all points of continuity.

We provide here proof of the first part of the theorem, since the second part can be established by the same arguments.

Proof Take some countable everywhere dense set of points $\mathscr{D}$ which does not contain the points of discontinuity of U. For each $n = 1, 2, \dots$ consider finite subsets $\mathscr{D}_n = \{t_k^n,\ k = 0, \dots, n+1,\ t_0^n = 0,\ t_{n+1}^n = T\}$ of ordered points $\{t_k^n\}$ such that $\mathscr{D}_n \subset \mathscr{D}_{n+1} \subset \mathscr{D}$ and $\bigcup_n \mathscr{D}_n = \mathscr{D}$. Now define a piecewise linear function $U^n(t)$ by

$$U^n(0) = U(0), \quad U^n(T) = U(T),$$

$$U^n(t_k^n) = U(t_k^n), \quad \text{for all} \quad t_k^n \in \mathscr{D}_n,$$

and linear between the points t_k^n. Then

$$\operatorname*{Var}_{[0,T]} U^n(t) = \sum_{k=1}^{n+1} \|U(t_k^n) - U(t_{k-1}^n)\| \leq \operatorname*{Var}_{[0,T]} U(t).$$

Each function $U^n(t)$ is absolutely continuous, so, by Helly's theorem, there exists a subsequence $\{U^{n_l}(t)\}$ that converges to some function $\widetilde{U}(t)$ of bounded variation. This function satisfies the inequality

$$\operatorname*{Var}_{[0,T]} \widetilde{U}(t) \leq \operatorname*{Var}_{[0,T]} U(t)$$

and, by construction, $\widetilde{U}(t) = U(t)$ for all $t \in \mathscr{D}$. However, the coincidence of these functions on the everywhere dense set $\mathscr{D}$ implies their coincidence at all points of continuity as well. □

A.1.2 Measures and Functions of Bounded Variation

The space of the right-continuous functions of bounded variation plays a key role in measure theory [204]. Consider the interval $[t_0, T]$ as a measurable set endowed with the σ-algebra $\mathscr{B}\{[t_0, T]\}$ of Borel subsets [204]. Let $\mu(A)$ be a σ-additive vector-valued measure defined on the σ-algebra $\mathscr{B}\{[t_0, T]\}$.

The measure $\mu(\cdot)$ has the following properties [204, 322, 409]:

1. There exists a constant $C < \infty$ such that $\|\mu(A)\| \leq C$ for all $A \in \mathscr{B}\{[t_0, T]\}$.
2. For any sequence of subsets $A_i \in \mathscr{B}\{[t_0, T]\}$ such that $A_i \cap A_j = \emptyset$ if $i \neq j$,

$$\mu\left(\sum_{i=1}^{\infty} A_i\right) = \sum_{i=1}^{\infty} \mu(A_i)$$

   i.e., $\mu(\cdot)$ is *σ-additive.*
3. There exists a constant $C' < \infty$ such that for any collection of subsets $\{A_i, i = 1, \ldots, N < \infty\}$, $A_i \in \mathscr{B}\{[t_0, T]\}$ such that $A_i \cap A_j = \emptyset$ if $i \neq j$,

$$\sum_{i=1}^{N} \|\mu(A_i)\| \leq C'.$$

4. The properties listed above imply that for any set $A \in \mathscr{B}\{[t_0, T]\}$ there exists

$$\sup_{\{A_i\}} \sum_{i=1}^{N} \|\mu(A_i)\| = \|\mu(A)\|,$$

where the supremum is taken over all (finite) collections of subsets $\{A_i \subseteq A, i = 1, \dots, N < \infty\}$ such that $A_i \cap A_j = \emptyset$ if $i \neq j$.

Definition A.3 The number $\|\mu(A)\|$ is called the *total variation* of the vector-valued measure μ on the set A..

5. If $A \subseteq B$, then $\|\mu(A)\| \leq \|\mu(B)\|$.
6. $\|\mu(A)\|$ is a non-negative scalar measure defined on $\mathscr{B}\{[t_0, T]\}$.
7. **Definition A.4** *A measure $\widetilde{\mu}$ is called an extension of the measure μ to a σ-algebra $\widetilde{\mathscr{B}}\{[t_0, T]\}$ such that $\mathscr{B}\{[t_0, T]\} \subset \widetilde{\mathscr{B}}\{[t_0, T]\}$, if $\widetilde{\mu}(A) = \mu(A)$ for every $A \in \mathscr{B}\{[t_0, T]\}$.*
8. **Definition A.5** A subset $A \subset [t_0, T]$ is said to be $\|\mu\|$*-measurable* if for any $\varepsilon > 0$ there exist two sets $B_1, B_2 \in \mathscr{B}\{[t_0, T]\}$ such that $B_1 \subseteq A \subseteq B_2$ and $\|\mu(B_2 \setminus B_1)\| \leq \varepsilon$.
9. The family $\mathscr{L}_{\|\mu\|}$ of all $\|\mu\|$-measurable subsets is a σ-algebra and the measure $\|\mu\|$ can be uniquely extended to $\mathscr{L}_{\|\mu\|}$.

Remark A.5 In what follows all measures are assumed to be extended and defined on the corresponding σ-algebra, completed by all the sets of measure zero.

Definition A.6 A function $f(\cdot)$ defined on $[t_0, T]$ is said to be:

1. *Borel measurable* if for any Borel subset $A \in \mathscr{B}(\mathbb{R}^1)$, the set $\{t : f(t) \in A\}$ belongs to $\mathscr{B}([t_0, T])$;
2. *V-measurable,* where V is a non-negative measure, if for any Borel subset $A \in \mathscr{B}(\mathbb{R}^1)$ the set $\{t : f(t) \in A\}$ belongs to $\mathscr{L}_V([t_0, T])$;
3. *Lebesgue measurable,* if it is V-measurable when $V(dt) = dt$ is the Lebesgue measure.

Remark A.6 In the above definition, the words "for any Borel subset" can be replace by "for any open or closed interval," as well as by "for any half-interval."

The following set of properties of measures concerns their classification and is necessary for the formulation of the fundamental Radon-Nikodym theorem.

Definition A.7 Let $\nu \geq 0$ be a σ-additive measure on the σ-algebra $\mathscr{L}_\nu$ of ν-measurable subsets of the segment $[t_0, T]$, and let μ be a vector-valued measure defined on $\mathscr{L}_\nu$. Then μ is said to be *concentrated* on a set $A \in \mathscr{L}_\nu$ if $\|\mu(E)\| = 0$ for every μ-measurable set $E \subseteq [t_0, T] \setminus A$. In this case the set A is called the *support* of the measure μ.

Definition A.8 Let ν, $\mathscr{L}_\nu$, $[t_0, T]$ and μ be as in Definition A.7. Then μ is said to be (with respect to the measure ν)

1. *Continuous*, if $\mu(\{\tau\}) = 0$ for every $\tau \in [t_0, T]$ such that the singleton $\{\tau\}$ is of ν-measure zero, i.e., $\nu(\{\tau\}) = 0$;
2. *Singular*, if μ is concentrated on a set of ν-measure zero;
3. *Discrete*, if μ is concentrated on a finite or countable set of points $D_\mu^{\mathrm{d}} = \{\tau \in [t_0, T] : \mu(\{\tau\}) \neq 0\}$;

4. *Absolutely continuous*, if $\nu(A) = 0$ implies $\|\mu(A)\| = 0$.

Theorem A.7 (Radon-Nikodym Theorem) *Let $\nu \geq 0$ be a σ-additive measure defined on the σ-algebra $\mathscr{L}_\nu$ of subsets of $[t_0, T]$, and let μ be a vector-valued measure defined on $\mathscr{L}_\nu$. Suppose μ is absolutely continuous with respect to ν.*

Then there exists a vector-function $\phi(t)$, $t \in [t_0, T]$, such that

1. $\displaystyle\int_{t_0}^{T} \|\phi(t)\| d\nu(t) < \infty$;
2. $\mu(A) = \displaystyle\int_A \phi(t) d\nu(t)$, *for any* $A \in \mathscr{L}_\nu$.

The function ϕ is unique up to its values on a set of ν-measure zero.

Definition A.9 The function ϕ is called the *Radon-Nikodym derivative* (of the measure μ with respect to the measure ν) and is denoted by $\phi(t) = \dfrac{d\mu}{d\nu}(t)$.

Theorem A.8 *Let λ be the Lebesgue measure on $[t_0, T]$. Suppose that the measure μ is absolutely continuous with respect to λ. Then the distribution function $\mu(t) = \mu([t_0, t])$ of μ admits a derivative $\dfrac{d\mu(t)}{dt}$ almost everywhere with respect to λ (λ-a.e.) and*

$$\frac{d\mu(t)}{dt} = \frac{d\mu}{d\lambda}(t), \quad \lambda\text{–a.e.}$$

Theorem A.9 *Let μ and ν be measures. Then the measure μ admits the following Lebesgue decomposition with respect to the measure ν :*

$$\mu = \mu^{\mathrm{a}} + \mu^{\mathrm{s}} + \mu^{\mathrm{d}},$$

where (with respect to ν)

- μ^{a} *is absolutely continuous;*
- μ^{s} *is continuous and singular;*
- μ^{d} *is discrete.*

The correspondence between functions of bounded variation and vector-valued measures on $[t_0, T]$ is specified by the following result.

Theorem A.10

1. *The distribution function $\mu(t) = \mu([t_0, t])$ of any vector-valued measure μ is a right-continuous function of bounded variation.*
2. *Any right-continuous function of bounded variation $f(\cdot)$ on $[t_0, T]$ generates exactly one vector-valued measure μ defined on half-intervals $(a, b] \subset [t_0, T]$ by the formula*

$$\mu((a, b]) = f(b) - f(a).$$

3. *For any interval* $[a, b] \subset [t_0, T]$,

$$\operatorname*{Var}_{[a,b]} \mu(t) \leq \|\mu([a, b])\|.$$

This correspondence provides a classification of measures and functions.

Definition A.10 Let g be a non-decreasing right-continuous function. A right-continuous function f defined on the interval $[t_0, T]$ is said to be *absolutely continuous* with respect to g, if for any $\varepsilon > 0$ there is a $\delta > 0$ such that

$$\sum_{k=1}^{n} \|f(b_k) - f(a_k)\| < \varepsilon$$

for every finite partition of the interval $[t_0, T]$ into pairwise-disjoint half-intervals $(a_k, b_k] \subset [t_0, T]$, $k = 1, \ldots, n$ satisfying

$$\sum_{k=1}^{n} \big(g(b_k) - g(a_k)\big) < \delta.$$

Remark A.7 If $g(t) = t - c$, we say that f is *absolutely continuous with respect to the Lebesgue measure,* simply *absolutely continuous.*

Let μ and ν be the measures generated by f and g, respectively. Then the absolute continuity of f with respect to g implies the absolute continuity of μ with respect to ν. The converse assertion is true if f and g are the distribution functions of μ and ν, respectively.

Definition A.11 A right-continuous function f of bounded variation defined on the interval $[t_0, T]$ is said to be *discrete* if there exist an at most countable set of points $D = \{\tau_i,\ i = 1, 2, \ldots\}$ and a corresponding sequence of real numbers $\{\alpha_i,\ i = 1, 2, \ldots\}$ such that

1. $\sum_{i=1}^{\infty} \|\alpha_i\| < \infty$;
2. $f(t) = \sum_{\{\tau_i \in D:\, \tau_i \leq t\}} \alpha_i.$

Thus, for a discrete function f we have

$$f(t-) = \sum_{\{\tau_i \in D:\, \tau_i < t\}} \alpha_i, \quad \Delta f(\tau_i) = f(\tau_i) - f(\tau_i-) = \alpha_i.$$

The corresponding discrete measure μ is defined by the formula

$$\mu(A) = \sum_{\{\tau_i \in A \cap D\}} \alpha_i$$

and is concentrated on the set D.

Finally, we give the definition of a function that is singular with respect to the Lebesgue measure (the general case is more complicated).

Definition A.12 A function f of bounded variation defined on the interval $[t_0, T]$ is said to be *singular* (with respect to the Lebesgue measure) if it is continuous and its derivative vanishes almost everywhere.

Remark A.8 A famous example of singular function is the *Cantor function* (see [204, 322]).

Remark A.9 In the general case, singularity is defined with respect to a concrete measure.

Definition A.13 A right-continuous function f of bounded variation defined on the interval $[t_0, T]$ is said to be *singular* with respect to a non-decreasing right-continuous function g, if the measure μ generated by f is singular with respect to the measure ν generated by g.

Any right-continuous function of bounded variation admits a Lebesgue decomposition analogous to the Lebesgue decomposition of the corresponding measure (see Theorem A.9).

Theorem A.11 *Let g be a non-decreasing right-continuous function. Any right-continuous function of bounded variation f defined on the interval $[t_0, T]$ admits a unique representation as*

$$f(t) = f^{\mathrm{a}}(t) + f^{\mathrm{d}}(t) + f^{\mathrm{s}}(t),$$

where (with respect to g)

- f^{a} *is absolutely continuous;*
- f^{d} *is discrete;*
- f^{s} *is continuous and singular.*

A.1.3 The Lebesgue-Stieltjes Integral

Definition A.14 Let $\mu(dt) \geq 0$ be a non-negative measure defined on the σ-algebra of Borel subsets of $\mathbb{R}^1$. The measure $\mu(dt)$ is generated by some *distribution function* $P(t)$ (i.e., $P(t)$ is a non-decreasing and right-continuous function and such that $f(-\infty) = 0$, $f(\infty) < \infty$) by the rule

$$\mu\{(a, b]\} = P(b) - P(a).$$

The *Lebesgue-Stieltjes integral* of the function f (with respect to P), denoted by

$$\int_a^b f(t)dP(t),$$

is defined as the Lebesgue integral

$$\int_a^b f(t)\mu(dt).$$

Let us introduce another important concept.

A *generalized measure* is, by definition, a measure that can be written in the form

$$\mu(dt) = \mu^+(dt) - \mu^-(dt),$$

where $\mu^+(dt)$ and $\mu^-(dt)$ are regular, scalar and non-negative measures. Accordingly, the distribution function of $\mu(dt)$ is the difference of two right-continuous and non-decreasing functions.

Definition A.15 *The Lebesgue-Stieltjes integral with respect to a generalized measure μ is defined as*

$$\int_a^b f(t)\mu(dt) = \int_a^b f(t)\mu^+(dt) - \int_a^b f(t)\mu^-(dt).$$

A vector-valued measure *has a vector-valued distribution function, the components of which are right-continuous functions of bounded variation. The Lebesgue-Stieltjes integral of a vector-valued function f with respect to a scalar measure is defined as the vector whose components are the Lebesgue-Stieltjes integrals of the respective components of the function.*

Remark A.10 The Lebesgue-Stieltjes integral over some measurable subset A is given by

$$\int_A f(t)dP(t) = \int_{-\infty}^{\infty} \mathbf{I}\{A\}(t)f(t)\mu(dt),$$

where $\mathbf{I}\{A\}$ stands for the *indicator function of the set* A, i.e.,

$$\mathbf{I}\{A\}(t) = \begin{cases} 1, & \text{if } t \in A \\ 0, & \text{if } t \notin A. \end{cases}$$

For example,

$$\int_{(a,b]} f(t)dP(t) = \int_{-\infty}^{\infty} \mathbf{I}\{(a,b]\}f(t)\mu(dt)$$

and

$$\int_{[a,b]} f(t)dP(t) = \int_{(a,b]} f(t)\mu(dt) + f(a)\mu\{a\} =$$
$$= \int_{(a,b]} f(t)dP(t) + f(a)\Delta P(a).$$

Thus, the Lebesgue-Stieltjes integral provides a general representation of any right-continuous function of bounded variation; this is a straightforward consequence of the decomposition theorem (see Theorem A.11).

Let P be a right-continuous function P of bounded variation. Then

$$P(t) = P(t_0) + \int_{t_0}^{t} \frac{dP(u)}{du}du + \sum_{t_0 \le \tau \le t} \Delta P(\tau) + P^{s}(t),$$

where the function $P^{s}(t)$ is continuous and singular with respect to the Lebesgue measure.

Consequently, the Lebesgue-Stieltjes integral of a function f (if it exists) admits the representation

$$\int_{t_0}^{T} f(u)dP(u) = \int_{t_0}^{T} f(u)\frac{dP(t)}{du}du + \sum_{t_0 \le \tau \le T} f(\tau)\Delta P(\tau) + \int_{t_0}^{T} f(u)dP^{s}(t).$$

The so-called Helly's first theorem establishes conditions under which passing to the limit in the Lebesgue-Stieltjes integral is allowed.

Theorem A.12 (First Helly Theorem) *Let $\{\Phi_n(t)\}$ be a sequence of functions of bounded variation on the segment $[t_0, T]$ that converges at each point $t \in [t_0, T]$ to a finite function $\Phi(t)$. If*

$$\operatorname*{Var}_{[t_0,T]} \Phi_n(t) \le C < \infty$$

for all $n = 1, 2, \ldots$, then $\Phi(t)$ has bounded variation, and for any continuous function $f(t)$,

$$\lim_{n} \int_{t_0}^{T} f(t)\, d\Phi_n(t) = \int_{t_0}^{T} f(t)\, d\Phi(t).$$

Remark A.11 In this theorem convergence everywhere can be replaced by convergence at the points of continuity.

We let $BV_m^+\{[t_0, T], X_0\}$ denote the space of right-continuous m-dimensional vector-functions of bounded variation defined on the interval $(-\infty, T]$, with $X(t) = X_0 = \text{const}$ for $t \in (-\infty, t_0)$, i.e., $X(t_0-) = X_0$. In the scalar case we denote this space by $BV^+\{[t_0, T], X_0\}$.

The metric properties of the space $BV_m^+\{[t_0, T], X_0\}$ are determined by the following lemma.

Lemma A.1 $BV_m^+\{[t_0, T], X_0\}$ *is a complete metric space with respect to the distance function (metric)*

$$\rho_T(f, g) = \operatorname*{Var}_{(-\infty,T]} \big(f(s) - g(s)\big).$$

Proof of Lemma A.1 Let $\{f_n(\cdot)\}$ be an arbitrary Cauchy sequence (see [204, 388]). Since

$$\|f_n(t) - f_k(t)\| \leq \operatorname*{Var}_{(-\infty,T]} \big(f_n(s) - f_k(s)\big),$$

for any $t \in [t_0, T]$, the sequence $f_n(t)$ converges to a function $f(t)$. Further, the variations of the functions $f_n(\cdot)$ are uniformly bounded, and since all the functions have the same values for $t < t_0$, they are also uniformly bounded. Hence, by Helly's Second Theorem, the limit function $f(t)$ has bounded variation. That is, for any Cauchy sequence $\{f_n(\cdot)\}$ with $f_n(\cdot) \in BV_m^+\{[t_0, T], X_0\}$ there exists a limit function $f(\cdot)$ with $\operatorname*{Var}_{(-\infty,T]} f(s) < \infty$ and such that $\lim_n \rho_T(f_n, f) = 0$ (see [204]). Indeed, since

$$\limsup_n \rho_T(f_n, f) \leq \limsup_{n,k} \rho_T(f_n, f_k) = 0,$$

the convergence in metric follows.

It remains to show that $f(\cdot)$ is right-continuous. It is readily seen that, for any $g, h \in BV_m^+\{[t_0, T], X_0\}$,

$$\sup_{(-\infty,T]} \|g(s) - h(s)\| \leq \rho_T(g, h).$$

Therefore, $f(t) = X_0$ when $t \in (-\infty, t_0)$ and

$$\lim_n \sup_{(-\infty,T]} \|f(s) - f_n(s)\| = 0. \tag{A.3}$$

Moreover, for any $\tau,\ t \in [0, T]$ with $\tau > t$ we have

$$\|f(\tau) - f(t)\| \leq \|f_n(\tau) - f_n(t)\| + \|f(t) - f_n(t)\| + \|f(\tau) - f_n(\tau)\|.$$

Consequently,

$$\lim_{\tau \downarrow t} \|f(\tau) - f(t)\| \le \lim_{n} \sup_{(-\infty, T]} \left[2\|f(s) - f_n(s)\|\right] = 0,$$

as needed. □

Remark A.12 If $X_0 = 0$, then $\|f\| = \operatorname*{Var}_{(-\infty,T]} f(t)$ has all the properties of a norm (see [204]), and the space $BV_m^+\{[t_0, T], 0\}$ is a complete normed space.

A.1.4 Some Results Concerning the Lebesgue-Stieltjes Integral

Let $f(\cdot)$ be a continuously differentiable scalar function of real variable. The Newton-Leibniz formula is well known from classical analysis (see, for example, [388]): for any X_0 and X_1 from $\mathbb{R}^1$

$$f(X_1) - f(X_0) = \int_{X_0}^{X_1} f'(x)dx. \tag{A.4}$$

Moreover, for any smooth curve $X(\cdot)$ defined on $[t_0, T]$, such that $X(t_0) = X_0$ and $X(T) = X_1$, we have

$$f(X_1) - f(X_0) = \int_{t_0}^{T} f'(X(t))dX(t).$$

In the above relations the integral is understood in the sense of Riemann, since

$$\int_{t_0}^{T} f'(X(t))dX(t) = \int_{t_0}^{T} f'(X(t))X'(t)dt$$

and the integrands are continuous.

These formulas allow the following generalization for the class of functions of bounded variation.

Theorem A.13 *Let $f(\cdot)$ be a continuously differentiable scalar function and let $X(\cdot)$ be a right-continuous function of bounded variation on the segment $[t_0, T]$, with $X(t_0-) = X_0$ and $X(T) = X_1$. Then*

$$f(X(T)) - f(X_0) = \int_{t_0}^{T} f'(X(t))dX^c(t) + Q(T), \tag{A.5}$$

where

$$Q(T) = \sum_{\tau \in D \cap [t_0, T]} \left[f(X(\tau)) - f(X(\tau-))\right], \tag{A.6}$$

and $X^{c}(\cdot)$ is the continuous component in the Lebesgue decomposition of the function $X(\cdot)$ (see Theorem A.11). The integral in (A.5) is understood in the sense of Lebesgue-Stieltjes, D is the set of points of discontinuity of the function $X(\cdot)$, and the series in (A.6) is absolutely convergent.

Remark A.13 Without loss of generality one can put $t_0 = 0$.

For the proof we need following lemmas.

Lemma A.2 *Under the assumptions of Theorem A.13, let $X(\cdot)$ be a continuous function of bounded variation. Then*

$$f\big(X(T)\big) - f(X_0) = \int_0^T f'(X(t))dX(t). \tag{A.7}$$

Proof of Lemma A.2 Let $0 = t_0 < t_1 < \cdots < t_n = T$ be a partition of the interval $[0, T]$. Then due to the continuity of the derivative $f'(\cdot)$ we have

$$\begin{aligned} f(X(T)) - f(X_0) &= \sum_{k=1}^{n} \big[f(X(t_k)) - f(X(t_{k-1}))\big] = \\ &= \sum_{k=1}^{n} f'\Big(X(t_{k-1}) + \xi_k\big(X(t_k) - X(t_{k-1})\big)\Big)\big[X(t_k) - X(t_{k-1})\big] = \\ &= \sum_{k=1}^{n} f'(X(t_{k-1}))\big[X(t_k) - X(t_{k-1})\big] + \delta_n, \end{aligned} \tag{A.8}$$

where $\xi_k \in [0, 1]$ and

$$\delta_n = -\sum_{k=1}^{n} \Big[f'(X(t_{k-1})) - f'\Big(X(t_{k-1}) + \xi_k\big(X(t_k) - X(t_{k-1})\big)\Big)\Big]\big[X(t_k) - X(t_{k-1})\big].$$

Since $f'(\cdot)$ is continuous and $X(\cdot)$ is a function of bounded variation,

$$|\delta_n| \leq \max_{1\leq k\leq n} \left|f'\big(X(t_{k-1})\big) - f'\Big(X(t_{k-1}) + \xi_k\big(X(t_k) - X(t_{k-1})\big)\Big)\right| \operatorname*{Var}_{[0,T]} X(t) \to 0,$$

as the partition of $[0, T]$ is refined so that $\max\limits_{1\leq k\leq n} |t_k - t_{k-1}| \to 0$. In this case we have

$$f(X(T)) - f(X_0) = \lim \left\{\sum_{k=1}^{n} f'(X(t_{k-1}))\big[X(t_k) - X(t_{k-1})\big]\right\}.$$

However, the right-hand side of this relation is, by definition (see [204]), the Riemann-Stieltjes integral of the continuous function $f'(\cdot)$, which coincides with the Lebesgue-Stieltjes integral of $f'(\cdot)$ (see [204] Theorem 1, Chap. 10, Sec. 36). The lemma is proved. □

To describe further properties of the Lebesgue-Stieltjes integral we resort to a discontinuous time transformation.

Consider on $[0, T]$ the increasing function

$$\Gamma(t) = t + \operatorname*{Var}_{[0,t]} X(u)$$

and its inverse function $\eta(s)$, defined on the interval $[0, T_1] = [0, \Gamma(T)]$ by the formula

$$\eta(s) = \inf\{u : \Gamma(u) > s\}.$$

Then

1. η is continuous and non-decreasing;
2. η is absolutely continuous and admits a derivative $\dot{\eta}(s) \in [0, 1]$ a.e. on $[0, T_1]$;
3. $\eta\big(\Gamma(t)\big) = t$ for all $t \in [0, T]$;
4. $\Gamma\big(\eta(s)\big) = s$ whenever $\eta(s)$ is a point of continuity point of Γ.

The proof of all these properties can be obtained arguing in much the same way as in the proof of Lemma 2.4. Let D_Γ (resp., D_X) be the set of points of discontinuity (jump points) of the function Γ (resp., X). Then $D_X = D_\Gamma$ and the interval $[0, T_1]$ admits the partition

$$[0, T_1] = \{s : \eta(s) \notin D_\Gamma\} \cup \bigcup_{\tau \in D_\Gamma} \big[\Gamma(\tau-), \Gamma(\tau)\big].$$

Define a function $Z(s)$ on $[0, T_1]$ by the formula

$$Z(s) = \begin{cases} X(\eta(s)), & \text{if } \eta(s) \notin D_\Gamma, \\ X(\tau-) + \dfrac{\Delta X(\tau)}{\Delta \Gamma(\tau)}\big(s - \Gamma(\tau-)\big), & \text{if } s \in \big[\Gamma(\tau_i-), \Gamma(\tau_i)\big]. \end{cases} \tag{A.9}$$

Lemma A.3 *The function $Z(s)$ is absolutely continuous, $Z(\Gamma(t)) = X(t)$, and for any Borel function $h(\cdot)$ defined on $[0, T]$ we have*

$$\int_0^T h(t)dX^{c}(t) = \int_0^{T_1} h\big(\eta(s)\big)\mathbf{I}\{s : \eta(s) \notin D_\Gamma\}\frac{dZ}{ds}(s)\,ds. \tag{A.10}$$

Proof of Lemma A.3 Suppose first that $s_1 < s_2$ and $s_1, s_2 \in \{s : \eta(s) \notin D_\Gamma\}$. Then there exist t_1 and t_2 such that $\eta(s_1) = t_1$, $\eta(s_2) = t_2$ and $\Gamma(t_1) = s_1,\ \Gamma(t_2) = s_2$. So we have

$$|Z(s_2) - Z(s_1)| = |X(t_2) - X(t_1)| \le |\Gamma(t_2) - \Gamma(t_1)| = s_2 - s_1. \tag{A.11}$$

Suppose now that $s_1 \in \left[\Gamma(t_1-), \Gamma(t_1)\right]$ for some $t_1 \in D_\Gamma$, and s_2 is as above. Then

$$|Z(s_2) - Z(s_1)| \leq |Z(s_2) - Z(\Gamma(t_1))| + |Z(\Gamma(t_1)) - Z(s_1)|.$$

Since $X(t_1) = Z(\Gamma(t_1))$, the first term on the right-hand side can be estimated by the difference $s_2 - \Gamma(t_1)$, and, as it follows from the definition of Z, the second term can be estimated by the difference $\Gamma(t_1) - s_1$. Hence, we again obtain inequality (A.11). The remaining cases can be treated in the same manner. The absolute continuity means that the function $Z(s)$ admits a derivative a.e. on $[0, T_1]$ and this derivative coincides a.e. with its Radon-Nikodym derivative of $Z(s)$ with respect to the Lebesgue measure.

The equality $Z(\Gamma(t)) = X(t)$ is an immediate consequence of the definition.

To prove the equality (A.10) it is sufficient to prove it for any interval indicator function $\mathbf{I}\{(a, b]\}(t)$ with $(a, b] \in [0, T]$ (see Remark 2.11). Setting $h(t) = \mathbf{I}\{(a, b]\}(t)$, we obtain

$$\int_0^T \mathbf{I}\{(a, b]\}(t) dX^c(t) = X^c(b) - X^c(a).$$

On the other hand,

$$\int_0^{T_1} \mathbf{I}\{s : \eta(s) \in (a, b]\} \mathbf{I}\{s :, \eta(s) \notin D_\Gamma\} \frac{dZ}{ds}(s)\, ds =$$

$$= \int_{\Gamma(a)}^{\Gamma(b)} \mathbf{I}\{s : \eta(s) \notin D_\Gamma\} \frac{dZ}{ds}(s)\, ds =$$

$$= Z(\Gamma(b)) - Z(\Gamma(a)) - \int_{\Gamma(a)}^{\Gamma(b)} \mathbf{I}\{s : \eta(s) \in D_\Gamma\} \frac{dZ}{ds}(s)\, ds =$$

$$= Z(\Gamma(b)) - Z(\Gamma(a)) - \sum_{\tau \in (a,b] \cap D_\Gamma} \int_{\Gamma(\tau-)}^{\Gamma(\tau)} \frac{dZ}{ds}(s) ds =$$

$$= Z(\Gamma(b)) - Z(\Gamma(a)) - \sum_{\tau \in (a,b] \cap D_\Gamma} \Delta X(\tau) =$$

$$= X(b) - X(a) - \sum_{\tau \in (a,b] \cap D_\Gamma} \Delta X(\tau) = X^c(b) - X^c(a).$$

This completes the proof. □

Proof of Theorem A.13 By Lemma A.2, the continuous function $Z(\cdot)$ satisfies

$$f(X(T)) - f(X_0) = f(Z(T_1)) - f(Z(0)) = \int_0^{T_1} f'(Z(s))dZ(s) =$$

$$= \int_0^{T_1} f'(Z(s))\mathbf{I}\{s : \eta(s) \notin D_\Gamma\}\frac{dZ}{ds}(s)\,ds + \sum_{\tau\in D_\Gamma} \int_{\Gamma(\tau-)}^{\Gamma(\tau)} f'(Z(s))\frac{dZ}{ds}(s)\,ds.$$

Next, by Lemma A.3,

$$\int_0^{T_1} f'\big(Z(s)\big)\mathbf{I}\{s : \eta(s) \notin D_\Gamma\}\frac{dZ}{ds}(s)ds = \int_0^{T} f'(X(t))dX^{c}(t),$$

because, by the definition of Z,

$$f'(Z(s))\mathbf{I}\{s : \eta(s) \notin D_\Gamma\} = f'(X(\eta(s)))\mathbf{I}\{s : \eta(s) \notin D_\Gamma\}.$$

The second term can be calculated directly. Indeed, since

$$\frac{dZ}{ds}(s) = \frac{\Delta X(\tau)}{\Delta\Gamma(\tau)}, \quad \text{if } s \in [\Gamma(\tau-), \Gamma(\tau)],$$

we deduce that

$$\int_{\Gamma(\tau-)}^{\Gamma(\tau)} f'(Z(s))Z'(s)ds = f(Z(\Gamma(\tau))) - f(Z(\Gamma(\tau-))) =$$

$$= f(X(\tau)) - f(X(\tau-)).$$

Finally, the absolute convergence of the series follows directly from the estimate

$$|f(X(\tau)) - f(X(\tau-))| \le \max_{x\in[X(\tau-),X(\tau)]} |f'(x)|\,|X(\tau) - X(\tau-)|,$$

the continuity of $f'(\cdot)$, and the boundedness of $X(\cdot)$ and of its variation. □

Remark A.14 In Theorem A.13 we cannot replace the assumption of continuous differentiability on f by the absolute continuity of f. The reason is that the composite function $f'(X(t))$ is not necessarily measurable with respect to $\|X^{c}\|(dt)$ even when $X^{c}(t)$ is absolutely continuous. Counterexamples can be found in [142]. However, in our case the function $f'(X(t)$ is Borel measurable, which ensures the existence of the integral.

Remark A.15 (Multi-Dimensional Case) Theorem A.13 remains valid in the case where $X(\cdot)$ is a vector-valued function.

Theorem A.14 *Let $f(\cdot)$ be a continuously differentiable function of m variables. Let $X(t) = \big(X_1(t), \dots, X_m(t)\big)$ be a vector-valued function of bounded variation defined on* $[0, T]$. *Then*

$$f\big(X(T)\big) - f\big(X(0-)\big) = \sum_{k=1}^{m} \int_0^T \frac{\partial f}{\partial x_k}(X(t))dX_k^{\mathrm{c}}(t) + Q(T), \tag{A.12}$$

where

$$Q(T) = \sum_{\tau \in D} \big[f\big(X(\tau)\big) - f\big(X(\tau-)\big)\big]. \tag{A.13}$$

Corollary A.1 (Integration by Parts Formula) *Under the assumptions of Theorem A.14, let $X = (Y, Z)$ be a two-component vector and $f(X) = f(Y, Z) = YZ$ be a real-valued function of two variables Y and Z. Then*

$$Y(T)Z(T) - Y(0-)Z(0-) = \int_0^T Y(t)dZ(t) + \int_0^T Z(t-)dY(t). \tag{A.14}$$

Indeed, the left-hand side of relation (A.12) is equal to

$$f(X(T)) - f(X(0-)) = Y(T)Z(T) - Y(0-)Z(0-), \tag{A.15}$$

while in view of (A.13) the right-hand side is a sum of two terms, namely

$$\int_0^T f'(X(t))dX^{\mathrm{c}}(t) = \int_0^T Y(t)dZ^{\mathrm{c}}(t) + \int_0^T Z(t)dY^{\mathrm{c}}(t) \tag{A.16}$$

and

$$\sum_{\tau \in D_X} \big[f(X(\tau)) - f(X(\tau-))\big] = \sum_{\tau \in D_Y \cup D_Z} [Y(\tau)Z(\tau) - Y(\tau-)Z(\tau-)]. \tag{A.17}$$

Moreover,

$$\int_0^T Z(t)dY^{\mathrm{c}}(t) = \int_0^T Z(t-)dY^{\mathrm{c}}(t),$$

since $Z(t) = Z(t-)$ $\|Y^{\mathrm{c}}\|$-a.e. (the measure $\|Y^{\mathrm{c}}(dt)\|$ is continuous, so the $\|Y^{\mathrm{c}}\|$-measure of the at most countable set D_Z is equal to zero). Now using the obvious relation

$$\begin{aligned} Y(\tau)Z(\tau) - Y(\tau-)Z(\tau-) &= Y(\tau)\big[Z(\tau) - Z(\tau-)\big] + \\ &+ Z(\tau-)\big[Y(\tau) - Y(\tau-)\big] = Y(\tau)\Delta Z(\tau) + Z(\tau-)\Delta Y(\tau), \end{aligned}$$

and combining (A.16) and (A.17) we finally deduce that

$$Y(T)Z(T) - Y(0-)Z(0-) = \int_0^T Y(t)dZ^c(t) + \sum_{\tau \in D_Z} Y(\tau)\Delta Z(\tau) +$$

$$+ \int_0^T Z(t-)dY^c(t) + \sum_{\tau \in D_Y} Z(\tau-)\Delta Y(\tau),$$

which coincides with (A.14). □

Note that equality (A.14) can be recast as

$$Y(T)Z(T) - Y(0-)Z(0-) = \int_{(0,T]} Y(t)dZ(t) + \int_{(0,T]} Z(t-)dY(t) +$$

$$+ Y(0)\big[Z(0) - Z(0-)\big] + Z(0-)\big[Y(0) - Y(0-)\big],$$

which yields another variant of the integration by parts formula:

$$Y(T)Z(T) - Y(0)Z(0) = \int_{(0,T]} Y(t)dZ(t) + \int_{(0,T]} Z(t-)dY(t). \tag{A.18}$$

Note also that equality (A.14) can be as well rewritten in the differential form

$$\begin{aligned} d[Y(t)Z(t)] &= Y(t)dZ(t) + Z(t-)dY(t) = \\ &= Y(t-)dZ(t) + Z(t-)dY(t) + \Delta Y(t)\Delta Z(t). \end{aligned} \tag{A.19}$$

Remark A.16 (A Matrix Form of the Integration by Parts Formula) Formula (A.14) remains valid when $Y(\cdot)$ and $Z(\cdot)$ are $n \times m$ and $m \times l$ matrix-valued functions, respectively. Indeed, by (A.14), the product of the elements $Y_{ik}(\cdot)$ and $Z_{kj}(\cdot)$ of these matrices satisfies

$$Y_{ik}(T)Z_{kj}(T) - Y_{ik}(0-)Z_{kj}(0-) = \int_0^T Y_{ik}(t)dZ_{kj}(t) + \int_0^T \big[dY_{ik}(t)\big]Z_{kj}(t-).$$

Summing here over k from 1 to m, we obtain an expression for the element $R_{ij}(\cdot)$ of the product matrix $R(\cdot) = Y(\cdot)Z(\cdot)$. The matrix analog of equation (A.14) reads

$$Y(T)Z(T) - Y(0-)Z(0-) = \int_0^T Y(t)dZ(t) + \int_0^T \big[dY(t)\big]Z(t-), \tag{A.20}$$

or, in differential form,

$$d\big[Y(t)Z(t)\big] = Y(t)dZ(t) + \big[dY(t)\big]Z(t-), \tag{A.21}$$

Lemma A.4 *Let $h(\cdot)$ be a right-continuous non-decreasing real-valued function on the segment $[0, T]$ such that $h(0-) = 0$. Then*

$$\int_0^t h^n(s-)\,dh(s) \le \frac{h^{n+1}(t)}{n+1} \le \int_0^t h^n(s)\,dh(s) \quad for\ n = 1, 2, \dots . \tag{A.22}$$

Remark A.17 Suppose $h(\cdot)$ is continuously differentiable. Then by standard calculations

$$\int_0^t h^n(s)dh(s) = \frac{h^{n+1}(t)}{n+1}.$$

Proof of Lemma A.4 By formula (A.14) and Theorem A.13,

$$\begin{aligned} h^{n+1}(t) &= \int_0^t h^n(s)dh(s) + \int_0^t h(s-)\,dh^n(s) = \\ &= \int_0^t h^n(s)\,dh(s) + \int_0^t h(s-)nh^{n-1}(s)\,dh^c(s) + \\ &+ \sum_{\tau\in D_h\cap[0,t]} h(\tau-)\big[h^n(\tau) - h^n(\tau-)\big]. \end{aligned}$$

Since h^c is continuous, it follows that

$$\int_0^t h(s-)nh^{n-1}(s)\,dh^c(s) = n\int_0^t h^n(s-)\,dh^c(s) = n\int_0^t h^n(s)\,dh^c(s). \tag{A.23}$$

Using the standard decomposition we obtain

$$\begin{aligned} &h(\tau-)\big[h^n(\tau) - h^n(\tau-)\big] = \\ &\quad = \Big[h(\tau-)h^{n-1}(\tau) + h^2(\tau-)h^{n-2}(\tau) + \dots + h^n(\tau-)\Big]\,\Delta h(\tau). \end{aligned} \tag{A.24}$$

Therefore, taking into account that $0 \le h(t-) \le h(t)$ and relations (A.23), (A.24), where we replace $h(\tau-)$ by $h(\tau)$, we easily obtain that

$$\begin{aligned} h^{n+1}(t) &\le \int_0^t h^n(s)\,dh(s) + n\int_0^t h^n(s)\,dh^c(s) + n\sum_{\tau\in D_h\cap[0,t]} h^n(\tau)\Delta h(\tau) = \\ &= (n+1)\int_0^t h^n(s)\,dh(s). \end{aligned}$$

This is just the assertion of the lemma for the right-hand inequality, while the left-hand one can be obtained in the same way if we replace $h(\tau)$ by $h(\tau-)$ in (A.23), (A.24). □

A.2 Linear Differential Equations with Measure

A.2.1 Some General Remarks and Definitions

Consider a linear differential equation with measure in the integral form

$$X(t) = X(t_0-) + \int_{t_0}^{t} \big[da(s)\big] X(s-) + b(t), \tag{A.25}$$

where $X(t) \in \mathbb{R}^m$, $da(\cdot)$ is a matrix-valued measure defined on $[t_0, T]$, with $T \leq \infty$, and $b(t) \in \mathbb{R}^m$ is a right-continuous vector-function of bounded variation on $[t_0, T]$ such that $b(t_0-) = 0$. As it follows from (A.25), its solution $X(t)$, if it exists, has to be a right-continuous function of bounded variation. For simplicity we transform (A.25) into an equation with a scalar measure. To this end we introduce the scalar function of bounded variation

$$\mu(t) = \operatorname*{Var}_{[t_0,t]} a(s) = \sum_{i,j=1}^{m} \operatorname*{Var}_{[t_0,t]} a_{ij}(s),$$

i.e., the sum of the variations of the elements of the matrix $a(s)$ for $s \in [t_0, t]$. Then for each of the elements a_{ij} the measure da_{ij} is absolutely continuous with respect to the measure $d\mu$:

$$da_{ij}(\cdot) \ll d\mu(\cdot).$$

Denote by $A_{ij}(\cdot)$ the Radon-Nikodym derivative of $da_{ij}(\cdot)$ with respect to measure $d\mu(\cdot)$ and by $A(\cdot)$ the matrix with the elements $A_{ij}(\cdot)$, $i, j = 1, \ldots, m$. Then

$$da(t) = A(t)d\mu(t), \quad t \in [t_0, T],$$

and the original Eq. (A.25) becomes

$$X(t) = X(t_0-) + \int_{t_0}^{t} A(s)X(s-)d\mu(s) + b(t). \tag{A.26}$$

Definition A.16 Equation (A.26) is called a *non-homogeneous linear differential equation with measure in normal form.*

Definition A.17 If $b(t) \equiv 0$, the corresponding equation

$$X(t) = X(t_0-) + \int_{t_0}^{t} A(s)X(s-)d\mu(s), \tag{A.27}$$

is called a *homogeneous linear differential equation with measure.*

Definition A.18 A right-continuous function of bounded variation $x = \varphi(t)$ is called a *solution* of Eq. (A.26) if it is defined on $[t_0, T]$ and, for any $t \in [t_0, T]$,

$$\varphi(t) = \varphi(t_0-) + \int_{t_0}^{t} A(s)\varphi(s-)d\mu(s) + b(t).$$

The notion of a solution of Eq. (A.27) is defined analogously.

The next result gives a property of these equations that follows directly from the definitions and properties of the integral.

Theorem A.15 *Any linear combination*

$$\varphi(t) = \sum_{k=1}^{n} \lambda_k \varphi_k(t), \quad n < \infty$$

of solutions of the non-homogeneous equation (A.27) with given

$$\varphi_k(t_0-) \quad \text{and} \quad b_k(t)$$

is a solution of the same non-homogeneous equation, corresponding to

$$X(t_0-) = \sum_{k=1}^{n} \lambda_k \varphi_k(t_0-) \quad \textit{and} \quad b(t) = \sum_{k=1}^{n} \lambda_k b_k(t).$$

A.2.2 Scalar Linear Differential Equation with Measure

Here we discuss the simplest differential equation with measure in the scalar case.

Theorem A.16 *The solution $\varphi_0(t)$ of the linear homogeneous equation*

$$X(t) = 1 + \int_{t_0}^{t} X(s-)d\mu(s), \quad \textit{with } X(t_0-) = 1, \quad \mu(t_0-) = 0, \tag{A.28}$$

is given by the formula

$$\varphi_0(t) = e^{\mu^c(t)} \prod_{\tau \in D_\mu \cap [t_0, t]} (1 + \Delta\mu(\tau)), \tag{A.29}$$

where, by convention, $\prod_{\emptyset} = 1$, and D_μ is the the set of points of discontinuity of the function $\mu(t)$.

Proof of Theorem A.16 We will show that the function $\varphi_0(\cdot)$ given by (A.29) is the solution of Eq. (A.28). Define

$$Y(t) = Y^{\mathrm{c}}(t) = \mu^{\mathrm{c}}(t) \quad \text{and} \quad Z(t) = \prod_{\tau \in D_\mu \cap [t_0, t]} \big(1 + \Delta\mu(\tau)\big). \tag{A.30}$$

Then clearly

$$dZ^{\mathrm{c}}(\cdot) = 0 \tag{A.31}$$

and

$$\Delta Z(\tau) = Z(\tau) - Z(\tau-) = Z(\tau-)\big[1 + \Delta\mu(\tau)\big] - Z(\tau-) = Z(\tau-)\Delta\mu(\tau). \tag{A.32}$$

Now consider the function of two variables $f(Y, Z) = Ze^Y$. We have

$$f\big(Y(t), Z(t)\big) = Z(t)e^{Y(t)} = \varphi_0(t). \tag{A.33}$$

Let us apply formulas (A.12) and (A.14) to the function $f(Y(t), Z(t))$. Then in view of the continuity of $e^{Y(t)}$ and relations (A.31) and (A.32), we have

$$\begin{aligned}
&Z(t)e^{Y(t)} - Z(t_0-)e^{Y(t_0-)} = \\
&\quad = \int_{t_0}^{t} Z(s)e^{Y(s)}dY^{\mathrm{c}}(s) + \sum_{\tau \in D_\mu \cap [t_0, t]} \Big[Z(\tau)e^{Y(\tau)} - Z(\tau-)e^{Y(\tau-)}\Big] = \\
&\quad = \int_{t_0}^{t} Z(s)e^{Y(s)}d\mu^{\mathrm{c}}(s) + \sum_{\tau \in D_\mu \cap [t_0, t]} e^{Y(\tau)}Z(\tau-)\Delta\mu(\tau) = \\
&\quad = \int_{t_0}^{t} Z(s-)e^{Y(s)}d\mu(s) = \int_{t_0}^{t} Z(s-)e^{Y(s-)}d\mu(s).
\end{aligned}$$

The last equality follows from the continuity of $Y(\cdot)$. As $Z(t_0-)e^{Y(t_0-)} = 1$, to complete the proof of the theorem it remains to verify the right-continuity, which follows immediately from (A.29) and the fact that the variation is bounded. Finally, note that

$$e^{|\mu(t)|} = \exp\left(\operatorname*{Var}_{[t_0, t]} \mu(s)\right)$$

is obviously a monotonically increasing function of bounded variation. At the same time, for any interval $[a, b] \subset [t_0, T]$,

$$|\varphi_0(b) - \varphi_0(a)| \le e^{|\mu(b)|} - e^{|\mu(a)|},$$

so the function $\varphi_0(t)$ is indeed of bounded variation. □

Corollary A.2 *In the scalar case, any solution $\varphi(t)$ of the equation*

$$X(t) = X_0 + \int_{t_0}^{t} A(s)X(s-)d\mu(s) \quad with \;\; X(t_0-) = X_0 \neq 0 \tag{A.34}$$

where $A(\cdot)$ is a μ-measurable function such that

$$\int_{t_0}^{T} |A(t)|d\mu(t) < \infty$$

is given by the formula

$$\varphi(t) = X_0\,\mathscr{E}(t, t_0), \tag{A.35}$$

where

$$\mathscr{E}(t, t_0) = \exp\left(\int_{t_0}^{t} A(s)d\mu^{c}(s)\right) \prod_{\tau\in D_\mu\cap[t_0,t]} (1 + A(\tau)\Delta\mu(\tau)), \tag{A.36}$$

with $\mathscr{E}(t_0-, t_0) = 1$.

Proof of Corollary A.2 Denote

$$Y(t) = \frac{X(t)}{X_0} \quad \text{and} \quad \nu(t) = \mu(t_0-) + \int_{t_0}^{t} A(s)d\mu(s).$$

In these notations, Eq. (A.34) takes on the form (A.28) and, in view of (A.36), formula (A.35), takes the form of (A.29). □

Now let us consider the non-homogeneous scalar case.

Theorem A.17 (Cauchy's Formula) *The solution $\varphi(t)$ of the non-homogeneous scalar equation*

$$X(t) = X_0 + \int_{t_0}^{t} A(s)X(s-)d\mu(s) + b(t), \quad with \;\; X(t_0-) = X_0 \tag{A.37}$$

is given by the formula

$$\varphi(t) = X_0\,\mathscr{E}(t, t_0) + \int_{t_0}^{t} \mathscr{E}(t, s+)db(s), \tag{A.38}$$

where

$$\mathscr{E}(t,s) = \exp\left(\int_s^t A(\tau)d\mu^c(\tau)\right) \prod_{\tau\in D_\mu\cap[s,t]} (1 + A(\tau)\Delta\mu(\tau)) \tag{A.39}$$

Remark A.18 Note that, in view of its definition, the function $\mathscr{E}(t,s)$ enjoys the following properties:

1. it is right-continuous with respect to t and left-continuous with respect to s;
2. it satisfies the differential equations with measure

$$d_t\mathscr{E}(t,s) = A(t)\mathscr{E}(t-,s)d\mu(t), \quad \mathscr{E}(s-,s) = 1$$

   and

$$d_s\mathscr{E}(t,s) = -\mathscr{E}(t,s+)A(s)d\mu(s), \quad \mathscr{E}(t,t+) = 1;$$

3. it satisfies the semigroup property: for any $t_0 \le s \le u \le t$,

$$\mathscr{E}(t,s) = \mathscr{E}(t,u)\mathscr{E}(u-,s);$$

4. if there exists $\tau \in [t_0,t]$ such that $1 + A(\tau)\Delta\mu(\tau) = 0$, then for any interval $[v,u]$ such that $\tau \in [v,u]$ we have $\mathscr{E}(u,v) = 0$.

Proof of Theorem A.17 In view of Theorem A.15, it suffices to show that the function

$$Z(t) = \int_{t_0}^t \mathscr{E}(t,s+)db(s) = \int_{[t_0,t]} \mathscr{E}(t,s+)db(s),$$

satisfies the equation

$$Z(t) = \int_{[t_0,t]} A(s)Z(s-)d\mu(s) + b(t). \tag{A.40}$$

The function $Z(t)$ is well defined, since $\mathscr{E}$ is a bounded Borel function, so the integral exists, and the integral is a Borel function of the upper integration limit (see Fubini's Theorem [204]). Now

$$Z(t-) = \int_{[t_0,t)} \mathscr{E}(t-,s+)db(s),$$

and substituting this expression in the integral on the right-hand side of (A.40) and switching the order of integration (see Fubini's Theorem [204]) we obtain

$$\int_{[t_0,t]} A(s)Z(s-)d\mu(s) = \int_{[t_0,t]} A(s)\Big[\int_{[t_0,s)} \mathscr{E}(s-,\tau+)db(\tau)\Big]d\mu(s) =$$

$$= \int_{[t_0,t]}\Big[\int_{(\tau,t]} A(s)\mathscr{E}(s-,\tau+)d\mu(s)\Big]db(\tau) = \int_{[t_0,t]} \mathscr{E}(t,\tau+)db(\tau) - b(t).$$

Here we used the fact that, due to the properties of $\mathscr{E}(t,s)$,

$$\int_{(\tau,t]} A(s)\mathscr{E}(s-,\tau+)d\mu(s) = \mathscr{E}(t,\tau+) - 1.$$

The proof is complete. □

A.2.3 Multi-Dimensional Case

In the multi-dimensional case we assume that there exists an $m \times m$ matrix-valued function $\mathscr{E}(t,s)$ with the same properties as in the scalar case (see Remark A.18), which satisfies for $t \geq s$ the differential equation

$$d_t\mathscr{E}(t,s) = A(t)\mathscr{E}(t-,s)d\mu(t), \quad \mathscr{E}(s-,s) = I_{(m\times m)}. \tag{A.41}$$

If the matrix-valued function $\mathscr{E}(t,s)$ is non-singular for any $t,s \in [t_0,T]$, then

$$d_s\mathscr{E}(t,s) = -\mathscr{E}(t,s+)A(s)d\mu(s), \quad \mathscr{E}(t,t+) = I_{(m\times m)}. \tag{A.42}$$

The existence of such a function will be proved later; here we merely note that, in contrast to the scalar case, it does not admit an explicit representation.

Theorem A.18 *Assume that the $m \times m$ matrix-valued function $A(\cdot)$ is μ-integrable, i.e.,*

$$\int_{[t_0,T]} \|A(t)\|d\mu(t) < \infty. \tag{A.43}$$

Let $\mathscr{E}(t,s)$ be the solution of Eq. (A.41). Then

$$\varphi(t) = \mathscr{E}(t,t_0)X_0 + \int_{[t_0,t]} \mathscr{E}(t,s+)db(s)$$

is the solution of the non-homogeneous Eq. (A.27) with $X(t_0-) = X_0 \in \mathbb{R}^m$.

Remark A.19 The proof mimics the proof in the scalar case and is omitted.

Now let us prove an existence and uniqueness theorem for the Eq. (A.41) in the multi-dimensional case.

Theorem A.19 *Suppose that the $m \times m$ matrix-valued function $A(\cdot)$ satisfies (A.43) where $\|\cdot\|$ is any of the matrix norms. Then there exists exactly one solution of the Eq. (A.41), which is right-continuous with respect to t and has bounded variation with respect to t.*

Proof of Theorem A.19 Define the operator $\mathscr{A}(g(\cdot))$ by the formula

$$\mathscr{A}(g)(t) = I_{(m\times m)} + \int_{[s,t]} A(s)g(u-, s)d\mu(u),$$

acting on matrix-valued functions $g(u, s)$ that are right-continuous with respect to u and of bounded variation, defined for $u \geq s$ and such that $g(u, s) = I_{(m\times m)}$ for $u < s$. As follows from the definition and condition (A.43), the operator $\mathscr{A}$ maps the space $BV^{+}_{m\times m}\{[s, T], I_{(m\times m)}\}$ into itself. Therefore, a solution of (A.41), if it exists, is a *fixed point* of $\mathscr{A}$ (see [204].)

To prove the existence and uniqueness of a fixed point of $\mathscr{A}$ in the complete metric space like $BV^{+}_{m\times m}\{[s, T], I_{(m\times m)}\}$ (see Lemma A.1), it suffices to show that some power of $\mathscr{A}$ is a *contraction mapping*, i.e., there exist $n \geq 1$ and $\alpha < 1$ such that

$$\rho_T(\mathscr{A}^n(g), \mathscr{A}^n(h)) \leq \alpha\rho_T(g, h), \quad \text{for all } g, h \in BV^{+}_{m\times m}\{[s, T], I_{(m\times m)}\}.$$

According to the definition,

$$\begin{aligned}\rho_t(\mathscr{A}(g), \mathscr{A}(h)) &= \operatorname*{Var}_{[s,t]}\left\| \int_{[s,u]} A(\tau)[g(\tau-, s) - h(\tau-, s)]d\mu(\tau)\right\| \leq \\ &\leq \rho_t(g, h)\int_{[s,t]} \|A(\tau)\|d\mu(\tau) = \rho_t(g, h)M(t),\end{aligned}$$

where $M(t)$ is a right-continuous function of bounded variation.

This last estimate implies that

$$\rho_t(\mathscr{A}^2(g), \mathscr{A}^2(h)) \leq \rho_t(g, h)\int_{[s,t]} M(s-)dM(s) \leq \rho_t(g, h)\frac{M^2(t)}{2},$$

and using (A.22) we conclude that

$$\rho_t(\mathscr{A}^n(g), \mathscr{A}^n(h)) \leq \rho_t(g, h)\frac{M^n(t)}{n!}.$$

Hence, for $n \geq 1$ we have

$$\rho_T(\mathscr{A}^n(g), \mathscr{A}^n(h)) \leq \rho_T(g, h)\frac{M^n(T)}{n!},$$

and so there exists n such that $\mathscr{A}^n$ is a contraction mapping.

This completes the proof of the existence and uniqueness of the matrix-valued right-continuous function $\bar{\mathscr{E}}(t,s)$ which is a solution of (A.41).

The existence of the solution of Eq. (A.42) follows from the semigroup property of $\mathscr{E}(t,s)$, namely that

$$\mathscr{E}(t,s) = \mathscr{E}(t,u)\mathscr{E}(u-,s), \tag{A.44}$$

for any $s \le u \le t$

In its turn, the semigroup property is a consequence of the relations

$$\begin{aligned}
\mathscr{E}(t,s) &= I_{(m\times m)} + \int_{[s,t]} A(\tau)\mathscr{E}(\tau-,s)d\mu(\tau) = \\
&= I_{(m\times m)} + \int_{[s,u)} A(\tau)\mathscr{E}(\tau-,s)d\mu(\tau) + \int_{[u,t]} A(\tau)\mathscr{E}(\tau-,s)d\mu(\tau) = \\
&= \mathscr{E}(u-,s) + \int_{[u,t]} A(\tau)\mathscr{E}(\tau-,s)d\mu(\tau),
\end{aligned}$$

$$\mathscr{E}(t,u)\mathscr{E}(u-,s) = \mathscr{E}(u-,s) + \int_{[u,t]} A(\tau)\mathscr{E}(\tau-,u)\mathscr{E}(u-,s)d\mu(\tau)$$

and the uniqueness of solutions.

The same arguments show that

$$\mathscr{E}(t,s) = \mathscr{E}(t,u+)\mathscr{E}(u,s). \tag{A.45}$$

Now to derive equation (A.42) it suffices to differentiate this relation with respect to u. Indeed, in view of formula (A.21),

$$\begin{aligned}
0 = d_u\mathscr{E}(t,s) &= d_u\mathscr{E}(t,u+)\mathscr{E}(u,s) = \\
&= d_u\mathscr{E}(t,u+)\mathscr{E}(u-,s) + \mathscr{E}(t,u+)A(u)\mathscr{E}(u-,s)d\mu(u).
\end{aligned}$$

Since $d_u\mathscr{E}(t,u+) = d_u\mathscr{E}(t,u)$ and the matrix-valued function $\mathscr{E}(u-,s)$ is non-singular thanks to the semigroup property and the non-singularity of $\mathscr{E}(t,s)$, Eq. (A.42) follows. In turn, (A.42) implies that the function $\mathscr{E}(t,s)$ has bounded variation with respect to the second argument. The terminal condition at the point t is a consequence of the relation

$$\mathscr{E}(t,t+) = \lim_{u\downarrow t} \mathscr{E}(u-,u) = I_{(m\times m)},$$

since $\mathscr{E}(t,s)$ is right-continuous with respect to the first argument and admits a right limit with respect to the second argument because its variation is bounded. □

Remark A.20 Assume that the matrix $\mathscr{E}(t, s)$ is nonsingular. Denote $\mathscr{G}(s, t) = \mathscr{E}^{-1}(t, s)$. Then the matrix-valued function $\mathscr{G}(s, t)$ satisfies the equation

$$d_t\mathscr{G}(s, t) = -\mathscr{G}(s, t)A(t)d\mu(t), \quad \mathscr{G}(s, s-) = I_{(m\times m)}. \tag{A.46}$$

Indeed, differentiating the identity $\mathscr{G}(s, t)\mathscr{E}(t, s) \equiv 0$ we obtain

$$d_t\mathscr{G}(s, t)\mathscr{E}(t-, s) + \mathscr{G}(s, t)A(t)\mathscr{E}(t-, s)d\mu(t) = 0,$$

and now (A.46) follows due to the invertibility of $\mathscr{E}(t-, s)$.

The matrix $\mathscr{E}(t, s)$ will be invertible if

$$\det\left[I_{(m\times m)} + A(\tau)\Delta\mu(\tau)\right] \neq 0, \quad \text{for all } \tau \in [s, t].$$

Since $\det(\cdot)$ is a continuous function and $\det I_{(m\times m)} = 1$, there exists $\varepsilon > 0$ such that

$$\det[I_{(m\times m)} + A] \neq 0 \quad \text{if} \quad \|A\| \leq \varepsilon.$$

This means that the singularity set

$$D^{\mathrm{s}} = \left\{\tau \in [t_0, T] : \det\left[I_{(m\times m)} + A(\tau)\Delta\mu(\tau)\right] = 0\right\}$$

is finite. Therefore, on the intervals that do not contain points from D^{s} (A.46) holds. Thus, if the inverse matrix exists, then one can use the formula

$$\mathscr{E}(t, s) = \mathscr{E}(t, t_0)\mathscr{E}^{-1}(s, t_0). \tag{A.47}$$

Another representation for $\mathscr{E}(t, s)$ is given by the following result.

Theorem A.20 *Assume that*

$$\det\left[I_{(m\times m)} + A(t)\Delta\mu(t)\right] \neq 0 \quad \textit{for all } t \in [t_0, T].$$

Then the matrix-function $\mathscr{E}(t, s)$ is invertible for any $t_0 \leq s \leq t \leq T$ and the inverse matrix-function $\widetilde{\mathscr{G}}(t, s) = \mathscr{E}^{-1}(t, s)$ is the solution to the equation

$$d\widetilde{\mathscr{G}}(t, s) = -\widetilde{\mathscr{G}}(t-, s)A(t)\left[I_{(m\times m)} + A(t)\Delta\mu(t)\right]^{-1}d\mu(t), \tag{A.48}$$

with the initial condition $\widetilde{\mathscr{G}}(s-, s) = I_{(m\times m)}$.

Proof of Theorem A.20 We only need to verify that the matrix-function

$$\widetilde{A}(t) = A(t)\left[I_{(m\times m)} + A(t)\Delta\mu(t)\right]^{-1}$$

is μ-integrable. As $\widetilde{A}(t) = A(t)$ a.e. with respect to the measure μ^{c}, is suffices to verify that there is a constant $C > 0$ such that

$$\left\| \left[I_{(m\times m)} + A(t)\Delta\mu(t) \right]^{-1} \right\| \leq C < \infty,$$

which is indeed true thanks to the assumption on the determinant in the statement of the theorem. Indeed, for some sufficiently small $\varepsilon > 0$ the above condition is met for $\Delta\mu(t) \leq \varepsilon$; however, the set of points where $\Delta\mu(t) > \varepsilon$ is finite and on it the condition is met thanks to the assumptions of the theorem. It follows that the solution of Eq. (A.48) exists and is unique. Then it remains only to verify the equality

$$d\widetilde{\mathscr{G}}(t,s)\mathscr{E}(t,s) = 0,$$

which is done by direct substitution in the integration by parts formula. □

Remark A.21 For a more detailed exposition of the theory of linear differential equations with measure we refer the reader to [165].

A.2.4 Generalized Gronwall-Bellman Inequality

The next result (the so-called generalized Gronwall-Bellman lemma) plays an important role in proving certain estimates for the solutions of differential equations with measure.

Lemma A.5 *Let $\mu(t)$ be a non-decreasing function of bounded variation on the interval $[t_0, T]$, with $\mu(t_0) = 0$. Let $h(t)$ and $f(t)$ be μ-integrable, non-negative, right-continuous, functions which admit left limits on $[t_0, T]$. Suppose that*

$$f(t) \leq h(t) + \int_{t_0}^{t} f(s-)d\mu(s) \quad \textit{for all } \; t \in [t_0, T]. \tag{A.49}$$

Then

$$f(t) \leq h(t) + \varphi_0(t) \int_{t_0}^{t} \varphi_0^{-1}(s)h(s-)d\mu(s), \tag{A.50}$$

where $\varphi_0(\cdot)$ is the function given by formula (A.29).

Proof of Lemma A.5 Denote

$$z(t) = \int_{t_0}^{t} f(s-)d\mu(s), \quad \text{and put} \quad z(t_0-) = 0.$$

Then inequality (A.49) takes on the form

$$f(t) \le h(t) + z(t). \tag{A.51}$$

By formula (A.14), the function $\varphi_0^{-1}(t)z(t)$ satisfies the relation

$$\begin{aligned}\varphi_0^{-1}(t)z(t) &= \int_{t_0}^{t} z(s-)d\varphi_0^{-1}(s) + \int_{t_0}^{t} \varphi_0^{-1}(s)dz(s) = \\ &= -\int_{t_0}^{t} z(s-)\varphi_0^{-1}(s)d\mu(s) + \int_{t_0}^{t} \varphi_0^{-1}(s)f(s-)d\mu(s).\end{aligned} \tag{A.52}$$

Here we used Eq. (A.46) for $\varphi_0^{-1}(\cdot)$ (see Remark A.20). Note that, by the definition of $\varphi_0(t)$, we have $\varphi_0(t) \ge 1$, so the inverse exists for any $t \ge t_0$. Multiplying inequality (A.51) by $\varphi_0^{-1}(t)$ and integrating its two sides with respect to the measure $d\mu(\cdot)$ we get

$$\begin{aligned}\int_{t_0}^{t} \varphi_0^{-1}(s)f(s-)d\mu(s) - \int_{t_0}^{t} \varphi_0^{-1}(s)z(s-)d\mu(s) &\le \\ \le \int_{t_0}^{t} \varphi_0^{-1}(s)h(s-)d\mu(s).\end{aligned} \tag{A.53}$$

Combining (A.52) and (A.53), we obtain the inequality

$$\varphi_0^{-1}(t)z(t) \le \int_{t_0}^{t} \varphi_0^{-1}(s)h(s-)d\mu(s),$$

which in view of (A.51) implies that

$$f(t) - h(t) \le z(t) \le \varphi_0(t)\int_{t_0}^{t} \varphi_0^{-1}(s)h(s-)d\mu(s).$$

This yields the claimed inequality (A.50). □

Remark A.22 Since the function $\varphi_0(t)$ is positive, we can use formula (A.47) and write

$$\varphi_0(t)\varphi_0^{-1}(s) = \mathscr{E}(t,s),$$

where $\mathscr{E}(\cdot)$ is given by formula (A.36).

Corollary A.3 *Under the conditions of Lemma A.5, suppose that*

$$f(t) \le h(t) + \int_{t_0}^{t} A(s)f(s-)d\mu(s), \tag{A.54}$$

where $A(t) \geq 0$ is a $d\mu(\cdot)$-integrable function on $[t_0, T]$. Then

$$f(t) \leq h(t) + \int_{t_0}^{t} \mathscr{E}(t,s)A(s)h(s-)d\mu(s). \tag{A.55}$$

Proof of Corollary A.3 Denote

$$\nu(t) = \mu(t_0-) + \int_{t_0}^{t} A(s)d\mu(s).$$

Then inequality (A.54) takes on the form (A.50) and formula (A.36) takes on the form (A.29). Hence, we infer that (A.55) takes the form of (A.50). The corollary is proved. □

The generalized Gronwall-Bellman lemma provides an estimate for the solutions of Eq. (A.26).

Theorem A.21 *The solution $\varphi(\cdot)$ of Eq. (A.26) with $\varphi(t_0-) = X_0$ satisfies the inequality*

$$\|\varphi(t)\| \leq \|X_0\|\mathscr{E}(t,t_0) + \|b(t)\| + \int_{t_0}^{t} \mathscr{E}(t,s)\|A(s)\|\,\|b(s-)\|\,d\mu(s), \tag{A.56}$$

where $\mathscr{E}(t,s)$ is defined by formula (A.39).

Proof of Theorem A.21 Equation (A.26) shows that the solution $\varphi(\cdot)$ satisfies the inequality

$$\|\varphi(t)\| \leq \|X_0\| + \|b(t)\| + \int_{t_0}^{t} \|A(s)\|\,\|\varphi(s-)\|d\mu(s),$$

where $\|A(\cdot)\|$ denotes any norm of the matrix $A(\cdot)$. By Lemma A.5,

$$\|\varphi(t)\| \leq \|X_0\| + \|b(t)\| + \int_{t_0}^{t} \mathscr{E}(t,s)\|A(s)\|\big(\|X_0\| + \|b(s-)\|\big)d\mu(s) =$$

$$= \|X_0\|\left[1 + \mathscr{E}(t,t_0) - 1\right] + \|b(t)\| + \int_{t_0}^{t} \mathscr{E}(t,s)\|A(s)\|\,\|b(s-)\|d\mu(s) =$$

$$= \|X_0\|\,\mathscr{E}(t,t_0) + \|b(t)\| + \int_{t_0}^{t} \mathscr{E}(t,s)\|A(s)\|\,\|b(s-)\|d\mu(s).$$

The theorem is proved. □

A.2.5 Matrix Equations

To conclude this section, we consider matrix linear differential equations that arise in filtering of discrete-continuous processes [291] and in observation control problems [279]:

$$\begin{aligned} X(t) = X(t_0-) + \int_{t_0}^{t} \big[A(s)X(s-) + X(s-)A^*(s)\big]d\mu(s) + \\ + \sum_{\tau\in D_\mu\cap[t_0,t]} A(\tau)X(\tau-)A^*(\tau)\big[\Delta\mu(\tau)\big]^2 + b(t), \end{aligned} \tag{A.57}$$

where $X(t)$ and $b(t)$ are $m \times m$ matrix-valued functions with components $X_{ij}(\cdot) \in BV^+\{[t_0, T], X_{ij}(t_0-)\}$ and $b_{ij}(\cdot) \in BV^+\{[t_0, T], 0\}$.

Clearly, the homogeneous equation corresponding to Eq. (A.57) reads

$$\begin{aligned} X(t) = X(t_0-) + \int_{t_0}^{t} \big[A(s)X(s-) + X(s-)A^*(s)\big]d\mu(s) + \\ + \sum_{\tau\in D_\mu\cap[t_0,t]} A(\tau)X(\tau-)A^*(\tau)\big[\Delta\mu(\tau)\big]^2. \end{aligned} \tag{A.58}$$

Note that the sum figuring in Eqs. (A.57) and (A.58) may be rewritten in integral form as

$$\sum_{\tau\in D_\mu\cap[t_0,t]} A(\tau)X(\tau-)A^*(\tau)[\Delta\mu(\tau)]^2 = \int_{t_0}^{t} A(s)X(s-)A^*(s)\Delta\mu(s)d\mu(s).$$

Theorem A.22 *Any solution $\varphi(\cdot)$ of the homogeneous matrix equation (A.58) with $X(t_0-) = X_0$ is given by the formula*

$$\varphi(t) = \mathscr{E}(t, t_0)X_0\mathscr{E}^*(t, t_0), \tag{A.59}$$

with $\mathscr{E}(t, t_0)$ defined by (A.41).

The solutions of the non-homogeneous equation (A.57) are described as follows.

Theorem A.23 *Any solution $\varphi(t)$ of the non-homogeneous matrix equation (A.57) with $X(t_0-) = X_0$ is given by the formula*

$$\varphi(t) = \mathscr{E}(t, t_0)X_0\mathscr{E}^*(t, t_0) + \int_{t_0}^{t} \mathscr{E}(t, s+)\big[db(s)\big]\mathscr{E}^*(t, s+), \tag{A.60}$$

with $\mathscr{E}(t, s)$ as above.

Remark A.23 The proof of these two results follows directly from the matrix integration by parts formula (see Remark A.16) and is omitted.

A.3 Nonlinear Differential Equations with Measure

A.3.1 Existence and Uniqueness of the Solution

Let $\mu(dt) = \mu^{\mathrm{c}}(dt) + \mu^{\mathrm{d}}(dt)$ be a vector-valued measure defined on the Borel subsets of the interval $[t_0, T]$ and such that $\|\mu([t_0, T])\| < \infty$, with a continuous component $\mu^{\mathrm{c}}(dt)$, a discrete component $\mu^{\mathrm{d}}(dt)$, and a distribution function $\mu(t) = \mu(\{[t_0, t]\}) \in BV_k^+\{[t_0, T], 0\}$. The function $\mu(t)$ admits the decomposition (see Theorem A.11)

$$\mu(t) = \mu^{\mathrm{c}}(t) + \sum_{\tau \in D_\mu \cap [t_0, t]} \Delta\mu(\tau), \qquad \Delta\mu(t_0) = 0,$$

with a continuous component $\mu^{\mathrm{c}}(t)$ and jumps $\Delta\mu(\tau) = \mu(\tau) - \mu(\tau-)$. As above, we denote by $D_\mu = \{\tau : \Delta\mu(\tau) \neq 0\}$ the set of jump time instants τ of the components of the vector-function $\mu(t)$.

Now consider on the space $BV_m^+\{[t_0, T], X_0\}$ the following nonlinear differential equation with measure in the integral form:

$$\begin{aligned} X(t) &= X_0 + \int_{t_0}^{t} F(X(s), s)d\mu^{\mathrm{c}}(s) + \sum_{\tau \le t} \Delta X(\tau) = \\ &= X_0 + X^{\mathrm{c}}(t) + X^{\mathrm{d}}(t), \end{aligned} \tag{A.61}$$

where $\Delta X(\tau) = \Psi(X(\tau-), \tau, \Delta\mu(\tau))$ and $\tau \in D_\mu^{\mathrm{d}}$. In this equation $\Psi(x, t, u)$ is an m-dimensional vector-function and $F(x, t)$ is a $m \times k$ matrix-function.

Remark A.24 Note that Eq. (A.61) turns into an ordinary differential equation when $\mu(t) \equiv t$ and into a difference equation when $\mu^{\mathrm{c}}(t) = 0$, $\Psi(x, t, 0) = 0$ and $D_\mu = \{\tau_i\}_{i=1,2,\ldots}$ is a monotonically increasing sequence of jump instants of $\mu(t)$. Indeed, in the last case we can put

$$X(\tau_i) = X_{i+1}, \quad X(\tau_i-) = X_i, \quad \Delta\mu(\tau_i) = u_i, \quad \Psi(X_i, u_i, \tau_i) = \Psi_i(X_i, u_i),$$

$$X_{i+1} - X_i = \Psi_i(X_i, u_i).$$

The next theorem gives conditions on $F(x, t)$ and $\Psi(x, t, u)$ that ensure the existence and uniqueness of the solution of Eq. (A.61).

Theorem A.24 *Let $F(x,t)$ and $\Psi(x,t,u)$ be Borel functions in the variables (x,t). Suppose that for any $x, y \in \mathbb{R}^m$, $u \in \mathbb{R}^k$ and $t \in [t_0, T]$ these functions satisfy Lipschitz and linear growth conditions ($d\mu^{c}(\cdot)$-a.e. with respect to $F(\cdot)$ and $d\mu^{d}(\cdot)$-a.e. with respect to $\Psi(\cdot)$), i.e.,*

$$\|F(x,t) - F(y,t)\| \le h_1(t)\,\|x - y\|, \tag{A.62}$$

$$\|F(x,t)\| \le h_2(t)\,\big(1 + \|x\|\big), \tag{A.63}$$

$$\sup_{[t_0,T]} \|F(0,t)\| < \infty, \tag{A.64}$$

$$\|\Psi(x,t,u) - \Psi(y,t,u)\| \le h_3(t)\,\|u\|\,\|x - y\|, \tag{A.65}$$

$$\|\Psi(x,t,u)\| \le h_4(t)\,\|u\|\,\big(1 + \|x\|\big), \tag{A.66}$$

where for $i = 1, 2, 3, 4$

$$\int_{t_0}^{T} h_i(s)\,dV(s) < \infty, \quad \textit{with} \quad V(s) = \operatorname*{Var}_{[t_0,s]} \mu(t) \tag{A.67}$$

(Here $\|\cdot\|$ stands for any of the matrix or vector norms and we use the simpler notation $\sum_{\tau \le t}$ for $\sum_{\tau \in D_\mu \cap [t_0,t]}$.)

Then for any initial condition $X(t_0-) = X_0$ Eq. (A.61) has exactly one solution.

Proof of Theorem A.24 Let us consider the operator Φ acting on the space $BV_m^+\{[t_0,T], X_0\}$ by the rule $\Phi\big(X(\cdot)\big) = Y(\cdot)$, where

$$Y(t) = X_0 + \int_{t_0}^{t} F(X(s), s)d\mu^{c}(s) + \sum_{\tau \le t} \Psi(X(\tau-), \tau, \Delta\mu(\tau)) \tag{A.68}$$

for $t \in [t_0, T]$, and $Y(t) = X_0$ for $t < t_0$. First, let us verify that Φ maps the space $BV_m^+\{[t_0,T], X_0\}$ into itself (see the proof of Theorem A.19). Indeed, the right-continuity of $Y(t)$ follows from its definition (A.68). Next, (A.62)–(A.66) and the obvious inequality

$$\operatorname*{Var}_{[t_0,t]} \int_{t_0}^{s} \varphi(\tau)d\mu(\tau) \le \int_{t_0}^{t} \|\varphi(\tau)\|\, d \operatorname*{Var}_{[t_0,\tau]} \mu(s) \tag{A.69}$$

which holds for any $d\mu(\cdot)$-integrable function $\varphi(\tau)$, imply that

$$
\begin{aligned}
\operatorname*{Var}_{[t_0,T]} Y(s) \le \|X_0\| + \int_{t_0}^{T} \|F(X(t),t)\|\, d \operatorname*{Var}_{[t_0,t]} \mu^{c}(s) \\
+ \sum_{\tau\le T} \|\Psi\big(X(\tau-),\tau,\Delta\mu(\tau)\big)\| \le \\
\le \|X_0\| + \int_{t_0}^{T} h_2(t)\big(1+\|X(t)\|\big)\, d \operatorname*{Var}_{[t_0,t]} \mu^{c}(s) + \\
\sum_{\tau\le T} h_4(\tau)\|\Delta\mu(\tau)\|\,\big(1+\|X(\tau-)\|\big) \le \\
\le \|X_0\| + \Big(1+\sup_{[t_0,T]}\|X(t)\|\Big)\int_{t_0}^{T} \big(h_2(t)+h_4(t)\big)\, dV(t) < \infty.
\end{aligned}
$$

Hence, $\Phi\big(X(\cdot)\big) = Y(\cdot) \in BV_m^{+}\{[t_0,T],X_0\}$, as needed.

Now we claim that some power of the operator Φ is a contraction. Indeed, take any two functions $X_1(\cdot),\ X_2(\cdot) \in BV_m^{+}\{[t_0,T],X_0\}$ and let $Y_i(\cdot) = \Phi\big(X_i(\cdot)\big)$, $i=1,2$. Then for any $t',t'' \in [t_0,T]$ with $t' < t''$, inequality (A.69) and the Lipschitz conditions (A.62) and (A.65) imply that

$$
\begin{aligned}
&\|\,\big[Y_2(t'')-Y_1(t'')\big]-\big[Y_2(t')-Y_1(t')\big]\,\| \le \\
&\quad\le \int_{t'}^{t''} \|F\big(X_2(s),s\big)-F\big(X_1(s),s\big)\|\, d \operatorname*{Var}_{[t',s]} \mu^{c}(t) + \\
&\quad+ \sum_{t'<\tau\le t''} \|\Psi\big(X_2(\tau-),\tau,\Delta\mu(\tau)\big)-\Psi\big(X_1(\tau-),\tau,\Delta\mu(\tau)\big)\| \le \\
&\quad\le \int_{t'}^{t''} h_1(s)\,\|X_2(s)-X_1(s)\|\, d \operatorname*{Var}_{[t_0,s]} \mu^{c}(t) + \\
&\quad+ \sum_{t'<\tau\le t''} h_3(\tau)\,\|\Delta\mu(\tau)\|\,\|X_2(\tau-)-X_1(\tau-)\| \le \\
&\quad\le \int_{t'}^{t''} \|X_2(s-)-X_1(s-)\| dM(s) \le \operatorname*{Var}_{[t',t'']}\big[X_2(s)-X_1(s)\big]\int_{t'}^{t''} dM(s),
\end{aligned}
$$

where $M(t)$ is a finite non-decreasing function of bounded variation which admits the decomposition

$$
M(t) = M^{c}(t) + M^{d}(t), \quad M(t_0) = 0, \tag{A.70}
$$

with

$$
dM^{c}(t) = h_1(t) d\|\mu^{c}(t)\|, \quad dM^{d}(t) = h_3(t) d\|\mu^{d}(t)\|.
$$

Summing up these increments for all $[t', t''] \in [t_0, t]$, we obtain

$$\rho_t\big(\Phi(X_2), \Phi(X_1)\big) = \rho_t(Y_2, Y_1) = \operatorname*{Var}_{(-\infty,t]}\big(Y_2(s) - Y_1(s)\big) \leq$$

$$\leq \int_{t_0}^{t} \rho_{s-}(X_2, X_1) dM(s) \leq \rho_t(X_2, X_1) \int_{t_0}^{t} dM(s) = \rho_t(X_2, X_1) M(t).$$

Then, in view of Lemma A.4, after n iterations we have

$$\rho_t\big(\Phi^n(X_2), \Phi^n(X_1)\big) \leq \rho_t(X_2, X_1) \int_{t_0}^{t} \int_{t_0}^{s_1-} \dots \int_{t_0}^{s_{n-1}-} dM(s_1) \dots dM(s_n) \leq$$

$$\leq \rho_t(X_2, X_1) \frac{1}{n!} \int_{t_0}^{t} M^n(s-) dM(s) \leq \rho_t(X_2, X_1) \frac{M^{n+1}(T)}{(n+1)!}.$$

Since $M(T) < \infty$, there exists n_0 such that for $n > n_0$, $\dfrac{M^{n+1}(T)}{(n+1)!} < 1$. Hence, the mapping Φ^n is a contraction. Therefore, by the fixed point theorem (see [204]) and Lemma A.1, Eq. (A.61) has a unique solution. The theorem is proved. □

A.3.2 Dependence on Parameters and Initial Conditions

As we mentioned in the Introduction, differential equations with measure can be used to describe approximately some real control processes with jump-like variables. Therefore, it is important to investigate the influence of small variations of the functions in the right-hand sides of such equations as well as the influence of small variations of the initial conditions. To this end, let us consider a differential equation with measure depending on a vector-parameter α:

$$X(t) = X_0 + \int_{t_0}^{t} F(X(s), s, \alpha) d\mu^c(s) + \sum_{\tau \leq t} \Delta X(\tau), \tag{A.71}$$

where $\Delta X(\tau) = \Psi\big(X(\tau-), \tau, \Delta\mu(\tau), \alpha\big)$.

Theorem A.25 *Suppose also that the functions $F(\cdot)$ and $\Psi(\cdot)$ in Eq. (A.71) depend continuously on the vector-parameter $\alpha \in A \subset \mathbb{R}^r$, where A is an open set. Suppose that the conditions (A.62)–(A.66) of Theorem A.24 hold uniformly with respect to α, and that the following Lipschitz conditions are satisfied for any $\alpha, \beta \in A$:*

$$\|F(x, t, \alpha) - F(x, t, \beta)\| \leq h_5(t)\, \|\alpha - \beta\|, \tag{A.72}$$

$$\|\Psi(x, t, u, \alpha) - \Psi(x, t, u, \beta)\| \leq h_6(t)\, \|u\|\, \|\alpha - \beta\|, \tag{A.73}$$

where $h_i(t)$, $i = 5, 6$, satisfy the condition (A.67). Then any solution $\varphi(t, \alpha)$ of Eq. (A.71) is continuous in α, uniformly in $t \in [t_0, T]$.

Proof of Theorem A.25 By Theorem A.24, for any fixed $\alpha \in A$ Eq. (A.71) has a unique solution. Let $\varphi(t, \alpha)$ and $\varphi(t, \beta)$ be the solutions of (A.71) for arbitrary $\alpha, \beta \in A$. Then,

$$
\begin{aligned}
\|\varphi(t,\alpha) - \varphi(t,\beta)\| \le \int_{t_0}^{t} \|F\big(\varphi(s,\alpha), s, \alpha\big) - F\big(\varphi(s,\beta), s, \beta\big)\| d \operatorname*{Var}_{[t_0,s]} \mu^{c}(\tau) + \\
+ \sum_{\tau \le t} \|\Psi\big(\varphi(\tau-,\alpha), \tau, \Delta\mu(\tau), \alpha\big) - \Psi\big(\varphi(\tau-,\beta), \tau, \Delta\mu(\tau), \beta\big)\| = \\
= \int_{t_0}^{t} \|F\big(\varphi(s,\alpha), s, \alpha\big) - F\big(\varphi(s,\alpha), s, \beta\big) + \\
+ F\big(\varphi(s,\alpha), s, \beta\big) - F\big(\varphi(s,\beta), s, \beta\big)\| d \operatorname*{Var}_{[t_0,s]} \mu^{c}(\tau) + \\
+ \sum_{\tau \le t} \|\Psi\big(\varphi(\tau-,\alpha), \tau, \Delta\mu(\tau), \alpha\big) - \Psi\big(\varphi(\tau-,\alpha), \tau, \Delta\mu(\tau), \beta\big) + \\
+ \Psi\big(\varphi(\tau-,\alpha), \tau, \Delta\mu(\tau), \beta\big) - \Psi\big(\varphi(\tau-,\beta), \tau, \Delta\mu(\tau), \beta\big)\| \le \\
\le \int_{t_0}^{t} h_5(s)\, \|\alpha - \beta\| d \operatorname*{Var}_{[t_0,s]} \mu^{c}(\tau) + \int_{t_0}^{t} h_1(s)\, \|\varphi(s,\alpha) - \varphi(s,\beta)\| d \operatorname*{Var}_{[t_0,s]} \mu^{c}(\tau) + \\
+ \sum_{\tau \le t} h_6(\tau)\, \|\Delta\mu(\tau)\|\, \|\alpha - \beta\| + \sum_{\tau \le t} h_3(\tau)\, \|\Delta\mu(\tau)\|\, \|\varphi(\tau-,\alpha) - \varphi(\tau-,\beta)\| = \\
= h(t) + \int_{t_0}^{t} \|\varphi(s-,\alpha) - \varphi(s-,\beta)\| dM(s).
\end{aligned}
$$

Here $M(t)$ is from (A.70) and and $h(t)$ is given by the formula

$$
h(t) = \|\alpha - \beta\| \int_{t_0}^{t} dM_1(s),
$$

with $dM_1(s) = dM_1^{c}(s) + dM_1^{d}(s)$ and

$$
dM_1^{c}(s) = h_5(s) d\|\mu^{c}(s)\|, \quad dM_1^{d}(s) = h_6(s) d\|\mu^{d}(s)\|,
$$

By Lemma A.5, it follows that

$$\|\varphi(t, \alpha) - \varphi(t, \beta)\| \leq$$
$$\leq \|\alpha - \beta\| \int_{t_0}^{t} dM_1(s) + \varphi_0(t) \int_{t_0}^{t} \varphi_0^{-1}(s) \|\alpha - \beta\| M_1(s-)\, dM(s) = \tag{A.74}$$
$$= \|\alpha - \beta\|\, f(t),$$

where

$$f(t) = M_1(t) + \varphi_0(t) \int_{t_0}^{t} \varphi_0^{-1}(s) M_1(s-)\, dM(s),$$

with

$$\varphi_0(t) = e^{M^c(t)} \prod_{\tau \leq t} \bigl(1 + \Delta M(\tau)\bigr).$$

The function $f(t)$ is obviously bounded: $f(t) \leq C$, for some constant C, So we conclude that

$$\sup_{t_0 \leq t \leq T} \|\varphi(t, \alpha) - \varphi(t, \beta)\| \leq \|\alpha - \beta\|\, C.$$

Passing to the limit (as $\|\alpha - \beta\| \to 0$) in both sides of this inequality, we obtain the assertion of the theorem. □

The next result provides sufficient conditions for differentiability of solutions of Eq. (A.71) with respect to a parameter.

Theorem A.26 *Under the conditions of Theorem A.25, suppose that the functions $F(\cdot)$ and $\Psi(\cdot)$ admit partial derivatives*

$$\frac{\partial F(x, t, \alpha)}{\partial \alpha_j}, \quad \frac{\partial F(x, t, \alpha)}{\partial x_i}, \quad \frac{\partial \Psi(x, t, u, \alpha)}{\partial \alpha_j}, \quad \frac{\partial \Psi(x, t, u, \alpha)}{\partial x_i},$$

$(i = 1, \ldots, m,\ j = 1, \ldots, r)$ that are continuous in (x, α). Let $\varphi(t, \alpha)$ be a solution of Eq. (A.71) for $\alpha \in A$, where $A \subset \mathbb{R}^r$ is an open convex set. Then not only $\varphi(t, \alpha)$, but also its partial derivatives $\xi^j(t, \alpha) := \partial \varphi(t, \alpha)/\partial \alpha_j$ are continuous in α.

Moreover, each of these partial derivatives satisfies the linear differential equation with measure

$$
\begin{aligned}
\xi^j(t,\alpha) = \sum_{l=1}^{k} \int_{t_0}^{t} \frac{\partial F^l\big(\varphi(s,\alpha),s,\alpha\big)}{\partial x} \xi^j(s,\alpha)\, d\mu_l^c(s) + \\
+ \sum_{\tau \le t} \frac{\partial \Psi\big(\varphi(\tau-,\alpha),\tau,\Delta\mu(\tau),\alpha\big)}{\partial x} \xi^j(\tau-,\alpha) + \\
+ \int_{t_0}^{t} \frac{\partial F\big(\varphi(s,\alpha),s,\alpha\big)}{\partial \alpha_j}\, d\mu^c(s) + \sum_{\tau \le t} \frac{\partial \Psi\big(\varphi(\tau-,\alpha),\tau,\Delta\mu(\tau),\alpha\big)}{\partial \alpha_j}
\end{aligned}
\tag{A.75}
$$

with $\xi^j(t_0-,\alpha) = 0$, $j = 1,\dots,r$. *Here* $F^l(x,s,\alpha)$ *is the* l*-th column of the matrix* $F(x,s,\alpha)$, $l = 1, \dots, k$.

We first prove an auxiliary proposition, known as Hadamard's Lemma [388].

Lemma A.6 *Let* $h(\alpha) = (h_1(\alpha),\dots,h_m(\alpha)$ *be a vector-function that is continuously differentiable in the vector argument* $\alpha = (\alpha_1,\dots,\alpha_r)$ *on an convex open set* $A \subset \mathbb{R}^r$. *Then there exists a matrix-valued function* $H(\alpha,\beta)$ *defined for* $\alpha,\ \beta \in A$, *with continuous components* $H_{ij}(\alpha,\beta)$ *such that*

$$
h(\alpha) - h(\beta) = H(\alpha,\beta)(\alpha-\beta) \quad \text{and} \quad H_{ij}(\alpha,\alpha) = \frac{\partial h_i}{\partial \alpha_j}(\alpha). \tag{A.76}
$$

Proof of Lemma A.6 For any $\alpha,\ \beta \in A$ and $\lambda \in [0,1]$, set $\gamma(\lambda) = \beta + \lambda(\alpha-\beta)$ and denote $f_i(\lambda) = h_i\big(\gamma(\lambda)\big)$, $i = 1,\dots,m$. Since A is convex, $\gamma(\lambda) \in A$ and so the function $f_i(\lambda)$ is defined for $\lambda \in [0,1]$.

In these notations, the difference $h_i(\alpha) - h_i(\beta)$ can be expressed as

$$
\begin{aligned}
h_i(\alpha) - h_i(\beta) = f_i(1) - f_i(0) = \int_0^1 \frac{df_i}{d\lambda}(\lambda)\,d\lambda = \\
= \int_0^1 \sum_{j=1}^{r} \frac{\partial h_i}{\partial \alpha_j}\big(\beta + \lambda(\alpha-\beta)\big) \frac{\partial \gamma_j}{\partial \lambda}(\lambda)\,d\lambda = \\
= \sum_{j=1}^{r} \left[\int_0^1 \frac{\partial h_i}{\partial \alpha_j}\big(\beta + \lambda(\alpha-\beta)\big)\,d\lambda \right] (\alpha_j - \beta_j).
\end{aligned}
$$

Denoting

$$
\int_0^1 \frac{\partial h_i}{\partial \alpha_j}\big(\beta + \lambda(\alpha-\beta)\big)\,d\lambda = H_{ij}(\alpha,\beta), \tag{A.77}
$$

we get

$$h_i(\alpha) - h_i(\beta) = \sum_{j=1}^{r} H_{ij}(\alpha, \beta)(\alpha_j - \beta_j).$$

Clearly, (A.76) is the matrix form of this equation. Moreover, it follows from (A.77) that $H_{ij}(\alpha, \alpha) = \partial h_i(\alpha)/\partial \alpha_j$. The lemma is proved. □

Proof of Theorem A.26 Fix an index j. Let $e = (0, \dots, 0, 1, 0, \dots, 0)^*$ be the standard r-dimensional vector (with the j-th component equal to 1 and all the other components equal to zero). Take arbitrary α and $\alpha + \varepsilon e$ in the set A, and let the functions $\varphi(t, \alpha)$ and $\varphi(t, \alpha + \varepsilon e)$ be corresponding solutions of Eq. (A.71). Then the difference $\Delta_\varepsilon \varphi(t, \alpha) := \varphi(t, \alpha + \varepsilon e) - \varphi(t, \alpha)$ satisfies

$$\Delta_\varepsilon \varphi(t, \alpha) = \int_{t_0}^{t} \big[F(\varphi(s, \alpha + \varepsilon e), s, \alpha + \varepsilon e) - F(\varphi(s, \alpha), s, \alpha)\big] d\mu^{c}(s) + \\ + \sum_{\tau \le t} \big[\Psi(\varphi(\tau-, \alpha + \varepsilon e), \tau, \Delta\mu(\tau), \alpha + \varepsilon e) - \Psi(\varphi(\tau-, \alpha), \tau, \Delta\mu(\tau), \alpha)\big]. \tag{A.78}$$

First we transform the integral term in (A.78). By Lemma A.6, the l-th column of the matrix $F(x, t, \alpha)$, $l = 1, \dots, k$ satisfies the relation

$$\begin{aligned} F^l\big(\varphi(s, \alpha + \varepsilon e), s, \alpha + \varepsilon e\big) - F^l\big(\varphi(s, \alpha), s, \alpha\big) = \\ = \big[F^l\big(\varphi(s, \alpha + \varepsilon e), s, \alpha + \varepsilon e\big) - F^l\big(\varphi(s, \alpha + \varepsilon e), s, \alpha\big)\big] + \\ + \big[F^l\big(\varphi(s, \alpha + \varepsilon e), s, \alpha\big) - F^l\big(\varphi(s, \alpha), s, \alpha\big)\big] = \\ = F^l_\alpha(s, \alpha, \varepsilon)\varepsilon + F^l_x(s, \alpha, \varepsilon)\Delta_\varepsilon \varphi(s, \alpha), \end{aligned} \tag{A.79}$$

where the vector-function $F^l_\alpha(s, \alpha, \varepsilon)$ and the matrix-function $F^l_x(s, \alpha, \varepsilon)$ have the components $(i, n = 1, \dots, m)$

$$F^l_{\alpha,i}(s, \alpha, \varepsilon) = \int_0^1 \frac{\partial F^l_i}{\partial \alpha_j}\big(\varphi(s, \alpha + \varepsilon e), s, \alpha + \lambda \varepsilon e\big) d\lambda$$

and

$$F^l_{x,i,n}(s, \alpha, \varepsilon) = \int_0^1 \frac{\partial F^l_i}{\partial x_n}\big(\varphi(s, \alpha) + \lambda \Delta_\varepsilon \varphi(s, \alpha), s, \alpha\big) d\lambda,$$

respectively, which are continuous with respect to α and ε.

Substituting the expression (A.79) in the integral term in (A.78), we get

$$\int_{t_0}^{t} \left[F\big(\varphi(s, \alpha + \varepsilon e), s, \alpha + \varepsilon e\big) - F\big(\varphi(s, \alpha), s, \alpha\big)\right] d\mu^{\mathrm{c}}(s) =$$

$$= \left[\sum_{l=1}^{k} \int_{t_0}^{t} F_{\alpha}^{l}(s, \alpha, \varepsilon) d\mu_l^{\mathrm{c}}(s)\right]\varepsilon + \left[\sum_{l=1}^{k} \int_{t_0}^{t} F_{x}^{l}(s, \alpha, \varepsilon)\Delta_{\varepsilon}\varphi(s, \alpha) d\mu_l^{\mathrm{c}}(s)\right]. \tag{A.80}$$

Now, as above, let $d\nu(\cdot)$ be the scalar measure with the distribution function

$$\nu(t) = \operatorname*{Var}_{[t_0,t]} \mu(s) = \sum_{l=1}^{k} \operatorname*{Var}_{[t_0,t]} \mu_l(s).$$

Let $N_l(\cdot)$ denote the Radon-Nikodym derivative of the measure $d\mu_l(\cdot)$ with respect to the measure $d\nu(\cdot)$ and $N(\cdot)$ denote the vector with the components $N_l(\cdot)$, $l = 1, \dots, k$. Then

$$d\mu^{\mathrm{c}}(t) = N(t)d\nu^{\mathrm{c}}(t), \quad d\mu^{\mathrm{d}}(t) = N(t)d\nu^{\mathrm{d}}(t), \quad t \in [t_0, T].$$

In these notations the difference (A.80) takes on the form

$$\varepsilon \int_{t_0}^{t} \left[\sum_{l=1}^{k} F_{\alpha}^{l}(s, \alpha, \varepsilon) N_l(s)\right] d\nu^{\mathrm{c}}(s) + \int_{t_0}^{t} \left[\sum_{l=1}^{k} F_{x}^{l}(s, \alpha, \varepsilon) N_l(s)\right] \Delta_{\varepsilon}\varphi(s, \alpha) d\nu^{\mathrm{c}}(s).$$

In much the same way we transform the difference in the second term in (A.78). We have the expression

$$\varepsilon \sum_{\tau \le t} \Psi_{\alpha}\big(\tau-, N(\tau)\Delta\nu(\tau), \alpha, \varepsilon\big) + \sum_{\tau \le t} \Psi_{x}\big(\tau-, N(\tau)\Delta\nu(\tau), \alpha, \varepsilon\big)\, \Delta_{\varepsilon}\varphi(\tau-, \alpha),$$

where the vector-function $\Psi_{\alpha}(\cdot)$ and the matrix-function $\Psi_x(\cdot)$ have continuous (in α and ε) components ($i, n = 1, \dots, m$)

$$\Psi_{\alpha,i}\big(\tau-, N(\tau)\Delta\nu(\tau), \alpha, \varepsilon\big) =$$

$$= \int_0^1 \frac{\partial \Psi_i}{\partial \alpha_j}\big(\varphi(\tau-, \alpha + \varepsilon e), \tau, N(\tau)\Delta\nu(\tau), \alpha + \lambda\varepsilon e\big)\, d\lambda$$

and

$$\Psi_{x,i,n}\big(\tau-, N(\tau)\Delta\nu(\tau), \alpha, \varepsilon\big) =$$

$$= \int_0^1 \frac{\partial \Psi_i}{\partial x_n}\big(\varphi(\tau-, \alpha) + \lambda\Delta_{\varepsilon}\varphi(\tau-, \alpha), \tau, N(\tau)\Delta\nu(\tau), \alpha\big)\, d\lambda,$$

respectively. Substituting the above expression for the transformed differences in (A.78) and dividing the result by ε, we get

$$\begin{aligned}\frac{\Delta_\varepsilon\varphi(t,\alpha)}{\varepsilon} &= h^\varepsilon(t,\alpha) + \int_{t_0}^{t} B^\varepsilon(s,\alpha)\frac{\Delta_\varepsilon\varphi(s,\alpha)}{\varepsilon}dv^{c}(s) + \\ &+ \sum_{\tau\le t}\Psi_x\big(\tau-, N(\tau)\Delta v(\tau), \alpha, \varepsilon\big)\frac{\Delta_\varepsilon\varphi(\tau-,\alpha)}{\varepsilon},\end{aligned} \tag{A.81}$$

where

$$h^\varepsilon(t,\alpha) = \int_{t_0}^{t}\left[\sum_{l=1}^{k} F_\alpha^l(s,\alpha,\varepsilon)N_l(s)\right]dv^{c}(s) + \sum_{\tau\le t}\Psi_\alpha\big(\tau-, N(\tau)\Delta v(\tau), \alpha, \varepsilon\big),$$

and

$$B^\varepsilon(s,\alpha) = \sum_{l=1}^{k} F_x^l(s,\alpha,\varepsilon)N_l(s).$$

Now note that $h^\varepsilon(t_0,\alpha) = 0$. Hence,

$$\frac{\Delta_\varepsilon\varphi(t_0,\alpha)}{\varepsilon} = 0.$$

Thus, (A.81) is a linear differential equation with a measure with respect to the vector-function $\Delta_\varepsilon\varphi(t,\alpha)/\varepsilon$, with null initial condition. Let $\eta(t,\alpha,\varepsilon)$ be a solution of this equation for $\varepsilon \neq 0$. Then, by Theorem A.24, this solution is unique, and the function

$$\eta(t,\alpha,\varepsilon) = \frac{\Delta_\varepsilon\varphi(t,\alpha)}{\varepsilon}, \tag{A.82}$$

is defined for ε in some neighborhood of $\varepsilon = 0$. By Theorem A.25, $\eta(t,\alpha,\varepsilon)$ depends continuously on α and ε. Therefore, letting $\varepsilon \to 0$ in (A.82), we obtain

$$\xi^j(t,\alpha) = \frac{\partial\varphi(t,\alpha)}{\partial\alpha_j} = \lim_{\varepsilon\to 0}\frac{\Delta_\varepsilon\varphi(t,\alpha)}{\varepsilon} = \lim_{\varepsilon\to 0}\eta(t,\alpha,\varepsilon) = \eta(t,\alpha,0), \tag{A.83}$$

with $\xi^j(t_0,\alpha) = \eta(t_0,\alpha,0) = 0$. Thus, the partial derivative $\partial\varphi(t,\alpha)/\partial\alpha_j$ exists and is continuous in α because so is $\eta(t,\alpha,0)$.

Now let us return to Eq. (A.75). To verify that it holds, we look at the Eq. (A.81). Since the functions $F(\cdot)$ and $\Psi(\cdot)$ satisfy Lipschitz conditions, all the functions under the integral sign in (A.81) are bounded by integrable functions that are independent on ε. By Lebesgue's bounded convergence theorem, we are allowed to let $\varepsilon \to 0$ in the integrals figuring in (A.81). So letting $\varepsilon \to 0$ in both sides of

(A.81) and using (A.83) and the continuity with respect to ε of all the integrands, we obtain Eq. (A.75). The theorem is proved. □

To conclude this subsection, let us show that the problem of the dependence of the solution on initial conditions can be reduced to the problem just settled.

Indeed, let $x(t) = \varphi(t, X_0)$ be the solution of Eq. (A.61) with $x(t_0-) = X_0$. Denoting $\alpha = X_0$, we introduce a new function $y(t, \alpha) = x(t) - \alpha$. Then $y(t, \alpha)$ satisfies the equation

$$Y(t) = \int_{t_0}^{t} F(Y(s) - \alpha, s)d\mu^c(s) + \sum_{\tau \le t} \Delta Y(\tau), \tag{A.84}$$

where $\Delta Y(\tau) = \Psi\big(Y(\tau-) - \alpha, \tau, \Delta\mu(\tau)\big)$ and $Y(t_0-) = 0$.

Thus, we obtained Eq. (A.84) with the vector-parameter α and the null initial condition. This equation and Eq. (A.61) are equivalent and the Lipschitz conditions (A.72) and (A.73) follow from (A.62) and (A.65). Hence, the next result is a direct corollary of Theorems A.25 and A.26.

Theorem A.27 *Suppose the functions $F(\cdot)$ and $\Psi(\cdot)$ satisfy the conditions of Theorem A.24. Then the solution $\varphi(t, X_0)$ of Eq. (A.61) depends continuously on the initial condition X_0, uniformly with respect to t.*

Moreover, if the partial derivatives

$$\frac{\partial F(x,t)}{\partial x_i}, \quad \frac{\partial \Psi(x,t,u)}{\partial x_i}, \quad i = 1, \dots, m$$

exists and are continuous in x, then not only $\varphi(t, X_0)$, but also its partial derivatives $\xi^i(t, X_0) = \partial\varphi(t, X_0)/\partial X_{0,i}$ are continuous in X_0 for the same values of t and X_0.

Each of these partial derivatives satisfies the linear differential equation with measure

$$\xi^i(t, X_0) = \sum_{l=1}^{k} \int_{t_0}^{t} \frac{\partial F^l\big(\varphi(s, X_0), s\big)}{\partial x} \xi^i(s, X_0)\, d\mu_l^c(s)$$
$$+ \sum_{\tau \le t} \frac{\partial \Psi\big(\varphi(\tau-, X_0), \tau, \Delta\mu(\tau)\big)}{\partial x} \xi^i(\tau-, X_0)$$

with $\xi^i(t_0-, X_0) = 0$, $i = 1, \dots, m$. Here $F^l(x, s)$ is the l-th column of the matrix $F(x, s)$, $l = 1, \dots, k$. □

A.3.3 *Riccati Differential Equation with Measure and a Generalized Kalman Filter*

This subsection discusses properties of solutions of the Riccati differential equation with a measure. This equation appears in a generalized Kalman filter, and consequently in linear-quadratic stochastic control problems. In particular, we already encountered this equation when solving problems of observation control. The issue here is that the equations of the classical Kalman filter for partially observable Gaussian processes (in discrete and continuous time) involve the inversion of certain matrices associated with the covariance matrix of the noises in the observations. It is known that in the case when these matrices are singular, inversion must be replaced with pseudo-inversion [244]. In discrete time, this replacement is always feasible. In continuous time, there exist sufficient conditions for the implementability of such a substitution [75]. In the case of equations with measure, the fact that the inversion operation can be replaced by pseudo-inversion requires a separate proof, see [291]. Moreover, in both the discrete and the continuous time cases, if the inverse (pseudo-inverse) matrices are ill conditioned, there arises the problem of constructing a regularized filter (in which these matrices would be well conditioned and could be numerically inverted with a prescribed accuracy). It is essential that the regularized filter should provide unbiased estimates that in addition are asymptotically close to optimal ones [75, 243, 291, 397].

The aim of this subsection is to obtain similar results in the case when a partially observable process is described by an Itô stochastic differential equation with measure. As already remarked, such a representation allows one to provide a unified description of continuous, discrete, and discrete-continuous processes. In addition, aiming at the application of filtering to stochastic control problems, we will also look at filtering for partially-observable linear controlled process with a control that is a functional (predictable) in the observations.

A rigorous formulation of the problem is as follows.

Consider a stochastic basis $(\Omega, F, \mathbf{F} = (F_t)_{0\le t\le T}, P)$ which satisfies the standard conditions [245]. Suppose that on $(\Omega, F, \mathbf{F})$ there are given an unobservable m-dimensional vector process $X = (X(t))_{0\le t\le T}$ and an observable k-dimensional vector process $Y = (Y(t))_{0\le t\le T}$, with respective representations

$$X(t) = X_0 + \int_{]0,t]} \big[da_1(s)\big]X(s-) + \eta(t), \tag{A.85}$$

and

$$Y(t) = Y_0 + \int_{]0,t]} \big[dA_1(s)\big]X(s-) + \zeta(t), \tag{A.86}$$

where $X(t-) = \lim X(s)$ as $s \uparrow t$, and X_0, Y_0 are independent Gaussian vectors with given mean values and covariance matrices; also, $\eta = (\eta(t))_{0\le t\le T}$

and $\zeta = (\zeta(t))_{0 \le t \le T}$ are independent Gaussian martingales (with respect to the family of σ-algebras $\mathbf{F}$) with regular paths (i.e., right-continuous and with left-limits P-a.s.), and with the quadratic characteristics (covariance matrices)

$$\langle \eta \rangle_t = \mathbf{E}\big(\eta(t)\eta^*(t)\big), \quad \langle \eta \rangle_0 = 0, \quad \langle \zeta \rangle_t = \mathbf{E}\big(\zeta(t)\zeta^*(t)\big), \quad \langle \zeta \rangle_0 = 0.$$

Here, as before, $\mathbf{E}$ stands for expectation and $*$ for transposition. The processes η, ζ and the initial conditions X_0, Y_0 are assumed to be independent.

The deterministic $m \times m$ and $k \times m$ matrix-functions $a_1(t)$ and $A_1(t)$ are right-continuous and of bounded variation for $t \in [0, T]$.

It is also assumed that there exists a $k \times m$ matrix-function $\rho(t)$ with regular components, such that the following consistency conditions are fulfilled (for $t \in [0, T]$):

$$dA_1^{\mathrm{c}}(t) = \big[d\langle \zeta \rangle_t^{\mathrm{c}}\big]\rho(t), \tag{A.87}$$

$$\operatorname{tr} \int_{[0,T]} \rho^*(t)\big[d\langle \zeta \rangle_t\big]\rho(t) < \infty. \tag{A.88}$$

Here and below, as before, the superscript c denotes the continuous components of a measure, and tr denotes the trace of a matrix.

Under the hypotheses listed above, the system (A.85) and (A.86) has a unique strong solution [255], and for any $T > 0$

$$\sup_{0 \le t \le T} |X(t)| < \infty, \qquad \sup_{0 \le t \le T} |Y(t)| < \infty \qquad (P\text{-a.s.}).$$

Our aim is to construct a Kalman filter for the estimation of the process X based on the observations Y, i.e., obtain the conditional mean $m(t) = \mathbf{E}\big(X(t)|F_t^Y\big)$, where $F_t^Y = \sigma\{Y(s),\ s \le t\}$ is the σ-algebra generated by observations of the process Y up to and including time t.

For this purpose we simplify the form the system (A.85) and (A.86) by reducing it to a system with a common scalar measure. Namely, we denote by $\operatorname{Var} a_1(t) = \sum_{i,j} \operatorname{Var} a_{1,ij}(t)$ the sum of the variations of the elements of the matrix $a_1(s)$ for $s \in [0, t]$, and we adopt a similar notation for the matrices $A_1(t)$, $\langle \zeta \rangle_t$ and $\langle \eta \rangle_t$.

Next, we introduce the function

$$\mu(t) = t + \operatorname{Var} a_1(t) + \operatorname{Var} A_1(t) + \operatorname{Var} \langle \zeta \rangle_t + \operatorname{Var} \langle \eta \rangle_t.$$

Let $b(t)$ be any of the elements of one of the matrices $a_1(t)$, $A_1(t)$, or $\langle \zeta \rangle_t$, $\langle \eta \rangle_t$. Then the measure $db(\cdot)$ is absolutely continuous with respect to the measure $d\mu(\cdot)$: $db(\cdot) \ll d\mu(\cdot)$.

Denote by $a_{ij}(t)$ the Radon-Nicodym derivative of the measure $da_{1,ij}(\cdot)$ with respect to $d\mu(\cdot)$, and by $a(t)$ the matrix with the elements $a_{ij}(t)$, $i, j = 1, \dots, m$. Then

$$da_1(t) = a(t)d\mu(t).$$

Analogously, we put

$$\begin{aligned} dA_1(t) &= A(t)d\mu(t), \\ d\langle\eta\rangle_t &= \alpha(t)d\mu(t), \\ d\langle\zeta\rangle_t &= \beta(t)d\mu(t), \\ dt &= \theta(t)d\mu^{\mathrm{c}}(t). \end{aligned} \tag{A.89}$$

In terms of these quantities, condition (A.87) says that the algebraic system

$$A(t) = \beta(t)\rho(t). \tag{A.90}$$

is consistent with respect to the elements of the matrix $\rho(t)$ for $d\mu^{\mathrm{c}}(\cdot)$-a.e. t. Next, in view of (A.89) and (A.90), condition (A.88) becomes

$$\mathrm{tr} \int_{[0,T]} A^*(t)\beta^+(t)A(t)d\mu(t) < \infty, \tag{A.91}$$

where the superscript "+" stands for pseudoinverse [244].

In the new notation, the system (A.85) and (A.86) reads

$$X(t) = X_0 + \int_{]0,t]} a(s)X(s-)d\mu(s) + \eta(t), \tag{A.92}$$

$$Y(t) = Y_0 + \int_{]0,t]} A(s)X(s-)d\mu(s) + \zeta(t). \tag{A.93}$$

As mentioned above, the output of the Kalman filter for the processes (A.92) and (A.93) requires a regularization of the problem and the rigorous justification of the corresponding limit transitions with respect to the regularization parameter. To achieve this, along with the observation process Y, we consider a regularized process (with non-degenerate noise) $Y^\varepsilon = (Y^\varepsilon(t))_{0\le t\le T}$ with the following regularization procedure:

$$Y_\varepsilon(t) = Y_0 + \int_{]0,t]} A(s)X(s-)d\mu(s) + \zeta(t) + \varepsilon hM(t). \tag{A.94}$$

Here h is a k-dimensional vector with unit components, $\varepsilon > 0$ is a parameter, and the process $M = (M(t))_{0\le t\le T}$ is a Gaussian martingale (with respect to the family of σ-algebras $\mathbf{F}$), which admits the decomposition $M = M^{\mathrm{c}} + M^{\mathrm{d}}$ into a continuous component $M^{\mathrm{c}} = (M^{\mathrm{c}}(t))_{0\le t\le T} = (W(t))_{0\le t\le T} = W$, which is a Wiener process independent of X_0, Y_0, η and ζ, and a discrete component $M^{\mathrm{d}} = (M^{\mathrm{d}}(t))_{0\le t\le T}$, which is a purely discontinuous martingale.

Thanks to Gaussianity, the jumps of the process M^{d} occur at determined instants τ_i, $i = 1, 2, \ldots$, which are also the jump instants of the measure $d\mu(\cdot)$.

The jumps of the process M^{d} are themselves independent zero-mean Gaussian random variables with covariance $\Delta\mu(\tau_i)$, where $\Delta\mu(\tau_i) = \mu(\tau_i) - \mu(\tau-_i)$, $i = 1, 2, \ldots$.

Denote (recall that $I_{(k\times k)}$ stands for the $k \times k$ identity matrix)

$$\zeta^{\varepsilon}(t) = \zeta(t) + \varepsilon h M(t) \tag{A.95}$$

and

$$d\langle\zeta^{\varepsilon}\rangle_t^{\mathrm{c}} = d\langle\zeta\rangle_t^{\mathrm{c}} + \varepsilon^2 I_{(k\times k)} dt. \tag{A.96}$$

Decomposing (by Lebesgue's theorem) the measure $d\mu(\cdot)$ into continuous and discrete components, $d\mu^{\mathrm{c}}(\cdot)$ and $d\mu^{\mathrm{d}}(\cdot)$, i.e., $d\mu(\cdot) = d\mu^{\mathrm{c}}(\cdot) + d\mu^{\mathrm{d}}(\cdot)$ we rewrite (A.96) (using (A.89)) as

$$d\langle\zeta^{\varepsilon}\rangle_t^{\mathrm{c}} = \big(\beta(t) + \varepsilon^2\theta(t) I_{(k\times k)}\big) d\mu^{\mathrm{c}}(t).$$

Therefore,

$$A(t) d\mu^{\mathrm{c}}(t) = A(t)\big(\beta(t) + \varepsilon^2\theta(t) I_{(k\times k)}\big)^{-1} d\langle\zeta^{\varepsilon}\rangle_t^{\mathrm{c}}. \tag{A.97}$$

Similarly, for the discrete components we have

$$d\langle\zeta^{\varepsilon}\rangle_t^{\mathrm{d}} = d\langle\zeta\rangle_t^{\mathrm{d}} + \varepsilon^2 I_{(k\times k)} d\mu^{\mathrm{d}}(t),$$

or

$$d\langle\zeta^{\varepsilon}\rangle_t^{\mathrm{d}} = \big(\beta(t) + \varepsilon^2 I_{(k\times k)}\big) d\mu^{\mathrm{d}}(t).$$

Therefore,

$$A(t) d\mu^{\mathrm{d}}(t) = A(t)\big(\beta(t) + \varepsilon^2 I_{(k\times k)}\big)^{-1} d\langle\zeta^{\varepsilon}\rangle_t^{\mathrm{d}}. \tag{A.98}$$

Now denote

$$H(t) = \theta(t)\mathbf{I}\{\Delta\mu(t) = 0\} + \mathbf{I}\{\Delta\mu(t) \neq 0\}.$$

Then, by (A.97) and (A.98),

$$A(t) d\mu(t) = A(t)\big(\beta(t) + \varepsilon^2 I_{(k\times k)} H(t)\big)^{-1} d\langle\zeta^{\varepsilon}\rangle_t,$$

and consequently (A.94) becomes

$$Y^{\varepsilon}(t) = Y_0 + \int_{]0,t]} A(s)\big(\beta(s) + \varepsilon^2 I_{(k\times k)} H(s)\big)^{-1}\big[d\langle\zeta^{\varepsilon}\rangle_s\big]X(s-) + \zeta^{\varepsilon}(t). \tag{A.99}$$

Formula (A.99) allows us to write the equations of the generalized Kalman filter [245] for the processes X and Y^{ε}. In accordance with the Kalman filtering theory, the evolution of the conditional expectation

$$\pi^{\varepsilon}(t) = \mathbf{E}\big(X(t)|F_t^{Y^{\varepsilon}}\big), \quad \text{with} \quad F_t^{Y^{\varepsilon}} = \sigma\{Y^{\varepsilon}(s), s \le t\}$$

is governed by the equation

$$\pi^{\varepsilon}(t) = \pi_0 + \int_{]0,t]} a(s)\pi^{\varepsilon}(s-)d\mu(s) + \int_{]0,t]} \Gamma^{\varepsilon}(s)d\nu^{\varepsilon}(s), \tag{A.100}$$

where

$$\begin{aligned}\Gamma^{\varepsilon}(t) = &\big(I_{(m\times m)} + a(t)\Delta\mu(t)\big)\gamma^{\varepsilon}(t-)A^*(t)\times\\ &\times\big[\beta(t) + \varepsilon^2 I_{(k\times k)}H(t) + A(t)\gamma^{\varepsilon}(t-)A^*(t)\Delta\mu(t)\big]^{-1},\end{aligned} \tag{A.101}$$

$$d\nu^{\varepsilon}(t) = dY^{\varepsilon}(t) - A(t)\pi^{\varepsilon}(t-)d\mu(t), \tag{A.102}$$

$$\begin{aligned}\gamma^{\varepsilon}(t) = \gamma_0 + &\int_{]0,t]} \big[a(s)\gamma^{\varepsilon}(s-) + \gamma^{\varepsilon}(s-)a^*(s)\big]d\mu(s) + \langle\eta\rangle_t +\\ &+ \sum_{\tau_i \le t} a(\tau_i)\gamma^{\varepsilon}(\tau_i-)a^*(\tau_i)\big(\Delta\mu(\tau_i)\big)^2 -\\ -\int_{]0,t]} &\Gamma^{\varepsilon}(s)\big[\beta(s) + \varepsilon^2 I_{(k\times k)}H(s) + A(s)\gamma^{\varepsilon}(s-)A^*(s)\Delta\mu(s)\big]\Gamma^{\varepsilon *}(s)d\mu(s).\end{aligned} \tag{A.103}$$

Here $\gamma_0 = \mathbf{E}\big[(X_0 - \pi_0)(X_0 - \pi_0)^*\big]$ is calculated by the normal correlation theorem [244], $\gamma^{\varepsilon}(t) = \mathbf{E}\big[\big(X(t) - \pi^{\varepsilon}(t)\big)\big(X(t) - \pi^{\varepsilon}(t)\big)^*\big]$, and the process $\nu^{\varepsilon} = \big(\nu^{\varepsilon}(t)\big)_{0\le t\le T}$ is a Gaussian martingale with regular paths with respect to the family $\mathbf{F}^{Y^{\varepsilon}} = (F_t^{Y^{\varepsilon}})_{0\le t\le T}$ of σ-algebras $F_t^{Y^{\varepsilon}} = \sigma\{Y^{\varepsilon}(s),\ s \le t\}$ (a so-called *renewal process*).

The procedure of passage to the limit in (A.103) when $\varepsilon \downarrow 0$ is defined by the following lemmas.

Lemma A.7 *The ε-parametric family $\gamma^{\varepsilon}(t)$ of solutions of Eq. (A.103) is uniformly bounded and has a uniformly bounded variation on the segment $[0, T]$ for sufficiently small $\varepsilon > 0$.*

Proof of Lemma A.7 Let $D(t)$ be a positive definite solution of the linear matrix equation

$$D(t) = \gamma_0 + \int_{]0,t]} \big[a(s)D(s-) + D(s-)a^*(s)\big]d\mu(s) + \langle\eta\rangle_t + \\ + \sum_{\tau_i \le t} a(\tau_i)D(\tau_i-)a^*(\tau_i)\big(\Delta\mu(\tau_i)\big)^2.$$

Then the matrix-function $\Delta_t^\varepsilon = D(t) - \gamma^\varepsilon(t)$ satisfies the equation

$$\Delta_t^\varepsilon = \int_{]0,t]} \big[a(s)\Delta_{s-}^\varepsilon + \Delta_{s-}^\varepsilon a^*(s)\big]d\mu(s) + \\ + \sum_{\tau_i \le t} a(\tau_i)\Delta_{\tau_i-}^\varepsilon a^*(\tau_i)\big(\Delta\mu(\tau_i)\big)^2 + f^\varepsilon(t), \tag{A.104}$$

where $f^\varepsilon(t)$ is a positive definite matrix, namely, the integral (nonlinear with respect to $\gamma^\varepsilon(t)$) term in the right-hand side of Eq. (A.103). The solution of Eq. (A.104) exists, is unique and is positive definite. Hence $\gamma^\varepsilon(t) \le D(t)$ (in the sense of positive definiteness of Δ_t^ε).

Next, for positive definite symmetric matrices D one always has that $\|D\| = \operatorname{tr} D$, so $\|\gamma^\varepsilon(t)\|$ is uniformly bounded. This implies that the variations of the functions

$$\int_{]0,t]} a(s)\gamma^\varepsilon(s-)d\mu(s), \quad \int_{]0,t]} \gamma^\varepsilon(s-)a^*(s)d\mu(s)$$

and

$$\sum_{\tau_i \le t} a(\tau_i)\gamma^\varepsilon(\tau_i-)a^*(\tau_i)(\Delta\mu(\tau_i))^2$$

are uniformly bounded.

The integral terms in (A.103) are positive definite and the integrals (denoted above by $f^\varepsilon(t)$) are uniformly bounded, since according to (A.103),

$$f^\varepsilon(t) = \int_{]0,t]} \big[a(s)\gamma^\varepsilon(s-) + \gamma^\varepsilon(s-)a^*(s)\big]d\mu(s) + \langle\eta\rangle_t + \\ + \sum_{\tau_i \le t} a(s)\gamma^\varepsilon(\tau_i-)a^*(\tau_i)\big(\Delta\mu(\tau_i)\big)^2 + \gamma_0 - \gamma^\varepsilon(t) < \infty.$$

Moreover, $f^\varepsilon(t)$ (as the integral of a non-negative definite function) is monotonically nondecreasing. Thus, the integral part of $f^\varepsilon(t)$ has uniformly bounded variation. This completes the proof of the lemma. □

Lemma A.8 *For each fixed $t \in [0, T]$, there exists the limit*

$$\lim_{\varepsilon \downarrow 0} \gamma^{\varepsilon}(t) = \gamma(t) = \mathbf{E}\big[\big(X(t) - m(t)\big)\big(X(t) - m(t)\big)^*\big].$$

Proof The argument is analogous to that in the proof of Lemma 10.5 in [244]. □

Corollary A.4 *The uniform boundedness of $\gamma^{\varepsilon}(t)$ and uniform boundedness of its variation imply that the limit function $\gamma(t)$ also enjoys these properties.*

Next, let us derive an equation for the limit function $\gamma(t)$.

The Lebesgue dominated convergence theorem [204] and Lemmas A.7 and A.8 imply that

$$\begin{aligned}
&\lim_{\varepsilon \downarrow 0} \int_{]0,t]} a(s)\gamma^{\varepsilon}(s-)d\mu(s) = \int_{]0,t]} a(s)\gamma(s-)d\mu(s),\\
&\lim_{\varepsilon \downarrow 0} \int_{]0,t]} \gamma^{\varepsilon}(s-)a^*(s)d\mu(s) = \int_{]0,t]} \gamma(s-)a^*(s)d\mu(s),\\
&\lim_{\varepsilon \downarrow 0}\sum_{\tau_i \le t} a(\tau_i)\gamma^{\varepsilon}(\tau_i-)a^*(\tau_i)\big(\Delta\mu(\tau_i)\big)^2 = \sum_{\tau_i \le t} a(\tau_i)\gamma(\tau_i-)a^*(\tau_i)\big(\Delta\mu(\tau_i)\big)^2.
\end{aligned} \tag{A.105}$$

Let us examine the nonlinear terms in Eq. (A.103). First of all, note that, in view of (A.90), the following equality holds ($d\mu^c(\cdot)$-a.e.):

$$\big(\beta(t) + \varepsilon^2\theta(t)I_{(k\times k)}\big)^{-1}A(t) = \beta^+(t)\big(I_{(k\times k)} + \varepsilon^2\theta(t)\beta^+(t)\big)^{-1}A(t). \tag{A.106}$$

Indeed, multiplying the left-hand side (l.h.s.) and right-hand side (r.h.s.) of (A.106) by $\beta(t) + \varepsilon^2\theta(t)I_{(k\times k)}$, using (A.90), and omitting the argument t, we have that

$$\text{l.h.s.} = A = \beta\rho$$

and

$$\begin{aligned}
\text{r.h.s.} &= \big(\beta + \varepsilon^2\theta I_{(k\times k)}\big)\beta^+\big(I_{(k\times k)} + \varepsilon^2\theta\beta^+\big)^{-1}A =\\
&= \big(\beta\beta^+ + \varepsilon^2\theta\beta^+\big)\big(I_{(k\times k)} + \varepsilon^2\theta\beta^+\big)^{-1}A =\\
&= \big(I_{(k\times k)} + \varepsilon^2\theta\beta^+\big)^{-1}\big(\beta\beta^+ + \varepsilon^2\theta\beta^+\big)A =\\
&= \big(I_{(k\times k)} + \varepsilon^2\theta\beta^+\big)^{-1}\big(\beta\beta^+A + \varepsilon^2\theta\beta^+A\big) =\\
&= \big(I_{(k\times k)} + \varepsilon^2\theta\beta^+\big)^{-1}\big(\beta\beta^+\beta\rho + \varepsilon^2\theta\beta^+\beta\rho\big) =\\
&= \big(I_{(k\times k)} + \varepsilon^2\theta\beta^+\big)^{-1}\big(\beta\rho + \varepsilon^2\theta\beta^+\beta\rho\big) =\\
&= \big(I_{(k\times k)} + \varepsilon^2\theta\beta^+\big)^{-1}\big(I_{(k\times k)} + \varepsilon^2\theta\beta^+\big)\beta\rho = \beta\rho,
\end{aligned}$$

Therefore, l.h.s. = r.h.s., i.e., (A.106) holds.

Next, from (A.106) it follows that ($d\mu^c(\cdot)$-a.e.)

$$\lim_{\varepsilon\downarrow 0}\left(\beta(t)+\varepsilon^2\theta(t)I_{(k\times k)}\right)^{-1}A(t)=\beta^+(t)A(t).$$

Thus, by the Lebesgue dominated convergence theorem [204],

$$\begin{aligned}\lim_{\varepsilon\downarrow 0}\int_{]0,t]}\gamma^\varepsilon(s)A^*(s)\big(\beta(s)+\varepsilon^2\theta(s)I_{(k\times k)}\big)^{-1}A(s)\gamma^\varepsilon(s)d\mu^c(s)=\\ =\int_{]0,t]}\gamma(s)A^*(s)\beta^+(s)A(s)\gamma(s)d\mu^c(s).\end{aligned}\tag{A.107}$$

Further, using (A.101) and (A.103) it is not difficult to show that

$$\begin{aligned}&\lim_{\varepsilon\downarrow 0}\Bigg(\gamma^\varepsilon(\tau_i-)A^*(\tau_i)\times\\ &\times\left[\beta(\tau_i)+\varepsilon^2I_{(k\times k)}+A(\tau_i)\gamma^\varepsilon(\tau_i-)A^*(\tau_i)\Delta\mu(\tau_i)\right]^{-1}A(\tau_i)\gamma^\varepsilon(\tau_i-)\Bigg)=\\ &=\gamma(\tau_i-)A^*(\tau_i)\big[\beta(\tau_i)+A(\tau_i)\gamma(\tau_i-)A^*(\tau_i)\Delta\mu(\tau_i)\big]^+A(\tau_i)\gamma(\tau_i-).\end{aligned}\tag{A.108}$$

Indeed, let us omit the arguments τ_i and τ_i-, denote $\gamma_\varepsilon=\gamma^\varepsilon(\tau_i-)\Delta\mu(\tau_i)$, and define on $\mathbb{R}^m\times\mathbb{R}^k$ the following functional of the variables x, y:

$$S_\varepsilon(x,y)=(x+A^*y)^*\gamma_\varepsilon(x+A^*y)+y^*(\beta+\varepsilon^2I_{(k\times k)})y.$$

Note that the functional $S_\varepsilon(x,y)$ is convex in the variable $y\in\mathbb{R}^k$. Consequently, it has a minimum (for any fixed $x\in\mathbb{R}^m$) on the element

$$y_\varepsilon(x)=-\big(A\gamma_\varepsilon A^*+\beta+\varepsilon^2I_{(k\times k)}\big)^{-1}A\gamma_\varepsilon x.$$

Therefore,

$$\begin{aligned}\lim_{\varepsilon\downarrow 0}\min_y S_\varepsilon(x,y)=\lim_{\varepsilon\downarrow 0}S_\varepsilon\big(x,y_\varepsilon(x)\big)=\\ =\lim_{\varepsilon\downarrow 0}x^*\big[\gamma_\varepsilon-\gamma_\varepsilon A^*\big(\beta+\varepsilon^2I_{(k\times k)}+A\gamma_\varepsilon A^*\big)^{-1}A\gamma_\varepsilon\big]x.\end{aligned}\tag{A.109}$$

On the other hand, for any $x\in\mathbb{R}^m$,

$$\begin{aligned}\min_y\lim_{\varepsilon\downarrow 0}S_\varepsilon(x,y)&=\min_y S_0(x,y)=\\ &=\min_y\left[(x+A^*y)^*\gamma_0(x+A^*y)+y^*\beta y\right]=\\ &=x^*\big[\gamma_0-\gamma_0A^*(\beta+A\gamma_0A^*)^+A\gamma_0\big]x,\end{aligned}\tag{A.110}$$

where $\gamma_0 = \lim \gamma_\varepsilon,\ \varepsilon \downarrow 0$, and the minimizing element has the form

$$y_0(x) = -(\beta + A\gamma_0 A^*)^+ A\gamma_0 x.$$

Now let us consider the obvious inequalities

$$S_\varepsilon\big(x, y_\varepsilon(x)\big) \le S_\varepsilon\big(x, y_0(x)\big),$$

$$\overline{\lim_{\varepsilon\downarrow 0}} S_\varepsilon\big(x, y_\varepsilon(x)\big) \le \overline{\lim_{\varepsilon\downarrow 0}} S_\varepsilon\big(x, y_0(x)\big) = \lim_{\varepsilon\downarrow 0} S_\varepsilon\big(x, y_0(x)\big) = S_0\big(x, y_0(x)\big). \tag{A.111}$$

On the other hand,

$$S_0\big(x, y_0(x)\big) \le S_0\big(x, y_\varepsilon(x)\big). \tag{A.112}$$

Suppose we could establish the *monotonicity* of the function γ_ε in the argument ε (a proof is given below, after the statement of Theorem A.28); then for any x, y we will have

$$S_0(x, y) \le S_\varepsilon(x, y)$$

and consequently (in view of (A.112))

$$S_0\big(x, y_0(x)\big) \le S_0\big(x, y_\varepsilon(x)\big) \le S_\varepsilon\big(x, y_\varepsilon(x)\big).$$

This implies that

$$S_0\big(x, y_0(x)\big) \le \underline{\lim}_{\varepsilon\downarrow 0} S_\varepsilon\big(x, y_\varepsilon(x)\big). \tag{A.113}$$

Comparing (A.111) and (A.113), we conclude that there exists the limit

$$\lim_{\varepsilon\downarrow 0} S_\varepsilon\big(x, y_\varepsilon(x)\big) = S_0\big(x, y_0(x)\big). \tag{A.114}$$

In view of (A.109) and (A.110), this implies (A.108).

The limit representations (A.105), (A.107), and (A.108) yield the following result.

Theorem A.28 *The matrix-function $\gamma(t)$ satisfies the generalized Riccati equation with measure*

$$\begin{aligned}\gamma(t) = \gamma_0 + \int_{]0,t]} \big[a(s)\gamma(s) + \gamma(s)a^*(s)\big]d\mu(s) + \langle\eta\rangle_t - \\ - \int_{]0,t]} \Gamma(s)\big[\beta(s) + A(s)\gamma(s-)A^*(s)\Delta\mu(s)\big]\Gamma^*(s)d\mu(s) + \\ + \sum_{\tau_i \le t} a(\tau_i)\gamma(\tau_i-)a^*(\tau_i)\big(\Delta\mu(\tau_i)\big)^2,\end{aligned} \tag{A.115}$$

where

$$\Gamma(t) = \big[I_{(m\times m)} + a(t)\Delta\mu(t)\big]\gamma(t-)A^*(t)\big[\beta(t) + A(t)\gamma(t-)A^*(t)\Delta\mu(t)\big]^+. \tag{A.116}$$

The most important question, the *uniqueness* of the solution of the Eq. (A.115), is settled below in Theorem A.29.

Now recall that we still need to prove the monotone dependence, assumed and used above, of γ^ε on ε. This is done in Lemma A.10 below, based in turn on the next result, Lemma A.9 (of independent interest), concerning the continuous dependence and differentiability of the solution to the generalized Riccati matrix equation with measure with respect to a parameter. Due to the specific matrix form of this equation, we prove Lemma A.9 directly.

Lemma A.9 *The solution $\gamma^\varepsilon(t)$ of Eq. (A.103) depends continuously on $\varepsilon > 0$ (in the topology defined by any matrix norm), uniformly in $t \in [0, T]$. In addition, it is differentiable with respect to ε.*

Proof of Lemma A.9 Let $\mathcal{T}$ denote the set of symmetric positive definite matrices. For $\gamma \in \mathcal{T}$ define the function

$$f(t, \gamma, \varepsilon) = \gamma A^*(t)\big[\beta(t) + \varepsilon^2 I_{(k\times k)}H(t) + A(t)\gamma A^*(t)\Delta\mu(t)\big]^{-1}A(t)\gamma.$$

Denote $\psi(\lambda) = f(t, \gamma_\lambda, \varepsilon)$, where

$$\gamma_\lambda = \gamma_1 + \lambda(\gamma_2 - \gamma_1), \quad \lambda \in [0, 1], \quad \gamma_i \in \mathcal{T},\ i = 1, 2.$$

Then

$$\begin{aligned}f(t, \gamma_2, \varepsilon) - f(t, \gamma_1, \varepsilon) = \psi(1) - \psi(0) = \int_{[0,1]} \frac{d\psi(\lambda)}{d\lambda}d\lambda = \\ = (\gamma_2 - \gamma_1)\int_{[0,1]} F_\lambda(t, \gamma_1, \gamma_2, \varepsilon)d\lambda + \\ + \left(\int_{[0,1]} F^*_\lambda(t, \gamma_1, \gamma_2, \varepsilon)d\lambda\right)(\gamma_2 - \gamma_1) - \\ - \left(\int_{[0,1]} F^*_\lambda(t, \gamma_1, \gamma_2, \varepsilon)(\gamma_2 - \gamma_1)F_\lambda(t, \gamma_1, \gamma_2, \varepsilon)d\lambda\right)\Delta\mu(t),\end{aligned} \tag{A.117}$$

where

$$
\begin{aligned}
&F_{\lambda}(t, \gamma_1, \gamma_2, \varepsilon) = \\
&\quad = A^*(t)\big[\beta(t) + \varepsilon^2 I_{(k\times k)} H(t) + A(t)\big(\gamma_1 + \lambda(\gamma_2 - \gamma_1)\big) A^*(t)\Delta\mu(t)\big]^{-1} \times \\
&\quad \times A(t)\big[\gamma_1 + \lambda(\gamma_2 - \gamma_1)\big]
\end{aligned}
$$

is a continuous function of γ_1, γ_2 and ε (in the norm topology).

Similarly, setting $\varphi(\lambda) = f(t, \gamma, \varepsilon_\lambda)$ with $\varepsilon_\lambda = \varepsilon_1 + \lambda(\varepsilon_2 - \varepsilon_1)$ and $\lambda \in [0, 1]$, we have

$$
\begin{aligned}
f(t, \gamma, \varepsilon_2) - f(t, \gamma, \varepsilon_1) &= \varphi(1) - \varphi(0) = \\
&= \int_{[0,1]} \frac{d\varphi(\lambda)}{d\lambda} d\lambda = -G(t, \gamma, \varepsilon_1, \varepsilon_2)(\varepsilon_2 - \varepsilon_1),
\end{aligned}
\tag{A.118}
$$

where

$$
\begin{aligned}
G(t, \gamma, \varepsilon_1, \varepsilon_2) = 2 \int_{[0,1]} &\gamma A^*(t) B_\lambda^{-1}(t, \gamma, \varepsilon_1, \varepsilon_2) \times \\
&\times \big[\varepsilon_1 + \lambda(\varepsilon_2 - \varepsilon_1)\big] I_{(k\times k)} H(t) B_\lambda^{-1}(t, \gamma, \varepsilon_1, \varepsilon_2) A(t)\gamma d\lambda
\end{aligned}
$$

is a continuous function of γ_1, ε_1 and ε_2 (in the norm topology), with

$$
B_\lambda(t, \gamma, \varepsilon_1, \varepsilon_2) = \beta(t) + \big[\varepsilon_1 - \lambda(\varepsilon_2 - \varepsilon_1)\big]^2 I_{(k\times k)} H(t) + A(t)\gamma A^*(t)\Delta\mu(t).
$$

Let us estimate the difference

$$
h(t, \varepsilon, \Delta\varepsilon) = \gamma^{\varepsilon+\Delta\varepsilon}(t) - \gamma^{\varepsilon}(t).
$$

From (A.103) we get

$$
\begin{aligned}
h(t, \varepsilon, \Delta\varepsilon) &= \int_{]0,t]} \big[a(s)h(s-, \varepsilon, \Delta\varepsilon) + h(s-, \varepsilon, \Delta\varepsilon)a^*(s)\big] d\mu(s) + \\
&+ \sum_{\tau_i \le t} a(\tau_i) h(s-, \varepsilon, \Delta\varepsilon) a^*(\tau_i)\big(\Delta\mu(\tau_i)\big)^2 - \\
&- \int_{]0,t]} \widetilde{a}(s)\Big[f\big(s, \gamma^{\varepsilon+\Delta\varepsilon}(s-), \varepsilon + \Delta\varepsilon\big) - f\big(s, \gamma^{\varepsilon}(s-), \varepsilon + \Delta\varepsilon\big) + \\
&+ f\big(s, \gamma^{\varepsilon}(s-), \varepsilon + \Delta\varepsilon\big) - f\big(s, \gamma^{\varepsilon}(s-), \varepsilon\big)\Big]\widetilde{a}^*(s) d\mu(s),
\end{aligned}
\tag{A.119}
$$

where $\widetilde{a}(s) = I_{(m\times m)} + a(s)\Delta\mu(s)$.

Using (A.117) and (A.118), the second integral in (A.119) is transformed as follows:

$$\begin{aligned}
&\int_{]0,t]} \widetilde{a}(s)\Big\{\Big[h(s-,\varepsilon,\Delta\varepsilon)\int_{[0,1]} F_\lambda\big(s,\gamma^\varepsilon(s-),\gamma^{\varepsilon+\Delta\varepsilon}(s-),\varepsilon+\Delta\varepsilon\big)d\lambda+\\
&\quad+\Big(\int_{[0,1]} F_\lambda^*\big(s,\gamma^\varepsilon(s-),\gamma^{\varepsilon+\Delta\varepsilon}(s-),\varepsilon+\Delta\varepsilon\big)d\lambda\Big)h(s-,\varepsilon,\Delta\varepsilon)\\
&\quad-\Delta\mu(s)\int_{[0,1]} F_\lambda^*\big(s,\gamma^\varepsilon(s-),\gamma^{\varepsilon+\Delta\varepsilon}(s-),\varepsilon+\Delta\varepsilon\big)\times\\
&\quad\times h(s-,\varepsilon,\Delta\varepsilon)F_\lambda\big(s,\gamma^\varepsilon(s-),\gamma^{\varepsilon+\Delta\varepsilon}(s-),\varepsilon+\Delta\varepsilon\big)d\lambda\Big]-\\
&\quad-G\big(s,\gamma^\varepsilon(s-),\varepsilon,\varepsilon+\Delta\varepsilon\big)\Delta\varepsilon\Big\}\widetilde{a}^*(s)d\mu(s).
\end{aligned} \tag{A.120}$$

Thanks to the uniform boundedness of $\gamma^\varepsilon(t)$ (see Lemma A.7), the functions $F_\lambda(\cdot)$ and $G(\cdot)$ are also uniformly bounded, and consequently

$$\begin{aligned}
\|h(t,\varepsilon,\Delta\varepsilon)\| &\le 2\int_{]0,t]} \|a(s)\|\,\|h(s-,\varepsilon,\Delta\varepsilon)\|d\mu(s)+\\
&+\sum_{\tau_i\le t}\|a(\tau_i)\|^2(\Delta\mu(\tau_i))^2\|h(\tau_i,\varepsilon,\Delta\varepsilon)\|+\\
&+K_1\int_{]0,t]}\|\widetilde{a}(s)\|^2\|h(s-,\varepsilon,\Delta\varepsilon)\|d\mu(s)+K_2\,\Delta\varepsilon\int_{]0,t]}\|\widetilde{a}(s)\|^2 d\mu(s),
\end{aligned}$$

where K_1, K_2 are constants. This inequality can be rewritten as

$$\|h(t,\varepsilon,\Delta\varepsilon)\| \le \int_{]0,t]}\|h(s-,\varepsilon,\Delta\varepsilon)\|d\widetilde{\mu}(s)+\alpha(t)\Delta\varepsilon,$$

where

$$\alpha(t)=K_2\int_{]0,t]}\|\widetilde{a}(s)\|^2 d\mu(s)$$

and

$$d\widetilde{\mu}(s)=\big(2\|a(s)\|+\|a(s)\|^2\Delta\mu(s)+K_1\|\widetilde{a}(s)\|^2\big)d\mu(s).$$

Applying the generalized Gronwall-Bellman inequality, we conclude that

$$\lim_{\Delta\varepsilon\downarrow 0}\sup_{t\in[0,T]}\|h(t,\varepsilon,\Delta\varepsilon)\|=0,$$

so the first assertion of the lemma is proved.

To prove the differentiability $\gamma^\varepsilon(t)$ with respect to ε, consider the expression

$$\kappa(t, \varepsilon, \Delta\varepsilon) = \frac{\gamma^{\varepsilon+\Delta\varepsilon}(t) - \gamma^\varepsilon(t)}{\Delta\varepsilon}.$$

By (A.119) and (A.120), when $\Delta\varepsilon \neq 0$ the function $\kappa(t, \varepsilon, \Delta\varepsilon)$ satisfies the following linear differential equation with null initial condition:

$$\begin{aligned}
z(t) = &\int_{]0,t]} \big[a(s)z(s-) + z(s-)a^*(s)\big]d\mu(s) + \\
&+ \sum_{\tau_i \le t} a(\tau_i)z(\tau_i)a^*(\tau_i)\big(\Delta\mu(\tau_i)\big)^2 - \\
&- \int_{]0,t]} \widetilde{a}(s)\Bigg[z(s-)\int_{[0,1]} F_\lambda\big(s, \gamma^\varepsilon(s-), \gamma^{\varepsilon+\Delta\varepsilon}(s-), \varepsilon + \Delta\varepsilon\big)d\lambda + \\
&+ \bigg(\int_{[0,1]} F_\lambda^*\big(s, \gamma^\varepsilon(s-), \gamma^{\varepsilon+\Delta\varepsilon}(s-), \varepsilon + \Delta\varepsilon\big)d\lambda\bigg)z(s-) - \\
&- \Delta\mu(s)\int_{[0,1]} F_\lambda^*\big(s, \gamma^\varepsilon(s-), \gamma^{\varepsilon+\Delta\varepsilon}(s-), \varepsilon + \Delta\varepsilon\big)z(s-) \times \\
&\times F_\lambda\big(s, \gamma^\varepsilon(s-), \gamma^{\varepsilon+\Delta\varepsilon}(s-), \varepsilon + \Delta\varepsilon\big)d\lambda\Bigg]\widetilde{a}^*(s)d\mu(s) + \\
&+ \int_{]0,t]} \Delta\mu(s)\widetilde{a}(s)G\big(s, \gamma^\varepsilon(s-), \varepsilon, \varepsilon + \Delta\varepsilon\big)\widetilde{a}^*(s)d\mu(s).
\end{aligned} \tag{A.121}$$

Let $z(t) = \rho(t, \varepsilon, \Delta\varepsilon)$ be the solution of this equation. Note that the solution exists in two different cases: for $\Delta\varepsilon > 0$, and for $\Delta\varepsilon = 0$.

Thanks the uniqueness of the solution of the Eq. (A.121), for $\Delta\varepsilon \neq 0$ we have that

$$\kappa(t, \varepsilon, \varepsilon) = \rho(t, \varepsilon, \Delta\varepsilon). \tag{A.122}$$

The function $\kappa(t, \varepsilon, \Delta\varepsilon)$ in (A.122) is continuous with respect to $\Delta\varepsilon$ and ε, so $\rho(t, \varepsilon, \Delta\varepsilon)$ also enjoys these properties. Letting $\Delta\varepsilon \to 0$ in both sides (A.122) we get

$$\frac{d\gamma^\varepsilon(t)}{d\varepsilon} = \lim_{\Delta\varepsilon\to 0} \frac{\gamma^{\varepsilon+\Delta\varepsilon}(t) - \gamma^\varepsilon(t)}{\Delta\varepsilon} = \lim_{\Delta\varepsilon\to 0} \rho(t, \varepsilon, \Delta\varepsilon) = \rho(t, \varepsilon, 0).$$

Thus, the derivative $d\gamma^\varepsilon/d\varepsilon$ exists and is continuous in ε (since $\rho(t, \varepsilon, 0)$ is continuous in ε). Lemma A.9 is proved. □

Finally, we are in the position to establish, as promised, the monotonicity of $\gamma^\varepsilon(t)$ in the parameter ε.

Lemma A.10 *For any fixed $t \in [0, T]$, the function $\varepsilon \mapsto \gamma^\varepsilon(t)$ is monotone for $\varepsilon > 0$.*

Proof of Lemma A.10 In view of Lemmas A.7 and A.9, we can pass from Eq. (A.103) to the following equation in variations, denoting $\varphi^\varepsilon(t) = d\gamma^\varepsilon(t)/d\varepsilon$ (the argument t in the integrand is omitted)

$$\varphi^\varepsilon(t) = \int_{]0,t]} \Big[a\varphi^\varepsilon + \varphi^\varepsilon a^* + \widetilde{a}\gamma^\varepsilon D_\varepsilon \varphi^\varepsilon D_\varepsilon \gamma^\varepsilon \widetilde{a}^* \Delta\mu + a\varphi^\varepsilon a^* \Delta\mu - \\ - \widetilde{a}\varphi^\varepsilon D_\varepsilon \gamma^\varepsilon \widetilde{a}^* - \widetilde{a}\gamma^\varepsilon D_\varepsilon \varphi^\varepsilon \widetilde{a}^* \Big] d\mu + \Phi_\varepsilon(t), \tag{A.123}$$

where

$$D_\varepsilon = A^* B_\varepsilon^{-1} A, \quad B_\varepsilon = \beta + A\gamma^\varepsilon A^* \Delta\mu + \varepsilon^2 I_{(k\times k)} H,$$

$$\widetilde{a} = I_{(m\times m)} + a\Delta\mu, \quad \Phi_\varepsilon(t) = 2\varepsilon \int_{]0,t]} H\widetilde{a}\gamma^\varepsilon A^* B_\varepsilon^{-2} A\gamma^\varepsilon \widetilde{a}^* d\mu.$$

Note that $\Phi_\varepsilon(t)$ is a symmetric non-negative definite matrix. Since Eq. (A.123) is linear in $\varphi^\varepsilon(t)$, the matrix $\varphi^\varepsilon(t)$ is also non-negative definite. This implies the claimed monotonicity of γ^ε with respect to ε. □

Next, we address the question of the *uniqueness* of the solution of equation (A.115).

Theorem A.29 *Suppose equation (A.115) admits a solution in the class of non-negative definite matrices. Then this solution is unique.*

The proof of Theorem A.29 is based on the following three lemmas.

Lemma A.11 *Let $f(\gamma)$ be the function defined on the set $\mathscr{T}$ of non-negative definite symmetric matrices by the formula*

$$f(\gamma) = \gamma - \gamma A^*(\beta + A\gamma A^* \Delta\mu)^+ A\gamma\Delta\mu,$$

where $\Delta\mu > 0$; γ, A and β are $m \times m$, $k \times m$ and $k \times k$ matrices, respectively; and $\beta \in \mathscr{T}$. Then

$$f(\gamma) = \lim_{\varepsilon \downarrow 0} \big[\gamma - \gamma A^*(\beta + A\gamma A^* \Delta\mu + \varepsilon I_{(k\times k)})^{-1} A\gamma\Delta\mu\big].$$

The proof of Lemma A.11 is similar to the proof of equality (A.108). □

Lemma A.12 *Let B be non-degenerate $k \times k$ matrix. Then under the assumptions of Lemma A.11, the differential equation*

$$\frac{d\gamma(\tau)}{d\tau} = -\gamma(\tau)A^*B^{-1}A\gamma(\tau), \quad \textit{with} \quad \gamma(0) = \gamma \in \mathcal{T},$$

has exactly one solution, given by

$$\gamma(\tau) = \gamma - \gamma A^*\big[B + A\gamma A^*\tau\big]^{-1}A\gamma\tau.$$

The proof of Lemma A.11 is similar to the proof of Theorem 1 from [265]. □

Lemma A.13 *Under the conditions of Lemmas A.11 and A.12, for any* $\gamma_1, \gamma_2 \in \mathcal{T}$ *it holds that*

$$\mathrm{tr}\big\{\big[f(\gamma_1) - f(\gamma_2)\big]^2\big\} \le \mathrm{tr}\big\{(\gamma_1 - \gamma_2)^2\big\}. \tag{A.124}$$

Proof of Lemma A.13 Consider the initial value problem for the matrix differential equation

$$\frac{d\gamma^{\varepsilon}(\tau)}{d\tau} = -\gamma^{\varepsilon}(\tau)A^*\big[\beta + \varepsilon^2 I_{(k\times k)}\big]^{-1}A\gamma^{\varepsilon}(\tau), \quad \text{with } \gamma^{\varepsilon}(0) = \gamma \in \mathcal{T}. \tag{A.125}$$

By Lemma A.12, its solution has the form

$$\gamma^{\varepsilon}(\tau) = \gamma - \gamma A^*\big[\beta + \varepsilon^2 I_{(k\times k)} + A\gamma A^*\tau\big]^{-1}A\gamma\tau. \tag{A.126}$$

Consider two solutions of this problem, $\gamma_1^{\varepsilon}(\tau)$ and $\gamma_2^{\varepsilon}(\tau)$, corresponding to different initial conditions γ_1 and γ_2. Then, in view of (A.125) and (A.126),

$$\begin{aligned}&\frac{d}{d\tau}\mathrm{tr}\big\{\big[\gamma_1^{\varepsilon}(\tau) - \gamma_2^{\varepsilon}(\tau)\big]^2\big\} = \\ &\quad = \mathrm{tr}\big\{-2\big[\gamma_1^{\varepsilon}(\tau) - \gamma_2^{\varepsilon}(\tau)\big]\big[\gamma_1^{\varepsilon}(\tau)B^{\varepsilon}\gamma_1^{\varepsilon}(\tau) - \gamma_2^{\varepsilon}(\tau)B^{\varepsilon}\gamma_2^{\varepsilon}(\tau)\big]\big\},\end{aligned} \tag{A.127}$$

where $B^{\varepsilon} = A^*\big[\beta + \varepsilon^2 I_{(k\times k)}\big]^{-1}A$.

The right-hand side of (A.127) can be easily rewritten as

$$-2\Big(\mathrm{tr}\big\{\big[\gamma_1^{\varepsilon}(\tau) - \gamma_2^{\varepsilon}(\tau)\big]^2\gamma_1^{\varepsilon}(\tau)B^{\varepsilon}\big\} + \mathrm{tr}\big\{\gamma_2^{\varepsilon}(\tau)\big[\gamma_1^{\varepsilon}(\tau) - \gamma_2^{\varepsilon}(\tau)\big]^2B^{\varepsilon}\big\}\Big). \tag{A.128}$$

Let us prove that the expression (A.128) is non-positive. For this purpose we need the following simple technical fact.

Lemma A.14 *For any three matrices* $P, Q, R \in \mathcal{T}$*, we have* $\mathrm{tr}\{PQR\} \ge 0$.

Remark For two matrices this result is well known [248].

Proof of Lemma A.14 We distinguish two cases: $R > 0$ and $R \ge 0$.

Case 1. Suppose the matrix R is positive definite. Then there is an orthogonal transformation S_1 that reduces R to diagonal form: $S_1^* R S_1 = \Lambda = \operatorname{diag}\{\lambda_1, \ldots, \lambda_m\}$, where $\lambda_i > 0$, $i = 1, \ldots, m$, are eigenvalues of R [30, 248]. Therefore,

$$\operatorname{tr}\{PQR\} = \operatorname{tr}\{S_1^* PQRS_1\} = \operatorname{tr}\{S_1^* PS_1 S_1^* QS_1 S_1^* RS_1\} = \operatorname{tr}\{P_1 Q_1 \Lambda\},$$

where we denoted $P_1 = S_1^* P S_1 \in \mathscr{T}$ and $Q_1 = S_1^* Q S_1 \in \mathscr{T}$. Next, let $S_2 : \mathbb{R}^m \to \mathbb{R}^m$ denote the non-degenerate transformation that sends each vector $x = (x_1, \ldots, x_m)^*$ into the component-wise "dilated" vector $y = (y_1, \ldots, y_m)^*$ with $y_i = \sqrt{\lambda_i} x_i$, $i = 1, \ldots, m$. This transformation converts the quadratic from $x^* \Lambda x = y^* y$ to a sum of squares, i.e., converts the matrix Λ to the unit matrix: $S_2^* \Lambda S_2 = I_{(m \times m)}$. Then we have

$$\begin{aligned}\operatorname{tr}\{P_1 Q_1 \Lambda\} &= \operatorname{tr}\{S_2^{-1} P_1 Q_1 \Lambda S_2\} = \operatorname{tr}\{S_2^{-1} P_1 (S_2^{-1})^* S_2^* Q_1 (S_2^*)^{-1} S_2^* \Lambda S_2\} = \\ &= \operatorname{tr}\{P_2 Q_2\} \geq 0,\end{aligned}$$

because $P_2 = S_2^{-1} P_1 (S_2^{-1})^* \in \mathscr{T}$, and $Q_2 = S_2^* Q_1 (S_2^*)^{-1} \geq 0$ due to the similarity Q_1 and Q_2.

Case 2. Now assume that $R \geq 0$. Consider the "regularized" matrix $R_\varepsilon = R + \epsilon I_{m \times m} > 0$, where $\varepsilon > 0$. By what we proved above, $\operatorname{tr}\{PQR_\varepsilon\} \geq 0$. Since the "trace" operator is continuous, letting $\varepsilon \downarrow 0$ in the last inequality yields the claim of the lemma. □

We continue the proof of the Lemma A.13.

The non-positivity of the expression (A.128) follows directly from Lemma A.14 (*Case* 1). The non-positivity of expression (A.128) implies that in (A.127)

$$\frac{d}{d\tau} \operatorname{tr}\left\{\left[\gamma_1^\varepsilon(\tau) - \gamma_2^\varepsilon(\tau)\right]^2\right\} \leq 0.$$

Thus, for $\Delta\mu > 0$ we have

$$\operatorname{tr}\left\{\left[\gamma_1^\varepsilon(\Delta\mu) - \gamma_2^\varepsilon(\Delta\mu)\right]^2\right\} \leq \operatorname{tr}\left\{(\gamma_1 - \gamma_2)^2\right\}.$$

Letting $\varepsilon \downarrow 0$ in this inequality and using Lemma A.11, we get (A.124). □

Back to the proof of Theorem A.29.

Let $\gamma_1(t)$ and $\gamma_2(t)$ be two non-negative definite and bounded solutions of Eq. (A.115). Then

$$z(t) = \operatorname{tr}\left\{\left[\gamma_1(t) - \gamma_2(t)\right]^2\right\} \geq 0. \tag{A.129}$$

By (A.115), the function $z(t)$ satisfies the equation

$$
\begin{aligned}
z(t) = \operatorname{tr} \int_{]0,t]} \Big[&(\gamma_1(s) - \gamma_2(s)) d(\gamma_1(s) - \gamma_2(s))^{\mathrm{c}} + \\
&+ \big(d(\gamma_1(s) - \gamma_2(s))^{\mathrm{c}}\big)(\gamma_1(s) - \gamma_2(s))\Big] + \sum_{\tau_i \le t} \Delta z(\tau_i),
\end{aligned} \tag{A.130}
$$

where $\Delta z(\tau_i) = z(\tau_i) - z(\tau_i -)$.

Omitting the time argument s, we write the integrand in (A.130) in the form

$$
\begin{aligned}
dz^{\mathrm{c}} = \operatorname{tr}\Big\{ 2(\gamma_1 - \gamma_2)\Big(&\big[a(\gamma_1 - \gamma_2) + (\gamma_1 - \gamma_2)a^*\big]d\mu^{\mathrm{c}} - \\
&- \gamma_1 A^* \beta^+ A \gamma_1 d\mu^{\mathrm{c}} + \gamma_2 A^* \beta^+ A \gamma_2 d\mu^{\mathrm{c}}\Big)\Big\}.
\end{aligned}
$$

The last two terms in this expression are of the form

$$
\Big\{\operatorname{tr}\big[-2(\gamma_1 - \gamma_2)(\gamma_1 B \gamma_1 - \gamma_2 B \gamma_2)\big]\Big\} d\mu^{\mathrm{c}}, \tag{A.131}
$$

where $B = A^* \beta^+ A \ge 0$. Note that the expression in the curly brackets in (A.131) coincides, up to notation, with the right hand-side of (A.127), and so, by Lemma A.14), is non-positive. Further,

$$
\begin{aligned}
&\Big|\operatorname{tr}\Big\{2(\gamma_1 - \gamma_2)\big[a(\gamma_1 - \gamma_2) + (\gamma_1 - \gamma_2)a^*\big]\Big\}\Big| = \\
&\quad = 2\Big|\operatorname{tr}\Big\{(\gamma_1 - \gamma_2)^2 (a + a^*)\Big\}\Big| \le 4\|a\| \operatorname{tr}\{(\gamma_1 - \gamma_2)^2\}.
\end{aligned} \tag{A.132}
$$

By virtue of (A.132) and the remark made above, the integral term in (A.130) obeys the estimate

$$
z^{\mathrm{c}}(t) \le 4 \int_{]0,t]} \|a(s)\| z(s) d\mu^{\mathrm{c}}(s).
$$

Next, in view of Lemma A.13, the discrete part in (A.130) is estimated as follows:

$$
\begin{aligned}
\Delta z(\tau_i) &= \operatorname{tr}\Big\{\big[\gamma_1(\tau_i) - \gamma_2(\tau_i)\big]^2 - \big[\gamma_1(\tau_i -) - \gamma_2(\tau_i -)\big]^2\Big\} = \\
&= \operatorname{tr}\Big\{\widetilde{a}(\tau_i)\big[f(\gamma_1(\tau_i -)) - f(\gamma_2(\tau_i -))\big]\widetilde{a}(\tau_i)^*\Big\}^2 - z(\tau_i -) \le \\
&\le \operatorname{tr}\Big\{\big[\widetilde{a}(\tau_i)\widetilde{a}(\tau_i)^*\big]^2 \big[\gamma_1(\tau_i -) - \gamma_2(\tau_i -)\big]^2\Big\} - z(\tau_i -) \le \\
&\le L(\tau_i) z(\tau_i -) \Delta\mu(\tau_i),
\end{aligned}
$$

where we denoted $\widetilde{a}(\tau_i) = I_{(m \times m)} + a(\tau_i)\Delta\mu(\tau_i)$, and the matrix $L(\tau_i)$ is such that

$$\sum_{\tau_i \le t} \|L(\tau_i)\| \Delta\mu(\tau_i) < \infty, \quad t \in [0, T].$$

This property of the matrix $L(\tau_i)$ follows from the boundedness of the variation of the measure $d\mu(\cdot)$ on $[0, T]$ and the condition

$$\sum_{\tau_i \le t} \|a(\tau_i)\| \Delta\mu(\tau_i) < \infty, \quad t \in [0, T],$$

which is required to ensure that Eq. (A.92) has a strong solution. Combining the estimates for the continuous and discrete parts in (A.130), we conclude that

$$z(t) \le 4 \int_{]0,t]} \|a(s)\| z(s) d\mu^{\mathrm{c}}(s) + \int_{]0,t]} \|L(s)\| z(s-) d\mu^{\mathrm{d}}(s).$$

Applying the generalized Gronwall-Bellman lemma shows that $z(t) \equiv 0$, which proves the assertion of Theorem A.29. □

We next present a generalized Kalman filter for a partially observed process $(X, Y) = (X(t), Y(t))_{0 \le t \le T}$, which is governed by the system (A.92) and (A.93), i.e., we derive an equation that is satisfied by the conditional expectation $m(t) = \mathbf{E}(X(t)|F_t^Y)$, where $F_t^Y = \sigma\{Y(s),\ s \le t\}$.

Theorem A.30 *The process $m = (m(t))_{0 \le t \le T}$ satisfies the equation*

$$\begin{aligned} m(t) = \pi_0 &+ \int_{]0,t]} a(s) m(s-) d\mu(s) + \\ &+ \int_{]0,t]} \Gamma(s) \big[dY(s) - A(s) m(s-) d\mu(s) \big], \end{aligned} \tag{A.133}$$

where $\Gamma(t)$ is defined by the formula (A.116), $Y(t)$ is subject to Eq. (A.93), and the process $\nu = (\nu(t))_{0 \le t \le T}$ with the Itô differential

$$d\nu(t) = dY(t) - A(t) m(t-) d\mu(t) \tag{A.134}$$

is a Gaussian martingale with regular paths with respect to the family $\mathbf{F}^Y = (F_t^Y)_{0 \le t \le T}$ of σ-algebras $F_t^Y = \sigma\{Y(s), s \le t\}$.

Proof of Theorem A.30 Define the process $\pi = (\pi(t))_{0 \le t \le T}$ as the solution of the equation

$$\pi(t) = \pi_0 + \int_{]0,t]} a(s) \pi(s-) d\mu(s) + \int_{]0,t]} \Gamma(s) \big[dY(s) - A(s) m(s-) d\mu(s) \big], \tag{A.135}$$

where $\pi_0 = \mathbf{E}(X_0|Y_0)$ and is calculated from a priori data by means the normal correlation theorem [244].

Denote $\lambda(t) = \mathbf{E}\big[\big(X(t) - \pi(t)\big)\big(X(t) - \pi(t)\big)^*\big]$. By Itô's formula [245], $\lambda(t)$ satisfies the equation

$$\begin{aligned}\lambda(t) &= \gamma_0 + \int_{]0,t]} \big[a(s)\lambda(s-) + \lambda(s-)a^*(s)\big]d\mu(s) + \langle\eta\rangle_t + \\ &+ \sum_{\tau_i \le t} a(\tau_i)\lambda(\tau_i-)a^*(\tau_i-)\big(\Delta\mu(\tau_i)\big)^2 - \\ &- \int_{]0,t]} \Gamma(s)\big[\beta(s) + A(s)\gamma(s-)A^*(s)\Delta\mu(s)\big]\Gamma^*(s)d\mu(s).\end{aligned} \tag{A.136}$$

Comparing (A.136) and (A.115) and taking into account the uniqueness of the solution of the Eq. (A.115) (Theorem A.29), we deduce that

$$\lambda(t) \equiv \gamma(t). \tag{A.137}$$

Further, set $\delta(t) = m(t) - \pi(t)$. Then

$$\begin{aligned}\lambda(t) &= \mathbf{E}\big[(X(t) - \pi(t))(X(t) - \pi(t))^*\big] = \\ &= \mathbf{E}\big[(X(t) - m(t) + \delta(t))(X(t) - m(t) + \delta(t))^*\big] = \\ &= \mathbf{E}\big[(X(t) - m(t))(X(t) - m(t))^*\big] + \mathbf{E}\big[\delta(t)\delta^*(t)\big] + \\ &+ \mathbf{E}\big[\delta(t)(X(t) - m(t))^*\big] + \mathbf{E}\big[(X(t) - m(t))\delta^*(t)\big].\end{aligned} \tag{A.138}$$

The last two terms in this equation vanish. Indeed,

$$\begin{aligned}\mathbf{E}\big[\delta(t)(X(t) - m(t))^*\big] &= \mathbf{E}\mathbf{E}\big[\delta(t)(X(t) - m(t))^* | F_t^Y\big] = \\ &= \mathbf{E}\delta(t)\mathbf{E}\big[(X(t) - m(t))^* | F_t^Y\big] = 0.\end{aligned} \tag{A.139}$$

From (A.138), taking (A.137) and (A.139) into account, we find that

$$\gamma(t) = \gamma(t) + \mathbf{E}[\delta(t)\delta^*(t)],$$

and so $\delta(t) = 0$ (P-a.e.). Therefore, $m(t) = \pi(t)$ (P-a.e.). Replacing $\pi(t)$ with $m(t)$ in (A.135), we get Eq. (A.133). The first assertion of Theorem A.30 is proved.

Let us prove the second assertion of Theorem A.30. It follows from the representation (A.134) that the process ν is $\mathbf{F}^Y$-adapted with regular paths. We claim that ν is a martingale. Indeed, (A.134) and the representation (A.93) show that, for $s \le t$,

$$\nu(t) - \nu(s) = \zeta(t) - \zeta(s) + \int_{]s,t]} A(u)\big[X(u-) - m(u-)\big]d\mu(u). \tag{A.140}$$

By the properties of conditional mathematical expectations,

$$\begin{aligned}\mathbf{E}\big[\zeta(t)-\zeta(s)|F_s^Y\big] &= \mathbf{E}\big(\mathbf{E}\big[\zeta(t)-\zeta(s)|F_s\big]|F_s^Y\big) = \\ &= \mathbf{E}\big[\zeta(s)-\zeta(s)|F_s^Y\big] = 0.\end{aligned} \tag{A.141}$$

Next, if $s < u$, then

$$\mathbf{E}\big(m(u-)|F_s^Y\big) = \mathbf{E}\big[\mathbf{E}\big(X(u-)|F_{u-}^Y\big)|F_s^Y\big] = \mathbf{E}\big(X(u-)|F_s^Y\big). \tag{A.142}$$

Thanks to the properties (A.141) and (A.142), the deterministic nature of the jumps of the measure $d\mu(\cdot)$ and the fact that the set of these jumps is at most countable, we have that

$$\begin{aligned}&\mathbf{E}\left(\int_{]s,t]} A(u)\big[X(u-)-m(u-)\big]d\mu(u)|F_s^Y\right) = \\ &= \int_{]s,t]} A(u)\mathbf{E}\big[X(u-)-m(u-)|F_s^Y\big]d\mu(u) = \\ &= \int_{]s,t]} A(u)\big[\mathbf{E}\big(X(u-)|F_s^Y\big)-\mathbf{E}\big(m(u-)|F_s^Y\big)\big]d\mu(u) = 0.\end{aligned} \tag{A.143}$$

Next, (A.140), (A.141), and (A.143) imply that $\mathbf{E}\big(\nu(t)-\nu(s)|F_s^Y\big) = 0$ or, due to the F_s^Y-measurability of $\nu(s)$, $\mathbf{E}\big(\nu(t)|F_s^Y\big) = \nu(s)$. So ν is indeed a martingale.

It remains to verify that ν is Gaussian. It suffices to prove that

$$\nu(t) \to \nu^{\varepsilon}(t), \quad \varepsilon \downarrow 0, \tag{A.144}$$

where ν^{ε} is given by (A.102). In fact, (A.102) and (A.134) yield the representation

$$\nu^{\varepsilon}(t)-\nu(t) = \big(Y^{\varepsilon}(t)-Y(t)\big) - \int_{]0,t]} A(s)\big[\pi^{\varepsilon}(s-)-m(s-)\big]d\mu(s). \tag{A.145}$$

Here

$$\mathbf{E}\big(Y^{\varepsilon}(t)-Y(t)\big)^2 = \varepsilon^2\langle M\rangle_t \to 0, \quad \varepsilon \downarrow 0. \tag{A.146}$$

Further,

$$\begin{aligned}\gamma_{ii}^{\varepsilon}(t) &= \mathbf{E}\big(X_i(t)-\pi_i^{\varepsilon}(t)\big)^2 = \mathbf{E}\big(X_i(t)-m_i(t)+m_i(t)-\pi_i^{\varepsilon}(t)\big)^2 = \\ &= \mathbf{E}\big(X_i(t)-m_i(t)\big)^2 + \mathbf{E}\big(m_i(t)-\pi_i^{\varepsilon}(t)\big)^2 + \\ &+ 2\mathbf{E}\big[\big(X_i(t)-m_i(t)\big)\big(m_i(t)-\pi_i^{\varepsilon}(t)\big)\big].\end{aligned} \tag{A.147}$$

The last term in this expression vanishes, because

$$\begin{aligned}
&\mathbf{E}\big[(X_i(t) - m_i(t))(m_i(t) - \pi_i^{\varepsilon}(t)\big] \\
&= \mathbf{E}\mathbf{E}\big[(X_i(t) - m_i(t))(m_i(t) - \pi_i^{\varepsilon}(t)|F_t^{Y,Y^{\varepsilon}}\big] = \\
&\quad = \mathbf{E}\big[\mathbf{E}\big(\mathbf{E}(X_i(t)|F_t^{Y,Y^{\varepsilon}}) - m_i(t))(m_i(t) - \pi_i^{\varepsilon}(t))\big] = 0.
\end{aligned}$$

The latter equality is a consequence of the independence of the processes X and M. Indeed,

$$\mathbf{E}\big(X_i(t)|F_t^{Y,Y^{\varepsilon}}\big) = \mathbf{E}\big(X_i(t)|F_t^{Y,M}\big) = \mathbf{E}\big(X_i(t)|F_t^{Y}\big) = m_i(t).$$

Consequently, in (A.147)

$$\gamma_{ii}^{\varepsilon}(t) = \gamma_{ii}(t) + \mathbf{E}\big(m_i(t) - \pi_i^{\varepsilon}(t)\big)^2, \tag{A.148}$$

and from (A.148) and Lemma A.8 it follows that

$$\mathbf{E}(\pi^{\varepsilon}(t) - m(t))^2 \to 0, \quad \varepsilon \downarrow 0. \tag{A.149}$$

Finally, in view of (A.145), relations (A.149) and (A.146) imply (A.144). Theorem A.30 is proved. □

To continue, we introduce the process $m^{\varepsilon} = \big(m^{\varepsilon}(t)\big)_{0\le t\le T}$ by

$$\begin{aligned}
m^{\varepsilon}(t) = \pi_0 + &\int_{]0,t]} a(s) m^{\varepsilon}(s-) d\mu(s) + \\
&+ \int_{]0,t]} \Gamma^{\varepsilon}(s)\big[dY(s) - A(s) m^{\varepsilon}(s-) d\mu(s)\big],
\end{aligned} \tag{A.150}$$

where $\Gamma^{\varepsilon}(s)$ is given by (A.101) and $Y(s)$ has the representation (A.93).

The process m^{ε} is called the *regularized* estimator of process X based on the observations Y.

Denote by

$$\widetilde{\gamma}^{\varepsilon}(t) = \mathbf{E}\big[\big(X(t) - m^{\varepsilon}(t)\big)\big(X(t) - m^{\varepsilon}(t)\big)^*\big]$$

the $m \times m$ covariance matrix of estimation errors of the process $X(t)$ corresponding to the regularized estimator $m^{\varepsilon}(t)$. Using Itô's formula, we get the representation

$$\begin{aligned}
\widetilde{\gamma}^{\varepsilon}(t) = \gamma_0 + &\int_{]0,t]} \big[\widetilde{\gamma}^{\varepsilon}(s-) a_{\varepsilon}^*(s) + a_{\varepsilon}(s)\widetilde{\gamma}^{\varepsilon}(s-)\big] d\mu(s) + \langle \eta \rangle_t + \\
&+ \sum_{\tau_i \le t} a_{\varepsilon}(\tau_i)\widetilde{\gamma}^{\varepsilon}(\tau_i-) a_{\varepsilon}^*(\tau_i)\big(\Delta\mu(\tau_i)\big)^2 + \int_{]0,t]} \Gamma^{\varepsilon}(s)\beta(s)\Gamma^{\varepsilon *}(s) d\mu(s),
\end{aligned} \tag{A.151}$$

where $a_{\varepsilon}(t) = a(t) - \Gamma^{\varepsilon}(t)A(t)$.

The following theorem, which establishes properties of the regularized estimator, is analogous to Theorem 10.4 in [244].

Theorem A.31 *The process $m^\varepsilon(t)$ introduced above enjoys the following properties (for any $t \in [0, T]$):*

(i) $\mathbf{E}\, m^\varepsilon(t) = \mathbf{E}\, X(t)$;
(ii) $\gamma_{ii}(t) \le \widetilde{\gamma}^\varepsilon_{ii}(t) \le \gamma^\varepsilon_{ii}(t), \quad i = 1, \dots, m$;
(iii) $\gamma(t) = \lim \widetilde{\gamma}^\varepsilon(t) = \lim \gamma^\varepsilon(t), \quad \varepsilon \downarrow 0$;
(iv) $\lim \mathbf{E}\big(m_i(t) - m_i^\varepsilon(t)\big)^2 = 0, \quad \varepsilon \downarrow 0, \quad i = 1, \dots, m.$

Proof of Theorem A.31 To prove (i), consider the process $\xi^\varepsilon = \big(\xi^\varepsilon(t)\big)_{0\le t\le T}$ with $\xi^\varepsilon(t) = X(t) - m^\varepsilon(t)$, which according to (A.92) and (A.150) is given by the equation

$$\xi^\varepsilon(t) = \xi_0 + \int_{]0,t]} a_\varepsilon(s)\xi^\varepsilon(s-)d\mu(s) + \eta(t) - \int_{]0,t]} \Gamma^\varepsilon(s)d\zeta(s),$$

where $\xi_0 = X_0 - \pi_0$, and $a_\varepsilon(s)$ is as in (A.151). Then

$$\mathbf{E}\,\xi^\varepsilon(t) = \int_{]0,t]} a_\varepsilon(s-)\mathbf{E}\,\xi^\varepsilon(s-)d\mu(s).$$

Therefore, $\mathbf{E}\,\xi^\varepsilon(t) = \mathbf{E}\,\big(X(t) - m^\varepsilon(t)\big) = 0$, so (i) is proved.

Next, consider the process $\vartheta^\varepsilon = \big(\vartheta^\varepsilon(t)\big)_{0\le t\le T}$ given by $\vartheta^\varepsilon(t) = m^\varepsilon(t) - \pi^\varepsilon(t)$. By (A.150) and (A.100), ϑ^ε satisfies the equation

$$\vartheta^\varepsilon(t) = \int_{]0,t]} a_\varepsilon(s)\vartheta^\varepsilon(s-)d\mu(s) - \varepsilon\int_{]0,t]} \Gamma^\varepsilon(s)h\,dM(s). \tag{A.152}$$

Since the martingale M does not depend on η, ζ, X_0, the processes ξ^ε and ϑ^ε are also independent. Consequently,

$$\mathbf{E}\,\xi^\varepsilon(t)\vartheta^\varepsilon(t) = \mathbf{E}\,\xi^\varepsilon(t)\mathbf{E}\,\vartheta^\varepsilon(t) = 0,$$

whence

$$\begin{aligned}\gamma^\varepsilon_{ii}(t) &= \mathbf{E}\big(X_i(t) - \pi_i^\varepsilon(t)\big)^2 = \mathbf{E}\big(X_i(t) - m_i^\varepsilon(t) + m_i^\varepsilon(t) - \pi_i^\varepsilon(t)\big)^2 = \\ &= \mathbf{E}\big(\xi_i^\varepsilon(t)\big)^2 + \mathbf{E}\big(\vartheta_i^\varepsilon(t)\big)^2 \ge \mathbf{E}\big(\xi_i^\varepsilon(t)\big)^2 = \widetilde{\gamma}^\varepsilon_{ii}(t),\end{aligned} \tag{A.153}$$

which, together with Lemma A.8, proves assertions (ii) and (iii).

Finally, we have

$$\begin{aligned}\gamma_{ii}(t) &= \mathbf{E}\,\big(X_i(t) - m_i(t)\big)^2 = \mathbf{E}\big(X_i(t) - m_i^\varepsilon(t) + m_i^\varepsilon(t) - m_i(t)\big)^2 = \\ &= \mathbf{E}\big(\xi_i^\varepsilon(t)\big)^2 + \mathbf{E}\big(m_i^\varepsilon(t) - m_i(t)\big)^2 = \widetilde{\gamma}^\varepsilon_{ii}(t) + \mathbf{E}\big(m_i^\varepsilon(t) - m_i(t)\big)^2,\end{aligned}$$

which in view of (A.153) and item (iii), proves item (iv). □

Next we address the important—from the point of view of practical applications—question of the rate of convergence of the regularized estimator m^{ε} as $\varepsilon \downarrow 0$. To estimate this rate, we estimate from above the functional

$$\begin{aligned} J^{\varepsilon}(t) = \mathbf{E}\big(m(t) - m^{\varepsilon}(t)\big)^2 = \mathbf{E}\big(m(t) - \pi^{\varepsilon}(t) + \pi^{\varepsilon}(t) - m^{\varepsilon}(t)\big)^2 \le \\ \le 2\mathbf{E}\big(m(t) - \pi^{\varepsilon}(t)\big)^2 + 2\mathbf{E}\big(\vartheta^{\varepsilon}(t)\big)^2. \end{aligned} \tag{A.154}$$

We start by evaluating the first term in the right-hand side of inequality (A.154). By the normal correlation theorem [244],

$$m(t) = \mathbf{E}X(t) + D_{XY}(t)D^{+}_{YY}(t)\big(Y(t) - \mathbf{E}Y(t)\big), \tag{A.155}$$

where $D_{XY}(t) = \operatorname{cov}\big(X(t), Y^{*}(t)\big)$ and $D_{YY}(t) = \operatorname{cov}\big(Y(t), Y^{*}(t)\big)$. Computing the conditional expectation in (A.155), we obtain

$$\begin{aligned} \pi^{\varepsilon}(t) = \mathbf{E}\big(m(t)|F^{Y^{\varepsilon}}_t\big) = \\ = \mathbf{E}X(t) + D_{XY}(t)D^{+}_{YY}(t)\big[\mathbf{E}\big(Y(t)|F^{Y^{\varepsilon}}_t\big) - \mathbf{E}Y(t)\big]. \end{aligned} \tag{A.156}$$

Note that $\mathbf{E}\big(Y(t)|F^{Y^{\varepsilon}}_t\big)$ is the best mean-square estimator of the process Y based on the observations Y^{ε}; in particular, it is better than the estimator equal to $Y^{\varepsilon}(t)$. Therefore, in view of (A.155) and (A.156), we have

$$\begin{aligned} &\mathbf{E}\big(m(t) - \pi^{\varepsilon}(t)\big)^2 = \operatorname{tr}\mathbf{E}\Big[\big(m(t) - \pi^{\varepsilon}(t)\big)\big(m(t) - \pi^{\varepsilon}(t)\big)^{*}\Big] = \\ &= \operatorname{tr}\Big[D_{XY}(t)D^{+}_{YY}(t)\mathbf{E}\Big[\big(Y(t) - \mathbf{E}\big(Y(t)|F^{Y^{\varepsilon}}_t\big)\big)\times \\ &\times \big(Y(t) - \mathbf{E}\big(Y(t)|F^{Y^{\varepsilon}}_t\big)\big)^{*}\Big]D^{+}_{YY}(t)D^{*}_{XY}(t)\Big] \le \\ &\le \operatorname{tr}\Big[D_{XY}(t)D^{+}_{YY}(t)\mathbf{E}\big[\big(Y(t) - Y^{\varepsilon}(t)\big)\big(Y(t) - Y^{\varepsilon}(t)\big)^{*}\big]D^{+}_{YY}(t)D^{*}_{XY}(t)\Big] = \\ &= \varepsilon^2\langle M\rangle_t \operatorname{tr}\big[D_{XY}(t)D^{+}_{YY}(t)D^{+}_{YY}(t)D^{*}_{XY}(t)\big], \end{aligned} \tag{A.157}$$

where, in accordance with the definition of M,

$$\langle M\rangle_t = t + \sum_{\tau_i \le t} \Delta\mu(\tau_i). \tag{A.158}$$

To calculate the covariance matrices $D_{XY}(t)$ and $D_{YY}(t)$, we consider the $(m + k)$-dimensional block-vector process

$$Z(t) = \begin{pmatrix} X(t) \\ Y(t) \end{pmatrix}.$$

By (A.92) and (A.93),

$$Z(t) = Z_0 + \int_{]0,t]} b(s)Z(s-)d\mu(s) + N(t),$$

where

$$b(t) = \begin{pmatrix} a(t) & {}_{m\times k} \\ A(t) & O_{k\times k} \end{pmatrix} \text{ is an } (m+k)\times(m+k)\text{-dimensional block-matrix,}$$

$$N(t) = \begin{pmatrix} \eta(t) \\ \zeta(t) \end{pmatrix} \text{ is a Gaussian martingale with}$$

$$\text{square characteristic} \quad \langle N\rangle_t = \int_{]0,t]} \begin{pmatrix} \alpha(s) & O_{m\times k} \\ O_{k\times m} & \beta(s) \end{pmatrix} d\mu(s),$$

where $O_{m\times k}$ is the $(m\times k)$-dimensional zero matrix.

Then the block-matrix

$$D(t) = \operatorname{cov}\big(Z(t), Z^*(t)\big) = \begin{pmatrix} D_{XX}(t) & D_{XY}(t) \\ D^*_{XY}(t) & D_{YY}(t) \end{pmatrix}$$

admits the representation

$$\begin{aligned} D(t) = D_0 &+ \int_{]0,t]} \big[D(s-)b^*(s) + b(s)D(s-)\big]d\mu(s) + \\ &+ \langle N\rangle_t + \sum_{\tau_i\le t} b(\tau_i)D(\tau_i)b^*(\tau_l)\big(\Delta\mu(\tau_l)\big)^2. \end{aligned} \tag{A.159}$$

Next we need to evaluate the second term in the right-hand side of inequality (A.154). Using the Itô formula for $q^\varepsilon(t) = \mathbf{E}\big(\vartheta^\varepsilon(t)\big)^2$, Eq. (A.152) and the Cauchy-Bunyakovsky-Schwarz inequality, we deduce that

$$q^\varepsilon(t) \le \int_{]0,t]} B_\varepsilon(s)q^\varepsilon(s-)d\mu(s) + \varepsilon^2 Q_\varepsilon(t), \tag{A.160}$$

where

$$B_\varepsilon(s) = 2\|a_\varepsilon(s)\|\Big[I\{\Delta\mu(s)=0\} + \Big(1 + \frac{1}{2}\|a_\varepsilon(s)\|\Big)\Delta\mu(s)\Big],$$

$$Q_\varepsilon(t) = \int_{]0,t]} \big(\Gamma^\varepsilon(s)h\big)^2 d\langle M\rangle_s.$$

Here $\|a\| = \sqrt{\operatorname{tr}(aa^*)}$.

By the generalized Gronwall-Bellman's lemma, inequality (A.160) implies that

$$q^{\varepsilon}(t) \leq \varepsilon^2\Big[\varphi_\varepsilon(t)\int_{]0,t]}\varphi_\varepsilon^{-1}(s)Q_\varepsilon(s-)B_\varepsilon(s)d\mu(s) + Q_\varepsilon(t)\Big], \tag{A.161}$$

where

$$\varphi_\varepsilon(t) = \exp\left\{\int_{]0,t]} B_\varepsilon(s)d\mu^c(s)\right\}\prod_{\tau_i\leq t}\big(1 + B_\varepsilon(\tau_i)\Delta\mu(\tau_i)\big).$$

Since $\|a_\varepsilon(t)\|$ and $\big(\Gamma^{\varepsilon}(t)h\big)^2$ are uniformly bounded for $t \in [0, T]$, inequality (A.161) gives

$$q^{\varepsilon}(t) \leq \varepsilon^2\langle M\rangle_t\, C_1, \quad \text{where } C_1 = \text{const}. \tag{A.162}$$

Finally, by (A.159), $\|D(t)\|$ is uniformly bounded, which together with (A.157), (A.162) and (A.158) implies that

$$\begin{aligned} J^{\varepsilon}(t) &= \mathbf{E}\big(m(t) - m^{\varepsilon}(t)\big)^2 \leq \varepsilon^2\langle M\rangle_t\, C_2 = \\ &= \varepsilon^2\Big(t + \sum_{\tau_i\leq t}\Delta\mu(\tau_i)\Big)C_2 \leq \varepsilon^2\big(t + C_3\big)C_2, \end{aligned}$$

where C_2 and C_3 are constants, with $C_3 = \underset{[0,T]}{\text{Var}}\ \mu(s)$.

A.3.4 Generalized Kalman Filter for Processes with Control

We consider again the stochastic basis $(\Omega, F, \mathbf{F} = (F_t)_{0\leq t\leq T}, P)$ defined in the previous section.

Let $\mathbf{D}^k$ be the space of $d\mu(\cdot)$-integrable and right-continuous k-dimensional vector-functions that admit left limits and are defined on the segment $[0, T]$.

We introduce the predictable σ-algebras

$$\mathscr{P}_t = \sigma\{]s,t] \times F : s < t,\ F \in F_s\}, \quad \mathscr{P} = (\mathscr{P}_t)_{0\leq t\leq T}$$

and

$$\widetilde{\mathscr{P}}_t = \sigma\{]s,t] \times G : s < t,\ G \in \mathscr{B}^k_s\}, \quad \widetilde{\mathscr{P}} = (\widetilde{\mathscr{P}}_t)_{0\leq t\leq T},$$

where $\mathscr{B}^k_s$ is the σ-algebra generated by cylindrical sets in the space $\mathbf{D}^k$ with bases on the segment $[0, s]$.

Denote by $\mathscr{U}^m$ the class of $\widetilde{\mathscr{P}}$-measurable m-dimensional vector-functions $u(t, g(\cdot)) = u(t, g)$ defined on $[0, T] \times \mathbf{D}^k$ and satisfying the Lipschitz condition on $t \in [0, T]$ with the constant $L > 0$, i.e.,

$$|u(t, g) - u(t, h)| \leq L\|g - h\|_t, \tag{A.163}$$

for any functions $g(\cdot),\ h(\cdot) \in \mathbf{D}^k$. Here

$$\|g\|_t = \text{ess sup}_{0\leq s\leq t} \|g(s)\| = \min\{\lambda \in \mathbb{R}^1 :\ |g(s)| \leq \lambda \text{ for } d\mu(\cdot)\text{-a.e. } s \leq t\}.$$

For any $\mathbf{F}$-adapted process Y with paths belonging to $\mathbf{D}^k$ the process $u(t, Y)$ will be $\mathscr{P}$-measurable [255]. Therefore, under the assumptions made above, the system of Itô stochastic differential equations

$$dX^u(t) = a(t)X^u(t-)d\mu(t) + u(t, Y^u)d\mu(t) + d\eta(t), \tag{A.164}$$

$$dY^u(t) = A(t)X^u(t-)d\mu(t) + d\zeta(t) \tag{A.165}$$

with $X(0) = X_0$ and $Y(0) = Y_0$ has a unique strong solution for any $T > 0$ [255].

The system (A.164) and (A.165) can be regarded as description of the evolution of the phase coordinates X^u of some controllable stochastic object and of the process Y^u of observation of these coordinates. In this interpretation, the class $\mathscr{U}^m$ plays the role of the class of admissible controls.

As remarked earlier, the problem of optimal control of the system (A.164) and (A.165) will not be considered here. The aim of this section is to derive the Kalman filter for this system. To this end, alongside the process (X^u, Y^u) from (A.164) and (A.165), we consider the process (X, Y) satisfying (A.92) and (A.93), i.e., when in (A.164), (A.165) the control $u(\cdot) \equiv 0$.

Put

$$\xi^u(t) = X^u(t) - X(t), \tag{A.166}$$

$$\psi^u(t) = Y^u(t) - Y(t). \tag{A.167}$$

Then

$$d\xi^u(t) = a(t)\xi^u(t-)d\mu(t) + u(t, Y^u)d\mu(t), \quad \xi^u(0) = 0, \tag{A.168}$$

$$d\psi^u(t-) = A(t)\xi^u(t-)d\mu(t), \quad \psi^u(0) = 0. \tag{A.169}$$

Let $\varphi(t, s)$ be the fundamental matrix corresponding to the matrix $a(t)$, i.e.

$$\varphi(t, s) = I_{(m\times m)} + \int_{]s,t]} a(v)\varphi(v-, s)d\mu(v). \tag{A.170}$$

Then (A.166), (A.167), (A.168), and (A.169) yield the representations

$$\xi^u(t) = \int_{]0,t]} \varphi(t,s)u(s,Y^u)d\mu(s)$$

and

$$Y^u(t) = Y(t) + \psi^u(t) = Y(t) + \int_{]0,t]} A(s)\xi^u(s-)d\mu(s),$$

or

$$Y^u(t) = Y(t) + \int_{]0,t]} A(s)\left[\int_{]0,s[} \varphi(s-,v)u(v,Y^u)d\mu(v)\right]d\mu(s). \tag{A.171}$$

The following theorem plays a key role in the derivation of the generalized Kalman filter for the controlled processes (X^u, Y^u). In the continuous time case, i.e., when $d\mu(t) = dt$, this result is well known [452].

Theorem A.32 *The family* $\mathbf{F}^{Y^u} = (F_t^{Y^u})_{0\le t\le T}$ *does not depend on the control; in particular,* $\mathbf{F}^{Y^u} = \mathbf{F}^{Y}$.

To prove this result we need the following lemma on the existence and uniqueness of the solution of a Volterra-type equation with measure. This technical lemma is of independent interest.

Lemma A.15 *Consider on* $\mathbf{D}^k$ *the operator* Φ *acting as*

$$(\Phi g)(t) = \int_{]0,t]} \varphi(t,s)u(s,g)d\mu(s), \tag{A.172}$$

where $u(\cdot) \in \mathscr{U}^m$, $\varphi(t,s)$ *satisfies (A.170), and* $A(\cdot)$ *is the* $k \times m$ *matrix defined above. Then for any function* $f(\cdot) \in \mathbf{D}^k$ *the integral equation*

$$g(t) = f(t) + \int_{]0,t]} A(s)(\Phi g)(s-)d\mu(s) \tag{A.173}$$

has exactly one solution (*up to modification on a set of* $d\mu(\cdot)$*-measure zero*) *in the space* $\mathbf{D}^k$.

Proof of Lemma A.15 The argument is based on the following properties of the integral operator figuring in the right-hand side of Eq. (A.173).

(1) First of all, note that if $g \in \mathbf{D}^k$, then

$$\Phi g \in \mathbf{D}^k. \tag{A.174}$$

Moreover, in view of (A.163), the operator Φ obeys the estimate

$$\|\Phi g - \Phi h\|_t \le C\mu(t)\|g - h\|_t, \quad C = \text{const.} \tag{A.175}$$

(2) For a given function $f \in \mathbf{D}^k$, define on $\mathbf{D}^k$ the operator $\mathbf{B}$ by the formula

$$(\mathbf{B}g)(t) = f(t) + \int_{[0,t]} A(s)(\Phi g)(s-)d\mu(s). \tag{A.176}$$

Note that

$$\mathbf{B}g \in \mathbf{D}^k \tag{A.177}$$

and

$$\|\mathbf{B}g - \mathbf{B}h\|_t \le C_1 \int_{[0,t]} \mu(s-)\|g - h\|_{s-}d\mu(s), \tag{A.178}$$

$$\|\mathbf{B}g - \mathbf{B}h\|_t \le C_1 \frac{\mu^2(t)}{2!}\|g - h\|_t, \quad C_1 = \text{const.}, \tag{A.179}$$

These inequalities are consequences of property (A.175) of the operator Φ. Indeed, using (A.175) and the obvious inequality

$$\int_{[0,t]} \mu^k(s-)d\mu(s) \le \frac{\mu^{k+1}(t)}{k+1}, \tag{A.180}$$

we have that

$$\|\mathbf{B}g - \mathbf{B}h\|_t = \|\int A(s)(\Phi g - \Phi h)(s-)d\mu(s)\|_t \le$$

$$\le C_1 \int_{[0,t]} \mu(s-)\|g - h\|_{s-}d\mu(s) \le C_1 \frac{\mu^2(t)}{2!}\|g - h\|_t,$$

as needed.

(3) From (A.178) and (A.179), one derives by induction an estimate for the iterated operator $\mathbf{B}^n$:

$$\|\mathbf{B}^n g - \mathbf{B}^n h\|_t \le C_1^n \int_{[0,t]} \frac{\mu^{2n-1}(s-)}{2^{n-1}(n-1)!}\|g - h\|_{s-}d\mu(s). \tag{A.181}$$

From this estimate and (A.180) it follows that

$$\|\mathbf{B}^n g - \mathbf{B}^n h\|_t \le \frac{K^n}{n!}\|g - h\|_t, \quad K = \text{const.} \tag{A.182}$$

Indeed, for $n = 1$ estimate (A.181) has the form (A.178). Using the (A.178), (A.179) and (A.180), we have for $n+1$ that

$$\|\mathbf{B}^{n+1}g - \mathbf{B}^{n+1}h\|_t = \|\mathbf{B}(\mathbf{B}^n g) - \mathbf{B}(\mathbf{B}^n h)\|_t \leq$$

$$\leq C_1 \int_{[0,t]} \mu(s-)\|\mathbf{B}^n g - \mathbf{B}^n h\|_{s-} d\mu(s) \leq$$

$$\leq C_1 \int_{[0,t]} \mu(s-)\left[C_1^n \int_{[0,s[} \frac{\mu^{2n-1}(r-)}{2^{n-1}(n-1)!}\|g-h\|_{r-} d\mu(r)\right] d\mu(s) \leq$$

$$\leq \frac{C_1^{n+1}}{2^{n-1}(n-1)!}\int_{[0,t]} \mu(s-)\|g-h\|_{s-}\left[\int_{[0,s[} \mu^{2n-1}(r-)d\mu(r)\right] d\mu(s) \leq$$

$$\leq \frac{C_1^{n+1}}{2^n n!}\int_{[0,t]} \mu^{2n+1}(s-)\|g-h\|_{s-} d\mu(s),$$

and so inequality (A.181) is proved.

Thus, continuing the sequence of inequalities in (A.181) we get

$$\|\mathbf{B}^n g - \mathbf{B}^n h\|_t \leq$$

$$\leq \frac{C_1^n}{2^{n-1}(n-1)!}\int_{[0,t]} \mu^{2n-1}(s-)\|g-h\|_{s-} d\mu(s) \leq$$

$$\leq \frac{C_1^n}{2^{n-1}(n-1)!}\|g-h\|_t \frac{\mu^{2n}(t)}{2^n n!} = \frac{C_1^n \mu^{2n}(t)}{2^n n!}\|g-h\|_t \leq$$

$$\leq \frac{K^n}{n!}\|g-h\|_t, \qquad \text{where } K = \frac{C_1\mu^2(T)}{2} = \text{const.}$$

Therefore, inequality (A.182) holds, which means that $\mathbf{B}$ is a contraction and hence, since the space $\mathbf{D}^k$ is complete with respect to our norm [204], the assertion of Lemma A.15 is proved. □

Remark A.25 Lemma A.15 remains valid if we consider (instead of functions with bounded variation) $d\mu(\cdot)$-equivalent square-integrable functions with the norm

$$\|g\|_t = \left(\int_{[0,t]} |g(s-)|^2 d\mu(s)\right)^{1/2}.$$

In this case the analogues of inequalities (A.181) and (A.182) have the form

$$\|\mathbf{B}^n g - \mathbf{B}^n h\|_t^2 \leq C_1^{2n}\int_{[0,t]} \frac{\mu^{3n-1}(s-)}{3^{n-1}(n-1)!}\|g-h\|_{s-}^2 d\mu(s),$$

and

$$\|\mathbf{B}^n g - \mathbf{B}^n h\|_t^2 \leq \frac{K^n}{n!}\|g - h\|_t^2,$$

respectively. The scheme of the proof of is the same.

Proof of Theorem A.32 Define the measurable map $\bar{\mathbf{B}} : (\mathbf{D}^k, \mathscr{B}_T^k) \to (\mathbf{D}^k, \mathscr{B}_T^k)$ by the formula

$$(\bar{\mathbf{B}}g)(t) = \int_{]0,t]} A(s)(\Phi g)(s-)d\mu(s). \tag{A.183}$$

In view of (A.172) and (A.183), Eq. (A.171) can be recast as

$$Y(t) = Y^u(t) - (\bar{\mathbf{B}}Y^u)(t).$$

Thus the inclusion $F_t^Y \subseteq F_t^{Y^u}$ follows.

To prove the opposite inclusion $F^{Y^u} \subset F_t^Y$, consider the recurrence relation

$$Y_{n+1}(t) = Y(t) + (\bar{\mathbf{B}}Y_n)(t), \quad n = 0, 1, 2, \ldots, \tag{A.184}$$

with $Y_0(t) \equiv 0$. One can verify by induction that the sequence $\{Y_n(t)\}_n$ is F_t^Y-measurable. Next, (A.184) and the proof of Lemma A.15, where the contractivity of the operator $\bar{\mathbf{B}}$ is established (see (A.179), where in our case the role of $g_n(t)$ is played by $Y_n(t)$)) imply the F_t^Y-measurability of $Y^u(t)$ and the convergence of the sequence $\{Y_n(t)\}_{n=1,2,\ldots}$ to the unique solution of the equation

$$Y^u(t) = Y(t) + (\bar{\mathbf{B}}Y^u)(t).$$

This shows that $F_t^{Y^u} \subseteq F_t^Y$, as desired. □

We now pass to the derivation of the Kalman filter for the system (A.164) and (A.165).

Theorem A.33 *The following holds:*

(1) *The process* $m^u = (m^u(t))_{0\leq t\leq T}$ *with* $m^u(t) = \mathbf{E}(X^u(t)|F_t^{Y^u})$ *satisfies the equation*

$$\begin{aligned} m^u(t) = \pi_0 + \int_{]0,t]} a(s)m^u(s-)d\mu(s) + \int_{]0,t]} u(s, Y^u)d\mu(s) + \\ + \int_{]0,t]} \Gamma(s)\big[dY^u(s) - A(s)m^u(s-)d\mu(s)\big], \end{aligned} \tag{A.185}$$

where $\Gamma(s)$ *is given by formula (A.116) and* $dY^u(s)$ *has the representation (A.165).*

(2) *The process* $\nu^u = (\nu^u(t))_{0\le t\le T}$ *with the Itô differential*

$$d\nu^u(t) = dY^u(t) - A(t)m^u(t-)d\mu(t) \tag{A.186}$$

is a Gaussian martingale with respect to the family of σ*-algebras* $\mathbf{F}^{Y^u}$.

(3) *Moreover, the following* σ*-algebras coincide for all* $t \in [0, T]$*:*

$$F_t^{Y^u} = F_t^Y = F_t^{\nu^u, Y_0}. \tag{A.187}$$

Proof Using (A.166), (A.167), (A.168), and (A.169), Theorem A.32, and the $\mathbf{F}^{Y^u}$-adaptivity of the processes $\xi^u = \big(\xi^u(t)\big)_{0\le t\le T}$, we deduce that

$$m(t) = \mathbf{E}(X(t)|F_t^Y) = \mathbf{E}(X^u(t)|F_t^Y) - \mathbf{E}(\xi^u(t)|F_t^Y) =$$

$$= \mathbf{E}(X^u(t)/F_t^{Y^u}) - \mathbf{E}(\xi^u(t)/F_t^{Y^u}) = m^u(t) - \xi^u(t).$$

But the process $m = \big(m(t)\big)_{0\le t\le T}$ satisfies the Eq. (A.133) and (A.134), which formally corresponds to the case $u(\cdot) \equiv 0$. Therefore, substituting $m(t) = m^u(t) - \xi^u(t)$ in (A.133), we find that

$$\begin{aligned} &m^u(t) - \xi^u(t) = \\ &= \pi_0 + \int_{]0,t]} a(s)m^u(s-)d\mu(s) - \int_{]0,t]} a(s)\xi^u(s-)d\mu(s) + \\ &+ \int_{]0,t]} \Gamma(s)\big[dY^u(s) - A(s)\xi^u(s-)d\mu(s) - A(s)m^u(s-)d\mu(s) \\ &\qquad + A(s)\xi^u(s-)d\mu(s)\big]. \end{aligned}$$

Eliminating here ξ^u by means of formula (A.168) we get (A.185). Assertion 1) of the theorem is proved.

Next, in view of the relations (A.166) and (A.167), the process ν^u defined by (A.186) coincides with the process $\nu = (\nu(t))_{0\le t\le T}$ that admits the representation

$$d\nu(t) = dY(t) - A(t)m(t-)d\mu(t). \tag{A.188}$$

Now note that the σ-algebras F_t^Y and F_t^{ν,Y_0} coincide. Indeed, the inclusion $F_t^{\nu,Y_0} \subseteq F_t^Y$ is a corollary of the representation (A.188) of the process ν, and the opposite inclusion $F_t^Y \subseteq F_t^{\nu,Y_0}$ follows from the representation

$$Y(t) = Y_0 + \nu(t) + \int_{]0,t]} A(s)\left[\varphi(s-,0)\pi_0 + \int_{]0,s[} \varphi(s-,v)\Gamma(v)d\nu(v)\right]d\mu(s),$$

which is a consequence of (A.188), (A.170), and (A.133). Thus, $F_t^Y = F_t^{\nu,Y_0}$. The equality $F_t^{Y^u} = F_t^Y$ is a result of Theorem A.32. Assertion (3) of the theorem is proved.

Finally, it follows from (A.187) and the coincidence of the processes ν^u and ν that the process ν^u is a Gaussian martingale with regular paths with respect to the family $\mathbf{F}^{Y^u}$. Assertion (2) of the theorem is proved. □

Corollary A.5 *The process m^u admits the stochastic Itô differential*

$$dm^u(t) = [a(t)m^u(t-)+u(t, Y^u)]d\mu(t)+\Gamma(t)d\nu(t), \quad m^u(0) = m_0. \tag{A.189}$$

A.4 Linear-Quadratic Gaussian Stochastic Control Problem

As an example of application of the Kalman filter for processes with controls, we discuss here the generalized (discrete-continuous) linear-quadratic Gaussian stochastic control problem for a partially observable system (X^u, Y^u) whose evolution is described by the Eqs. (A.164) and (A.165).

The objective is to find an admissible control $u^0(\cdot) \in \mathscr{U}^m$ which minimizes the criterion

$$\begin{aligned} &J(u(\cdot, Y^u)) = \\ &= \mathbf{E}\Big\{ \int_{[0,T]} \Big[X^u(t-)^* K_t X^u(t-) + u(t, Y^u)^* R_t u(t, Y^u)\Big] d\mu(t) \\ &\quad + X^u(T)^* K X^u(T)\Big\}, \end{aligned} \tag{A.190}$$

where K, K_t and R_t are symmetric $d\mu(\cdot)$-integrable and $d\mu(\cdot)$-a.e. non-negative definite matrix functions of appropriate dimensions. We assume that $K_T = 0$, $R_T = 0$, and the (possibly, degenerate) matrix R_t satisfies the condition

$$\operatorname{tr} \int_{[0,T]} R_t^+ d\mu(t) < \infty. \tag{A.191}$$

In addition, we assume that the matrix algebraic equation

$$R_t v + I_{(m\times m)} = 0 \tag{A.192}$$

is solvable $d\mu^c(\cdot)$-a.e. with respect to $v \in \mathbb{R}^{m\times m}$.

To solve this problem (called in the literature a Linear Quadratic Gaussian (LGQ) problem), we will follow the standard procedure [386], the first step of which is to reduce the initial control problem for a partially observable system (the so-called control problem with incomplete data) to a completely conditional-mean observable control system.

Thanks to properties of conditional expectations, we get

$$\begin{aligned}\mathbf{E}\big[X^u(t-)^* K_t X(t)\big] &= \mathbf{E}\mathbf{E}\big[X^u(t-)^* K_t X(t-)|\mathscr{F}_{t-}^{Y^u}\big] = \\ &= \mathbf{E}\mathbf{E}\big[\big(X^u(t-) - m^u(t-) + m^u(t-)\big)^* K_t\big(X^u(t-) - m^u(t-) \\ &\quad + m^u(t-)\big)|\mathscr{F}_{t-}^{Y^u}\big] = \\ &= \mathbf{E}\big[m^u(t-)^* K_t m^u(t-)\big] + \mathrm{tr}\big[K_t^{1/2}\gamma(t-)K_t^{1/2}\big].\end{aligned}$$

Therefore, the criterion (A.190) takes on the form

$$\begin{aligned}J(u(\cdot, Y^u)) = \mathbf{E}\Bigg\{ &\int_{[0,T]} \mathrm{tr}\big[K_t^{1/2}\gamma(t-)K_t^{1/2}\big]d\mu(t) + m^u(T)^* K m^u(T) + \\ &+ \int_{[0,T]} \big[m^u(t-)^* K_t m^u(t-) + u(t, Y^u)^* R_t u(t, Y^u)\big]d\mu(t)\Bigg\}.\end{aligned} \tag{A.193}$$

Since the first term in (A.193) is control-independent, we can replace $J(u(\cdot, Y^u))$ by the criterion

$$\begin{aligned}\widehat{J}(u(\cdot, Y^u)) = \mathbf{E}\Bigg\{ &m^u(T)^* K m^u(T) + \\ &+ \int_{[0,T]} \big[m^u(t-)^* K_t m^u(t-) + u(t, Y^u)^* R_t u(t, Y^u)\big]d\mu(t)\Bigg\}.\end{aligned} \tag{A.194}$$

Thus, we arrived at a control problem (already with completely observable data) for the stochastic process (A.189) with the criterion (A.194) to be minimized by a control $u(\cdot) \in \mathscr{U}^m$. It is natural to regard this problem on the stochastic basis $(\Omega, \mathscr{F}, \mathbf{F}^Y = (\mathscr{F}_t^Y)_{0\le t\le T}, P)$, where, as we noted above, the process ν is a Gaussian martingale.

In [386] there is given the solution of the generalized (discrete-continuous) LQG control problem with a singular criterion (degenerate matrix R_t in (A.194)) for the case of a completely observable control process. An application of these results to the linear-quadratic problem for partially observable hybrid systems gives the following theorem.

Theorem A.34 *Consider the partially observable system (X^u, Y^u) described by the Eqs. (A.164) and (A.165), with the performance criterion (A.190) and the conditions (A.90), (A.91), (A.191), and (A.192). Then the optimal control $u^0(\cdot) \in \mathscr{U}^m$ that minimizes (A.190) has the form*

$$u^0(t, Y^0) = \begin{cases} -R_t^+ q_t m^0(t), & t \in \mathscr{T}^{\mathrm{c}}, \\ -r_t^+ q_t \big[I_{(m\times m)} + a(t)\Delta\mu(t)\big] m^0(t-), & t \in \mathscr{T}^{\mathrm{d}}, \end{cases}$$

where one denotes: $r_t = R_t + q_t \Delta\mu(t)$, $\mathscr{T}^{\mathrm{d}} = \{t \in [0, T] : \Delta\mu(t) > 0\}$ *and* $\mathscr{T}^{\mathrm{c}} = [0, T] \setminus \mathscr{T}^{\mathrm{d}}$ *(the set* $\mathscr{T}^{\mathrm{d}}$ *is at most countable); the superscript "0" labels the variables corresponding to the optimal control. The process* $m^0 = (m^0(t))_{0\le t\le T}$ *with* $m^0(t) = \mathbf{E}(X^0(t)|F_t^{Y^0})$ *satisfies the Eq. (A.185) with* $\Gamma(t)$ *given in (A.116) and* $\gamma(t)$ *given in (A.133). The deterministic matrix-function* q_t *satisfies the Riccati matrix differential equation with measure*

$$\begin{aligned} - dq_s &= \Big[a^*(s)q_s + q_s a(s) + K_s - \big(I_{(m\times m)} + a(s)\Delta\mu(s)\big)^* \times \\ &\times q_s (R_s + q_s \Delta\mu(s))^+ q_s (I_{(m\times m)} + a(s)\Delta\mu(s))\Big] d\mu(s) \\ &\quad + a^*(s) q_s a(s)\Delta\mu(s) d\mu(s) \end{aligned}$$

and the terminal condition $q_T = K$.

In conclusion, we note that the above result is applicable to a wide class of hybrid (discrete-continuous) linear control systems for which the well-known separation "principle" is valid; this principle allows one to reduce the control problem with incomplete data to two other problems: a filtering problem (finding an estimate of an unobservable process and the covariance matrix of the estimation errors, which play the role of the so-called sufficient statistics) and the control problem based already on complete data, with actually sufficient statistics [133].

A.5 Example

In this section we give the explicit form of the filter in a case with an infinite number of observation instants. Specifically, we consider a ball bouncing on a horizontal elastic conveyor belt with initial velocity $V_0 = (V_{x0}, V_{y0})$. We assumes that the belt is not ideally elastic; then the vertical component of the velocity, V_y, diminishes after each impact at the instants $\tau_i, i = 1, 2, \ldots$ and tends to zero. As a consequence, the time intervals $[\tau_{i-1}, \tau_i]$ between impacts shrink and tend to zero as well. It is readily seen that there exists the limit $\lim \tau_n = \tau$, $n \to \infty$. After the instant τ, the vertical velocity component vanishes and the ball has only a horizontal velocity component $V_x \neq 0$. Suppose one is able to observe the coordinates $x(\tau_i)$ of impact points with some additive Gaussian noise, and we want to "calm" the ball, i.e., reduce the horizontal component V_x, by moving the belt forward or backward, in a controlled manner, at the impact instants τ_i.

A mathematical model of this mechanical system consists of

An unobservable process:

$$dV_x(t) = -k\widehat{V}_x(t-)\,d\mu(t), \quad \text{with } V_x(0-) = V_{x0}.$$

An observations process:

$$dx(t) = V_x(t-)\,d\mu(t) + d\zeta(t), \quad \text{with } x(0-) = x_0.$$

In this model the control law appears in the explicit form $u(t,x) = -k\widehat{V}_x(t-)$, where k =const and $\widehat{V}_x(t)$ is an $\mathscr{F}_t^x$-measurable estimate of the process $V_x(t)$ (for example, $\widehat{V}_x(t) = m(t) = \mathbf{E}(V_x(t)|\mathscr{F}_t^x)$).

Let V_{x0} and x_0 be independent Gaussian quantities with given mean $\bar{V}_{x0}$ and $\bar{x}_0$, and with covariance $d_{VV} = \operatorname{cov}(V_{x0}, V_{x0})$ and $d_{xx} = \operatorname{cov}(x_0, x_0)$, respectively. The measure $\mu(t)$ has the form

$$\mu(t) = \sum_{\tau_i \le t} \Delta\mu(\tau_i) + (t-\tau)\mathbf{I}\{t \ge \tau\}, \quad \text{with } \Delta\mu(\tau_i) = \tau_i - \tau_{i-1},\ \tau_0 = 0.$$

Suppose the noise affecting the observations is governed by the equation

$$\zeta(t) = \sum_{\tau_i \le t} \Delta\zeta(\tau_i) + \sqrt{\beta}\, W(t)\mathbf{I}\{t \ge \tau\},$$

where $\Delta\zeta(\tau_i)$ $(i = 1, 2, \ldots)$ are independent Gaussian quantities with zero mean and covariance $\beta\Delta\mu(\tau_i)$, β =const, $W(t)$ is the standard Wiener process independent of V_{x0}, x_0, and $\Delta\zeta(\tau_i)$, $i = 1, 2, \ldots$.

For this model the equations of the Kalman filter read

$$m(t) = \bar{V}_{x0} - k\sum_{\tau_i \le t} m(\tau_{i-1})\Delta\mu(\tau_i) - k\int_{[0,t]} \mathbf{I}\{t \ge \tau\}m(s)ds +$$

$$+\sum_{\tau_i \le t} \Gamma(\tau_i)\big[\Delta x(\tau_i) - m(\tau_i)\Delta\mu(\tau_i)\big] + \int_{[0,t]} \mathbf{I}\{t \ge \tau\}\Gamma(s)\big[dx(s) - m(s)ds\big],$$

where

$$\Gamma(\tau_i) = \frac{\gamma(\tau_{i-1})}{\beta + \gamma(\tau_{i-1})\Delta\mu(\tau_i)}, \quad \Gamma(t) = \gamma(t)\,\beta^{-1}, \ \text{ for } t \ge \tau,$$

and

$$\gamma(t) = d_{VV} - \sum_{\tau_i \le t} \frac{\gamma(\tau_{i-1})}{\beta + \gamma(\tau_{i-1})\Delta\mu(\tau_i)}\Delta\mu(\tau_i) - \frac{1}{\beta}\int_{[0,t]} \mathbf{I}\{t \ge \tau\}\gamma^2(s)ds.$$

Bibliography

1. V. Acary, B. Brogliato, *Numerical Methods for Nonsmooth Dynamical Systems*. Applications in Mechanics and Electronics (Springer, Berlin, 2008)
2. J. Ahn, D.E. Stewart, Existence of solutions for a class of impact problems without viscosity. SIAM J. Math. Anal. **38**(1), 37–63 (2006)
3. M.A. Aizerman, *Classical Mechanics* (in Russian) (Nauka, Moscow, 1980)
4. C. Altafini, R. Frezza, Motion on submanifolds of noninvariant holonomic constraints for kinematic control systems evolving on a matrix Lie group. Syst. Control Lett. **50**, 241–250 (2003)
5. T.F. Anan'ina, On optimization of a controlled system in the class of generalized controls. Differ. Uravnen. **11**(4), 595–603 (1975)
6. P. Antosik, J. Ligeza, Products of measures and functions of finite variation, in *Proceedings of the Conference on Generalized Functions and Operational CalculusVarna, 1975, Sofia* (1979), pp. 20–26
7. Y. Aoustin, A. Formalsky, C. Chevallereau, Virtual quadruped: mechanical design, control, simulation and experimentation. J. Math. Sci. **147**(2), 6552–6568 (2007)
8. A.V. Arutyunov, D.Y. Karamzin, F.L. Pereira, A nondegenerate maximum principle for the impulse control problem with state constraints. SIAM J. Control Optim. **43**, 1812–1843 (2005)
9. A.V. Arutyunov, D.Y. Karamzin, F.L. Pereira, State constraints in impulsive control problems: Gamkrelidze-like conditions of optimality. J. Optim. Theory Appl. **166**, 440–459 (2015)
10. A.V. Arutyunov, D.Y. Karamzin, F.L Pereira, Conditions of the jump absence in solution of adjoint system of maximum principle in optimal control problem with phase constraints. Proce. Steklov Inst. Math. **292**(suppl. 1), 27–35 (2016)
11. A.V. Arutyunov, D.Y. Karamzin, F.L. Pereira, *Optimal Impulsive Control. The Extension Approach*. Springer Lecture Notes in Control and Information Sciences (Springer, Cham, 2019)
12. A.A. Aslanyan, Optimality conditions in the control problem of systems with impulse actions. Dokl. Akad. Nauk Ukr. SSR Ser. Fiz.-Mat. Tekh. Nauki **11**, 3–6 (1982)
13. A.A. Aslanyan, Maximum principle for discontinuous dynamical systems. Teor. Funkts. Funkts. Anal. Prilozh. **37**, 132–137 (1982)
14. V.K. Astashev, V.I. Babitsky, *Ultrasonic Processes and Machines*. Dynamics, Control and Applications (Springer, Berlin, 2007)
15. M. Athans, The matrix minimum principle. Inform. Control **11**, 562–606 (1968)

B. Miller, E. Rubinovich, *Optimization of Dynamical Systems with Impulse Controls and Shocks*, Systems & Control: Foundations & Applications,
https://doi.org/10.1007/978-3-031-64124-4

16. J.P. Aubin, J. Lygeros, M. Quincampoix, S. Sastry, N. Seube, Impulse differential inclusions: a viability approach to hybrid systems. IEEE Trans. Autom. Control **47**(1), 2–20 (2002)
17. V.I. Babitsky, *Theory of Vibro-Impact Systems and Applications* (Springer, Berlin, 1998). (Revised translation from Russian), Nauka, Moscow (1978)
18. P. Ballard, The dynamics of discrete mechanical systems with perfect unilateral constraints. Arch. Rational Mech. Anal. **154**, 199–274 (2000)
19. P. Ballard, Formulation and well-posedness of the dynamics of rigid-body systems with perfect unilateral constraints. Philos. Trans. R. Soc. Lond. **A 359**, 2327–2346 (2001)
20. P. Ballard, Dynamics of rigid bodies with unilateral or frictional constraints, in *Advances in Mechanics and Mathematics* (Kluwer Academic, Dordrecht, 2002), pp. 3–87
21. P. Ballard, S. Basseville, Existence and uniqueness for dynamical unilateral contact with Coulomb friction: a model problem. Math. Model. Numer. Anal. **39**(1), 59–77 (2005)
22. G. Barles, An approach of deterministic control problems with unbounded data. Ann. Inst. Henry Poincaré **7**, 235–258 (1990)
23. E.N. Barron, Remarks on a simple optimal control problem with monotone control functions. J. Optim. Theory Appl. **41**, 573–586 (1983)
24. E.N. Barron, R.Jensen, A.G. Malliaris, Minimizing a quadratic payoff with monotone controls. Math. Oper. Res. **12**(2), 297–308 (1987)
25. E.N. Barron, R. Jensen, J.L. Menaldi, Optimal control and differential games with measures. Nonlinear Anal. Theory Methods Appl. **21**, 241–268 (1993)
26. M.V. Basin, Filtering of stochastic Ito-Voltera processes with respect to discrete-continuous observations. Autom. Remote Control **53**(8), 1183–1192 (1992)
27. M.V. Basin, Y.V. Orlov, Guaranteed estimation of linear dynamic systems with respect to discrete-continuous observations. Autom. Remote Control **53**(3), 349–357 (1992)
28. J. Bastien, M. Schatzman, C.H. Lamarque, Study of some reological models with finite number of degrees of freedom. Eur. J. Mech. A/Solids **19**, 277–307 (2000)
29. G. Bayada, M. Chambat, A. Lakhal, A dynamic contact problem with regularized friction law. Application to impact problems. Math. Comput. Model. **28**(4–8), 67–85 (1998)
30. R. Bellman, *Introduction to Matrix Analysis*. Classics in Applied Mathematics (McGraw-Hill, New York, 1970)
31. A. Bensoussan, J.L. Lions, Nouvelles methodes en contrôle impulsionnel. Appl. Math. Optim. **1**(4), 289–312 (1975)
32. A. Bensoussan, S. Sethy, R. Vickson, N. Derzko, Stochastic production planning with production constraints. SIAM J. Control Optim. **22**, 920–935 (1984)
33. J. Bentsman, B.M. Miller, Control of dynamic systems with unilateral constraints and differential equations, with measures, in *Preprints of the 4th IFAC Nonlinear Control Systems Design Symposium 1998 (NOLCOS'98), University of Twente, Enschede, July 1–3, 1998*, ed. by H.J.C. Huijberts, H. Nijmeijer, A.J. van der Schaft, J.M.A. Scherpen, vol. 2 (University of Twente, Enschede, 1998), pp. 411–416
34. J. Bentsman, B.M. Miller, Dynamical systems with controllable singularities: multi-scale and limit representation and optimal control, in *Proceedings of the 40th IEEE Conference on Decision and Control (CDC'01), Orlando, December, 2001* (2001), pp. 3681–3686
35. J. Bentsman, B.M. Miller, Mechanical systems with unilateral constraints: controlled singularities approach, in *Proceedings of the 40th IEEE Conference on Decision and Control (CDC'01), Orlando, December, 2001* (2001), pp. 3692–3697
36. J. Bentsman, B.M. Miller, Dynamical systems with controlled singularities: physically-based representation and control-oriented modeling, in *Proceedings of the 42th IEEE Conference on Decision and Control (CDC'03), Maui, December, 2003* (2003), pp. 6351–6356
37. J. Bentsman, B.M. Miller, Dynamical systems with active singularities of elastic type: a modeling and controller synthesis framework. IEEE Trans. Autom. Control **52**(1), 39–55 (2007)
38. J. Bentsman, B.M. Miller, E.Y. Rubinovich, Modeling and control of dynamical systems with active singularities and sensing in a singular motion phase, in *Proceedings of the 16th*

IFAC World Congress, Prague, July 4-8, IFAC Proceedings, vol. 38, no. 1 (2005), pp. 477–482

39. J. Bentsman, B.M. Miller, E.Y. Rubinovich, K. Zheng, Hybrid dynamical systems with controlled discrete transitions. Nonlinear Anal. Hybrid Syst. **1**, 466–481 (2007)
40. J. Bentsman, B.M. Miller, E.Y. Rubinovich, Dynamical systems with active singularities: input/state/output modeling and control. Automatica **44**, 1741–1752 (2008)
41. J. Bentsman, B.M. Miller, E.Y. Rubinovich, S.K. Mazumder, Optimal control of dynamical systems with active singularities under single and multi-impact sequences: a ball/racket system example, in *Proceedings of the 48th IEEE Conference on Decision and Control (CDC'09), Shanghai, December 16–18, 2009* (2009), pp. 5033–5038
42. J. Bentsman, B.M. Miller, E.Y. Rubinovich, S.K. Mazumder, Modeling and control of systems with active singularities under energy constraints: single- and multi-impact sequences. IEEE Trans. Autom. Control **57**(7), 1854–1859 (2012)
43. A. Bensoussan, J.L. Lions, *Control Impulsionnel et Inequations Quasivariationnelles* (Bordas, Paris, 1982)
44. J. Bentsman, B.M. Miller, Control of distributed systems with generalized inputs. Autom. Remote Control **58**(7), 1119–1131 (1997)
45. F. Bernardin, M. Schatzman, C.H. Lamarque, A stochastic differential equation from friction mechanics. C. R. Acad. Sci. Paris Ser. I **338**, 837–842 (2004)
46. F. Bernardin, M. Schatzman, C.H. Lamarque, Second-order multivalued stochastic differential equations on Riemannian manifolds. Proc. R. Soc. Lond. **A 460**, 3095–3121 (2004)
47. A. Blaquiere, Impulsive optimal control with finite or infinite time horizon. J. Optim. Theory Appl. **46**(4), 431–439 (1985)
48. K.F. Bohringer, K. Goldberg, M. Cohn, R. Howe, A. Pisano, Parallel microassembly with electrostatic force fields, in *Procedings of 1998 IEEE International Conference on Robotics and Automation*, vol. 2 (1998), pp. 1204–1211
49. K.F. Bohringer, B.R. Donald, R. Mihailovich, N.C. MacDonald, A theory of manipulation and control for microfabricated actuator arrays, in *Procedings of IEEE Micro Electro Mechanical Systems* (1994), pp. 102–107
50. K.F. Bohringer, B.R. Donald, N.C. MacDonald, What programmable vector fields can (and cannot) do: force field algorithms for MEMS and vibratory plate parts feeders, in *Proceedings of 1996 IEEE International Conference on Robotics and Automation*, vol. 1 (1996), pp. 822–829
51. J. Boldrini, M. Costa, Therapy burden, drug resistance, and optimal treatment regimen for cancer chemotherapy. IMA J. Math. Appl. Med. Biol. **17**, 33–51 (2000)
52. E.A. Bolotov, *About the Motion of Matherial Plane Figure with Friction*. Sbornik Mathematics, vol. 25, no. 4 (University Press, 1906), pp. 562–708
53. V.G. Boltyanskii, *Optimal Control of Discrete Systems* (Wiley, New York, 1978)
54. F. Bornemann, *Homogenization in Time of Singularly Perturbed Mechanical Systems*. Lecture Notes in Mathematics 1687 (Springer, Berlin, 1998)
55. V.N. Brendelev, On realization of constraints in nonholonomic mechanics. J. Appl. Math. Mech. **45**, 351–355 (1981)
56. A. Bressan, Incompatibilità dei teoremi di existenza e di Unicita del moti per un tipo molto comune e regolare di sistemi meccanici. Ann. della Scuola Normale Super. di Pisa Serie III **14**, 333–348 (1960)
57. A. Bressan, On differential systems with impulsive controls. Rend. Sem. Math. Univ. Padova **78**, 227–236 (1987)
58. A. Bressan, On the application of control theory to certain problems of Lagrangian systems, and hyper-impulsive motion for these, I, II. Atti. Acc. Lincei Rend. Fis. **LXXXII**, 91–105, 107–118 (1988)
59. A. Bressan, Hiper-impulsive motion and controllizable coordinates for Lagrangian systems. Atti. Acc. Lincei Rend. Fis. **XIX**, 197–246 (1989)

60. A. Bressan, M. Motta, A class of mechanical systems with some coordinates as controls. A reduction of certain optimization problems for them. Solution methods. Atti. Acc. Lincei Rend. **II**, 7–30 (1993)
61. A. Bressan, M. Motta, Some optimization promlems with a monotone impulsive character. Approximation by means of structural discontinuities. Atti. Acc. Lincei Rend. **II**, 31–52 (1994)
62. A. Bressan, M. Motta, Structural discontinuities to approximate some optimization problems with a nonmonotone impulsive character. Rend. Mat. Acc. Lincei **VI**, 93–109 (1995)
63. A. Bressan, F. Rampazzo, On differential systems with vector-valued impulsive controls. Boll. Un. Mat. Ital. **(7)**(2-B), 641–656 (1988)
64. A. Bressan, F. Rampazzo, Impulsive control systems with commutative vector fields. J. Optim. Theory Appl. **71**(1), 67–83 (1991)
65. A. Bressan, F. Rampazzo, On differential systems with quadratic impulses and their applications to Lagrangian mechanics. SIAM J. Control Optim. **31**, 1206–1220 (1993)
66. A. Bressan, F. Rampazzo, Impulsive control systems without commutativity Assumptions. J. Optim. Theory Appl. **81**(3), 435–457 (1994)
67. A. Bressan, K. Han, F. Rampazzo, On the control of non-holonomic systems by active constraints. Discrete Contin. Dynam. Syst. **33**, 3329–3353 (2013)
68. R.W. Brockett, Lie algebras and Lie groups in control theory, in *Geometric Methods in System Theory*, ed. by D.Q. Mayne, R.W. Brockett. Proceedings of the NATO Advanced Study Institute Held at London, Aug. 27–Sept. 7, 1993 (D. Reidel Publishing Company, Dordrecht, 1973), pp. 43–82
69. B. Brogliato, S.I. Niculescu, P. Orhant, On the control of finite-dimensional mechanical systems with unilateral constraint. IEEE Trans. Autom. Control **42**(2), 200–215 (1997)
70. B. Brogliato, *Nonsmooth Impact Mechanics. Models, Dynamics and Control*. Communications and Control Engineering Series (Springer, London, 1999)
71. B. Brogliato, A. Zavala Rio, On the control of complementary-slackness juggling mechanical systems. IEEE Trans. Autom. Control **45**(2), 235–246 (2000)
72. B. Brogliato, S. Niculescu, M. Monteiro-Marques, On tracking control of a class of complementary-slackness hybrid mechanical systems. Syst. Control Lett. **39**, 255–266 (2000)
73. B. Brogliato, Some perspectives on the analysis and control of complementary systems. IEEE Trans. Autom. Control **48**(6), 918–935 (2003)
74. B. Brogliato, R. Lozano, B. Maschke, O. Egeland, *Dissipative Systems. Analysis and Control, Theory and Applications* (Springer, London, 2007)
75. A.A. Butov, Optimal linear filtering in noise degeneration in observations. Autom. Remote Control **41**(11), 1506–1511 (1981)
76. A.A. Butov, N.A. Kuznetsov, R.S. Liptser et al., Generalized observation control in problems of stochastical optimization, in *Preprints of VIII IFAC Congress, August 24–28, 1981, Kyoto, Control Science and Technology for the Progress of Society*, vol. 6 (1981), pp. 851–856
77. G. Buttazzo, D. Percivale, On the approximation of the elastic bounce problem on riemannian manifolds. J. Differ. Equ. **47**, 227–245 (1983)
78. F. Camilli, M. Falcone, Analysis and approximations for the impulse control problem on the infinite time interval. Autom. Remote Control **58**(7), 1203–1215 (1997)
79. H. Cartan, *Calcul Différentiel. Formses Différentielles* (Hermann, Paris, 1967)
80. W.L. Chan, S.K. Ng, Variational control problems for linear differential systems with Stieltjes boundary conditions. J. Aust. Math. Soc. **20**(Ser. B), 434–445 (1978)
81. C.C. Chang, R. Huston, Collision of multiboby systems. Comput. Mech. **27**, 436–444 (2001)
82. W.L. Chang, P.T. Liu, Maximum principle for control systems described by measure functional differential equations. J. Optim. Theory Appl. **15**(5), 517–531 (1975)
83. C.C. Chang, S.T. Peng, Impulsive motion of multibody systems. Multibody Syst. Dynam. **17**, 47–70 (2007)
84. A. Chentsov. *Finitely Additive Measures and Relaxations of Extremal Problems* (Springer, Berlin, 1996)

85. A.G. Chentsov, S.I. Morina, *Extension and Relaxation* (Kluwer Academic, Dordrecht, 2002)
86. F.L. Chernous'ko, The motion of a three-link system along a plane. J. Appl. Math. Mech. **66**(1), 3–9 (2002)
87. F.L. Chernous'ko, V.B. Kolmanovskii, *Optimal Control at Random Perturbations* (in Russian). (M. Nauka, Moscow, 1978)
88. M.B. Chiarolla, U.G. Haussmann, Optimal control of inflation: central bank problem. SIAM J. Control Optim. **36**(3), 1099–1132 (1998)
89. F.N. Chowdhury, J.L. Aravena, A modular methodology for fast fault detection and classification in power systems. IEEE Trans. Control Syst. Technol. **6**(5), 623–634 (1998)
90. B.E. Chuprun, Problems with restriction on control variation. Autom. Remote Control **3**, 24–37 (1975)
91. B.E. Chuprun, Problems with constraints on changes of vector control. Autom. Remote Control **12**, 20–31 (1976)
92. B.E. Chuprun, Problems with bounded or minimized variation of the vector function of the system coordinates. Autom. Remote Control **59**(7), 950–956 (1998)
93. F.H. Clarke, *Optimization and Nonsmooth Analysis* (Wiley, New York, 1983)
94. F.H. Clarke, R.B. Vinter, Applications of optimal multiprocesses. SIAM J. Control Optim. **27**(5), 1048–1071 (1989)
95. F.H. Clarke, R.B. Vinter, Optimal multiprocesses. SIAM J. Control Optim. **27**(5), 1072–1091 (1989)
96. E.A. Coddington, N. Levinson, *Theory of Ordinary Differential Equations* (McGeaw-Hill, New York, 1955)
97. A.J. Coldman, J.M. Murray, Optimal control for a stochastic model of cancer chemotherapy. Math. Biosci. **168**, 187–200 (2000)
98. G. Dal Maso, F. Rampazzo, On systems of ordinary differential equations with measures as control. Differ. Integral Equ. **4**(4), 739–765 (1991)
99. P.C. Das, R.R. Sharma, On optimal control for measure delay-differential equations. SIAM J. Control **9**(1), 43–61 (1971)
100. G. Dal-Maso, M. Negri, D. Percivale, Linearized elasticity as Γ-limit of finite elasticity. Set-Valued Anal. **10**, 165–183 (2002)
101. H. Dankowicz, X. Zhao, Local analysis of co-dimension-one and co-dimension-two grazing bifurcations in impact microactuators. Physica **D 202**, 238–257 (2005)
102. H. Dankowicz, F. Svahn, On the stabilizability of near-grazing dynamics in impact oscillators. Int. J. Robust Nonlinear Control **17**, 1405–1429 (2007)
103. A. Daryin, A. Malakaeva, Numerical methods for linear impulse feedback problems. J. Comput. Syst. Sci. Int. **47**(2), 207–213 (2008)
104. A.N. Daryin, A.B. Kurzhanskii, A.V. Seleznev, The dynamic programming method in impulsive control synthesis. Differ. Equ. **41**(11), 1566–1576 (2005)
105. A.N. Daryin, I.A. Digailova, A.B. Kurzhanski, On the problem of impulse measurement feedback control. Proc. Steklov Inst. Math. **268**(suppl. 1), 71–84 (2010)
106. M.H.A. Davis, *Linear Estimation and Stochastic Control* (Chapman and Hill, London, 1977)
107. G. De Ans, D. Gottlieb, P. Kokotovic, Optimal control of bacterial growth. Automatica **8**, 729–736 (1972)
108. M.V. Deryabin, The realization of unilateral constraints. J. Appl. Math. Mech. **58**(6), 1079–1083 (1994)
109. M.V. Deryabin, V.V. Kozlov, A theory of systems with unilateral constraints. J. Appl. Math. Mech. **59**(4), 505–512 (1995)
110. A.V. Dmitruk, V.A. Dykhta, Quadratic necessary and sufficient optimality conditions for singular and impulsive processes, in *IMACS/IFAC International Workshop, Methods and Software for Automatic Control Systems*. Summaries of papers, Irkutsk (1991), pp. 14–15
111. J.R. Dorron, G. Ferreyra, A multistage, multicontrol problem with unbounded controls. SIAM J. Control Optim. **32**, 1322–1331 (1994)
112. A.Y. Dubovitskii, A.A. Milyutin, Extremum problems in the presence of restrictions. USSR Comput. Math. Math. Phys. **5**, 1–80 (1965)

113. A.Y. Dubovitskii, A.A. Milyutin, Necessary conditions for a weak extremum in optimal control problems with mixed constraints of the inequality type. USSR Comput. Math. Math. Phys. **8**(4), 24–98 (1968)
114. F. Dufour, B.M. Miller, Generalized solutions in nonlinear stochastic control problems. SIAM J. Control Optim. **40**, 1724–1745 (2002)
115. F. Dufour, B.M. Miller, Singular stochastic control problems. SIAM J. Control Optim. **43**, 708–730 (2004)
116. F. Dufour, B.M. Miller, Maximum principle for singular stochastic control problems. SIAM J. Control Optim. **45**(2), 668–698 (2006)
117. F. Dufour, B.M. Miller, Necessary conditions for optimal singular stochastic control problems. Stoch. Int. J. Probab. Stoch. Process. **79**(5), 469–504 (2007)
118. V.A. Dykhta, Conditions of local minimum for singular modes in systems with linear control. Autom. Remote Control **42**(12), 1583–1587 (1981)
119. V.A. Dykhta, Impulse-trajectory extension of degenerated optimal control problems. IMACS Ann. Comput. Appl. Math. **8**, 103–109 (1990)
120. V.A. Dykhta, Variational maximum principle for impulse and particular regimes in problem of optimization, linear with respect to control. Izv. Vyssh. Uchebn. Zaved Mat. **11**, 89–91 (1991)
121. V.A. Dykhta, A variational maximum principle for pulse and singular regimes in an optimization problem that is linear with respect to control. Soviet Math. (Iz. VUZ) **35**(11), 89–91 (1991)
122. V.A. Dykhta, Impulse-trajectory extention of the optimal control problems, in *Development and Application of the Lyapunov Function Method* (Nauka, Novosibirsk, 1992), pp. 170–182
123. V.A. Dykhta, A variational maximum principle and quadratic conditions of optimality for impulsive and singular processes. Sib. Math. J. **35**(1), 70–82 (1994)
124. V.A. Dykhta, Second order necessary optimality conditions for impulsive control problem and multiprocesses, in *Proceedings of III International IFAC Workshop on Singular Solutions and Perturbations in Control Systems (SSPCS'97), Pereslavl-Zalessky, Russia, July 7–11, 1997*, ed. by V.I. Gurman, B.M. Miller, M.G. Dmitriev (Elsevier, Amsterdam, 1997), pp. 97–101
125. V.A. Dykhta, Impulsive optimal control in models of economics and quantum electronics. Autom. Remote Control **60**(11), 1603–1613 (1999)
126. V.A. Dykhta, N. Antipina, Application of variational maximum principle to optimization models with discontinuous trajectories, in *Proceedings of 5-th IFAC Symposium "Nonlinear Control Systems", St.-Petersburg, Russia, July, 2001. Proceedings* (2001), pp. 614–618
127. V.A. Dykhta, G.A. Kolokol'nikova, Minimum conditions on the set of sequences in a degenerate variational problem. Math. Not. **34**(5), 859–863 (1983)
128. V.A. Dykhta, O.N. Samsonyuk, *Optimal Impulse Control with Applications* (in Russian) (Fizmathlit, Moscow, 2000)
129. V.A. Dykhta, O.N. Samsonyuk, Maximum principle for non-smooth optimal impulse problems, in *Proceedings of 5-th IFAC Symposium "Nonlinear Control Systems", St.-Petersburg, Russia, July, 2001. Proceedings* (2001), pp. 1381–1385
130. M.C. Ferris, J.S. Pang, Engineering and economic applications of complementary problems. SIAM Rev. **39**(4), 669–713 (1997)
131. R.C. Fetecau, J.E. Marsden, M. Ortiz, M. West, Nonsmooth lagrangian mechanics and variational collision integrators. SIAM J. Appl. Dynam. Syst. **2**(3), 381–416 (2003)
132. T.F. Filippova, Trajectory tubes to impulsive control systems, in *Proceedings of LSSC 2005*, ed. by I. Lirkov, S. Margenov, J. Warsniewski. Lecture Notes in Computer Science, vol. 3743 (Springer, Berlin, 2006), pp. 223–230
133. W. Fleming, R. Rishel, *Deterministic and Stochastic Optimal Control* (Springer, New York, 1975)
134. A.M. Formalskii, *Modelling of Antropomorphic Mechanisms* (in Russain). (Nauka, Moscow, 1982)

135. L. Fridman, Y. Aoustin, F. Plestan, Decomposition of existence and stability analysis of periodic solutions of systems with impacts: application to bipedal walking robot, in *Proceedings of the 47th IEEE Conference on Decision and Control Cancun, Mexico, Dec. 9–11* (2008), pp. 5238–5243
136. P. Gajardo, C.H. Ramirez, A. Rapaport, Minimal time sequential batch reactors with bounded and impulse controls for one or more species. SIAM J. Control Optim. **47**(6), 2827–2856 (2008)
137. F.R. Gantmacher, *Theory of Matrices* (Chelsea Publication, New York, 1959)
138. A. Galka, J. Telega, Dynamic contact problems for systems with finite degrees of freedom: the case of piecewise regular constraints. Math. Comput. Model. **28**(4–8), 479–495 (1998)
139. A.A. Galyaev, Optimal pulse control of dynamic systems in the shock phase. Autom. Remote Control **67**(1), 65–77 (2006)
140. A.A. Galyaev, Impact of a system of material points against an absolutely rigid obstacle: a model for its impulsive action. Autom. Remote Control. **67**(6), 856–867 (2006)
141. A.A. Galyaev, B.M. Miller, E.Y. Rubinovich, *Optimal Impulsive Control of Dynamical System in an Impact Phase*. Lecture Notes in Applied and Computational Mechanics, vol. 27 (Springer, Berlin, 2006), pp. 385–386
142. B.R. Gelbaum, J.M.H. Olmsted, *Counterexamples in Analysis* (Holden-Day, San Francisco, 1964)
143. F. Génot, B. Brogliato, New Results on é Paradoxes. INRIA, Report No. 3366 (1998)
144. F. Génot, B. Brogliato, New results on painlevé paradoxes. Eur. J. Mech. A/Solids **18**(18), 653–678 (1999)
145. W.M. Getz, D.H. Martin, Optimal control systems with state variable jump discontinuities. J. Optim. Theory Appl. **31**(2), 195–205 (1980)
146. I.V. Girsanov, *Lectures on Mathematical Theory of Extremum Problems* (Springer, Berlin, 1972)
147. B. Goh, Optimal singular control for multi-input singular Bolza problem. SIAM J. Control Optim. **4**(2), 309–325 (1966)
148. W. Goldsmith, Bouncing from one collision to the next. Appl. Mech. Rev. **52**, R27–R43 (1999)
149. V.K. Gorbunov, On optimizing processes with controlled discontinuities. Nauch. Tr. Tashkent. Univ. **455**, 19–25 (1975)
150. V.K. Gorbunov, G.U. Nurakhunova, Processes with controlled discontinuities of phase trajectories and simulation of production with moving basic funds. Izv. Akad. Nauk SSSR, Tekh. Kibern. **6**, 55–61 (1975)
151. F.N. Grigor'ev, On data processing control in discrete time systems. Autom. Remote Control **43**(9), 1149–1155 (1982)
152. F.N. Grigor'ev, N.A. Kuznetsov, Control of the observation process in continuous systems. Probl. Control Inform. Theory **6**(3), 181–201 (1977)
153. F.N. Grigor'ev, N.A. Kuznetsov, A.P. Serebrovskii, *Observation Control in Automatic Systems* (Nauka, Moscow, 1986)
154. S.S. Grigoryan, The solution to the painleve paradox for dry friction. Dokl. Phys. **46**(7), 499–503 (2001)
155. J.W. Grizzle, G. Abba, F. Plestan, Asymptotically stable walking for biped robots: analysis via systems with impulse effects. IEEE Trans. Autom. Control **46**(1), 51–64 (2000)
156. J. Groh, On the optimal control for nonlinear equations with impulses. Math. Nach. **104**, 41–48 (1981)
157. V.I. Gurman, On optimal controlled processes with unbounded derivatives (in Russian). Avtom. Telemekh. **12**, 14–21 (1972)
158. V.I. Gurman, *Degenerate Problems of Optimal Control* (in Russian). (Nauka, Moscow, 1977)
159. V.I. Gurman, *Extension Principle in Control Problems* (in Russian). (Nauka, Moscow, 1985)
160. V.I. Gurman, V.A. Dykhta, Singular problems of optimal control and the method of multiple maxima. Autom. Remote Control **38**(3), 343–350 (1977)

161. V.I. Gurman, V.A. Dykhta, G.N. Konstantinov, et al., The extension principle in optimal control problems. Izv. Akad. Nauk SSSR, Tekh. Kibern. **2**, 200–213 (1983)
162. U. Haussmann, W. Suo, Singular optimal stochastic control I, II. SIAM J. Control Optim. **33**(3), 916–936, 937–959 (1995)
163. O. Hâjek, Review of the book [339]. Bull. Am. Math. Soc. **12**(2), 272–279 (1985)
164. W.P.M.H. Heemels, J.M. Schumacher, S. Weiland, Linear complementary systems. SIAM J. Appl. Math. **60**(4), 1234–1269 (2000)
165. T.H. Hildebrandt, On systems of linear differentio-Stieltjes integral equations. Ill. J. Math. **3**(3), 352–373 (1959)
166. R.L. Hipwell, R.S. Muller, A.P. Pisano, Characterization of thin-film impact microactuators, in *Proceedings of ASME, DSC-62, ASME International Congress and Exposition, November 1997* (1997), pp. 87–91
167. J. Hong, Optimal substrate feeding policy for fed batch fermentation with substrate and product inhibition kinetics. Biotechnol. Bioengng. **28**, 1421–1431 (1986)
168. T. Huillet, A. Monin, G. Montseny, G. Salut, Generalized inputs to non-linear systems, in *New Trends in Nonlinear Control Theory, Proceedings of the International Conference Nonlinear System, Nantes, June 13–17, 1988* (1989), pp. 414–428
169. T. Huillet, A. Monin, G. Salut, Representations exponentielles des systemes analytiques generaux. AP II, Autom. Prod. Inf. Ind. **24**, 37–56 (1990)
170. T. Huillet, A. Monin, G. Montseny, G.Salut, Reversibilite des systemes non lineares a commande ponctuelle. AP II Autom. Prod. Inf. Ind. **24**, 57–82 (1990)
171. Y. Hurmuzglu, An energy-based coefficient of restitution for planar impacts of slender bar with massive external surfaces. Trans. ASME J. Appl. Mech. **65**, 952–962 (1998)
172. A.D. Ioffe, V.M. Tikhomirov, An extension of variation problems. Tr. Mosk. Mat. Obshch. **18**, 187–246 (1968)
173. A.D. Ioffe, V.M. Tikhomirov, *Theory of Extremal Problems* (North-Holland, Amsterdam, 1979)
174. R.L. Irvine, L.H. Ketchum, Sequencing batch reactors for biological wastewater treatment. Crit. Rev. Environ. Control **18**, 255–294 (1989)
175. K.V. Isaev, Optimal control of discrete-continuous systems. Kibern. Vychisl. Tek. **79**, 45–51 (1988)
176. A.P. Ivanov, On the correctness of basic problems of dynamics in systems with friction. J. Appl. Math. Mech. **50**(5), 712–716 (1986)
177. V.K. Ivanov, *Mathematical Modelling and Optimization of the Tumors Radio-therapy* (Energoatomizdat, Moscow, 1986)
178. A.P. Ivanov, A constructive model of impact with friction. J. Appl. Math. Mech. **52**(6), 700–704 (1988)
179. A.P. Ivanov, Collision-free motion in systems with non-retaining constraints. J. Appl. Math. Mech. **56**(1), 1–12 (1992)
180. A.P. Ivanov, Energetics of a collision with friction. J. Appl. Math. Mech. **56**(4), 527–534 (1992)
181. A.P. Ivanov, The dynamics of systems near to grazing collision. J. Appl. Math. Mech. **58**(3), 437–444 (1994)
182. A.P. Ivanov, On multiple impact. J. Appl. Math. Mech. **59**(6), 887–902 (1995)
183. A.P. Ivanov, *Dynamic of Systems with Mechanical Collisions* (International Educqtion Programm, Moscow, 1997)
184. A.P. Ivanov, The problem of constrained impact. J. Appl. Math. Mech. **61**(3), 341–353 (1997)
185. A.P. Ivanov, On discontinuous motions in systems with unilateral constraints. J. Appl. Math. Mech. **62**(3), 351–358 (1998)
186. A.P. Ivanov, The stability of mechanical systems subjected to impulsive actions. J. Appl. Math. Mech. **65**(4), 617–629 (2001)
187. A.P. Ivanov, Singularities in the dynamics of systems with non-ideal constraints. J. Appl. Math. Mech. **67**(2), 185–192 (2003)

188. A.P. Ivanov, The properties of solutions of the fundamental problem of dynamics in systems with non-ideal constraints. J. Appl. Math. Mech. **69**(3), 338–350 (2005)
189. A.P. Ivanov, *Basics of the Systems with Friction*. S.E.C. Regular and Chaiotic Dynamics (Institute of Computer Research, Igevsk, 2011)
190. A.P. Ivanov, The equilibrium of systems with dry friction. J. Appl. Math. Mech. **79**(3), 217–228 (2015)
191. V.K. Ivanov, B.M. Miller, P.I. Kitsul, A. M. Petrovskii, A mathematical model of control of the healing of an organism damaged by a malignant growth, in *Biological Aspects of Control Theory* (in Russian), vol. 8 (Institue of Control Sciences, Moscow, 1976), pp. 15–22
192. V.V. Ivashkin, Energetically optimal transitions from hyperbolic orbit in the absence of constraints on the transition time (in Russian). Kosm. Issled. **4**(1), 17–25 (1966)
193. V.V. Ivashkin, *Optimization of Space Maneuvers Under Constraints on Planets Distance* (in Russian). (Nauka, Moscow, 1975)
194. M. Jean, J.J. Moreau, *Dynamics of Elastic or Rigid Bodies with Frictional Contact: Numerical Methods*, vol. 124 (Publications of Laboratory of Mechanics and Acoustics, Marseille, 1991)
195. K. Johansson, J. Lygeros, S. Sastry, M. Egerstedt, Simulation of Zeno hybrid automata, in *Proceedings of the 38th IEEE Conference on Decision and Control 1999, CDC'99, Phoenix, Arizona, December* (1999), pp. 3538–3543
196. K. Johansson, M. Egerstedt, J. Lygeros, S. Sastry, On the regularization of Zeno hybrid automata. Hybrid control systems. Syst. Control Lett. **38**(3), 141–150 (1999)
197. D.Y. Karamzin, Necceaary conditions of the minimum in an impulse optimal control problems. J. Math. Sci. **139**(6), 7087–7150 (2006)
198. A. Khalanai, D. Veksler, *Qualitative Theory of Systems with Impulses* (in Russain, Engl. Translaiton). (Mir Publishers, Moscow, 1971)
199. P.V. Kharlamov, A critique of some mathematical models of mechanical systems with differential constraints. J. Appl. Math. Mech. **56**(4), 584–594 (1992)
200. A. Klabring, J.S. Pang, Existence of solution to discrete semicoecitive frictional problems. SIAM J. Optim. **8**(2), 414–442 (1998)
201. R. Klotzler, On a distributional version of Pontryagin's maximum principle. Optimization **19**(3), 413–419 (1988)
202. J. Kokavecz, P. Heszler, Z. Tóth, Á. Mechler, Effect of step function-like perturbation on intermittent contact mode sensors: a response analysis. Appl. Surface Sci. **210**, 123–127 (2003)
203. J. Kokavecz, O. Marti, P. Heszler, Á. Mechler, Imaging bandwidth of the tapping mode atomic force microscope probe. Phys. Rev. **B 73**, 155403-(1-8) (2006)
204. A.N. Kolmogorov, S.V. Fomin, *Introductory Real Ananlysis* (Prentice-Hall, Englewood Cliffs, 1970)
205. G.A. Kolokol'nikova, Optimality condition of impulse regimes in degenerate problems of optimal control, in *Numerical Methods in Mathematical Physics* (Moscow State University, Moscow, 1986), pp. 83–85
206. G.A. Kolokol'nikova, Second-order necessary conditions for the optimality of singular and pulse modes. Autom. Remote Control **51**(6), 760–767 (1990)
207. V.N. Koshelev, S.F. Morozov, Sufficient conditions of existence of discontinuous solutions for simplest integral of variational calculus. I, II (in Russsian). Izv. Vyssh. Uchebn. Zaved. Mat. 11, 21–30 (1967); no. 12, 38–46
208. V.N. Koshelev, S.F. Morozov, Existence theorems of discontinuous solutions in space variational problems (in Russsian). Izv. Vysh. Uchebn. Zaved. Mat. **5**, 47–52 (1970)
209. V.N. Koshelev, S.F. Morozov, Necessary conditions of extremum in variational problems in nonparametric form on totality of discontinuous functions (in Russsian). Izv. Vysh. Uchebn. Zaved. Mat. **12**, 37–46 (1970)
210. M. Koska, On some extremum problem in the family of non-decreasing functions. Acta VL Folia Math. **1**, 57–75 (1984)

211. A.M. Kovalev, N.V. Kravchenko, V.N. Nespirnyi, Problems of control and stabilization of impulsive-control dynamic systems with application to nonholonomic mechanics. Autom. Remote Control **68**(8), 1444–1458 (2007)
212. V.V. Kozlov, A constructive method for justifying the theory of systems with nonretaining constraints. J. Appl. Math. Mech. **52**(6), 691–699 (1988)
213. V.V. Kozlov, The problem of realizing constraints in dynamics. J. Appl. Math. Mech. **56**(4), 594–600 (1992)
214. V.V. Kozlov, A.I. Neishtadt, Realization of holonomic constraints. J. Appl. Math. Mech. **54**(5), 705–708 (1990)
215. V.V. Kozlov, D.V. Treschev, *Billiards. A Genetic Introduction to the Dynamics of Systems with Impacts*. Translations of Mathematical Monographs, vol. 89. (American Mathematical Society, Providence, 1991)
216. M.A. Krasnoselskii, A.V. Pokrovskii, Vibrostable differential equations with a continuous right-hand side. Proc. Moscow Math. Soc. **27**, 739–765 (1972)
217. M.A. Krasnoselskii, A.V. Pokrovskii, *Systems with Hysteresis* (Springer, Berlin, 1989)
218. N.N. Krasovskii, *Theory of Motion Control* (in Russian). (Nauka, Moscow, 1968)
219. V.F. Krotov, Discontinuous solutions of variation problems I, II (in Russian). Izv. Vysh. Uchebn. Zaved. Mat. **5**, 89–98 (1960); No. 2, 75–89 (1961)
220. V.F. Krotov, *Global Methods in Optimal Control* (Marcel Dekker, New York, 1996)
221. V.F. Krotov, V.I. Gurman, *Methods and Problems of the Optimal Control* (in Russian). (Nauka, Moscow, 1973)
222. V.F. Krotov, V.I. Gurman, V.Z. Bukreev, *New Variational Methods in Flight Dynamics*, Issue 657 NASA Technical Translation (Publisher Israel Program for Scientific Translations, Jerusalem, 1971)
223. J. Kučera, Solution in large of control problem $\dot{x} = (A(1-u)+Bu)x$. Chech. Math. J. **16**(91), 600–623 (1966)
224. J. Kučera, Solution in large of control problem $\dot{x} = (Au+Bv)x$. Chech. Math. J. **17**(92), 91–96 (1967)
225. P. Kundur, *Power System Stability and Control* (McGraw-Hill, New York, 1994)
226. A.B. Kurzhanski, On synthesizing impulse controls and the theory of fast controls. Proc. Steklov Inst. Math. **268**, 207–221 (2010)
227. A.B. Kurzhanski, A.N. Daryin, *Dynamic Programming for Impulse Feedback and Fast Controls. The Linear Systems Case*. Lecture Notes in Control and Information Sciences, vol. 468 (Springer, Berlin, 2020)
228. A.B. Kurzhanski, P. Varaiya, *Dynamics and Control of Trajectory Tubes. Theory and Computation* (Springer, Cham, 2014)
229. J. Kurzweil, Generlized ordinary differential equations. Chech. Math. J. **8**(3), 360–388 (1958)
230. J. Kurzweil, Linear differential equations with distributions as coefficients. Bull. Acad. Polon.Sci. **7**(9), 557–560 (1959)
231. H.J. Kushner, On the optimal timing of observation for linear control systems with unknown initial states. IEEE Trans. Autom. Control **AC-9**, 144–145 (1964)
232. H.J. Kushner, K.M. Ramachandran, Optimal and approximately optimal control policies for queues in heavy traffic. SIAM J. Control Optim. **27**, 1293–1318 (1989)
233. N.A. Kuznetsov, R.S. Liptser, A.P. Serebrovskii, Optimal control and information processing in continuous time (a linear system and quadratic functional). Autom. Remote Control **41**(10), 1369–1374 (1981)
234. K.H. Kyung, Impulse control: boolean programming and numerical algorithms. IEEE Trans. Autom. Control **51**(8), 1230–1248 (2006)
235. V. Laksmikantham, D.D. Bainov, P.S. Simeonov, *Theory of Impulsive Differential Equations* (World Scientific, Singapore, 1989)
236. D. Lawden, *Optimal Trajectories for Space Navigation* (Butterworth, London, 1963)
237. E.B. Lee, L. Markus, *Foundations of Optimal Control Theory* (Wiley, New York, 1967)

238. A.P. Lee, A.P. Pisano, M.G. Lim, Impact, friction, and wear testing of microsamples of polycrystalline silicon, in *Proceedings of Matherials Research Society Symposium "Smart Materials Fabrication and Materials for Micro-Electro-Mechanical Systems"*, vol. 276 (1992), pp. 67–78
239. R.I. Leine, N. van de Wouw, Stability properties of equilibrium sets of non-linear mechanical systems with dry friction and impact. Nonlinear Dynam. **51**, 551–583 (2008)
240. R.I. Leine, N. van de Wouw, *Stability and Convergence of Mechanical Systems with Unilateral Constraints*, ed. by F. Pfeiffer, P. Wriggers. Lecture Notes in Applied and Computational Mechanics, vol. 36 (Springer, Berlin, 2008)
241. H. Leiva, B. Lehman, Approximate controllability of special classes of bounded abstract nonautonomous evolution equations, in *Proceedings of 32th IEEE CDC. San-Antonio, December 1993*, vol. 3 (1993), pp. 2874–2875
242. S. Lenhart, J.T. Workman, *Optimal Control Applied to Biological Models* (Chapman & Hall, Boca Raton, 2007)
243. R.S. Liptser, Equations of near-optimal Kalman filter with a singular matrix of noise covariations in observations. Autom. Remote Control **35**(1), 29–35 (1974)
244. R.S. Liptser, A.N. Shiryaev *Statistics of Random Processes* (Springer, New-York, 1978)
245. R.S. Liptser, A.N. Shiryaev, *Theory of Martingales* (Kluver Aacdemic, Dordrecht, 1989)
246. R.W. Longman, C.A. Cooper, Optimal selection of observation times in the linear-quadratic gaussian control problem. J. Optim. Theory Appl. **39**(1), 47–58 (1983)
247. L. Mailleret, F. Grognard, Global stability and optimisation of a general impulsive biological control model. Math. Biosci. **221**, 91–100 (2009)
248. M. Marcus, H. Minc, *A Survey of Matrix Theory and Matrix Inequalities* (Allyn and Bacon, Boston, 1964)
249. V.V. Malyshev, M.N. Krasilshchikov, V.I. Karlov, *Optimization of Observation and Control Processes*. AIAA Education Series (AIAA Education, Reston, 1992)
250. L.F. Martins, H.J. Kushner, Routing and singular control for queueing networks in heavy traffic. SIAM J. Control Optim. **28**, 1209–1233 (1990)
251. D.D. Mazon, V.D. Dickerson, D.V. Smith, A variational method of optimal staging. AIAA J. **3**(11), 2007–2012 (1965)
252. J. Mazumdar, *An Introduction to Mathematical Physiollogy and Biology* (Cambridge University Press, Cambridge, 1989/1999)
253. N.H. McClamroch, Singular systems of differential equations as dynamic model for constrained robot systems, in *Proceedings of IEEE International Conference on Robotics and Automation, San Francisco, April 8–10* (1986), pp. 21–28
254. N.H. McClamroch, A singular perturbation approach to modeling and control of manipulators constrained by a stiff environment, in *Proceedings of IEEE Conference on Decision and Control, Tampa*, vol. 3, (1989), pp. 2407–2411
255. A.V. Melnikov, On the theory of stochastic equations in components of semimartingales. Math. USSR-Sb. **38**(3), 381–394 (1981)
256. L. Menini, A. Tornambe, Control of (otherwise) uncontrollable linear mecahnical systems through non-smooth impacts. Syst. Control Lett. **49**, 311–322 (2003)
257. P.A. Meyer, *Probabilités et Potentiels* (Herman, Paris,1966)
258. B.M. Miller, On one problem of impulse nonlinear control. Autom. Remote Control **37**(6), 865–873 (1976)
259. B.M. Miller, On one class of impulse control problems, in *Algorithms and Models of Control in Technical and Organization Systems* (in Russian), vol. 10 (Institute of Control Sciences, Moscow, 1976), pp. 27–30
260. B.M. Miller, On synthesis of the optimal control in systems described by differential equations with measure. Dissertation Candidat of Sceinces, Mathematics. Moscow, Institute of Physics and Technology (1977)
261. B.M. Miller, Stability of solutions of ordinary differential equations with measure. Uspekhi Mat. Nauk **2**, 198 (1978)

262. B.M. Miller, Nonlinear sampled-data control of processes described by ordinary differential equations. I, II. Autom. Remote Control **39**(1), 57–67 (1978); No. 3, 338–344
263. B.M. Miller, An optimality condition in control of a system which is described by a differential equation with a measure. Autom. Remote Control **43**(6), 752–761 (1982)
264. B.M. Miller, A sufficient optimality condition in linear control of plants described by differential equations with a measure. Autom. Remote Control **45**(9), 1171–1180 (1984)
265. B.M. Miller, Optimal control of observations in the filtering of diffusion processes I, II. Autom. Remote Control **46**(2), 207–214 (1985); No. 6, 745–754
266. B.M. Miller, Optimization of dynamic systems with a generalized control. Autom. Remote Control **50**(6), 733–742 (1989)
267. B.M. Miller, Generalized optimization in observation control problems. Autom. Remote Control **52**(10), 1397–1404 (1991)
268. B.M. Miller, Generalized solutions of discrete-continuous and impulsive systems, in *Generalized Functions in Control Problems and Differential Equations* (in Russian). (Ural Branch, Russian Academy of Sciences, Sverdlovsk, 1992), pp. 49–58
269. B.M. Miller, Conditions for optimality in generalized control problems. I. Necessary conditions for optimality. Autom. Remote Control **53**(3), 362–370 (1992). "II: Sufficient conditions of optimality", *Automation and Remote Control*, **53**, No. 4, 505–513 (1992)
270. B.M. Miller, Method of discontinuous time change in problems of control for impulse and discrete-continuous systems. Autom. Remote Control **54**(12), 1727–1750 (1993)
271. B.M. Miller, Observation control for discrete-continuous stochastic systems, in *Proceedings of IEEE International Workshop on Information Theory, Moscow, July 3–8* (Moscow Institute of Physics and Technology, Moscow, 1994), pp. 66–70
272. B.M. Miller, Generalized solutions of nonlinear optimization problems with impulse controls. I. Existence of solutions. Autom. Remote Control **56**(4), 505–516 (1995). II. Representation of solutions by means of differential equations with a measure, **56**, No. 5, 657–669
273. B.M. Miller, The generalized solutions of nonlinear optimization problems with impulse control. SIAM J. Control Optimi. **34**, 1420–1440 (1996)
274. B.M. Miller, The generalized solutions of ordinary differential equations in the impulse control problems. J.Math. Syst. Est. Control **6**, 415–435 (1996)
275. B.M. Miller, Representation of robust and non-robust solutions of nonlinear discrete-continuous systems, in *Proccedings of International Workshop on Hybrid and Real-Time Systems (HART'97), Grenoble, March 26–28*, ed. by O. Maler. Lecture Notes in Computer Science, vol. 1201 (Springer, Berlin, 1997), pp. 228–239
276. B.M. Miller, Observation control problems for discrete-continuos stochastic systems, in *Proceedings of AIICSR'97, Smolenice Castle, Sept 10–14*, ed. by I. Plander (World Scientific, Singapore, 1997), pp. 115–126
277. B.M. Miller, Representation of generalized solutions for differential equations with affine dependence on unbounded controls, in *Proceedings of III International IFAC Workshop on Singular Solutions and Perturbations in Control Systems (SSPCS'97), Pereslavl-Zalessky, July 7–11*, ed. by V.I. Gurman, B.M. Miller, M.G. Dmitriev (Elsevier, Amsterdam, 1997), pp. 43–48
278. B.M. Miller, Optimization of generalized solutions of nonlinear hybrid (discrete-continuos) systems, in *Proceeings of First International Workshop. Hybrid Systems: Computation and Control, (HSCC'98), Berkeley, April 13–15*, ed. by T.A. Henzinger, S. Sastry. Lecture Notes in Computer Science, vol. 1386 (Springer, Berlin, 1998), pp. 334–345
279. B.M. Miller, Observation control for discrete-continuous stochastic systems. IEEE Trans. Autom. Control **AC-45**(5), 993–998 (2000)
280. B.M. Miller, Optimal control problems in discrete-continuous (hybrid) systems with phase constraints. Int. J. Hybrid Syst. **1**(1), 1–18 (2001)
281. B.M. Miller, Generalized solutions in optimal control problems with state constraints, in *Proceedings of IFAC Workshop GSCP-04, Pereslavl-Zalessky, September 21–29*, ed. by Y.L. Sachkov (Fizmatlit, Moscow, 2004), pp. 176–184

282. B.M. Miller, Controllable systems with impacts. J. Math. Sci. **199**(5), 571–582 (2014)
283. B.M. Miller, J. Bentsman, A new approach to control of dynamic syatems with unilateral constraints, in *Proceedings of 14th IFAC World Congress, Beijing* (1999), pp. 199–203
284. B.M. Miller, J. Bentsman, Generalized solutions in dynamic systems with unilateral constraints, in *Proccedings of NOLCOS'01, 5th IFAC Symposium on Nonlinear Control Systems, Saint-Petersbourg* (2001), pp. 625–628
285. B.M. Miller, J. Bentsman, The singularity opening approach to control of mechanical systems with constraints, in *Proccedings of 2nd IFAC Workshop on Lagrangian and Hamiltonian Methods for Nonlinear Control, LHMNLC'03, April 3–5, 2003, Seville* (2003), pp. 293–298
286. B.M. Miller, J. Bentsman, Optimal control problems in hybrid systems with active singularities. Nonlinear Anal. **65**, 999–1017 (2006)
287. B.M. Miller, J. Bentsman, Representation of motion of controlled dynamic systems with unilateral constraints, in *Proceedings of the 45th IEEE Conference on Decision and Control, San Diego, December 13–15* (2006), pp. 4787–4792
288. B.M. Miller, J. Bentsman, Generalized solutions in systems with active unilateral constraints. Nonlinear Anal. Hybrid Syst. **1**, 510–526 (2007)
289. J.K. Mills, C. Nguyen, Robotic manipulator collision: modeling and simulation. Trans. ASME J. Dynam. Syst. Meas. Control **114**(4), 650–659 (1993)
290. B.M. Miller, E.Y. Rubinovich, Optimal impulse control problem with constrained number of impulses. Math. Comput. Simul. **34**, 23–49 (1992)
291. B.M. Miller, E.Y. Rubinovich, Regularization of a generalized Kalman filter. Math. Comput. Simul. **39**, 87–108 (1995)
292. B.M. Miller, E.Y. Rubinovich, *Impusive Control in Continuous and Discrete-Continuous Systems* (Kluwer Academic/Plenum Publishres, New York, 2003)
293. B.M. Miller, E.Y. Rubinovich, Kalman filter for controlled hybrid systems. Syst. Control Lett. **50**, 39–50 (2003)
294. B.M. Miller, E.Y. Rubinovich, *Optimization of Dynamical Systems with Impulse Controls* (in Russian). (Nauka, Moscow, 2005)
295. B.M. Miller, E.Y. Rubinovich, Impulsive control with impulsive actions of two types. Autom. Remote Control **70**(11), 1795–1813 (2009)
296. B.M. Miller, E.Y. Rubinovich, Discontinuous solutions in the optimal control problems and their representation by singular space-time transformations. Autom. Remote Control **74**(12), 1969–2006 (2013)
297. B.M. Miller, E.Y. Rubinovich, Dynamical systems with discontinuous solutions and problems with unbounded derivatives. Bullet. Irkutsk State Uni. Math. **19**, 136–149 (2017)
298. B.M. Miller, E.Y. Rubinovich, *Optimization of Dynamical Systems with Impulse Controls and Shocks* (in Russain) (URSS, Lenand, Moscow, 2019)
299. B.M. Miller, E.Y. Rubinovich, Numerical analysis of shock interactions with the example of Painlevé paradox with a "Slanted" Fall of a Rod. Autom. Remote Control **80**(10), 1835–1846 (2019)
300. B.M. Miller, W.J. Runggaldier, Optimization of observations: a stochastic control approach. SIAM J. Control Optim. **35**(3), 1030–1052 (1997)
301. B.M. Miller, K.V. Stepanyan, Observation control in tracking problem for moving object, in *Proceedings of the IFAC Workshop on Motion Control, October 9–11, Munich* (1995), pp. 245–252
302. B.M. Miller, K.V. Stepanyan, Observation control problem for discrete-continuous systems as a singular control problem, in *Proceedings of III International IFAC Workshop on Singular Solutions and Perturbations in Control Systems (SSPCS'97), Pereslavl-Zalessky, July 7–11*, ed. by V.I. Gurman, B.M. Miller, M.G. Dmitriev (Elsevier, Amsterdam, 1997), pp. 180–185
303. B.M. Miller, K.V. Stepanyan, Optimal linear filtering in systems with noise in observations dependent on signal and estimate. Autom. Remote Control **59**(11), 1609–1620 (1998)

304. B.M. Miller, K.V. Stepanyan, Discrete-continuous observation control for systems with state dependent noise, in *Proceedings of 5th IFAC Symposium on Nonlinear Control Systems, NOLCOS'01, Saint-Petersbourg, July 4–6* (2001), pp. 1191–1196
305. B.M. Miller, K.V. Stepanyan, Observation control for discrete-continuous (hybrid) stochastic systems with the estimate dependent noise. Int. J. Hybrid Syst. **2**, 337–368 (2003).
306. B.M. Miller, E.Y. Rubinovich, J. Bentsman, Spatiotemporal singular transformation in dynamical systems with impacts and its use in obtaning generalized solution of painlevé problem, in *Preprints of the 18th IFAC World Congress, Milano, August 28 - September 2* (2011), pp. 3463–3473
307. B.M. Miller, E.Y. Rubinovich, J. Bentsman, Singular space-time transformations. Toward one method for solving the painlevé problem. J. Math. Sci. **219**(2), 208–219 (2016)
308. M.D.P. Monteiro-Marques, *Differential Inclusions in Nonsmooth Mechanical Problems: Shocks and Dry Friction* (Birkhauser, Boston, 1993)
309. I.C. Morarescu, B. Brogliato, Passivity-based tracking control of multiconstraint complementarity Lagrangian systems. INRIA Report 6483 (2008)
310. J.J. Moreau, Bounded variation in time, in *Topics in Nonsmooth Mechanics,* ed. by J.J. Moreau, P.D. Panagiotopolous, G. Strang (Birkhauser Verlag, Basel, 1988), pp. 1–74
311. J.J. Moreau, Unilateral contacts and dry friction in finite freedom dynamics, in *Nonsmooth Mechanics and Applications*. CIMS Course and Lectures, vol. 302 (Springer, Wien, 1988), pp. 1–82
312. S.I. Morina, A.G. Chentsov, An extension of control problem in the class of vector finitely additive measures. Autom. Remote Control **58**(7), 1235–1244 (1997)
313. M. Motta, Viscosity solutions of HJB equations with unbounded data and characteristic points. Appl. Math. Optim. **49**, 1–26 (2004)
314. M. Motta, F. Rampazzo, Space-time trajectories of nonlinear systems driven by ordinary and impulsive control. Differ. Integral Equ. **8**(2), 269–288 (1995)
315. M. Motta, F. Rampazzo, Dynamic programming for nonlinear systems driven by ordinary and impulsive control. SIAM J. Control Optim. **34**(1), 199–225 (1996)
316. M. Motta, F. Rampazzo, Nonlinear system with unbounded controls and state constraints: a problem of proper extension. Nonlinear Differ. Equ. Appl. **3**, 191–216 (1996)
317. M. Motta, C. Sartori, Normality and nondegeneracy of the maximum principle in optimal impulsive control under state constraints. JOTA **185**, 44–71 (2020). https://doi.org/10.1007/s10957-020-01641-w
318. J.M. Murray, Existence theorems for optimal control and calculus of variations problems where the state can jump. SIAM J. Control Optim. **24**(3), 412–438 (1989)
319. J.M. Murray, Optimal Control for Cancer Chemotherapy with General Growth and Loss Functions. Math. Biosci. **98**, 273–287 (1990)
320. J.M. Murray, Some optimal control problems in cancer chemotherapy with a toxicity limit. Math. Biosci. **100**, 49–67 (1990)
321. A.D. Myshkis, Stability of solutions of differential equations under generalized pulse perturbations. Autom. Remote Control **68**(10), 1844–1851 (2007)
322. I.G. Natanson, *Theory of Functions of a Real Variable*, vols. 1–2. (Frederick Ungar, New York, 1955/1960)
323. Y.I. Neimark, Mathematical modelling as the science and the art (in Russian). (Nizhny Novgorod State University, Nizhny Novgorod, 2010)
324. Y.I. Neimark, N.A. Fufaev, The Penlevé paradoxes and the Dynamics of a brake shoe. J. Appl. Math. Mech. **59**(3), 343–352 (1995)
325. Y.I. Neimark, V.N. Smirnova, Singularly perturbed problems and the painvelé paradox. Differ. Equ. **36**(11), 1639–1646 (2000)
326. Y. I. Neimark, V.N. Smirnova, Contrast structures, limit dynamics, and the painlevé paradox. Differ. Equ. **37**(11), 1580–1588 (2001)
327. Y.I. Neimark, V.N. Smirnova, Specific properties of jumps in singularly perturbed cauchy problems. Differ. Equ. **40**(11), 1649–1652 (2004)

328. L. Neustadt, Optimization, a moment problem, and nonlinear programming. J. SIAM. Ser. A Control **2**(1), 33–53 (1964)
329. L. Neustadt, A general theory of minimum-fuel space trajectories. J. SIAM. Ser. A Control **3**(2), 317–356 (1965)
330. D.E. Okhotsimsky, On the theory of rocket propulsion (in Russian). Prikl. Math. Mech. **10**(2), 251–272 (1946)
331. D.E. Okhotsimsky, T.M. Eneyev, Some variational problems related to the launch of an artificial satellite of the earth (in Russian). Uspekhi Fizicheskih Nauk **XIII**(1a), 5–32 (1957)
332. N.M. Olgac, C.A. Cooper, R.W. Longman, The impact of costly observation and observation delay in stochastic optimal control problems. Int. J. Control **41**(3), 769–785 (1984)
333. Y.V. Orlov, *Theory of Optimal Systems with Generalized Controls* (in Russian). (Nauka, Moscow, 1988)
334. L.V. Ovsiannikov, *Group Analysis of Differential Equations* (Academic Press, New York, 1982)
335. P. Painlevé, *Leçons sur le Frottement* (Hermann, Paris, 1895)
336. P. Panagiotopolos, C. Glocker, Inequality constraints with elastic impacts in deformable bodies. The convex case. Arch. Appl. Mech. **70**, 349–365 (2000)
337. A.I. Panasyuk, J. Bentsman, Application of quasidifferential equations to the description of discontinuous processes. Differ. Equ. **33**(10), 1346–1355 (1997)
338. Y.G. Panovko, *Introduction to the Theory of mechanical shock* (in Russian). (Nauka, Moscow, 1977)
339. S.G. Pandit, S.G. Deo, *Differential Systems Involving Impulses* Lecture Notes in Mathematics, vol. 954 (Springer, Berlin, 1982)
340. L. Paoli, An existence result for vibrations with unilatral constraints: case of non-smooth set of constraints. Math. Models Methods Appl. Sci. **10**(6), 815–831 (2000)
341. L. Paoli, Problèmes de vibro-impact: étude de la dépendance par rapport aux données. C. R. Acad. Sci. Paris **I 339**, 27–32 (2004)
342. L. Paoli, Continuous dependence on data for vibro-impact problems. Math. Models Methods Appl. Sci. **15**(1), 53–93 (2005)
343. L. Paoli, An existence result for non-smooth vibro-impact problems. J. Differ. Equ. **211**, 247–281 (2005)
344. L. Paoli, Time-stepping approximation of rigid-body dynamics with perfect unilateral constraints. I: the inelastic impact case. Arch. Ration. Mech. Anal. **198**(2), 457–503 (2010)
345. L. Paoli, Time-stepping approximation of rigid-body dynamics with perfect unilateral constraints. II: the partially elastic impact case. Arch. Ration. Mech. Anal. **198**(2), 505–568 (2010)
346. L. Paoli, M. Schatzman, Mouvement à un nombre fini de degrés de libérte avec contraintes unilatérales: cas avec perte d'énergie. Modél. Math. Anal. Num. **27**, 673–717 (1993)
347. L. Paoli, M. Schatzman, Resounance in impact problems. Math. Comput. Model. **28**(4–8), 385–406 (1998)
348. L. Paoli, M. Schatzman, Approximation et existence en vibro-impact. C. R. Acad. Sci. Paris **329**, 1103–1107 (1999)
349. L. Paoli, M. Schatzman, *Non Convex Aspects of Dynamics with Impact. From Convexity to Nonconvexity*. Nonconvex Optimization and Its Applications, vol. 55 (Kluwer Academic, Dordrecht, 2001), pp. 207–222
350. L. Paoli, M. Schatzman, Penalty approximation for non-smooth constraints in vibroimpact. J. Differ. Equ. **177**, 375–418 (2001)
351. L. Paoli, M. Schatzman, A numerical scheme for impact problems I: the one-dimencional case. SIAM J. Numer. Anal. **40**(2), 702–733 (2002)
352. L. Paoli, M. Schatzman, A numerical scheme for impact problems II: the multidimencional case. SIAM J. Numer. Anal. **40**(2), 734–768 (2002)
353. L. Paoli, M. Schatzman, Penalty approximation for dymanical systems submitted to multiple non-smooth constraints. Multibody Syst. Dynam. **8**, 347–366 (2002)

354. L. Paoli, M. Schatzman, Numerical simulation of the dynamics of an impacting bar. Comput. Methods Appl. Mech. Eng. **196**, 2839–2851 (2007)
355. M. Payr, C. Glocker, Oblique frictional impact of a bar: analysis and comparison of different impact laws. Nonlinear Dynam. **41**, 361–383 (2005)
356. F. Pereira, G. Silva, Necessary conditions of optimality for vector-valeud impulsive control problems. Syst. Control Lett. **40**, 205–215 (2000)
357. F. Pereira, G. Silva, R. Vinter, Necessary conditions of optimality for impulsive differential inclusions, in *Proceedings of 37-th IEEE CDC, Tampa, December 1998*, vol. 1 (1998), pp. 16–18
358. F. Pereira, A. Matos, G. Silva, Hamilton-Jacobi conditions for an impulsiev control problems, in *Proceedings of 5-th IFAC Symposium on Nonlinear Control Systems, St.-Petersburg, July* (2001), pp. 1375–1380
359. D. Percivale, Uniqueness in the elastic bounce problem. J. Differ. Equ. **56**, 206–215 (1985)
360. D. Percivale, Uniqueness in the elastic bounce problem, II. J. Differ. Equ. **90**, 304–315 (1991)
361. F. Pfeifer, *Mechanical System Dynamics*. Lecture Notes in Applied and Computational Mechanics, 1st edn. (Springer, Berlin, 2005). Corrected Second Printing (2008)
362. F. Pfeifer, C. Glocker, *Multi-Body Dynamics with Unilateral Constraints* (Wiley, New York, 1996)
363. F. Pfeiffer, C. Glocker, Contacts in multibody systems. J. Appl. Math. Mech. **64**(5), 773–782 (2000)
364. A.F. Philippov, On some questions of the theory of optimal control. Vestn. Mosk. Univ. Ser. Mat. Mekh. **2**, 25–32 (1959)
365. A.P. Phillipov, *Differential Equations with Discontinuous Right-Hand Side* (Kluwer, Dordrecht, 1988)
366. J.G. Pierce, A. Schumitzky, Optimal impulsive control of compartment models, I: qualitative aspects; II: algorithm. J. Optim. Theory Appl. **18**, 537–554 (1976); **26**, 581–599 (1978)
367. V.N. Pilipchuk, Calculation of mechanical systems with pulsed excitation. J. Appl. Math. Mech. **60**(2), 217–226 (1996)
368. V.N. Pilipchuk, Some remarks on non-smooth transformations of space and time for vibrating systems with rigid barriers. J. Appl. Math. Mech. **66**(1), 31–37 (2002)
369. J.F. Pommaret, *Systemes of Partial Differential Equations and Lie Pseudogroups* (Gordon and Breach Science Publishers, New York, 1978)
370. K. Popp, Non-smooth mechanical systems. J. Appl. Math. Mech. **64**(2), 765–772 (2000)
371. A.J. Pritchard, Introduction to semigroup theory, in *Analysis and Optimization of Systems: State and Frequency Domain Approach for Infinite-Dimensional Systems. Proceedings of 10-th Intenational Conference, Sophia-Antipolis, France, 1992*. Lecture Notes Control Inform. Sciences, vol. 185 (Springer, Berlin, 1992), pp. 3–22
372. A.I. Propoi, *Elements of the Theory of Optimal Discrete Processes* (in Russian). (Nauka, Moscow, 1973)
373. I. Radchenko, Discrete modeling of dynamic states of contact systems with arbitrary number of elastic bodies. Math. Comput. Model. **28**(4–8), 449–454 (1998)
374. K. Ragulskis, K.R. Bansevicious, R. Barauskas, G. Kulvetis, *Vibrimotors for Precision Microrobots* (Hemisphere Publishers, New York, 1988)
375. F. Rampazzo, Differential games with unbounded versus bounded controls. SIAM J. Control Optim. **36**(3), 814–839 (1998)
376. F. Reisz, B.S. Nagy, *Leçons D'Analyse Functionelle* (Akadémiai Kiadó, Budapest, 1972)
377. R.W. Rishel, An extended Pontriagin principle for control systems whose control laws contain measures. J. SIAM. Ser. A Control **3**(2), 191–205 (1965)
378. R.W. Rishel, An example concerning rockets capable of impulsive thrust. SIAM J. Control **4**(4), 740–744 (1966)
379. R.T. Rockafellar, *Convex Analysis* (Princeton University Press, Princeton, 1970)

380. R.T. Rockafellar, Dual problems of lagrange arcs of bounded variation, in *Calculus of Variation and Control Theory*, ed. by D.L. Russel (Academic Press, New York, 1976), pp. 155–192
381. R.T. Rockafellar, Moreau's proximal mapping and convexity in Hamilton-Jacobi theory, in *Nonsmooth Mechanics and Analysis Theoretical and Numerical Advances*, ed. by P. Alart, O. Maisonneuve, T. Rockafellar (Springer, Berlin, 2006), pp. 3–12
382. J.W. Rogers, L.M. Phinney, Process yields for laser repair of aged stiction-failed MEMS devices. J. Microelectromech. Syst. **10**(2), 280–285 (2001)
383. J.W. Rogers, L.M. Phinney, Nanosecond laser repair of adhered MEMS structures. J. Heat Trans. **124**(2), 394–396 (2002)
384. H. Rubin, P. Ungar, Motion under a strong constraining force. Commun. Pure Appl. Math. **10**(1), 67–87 (1957)
385. E.Y. Rubinovich, Trajectory-wise control of observations in digital stochastic problems. Autom. Remote Control **41**(3), 365–372 (1980)
386. E.Y. Rubinovich, The generalized linear-quadratic stochastic control from incomplete data. Autom. Remote Control **58**(7), 1243–1262 (1997)
387. E.Y. Rubinovich, Minimax generalized linear-quadratic stochastic control problem with incomplete information, in *Proceedings of III International IFAC Workshop on Singular Solutions and Perturbations in Control Systems (SSPCS'97), Pereslavl-Zalessky, July 7–11*, ed. by V.I. Gurman, B.M. Miller, M.G. Dmitriev (Elsevier, Amsterdam, 1998), pp. 186–191
388. W. Rudin, *Real and Complex Analysis* (McGraw-Hill, New York, 1966)
389. V.A. Samsonov, The Dynamics of a Brake shoe and "Impact Generated by Friction". J. Appl. Math. Mech. **69**, 816–824 (2005)
390. A.V. Sarychev, Integral representation of trajectories of a controllable system with a generalized right-hand side. Differ. Equ. **24**(9), 1021–1031 (1988)
391. M. Schatzman, A class of nonlinear differential equations of second order in time. Nonlinear Anal. TMA **2**, 355–373 (1978)
392. M. Schatzman, Uniqueness and continuous dependence on data for one-dimensional impact problems. Math. Comput. Model. **28**(4–8), 1–18 (1998)
393. M. Schatzman, Penalty method for impact in generalized coordinates. Phil. Trans. R. Soc. A **359**, 2429–2446 (2001)
394. M. Schatzman, Penalty approximation of Painlevé problem, in *Nonsmooth Mechanics and Analysis, Theoretical and Numerical Advances*, vol. 12 (Springer, New York, 2006), pp. 129–143
395. W.W. Schmaedeke, Optimal control theory for nonlinear vector differential equation containing measures. J. SIAM. Ser. A Control **3**(2), 231–280 (1965)
396. J. Schumacher, Complemetary systems in optimization. Math. Program Ser. B **101**(1), 263–295 (2004)
397. A.P. Serebrovskii, Design of suboptimal control in a problem with a degenerate payoff for the control signal. Autom. Remote Control **39**(5), 640–644 (1978)
398. A.N. Sesekin, Discontinuous solutions of integral equations and their estimates, in *Generalized Functions and Differential Equations* (in Russian). (Ural. Nauchn. Tsentr Akad. Nauk SSSR, Sverdlovsk, 1985), pp. 62–68
399. A.N. Sesekin, Nonlinear differential equations containing product of discontinuous functions by generalized functions, in *Generalized Functions and Differential Equations* (in Russian). (Ural. Nauchn. Tsentr Akad. Nauk SSSR, Sverdlovsk, 1985), pp. 48–61
400. A.N. Sesekin, Continuous dependence on the right-hand sides and stability of approximable solutions of differential equations containing products of discontinuous and generalized functions (in Russian). Differ. Uravn. **22**(11), 2009–2011 (1986)
401. A.N. Sesekin, Optimal control of a linear system with a scarce resource, in *Nonlinear Problems in Generalized Functions* (in Russian). (Ural. Branch, Russian Academy of Sciences, Sverdlovsk, 1988), pp. 77–84
402. A.N. Sesekin, Nonlinear differential equations in a class of functions of bounded variation. Differ. Equ. **25**(11), 1356–1361 (1989)

403. A.N. Sesekin, On sets of discontinuous solutions of nonlinear differential equations. Soviet Math. (Iz. VUZ) **38**(6), 81–87 (1989)
404. A.N. Sesekin, Discontinuous solutions of ordinary differential equations with heredity of neutral type, in *Interschool Collection of Scientific Works, Boundary-value Problems* (in Russian). (Permskii Politekhn. Institute, Perm, 1991), pp. 107–112
405. A.N. Sesekin, On the best trajectory in an integral funnel of discontinuous solutions, in *Generalized Functions and Differential Equations* (in Russian). (Ural. Branch, Russian Academy of Sciences, Sverdlovsk, 1992), pp. 78–84
406. A.N. Sesekin, Impulse extension in the problem of optimizing the energy functional. Autom. Remote Control **53**(8), 1174–1182 (1992)
407. A.N. Sesekin, Properties of the attainability set of a dynamical system with impulse control. Autom. Remote Control **55**(2), 190–195 (1994)
408. P. Shi, The restitution coefficient for a linear elastic rod. Math. Comput. Model. **28**(4–8), 427–435 (1998)
409. G.E. Shilov, B.L. Gurevich, *Integral, Measure and Derivative: A Unified Approach* (Prentice-Hall, Englewood Cliffs, 1966)
410. G.N. Silva, R.V. Vinter, Measure differential inclusions. J. Math. Anal. Appl. **202**, 727–746 (1996)
411. G.N. Silva, R.V. Vinter, Necessary conditions for optimal impulsive control problems. SIAM J. Control Optim. **35**(6), 1829–1846 (1997)
412. P. Song, P. Kraus, V. Kumar, P. Dupont, Analysis of rigid-body dynamic models for simulation of systems with frictional contacts. Trans. ASME **68**, 118–128 (2001)
413. M. Spong, Impact controllability of an air hockey puck. Syst. Control Lett. **42**, 333–345 (2001)
414. D. Stainovici, Y. Hurmuzglu, A critical study of the applicability of rigid-body collision theory. Trans. ASME J. Appl. Mech. **63**, 307–316 (1996)
415. D.E. Stewart, Convergence of a time-stepping scheme for rigid-body dynamics and resolution of painleve's problem. Arch. Ration. Mech. Anal. **145**, 215–260 (1998)
416. D.E. Stewart, Rigid-body dynamics with friction and impact. SIAM Rev. **42**(1), 3–39 (2000)
417. D.E. Stewart, Reformulations of measure differential inclusions and their closed graph property. J. Differ. Equ. **175**, 108–129 (2001)
418. D.E. Stewart, Convolution complementarity problems with application to impact problems. IMA J. Appl. Math. **71**, 92–119 (2006)
419. S. Straszewicz, Uber exponierte Punkte abgeschlossener Punktmengen. Fund. Math. **24**, 139–143 (1935)
420. W.J. Stronge, Rigid body collisions with friction. Proc. Roy. Soc. Lond. Ser. A **431**, 168–181 (1990)
421. C. Studer, *Numerics of Unilateral Contacts and Friction. Modeling and Numerical Time Integration in Non-Smooth Dynamics* (Springer, Berlin, 2009)
422. H. Sussmann, On the gap between deterministic and stochastic ordinary differntial equations. Ann. Probab. **6**(1), 19–41 (1978)
423. H. Sussmann, A continuation method for nonholonomic path-finding problem, in *Proceedings of 32-th IEEE CDC, San-Antonio, December 1993. Proceedings*, vol. 3 (1993), pp. 2718–2723
424. G.W. Swan, General applications of optimal control theory in cancer chemotherapy. IMA J. Math. Appl. Med. Biol. **5**, 303–316 (1988)
425. G.W. Swan, Role of optimal control theory in cancer chemotherapy. Math. Biosci. **101**, 237–284 (1990)
426. J. Sztajnic, Transversality conditions for control with bounded variation. Acta VL Folia Math. **1**, 131–149 (1984)
427. M. Taylor, Grazing rays and reflection of singularities of solution to wave equations. Commun. Pure Appl. Math. **XXVIV**, 1–38 (1976)
428. J.S. Thorp, C. E. Seyler, A.G. Phadke, Electromechanical wave propagation in large electric power systems. IEEE Trans. Circuits Syst. I Fundam. Theory Appl. **45**(6), 614–622 (1998)

429. A. Tornambe, Modeling and control of impact mechanical systems: theory and experimental results. IEEE Trans. Automa. Control **AC-44**(2), 294–309 (1999)
430. Y.Z. Tsypkin, *Theory of Linear Impulse Systems* (in Russian). (Fizmatgiz, Moscow, 1963)
431. Y.Z. Tsypkin, Y.S. Popkov, *Theory of Nonlinear Impulse Systems* (in Russian). (Nauka, Moscow, 1973)
432. K. Uchino, Piezoelectric ultrasonic motors: overview. Smart Mater. Struct. **7**, 273–285 (1998)
433. K. Uchino, *Piezoelectric Actuators and Ultrasonic Motors* (Kluwer Academic, Boston, 1997)
434. A. van der Schaft, J.M. Schumacher, Complementary modeling of hybrid systems. IEEE Trans. Automat. Control **43**,(4), 483–490 (1998)
435. A. van der Schaft, J.M. Schumacher, Modelling and analysis of hybrid dynamical systems, in *Advances in the Control of Nonlinear Systems*. Lecture Notes in Control and Information Sciences (Springer, London, 2000). **264**, 195–224 (2001)
436. C. Vârsan, On decomposition and integral representation of solutions for affine control systems. Syst. Control Lett. **22**, 53–59 (1994)
437. I.B. Vapnyarskii, An existence theorem for optimal control in the Boltz problem, some of its applications and the necessary conditions for the optimality of moving and singular systems. USSR Comput. Math. Math. Phys. **7**(2), 22–54 (1967)
438. V.G. Veretennikov, V.A. Sinitsin, Discontinuous variational problem of optimizing control processes. J. Appl. Math. Mech. **36**(2), 333–336 (1972)
439. Y.M. Volin, G.M. Ostrovskii, The maximum principle for discontinuous systems and its application to problems with phase restrictions. Izv. Vyssh. Uchebn. Zaved. Radiofiz. **12**(11), 1609–1621 (1969)
440. R.V. Vinter, G. Pappas, A maximum principle for nonsmooth optimal control problems with state constraints. J. Math. Anal. Appl. **89**, 212–232 (1982)
441. R.V. Vinter, F.M.F.L. Pereira, A maximum principle for optimal processes with discontinuous trajectories. SIAM J. Control Optim. **26**(1), 205–229 (1988)
442. R.V. Vinter, G.N. Silva, Optimal impulsive control problems with state constraints, in *Proceedings of 32-th IEEE CDC, San Antonio, December 1993. Proceedings*, vol. 4 (1993), pp. 3811–3812
443. S. Walazak, On some control problems. Acta VL Folia Math. **1**, 187–196 (1984)
444. X. Wang, S.N. Balakrishnan, Optimal Neuro-controller synthesis for impulse-driven system, in *Proceedings of the 2007 American Control Conference, New York City, July 11–13* (2007), 1239–1244
445. J. Warga, Variational problems with unbounded controls. J. SIAM. Ser. A Control **3**(2), 428–438 (1965)
446. J. Warga, *Optimal Control of Differential and Functional Equations* (Academic Press, New York, 1972)
447. J. Warga, Q.J. Zhu, The equivalence of extremals in different representations of unbounded control problems. SIAM J. Control Optim. **32**, 1151–1169 (1994)
448. J. Wei, E. Norman, On global representation of the solutions of linear differential equations as a product of exponentials. Proc. Am. Math. Soc. **15**, 327–334 (1964)
449. Y. Wen-li, C. Bin, L. Cai-shan, X. Jian, Sliding state stepping algorithm for solving impact problems of multi-rigid-body system with joint friction. Appl. Math. Mech. **28**(12), 1621–1627 (2007)
450. P. Wolenski, S. Zabic, A differential solution concept for impulsive systems. Dynam. Contin. Discrete Impuls. Syst. Ser. A Math. Anal. **13**(B), 199–210 (2006)
451. P. Wolenski, S. Zabic, A sampling method and approximation results for impulsive systems. SIAM J. Control Optim. **46**(3), 983–998 (2007)
452. W.M. Wonham, On the separation theorem of stochastic control. SIAM J. Control **6**(2), 312–326 (1968)
453. M. Wösle, F. Pfeifer, Dynamics of spatial structure-varing rigid mutibody systems. Arch. Appl. Mech. **69**, 265–285 (1999)

454. T. Yang, *Impulsive Control Theory*. Lecture Notes in Control and Information Sciences, vol. 272 (Springer, Berlin, 2001)
455. J. Yong, Infinite-dimensional Volterra-Stieltjes evolution equations and related optimal control problems. SIAM J. Control Optim. **31**(3), 539–568 (1993)
456. L.C. Young, *Lectures on Variational Calculus and the Theory of Optimal Control* (W. B. Saunders Company, Philadelphia, 1969)
457. K. Yunt, C. Glocker, Modeling and optimal control of hybrid rigidbody mechanical systems, in *Proceedings of HSCC 2007*, ed. by A. Bemporad, A. Bicchi, G. Buttazzo. Lecture Notes in Computer Science, vol. 4416 (Springer, Berlin, 2007), pp. 614–627
458. K. Yunt, C. Glocker, A combined continuation and penalty method for the determination of optimal hybrid mechanical trajectories, in *Proceedings of IUTAM Symposium on Dynamics and Control of Nonlinear Systems with Uncertainty*, ed. by H.Y. Hu, E. Kreuzer (Springer, Berlin, 2007), pp. 187–196
459. R. Zander, M. Förg, H. Ulbrich, Impacts on beam structures: interactions of wave propagation and global dynamics, in *Multiscale Problems in Multibody System Contacts*, ed. by P. Eberhard (Springer, Berlin, 2007), pp. 327–338
460. A. Zavala-Rio, B. Brogliato, Direct adaptive control desigh for one-degree-of-freedom complementary-slackness jugglers. Automatica **37**, 1117–1123 (2001)
461. S.T. Zavalishchin, A supplement to Lawden's theory. J. Appl. Math. Mech. **53**(5), 572–578 (1989)
462. S.T. Zavalishchin, Time-optimal interorbital flight, in *Generalized Functions in Control Problems and Differential Equations* (in Russian). (Ural Branch, Russain Academy of Sciences, Sverdlovsk, 1992), pp. 17–29
463. S.T. Zavalishchin, Impulse dynamic systems and applications to mathematical economics. Dynam. Syst. Appl. **3**(3), 443–450 (1994)
464. S.T. Zavalishchin, A.N. Sesekin, *Impulsive Processes. Models and Applications* (in Russian). (Nauka, Moscow, 1991)
465. S.T. Zavalishchin, A.N. Sesekin, *Dynamic Impulse Systems. Theory and Applications* (Kluwer Academic, Dordrecht, 1997)
466. S.T. Zavalishchin, V.I. Sukhanov, *Applied Problems of Synthesis and Design of Control Algorithms* (in Russian). (Nauka, Moscow, 1985)
467. S.T. Zavalishchin, A.N. Sesekin, S.E. Drozdenko, *Dynamic Systems with Impulse Structure* (in Russian). (Sredn.-Ural. Book Publishers, Sverdlovsk, 1983)
468. J. Zhang, K. Johansson, J. Ligeros, S. Sastry, Zeno hybrid systems. Hybrid systems in control. Int. J. Robust Nonlinear Control **11**(5), 435–451 (2001)
469. X. Zhao, H. Dankowicz, Control of impact microactuators for precise positioning. Trans. ASME J. Comput. Nonlinear Dynam. **1**, 65–70 (2006)
470. X. Zhao, H. Dankowicz, Unfolding degenerate grazing dynamics in impact actuators. Nonlinearity **19**, 399–418 (2006)
471. X. Zhao, H. Dankowicz, C. Reddy, A. Nayfeh, Modeling and simulation methodology for impact microactuators. J. Micromech. Microeng. **14**, 775–784 (2004)
472. Z. Zhao, B. Chen, C. Liu, H. Jin, Impact model resolution of Painlevé paradox. Acta Mech. Sin. **20**(6), 649–660 (2004)
473. Z. Zhao, C. Liu, B. Chen, The Painlevé paradox studied at a 3D slender rod. Multibody Syst. Dynam. **19**, 323–343 (2008)
474. V.F. Zhuravlev, Equations of motion of mechanical systems with ideal onesided links. J. Appl. Math. Mech. **42**(5), 839–847 (1978)

Index

B. Miller, E. Rubinovich, *Optimization of Dynamical Systems with Impulse Controls and Shocks*, Systems & Control: Foundations & Applications, https://doi.org/10.1007/978-3-031-64124-4

GPSR Compliance
The European Union's (EU) General Product Safety Regulation (GPSR) is a set of rules that requires consumer products to be safe and our obligations to ensure this.

If you have any concerns about our products, you can contact us on

ProductSafety@springernature.com

In case Publisher is established outside the EU, the EU authorized representative is:

Springer Nature Customer Service Center GmbH
Europaplatz 3
69115 Heidelberg, Germany

www.ingramcontent.com/pod-product-compliance
Lightning Source LLC
Chambersburg PA
CBHW071232230926
51849CB00001B/81
9783031641237